**Europe's Changing Woods and Forests
From Wildwood to Managed Landscapes**

———————————

To the memory of Oliver Rackham, 1939–2015

Europe's Changing Woods and Forests
From Wildwood to Managed Landscapes

———————

Edited by

Keith J. Kirby

Department of Plant Sciences, Oxford, UK

and

Charles Watkins

School of Geography, Nottingham, UK

CABI is a trading name of CAB International

CABI	CABI
Nosworthy Way	38 Chauncy Street
Wallingford	Suite 1002
Oxfordshire OX10 8DE	Boston, MA 02111
UK	USA
Tel: +44 (0)1491 832111	Tel: +1 800 552 3083 (toll free)
Fax: +44 (0)1491 833508	E-mail: cabi-nao@cabi.org
E-mail: info@cabi.org	
Website: www.cabi.org	

© CAB International 2015. All rights reserved. No part of this publication may be reproduced in any form or by any means, electronically, mechanically, by photocopying, recording or otherwise, without the prior permission of the copyright owners.

A catalogue record for this book is available from the British Library, London, UK.

Library of Congress Cataloging-in-Publication Data

Europe's changing woods and forests : from wildwood to managed landscapes / editors: Keith J. Kirby, Department of Plant Sciences, Oxford, Charles Watkins, School of Geography, Nottingham.
    pages cm
  Includes bibliographical references and index.
  ISBN 978-1-78064-337-3 (hbk : alk. paper) 1. Forest ecology--Europe--History. 2. Forest management--Europe--History. 3. Forests and forestry--Europe--History. I. Kirby, K. J., editor.

 SD177.E858 2015
 634.9'2094--dc23

2014041667

ISBN-13: 978 1 78064 337 3

Commissioning editors: Vicki Bonham and Nicki Dennis
Editorial assistant: Emma McCann
Production editors: Lauren Povey and Emma Ross

Typeset by SPi, Pondicherry, India.
Printed and bound in the UK by CPI Group (UK) Ltd, Croydon, CR0 4YY.

Contents

Contributors		xv
Preface		xvii
Acknowledgements		xix

PART I. INTRODUCTION AND OVERVIEW

1 Overview of Europe's Woods and Forests — 3
Keith J. Kirby and Charles Watkins

1.1	Introduction	3
1.2	The Current State and Composition of European Woods and Forests	4
	1.2.1 European forests in a global context	4
	1.2.2 Variation in forest cover across the Continent	6
	1.2.3 Variation in forest composition	8
1.3	Forestry Policy and Cooperation at a European Level	8
	1.3.1 Forestry policy	8
	1.3.2 Conservation measures	9
	1.3.3 Landscape and amenity conservation	11
	1.3.4 Certification as an approach to sustainable forestry management	14
	1.3.5 Forest research cooperation across Europe	14
1.4	Conclusion	14
	References	14

2 Methods and Approaches in the Study of Woodland History — 18
Charles Watkins

2.1	Introduction	18
2.2	Oral History	19
2.3	Photographs and Drawings	21
2.4	Biological Indicators	22
2.5	Historical Records	23
2.6	Preserved Wood and Dendrochronology	24

v

2.7	Lidar and GIS	26
2.8	Applying Archaeological Insights to Ecological Issues	26
2.9	Pollen and Charcoal Analysis	27
2.10	Genetic and Molecular Markers	28
2.11	Conclusion	29
References		29

3 The Forest Landscape Before Farming 33
Keith J. Kirby and Charles Watkins

3.1	Where to Begin?	33
3.2	A Cold Open Continent	34
3.3	Trees Spread Back After the Ice	34
	3.3.1 Forming a canopy	35
	3.3.2 The wood beneath the trees	36
	3.3.3 Molecular markers for recolonization routes	36
3.4	A Holey Blanket of Trees	37
3.5	The Role of Large Herbivores, Particularly Bison, Wild Horse and Aurochs	38
3.6	People in the Landscape: The Trees in Retreat	41
References		41

4 Evolution of Modern Landscapes 46
Keith J. Kirby and Charles Watkins

4.1	Introduction	46
4.2	The Emergence of Woodland Management	46
4.3	Changes in Forest Extent and Distribution	47
	4.3.1 Reductions in forest cover	47
	4.3.2 Increases as well as decreases	48
	4.3.3 Patterns of clearance and survival	48
	4.3.4 The ecological consequences of a patchy landscape	49
4.4	Changes in Structure and Composition Through Management	49
4.5	Deliberate Modification of the Tree and Shrub Composition of Forests	50
4.6	Other Species Gains and Losses	51
4.7	Changes to the Fire Regime	51
4.8	Changes to the Forest Soil	52
4.9	Forests and Atmospheric Pollution	52
4.10	Climate Change	53
4.11	Conclusion	54
References		54

PART II. THE VARIETY OF MANAGEMENT ACROSS EUROPEAN WOODS AND FORESTS

5 Wood-pastures in Europe 61
Tibor Hartel, Tobias Plieninger and Anna Varga

5.1	Introduction	61
5.2	Wood-pasture: A Multi-purpose System	63
5.3	Historical Development of Wood-pastures in Europe	65
	5.3.1 Forest grazing and pasturing in ancient times	65
	5.3.2 Driving the livestock out of the forest (18th–19th centuries)	66
	5.3.3 New recognition for wood-pastures?	68
5.4	National Inventories of Wood-pastures	69

	5.5	Wood-pastures as Multifunctional Landscape Elements: Past and Present	69
	5.6	Threats to Wood-pastures	69
		5.6.1 Management changes	71
		5.6.2 Policy mismatch	71
		5.6.3 Decline of old, hollowing or dying trees	72
		5.6.4 Lack of regeneration	72
	5.7	Conclusion	72
	Acknowledgements		72
	References		73
6	**Coppice Silviculture: From the Mesolithic to the 21st Century**		**77**
	Peter Buckley and Jenny Mills		
	6.1	Introduction	77
	6.2	The Physiological and Evolutionary Significance of Coppice	77
	6.3	Historic Development of Coppice Silviculture	78
	6.4	The Rise and Fall of Coppice as an Industrial Resource	80
	6.5	Surviving and Neglected Coppice in Europe: The Extent of the Forest Estate	81
	6.6	Coppice Silviculture	82
		6.6.1 Cutting methods	83
		6.6.2 Time of cutting	84
	6.7	Conversion to High Forest	84
		6.7.1 Coppice versus high forest yields	85
	6.8	Reinstating Coppice Management	85
	6.9	Future Drivers of Change	86
	References		87
7	**High Forest Management and the Rise of Even-aged Stands**		**93**
	Peter Savill		
	7.1	Introduction	93
	7.2	Changing from Coppice to High Forest Systems	93
	7.3	The Need for New Administrative Tools	94
	7.4	Silvicultural Systems	95
	7.5	The Rise of Plantations	98
	7.6	Increased Use of Conifers and Introduced Species	99
	7.7	How Forestry is Changing	101
	7.8	Future High Forest and Natural Forest Structures	103
	References		103
8	**Close-to-nature Forestry**		**107**
	Matthias Bürgi		
	8.1	Introduction	107
	8.2	Roots and Prerequisites	109
	8.3	Developments in the 20th Century	110
	8.4	Ecological Implications	112
	8.5	Conclusion	113
	References		113
9	**The Impact of Hunting on European Woodland from Medieval to Modern Times**		**116**
	John Fletcher		
	9.1	Introduction	116
	9.2	Early Impacts of Hunting	116

	9.3	Meat or Merit?	117
	9.4	Medieval Hunting Reserves	117
	9.5	Early Modern Hunting Parks in Europe	118
	9.6	Hunting and the Wider Landscape	119
	9.7	Modern Hunting	121
		9.7.1 The influence of driven pheasant shoots on British woodland	122
		9.7.2 The influence of modern hunting enclosures on Spanish woodland	123
	9.8	Conclusion	124
	References	124

PART III. HOW PLANTS AND ANIMALS HAVE RESPONDED TO THE CHANGING WOODLAND AND FOREST COVER

10 The Flora and Fauna of Coppice Woods: Winners and Losers of Active Management or Neglect? — 129
Peter Buckley and Jenny Mills

	10.1	Introduction	129
	10.2	The Diversity of Coppice	129
		10.2.1 Plants	130
		10.2.2 Birds	131
		10.2.3 Invertebrates	131
		10.2.4 Dead wood and associated species	133
		10.2.5 Mammals	133
	10.3	Impacts of Deer Browsing on Flora and Fauna in Coppice	134
	10.4	Conservation Strategies	135
	10.5	Short-rotation Coppice	135
	10.6	Conclusion	136
	References	136

11 The Importance of Veteran Trees for Saproxylic Insects — 140
Juha Siitonen and Thomas Ranius

	11.1	Introduction	140
	11.2	What Are Saproxylic Species?	140
	11.3	Veteran Trees in Past and Present Landscapes	141
	11.4	Important Structures and Associated Species in Old Trees	141
		11.4.1 Microhabitat diversity	141
		11.4.2 Tree cavities and their invertebrates	142
		11.4.3 Other microhabitats	143
	11.5	Effects of Environmental Factors on the Invertebrate Fauna	144
		11.5.1 Effects of tree characteristics on species assemblages	144
		11.5.2 Effects of surrounding landscape on species assemblages	145
		11.5.3 Catering for the needs of the adults as well as the larvae	146
		11.5.4 Survey methods	146
	11.6	Current Situation in Europe	147
	11.7	How to Preserve the Specialized Saproxylic Species?	147
		11.7.1 Management for increasing habitat amount and quality	147
		11.7.2 Management for securing spatio-temporal continuity	148
	11.8	Future Prospects	150
	References	150

12	**The Changing Fortunes of Woodland Birds in Temperate Europe**	**154**
	Shelley A. Hinsley, Robert J. Fuller and Peter N. Ferns	
	12.1 Introduction	154
	12.2 The Birds of the Early Holocene	154
	12.3 The Birds of the Wildwood: Alternative Models of Forest Dynamics	155
	12.3.1 Largely closed forest – the 'closed canopy' scenario	158
	12.3.2 Open mosaic landscape – the 'wood-pasture' scenario	159
	12.3.3 Forest-dominated, but more varied – the 'closed but varied' scenario	159
	12.4 Fragmentation of the Wildwood	159
	12.5 Effects of the Historical Emergence of Management	161
	12.6 The Age of Managed Pasture Woods and Coppice	163
	12.7 The Shift Towards High Forest	164
	12.8 Woodland Birds Today	164
	12.8.1 Population trends	164
	12.8.2 Influences of agriculture	165
	12.8.3 Forestry intensification	165
	12.8.4 Birds and afforestation	166
	12.9 Recent Trends	167
	12.10 Conclusion	168
	References	168
13	**Evolution and Changes in the Understorey of Deciduous Forests: Lagging Behind Drivers of Change**	**174**
	Martin Hermy	
	13.1 Introduction	174
	13.2 Background	174
	13.3 What Sorts of Plants Occur in Forests?	175
	13.4 Comparing Ancient and Recent Forests	177
	13.5 Colonization of New Forests	178
	13.6 Dispersal and Recruitment Limitation	179
	13.7 Changing Ancient Forests	181
	13.7.1 Management effects	181
	13.7.2 Effects of environmental changes	182
	13.7.3 Effects of grazing	183
	13.7.4 Effects of invasive non-native species	183
	13.8 Conserving and Expanding Forests: Does It Work?	183
	References	186
14	**Gains and Losses in the European Mammal Fauna**	**193**
	Robert Hearn	
	14.1 Introduction	193
	14.2 Aurochs	193
	14.3 Carnivores	195
	14.3.1 The wolf	195
	14.3.2 The brown bear	196
	14.3.3 The lynx	197
	14.4 The Beaver	197
	14.5 A Species that Has Done too Well: The American Grey Squirrel	199

	14.6	The Decline and Rise of Wild Boar and Deer	200
		14.6.1 Wild boar	200
		14.6.2 Deer	201
	14.7	Conclusion	202
	References		202

15 The Curious Case of the Even-aged Plantation: Wretched, Funereal or Misunderstood? 207
Chris P. Quine

	15.1	Introduction	207
	15.2	What is an Even-aged Plantation?	207
	15.3	A Brief Historical Overview of Atlantic Spruce Forests	209
		15.3.1 The dominance of Sitka spruce	210
		15.3.2 Breaking up the conifer blanket	210
	15.4	Species Composition of Spruce Plantations	211
	15.5	Ecological Implications of Stand Dynamics	212
		15.5.1 Precursors – the creation of woodland through afforestation (Stage 0)	213
		15.5.2 Stand initiation (Stage 1)	214
		15.5.3 The impact of stand development – canopy closure and mortality (Stages 2 and 3)	214
		15.5.4 Prolonging the rotation and developing multiple storeys (Stage 4)	215
		15.5.5 Resetting the woodland through disturbance	216
	15.6	Forest Design	216
	15.7	The Landscape Setting	217
	15.8	Where Next?	217
	15.9	Conclusion	218
	References		218

PART IV. A VARIETY OF WOODLAND HISTORIES

16 Historical Ecology in Modern Conservation in Italy 227
Roberta Cevasco and Diego Moreno

	16.1	Introduction	227
	16.2	Background	227
	16.3	The Spread of a Historical Ecological Approach in European Conservation Thinking	230
		16.3.1 Forestry versus woodmanship	230
		16.3.2 Woodland or land bearing trees	231
		16.3.3 The need for an interdisciplinary approach	231
		16.3.4 The role of historical ecology	232
	16.4	Integrating Historical and Local Knowledge into Management Strategies	232
		16.4.1 An introduction to the case studies	234
		16.4.2 Trees and woodlands producing leaf fodder	234
		16.4.3 Trees, woodland and soil fertility	236
		16.4.4 The collection of litter	237
		16.4.5 Trees invading bogs: an experiment in applied historical ecology	238
	16.5	Conclusion	239
	References		239

17	Białowieża Primeval Forest: A 2000-year Interplay of Environmental and Cultural Forces in Europe's Best Preserved Temperate Woodland		243
	Małgorzata Latałowa, Marcelina Zimny, Bogumiła Jędrzejewska and Tomasz Samojlik		
	17.1	Introduction	243
	17.2	Previous Studies	243
	17.3	A New Palaeoecological Record for Białowieża Primeval Forest	245
		17.3.1 Methods	245
		17.3.2 Results	246
	17.4	Archaeological Evidence	250
	17.5	Archival Studies	252
		17.5.1 The royal forest of Polish kings	252
		17.5.2 Under Russian rule	253
		17.5.3 World War I to the present	254
		17.5.4 Changes in land-use extent and character	255
	17.6	Dendrochronological Analyses of Fire Dynamics	255
	17.7	Interplay of Natural and Cultural Forces	257
		17.7.1 The Iron Age	257
		17.7.2 The Migration Period, medieval and early modern times	257
		17.7.3 The 17th and 18th centuries	258
		17.7.4 The 19th to mid-20th centuries	258
		17.7.5 The recent decades	259
	17.8	The Role of Large Herbivores in Shaping Białowieża National Park	259
	17.9	Conclusion	259
	Acknowledgements		260
	References		260
18	Woodland History in the British Isles – An Interaction of Environmental and Cultural Forces		265
	George F. Peterken		
	18.1	Introduction	265
	18.2	Outline of British Woodland History	265
	18.3	Historical Stages and Processes of Change	266
	18.4	Regions	267
	18.5	Pre-Neolithic Wildwood	268
	18.6	Exploited Wildwood	269
	18.7	Traditional Woodland Management	269
	18.8	Parks, Forests and Wooded Commons	272
	18.9	Improved Traditional Management	273
	18.10	Plantations	275
	18.11	Revival and Restoration of Native Woodland	275
	18.12	Some Consequences of Differences in Regional History	276
	References		277
19	Forest Management and Species Composition: A Historical Approach in Lorraine, France		279
	Xavier Rochel		
	19.1	Introduction	279
	19.2	The Study of Forest History in France	279
	19.3	Historical Forest Uses and Their Consequences for Forest Management	280
	19.4	The Making of the Technical and Legislative Framework	283

19.5	Consequences of Forestry Policies for Forest Composition in the Woodlands of Lorraine	285
19.6	The Modern Forest – Conclusion	288
References		288

20 Barriers and Bridges for Sustainable Forest Management: The Role of Landscape History in Swedish Bergslagen — 290
Per Angelstam, Kjell Andersson, Robert Axelsson, Erik Degerman, Marine Elbakidze, Per Sjölander and Johan Törnblom

20.1	Introduction	290
20.2	The European Scale	290
20.3	The Regional Scale	291
20.4	Bergslagen – An Introduction	292
20.5	Forests, Forest Ownership and Land-use Dynamics	292
20.6	Barriers to Sustainability	294
	20.6.1 Ecological sustainability	294
	20.6.2 Economic sustainability	297
	20.6.3 Social and cultural sustainability	298
20.7	Bridges Towards Sustainable Forest Management	299
20.8	Discussion	300
	20.8.1 From forest history to history of forest landscapes	300
	20.8.2 Landscapes with different histories: using space for time substitution	301
References		301

PART V. LESSONS FROM THE PAST FOR THE FUTURE?

21 The Development of Forest Conservation in Europe — 309
James Latham

21.1	Introduction	309
21.2	Why Conserve Forests?	309
	21.2.1 As a spiritual place	309
	21.2.2 As a place for the chase	310
	21.2.3 As a source of raw materials and a barrier against the elements	311
	21.2.4 For a new form of communing with the forests	312
21.3	Type and Extent of Protected Forest Areas	312
21.4	Selection of Protected Areas	315
21.5	Developing a European Perspective	315
21.6	Forest Protection and Conservation as Part of Land-use Practice	318
21.7	Rewilding and Forest Conservation	321
21.8	From the Past to the Future	321
	21.8.1 Conservation for people?	321
	21.8.2 What sorts of woods and forests will be conserved in future?	322
References		323

22 The UK's Ancient Woodland Inventory and its Use — 326
Emma Goldberg

22.1	Introduction	326
22.2	Developing the Ancient Woodland Concept	327
22.3	The Creation of the Ancient Woodland Inventory	328

22.4	Developing and Using the Inventories	329
	22.4.1	England: the 'Red Queen' dilemma	329
	22.4.2	Wales	331
	22.4.3	Scotland	331
	22.4.4	Northern Ireland	332
22.5	Testing the Limits of the English Inventories	332
	22.5.1	Uncertain evidence	332
	22.5.2	What is a wood?	332
	22.5.3	How small can an ancient wood be?	333
22.6	Conclusion	334
	References	335

23 Tree and Forest Pests and Diseases: Learning from the Past to Prepare for the Future — 337
Clive Potter

23.1	Introduction	337
23.2	Background	338
	23.2.1 Dutch elm disease, ramorum blight and ash dieback	339
23.3	The Dutch Elm Disease Outbreak	339
23.4	'Sudden Oak Death' (Ramorum Blight) in the UK	341
23.5	A Landscape Without Ash?	343
23.6	The Lessons from History	344
	References	345

24 Reflections — 347
Charles Watkins and Keith J. Kirby

24.1	Introduction	347
24.2	Ways of Exploring and Understanding Woodland Histories	347
24.3	Issues for the Future Historian	348
24.4	From Cultural Landscapes Back to Wildwood?	349
24.5	Europe's Woods and Forests: The Future?	349

Index — 351

Contributors

Kjell Andersson deceased, formerly of School for Forest Management, Swedish University of Agricultural Sciences, Skinnskatteberg, Sweden.
Per Angelstam, School for Forest Management, Swedish University of Agricultural Sciences, PO Box 43, 73921 Skinnskatteberg, Sweden. E-mail: per.angelstam@slu.se
Robert Axelsson, School for Forest Management, Swedish University of Agricultural Sciences, PO Box 43, 73921 Skinnskatteberg, Sweden. E-mail: robert.axelsson@slu.se
Peter Buckley, Peter Buckley Associates, 8 Long Row, Mersham, Ashford, Kent, TN25 7HD, UK. E-mail: peterbuckleyassociates@gmail.com
Matthias Bürgi, Research Unit for Landscape Dynamics, Swiss Federal Research Institute WSL, Zürcherstrasse 111, CH-8903 Birmensdorf, Switzerland. E-mail: matthias.buergi@wsl.ch
Roberta Cevasco, Centro per l'Analisi Storica del Territorio (CAST), Università del Piemonte Orientale, Alessandria, Italy. E-mail: robcev@gmail.com
Erik Degerman, Department of Aquatic Resources, Institute of Freshwater Research, Swedish University of Agricultural Sciences (SLU), Pappersbruksallén 22, SE-702 15 Örebro, Sweden. E-mail: erik.degerman@slu.se
Marine Elbakidze, School for Forest Management, Swedish University of Agricultural Sciences, PO Box 43, 73921 Skinnskatteberg, Sweden. E-mail: marine.elbakidze@slu.se
Peter N. Ferns, Cardiff School of Biosciences (BIOSI2), Cardiff University, Cardiff, CF10 3AX, UK. E-mail: FernsPN@cardiff.ac.uk
John Fletcher, Harthill, Reediehill Deer Farm, Auchtermuchty, Fife, KY14 7HS, Scotland, UK. E-mail: Tjohn.fletcher@virgin.net
Robert J. Fuller, British Trust for Ornithology, The Nunnery, Thetford, Norfolk, IP24 2PU, UK. E-mail: rob.fuller@bto.org
Emma Goldberg, Natural England, Suite D, Unex House, Bourges Boulevard, Peterborough PE1 1NG, UK. E-mail: emma.goldberg@naturalengland.org.uk
Tibor Hartel, Department of Environmental Studies, Faculty of Sciences and Arts, Sapientia Hungarian University of Transylvania, Cluj-Napoca, Romania. E-mail: hartel.tibor@gmail.com
Robert Hearn, Laboratorio di Archeologia e Storia Ambientale (LASA), Dipartimento di Antichità, Filosofia, Storia, Geografia (DAFIST)/Dipartimento di Scienze della Terra, dell'Ambiente e della Vita(DISTAV), Università degli Studi di Genova, Via Balbi 6, 16126, Genoa, Italy. E-mail: robert.hearn@edu.unige.it; roberthearn3@me.com

Martin Hermy, Division of Forest, Nature and Landscape, Department of Earth and Environmental Sciences, University of Leuven (KU Leuven), Celestijnenlaan 200E, B-3001 Leuven, Belgium. E-mail: martin.hermy@ees.kuleuven.be

Shelley A. Hinsley, CEH Wallingford, Maclean Building, Crowmarsh Gifford, Wallingford, Oxfordshire, OX10 8BB, UK. E-mail: sahi@ceh.ac.uk

Bogumiła Jędrzejewska, Mammal Research Institute, Polish Academy of Sciences, ul. Waszkiewicza 1c, 17-230 Białowieża, Poland. E-mail: bjedrzej@ibs.bialowieza.pl

Keith J. Kirby, Department of Plant Sciences, University of Oxford, South Parks Road, Oxford, OX1 3RB, UK. E-mail: keith.kirby@bnc.oxon.org

Małgorzata Latałowa, Laboratory of Palaeoecology and Archaeobotany, Department of Plant Ecology, University of Gdańsk, ul. Wita Stwosza 59, 80-308 Gdańsk, Poland. E-mail: m.latalowa@ug.edu.pl

James Latham, Bryn Ffynnon, Llanddona, Anglesey, LL58 8UG, UK. E-mail: jimstardrift@googlemail.com

Jenny Mills, Peter Buckley Associates, 8 Long Row, Mersham, Ashford, Kent, TN25 7HD, UK. E-mail: peterbuckleyassociates@gmail.com

Diego Moreno, Laboratorio di Archeologia e Storia Ambientale (LASA), Dipartimento di Antichità, Filosofia, Storia, Geografia (DAFIST)/Dipartimento di Scienze della Terra, dell'Ambiente e della Vita (DISTAV), Università di Genova, Via Balbi 6, 16126, Genoa, Italy. E-mail: diego.moreno@unige.it

George F. Peterken, Beechwood House, St Briavels Common, Lydney, Gloucestershire, GL15 6SL, UK. E-mail: gfpeterken@tiscali.co.uk

Tobias Plieninger, Section for Landscape Architecture and Planning, Department of Geosciences and Natural Resource Management, University of Copenhagen, Rolighedsvej 23, 1958 Frederiksberg C, Denmark. E-mail: tobias.plieninger@ign.ku.dk

Clive Potter, Centre for Environmental Policy, Imperial College London, South Kensington Campus, 14 Princes Gardens, London, SW7 2AZ, UK. E-mail: c.potter@imperial.ac.uk

Chris P. Quine, Forest Research, Centre for Ecosystems, Society and Biosecurity, Northern Research Station, Roslin, Midlothian, EH25 9SY, UK. E-mail: chris.quine@forestry.gsi.gov.uk

Thomas Ranius, Department of Ecology, Swedish University of Agricultural Sciences, PO Box 7044, SE-75007 Uppsala, Sweden. E-mail: thomas.ranius@slu.se

Xavier Rochel, Département de Géographie, Université de Lorraine, BP 13397, 54015 Nancy Cedex, France. E-mail: xavier.rochel@univ-lorraine.fr

Tomasz Samojlik, Mammal Research Institute, Polish Academy of Sciences, ul. Waszkiewicza 1c, 17-230 Białowieża, Poland. E-mail: samojlik@ibs.bialowieza.pl

Peter Savill, Department of Plant Sciences, University of Oxford, South Parks Road, OX1 3RB, UK. E-mail: peter.savill@plants.ox.ac.uk

Juha Siitonen, Natural Resources Institute Finland, PO Box 18, FI-01301 Vantaa, Finland. E-mail: juha.siitonen@metla.fi

Per Sjölander, Academy North, Development Unit, SE-923 81 Storuman, Sweden. E-mail: per.sjolander@vilhelmina.se

Johan Törnblom, School for Forest Management, Swedish University of Agricultural Sciences, PO Box 43, 73921 Skinnskatteberg, Sweden. E-mail: johan.tornblom@slu.se

Anna Varga, Centre for Ecological Research, Institute of Ecology and Botany, Hungarian Academy of Sciences, Vácrátot, Hungary. E-mail: varga.anna@gmail.com

Charles Watkins, School of Geography, University Park, Nottingham, NG7 2RD, UK. E-mail: charles.watkins@nottingham.ac.uk

Marcelina Zimny, Laboratory of Palaeoecology and Archaeobotany, Department of Plant Ecology, University of Gdańsk, ul. Wita Stwosza 59, 80-308 Gdańsk, Poland. E-mail: m.zimny@ug.edu.pl

Preface

In the summer of 1996 we organized a conference on *Advances in Forest and Woodland History* supported by the International Union of Forest Research Organizations (IUFRO), the British Ecological Society and the School of Geography at University of Nottingham. Papers at that conference and from others in the 1980s and 1990s demonstrated the strength and breadth of research in woodland history, historical ecology and associated disciplines at the time. Since then, a wealth of other work has led to significant changes in our knowledge of European woodland history. Some of the research has been carried out within a single discipline, such as ornithology, entomology or archaeology. Other research is consciously interdisciplinary and attempts to bring together historical, cultural and technical knowledge. Both types of approach have produced significant insights that are reported in this book and have broad implications for woodland history, policy and management.

In this book, we have brought together information about Europe's forests and how they have developed since the last Ice Age. The first part (Chapters 1–4) gives an overview of Europe's woods and forests in space and over time; the second part (Chapters 5–9) looks at how they have been managed; the third part (Chapters 10–15) deals with how plants and animals have responded to our altering of the forest cover; the fourth part (Chapters 16–20) illustrates a range of different histories for locations from Italy to Sweden, from Poland to the UK; while the final part (Chapters 21–24) explores some of the issues around woodland conservation, now and in the future.

Keith J. Kirby and Charles Watkins
24 September 2014

Acknowledgements

We are grateful to the publishers for proposing that we produce a reworking of thoughts on European forest history building on the ideas from our 1998 volume, *The Ecological History of European Forests*, and we are indebted to our co-authors for the time and effort that they have put into this venture. We thank various colleagues who provided material, comments and advice on different sections, notably Alice Broome, Althea Davies, Julian Evans, Frank Götmark, Phil Grice, Carl Griffin, Jeanette Hall, Ralph Harmer, Dawn Isaac, Rob Jarman, Gary Kerr, Roger Key, Rob Marrs, Fraser Mitchell, Tony Mitchell-Jones, John O'Halloran, Suzanne Perry, Andy Poore, Firini Saratsi, Richard Tipping, Peter Thomas and Judith Tsouvalis. The School of Geography in Nottingham and the Department of Plant Sciences in Oxford provided us with support and encouragement to carry out this work.

We note with deep regret the death of Kjell Andersson, who co-authored Chapter 20, just before the end of the project.

Part I

Introduction and Overview

Within Part 1 of the book, the first chapter provides an overview of the current extent and composition of European woods and forests. This overview is followed (in Chapter 2) by a look at the methods used to study woodland history, including the exciting insights into past species movements that have been made possible by recent molecular genetic techniques.

We then look at how this variation in the extent and character of European woods and forests has arisen. What might the 'natural' forest cover across Europe – the wildwood of the title – have been prior to the development of farming in the Neolithic period (Chapter 3)? This is an area of active debate following the challenge to the conventional views of natural forest that have been made by Frans Vera.[1] In Chapter 4 the ecological consequences of increasing levels of human intervention in European woodland are described in terms of (for most of history) reductions in forest cover, changing tree composition and alterations to the other wildlife it contains.

---

[1] Vera, F.W.M. (2000) *Grazing Ecology and Forest History*. CAB International, Wallingford, UK.

1 Overview of Europe's Woods and Forests

Keith J. Kirby[1]* and Charles Watkins[2]

[1]*Department of Plant Sciences, University of Oxford, Oxford, UK;*
[2]*School of Geography, University of Nottingham, Nottingham, UK*

1.1 Introduction

Europe's trees and woods range from Mediterranean olive groves to extensive forests of pine and spruce in Scandinavia, from tall lime trees in the forests of Poland to scrubby oaks barely overtopping the heather on Atlantic cliffs. Some contain beautiful orchids, strange beetles or wild wolves. These patterns reflect variations in past and present climates and soil conditions; the natural environment sets limits on what can live where. However, people have also been living in Europe for thousands of years. Since the last Ice Age, our ancestors have shaped the distribution, composition and structure of woods and forests (Williams, 2006). There is less forest now and it is more fragmented than in the distant past; in many countries the proportion of conifers to broadleaves has increased; some animals are now extinct, such as the wild ox, while others, such as the grey squirrel, have been introduced and become pests.

In this book, we explore the history and ecology of European woods and forests, and how the interplay of environmental and human factors has created different wooded landscapes across the continent. Unless we understand how the current patterns formed, we cannot expect to address future challenges to their management and conservation.

We have tried to cover the full spectrum of woodland cover – from dense closed-canopy plantations (Savill *et al.*, 1997) to wood-pastures where trees occur in more open park-like landscapes maintained by grazing (Rotherham, 2013; Hartel and Plieninger, 2014) (Box 1.1). Where one category stops and another begins is not always clear-cut, so there is a fuzziness to the boundaries of the different definitions. We have, though, excluded the forests of the Russian Federation, which are so vast and distinct, historically, geographically and ecologically, that they deserve separate treatment (Teplyakov *et al.*, 1998).

On the whole, we do not deal with the changing use and processing of the wood and timber (Peck, 2001; Owende, 2004), except where this has implications for the forest itself. We have also not considered, except in passing, other types of landscape change, for example the losses of species-rich grassland or heath through agricultural intensification (Meeus *et al.*, 1990; Henle *et al.*, 2008; Peterken, 2013).

Europe's woods and forests and their history can be grouped into broad geographic zones – Mediterranean, temperate broadleaved and boreal coniferous. Nevertheless, within

---

*E-mail: keith.kirby@bnc.oxon.org

© CAB International 2015. *Europe's Changing Woods and Forests: From Wildwood to Managed Landscapes* (eds K.J. Kirby and C. Watkins)

> **Box 1.1.** Woods, forests and trees.
>
> - 'Wood', 'woodland' and 'forest' are all used generally to describe tree-covered lands. Wood tends to be used where relatively small discrete areas of land are involved; woodland and forest are used for more extensive tracts.
> - 'Forest' is also used in a more specialized sense in some chapters where it refers to land subject to Forest Law, particularly in the medieval period. Forest law was primarily concerned with regulating hunting and the land to which it applied might or might not be covered by trees, i.e. not all of it was forest in the modern sense.
> - 'Wood-pasture' refers to landscapes where grazing by domestic stock or deer has created or maintained a relatively open tree cover. This includes parks whose boundaries are often marked by walls or fences as well as less well-defined areas with scattered trees.
> - 'Coppice' refers to the practice of repeatedly cutting trees close to ground level, resulting in the regrowth of multiple stems from the stump, which can be harvested again when they have regrown, usually after intervals of between 5 and 30 years. 'Pollarding' is a similar process, but the cut is made at 2–3 m above the ground so that the regrowth is out of the reach of browsing animals.
> - 'Plantations' are areas where the majority of trees have been planted. The stands may be created within existing woodland or on previously open ground, and are often referred to as 'planted forests'. The trees may be native to the area or introductions; they may be planted in large even-aged blocks or as wide-spaced individual stems (Evans, 2009). Old plantations may be difficult to distinguish from naturally regenerated stands.
> - 'Ancient woods' (or ancient forests) are those where there has been continuous woodland cover since a set threshold date, often around 1800, but sometimes earlier. These might, however, be on land that was open at some time before this date, so they are not necessarily primary. They have also usually been cut over or managed at some time.
> - 'Ancient trees' are old for their species with features such as cavities or a hollow trunk, bark loss over sections of the trunk and a large quantity of dead wood in the canopy. The broader term, 'veteran trees', also includes younger individuals that have developed similar characteristics, perhaps due to adverse growing conditions or injury.

each zone, there are many variations on a theme, sometimes even between the history of one wood and the next, as demonstrated in surveys of woods in eastern England or for different Mediterranean landscapes (Grove and Rackham, 2001; Rackham, 2003). This variation in what has happened in particular places and regions is important. Both general trends and local differences are reflected, not just in the trees and shrubs, but in the smaller plants and in the insects, bird and mammals that live in the forests; all of these contribute to our rich cultural and biological heritage.

1.2 The Current State and Composition of European Woods and Forests

1.2.1 European forests in a global context

Just over a third (34%) of Europe's land surface is wooded, but this makes only a minor contribution (5%) to the world's forests, according to the *Global Forest Resources Assessment 2010* of the Food and Agriculture Organization of the United Nations (FAO, 2010) (Fig. 1.1). This FAO report focuses on forest land use, not land cover. Forest land use is defined as areas with tree cover, or where management or natural processes will ultimately restore tree cover, and the predominant use is forestry. Areas are included if they span more than 0.5 ha with trees higher than 5 m and a canopy cover of more than 10%, or trees able to reach these thresholds *in situ*. Land that is predominantly under agricultural or urban land use is not included. In some cases, forest land use may include land temporarily without tree cover, for example during cycles of shifting cultivation, forest plantations and even-aged forest management.

A recent alternative analysis has sought to address these problems by using Landsat data to quantify the extent of land where tree cover is greater than 25% (Hansen *et al.*, 2013). There is a strong correlation for European

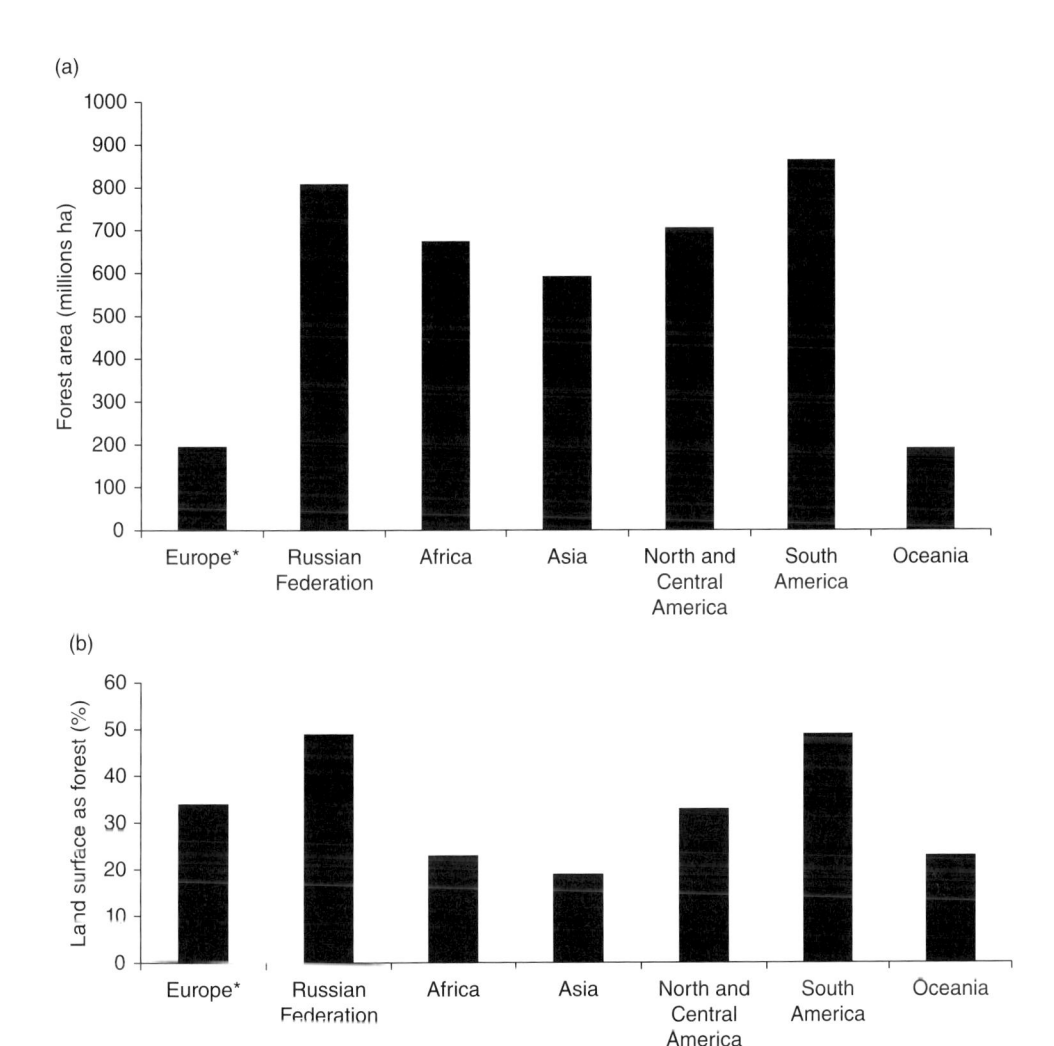

Fig. 1.1. European forest cover (*excluding the Russian Federation) compared with that of other continents: (a) total area (millions ha); (b) % land surface as forest. (From FAO, 2010.)

countries between the estimates based on land use and on land cover by forest. The FAO data for the Iberian peninsula were, however, higher than the remotely sensed estimates (Spain 36% versus 23%; Portugal 38% versus 28%), reflecting the ambiguities around the inclusion (or not) of the extensive wood-pasture areas with relatively low tree cover in these countries.

Europe's forest cover increased during the period 2000–2010, as it had done in the previous decade, although there was some loss of other wooded land between 1990 and 2000 (FAO, 2010). European forests are not just growing in extent but there has also been an increase in the growing stock (expressed as m^3 ha^{-1} of wood). This is despite a steady increase of 1.5% per annum (1990–2005) in the amount of wood harvested from European forests. Increasing nitrogen supply seems to have been the major cause of the changes observed during the 20th century (increased atmospheric deposition, but also improved soil nitrogen availability on sites that had formerly been degraded). Future changes in forest growth are more likely to be caused by increasing atmospheric carbon dioxide and increasing temperatures (particularly in northern latitudes) (Kahle *et al.*, 2008).

Compared with other continents, Europe has the lowest percentage (<3%) of forests classed as primary, which are defined by FAO as consisting of native species, where there are no clearly visible indications of human activities, and the ecological processes have not been significantly disturbed. It should be noted though that, in Europe, determining whether any areas have never been disturbed by humans (directly or indirectly by grazing or deliberate fire) such that they may be termed primary (or 'primeval', or 'virgin' forest) is very difficult because evidence for woodland management stretches back thousands of years.

A more useful distinction in Europe is between woods where there has at least been continuity of tree cover (even if managed) for several 100 years, from other woods that have developed on abandoned farmland within the last two centuries. The former have been termed 'ancient woodland', a concept developed by Peterken (1977) and Rackham (1976) in England, but which has since been adopted more widely across the Continent (Tack and Hermy, 1998; Wulf, 2004; Goldberg, Chapter 22).

The way that woods are treated (woodpasture, coppice, various forms of high forest, etc.; Matthews, 1991) reflects the products that a particular society values at that time. In Europe, more of the harvested wood tends to be used for industrial purposes and less goes for wood fuel than elsewhere. Worldwide, the emphasis in production forestry is shifting from natural/semi-natural stands towards plantations and Europe has the highest proportion (>20%) of planted forest, albeit these are predominantly of native trees (Evans, 2009; FAO, 2010). Nonetheless, non-wood forest products remain at least locally important, including nuts, mushrooms, berries and honey, Christmas trees and cork. Many countries have a long tradition of hunting and extensive areas of woodland are managed for this purpose (Fletcher, Chapter 9). Grazing by livestock also remains a major use of forests in parts of Europe (Hartel and Plieninger, 2014).

The areas designated for biodiversity conservation have tended to increase (Latham, Chapter 21). There is also some indication that the amounts of dead wood, a critical resource for biodiversity (European Environment Agency, 2010), have been increasing, although in some Mediterranean areas more dead wood may also increase the fire risk. In mountainous regions of Europe, the protective value of forests against avalanches has also long been recognized (Schneebeli and Bebi, 2004) and about 22 million ha of forests are reported as designated for the protection of soil and water (FAO, 2010). This probably underestimates the area of woodland and forest where such protection is important, because some protective areas may be listed as managed for production or for biodiversity conservation.

European forests have always been subject to a wide variety of periodic disturbances (Peterken, 1996; Schelhaas et al., 2003; Lindner et al., 2010; Thomas and McAlpine, 2010; Waller, 2013). FAO (2010) records an increase in the area affected by tree diseases in Europe since 1990. Fire is a major factor in the Mediterranean region with 300–600 thousand ha typically being burnt each year, whereas in the rest of Europe it is only 20–80 thousand ha. Catastrophic storms tend to occur every 5–10 years. These disturbances may interact: major storms in 2005 and 2007 in spruce stands in southern Sweden resulted in increased populations of the spruce bark beetle, *Ips typographus*; storms in central Europe in 2004/05 were also followed by severe bark beetle outbreaks. The frequency, if not the magnitude, of major disturbances is likely to increase with climate change (Seidl et al., 2011).

1.2.2 Variation in forest cover across the Continent

Forest extent varies considerably across the Continent (Päivinen et al., 2001; Schuck et al., 2002; FAO, 2010; Kempeneers et al., 2011; Hansen et al., 2013). Of the 196 million ha of forest in Europe (excluding the Russian Federation) nearly half (84 million ha) is in just four countries – Sweden, Finland, Spain (including *dehesas*) and France (FAO, 2010) (Fig. 1.2a). Nine countries (plus a further 14 very small states and territories) have less than 1 million ha of forest. Five countries

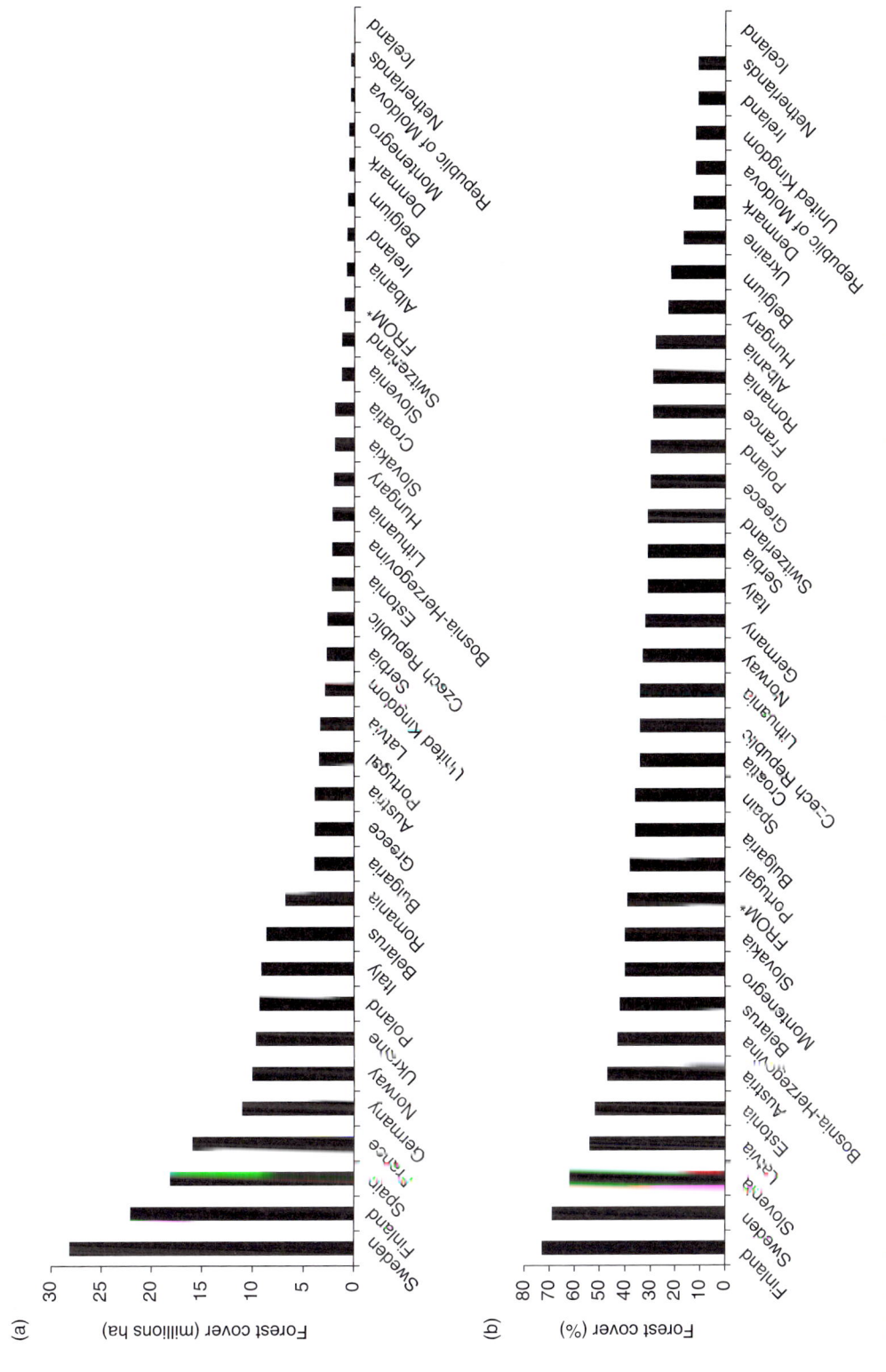

Fig. 1.2. Forest cover by country: (a) by extent (millions ha); (b) as % land surface. *, the former Yugoslav Republic of Macedonia. Small states and territories have been excluded. (Based on FAO, 2010.)

(Finland, Sweden, Slovenia, Latvia and Estonia) have more than 50% of their land surface as forest. Six countries (apart from the very small states and territories) have less than 17% forest cover, i.e. less than half the European average (Fig. 1.2b). For Iceland, this reflects a lack of suitable conditions for tree growth, but in Denmark, the Republic of Moldova, the UK, Ireland and the Netherlands it is largely due to a long history of forest clearance. There is a negative correlation between population density and forest cover; historically, higher population density has tended to mean greater demands on land to produce food (FAO, 2010).

Low forest cover may lead to interest in afforestation and the use of more productive (often introduced) species. So Denmark, the UK, Ireland and the Netherlands feature among the countries with the highest proportion of planted forests and use of introduced tree species. In both Ireland and the UK, forest cover has increased substantially over the last century (Quine, Chapter 15). Overall, Europe (as a continent) has the lowest proportion (about 44%) of forests in public versus private ownership, despite the predominance of publically owned forests in many Eastern European states. By contrast, in France, Slovenia, Sweden, Austria, Norway and Portugal, more than 70% of the forests are in private ownership.

1.2.3 Variation in forest composition

Much of Europe has a potential natural vegetation of mixed broadleaved and coniferous temperate forest with conifer-dominated forests to the north, and thermophilous mixed deciduous forests and sclerophyllous forest and scrub to the south around the Mediterranean Sea (Bohn *et al.*, 2000). The growing stock of European forests is split 60:40 conifers to broadleaves, with the broadleaves predominating in the south and particularly in central Europe, with more extensive coniferous areas to the north (Plates 1 and 2).

Descriptive accounts of the variation in forest vegetation can be found, for example, in Polunin and Walters (1985) and Ellenberg (1988) for Central Europe, in Scarascia-Mugnozza *et al.* (2000) and Fady and Medail (2004) for Mediterranean forests and in Arnborg (1990) for Swedish boreal forests. Rodwell (1991) is an example of a more detailed account just for the UK. Separate country-based classifications, such as this last, have been brought together and harmonized in different ways, for example the EUNIS (European Nature Information System) Habitat Classification (Davies *et al.*, 2004), the overview of phytosociological alliances presented by Rodwell *et al.* (2002) and that developed for reporting purposes under the Ministerial Conferences on the Protection of Forests in Europe (MCPFE) (Table 1.1).

Data from the International Co-operative Programme (ICP) on Assessment and Monitoring of Air Pollution Effects on Forests plots (Packalen and Maltamo, 2002), classified according to the forest types described in Table 1.1, illustrate the transition from mainly broadleaved evergreen woodland in the Mediterranean countries, through beech and mixed deciduous woodland in central and western Europe to boreal forests in Fenno-Scandinavia (European Environment Agency, 2007) (Table 1.2).

1.3 Forestry Policy and Cooperation at a European Level

1.3.1 Forestry policy

The coordination of European forestry policy comes under the MCPFE (2014), whose resolutions serve as a framework for action by the participating countries. The Helsinki Declaration in 1993 focused on biodiversity and climate change; at the Lisbon (1998) and Vienna (2003) meetings, more emphasis was placed on the socio-economic and cultural aspects of forest management; at Warsaw (2007) the resolutions concerned forests for energy and in water management. At the meeting in June 2011 in Oslo, ministers responsible for forests in Europe signed a mandate for negotiating a 'Legally Binding Agreement on Forests in Europe'.

Within the European Union (EU), unlike the situation for farming, there is no specific provision for a common forestry policy. Forestry policies remain nationally distinct and may also differ between regions within member states. There are, however, a number of areas where the European Union does have competence that relates to forests and forestry, including environmental policy, common agricultural policy (e.g. support for the afforestation of farmland), internal markets and trade (including efforts to combat illegal logging), and renewable energy.

In 1989, a standing forestry committee was set up (European Commission, 2013a) and in 2013, a forest strategy was agreed (European Commission, 2013b) with the following objectives for 2020:

> To ensure and demonstrate that all forests in the EU are managed according to sustainable forest management principles and that the EU's contribution to promoting sustainable forest management and reducing deforestation at global level is strengthened, thus contributing to balancing various forest functions, meeting demands, and delivering vital ecosystem services; and providing a basis for forestry and the whole forest-based value chain to be competitive and viable contributors to the bio based economy.

In May 2014, this interest in forestry was reiterated by the Council of the European Union (2014), when it welcomed the new EU Forest Strategy published by the Commission in September 2013 by stating that:

> as the forest sector is affected by an increasing number of EU policy initiatives, such as those dealing with energy and climate policy, the forest sector's contribution to preparing these initiatives needs to be strengthened. The ministers acknowledge that the new EU Forest Strategy should enhance coordination and facilitate the coherence of forest-related policies and should allow for synergies with other sectors that influence forest management and offer the key reference in EU forest-related policy development.

1.3.2 Conservation measures

More specific commitments on the conservation of species and habitats are included in the Bern Convention (1979) and, for the EU, in the Habitats and Species Directive (1992) and the Birds Directive (1979) (Latham, Chapter 21).

The Bern Convention (Council of Europe, 2013) aims are to conserve wild flora and fauna and their natural habitats and to promote European cooperation in this area. Article 4 paragraph 1 states that:

> Each Contracting Party shall take appropriate and necessary legislative and administrative measures to ensure the conservation of the habitats of the wild flora and fauna species, especially those specified in Appendices I and II, and the conservation of endangered natural habitats.

Among the species listed are a number that use woods and forests, from wide-ranging species such as the wolf (*Canis lupus*) to bat species that forage in woods but may roost in just one particular tree.

The Birds Directive (European Commission, 2013d) covers activities that directly threaten birds (other than pest species). Its provisions include restrictions on the destruction of their nests, which may have implications for when forestry management operations can be carried out, even within woods managed primarily for production. Thus, in England, concerns about the disturbance of the roosting or resting sites for birds and other European protected species have led to the development of specific guidance on woodland management (Forestry Commission, 2013).

The Birds Directive and the Habitats and Species Directive (European Commission, 1992, 2013c) are built around two pillars: the Natura 2000 Network of protected sites and a strict system of species protection. These two directives seek to protect over 1000 animals and plant species and over 200 'habitat types' (covering forests, meadows, wetlands, etc.) that are deemed to be of European importance. These include 69 forest types, which make up about half of the total area within the Natura series (Latham, Chapter 21). The species identified as in need of protection include large charismatic beasts such as the European bison (*Bison bonasus*), most bear and lynx populations (*Ursus arctos*, *Lynx lynx*) and the wolverine (*Gulo gulo*), but also many invertebrates

Table 1.1. Main European forest types. (From European Environment Agency, 2007.)

Type[a]	Description
Broadleaved evergreen forests	The sclerophyllous forests and scrubs of the Mediterranean region and the laurel-like forests of the Canary Islands and Azores. Water availability is a key limiting factor. The structure and occurrence of these forests has been much modified by the long history of human occupation of the Mediterranean basin.
Coniferous forests of the Mediterranean, Anatolian and Macaronesian regions	A large group of coniferous forests, pines (*Pinus* spp.), fir (*Abies* spp.) and juniper (*Juniperus* spp.), mainly on dry, nutrient-poor sites, from coastal regions to high mountains. Some pine forests in this category are adapted to fire (*P. halepensis*, *P. canariensis*) but repeated anthropogenic fires are a major factor leading to forest degradation.
Thermophilous deciduous forests	Milder climates of the Mediterranean region allow the development of mixed deciduous and semi-deciduous species, mainly of oaks, *Quercus* spp., but with maples, *Acer* spp., *Ostrya* sp., ashes, *Fraxinus* spp. and *Carpinus betulus* as frequent associates. Many were managed as coppice and sweet chestnut, *Castanea sativa*, has been widely introduced to woods and as new plantations. Often the coppice has now been abandoned and the stands are developing into high forest. The distribution of these forests tends to be limited to the north and at higher altitudes by temperature and to the south and on lower slopes by drought.
Beech forests	Beech forest (mainly *Fagus sylvatica*, but with *F. orientalis* in the eastern and southern parts of the Balkan peninsula) occurs widely across lowland and submontane Europe on soils ranging from thin limestone to deep acidic free-draining types. The forest is limited by low winter temperatures and short growing seasons at its northern and eastern edges, while water deficits become more critical to the south.
Mesophytic deciduous forests	Found on richer soils than the previous type, often with mixed canopy of hornbeam (*Carpinus betulus*), oak, ash, lime (*Tilia* spp.) and maples (*Acer* spp.). Much has been converted to productive farmland and the remaining areas may be intensively managed as a consequence. However, they remain widespread and highly variable often with rich understoreys.
Flood plain forests	High water tables and occasional flooding determine the appearance of this forest type, which must once have been widespread along the main European river channels. The forests are often species rich and multilayered with combinations of alders (*Alnus* spp.), birches (*Betula* spp.), poplars (*Populus* spp.), willows (*Salix* spp.), ash and elm (*Ulmus* spp.). Clearance to create farmland, management of the rivers (dams, canalization) and drainage of wetlands have all contributed to loss of large areas of these forests.
Mountain beech forests	Conifers (*Picea abies*, *Abies* spp.) are a much stronger component than in the other beech type. Birches may also be locally important. The range of these forests is centred on central European mountains, tending to occur at higher levels the further south they are. Traditionally, the forests were intensively managed for fuelwood by coppicing, but most have been turned into high forest.
Alpine coniferous forests	There is a similar harsh climate to the boreal zone, but with a different light regime and day length. These forests show a disjunct and patchy distribution because of their restriction to the highest elevations of European mountain ranges. Forest composition varies broadly with the altitudinal vegetation belts; as well as spruce (*Picea abies*) and Scots pine (*Pinus sylvestris*), larch (*Larix decidua*), *Pinus cembra* and *Pinus mugo* may also be abundant. Traditional pastoral farming modified the distribution and structure of these forests, but over wide areas has now been abandoned.
Acidophilous oak and oak–birch forests	This is found on nutrient-poor soils in the temperate deciduous forest zone. Oaks (*Quercus robur*, *Q. petraea*) predominate with birches (*Betula* spp.). Often, these forests were managed for coppice, sometimes with some grazing, but they have tended to be converted to broadleaved high forest, to conifer plantations, or simply abandoned, in recent decades.

Continued

Table 1.1. Continued.

Type[a]	Description
Non-riverine alder, birch and aspen forests	These communities occur in specific ecological conditions, for example mountain birch formations, and as pioneer stages of forest succession.
Hemiboreal forests and nemoral coniferous and mixed broadleaved–coniferous forests	Mixed forests where boreal coniferous species coexist with temperate broad-leaved trees such as oaks, ash, elms and limes. The coniferous species tend to be on the poorer soils and the broadleaves on the richer ones. The extent and composition of these forests have been much affected by human activity.
Mire and swamp forests	These occur on waterlogged, peaty soils, mainly in the boreal zone. Micro-topographic variations lead to a variety of nutrient and water regimes. Spruce and Scots pine build up mire forests, while species of alder (*Alnus* spp.), birch, oak and poplar dominate the deciduous swamp forests. Many areas have been drained to convert them to farmland or to allow more productive coniferous crops to be grown.
Boreal forests	Dominant in northern Europe, the climatic conditions mean that two conifers, spruce and Scots pine, predominate. Deciduous trees such as birches tend to occur most often in the early stages of succession. Under natural conditions, forest fires may have been the major disturbance type in many areas, but much of the area is now managed as even-aged stands for commercial forestry.
Plantations and self-sown exotic tree species stands	These have usually been established and are intensively managed for wood production or for the rehabilitation of degraded land. While some plantations may be of native species, introduced species are often used, some of which, such as *Robinia pseudoacacia* and *Ailanthus altissima*, have subsequently proved to be very invasive.

[a]The order reflects broadly the distribution of these types across the continent from the Mediterranean to boreal zones, with plantations added separately at the end.

such as the hermit beetle (*Osmoderma eremita*), the violet click beetle (*Limoniscus violaceus*) and the Kerry slug (*Geomalacus maculosus*).

1.3.3 Landscape and amenity conservation

The visual and social aspects of forests are also more widely recognized than in the past; landscape conservation and planning often rest alongside that for biodiversity (Ward Thompson, 2004). The Council of Europe (2000) has agreed a European Landscape Convention, specifically:

- to recognize landscapes in law as an essential component of people's surroundings, an expression of the diversity of their shared cultural and natural heritage, and a foundation of their identity;
- to establish and implement landscape policies aimed at landscape protection, management and planning;
- to establish procedures for the participation of the general public, local and regional authorities, and other parties with an interest in the definition and implementation of the landscape policies mentioned in the bullet point above; and
- to integrate landscape into its regional and town planning policies and in its cultural, environmental, agricultural, social and economic policies, as well as in any other policies with possible direct or indirect impact on landscape.

The convention is particularly relevant for afforestation and deforestation, because these can create significant changes to the landscape. Hence, these activities are also included under European requirements on Environmental

Table 1.2. Distribution of the main forest types across selected European countries, based on percentage of International Co-operative Programme (ICP) Assessment and Monitoring of Air Pollution Effects on Forests plots assigned to each type. (From European Environment Agency, 2007.)

Forest type[a]/ Country[b]	Broad-leaved evergreen forests	Coniferous forests of the Mediterranean region	Thermophilous deciduous forests	Beech forests	Mesophytic deciduous forests	Flood plain forests	Mountain beech forests	Alpine coniferous forests	Acidophilous oak and oak-birch forests	Non-riverine alder, birch and aspen forests	Hemiboreal forests	Mire and swamp forests	Boreal forests	Plantations and self-sown exotic tree species stands
Portugal	48	29	4		1					1				18
Spain	26	43	9		2		2	3	2					12
Greece	16	43	19	2	9		10							1
Italy	4	4	40		2		16	23	1	3				6
Serbia		1	35	23	11	5	11	1		1	2			12
Bulgaria		17	17	12	5		7	26			8			8
Croatia	2	6	14	20	15	19	11	6	2		1			2
Switzerland			6	6	8		13	50			15			2
France	4	10	14	7	24	1	5	9	6	2	4			15
Hungary			19	7	21	5				7	5			36
Romania			10	22	21		21	16		2	1			6
Moldova			10		80									10
Austria				1	3	1	6	65			24			
Slovenia			5	21	2	2	29	19		2	12			7
Slovakia			1	25	16		10	39			5			4
Denmark				30	10									60
UK				14	16	1			4		4			61
Belgium					10				10					80
Netherlands					27				9					64
Ireland					[c]				[c]					100
Germany				12	8	1	6	4	1	3	51			14
Czech Republic				4	9	1	4	1	1	3	69			9
Poland					7	1	2	5	1	5	75			1
Lithuania			2	2					17	76	13	5	
Belarus					1	1				14	62		8	

12 K.J. Kirby and C. Watkins

Estonia				3	77	12	7
Latvia	1			22	59		19
Finland				6	3	4	88
Sweden		1	1	6	39		50
Norway				27	4		68	1

^aThe order of forest types reflects broadly the distribution of these types across the continent from the Mediterranean to Boreal zones; with plantations added separately at the end.
^bCountries are ordered according to the broad composition of their recorded forest types.
^cForest types certainly present but not represented in the ICP Monitoring Plots because of the scarcity of semi-natural woodland.

Impact Assessment (http://ec.europa.eu/environment/eia/eia-legalcontext.htm).

1.3.4 Certification as an approach to sustainable forestry management

Implementation of the above strategies, resolutions and commitments remains the responsibility of the individual countries through their particular forestry legislation and forest services. However, paralleling government actions on harmonizing approaches to sustainable forest management and conservation has been the development of independent, voluntary certification schemes (Bass, 2004). These emerged from the 1992 Conference on Biodiversity – 'the Earth Summit' – held in Rio de Janeiro. The two major approaches that have developed are the Forest Stewardship Council with its principles and criteria for sustainable management (Forest Stewardship Council, 2013) and the Programme for the Endorsement of Forest Certification (PEFC, 2013). Timber that is harvested from forests run according to the principles and standards approved by these bodies can be marked to show that has it been sustainably produced. The extent to which these different certification systems have been taken up across Europe varies in both the state and private sectors. In general, there is more incentive for large-scale producers (including state forest services) to take up such systems than owners of small areas of woodland.

1.3.5 Forest research cooperation across Europe

There has been an increasing amount of joint working in many fields of forest research across Europe in the last three decades. Programmes organized under the European Cooperation in Science and Technology programme (COST, 2013), are one of the longest running European frameworks supporting cooperation among researchers across Europe. COST actions have included reviews of protected areas, the role of trees in human health and well-being, climate change and its implications for silviculture, and developing strategies to deal with ash dieback (caused by *Hymenoscyphus fraxineus*). The European Forest Institute (EFI), established at Joensuu, Finland, in 1993, now has over 100 associated organizations spread across 36 countries (http://www.efi.int/portal/home/) with a particular focus on the use of science and evidence to develop forest policy.

Another relevant research grouping has been the 'Forest History and Traditional Knowledge' chapter within International Union of Forest Research Organizations (IUFRO, 2013). Conferences organized under the auspices of IUFRO have brought together studies of forest history using a wide range of techniques, from oral histories to deep-time palaeoecology (Kirby and Watkins, 1998; Honnay *et al.*, 2004; Saratsi *et al.*, 2009; Parrotta and Trosper, 2012; Watkins, Chapter 2).

1.4 Conclusion

The variation in European woods and forests is very apparent, but equally there are many benefits from looking at how the forest cover of different areas has evolved or been treated. This will become even more necessary as the impacts of climate change start to bite (Lindner *et al.*, 2010). Woods in central Europe will not suddenly turn into copies of those currently found in the Mediterranean zone, but there will be lessons that can be learnt from such areas. We hope that the remaining chapters in this volume can contribute to that learning process.

References

Arnborg, T. (1990) Forest types of northern Sweden. *Vegetatio* 90, 1–13.
Bass, S. (2004) Certification. In: Burley, J., Evans, J. and Youngquist, J.A. (eds) *Encyclopedia of Forest Sciences*. Elsevier, Oxford, UK, pp. 1350–1357.
Bohn, U., Gollub, G. and Hettwer, C. (2000) *Map of the Natural Vegetation of Europe*. Federal Agency for Nature Conservation, Bonn, Germany.

COST (2013) Forests, their Products and Services (FPS): Actions. European Cooperation in Science and Technology, Brussels. Available at: http://www.cost.eu/domains_actions/fps/Actions (accessed 9 January 2013).

Council of Europe (2000) *European Landscape Convention and Reference Documents*. Available at: http://www.coe.int/t/dg4/cultureheritage/heritage/Landscape/Publications/Convention-Txt-Ref_en.pdf (accessed 31 July 2014).

Council of Europe (2013) *Convention on the Conservation of European Wildlife and Natural Habitats*. Available at: http://www.coe.int/t/dg4/cultureheritage/nature/bern/default_en.asp (accessed 9 January 2014).

Council of the European Union (2014) *New EU Forestry Strategy: Conclusions Adopted by the Council*. Available at: http://www.consilium.europa.eu/uedocs/cms_data/docs/pressdata/en/agricult/142685.pdf (accessed July 2014).

Davies, C.E., Moss, D. and Hill, M.O. (2004) *EUNIS Habitat Classification Revised 2004*. European Environment Agency, Copenhagen. Available at: http://www.eea.europa.eu/themes/biodiversity/eunis/eunis-habitat-classification/documentation/eunis-2004-report.pdf/download (accessed 14 November 2014).

Ellenberg, H. (1988) *The Vegetation Ecology of Central Europe*. Cambridge University Press, Cambridge, UK.

European Commission (1992) *Directive on the Conservation of Natural Habitats and Wild Fauna and Flora: The Habitats Directive, 92/43/EEC*. European Commission, Brussels.

European Commission (2013a) Standing Forestry Committee. European Commission, Brussels. Available at: http://ec.europa.eu/agriculture/committees/forestry_en.htm (accessed 9 January 2014).

European Commission (2013b) The New EU Forest Strategy. European Commission, Brussels. Available at: http://ec.europa.eu/agriculture/forest/strategy/index_en.htm (accessed 9 January 2014).

European Commission (2013c) The Habitats Directive. European Commission, Brussels. Available at: http://ec.europa.eu/environment/nature/legislation/habitatsdirective/ (accessed 9 January 2013).

European Commission (2013d) The Birds Directive. European Commission, Brussels. Available at: http://ec.europa.eu/environment/nature/legislation/birdsdirective/index_en.htm (accessed 9 January 2014).

European Environment Agency (2007) *European Forest Types: Categories and Types for Sustainable Forest Management Reporting and Policy*. EEA Technical Report No 9/2006, European Environment Agency, Copenhagen. Available at: http://www.env-edu.gr/Documents/European%20forest%20types.pdf (accessed 14 November 2014).

European Environment Agency (2010) Forest: deadwood (SEBI 018) – Assessment published 2010. European Environment Agency, Copenhagen. Available at: www.eea.europa.eu/data-and-maps/indicators/forest-deadwood/forest-deadwood-assessment-published-may-2010 (accessed 8 January 2014).

Evans, J. (ed.) (2009) *Planted Forests – Uses, Impacts and Sustainability*. Food and Agriculture Organization of the United Nations (FAO), Rome and CAB International, Wallingford, UK.

Fady, B. and Medail, F. (2004) Mediterranean forest systems. In: Burley, J., Evans, J. and Youngquist, J.A. (eds) *Encyclopedia of Forest Sciences*. Elsevier, Oxford, UK, pp. 1403–1414.

FAO (2010) *Global Forest Resources Assessment 2010*. Food and Agriculture Organization of the United Nations, Rome.

Forest Stewardship Council (2013) FSC Principles and Criteria: International Guidelines to forest management. FSC International Center, Bonn, Germany. Available at: https://ic.fsc.org/principles-and-criteria.34.htm (accessed 9 January 2014).

Forestry Commission (2013) *Safeguarding European Protected Species*. Forestry Commission England, Bristol, UK. Available at: http://www.forestry.gov.uk/england-protectedspecies (accessed 9 January 2013).

Grove, A.T. and Rackham, O. (2001) *The Nature of Mediterranean Europe: An Ecological History*. Yale University Press, New Haven, Connecticut.

Hansen, M.C., Potapov, P.V., Moore, R., Hancher, M., Turubanova, S.A., Tyukavina, A., Thau, D., Stehman, S.V., Goetz, S.J., Loveland, T.R. *et al.* (2013) High-resolution global maps of 21st-century forest cover change. *Science* 342, 850–853.

Hartel, T. and Plieninger, T. (eds) (2014) *European Wood-Pastures in Transition: A Social–Ecological Approach*. Earthscan from Routledge (imprint of Taylor & Francis), Abingdon, UK.

Henle, K., Alard, D., Clitherow, J., Cobb, P., Firbank, L., Kull, T., Mccracken, D., Moritz, R.F.A., Niemela, J., Rebane, M. *et al.* (2008) Identifying and managing the conflicts between agriculture and biodiversity conservation in Europe – a review. *Agriculture, Ecosystems and Environment* 124, 60–71.

Honnay, O., Verheyen, K., Bossuyt, B. and Hermy, M. (eds) (2004) *Forest Biodiversity – Lessons from History for Conservation.* CAB International, Wallingford, UK.

IUFRO (2013) 9.03.00 – Forest History and Traditional Knowledge: Activities and Events. International Union of Forest Research Organizations, Vienna. Available at: http://www.iufro.org/science/divisions/division-9/90000/90300/ (accessed 14 November 2014).

Kahle, H.-P., Karjalainen, T., Schuck, A., Agren, G.I., Kellomaki, S., Mellert, K., Prietzel, J., Rehfuess, K.-E. and Spiecker, H. (eds) (2008) *Causes and Consequences of Forest Growth Trends in Europe – Results of the Recognition Project.* European Forest Institute Research Report 21. Brill, Leiden, The Netherlands.

Kempeneers, P., Sedano, F., Seebach, L., Strobl, P. and San-Miguel-Ayanz, J. (2011) Data fusion of different spatial resolution remote sensing images applied to forest type mapping. *IEEE Transactions on Geoscience and Remote Sensing* 49, 4977–4986.

Kirby, K.J. and Watkins, C. (eds) (1998) *The Ecological History of European Forests.* CAB International, Wallingford, UK.

Lindner, M., Maroschek, M., Netherer, S., Kremer, A., Barbati, A., Garcia-Gonzalo, J., Seidl, R., Delzon, S., Corona, P., Kolström, M. *et al.* (2010) Climate change impacts, adaptive capacity, and vulnerability of European forest ecosystems. *Forest Ecology and Management* 259, 698–709.

Matthews, J.D. (1991) *Silvicultural Systems.* Oxford University Press, Oxford, UK.

MCPFE (2014) Ministerial Conferences. Forest Europe (The Ministerial Conference on the Protection of Forests in Europe, Liaison Unit, (currently) Madrid. Available at: www.foresteurope.org (accessed 9 January 2014).

Meeus, J.H.A., Wijermans, M.P. and Vroom, M.J. (1990) Agricultural landscapes in Europe and their transformation. *Landscape and Urban Planning* 18, 289–352.

Owende, P.M.O. (2004) Wood delivery. In: Burley, J., Evans, J. and Youngquist, J.A. (eds) *Encyclopedia of Forest Sciences.* Elsevier, Oxford, UK, pp. 269–279.

Packalen, P. and Maltamo, M. (2002) *Evaluation of the Suitability of ICP Level I Data to Support Forest Biodiversity Monitoring.* Joint Research Centre, European Commission, Brussels.

Päivinen, R., Lehikoinen, M., Schuck, A., Häme, T., Väätäinen, S., Kennedy, P. and Folving, S. (2001) *Combining Earth Observation Data and Forest Statistics.* EFI Research Report 14. Jointly published by European Forest Institute, Joensuu, Finland and Joint Research Centre, European Commission, Brussels.

Parrotta, J.A. and Trosper, R.L. (eds) (2012) *Traditional Forest-Related Knowledge: Sustaining Communities, Ecosystems and Biocultural Diversity.* Springer, New York.

Peck, T. (2001) *The International Timber Trade.* Woodhead Publishing, Cambridge, UK.

PEFC (2013) PEFC International Standards. PEFC (Programme for the Endorsement of Forest Certification) International, Geneva, Switzerland. Available at: http://www.pefc.org/standards/technical-documentation/pefc-international-standards-2010 (accessed 14 November 2014).

Peterken, G.F. (1977) Habitat conservation priorities in British and European woodland. *Biological Conservation* 11, 223–236.

Peterken, G.F. (1996) *Natural Woodland.* Cambridge University Press, Cambridge, UK.

Peterken, G.F. (2013) *Meadows.* British Wildlife Publishing, Gillingham, UK.

Polunin, O. and Walters, M. (1985) *A Guide to the Vegetation of Europe.* Oxford University Press, Oxford, UK.

Rackham, O. (1976) *Trees and Woodland in the British Landscape.* Dent, London.

Rackham, O. (2003) *Ancient Woodland: Its History, Vegetation and Uses in England*, revised edn. Castlepoint Press, Dalbeattie, UK.

Rodwell, J.S. (1991) *British Plant Communities: 1 Woodlands and Scrub.* Cambridge University Press, Cambridge, UK.

Rodwell, J.S., Schaminee, J., Mucina, L., Pignatti, S., Dring, J. and Moss, D. (2002) *The Diversity of European Vegetation: An Overview of Phytosociological Alliances and Their Relationships to EUNIS Habitats.* Ministerie van Landbouw, Natuurbeheer en Visserij/European Environment Agency, Wageningen, The Netherlands. Available at: http://forum.eionet.europa.eu/nrc-nature-and-biodiversity-interest-group/library/eunis_classification/eunis_meeting_2011/diversity_vegetationpdf/download/1/The%20Diversity%20of%20European%20Vegetation.pdf (accessed 14 November 2014).

Rotherham, I.D. (ed.) (2013) *Trees, Forested Landscapes and Grazing Animals.* Routledge, Abingdon, UK.

Saratsi, E., Bürgi, M., Johann, E., Kirby, K.J., Moreno, D. and Watkins, C. (eds) (2009) *Woodland Cultures in Time and Space.* Embryo Publications, Athens.

Savill, P.S., Evans, J., Auclair, D. and Falck, J. (1997) *Plantation Silviculture in Europe.* Clarendon Press, Oxford, UK.

Scarascia-Mugnozza, G., Oswald, H., Piussi, P. and Radoglou, K. (2000) Forests of the Mediterranean region: gaps in knowledge and research needs. *Forest Ecology and Management* 132, 97–109.

Schelhaas, M.-J., Nabuurs, G.-J. and Schuck, A. (2003) Natural disturbances in the European forests in the 19th and 20th centuries. *Global Change Biology* 9, 1620–1633.

Schneebeli, M. and Bebi, P. (2004) Snow and avalanche control. In: Burley, J., Evans, J. and Youngquist, J.A. (eds) *Encyclopedia of Forest Sciences*. Elsevier, Oxford, UK, pp. 397–402.

Schuck, A., Van Brusselen, J., Päivinen, R., Häme, T., Kennedy, P. and Folving, S. (2002) *Compilation of a Calibrated European Forest Map Derived from NOAA-AVHRR Data*. EFI Internal Report 13, European Forest Institute, Joensuu, Finland. Available at: http://www.efi.int/files/attachments/publications/ir_13_bw.pdf (accessed 14 November 2014).

Seidl, R., Schelhaas, M.-J. and Lexer, M.J. (2011) Unraveling the drivers of intensifying forest disturbance regimes in Europe. *Global Change Biology* 17, 2842–2852.

Tack, G. and Hermy, M. (1998) Historical ecology of woodlands in Flanders. In: Kirby, K.J. and Watkins, C. (eds) *The Ecological History of European Forests*. CAB International, Wallingford, UK, pp. 283–292.

Teplyakov, V.K., Kuzmichev, Y.P., Baumgartner, D.M. and Everett, R.L. (1998) *A History of Russian Forestry and Its Leaders*. Washington State University with Federal Forest Service of Russia and Pacific Northwest Station, USDA Forest Service, Pullman, Washington, Also available via DIANE Publishing, Darby, Pennsylvania, at. http://books.google.co.uk/books?id=QR7gGlXFWnYC&printsec=frontcover&redir_esc=y#v=onepage&q&f=false (accessed 14 November 2014).

Thomas, P.A. and McAlpine, R. (2010) *Fire in the Forest*. Cambridge University Press, Cambridge, UK.

Waller, M. (2013) Drought, disease, defoliation and death: forest pathogens as agents of past vegetation change. *Journal of Quaternary Science* 28, 336–342.

Ward Thompson, C. (2004) Forest amenity planning approaches. In: Burley, J., Evans, J. and Youngquist, J.A. (eds) *Encyclopedia of Forest Sciences*. Elsevier, Oxford, UK, pp. 478–486.

Williams, M. (2006) *Deforesting the Earth*. The University of Chicago Press, Chicago, Illinois.

Wulf, M. (2004). Relative importance of habitat quality and forestry continuity for the floristic composition of ancient, old and recent woodland. In: Honnay, O., Verheyen, K., Bossuyt, B. and Hermy, M. (eds) *Forest Biodiversity – Lessons from History for Conservation*. CAB International, Wallingford, UK, pp. 67–79.

2 Methods and Approaches in the Study of Woodland History

Charles Watkins*
School of Geography, University of Nottingham, Nottingham, UK

2.1 Introduction

There have been very significant advances in the study of forest and woodland history over the last 20 years. This chapter explores the wide range of methods and approaches to woodland history that have been used, with examples taken from across Europe. Interdisciplinary research undertaken by archaeologists, ecologists, geographers and historians is increasingly the norm and there has also been valuable international collaboration. Studies in woodland history span a vast range in time and space, with different methods providing varying degrees of precision in each dimension.

Much research draws on well-established methods developed in the 1970s and 1980s and associated with historical ecology; these include surveys of the flora and fauna and relating these to archaeological features, documentary and map evidence and field observation of the form of the trees themselves (Peterken, 1974; Rackham, 1976, 1980; Moreno, 1990; Watkins, 1990; Wulf, 1997) (Plates 3 and 4). Palaeoecological studies have tended to address regional changes over periods of hundreds or thousands of years, though smaller spatial scales and shorter time intervals are increasingly explored. Oral history may look at what has happened to a single site over the last few decades. The outcomes from these studies are found spread across ecological, historical, geographical and forestry journals as well as in the proceedings of conferences and symposia (Agnoletti and Anderson, 2000a,b; Honnay *et al.*, 2004; Saratsi *et al.*, 2009).

There has, moreover, been a massive increase in the ease of access and availability of documentary evidence, aerial photographs, landscape photographs, manuscripts and rare published material through digitization. This Web availability, and the associated speed of searching for relevant historical woodland documentation, has allowed an improvement in scholarly resolution. It has also improved the potential for searching for relevant research publications outside one's own immediate discipline.

A major advance in the molecular genetic analysis of trees has provided hitherto unknown understandings of the origin and movement of species over the last 20,000 years. The development of Lidar (light detection and ranging) has enabled the survey and analysis of archaeological features within woods and of the dynamic structural patterns of woodland. Advances in geographical information

---

*E-mail: charles.watkins@nottingham.ac.uk

systems (GIS) have improved the ability to map and analyse woodland change. Finally, there has been an increase in emphasis on place-based specific change and a realization that oral histories are vitally important for understanding woodland change in the 20th century.

2.2 Oral History

There has been increasing interest in gaining knowledge about the history of land use, agriculture and forestry through interviewing older people who have managed the land (Riley, 2004). George Ewart Evans (1956) pointed out that there is an imperative to ask older people about their experiences as 'once this knowledge is under the soil, no amount of digging will ever again recover it'. The value of oral histories and interviews that take place within the places and landscapes being discussed has been stressed by Fish *et al.* (2003) and Riley (2007). Ruth Tittensor (2009) used this approach to uncover the social and ecological history of the creation of around 6000 ha of coniferous forest at Whitelee near Glasgow, Scotland, in the second half of the 20th century.

Oral history is particularly useful in understanding the fine detail of how landscapes were used. Saratsi (2003) interviewed older residents in the Zagori region of northwestern Greece to examine how the woods were used for firewood, fodder and timber in the mid-20th century. In this region, there are many old pollarded or shredded trees growing along the edges of fields, now largely abandoned, or in narrow strips of woodland, known locally as 'kladera', which were especially used for the provision of winter fodder. Eight species of oak are found in the area and the interviews revealed that all were used in one way or other for feeding animals. If the deciduous oaks are cut in the summer, the branches retain their leaves when dried, which makes them easier to store. The different species of oak had characteristics that were well known to the people who fed the animals.

A 70-year-old man said 'Here we have Tzero (*Quercus cerris*), we have Drios (*Quercus frainetto*), we have Douskou (*Quercus robur*) we have Granitsa (*Quercus petraea*). … Tzero, we did not cut it very much because it crushes into bits easily, and was also a little sharp'. A 90-year-old woman recalled that the evergreen prickly oak was used differently: 'we did not store it indoors, we used to go even when it snowed…to bring a branch for the goats to eat'. The branches of other species of tree, which goats preferred, such as the field maple (*Acer campestre*), hop hornbeam (*Ostrya carpinifolia*) and hornbeam (*Carpinus betulus*) did not retain their leaves once cut, so were only used as fresh leaf fodder in the spring. The species most favoured for young goats were the flowering branches of the lime tree (*Tilia alba*), locally called lipanthia; a man from Micro Papigo noted that 'we used to cut it a lot but the problem was where could we find it'. The selection of the correct type of tree for pollarding and shredding of trees required detailed local knowledge.

In another study, Arvanitis (2011) used the oral history of the Psiloritis Mountains of central Crete to show the relative importance of different tree species for the provision of leaf fodder for shepherds (Fig. 2.1). The most important leaf fodder trees were the maple (*Acer criticum*) and the kermes oak (*Q. coccifera*), which are the most common trees growing in the Psiloritis Mountains. Other trees, such as *Q. ilex* and *Phillyrea latifolia* were not used because shepherds considered that their animals did not like to eat them.

One shepherd remembered that 'Many times when someone had a few goats, 100 animals, he could cut now and then kermes oak in the end of the summer, autumn, and that was good for the trees'. Indeed, the dominance of shrubby maples and kermes oaks in the area may be due to careful protection by shepherds over many centuries as it was in their interests to ensure that there was a regular supply of fodder. A different (78-year-old) shepherd pointed out that the best time to cut the foliage was linked with the life cycle of their stock: 'When the tree had both leaves and fruits, we cut it in order to fatten the young goats'. The kermes oak was shredded from August onwards, when there was much less grass, and especially in September and early October when the oak

Fig. 2.1. Browsed oak, Zaros, Crete, 2010.

branches bore acorns as well as leaves. Later, the acorns would be threshed from the trees to provide feed for the goats, and in October and November the acorns would fall down from the branches themselves, especially in heavy rain.

Leaf fodder in Crete, as in many parts of the Mediterranean, was an important supplement to grasses and other herb species, and still provides essential nourishment, especially towards the end of the long hot Cretan summers. Although the practice is dying out now, the testimony of retired shepherds and farmers illustrates the specialized knowledge and techniques used to make use of leaf fodder.

Modern foresters are beginning to realize that an understanding of these ancient practices may be of vital importance in the development of management strategies for the conservation of old trees and their associated habitats. Gimmi and Bürgi (2007) compared the value of oral history and old forest management plans to assess the extent of non-timber forest uses in the Swiss Rhone Valley. In the last decades of the 20th century, the woodland in this area has shown a decline in *Pinus sylvestris* and an increase in *Q. pubescens*. The management plans were poor at giving precise information about traditional practices, so oral histories were used to explore changes in management practices that might have encouraged these trends. Gimmi and Bürgi interviewed 12 people who had been involved in such practices, including the management of herds of goats and, more especially, the collection of forest litter for agricultural purposes.

The leaves and needles of trees were collected, mainly by women and children, with the use of rakes and either stored in heaps in the forest to be moved later, or taken straight away to cattle barns. Here, the litter was used in place of straw as a means of binding the manure together; the increasing use of indoor housing of cattle meant that this practice had become more common in the early 20th century. The interviews showed that litter from spruce, pine and birch was preferred, while larch litter decomposed too rapidly to be of much use. Sometimes, some of the upper layer of soil was carried away too. The 'continuous removal of the litter cover and parts of the upper soil led to excellent conditions for pine regeneration'. As well as regenerating well, pine was less damaged by browsing animals than oak, meaning that pines were favoured overall. Now that this practice has ceased, the oak trees are becoming more dominant. These insights have been helpful in devising the current management strategies for the area.

2.3 Photographs and Drawings

The greater availability through digitization of historical images such as photographs, postcards and topographical drawings means that these are increasingly used to explore changes in the extent of woodland. Some sites have been part of an established tourist itinerary for many years and so there are many surviving photographs.

Proctor *et al* (1980) used the large number of photographs taken by tourists and ecologists of Wistman's Wood on Dartmoor in south-western England to examine how it had changed since 1880. The scenes were located by reference to distinctive trees and rocks. The received wisdom that the woodland was fairly static was incorrect. Instead, it had undergone profound changes: its area had increased and individual trees, celebrated for their stunted and windblown appearance, were increasing in size and the younger growth showed more conventional stem forms. Wider landscape-scale changes can also be detected, as with Berto Giuffa's collection of 15,000 photographs, taken between 1933 and 1997, of the Val d'Aveto in the Italian Apennines (north-western Italy). Many photographs show the distribution of trees and woods before the abandonment of traditional agriculture in this area in the 1950s and 1960s, and the subsequent natural regeneration of secondary woodland (Gemignani, 2007).

Before photography became popular, people relied on sketching to capture their views of an area. Such 18th and 19th century topographical drawings can also be used to examine woodland change if they can be precisely dated and located. Piana *et al.* (2012) examined drawings by Elizabeth Fanshawe of the Ligurian coastline and mountains of north-western Italy. Figure 2.2, for example, shows the view looking down from the Bracco Pass to Sestri Levante and towards the Portofino. The drawing shows terraced slopes in the middle distance with olive groves, recognizable by the particular rounded

Fig. 2.2. Sestri from the ascent of the Bracca by Elizabeth Fanshawe, 19 November 1829.

shape of the trees. Behind this there are two hills, and the one to the right appears to have a group of evergreen trees above a small group of houses. Fieldwork has confirmed that evergreen oaks (*Q. ilex*) grow today on this site above the hamlet of Casa Ginestra. The foreground shows other distinct types of trees, including a prominent group of stone pines (*Pinus pinea*) with their characteristic umbrella shape. This is not native to the region and was planted for its valuable seed kernels, for shade and for its bark, which was ground to obtain the black tannin used to help preserve fishing nets.

Thus, topographical drawings can be a valuable addition to the range of resources (historical maps, land registers and present-day field surveys) available for the study of woodland history. The placement and localization of such topographical drawings is helped by the use of novel landscape visualization techniques which identify precise locations of specific views and features, as with Edward Lear's views of the English Lake District made on his tour of 1836 (Priestnall and Cowton, 2009).

2.4 Biological Indicators

Close examination of the form and distribution of individual trees and groups of trees can provide an understanding of the history of a particular site, as illustrated by Koop and Hilgen (1987) at the Forest of Fontainebleau in France, and by Tubbs (1986) for the New Forest in southern England (Plate 3). Repeated monitoring of tree and shrub species from permanent transects, for example at Lady Park Wood in the Forest of Dean, Monmouthshire, Wales (Peterken and Jones, 1987; Peterken and Mountford, 1997), creates a picture of forest dynamics that can be applied elsewhere and to other times.

Hæggström (2000) examined 84 stools of *Corylus avellana* on Nåtö Island, Åland Islands, south-west Finland. Nåtö Island means 'nut island' in Swedish and the trees have been regularly coppiced at least since the 19th century for the construction of hoops for salt herring barrels. The *C. avellana* coppice is mixed with pollarded *Fraxinus excelsior* and *Betula pubescens*. The girth of each stool was measured and cores were taken from a sample of 169 stems to determine the age of the stems and the stools. The oldest individual stem was 60 years. However, the 20 largest stools examined were estimated to be between 647 and 990 years old, with five stools being aged at over 900 years.

There has been a major expansion of studies of the links between tree and woodland continuity and the distribution of various groups of species (see also Siitonen and Ranius, Chapter 11; Hermy, Chapter 13). A study of click beetles in one of the largest Czech ancient oak and beech wood-pastures found, for example, that the majority of species sampled 'preferred solitary trees in fully sun-exposed habitats'. The study was undertaken at Lány Game Park, which has been a hunting park from the Middle Ages onwards, and emphasizes the importance of long-term openness in forests for the 'maintenance of high species richness of click beetles' and other organisms (Horák and Rébl, 2013). A study of flightless saproxylic weevils in the Weser-Ems region of north-west Germany confirmed that at least seven species were restricted to ancient woodland. Samples taken from litter in both ancient and recent woodland were supplemented with data from institutional and private collections made by entomologists in the 20th century. Suitable habitat in the form of many ground-level shoots following coppicing is thought to 'increase the available habitat for weevils' (Buse, 2012).

Research in Britain and Germany has shown a link between lichen richness and old woodland, and particularly old trees (Rose, 1988; Kirby et al., 1995). Manegold et al. (2009) examined the growth of mosses and lichens in the Feldberg region of the Black Forest (south-west Germany) in woodland of different origins: 'ancient', present for more than 300 years; 'old', formed between 1772 and 1900, and 'new', established after 1900. No moss species were restricted to ancient forests and generally the more recent woodland had the more diverse mosses. In contrast, ancient forests had a higher cover rate of crustose and fruticose lichens with larger thalli.

The timbers found in vernacular buildings have long been recognized as sources for the study of woodland history (Rackham, 1980), but more recently the lichens and mosses preserved on such timbers have been looked at as evidence of environmental and landscape change. Yahr et al. (2011) studied untreated roof timbers of 78 buildings in southern England. The epiphytes were often hidden by centuries of dust and cobwebs, but if the timbers were carefully swept with soft brushes it was possible to identify lichen fruit bodies and tiny moss stems. Epiphytes were present in over half of the buildings surveyed. This suggested a significant loss of epiphyte species during industrialization in southern England, with distinct regional signatures in the loss patterns influenced by the pollution regime and the extent of ancient woodland (Ellis et al., 2011).

2.5 Historical Records

The value placed on trees and woods and the products from them means that they regularly feature in historic documents from major surveys such as the Domesday Book in England (1086) to monastic accounts, records of land or timber sales, wills and disputes over land ownership (Rackham, 1980; Smout et al., 2005; Cevasco, 2008). Such documents provide a contemporary record of how woods and woodland management were perceived, although they need to be interpreted in context, preferably in conjunction with other sources, as for example Wright's (2003) use of historical records and archaeological evidence to examine the changing pattern of ancient woodland in eastern Hertfordshire, UK.

French monastic records indicate that specific contracts for wood cutting started to become common in the early 13th century. An example illustrating the division of woodland into sections is a sale by the Benedictines of Molesme, approved by Countess Blanche of Champagne in 1219, of a thousand arpents (400 ha) of woodland near Jeugny, south of Troyes (Keyser, 2009). The large-scale nature of this contract is shown by the fact that the two purchasers, Girard Judas and Guillaume de Vaudes, gained a 10-year lease and had to cut 100 arpents (40 ha) of woodland a year. In 1217, Blanche approved contracts where 'the countess sold cutting rights' in two small forests, or 'forestellas', over a 6-year period 'on condition that the merchants "cut each tree only once so that it grows back quickly"'. Another contract specified that 400 arpents (162 ha) would be cut over a 10-year period in the forest of Gault, and adjoining parcels had to be felled in sequence.

Restrictions on the grazing of freshly coppiced areas for a period of between 4 to 6 years after cutting were commonplace by the mid-13th century. In 1271, the Grand Jours of Troyes, the high court of Champagne, 'upheld against the community of Chaource a customary exclusion of pasturage (vaine-paturage) for 5 years after cutting', which allowed the woodland 'to defend itself'. This exclusion period was extended to 6 or 7 years for woodland on poor soils where the coppice regrowth was likely to be less rapid – an early development in sustainable woodland management.

Kottler *et al.* (2005) studied woodland change in the 20th century on a large private estate, Thoresby, in Sherwood Forest, English Midlands, using a combination of aerial photographs, oral history, historical maps, estate documents and field survey. This woodland landscape was dynamic and had undergone revolutionary changes through interventions by land agents and landowners in response to changing social, economic and government policy pressures. Further upheaval came about through the impact of outside interests: the creation of an army camp in 1942, the use of land as a military training area and the expansion of a coal mine and its spoil tip, which encroached on former heathland, ancient woodland and plantations.

Different species went rapidly in and out of fashion at Sherwood. In the early 20th century, birch was highly valued as an essential part of the picturesque landscape ensemble. However, grazing stock and rabbits halted regeneration and attempts were made to plant the species at Birklands. This proved unnecessary in the longer term and the cessation of grazing after the Second World War resulted in large scale natural regeneration of birch, particularly on the sandy soils disturbed by military manoeuvres. Indeed, the loss of large areas of wood-pasture and heathland meant that birch then came to be seen as a threat to the characteristic open, grazed landscape. The ancient oaks of Sherwood had been preserved until the 1940s for their historic and aesthetic interest, but large numbers were felled or burnt in response to the pressure to grow commercial timber in the postwar period. Under what was then seen as good forestry practice, large areas of *Quercus* and *Betula* woodland were replaced by plantations of *P. sylvestris* and *P. nigra*. Times change, and now the estate's management policy is to conserve all ancient trees on the estate and the pine plantations have come to be at risk from disease.

2.6 Preserved Wood and Dendrochronology

The archaeological study of preserved wood has increased knowledge both about the use of wood in the past and about the forests from which it was obtained. There have been great advances in techniques for conserving wood once it has been excavated and in interpreting aspects of woodland history (Brennan and Taylor, 2003; Taylor, 2010). In the Somerset levels in south-west England, considerable quantities of wood were discovered in the 1970s, including ash planks which had been split from large trees. These and other finds were identified as parts of a Neolithic trackway, built as a raised path across a reed swamp (Coles and Coles, 1986). The wood used gives an indication of trees growing in the Somerset Levels in the Neolithic period: mainly oak (*Quercus* spp.), elm (*Ulmus* spp.), lime (*Tilia* spp.) and ash (*Fraxinus excelsior*), with an undergrowth of hazel (*Corylus avellana*) and holly (*Ilex aquifolium*), and some alder (*Alnus glutinosa*), willow (*Salix* spp.) and poplar (*Populus* spp.). The lime trees, cut to produce long and straight planks, must have been tall and straight, with few side branches, which suggests that they had grown fairly close together in dense woodland.

The archaeologists were impressed with the quality of woodworking skills indicated by the finds. Oak was the most frequent species used for planks and 'the trunks were converted into planks by splitting with wedges, either of stone or seasoned oak'. Most of 'the splits were radial, exploiting the tendency of oak to split along its rays', but smaller oak trunks were cut 'at right angles to the rays, more or less around the rings'. Some of the large oak planks indicated that the oak trunks used were up to 5 m long and 1 m in diameter. Dendrochronological studies indicated that most of the oak and ash found in the track had been felled in the same year, although the tree ages ranged from 400 years down to just over 100 years. Large numbers of the hazel rods used were 7 years old, indicating that areas of hazel were being deliberately coppiced for their production.

Six similar Neolithic trackways have been investigated at Campemoor in north-west Germany that are dated to 5000–3000 BC. One, dated to 2925 BC, consisted of very well-preserved pine timbers in the context of 'natural sub-fossil tree remains of a surrounding pine forest' (Leuschner et al., 2007). Comparison of the tree rings in the pines used to construct the track and those in the surrounding wood indicated that the building of the track and the death of standing trees were both associated with a change in climate to generally wetter conditions.

Trees were frequently used to mark borders, routes and property boundaries, and references to boundary trees in early medieval charters and monastic records are a useful source (Hooke, 2010; Balzaretti, 2013). The trees themselves might be marked and the marks may still be visible. Östlund et al. (2002) argue that the boreal forests of Scandinavia are an archive of trees that were modified by 'ancient ethnic groups' for a variety of purposes until the end of the 19th century and that culturally modified trees can provide insights into past land-use patterns. A star-shaped symbol discovered on a waterlogged tree at Čelákovice in the Czech Republic was recently dated to the 6th to 9th centuries AD (Dreslerová and Mikulás, 2010). The tree trunk was found when a large number of oak tree trunks were being excavated from a gravel pit on the banks of the River Elbe. The carving 'had been cut into the tree trunk but its image survived in relief that was carried on the new growth which formed as the tree repaired its wound and filled in the original cuts'. The precarious survival of the mark is remarkable and its meaning remains obscure, but it was most probably used to mark a boundary or route.

Dendrochronology can be used to examine aspects of the history of stands of living and dead trees, as in the case of the disturbance history of a rare old-growth stand dominated by *Fagus sylvatica* with *Picea abies* and *Abies alba* growing on the southern slopes of Dürrenstein in the northern limestone alps of Lower Austria (Splechtna et al., 2005). Samples were taken from four 100 m^2 permanent sample plots and two cores from bark to pith were taken from 'non-suppressed' trees. The results suggested that the disturbance history of the forest studied was characterized 'by a relatively low-severity disturbance regime during the 19th century, but higher severity in the 18th century and a strong increase in the 20th century'. The frequent periods of establishment of young trees were correlated with powerful storms that affected the area in 1966, 1976 and 1990, and the results suggested that episodic disturbances were driving the dynamics of the gap-phase forest.

A study of the ancient oaks of Sherwood Forest used dendrochronology to estimate the date of death of standing dead trees and the likely longevity of surviving ancient trees (Watkins et al., 2004). Previous work had shown that around 35% of the cohort of ancient oaks in the study area at Buck Gates was dead, as were large sections of many living oaks. Accurate tree ring measurements could be made from 80 samples and these measurements were related to the East Midland Tree-ring Chronology. The earliest measured ring was ascribed to the year 1415 and 43% of the trees contained wood dated back to 1765 or earlier. The great majority of the dead trees still detectable in the study area had died within the previous 150 years, which implies that trees which had died more than 150 years ago are likely to be completely rotted or have been removed. Such dendrochronological analysis provides useful data for managers who wish to maintain a succession of age classes of oaks to replace the current cohorts of veteran trees.

Ancient trees are increasingly being valued as repositories of data on environmental conditions. The examination of tree rings provides indications of fluctuations in temperature and likely regional histories of climate change or human occupation. At Loch Katrine, Mills *et al.* (2013) cored ancient trees in the wood-pastures on the south eastern shores of the lake. No one now lives in the area, but it was once important for cattle and sheep farming, and its significance for hunting was captured in Walter Scott's poem *The Lady of the Lake* (1810). The cores from two ash trees showed that both were of late 17th century origin and that one had been regularly pollarded in the 18th century. Several of the alder and oak trees from the 19th century had originally been coppiced, but the coppice regrowth had then fused to form what appeared to be a single stem.

2.7 Lidar and GIS

Lidar, or airborne laser scanning, has been increasingly used for examining landscape topography. It allows the precise mapping of small variations in height and so picks up features such as minor earthworks, ditches, terraces, charcoal platforms and ridges and furrows. In woodland areas, it can penetrate the woodland canopy to give a picture of the ground surface under the trees, which is difficult if not impossible using conventional archaeological surveys and aerial photographs.

Devereux *et al.* (2005) used Lidar at Welshbury Hill in the Forest of Dean, a hill fort which has three distinct ditches and banks, surrounded by an area characterized by hut and charcoal platforms and a network of boundaries from a possible Bronze Age field system. All are now covered by trees, but the application of a 'vegetation-removal algorithm' to the Lidar output allowed the creation of a digital elevation model which revealed small and subtle linear features such as field system boundaries and charcoal platforms.

In Sherwood Forest, Lidar has been used to identify and map a small site known as Thynghowe, which consists of a mound about 0.7 m high and 8 m in diameter where there are also three parish boundary stones. There are also references to the site as Thynghowe in 1334 and 1608, which may point to a mid-10th century origin as a Norse 'thing' or meeting place under the Danelaw (Gaunt, 2011; Mallett *et al.*, 2012).

Crutchley (2009) compared Lidar results with those from standard aerial photography taken over many years in Savernake Forest, southern England. A large number of features which had previously not been recorded were identified, although several of these features were visible on the aerial photographs, but had been missed. The Lidar results also picked up previously unknown sites, including a possible Iron Age temple complex.

The spread of GIS has made it a lot easier to bring together historical maps, cadastral surveys, land registers and aerial photographs to quantify changes in landscape over the last 200 years. Bender *et al.* (2005), for example, analysed changes in the landscape of southern Germany – in Brotjacklriegel in Bavaria and in Wüstenstein in the Franconian Alb. Both studies confirmed a substantial increase in woodland between 1850 and 2000. In Bavaria, the decline in agriculture and rural depopulation has been combined with commercial afforestation and the spread by natural regeneration of semi-natural woods. In the Franconian Alb, the long-established process of afforesting areas of rough pasture with pine forests has more recently been combined with the afforestation of meadows in the valleys; the latter is a threat to biodiversity because these open grasslands are ecologically very rich.

2.8 Applying Archaeological Insights to Ecological Issues

Dambrine *et al.* (2007) pointed out that the lack of evidence from artefacts brought to the surface by ploughing and the greater difficulty of interpreting aerial photographs had led to the 'questionable impression that ancient forests had never been cultivated before the historical period'. For example, the large forest of Tronçais in Berry, France, is characterized by extensive stands of *Q. petraea* on acid sandy soils. The earliest map of 1665 shows the forest to have similar boundaries to today's forest, though in the 17th century much of the forest consisted of wood-pasture which was only converted to high forest in the 18th century. However, a detailed archaeological survey identified 108 Roman settlements dating from the first to the fourth centuries AD, highlighting that the area must have been much more open during that period. Three large sites and seven smaller sites, all today consisting of even-aged high forest dominated by *Q. petraea* with *C. betulus*, were used as case studies. The vegetation and soils of these sites were surveyed and sampled. The results showed that forest composition was still strongly influenced by Roman cultivation over 1500 years since the settlements were abandoned: this impact had not been disguised by several centuries of intensive woodland management. The conclusion was that 'any type of agriculture involving long-term fertility transfers and accumulation' can create long-lasting changes because of the 'conservative character of forest biogeochemical cycles'.

Verheyen *et al.* (2001) provide another example of why human land use influences have to be taken into account when considering present-day woodland ecology. Their study site was of Ename Wood in Flanders, a 62 ha remnant of a larger 145 ha wood which was owned from 1278 to 1795 by the Abbey of Ename. Most of the wood is on sandy loams with a ground flora characterized by *Hyacynthoides non-scripta*. The land-use changes between 1278 and 1990 were quantified using historical documents and maps, and 18 land-use maps were constructed and analysed with the help of GIS. Parts of the wood were managed as coppice with standards for hundreds of years but much of it has more recently been managed as high forest. In 1278, around a third of the wood was used for wood production and the remainder was wood-pasture. Through the centuries, the wood-pasture was converted to either arable land or woodland and this polarization increased in the second half of the 18th century. Verheyen *et al.* (2001) took soil samples to examine its chemical properties in sections of the wood that had had different land-use histories. The impact of pastoral land use hundreds of years previously remained evident in terms of the depletion in soil potassium, magnesium and calcium.

2.9 Pollen and Charcoal Analysis

The precise analysis of pollen and charcoal remains has demonstrated the importance of humans in affecting woodland change over thousands of years. For example, Hannon *et al.* (2000) studied the pollen, macrofossils and charcoal found in 130 cm of sediment cored from a small, wet hollow in the wood of Suserop Skov on the island of Sjælland, in eastern Denmark. From this, they reconstructed the history of the stand over 6000 years.

Five phases were identified in the vegetation history of the site. From 4200 to 3200 BC, the area was characterized by *Tilia*, *Quercus*, *Ulmus* and *Corylus* forest as represented in pollen and macrofossils. There was evidence of fire, but this was probably of natural origin. In the second phase, 3200–2700 BC, there was a decline in *Ulmus* and *Pinus* pollen with increased values for *Corylus*, *Fraxinus* and *Acer*, interpreted as possibly linked to an increase in cattle browsing. The third phase, of *Tilia*, *Quercus*, *Pine* forest (2700–600 BC) showed a decline in woody diversity and the reappearance of fire, now linked to human agricultural activities, while the fourth phase (600 BC–900 AD) was characterized by a remarkable decline in *Tilia* and a sustained increase in grass and cereal pollen. The evidence suggests that the site was now surrounded by open ground with scattered trees and the area was heavily disturbed by humans. The final phase (900 AD–present) showed the establishment of *Fagus* and *Quercus* forest with *F. sylvatica* as the shade-tolerant dominant species. Overall, the loss of *Tilia cordata* and *T. platyphyllos* around 2500 years ago was the most significant woodland change and this was probably brought about by human activity, although the decline is contemporaneous with an abrupt change in the climate. A similar pattern of loss of *Tilia* and later increase in *Fagus* was found at another site in eastern Denmark at Gribskov (Overballe-Petersen *et al.*, 2012).

Robin *et al.* (2014) used soil charcoal in the eastern part of the Harz Mountains of northern Germany to study the occurrence of fires and examine the development of forest. They examined a minimum of four vertical sequences from soil profiles or trenches at three field sites. All charcoal more than 1 mm in size was extracted from the soil samples, woody species were identified and 24 single charcoal pieces were radiocarbon dated.

Despite the 'stratigraphical, chronological and taxonomical limits of soil charcoal analysis', the results identified 'previously undocumented fire events and the related burnt forest types'. Formerly, the earliest human impact on the lower parts of the mountain range was mining activity in *c*.700–800 AD, but this study indicated that humans had a significant effect on the landscape through mid-Holocene fires probably associated with a Neolithic settlement 10 km away which contained much oak that had been used for construction and fuel.

Another source of information can be plant macrofossils such as fruits, needles and bud scales, but these are often only found in low concentrations. Ammann *et al.* (2014) demonstrated the value of the analysis of stomata derived from conifer needles as a proxy for the local presence of conifers from a study of 13 sites on a transect across the Alps. They concluded that, when considered together with pollen analysis, 'stomata and plant macrofossils may contribute to a better assessment of the times of apparent establishment, expansion, decline and extinction' of conifer species.

2.10 Genetic and Molecular Markers

Advances in knowledge and the interpretation of genetic and molecular markers have enabled the use of such markers to be among the latest techniques to be applied to unravelling woodland history (see also Kirby and Watkins, Chapter 3). Analysis of samples taken from the English elm (*U. procera*) and the wych elm (*U. glabra*) (Gil *et al.*, 2004) in Spain, France, Greece, Italy and Britain established the variability of chloroplast DNA and likely genetic lineages. A clone of *U. procera* common to Italy, Spain and Britain was identified. This elm only rarely produces seed, but spreads by suckers very well and was once one of the commonest hedgerow trees in England. It had long been thought that the tree was introduced and the genetic research confirmed that the elm originated in Italy. It was most likely introduced by the Romans as the clone is the same as that of the Atinian elm from Latium recommended by Columella (*c*.50 AD) in *De Re Rusticus* as vine supports.

Research on *Castanea sativa* also confirms the classical understanding of its history. Mattioni *et al.* (2009) found three 'distinct gene pools' for the European *C. sativa*: one for Italy and Spain, one for Greece and West Turkey, and one for East Turkey. They conclude that the Turkish region was 'the centre of origin of the European chestnut', which ties in well with the belief of the Greeks and the Romans that the chestnut 'came to Europe from the East' (Meiggs, 1982).

Vendramin *et al.* (2008) carried out genetic analysis of *P. pinea*, which is widespread across Mediterranean Europe from Spain through to Lebanon. It is a genetically depauperate species and the work of these authors raises questions about the belief that 'genetic diversity is essential and worthy of conservation'. The study of 34 populations, using 12 paternally inherited chloroplast microsatellites, found 'an almost complete lack of chloroplast-microsatellite variation across the full species range'. Vendramin *et al.* consider that *Pinus pinea* is 'truly exceptional among widespread, sexually reproducing plant species for its low level of genetic diversity'. Perhaps at some stage in its history the species suffered a wide-ranging decline and became limited to a single, geographically circumscribed, population. This prolonged demographic bottleneck may have been exacerbated by the seed dispersal mechanism of this species, which contrary to most pines, depends on mammals and birds rather than wind. The 'scarcity of seed dispersers during critical periods of the species history' may then have reduced the ability of the tree to move to new territories and made it 'more susceptible to range contractions', eventually influencing its genetic diversity. The subsequent expansion of the species began about 3000 years ago when humans began to plant and cultivate the tree because of the value of its pine nuts. Puglisi *et al.* (2000) examined the genetic structure of nine 'natural' populations of this pine growing in Apulia, Calabria and Basilicata in Italy. The data from this analysis was combined with charcoal data from archaeological sites and it was concluded that there was a strong possibility that these populations of trees, which had a higher genetic variation than other Italian populations, could have been introduced by Greek colonists to southern Italy.

Human influence on species distribution was also investigated by Cottrell *et al.* (2005) in their major study of the postglacial migration of *Populus nigra* using chloroplast DNA. They concluded that there were two glacial refugia for this species, one in Spain and one in Italy and the Balkans. The Pyrenees formed a significant barrier to the spread of the species as very few of the haplotypes found in Spain are found elsewhere in Europe. However, haplotypes from both the eastern and western refugia are found in Britain. There are three clones from the western refugia and 20 from the east, which suggests that the three clones from the western refugia may have been introduced by humans; *P. nigra* is very readily grown from small cuttings and can easily be transported long distances.

2.11 Conclusion

Both well-established and novel techniques are being applied to the study of woodland history, from molecular genetics to remote sensing. These have transformed our understanding of the historical geography of the distribution and spread of tree species and their interrelationship with human activities. Advances in woodland archaeology demonstrate the complex interplay between humans and woodland.

There must still be care though in the interpretation of evidence and results. Wood and charcoal from archaeological sites may reflect selection of particular species, not the composition of former woodland, and pollen analyses have in the main not been particularly helpful in identifying woodland management. The increase in the amount of historical ecological research allows for valuable national and regional comparisons but, at the same time, an increased awareness of the distinctive histories of individual sites and species. The changing fortunes of woods and forests, and the different ways in which they are treated, must be understood in the context of particular places, societies and periods.

References

Agnoletti, M. and Anderson, S. (eds) (2000a) *Forest History. International Studies on Socio-Economic and Forest Ecosystem Change*. IUFRO Research Series 2, Report No. 2 of the IUFRO Task Force on Environmental Change. CAB International, Wallingford, UK in association with International Union of Forestry Research Organizations, Vienna.

Agnoletti, M. and Anderson, S. (eds) (2000b) *Methods and Approaches in Forest History*. IUFRO Research Series 3, Report No. 3 of the IUFRO Task Force on Environmental Change. CAB International, Wallingford, UK in association with International Union of Forestry Research Organizations, Vienna.

Ammann, B., van der Knaap, W.O., Lang, G., Gaillard, M.-J., Kaltenrieder, P., Rösch, M., Finsinger, W., Wright, H.E. and Tinner, W. (2014) The potential of stomata analysis in conifers to estimate presence of conifer trees: examples from the Alps. *Vegetation History and Archaeobotany* 23, 249–264.

Arvanitis, P. (2011) Traditional forest management in Psiloritis, Crete, c.1850–2011: integrating archives, oral history and GIS. PhD thesis, University of Nottingham, Nottingham, UK.

Balzaretti, R. (2013) *Dark Age Liguria*. Bloomsbury, London.

Bender, O., Boehmer, H.J., Jens, D. and Schumacher, K.P. (2005) Using GIS to analyse long-term cultural landscape change in Southern Germany. *Landscape and Urban Planning* 70, 111–125.

Brennan, M. and Taylor, M. (2003) The survey and excavation of a Bronze Age timber circle at Holme-next-the-Sea, Norfolk. *Proceedings of the Prehistoric Society* 69, 1–84.

Buse, J. (2012) 'Ghosts of the past': flightless saproxylic weevils (Coleoptera: Curculionidae) are relict species in ancient woodlands. *Journal of Insect Conservation* 16, 93–102.

Cevasco, R. (2008) *Memoria Verde. Nuovi Spazi per la Geografia*. Diabasis, Reggio Emilia, Italy.

Coles, J. and Coles, B. (1986) *Sweet Track to Glastonbury: The Somerset Levels in Prehistory*. Thames and Hudson, London.

Cottrell, J.E., Krystufek, V., Tabbener, H.E., Milner, A.D., Connolly, T., Sing, L., Fluch, S., Burg, K., Lefèvre, F., Achard, P. *et al.* (2005) Postglacial migration of *Populus nigra* L.: lessons learnt from chloroplast DNA. *Forest Ecology and Management* 206, 71–90.

Crutchley, S. (2009) Ancient and modern: combining different remote sensing techniques to interpret historic landscapes. *Journal of Cultural Heritage* 10(Supplement 1), e65–e71.

Dambrine, E., Dupouey, J.-L., Laüt, L., Humbert, L., Thinon, M., Beaufils, T. and Richard, H. (2007) Present forest biodiversity patterns in France related to former Roman agriculture. *Ecology* 88, 1430–1439.

Devereux, B.J., Amable, P., Crow, P. and Cliff, A.D. (2005) The potential of airborne Lidar for detection of archaeological features under woodland canopies. *Antiquity* 79, 648–660.

Dreslerová, D. and Mikulás, R. (2010) An early medieval symbol carved on a tree trunk: pathfinder or territorial marker? *Antiquity* 84, 1067–1075.

Ellis, C.J., Yahr, R. and Coppins, B.J. (2011) Archaeobotanical evidence for a massive loss of epiphyte species richness during industrialisation in southern England. *Proceedings of the Royal Society B: Biological Sciences* 278, 3482–3489.

Evans, G.E. (1956) *Ask the Fellows Who Cut the Hay*. Routledge, London.

Fish, R., Seymour, S. and Watkins, C. (2003). Conserving English landscapes. Land managers and agri-environmental policy. *Environment and Planning A* 35, 19–41.

Gaunt, A. (2011) *A Topographic Earthwork Survey of Thynghowe. Hanger Hill, Nottinghamshire*. Publication NCA-016, Nottinghamshire Community Archaeology, Nottinghamshire County Council, Nottingham, UK.

Gemignani, C.A. (2007) *Adalberto 'Berto' Giuffra un Fotografo di Montagne*. Comune di Santa Stefano d'Aveto, Genova, Italy.

Gil, L., Fuentes-Utrilla, P., Soto, Á., Cervera, M. and Collada, C. (2004) English elm is a 2,000-year-old Roman clone. *Nature* 431, 1053.

Gimmi, U. and Bürgi, M. (2007) Using oral history and forest management plans to reconstruct traditional non-timber forest uses in the Swiss Rhone Valley (Valais) since the late nineteenth century. *Environment and History* 13, 211–246.

Hæggström, C.-A. (2000) The age and size of hazel (*Corylus avellana* L.) stools of Nåtö Island, Åland Islands, SW Finland. In: Agnoletti, M. and Anderson, S. (eds) *Methods and Approaches in Forest History*. IUFRO Research Series 3, Report No. 3 of the IUFRO Task Force on Environmental Change. CAB International, Wallingford, UK in association with International Union of Forestry Research Organizations, Vienna, pp. 67–77.

Hannon, G.E., Bradshaw, R. and Emborg, J. (2000) 6000 years of forest dynamics in Suserup Skov, a semi-natural Danish woodland. *Global Ecology and Biogeography* 9, 101–114.

Honnay, O., Verheyen, K., Bossuyt, B. and Hermy, M. (eds) (2004) *Forest Biodiversity: Lessons from History for Conservation*. IUFRO Research Series 10. CAB International, Wallingford, UK in association with International Union of Forestry Research Organizations, Vienna.

Hooke, D. (2010) *Trees in Anglo-Saxon England*, Boydell Press, Woodbridge, UK.

Horák, J. and Rébl, K. (2013) The species richness of click beetles in ancient pasture woodland benefits from a high level of sun exposure. *Journal of Insect Conservation* 17, 307–318.

Keyser, R. (2009) The transformation of traditional woodland management: commercial sylviculture in medieval Champagne. *French Historical Studies* 32, 353–384.

Kirby, K.J., Thomas, R.C., Key, R.S., Mclean, I.F.G. and Hodgetts, N. (1995) Pasture-woodland and its conservation in Britain. *Biological Journal of the Linnean Society* 56(Supplement s1), 135–153.

Koop, H. and Hilgen, P. (1987) Forest dynamics and regeneration mosaic shifts in unexploited beech (*Fagus sylvatica*) stands at Fontainebleau (France). *Forest Ecology and Management* 20, 135–150.

Kottler, D., Watkins, C. and Lavers, C. (2005) The transformation of Sherwood Forest in the twentieth century: the role of private estate forestry. *Rural History* 16, 95–110.

Leuschner, H.H., Bauerochse, A. and Metzler, A. (2007) Environmental change, bog history and human impact around 2900 BC in NW Germany – preliminary results from a dendroecological study of a sub-fossil pine woodland at Campemoor, Dummer Basin. *Vegetation History and Archaeobotany* 16, 183–195.

Mallett, L., Reddish, S., Baker, J., Brookes, S. and Gaunt, A. (2012) Community archaeology at Thynghowe, Birklands, Sherwood Forest. *Transactions of the Thoroton Society of Nottinghamshire* 116, 53–71.

Manegold, M., Pfab, H. and Bogenrieder, A. (2009) Indicator value of lichens and mosses for land use history in mountainous ancient woodlands. In: Saratsi, E., Bürgi, M., Johann, E., Kirby, K.J., Moreno, D. and Watkins, C. (eds) *Woodland Cultures in Time and Space – Tales from the Past, Messages for the Future. Proceedings of Conference in Thessaloniki, Greece in 2007, Organized by IUFRO Division 6.07.04*. Embryo Publications, Athens, pp. 178–186.

Mattioni, C., Cherubini, M., Taurchini, D., Villani, F. and Martin, M.A. (2009) Genetic diversity in European chestnut populations. In: Bounous, G. (Convenor) *Proceedings of the 1st European Congress on Chestnut,*

Castanea 2009, Cuneo-Torino, Italy, October 13–16, 2009. International Society for Horticultural Science, Leuven, Belgium. *Acta Horticulturae* 866, 163–167.

Meiggs, R. (1982) *Trees and Timber in the Ancient Mediterranean World*. Oxford University Press, Oxford, UK.

Mills, C., Quelch, P. and Stewart, M. (2013) Historic woodland survey at South Loch Katrine. In: *Managing the Historic Environment Case Study*. Forestry Commission Scotland, Edinburgh, UK, pp. 16–22. Available at: http://scotland.forestry.gov.uk/images/corporate/pdf/Managing-the-historic-environment-case-studies.pdf (accessed 20 November 2014).

Moreno, D. (1990) *Dal Documento al Terreno. Storia e Archeologia dei Sistemi Agro-silvo-pastorali*. Il Mulino, Bologna, Italy.

Östlund, L., Zackrisson, O. and Hörnberg, G. (2002) Trees on the border between nature and culture: culturally modified trees in boreal Sweden. *Environmental History* 7, 48–68.

Overballe-Petersen, M.V., Nielsen, A.B., Hannon, G.E., Halsall, K. and Bradshaw, R.H. (2012) Long-term forest dynamics at Gribskov, eastern Denmark with early-Holocene evidence for thermophilous broadleaved tree species. *The Holocene* 23, 243–254.

Peterken, G.F. (1974) A method for assessing woodland flora for conservation using indicator species. *Biological Conservation* 6, 239–245.

Peterken, G.F. and Jones, E.W. (1987) Forty years of change in Lady Park Wood: the old-growth stands. *Journal of Ecology* 75, 477–512.

Peterken, G.F. and Mountford, E.P. (1997) Effects of drought on beech in Lady Park Wood, an unmanaged mixed deciduous woodland. *Forestry* 69, 125–136.

Piana, P., Balzaretti, R., Moreno, D. and Watkins, C. (2012) Topographical art and landscape history: Elizabeth Fanshawe (1779–1856) in early nineteenth-century Liguria. *Landscape History* 33, 65–82.

Priestnall, G. and Cowton, J. (2009) Putting landscape drawings in their place: virtual tours in an exhibition context. In: *E-Science Workshops, 2009 5th IEEE International Conference on 9–11 Dec. 2009*. Institute of Electrical and Electronics Engineers, Washington, DC. Available at: http://ieeexplore.ieee.org/stamp/stamp.jsp?tp=andarnumber=5407975 (accessed 17 November 2014).

Proctor, M.C.F., Spooner, G.M. and Spooner, M.F. (1980) Changes in Wistman's Wood, Dartmoor: photographic and other evidence. *Reports and Transactions of the Devonshire Association for the Advancement of Science* 112, 43–79.

Puglisi, S., Fiorentino, G., Lovreglio, R. and Leone, V. (2000) Integration between genetic and archaeobotanical data in a study on the evolutionary history of *Pinus halepensis* Mill. populations in southern Italy. In: Agnoletti, M. and Anderson, S. (eds) *Methods and Approaches in Forest History*. IUFRO Research Series 3, Report No. 3 of the IUFRO Task Force on Environmental Change. CAB International, Wallingford, UK in association with International Union of Forestry Research Organizations, Vienna, pp. 221–231.

Rackham, O. (1976) *Trees and Woodland in the British Landscape*. Dent, London.

Rackham, O. (1980) *Ancient Woodland: Its History, Vegetation and Uses in England*. Arnold, London.

Riley, M. (2004) Ask the fellows who cut the hay: farm practices, oral history and nature conservation. *Oral History* 32, 42–51.

Riley, M. (2007) Reconsidering conceptualisations of farm conservation activity: the case of conserving hay meadows. *Journal of Rural Studies* 22, 337–353.

Robin, V., Bork, H.-R., Nadeau, M.-J. and Nelle, O. (2014) Fire and forest history of central European low mountain forest sites based on soil charcoal analysis: the case of the eastern Harz. *The Holocene* 24, 35–47.

Rose, F. (1988) Phytogeographical and ecological aspects of *Lobarion* communities in Europe. *Botanical Journal of the Linnean Society* 96, 69–79.

Saratsi, E. (2003) Landscape history and traditional management practices in the Pindos Mountains, Northwest Greece, c.1850–2000. PhD thesis, University of Nottingham, Nottingham, UK.

Saratsi, E., Bürgi, M., Johann, E., Kirby, K.J., Moreno, D. and Watkins, C. (eds) (2009) *Woodland Cultures in Time and Space – Tales from the Past, Messages for the Future. Proceedings of Conference in Thessaloniki, Greece in 2007, Organized by IUFRO Division 6.07.04*. Embryo Publications, Athens.

Smout, T.C., MacDonald, A.R. and Watson, F. (2005) *A History of the Native Woodlands of Scotland 1500–1920*. Edinburgh University Press, Edinburgh, UK.

Splechtna, B.E., Gratzer, G. and Black, B.A. (2005) Disturbance history of a European old-growth mixed-species forest – a spatial dendro-ecological analysis. *Journal of Vegetation Science* 16, 511–522.

Taylor, M. (2010) Big trees and monumental timbers. In: Pryor, F. and Bamforth, M. (eds) *Flag Fen, Peterborough: Excavation and Research 1995–2007*. Oxbow Books, Oxford, UK, pp. 90–97.

Tittensor, R. (2009) *From Peat Bog to Conifer Forest. An Oral History of Whitelee, Its Community and Landscape*. Packard, Chichester, UK.

Tubbs, C. (1986) *The New Forest*. Collins, London.

Vendramin, G.G., Fady, B., González-Martínez, S.C., Hu, F.S., Scotti, I., Sebastiani, F., Soto, Á. and Petit, R.J. (2008) Genetically depauperate but widespread: the case of an emblematic Mediterranean pine. *Evolution* 62, 680–688.

Verheyen, K., Bossuyt, B., Hermy, M. and Tack, G. (2001) The land use history (1278–1990) of a mixed hardwood forest in western Belgium and its relationship with chemical soil characteristics. *Journal of Biogeography* 26, 1115–1128.

Watkins, C. (1990) *Woodland Management and Conservation*. David and Charles, Newton Abbot, UK.

Watkins, C., Lavers, C. and Howard, R. (2004) The use of dendrochronology to evaluate dead wood habitats and management priorities for the ancient oaks of Sherwood Forest. In: Honnay, O., Verheyen, K., Bossuyt, B. and M. Harmy, M. (eds) *Forest Biodiversity: Lessons from History for Conservation*. IUFRO Research Series 10. CAB International, Wallingford, UK in association with International Union of Forestry Research Organizations, Vienna, pp. 247–268.

Wright, L.W. (2003) Woodland continuity and change: ancient woodland in eastern Hertfordshire. *Landscape History* 25, 67–78.

Wulf, M. (1997) Plant species as indicators of ancient woodland in north-western Germany. *Journal of Vegetation Science* 8, 635–642.

Yahr, R., Coppins, B.J. and Ellis, C.J. (2011) Preserved epiphytes as an archaeological resource in post-medieval vernacular buildings. *Journal of Archaeological Science* 38, 1191–1198.

3 The Forest Landscape Before Farming

Keith J. Kirby[1]* and Charles Watkins[2]
[1]*Department of Plant Sciences, University of Oxford, Oxford, UK;*
[2]*School of Geography, University of Nottingham, Nottingham, UK*

3.1 Where to Begin?

Trees have spread back and forth across Europe many times, including species that we now think of as quite exotic (Watts, 1988). By the first half of the Pleistocene (about 2 million years ago), the flora was much more like that of today but, even so, non-native genera such as *Tsuga* and *Pterocarya* turn up in British pollen samples (Ingrouille, 1995).

During the Pleistocene era, there was a sequence of warm and cold phases. At the start of a warming period much of the landscape would have been composed of young, immature soils disturbed by periods of freezing and thawing, and supporting low shrub and herb communities with arctic–alpine species (cryocratic phase). Later, both vegetation and soils developed (protocratic phase) to the point where, through central Europe, deciduous and mixed tree cover would be expected (mesocratic phase). With subsequent climatic cooling, and further soil leaching and podzol development, there might be a shift towards heath or moorland development, or towards more conifer-dominated landscapes (telocratic phase) (Watts, 1988). As conditions became colder, forest species were restricted to 'refugia', such as in southern Europe, until the next interglacial period.

However, each period of warming and cooling was different and vegetation development and faunal spread did not follow the same fixed track each time. Stewart (2008), for example, found increasing differences between the mammal fauna of the British Isles in the Holocene and successively older interglacial faunas. The magnitude of the changes and their timings varied across the Continent, and in this latest warming phase (the Holocene), an important factor has been the influence of humans and the absence or scarcity of megafauna such as mammoths, woolly rhinoceros and various horse/ass species.

After a brief look at the vegetation and fauna of the last Ice Age, our starting point is the development of the modern landscape after this glacial period. We look at the spreading back of different groups of plants and animals, the balance between open and closed woodland landscapes, the role of grazing and fire, and the influence of humans in the time before farming became widespread. An important theme is that conditions varied across time and across the continent at both small and large scales; there is not one 'natural state' that can be used as a reference for what conditions were like prior to the Neolithic age.

*E-mail: keith.kirby@bnc.oxon.org

© CAB International 2015. *Europe's Changing Woods and Forests: From Wildwood to Managed Landscapes* (eds K.J. Kirby and C. Watkins)

3.2 A Cold Open Continent

During the last glacial maximum, about 25,000–18,000 years BP, there were extensive ice sheets over much of northern Europe (Böse et al., 2012) (Plate 1); glaciers also occurred further south, for instance in the Alps and the Pyrenees, although their retreat may have started earlier (Moreno et al., 2010). Sea levels were much lower, exposing parts of the Atlantic continental shelf and creating land bridges, such as between Britain and the rest of Europe.

Much of Europe south of the ice was covered by large areas of 'mammoth steppe' – mixtures of grasses, herbs and low shrubs – that stretched from Spain to Canada, from the Arctic islands to China (Tarasov et al., 2000; Zimov et al., 2012; Willerslev et al., 2014). It was a highly productive environment supporting a large (in every sense) fauna which, in turn, influenced the structure and composition of the vegetation.

The relative contribution of climate change and hunting to the loss of this megafauna is debated. One model is that the spread of trees and shrubs, as the climate warmed, lessened the available food for the large herbivores, leading to reductions in their populations. An alternative is that human hunting pressure reduced the herbivore levels to the point where this triggered vegetation change and the spread of woody species (Stewart, 1993; Allen et al., 2010; Ukkonen et al., 2011; Zimov et al., 2012).

Similar loss of megafauna happened in North and South America and, at earlier periods, in Africa and Australia, where the evidence for a human cause is stronger. However, even if humans were only a contributory factor in megafaunal extinction in Europe, this represents a major anthropogenic modification of the landscape. The loss of these animals changed the nature of the grazing and browsing pressure on the vegetation and, because large animals tend to move longer distances than smaller ones, the dispersal patterns of seeds and nutrients (via dung) were also changed (Corlett, 2013; Doughty et al., 2013).

The mammoth steppe may have been largely treeless, but there is increasing evidence for survival within it in of some trees and perhaps small patches of woodland, even during the last glacial maximum (Stewart and Lister, 2001; Provan and Bennett, 2008). Kullman (2008) suggests an analogy with present-day south-western Greenland, where small isolated birchwoods occur at the inner fjords, close to inland ice, while the outer reaches support arctic vegetation.

South of the steppe zone, around the Mediterranean and Black seas, pollen analysis suggests the presence of more extensive scrub and woodland (Elenga et al., 2000; Tarasov et al., 2000). For example, González-Sampériz et al. (2010) conclude that the Iberian peninsula during the Pleistocene was a highly variable mix of pine woodland, deciduous and mixed forests, savannah-like parkland, and scrub, as well as steppes and grassland.

3.3 Trees Spread Back After the Ice

Conditions started to become warmer from about 16,000 years BP onward, apart from a short cold spell (12,800 and 11,500 years BP) known as the Younger Dryas. Pollen remains suggest that there was a fairly rapid response from the vegetation to the changing climate (Huntley, 1988; Williams et al., 2002). A similar picture emerges for at least some groups of insects (Coope and Wilkins, 1994).

Species range changes would have tracked changing environmental conditions in an individual way, rather than as a whole assemblage. Major environmental changes might be reflected in synchronous vegetation shifts across a region or the continent, but other developments would be site and species specific (Giesecke et al., 2011). In some places, at some times, the plant and animal communities would have looked familiar to us today, while other communities might have no modern analogues. Huntley (1990) suggests that 'plant communities' have generally persisted for periods of less than a 1000 years before changing conditions led to a reassortment of their constituent species.

We consider first the recolonization of Europe by trees, then what is known about the spread of other woodland plants, both largely based on pollen studies, before considering some of the more recent insights from molecular genetic research.

3.3.1 Forming a canopy

As conditions warmed, trees species spread out from the glacial refugia through the rest of the Continent (Huntley and Birks, 1983). The most important sources were thought to have been in the Iberian peninsula, Italy and the western Balkan mountains, but the early appearance of trees in Scandinavia points to contributions from more northerly small refugia (Bennett et al., 1991; Kullman, 2008).

Various factors may have contributed to the differences between species in their apparent rate of spread. Deciduous oaks (*Quercus robur*/*Q. petraea*) were established through southern Norway and Finland by about 7000 years BP, but beech (*Fagus sylvatica*) seems to have spread much more slowly. Even by 6000 years BP it appears to have been only a minor component of the tree cover in north-west Europe (Huntley and Birks, 1983). Small populations of species probably established well ahead of the main 'front'. For instance, the main expansion of beech in Britain was about 3000 years ago, but there is enough pollen evidence to suggest a low level of beech presence from several millennia before this (Rackham, 2003). Later arrivals might have taken longer to become abundant because they had to contend with competition from the tree species already established.

The climate conditions had to have improved sufficiently for the species to be able to survive – species that depend on warmer conditions appear later in the record. Light, wind-dispersed seeds such as those of Scots pine (*Pinus sylvestris*) and birch (*Betula* spp.) are likely to have spread faster and further than the heavier fruits and seeds of trees such as oak and beech. Among heavy-seeded species, those with seeds dispersed by birds might be expected to spread faster than species moved by small mammals, or which appear not to have any specialist dispersal agent. For example, the faster spread back of oaks compared with beech to north-western Europe may be because jays (*Garrulus glandarius*) carry acorns into open areas as part of their hoarding behaviour (Bossema, 1979).

There has to be sufficient continuity of suitable conditions along the way for the species to establish, grow and produce the seeds for the next jump. Small gaps may be crossed by occasional long-distance dispersal, but larger ones may prove to be barriers to species spread. Sycamore (*Acer pseudoplatanus*) occurs in northern France, but appears not to have reached Britain naturally, presumably because its northward movement was blocked by the English Channel; once introduced, it became widely naturalized in Britain and Ireland. The spread of spruce (*Picea abies*) into parts of northern Fenno-Scandinavia may have been limited by zones of unsuitable geology and soils (Sutinen et al., 2007); in the Alps, the slower spread of spruce than pine may have been because spruce needed soils that had developed a higher water-holding capacity (Henne et al., 2011).

Svenning and Skov (2004) used the climatic requirements for 55 tree species to model a potential range, which they then compared with the actual areas where the species were considered native. Many trees seem to be filling less than 50% of their potential climatic range, with the species showing the greatest discrepancies being concentrated in southern Europe. Species such as hornbeam (*Carpinus betulus*), sweet chestnut (*Castanea sativa*), walnut (*Juglans regia*), hop hornbeam (*Ostrya carpinifolia*), plane (*Platanus orientalis*) and Turkey oak (*Q. cerris*) are wholly or largely missing from the Iberian peninsula, despite it having conditions suitable for these species. Barriers such as the Mediterranean Sea and the Pyrenees may have blocked postglacial spread from refugia further east or south (Svenning et al., 2010). Dispersal limitations thus seem still to be important.

By about 6000–4000 years BP, trees had reached their greatest extension northward and also in terms of elevation of the treeline (Huntley, 1988), even to beyond current limits. The modern pattern of broad forest zones – sclerophyllous forest and scrub in the south, broadleaved deciduous forests through central Europe, boreal and sub-boreal conifer/conifer–deciduous mixes in the north – was largely established. Range shifts continued, however, for individual species; examples include the westward spread of spruce from Russia through Fenno-Scandinavia, the spread of beech (*F. sylvatica*) into north-west Europe, and a similar spread of hornbeam (*C. betulus*) accompanied by some decline in parts of its southern range (Huntley, 1988).

3.3.2 The wood beneath the trees

Pollen records tend to be less complete for plants other than trees and shrubs. Some woodland herbs produce little pollen and so may not be very visible in the record; in other cases the pollen may be assignable only to broad categories such *Poaceae* (grasses) or *Cyperaceae* (sedge and sedge-like plants). Even if species were present in abundance under the trees, their flowering, and hence pollen production, may have been limited by shade.

The spread of what we now think of as 'woodland flora' may not have required prior establishment of trees. Species regularly associated with woodland habitats under current conditions may also occur on cliffs or in meadows (Peterken, 2013). Their spread could have been among the open postglacial vegetation initially, and only later did the association with woodland develop. Huntley's (1990) point about possible non-analogue assemblages applies also to the ground flora (and to the fauna).

Svenning *et al.* (2008) looked for evidence of postglacial migratory lags in species distributions that were similar to those found for some trees and shrubs by comparing modelled climatic space and current distribution for 39 woodland ground flora species. Ferns such as *Athyrium filix-femina*, *Dryopteris filix-mas* and bracken (*Pteridium aquilinum*), and herbs such as *Anemone nemorosa*, *Moehringia trinervia* and *Ranunculus auricomus*, were widespread across northern Europe, so not apparently limited by dispersal rate. These authors did, however, find evidence for migratory lag in other species, such as the ferns *Polystichum aculeatum* and *Thelypteris limbosperma*, and the herbs *Lunaria rediviva* and *R. lanuginosus*, even after controlling for the direct effects of climate and soils as limiting factors on species distributions.

An intriguing question is why more woodland ground flora species do not show a migratory lag effect. Under current conditions, various studies have recorded rates of spread of only a few metres a year (Brunet and Von Oheimb, 1998; Rackham, 2003; Brunet, 2007), which contributes to their association with ancient woodland (Peterken, 1974; Hermy, Chapter 13). Even an average rate of spread of 100 m a year for 5000 years equates to just 500 km, yet we presume that some of these species must have reached northern Europe and spread into Britain by about 7000 years ago, before the land bridge was severed.

Some refugia might have been closer; there is evidence that cold-adapted species survived further north than previously thought (Bhagwat and Willis, 2008; Provan and Bennett, 2008). In addition, rare, long-distance dispersal events may have been important, but are difficult to detect (Cain *et al.*, 2000; Nathan *et al.*, 2003). The contribution of large mammals such as boar (*Sus scrofa*), various species of deer and the now extinct aurochs (*Bos primigenius*) to such long-distance dispersal also needs further consideration (Heinken *et al.*, 2002; Schmidt *et al.*, 2004; Picard and Baltzinger, 2012).

3.3.3 Molecular markers for recolonization routes

Molecular genetic markers have provided a new way of exploring the possible location of Ice Age refugia and recolonization routes. There is a tendency for an increase in genetic diversity from northern to southern Europe (where the main supposed refugia were) in terms of numbers of species, extent of subspecific division and the allelic variation of populations (Hewitt, 1999; Svenning *et al.*, 2011). Populations of species in northern Europe often show markers in their genetic make-up which suggest that they have been largely derived by migration from just one of these refugia; or they may show mixtures and hybrid zones suggesting that individuals from more than one refugia were involved in recolonization. This could lead to the highest levels of genetic diversity being in the mid-latitudes where spreading populations meet, rather than in the refugia zones themselves (Petit *et al.*, 2003).

Based on analyses of the genetic diversity of different species, Hewitt (1999) suggested three different general patterns of postglacial recolonization of northern Europe. The first is that typified by the grasshopper (*Chorthippus parallelus*), for which most of Europe seems to have been repopulated from a Balkan refugium, with very limited contribution from the populations surviving in the Italian or Iberian peninsulas. The second, the hedgehog (*Erinaceus*) type, involves spread from all three refugia, although different parts of northern Europe may owe more or less to one particular refugium – so hedgehogs in Britain are more linked to the Iberian population, whereas those in Scandinavia show more affinities with the Italian population. The third pattern – the brown bear (*Ursus arctos*) type – involves colonization from the Iberian peninsula in the west and from eastern refugia, but with only limited input from the populations from Italian or Balkan refuges.

This technique has now been applied to a wide range of species. The crab apple (*Malus sylvestris*) in western Europe is probably derived from refugia in the Iberian peninsula; a population in the Carpathian Mountains spread north through eastern Europe; with a Balkan population contributing little (Cornille *et al.*, 2013). For Scots pine, the main spread seems to have been from a refugium east of the Alps, in the Hungarian plain and Danube region, with the Iberian and Italian populations showing little expansion (Cheddadi *et al.*, 2006). Korsten *et al.* (2009) propose that the Ural Mountains might have been an important refuge area for species such as the common and pygmy shrews (*Sorex araneus*, *S. minutus*), which then spread extensively to the north-west into Europe and north-east into Asia. For other species, such as the badger (*Meles meles*), the Urals seem to have been a barrier to population spread from the west, leading to distinct genetic differences on either side of the range. Some wide-ranging carnivores, such as the pine marten (*Martes martes*) (Davison *et al.*, 2001), show very little genetic population structuring, suggesting expansion from a single refugium, but with little indication as to where that was.

There may have been different colonization routes even within a small geographic region, with, for example, oak coming into Denmark both from the east and the west (Jøhnk and Siegismund, 1997; Brewer *et al.*, 2002). However, interpreting patterns based on expansion from a small number of southern refugia is complicated by possible contributions from northerly refugia; for instance, Scots pine in England seems to have colonized from the Continent, but the populations in Scotland appear to have two separate origins, one in Ireland and another whose location is currently unknown (Stewart and Lister, 2001).

3.4 A Holey Blanket of Trees

Tree cover probably reached its maximum across Europe about 7000 years BP, limited in the north particularly by temperature, but in the south more by water availability. The composition of the tree cover would have varied in response to climatic variations across the continent as in models of current potential natural vegetation (Bohn *et al.*, 2000). Under any one set of climatic conditions, further variations would occur according to soil differences, competition between tree species and the effects of various types of disturbance (grazing, fire, windthrow, disease, etc.) as in near-natural forests today (Peterken, 1996). The tree species themselves modified the local environmental conditions through the shade that they cast, the nature of their leaf litter and dead wood, and their below-ground interactions (Ellenberg, 1988) leading, in turn, to more-or-less favourable conditions for other species of plant and animal.

By analogy with relatively undisturbed areas of North America and northern Asia, much of the continent might have been covered by 'old-growth' forest with many large trees and much dead wood. The Białowieża Forest in Poland (Faliński, 1986) has often been seen as a model for the pre-Neolithic landscape through much of central Europe, but its mixed composition and mixed-age structure are only one possible outcome of the interplay of natural processes.

In addition, the forest has more of a history of human influence than is often appreciated (Latałowa *et al.*, Chapter 17).

Various disturbance factors would have created gaps and opportunities for young stands to develop, leading to a dynamic and variable forest structure. In southern Europe and in the boreal zone fire could kill off large patches giving either rise to young even-aged regenerating stands or to open woodland and scrub formations (Bradshaw and Hannon, 1992; Angelstam, 1998; Grove and Rackham, 2001). Major windstorms periodically blow down modern forests (Gardiner *et al.*, 2010), and the remnants of such stands blown down *c.*6000 years ago have been found off the coast of Britain (Allen, 1992). Dutch elm disease has been present for millennia and contributed to the prehistoric decline of elms (*Ulmus* spp.) across Europe (Parker *et al.*, 2002; Potter, Chapter 23). In mountainous areas, avalanches and landslides would create open swathes where young growth might subsequently develop (Bebi *et al.*, 2009). Along rivers and streams, floods, but also beavers (*Castor fiber*), would modify the pattern of the landscape (Hughes, 2003; Coles, 2006).

More permanent open areas probably existed within the forest zone, for example on mobile dunes, at high altitudes or on the wettest bogs. In the past, the presence of open ground was often difficult to detect in pollen records, which were the main source for reconstructing landscape structure and composition. More recent work has refined our ability to reconstruct landscapes through samples from small basins or moss tussocks, which largely represent the vegetation in the immediate vicinity of the sample point (Hannon *et al.*, 2000; Bunting, 2002); such reconstructions have also been aided by the use of correction factors that allow for different levels of pollen production and spread between species (Nielsen *et al.*, 2012). These studies have helped to make the open vegetation component of the landscape more visible.

Across Denmark and Germany, Nielsen *et al.* (2012) estimated 'openness' to be about 10–30% over much of the landscape between about 8000 and 4000 BP, but with 30–40% through east Jutland. This latter area has the sandiest soils and is closest to the coast; it also showed the highest levels of heather (*Calluna vulgaris*), suggesting the start of the development of open heathland. From 4000 years BP onward, 'openness' estimates increased across the whole landscape, along with an increase in indicators of possible early farming (*Rumex* spp., *Artemisia*, *Plantago* spp.).

Fyfe *et al.* (2013), applying the same approach to pollen records from Britain and Ireland, found similarly strong regional variation. In Scotland, the maximum tree and shrub cover was reached by about 6700 years BP. This was 60–90% in the south and the east, but dropped to about 50% further north on the mainland, and to even lower levels on the Orkney Islands and some of the Western Isles. In England, the pre-Neolithic woodland cover estimates (*c.*6000 years BP) varied from 60–90% in the south and east to 30–40% in the uplands of Dartmoor and north-west England. In Ireland, the highest levels of tree and shrub cover (>80%) occurred earlier (*c.*8000 years BP) in three of four data sets; in a fourth, it was lower, at 40–50%. Heather, grass and/or other herbaceous cover increased as the tree and shrub cover declined. These data are consistent with a trend towards increasing development of bog under the more Atlantic climate of Britain and Ireland compared with the Continent, with an earlier shift perhaps towards the teleocratic phase in the interglacial cycle (Watts, 1988). The authors note that their data sets are biased towards uplands and wetlands, with no data from the bulk of central England where tree cover might be expected to be highest; this may also contribute to the higher estimates of openness compared with those found on the Continent.

3.5 The Role of Large Herbivores, Particularly Bison, Wild Horse and Aurochs

The balance between open and closed tree cover before the widespread adoption of farming in the Neolithic age has been a key recent debate. How much of the landscape was dense forest with only limited openings; how much was largely open, interspersed by scattered trees, patches of woodland and occasional larger blocks (Rackham, 1998); and what role might large herbivores have played in creating and maintaining open landscapes (Vera, 2000, 2009)? The question is of interest because it may influence how we view species and habitats that we seek to conserve, such as wood-pastures, meadows or heaths.

The biggest herbivores, the mammoth (*Mammuthus primigenius*) and woolly rhinoceros (*Coelodonta antiquitatis*), which had helped create and maintain the open steppe conditions, did not survive as part of most European ecosystems into the Holocene. However, grazers and browsers, including red deer (*Cervus elaphus*), roe deer (*Capreolus capreolus*), wild ox (or aurochs), bison (*Bison bonasus*), wild horse (*Equus ferus*), elk (*Alces alces*) and reindeer (*Rangifer tarandus*), remained widespread, along with wild boar, which include much plant material in their diet.

Thus, an alternative scenario proposed by Frans Vera (2000) is that the wild ox, bison, and the wild horse in particular, were the main driving forces determining the character of the vegetation and landscape structure. Instead of the closed forest as the natural state for much of the European landscape, Vera proposes that wood-pastures such as the New Forest (southern England) represent a better analogue. The closed forest model, he argues, is based on the overemphasis of the tree and shrub component in early palynological studies, a misinterpretation of early documentary references to forest and woods, and studies of forest succession in the absence of grazing. Instead, the landscape would have been a shifting mosaic of open areas, patches of regeneration, patches of mature trees or single open-grown trees (Fig. 3.1).

Key elements of Vera's argument are that the landscape must have been relatively open to allow for hazel (*Corylus avellana*) to flower in abundance – flowering of this species is much reduced in the shade – and also to allow for the high levels of oak (*Quercus* spp.) pollen found: oak is outcompeted under closed forest conditions in ungrazed modern forest reserves by more shade-tolerant species, particularly beech. Regeneration of oak and hazel, but also of other trees and shrubs, could still have occurred, despite grazing by the wild ox, bison, wild horse, deer, etc., through trees establishing in the shelter of thorny shrubs such as hawthorn (*Crataegus monogyna*), blackthorn (*Prunus spinosa*) or bramble (*Rubus fruticosus*). Such associational resistance to grazing has indeed been described by various authors (Olff *et al.*, 1999; Bakker *et al.*, 2004; Uytvanck *et al.*, 2008).

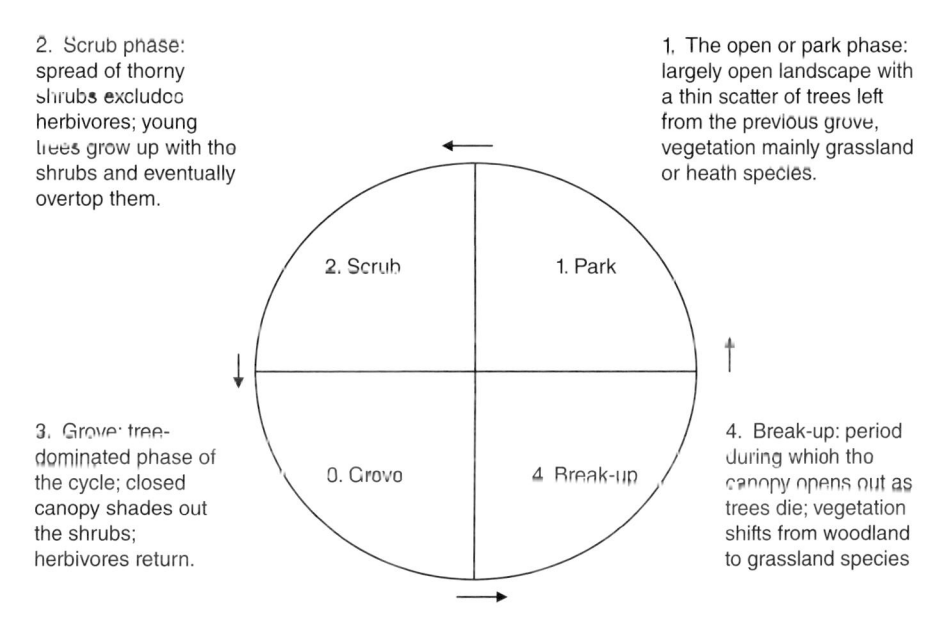

2. Scrub phase: spread of thorny shrubs excludes herbivores; young trees grow up with the shrubs and eventually overtop them.

1. The open or park phase: largely open landscape with a thin scatter of trees left from the previous grove, vegetation mainly grassland or heath species.

3. Grove: tree-dominated phase of the cycle; closed canopy shades out the shrubs; herbivores return.

4. Break-up: period during which the canopy opens out as trees die; vegetation shifts from woodland to grassland species

Fig. 3.1. A representation of Vera's shifting mosaic cycle. (From Kirby, 2004.)

There is agreement that the pre-Neolithic landscape contained more trees than previous interglacials, that there were large herbivores present and that there were open areas scattered through the landscape. The key differences between Vera's hypothesis and the 'closed forest' can then be posed as follows:

- How much of the landscape was predominantly open with trees (singly and in patches of varying sizes), versus predominantly woodland cover with gaps?
- In which areas/under what circumstances were the populations of the herbivores such that they would create and maintain the half-open shifting mosaic that Vera proposes, without it becoming either a largely open landscape or a largely wooded one?

Recent pollen models have increased the estimates of landscape openness at a regional level based on a simple comparison of the ratio of arboreal to non-arboreal pollen (Nielsen et al., 2012). However, there is not yet the evidence for the sort of widespread shifts from open to tree cover, and then back to open conditions, that Vera's model requires.

Large herbivores did spread across Europe under the open steppe-like conditions that developed in the early Holocene. None the less, direct evidence for large herds of wild ox, bison or horse in north-west Europe is lacking. A possible surrogate for the herbivore populations is the abundance of dung beetles, which do appear to suggest low levels of large herbivores in the pre-Neolithic landscape (Whitehouse and Smith, 2010; Sandom et al., 2014), which is more consistent with the 'closed forest' model. The use of dung fungal spores might also help to address this issue in future (Burney et al., 2003; Davis and Shafer, 2006; Gill et al., 2009). Innes and Blackford (2003) illustrate the use of dung fungal assemblages to suggest concentrations of animals in temporary clearances created by fire in the North York Moors (northern England).

Perhaps as the tree cover spread, so the shading out of the ground layers reduced the available herbage, leading to declines in herbivore populations and with large numbers confined to specific situations. The fossil record for the wild horse suggests this pattern: Sommer et al. (2011) conclude that this species was almost absent in the central European lowlands in the middle Holocene when tree cover appears to have been most extensive. The geographic range of the European wild ass, *Equus hydruntinus*, also became highly fragmented into discrete subpopulations during the Holocene, surviving in regions of open habitat, before the species was progressively lost from the whole Continent (Crees and Turvey, 2014). Hall (2007) suggests that in Britain the aurochs was mainly a species of low-lying flat, fertile ground, where flooding could have kept the landscape more open, consistent with Van Vuure's interpretation of this species being associated with marshes and fens (Van Vuure, 2005).

Hodder et al. (2005, 2009) set out arguments that have developed around the shifting landscape mosaic hypothesis. The activity of large herbivores would not necessarily produce 'a half open landscape' (Kirby, 2004; Samojlik and Kuijper, 2013); the pollen record for regions with differing levels of large herbivores do not show different patterns (Mitchell, 2005); beech may not have been such a competitor with oak in the past because it took longer to spread and so was not very widespread in the landscape 6000 years ago (Huntley and Birks, 1983). Medieval and early modern written sources do often describe open woodland, but this says very little about the pre-Neolithic state of the landscape, because this had long since been altered (Szabó, 2009). The experience of predator reintroduction in Yellowstone National Park (Wyoming, USA) suggests another mechanism by which herbivores might not necessarily produce a half-open landscape; they might have been restricted in their ability to control regeneration (Ripple and Beschta, 2003, 2012; but see also Theuerkauf and Rouys, 2008; Kauffman et al., 2010).

The arguments seem to shifting back towards a predominantly wooded landscape, but with higher levels of openness (Nielsen et al., 2012; Fyfe et al., 2013) than might have been envisaged 30 years ago. The presence of some oaks grown in relatively open conditions is indicated by the occurrence of saproxylic beetles now found predominantly in wood-pastures (Buckland, 2005). Equally, patches of dense, close-grown trees are known to have still been present around 4000 years ago because their remains survive. Dugout canoes require relatively straight knot-free trunks, and have been found across Europe; one in Dublin Museum (dated to 4500 years BP) is 15 m long (Plate 4). Another vessel, the 'Dover Boat', dated to about 3500 years BP, was made from four oak planks (Clark, 2004), which came from oak trees at least a metre in diameter and about 350 years old when felled. The planks show only small knots, indicating an absence of large side branches (hence growth in closed conditions) in the lower 11 m of trunk used.

Large herbivores probably did determine the composition and structure of the landscape in some places and at some times, but elsewhere, the key disturbance factor might have been disease, fire, wind, etc. No one type of landscape or one type of disturbance factor would dominate in as varied an environment as Europe. Across the continent and across time, the tree and forest cover has changed and will continue to change in response to a wide variety of different factors, and almost from the beginning of the Holocene, people have been one of those factors.

3.6 People in the Landscape: The Trees in Retreat

It is often assumed that human influence on the landscape in the early Holocene was limited and that, therefore, the landscape could be regarded as 'natural' (whether your preferred vision for this period is closed forest or a half-open landscape). However, humans were even then a force in the landscape; they seem to have contributed to the elimination of the Pleistocene megafauna, and they may have created or maintained open areas, perhaps through fire, so that they could hunt prey animals more easily (Innes and Blackford, 2003; Simmons, 2003).

Farming spread across Europe from about 7000 years BP (Ammerman and Cavalli Sforza, 1971; Pinhasi et al., 2005), and human populations and their domestic stock started to become the dominant driver of forest change. Gregg (1988) suggests that a settlement of about 30 people might have required about 600 ha to meet their needs: about 50 ha of cleared land (pasture, meadow and arable), and up to 550 ha of 'forest browsing', depending on the density of the tree cover and hence the available fodder for livestock. The pollen record and other sources point to increasing areas of open landscapes such as heathland and calcareous grassland (Webb, 1998; Poschlod and WallisDeVries, 2002). The landscape was becoming cultural in character, and some now refer to the current era as the Anthropocene to reflect the degree of human influence. Woodland 'history' had begun.

References

Allen, J.R.L. (1992) Trees and their response to wind: mid-Flandrian strong winds, Severn Estuary and Inner Bristol Channel, Southwest Britain. *Philosophical Transactions of the Royal Society B: Biological Sciences* 338, 335–364.

Allen, J.R.M., Hickler, T., Singarayer, J.S., Sykes, M.T., Valdes, P.J. and Huntley, B. (2010) Last glacial vegetation of northern Eurasia. *Quaternary Science Reviews* 29, 2604–2618.

Ammerman, A.J. and Cavalli-Sforza, L.L. (1971) Measuring the rate of spread of early farming in Europe. *Man* 6, 674–688.

Angelstam, P.K. (1998) Maintaining and restoring biodiversity in European boreal forests by developing natural disturbance regimes. *Journal of Vegetation Science* 9, 593–602.

Bakker, E.S., Olff, H., Vandenberghe, C., De Maeyer, K., Smit, R., Gleichman, J.M. and Vera, F.W.M. (2004) Ecological anachronisms in the recruitment of temperate light-demanding tree species in wooded pastures. *Journal of Applied Ecology* 41, 571–582.

Bebi, P., Kulakowski, D. and Rixen, C. (2009) Snow avalanche disturbances in forest ecosystems—state of research and implications for management. *Forest Ecology and Management* 257, 1883–1892.

Bennett, K.D., Tzedakis, P.C. and Willis, K.J. (1991) Quaternary refugia of north European trees. *Journal of Biogeography* 18, 103–115.

Bhagwat, S.A. and Willis, K.J. (2008) Species persistence in northerly glacial refugia of Europe: a matter of chance or biogeographical traits? *Journal of Biogeography* 35, 464–482.

Bohn, U., Gollub, G. and Hettwer, C. (2000) *Map of the Natural Vegetation of Europe*. Federal Agency for Nature Conservation, Bonn, Germany.

Böse, M., Lüthgens, C., Lee, J.R. and Rose, J. (2012) Quaternary glaciations of northern Europe. *Quaternary Science Reviews* 44, 1–25.

Bossema, I. (1979) Jays and oaks: an eco-ethological study of a symbiosis. *Behaviour* 70, 1–117.

Bradshaw, R. and Hannon, G. (1992) The disturbance dynamics of Swedish boreal forest. In: Teller, A., Mathy, P. and Jeffers, J.N.R. (eds) *Responses of Forest Ecosystems to Environmental Changes. European Symposium on Terrestrial Ecosystems: Forests and Woodland*. Commission of the European Communities, European Science Foundation and Consiglio Nazionale Delle Ricerche (Italy). Elsevier Applied Science, London, pp. 528–535.

Brewer, S., Cheddadi, R., de Beaulieu, J.L. and Reille, M. (2002) The spread of deciduous *Quercus* throughout Europe since the last glacial period. *Forest Ecology and Management* 156, 27–48.

Brunet, J. (2007) Plant colonization in heterogeneous landscapes: an 80-year perspective on restoration of broadleaved forest vegetation. *Journal of Applied Ecology* 44, 563–572.

Brunet, J. and Von Oheimb, G. (1998) Migration of vascular plants to secondary woodlands in southern Sweden. *Journal of Ecology* 86, 429–438.

Buckland, P.C. (2005) Palaeoecological evidence for the Vera hypothesis. In: Hodder, K.H., Bullock, J.M., Buckland, P.C. and Kirby, K.J. *Large Herbivores in the Wildwood and in Modern Naturalistic Grazing Systems*. English Nature Research Reports No. 648, English Nature, Peterborough, UK, pp. 62–116.

Bunting, M.J. (2002) Detecting woodland remnants in cultural landscapes: modern pollen deposition around small woodlands in northwest Scotland. *The Holocene* 12, 291–301.

Burney, D.A., Robinson, G. and Burney, L.P. (2003) *Sporormiella* and the late Holocene extinctions in Madagascar. *Proceedings of the National Academy of Sciences of the United States of America* 100, 10800–10805.

Cain, M.L., Milligan, B.G. and Strand, A.E. (2000) Long-distance seed dispersal in plant populations. *American Journal of Botany* 87, 1217–1227.

Cheddadi, R., Vendramin, G.G., Litt, T., François, L., Kageyama, M., Lorentz, S., Laurent, J.-M., de Beaulieu, J.-L., Sadori, L., Jost, A. and Lunt, D. (2006) Imprints of glacial refugia in the modern genetic diversity of *Pinus sylvestris*. *Global Ecology and Biogeography* 15, 271–282.

Clark, P. (2004) *The Dover Bronze Age Boat*. English Heritage, Swindon, UK.

Coles, B.J. (2006) *Beavers in Britain's Past*. Oxbow, Oxford, UK.

Coope, G.R. and Wilkins, A.S. (1994) The response of insect faunas to glacial–interglacial climatic fluctuations. *Philosophical Transactions of the Royal Society B: Biological Sciences* 344, 19–26.

Corlett, R.T. (2013) The shifted baseline: prehistoric defaunation in the tropics and its consequences for biodiversity conservation. *Biological Conservation* 163, 13–21.

Cornille, A., Giraud, T., Bellard, C., Tellier, A., Le Cam, B., Smulders, M.J.M., Kleinschmit, J., Roldan-Ruiz, I. and Gladieux, P. (2013) Postglacial recolonization history of the European crab-apple (*Malus sylvestris* Mill.), a wild contributor to the domesticated apple. *Molecular Ecology* 22, 2249–2263.

Crees, J.J. and Turvey, S.T. (2014) Holocene extinction dynamics of *Equus hydruntinus*, a late-surviving European mega-faunal mammal. *Quaternary Science Reviews* 91, 16–29.

Davis, O.K. and Shafer, D.S. (2006) *Sporormiella* fungal spores, a palynological means of detecting herbivore density *Palaeogeography, Palaeoclimatology, Palaeoecology* 237, 40–50.

Davison, A., Birks, J.D.S., Brookes, R.C., Messenger, J.E. and Griffiths, H.I. (2001) Mitochondrial phylogeography and population history of pine martens *Martes martes* compared with polecats *Mustela putorius*. *Molecular Ecology* 10, 2479–2488.

Doughty, C.E., Wolf, A. and Malhi, Y. (2013) The impact of large animal extinctions on nutrient fluxes in early river valley civilizations. *Ecosphere* 4(12), art148.

Elenga, H., Peyron, O., Bonnefille, R., Jolly, D., Cheddadi, R., Guiot, J., Andrieu, V., Bottema, S., Buchet, G., de Beaulieu, J.L. *et al*. (2000) Pollen-based biome reconstruction for southern Europe and Africa 18,000 yr BP. *Journal of Biogeography* 27, 621–634.

Ellenberg, H. (1988) *The Vegetation Ecology of Central Europe*. Cambridge University Press, Cambridge, UK.

Falinski, J.B. (1986) *Vegetation Dynamics in Temperate Lowland Primeval Forests*. W. Junk, Dordrecht, The Netherlands.

Fyfe, R.M., Twiddle, C., Sugita, S., Gaillard, M.-J., Barratt, P., Caseldine, C.J., Dodson, J., Edwards, K.J., Farrell, M., Froyd, C. et al. (2013) The Holocene vegetation cover of Britain and Ireland: overcoming problems of scale and discerning patterns of openness. *Quaternary Science Reviews* 73, 132–148.

Gardiner, B., Blennow, K., Carnus, J.-M., Lindner, M., Marzano, M., Nicoll, B., Orazio, C., Peyron, J.-L., Reviron, M.-P., Schelhaas, M.J. et al. (2010) *Destructive Storms in European Forests: Past and Forthcoming Impacts.* European Forest Institute, Joensuu, Finland.

Giesecke, T., Bennett, K.D., Birks, H.J.B., Bjune, A.E., Bozilova, E., Feurdean, A., Finsinger, W., Froyd, C., Pokorný, P., Rösch, M. et al. (2011) The pace of Holocene vegetation change – testing for synchronous developments. *Quaternary Science Reviews* 30, 2805–2814.

Gill, J.L., Williams, J.W., Jackson, S.T., Lininger, K.B. and Robinson, G.S. (2009) Pleistocene mega-faunal collapse, novel plant communities and enhanced fire regimes in North America. *Science* 326, 1100–1103.

González-Sampériz, P., Leroy, S.A.G., Carrión, J.S., Fernández, S., García-Antón, M., Gil-García, M.J., Uzquiano, P., Valero-Garcés, B. and Figueiral, I. (2010) Steppes, savannas, forests and phytodiversity reservoirs during the Pleistocene in the Iberian peninsula. *Review of Palaeobotany and Palynology* 162, 427–457.

Gregg, S.A. (1988) *Foragers and Farmers: Population Interactions and Agricultural Expansion in Prehistoric Europe.* University of Chicago Press, Chicago, Illinois.

Grove, A.T. and Rackham, O. (2001) *The Nature of Mediterranean Europe: An Ecological History.* Yale University Press, New Haven, Connecticut.

Hall, S.J.G. (2007) A comparative analysis of the habitat of the extinct aurochs and other prehistoric mammals in Britain. *Ecography* 31, 187–190.

Hannon, G.E., Bradshaw, R. and Emborg, J. (2000) 6000 years of forest dynamics in Suserup Skov, a semi-natural Danish woodland. *Global Ecology and Biogeography* 9, 101–114.

Heinken, T., Hanspach, H., Raudnitschka, D. and Schaumann, F. (2002) Dispersal of vascular plants by four species of wild mammals in a deciduous forest in NE Germany. *Phytocoenologia* 32, 627–643.

Henne, P.D., Elkin, C.M., Reineking, B., Bugmann, H. and Tinner, W. (2011) Did soil development limit spruce (*Picea abies*) expansion in the Central Alps during the Holocene? Testing a palaeobotanical hypothesis with a dynamic landscape model. *Journal of Biogeography* 38, 933–949.

Hewitt, G. (1999) Post-glacial re-colonization of European biota. *Biological Journal of the Linnean Society* 68, 87–112.

Hodder, K.H., Bullock, J.M., Buckland, P.C. and Kirby, K.J. (2005) *Large Herbivores in the Wildwood and Modern Naturalistic Grazing Systems.* English Nature Research Reports No. 648, English Nature, Peterborough, UK.

Hodder, K.H., Buckland, P.C., Kirby, K.J. and Bullock, J.M. (2009) Can the pre-Neolithic provide suitable models for re-wilding the landscape in Britain? *British Wildlife* 20(supplement), 4–15.

Hughes, F.M.R.E. (2003) *The Flooded Forest: Guidance for Policy Makes and River Managers on the Restoration of Floodplain Forests.* Department of Geography, University of Cambridge, Cambridge, UK.

Huntley, B. (1988) Europe. In: Huntley, B. and Webb, T. III (eds) *Vegetation History.* Kluwer, Dordrecht, The Netherlands.

Huntley, B. (1990) European vegetation history: palaeovegetation maps from pollen data – 13 000 yr BP to present. *Journal of Quaternary Science* 5, 103–122.

Huntley, B. and Birks, H.J.B. (1983) *An Atlas of Past and Present Pollen Maps for Europe: 0–13,000 years ago.* Cambridge University Press, Cambridge, UK.

Ingrouille, M. (1995) *Historical Ecology of the British Flora.* Chapman and Hall, London.

Innes, J.B. and Blackford, J.J. (2003) The ecology of Late Mesolithic woodland disturbances: model testing with fungal spore assemblage data. *Journal of Archaeological Science* 30, 185–194.

Johnk, N. and Siegismund, H.R. (1997) Population structure and post-glacial migration routes of *Quercus robur* and *Quercus petraea* in Denmark, based on chloroplast DNA analysis. *Scandinavian Journal of Forest Research* 12, 130–137.

Kauffman, M.J., Brodie, J.F. and Jules, E.S. (2010) Are wolves saving Yellowstone's aspen? A landscape-level test of a behaviourally mediated trophic cascade. *Ecology* 9, 2742–2755.

Kirby, K.J. (2004) A model of a natural wooded landscape in Britain as influenced by large herbivore activity. *Forestry* 77, 405–420.

Korsten, M., Ho, S.Y.W., Davison, J., Pähn, B., Vulla, E., Roht, M., Tumanov, I.L., Kojola, I., Andersone-Lilley, Z., Ozolins, J. et al. (2009) Sudden expansion of a single brown bear maternal lineage across northern continental Eurasia after the last Ice Age: a general demographic model for mammals? *Molecular Ecology* 18, 1963–1979.

Kullman, L. (2008) Early postglacial appearance of tree species in northern Scandinavia: review and perspective. *Quaternary Science Reviews* 27, 2467–2472.

Mitchell, F.J.G. (2005) How open were European primeval forests? Hypothesis testing using palaeoecological data. *Journal of Ecology* 93, 168–177.

Moreno, A., Valero-Garcés, B.L., Jimenez-Sanchez, M., Dominquez-Cuesta, M.J., Mata, M.P., Navas, A., González-Sampériz, P., Stoll, H., Farias, P., Morellon, M. *et al.* (2010) The last deglaciation in the Picos de Europa National Park (Cantabrian Mountains, northern Spain). *Journal of Quaternary Science* 25, 1076–1091.

Nathan, R., Perry, G., Cronin, J.T., Strand, A.E. and Cain, M.L. (2003) Methods for estimating long-distance dispersal. *Oikos* 103, 261–273.

Nielsen, A.B., Giesecke, T., Theuerkauf, M., Feeser, I., Behre, K.-E., Beug, H.-J., Chen, S.-H., Christiansen, J., Dörfler, W., Endtmann, E. *et al.* (2012) Quantitative reconstructions of changes in regional openness in north-central Europe reveal new insights into old questions. *Quaternary Science Reviews* 47, 131–149.

Olff, H., Vera, F.W.M., Bokdam, J., Bakker, E.S., Gleichman, J.M., De Maeyer, K. and Smit, R. (1999) Shifting mosaics in grazed woodland driven by the alternation of plant facilitation and competition. *Plant Biology* 1, 127–137.

Parker, A.G., Goudie, A.S., Anderson, D.E., Robinson, M.A. and Bonsall, C. (2002) A review of the mid-Holocene elm decline in the British Isles. *Progress in Physical Geography* 26, 1–45.

Peterken, G.F. (1974) A method for assessing woodland flora for conservation using indicator species. *Biological Conservation* 6, 239–245.

Peterken, G.F. (1996) *Natural Woodland*. Cambridge University Press, Cambridge, UK.

Peterken, G.F. (2013) *Meadows*. British Wildlife Publishing, Gillingham, UK.

Petit, R.J., Aguinagalde, I., de Beaulieu, J.L., Bittkau, C., Brewer, S., Cheddadi, R., Ennos, R., Fineschi, S., Grivet, D., Lascoux, M. *et al.* (2003) Glacial refugia: hotspots but not melting pots of genetic diversity. *Science* 300, 1563–1565.

Picard, M. and Baltzinger, C. (2012) Hitch-hiking in the wild: should seeds rely on ungulates? *Plant Ecology and Evolution* 145, 24–30.

Pinhasi, R., Fort, J. and Ammerman, A.J. (2005) Tracing the origin and spread of agriculture in Europe. *PLoS Biology* 3(12): e410.

Poschlod, P. and WallisDeVries, M.F. (2002) The historical and socioeconomic perspective of calcareous grasslands – lessons from the distant and recent past. *Biological Conservation* 104, 361–376.

Provan, J. and Bennett, K.D. (2008) Phylogeographic insights into cryptic glacial refugia. *Trends in Ecology and Evolution* 23, 564–571.

Rackham, O. (1998) Savanna in Europe. In: Kirby, K.J. and Watkins, C. (eds) *The Ecological History of European Forests*. CAB International, Wallingford, UK, pp. 1–24.

Rackham, O. (2003) *Ancient Woodland: Its History, Vegetation and Uses in England*, rev. edn. Castlepoint Press, Dalbeattie, Scotland.

Ripple, W.J. and Beschta, R.L. (2003) Wolf reintroduction, predation risk, and cottonwood recovery in Yellowstone National Park. *Forest Ecology and Management* 184, 299–313.

Ripple, W.J. and Beschta, R.L. (2012) Trophic cascades in Yellowstone: the first 15 years after wolf introduction. *Biological Conservation* 145, 205–213.

Samojlik, T. and Kuijper, D. (2013) Grazed wood-pasture versus browsed high forest: impact of ungulates on forest landscapes from the perspective of the Białowieża Primeval Forest. In: Rotherham, I.D. (ed.) *Trees, Forested Landscapes and Grazing Animals: A European Perspective on Woodlands and Grazed Treescapes*. Earthscan from Routledge (imprint of Taylor & Francis), Abingdon, UK, pp. 143–162.

Sandom, C.J., Ejrnæs, R., Hansen, M.D.D. and Svenning, J.-C. (2014) High herbivore density associated with vegetation diversity in interglacial ecosystems. *Proceedings of the National Academy of Sciences of the United States of America* 111, 4162–4167.

Schmidt, M., Sommer, K., Kriebitzsch, W.-U., Ellenberg, H. and Oheimb, G. (2004) Dispersal of vascular plants by game in northern Germany. Part I: roe deer (*Capreolus capreolus*) and wild boar (*Sus scrofa*). *European Journal of Forest Research* 123, 167–176.

Simmons, I.G. (2003) *Moorlands of England and Wales – An Environmental History 8000 BC–AD 2000*. Edinburgh University Press, Edinburgh, UK.

Sommer, R.S., Benecke, N., Lougas, L., Oliver, N. and Schmolcke, U. (2011) Holocene survival of the wild horse in Europe: a matter of open landscape. *Journal of Quaternary Science* 26, 805–812.

Stewart, A.J. (1993) The failure of evolution: late quaternary mammalian extinctions in the holarctic. *Quaternary International* 19, 101–107.

Stewart, J.R. (2008) The progressive effect of the individualistic response of species to Quaternary climate change: an analysis of British mammalian faunas. *Quaternary Science Reviews* 27, 2499–2508.

Stewart, J.R. and Lister, A.M. (2001) Cryptic northern refugia and the origins of the modern biota. *Trends in Ecology and Evolution* 16, 608–613.

Sutinen, R., Aro, I., Herva, H., Muurinen, T., Piekkari, M. and Timonen, M. (2007) Macrofossil evidence dispute ubiquitous birch–pine–spruce succession in western Finnish Lapland. In: Johansson, P. and Serala, P. (eds) *Applied Quaternary Research in the Central Part of Glaciated Terrain. Proceedings of the INQUA Peribaltic Group Field Symposium 2006, Oulanka Biological Research Station, Finland, September 11–15*. Special Paper 46, Geological Survey of Finland, Espoo, Finland, pp. 93–98.

Svenning, J.C. and Skov, F. (2004) Limited filling of the potential range in European tree species. *Ecology Letters* 7, 565–573.

Svenning, J.C., Normand, S. and Skov, F. (2008) Postglacial dispersal limitation of widespread forest plant species in nemoral Europe. *Ecography* 31, 316–326.

Svenning, J.C., Fitzpatrick, M.C., Normand, S., Graham, C.H., Pearman, P.B., Iverson, L.R. and Skov, F. (2010) Geography, topography, and history affect realised to potential tree species richness patterns in Europe. *Ecography* 33, 1070–1080.

Svenning, J.C., Fløjgaard, C. and Baselga, A. (2011) Climate, history and neutrality as drivers of mammal beta diversity in Europe: insights from multiscale deconstruction. *Journal of Animal Ecology* 80, 393–402.

Szabó, P. (2009) Open woodland in Europe in the Mesolithic and in the middle ages: can there be a connection? *Forest Ecology and Management* 257, 2327–2330.

Tarasov, P.E., Volkova, V.S., Webb, T. III, Guiot, J., Andreev, A.A., Bezusko, L.G., Bezusko, T.V., Bykova, G.V., Dorofeyuk, N.I., Kvavadze, E.V. et al. (2000) Last glacial maximum biomes reconstructed from pollen and plant macrofossil data from northern Eurasia. *Journal of Biogeography* 27, 609–620.

Theuerkauf, J. and Rouys, S. (2008) Habitat selection by ungulates in relation to predation risk by wolves and humans in Białowieża Forest, Poland. *Forest Ecology and Management* 256, 1325–1332.

Ukkonen, P., Aaris-Sørensen, K., Arppe, P.U., Daugnora, L., Lister, A.M., Lõugas, L., Seppä, H., Sommer, R.S., Stuart, A.J., Wojtal, P. and Zupins, I. (2011) Woolly mammoth (*Mammuthus primigenius* Blum.) and its environment in northern Europe during the last glaciation. *Quaternary Science Reviews* 30, 693–712.

Uytvanck, J.V., Maes, D., Vandenhaute, D. and Hoffmann, M. (2008) Restoration of woodpasture on former agricultural land: the importance of safe sites and time gaps before grazing for tree seedlings. *Biological Conservation* 141, 78–88.

Van Vuure, C. (2005) *Retracing the Aurochs*. Pensoft, Sofia-Moscow, Russia.

Vera, F.W.M. (2000) *Grazing Ecology and Forest History*. CAB International, Wallingford, UK.

Vera, F.W.M. (2009) Large-scale nature development – the Oostvaardersplassen. *British Wildlife* 20(supplement), 28–38.

Watts, W.A. (1988) Europe. In: Huntley, B. and Webb, T. (eds) *Vegetation History*. Kluwer, Dordrecht, The Netherlands.

Webb, N.R. (1998) The traditional management of European heathlands. *Journal of Applied Ecology* 35, 987–990.

Whitehouse, N.J. and Smith, D. (2010) How fragmented was the British Holocene wildwood? Perspectives on the 'Vera' grazing debate from the fossil beetle record. *Quaternary Science Reviews* 29, 539–553.

Willerslev, E., Davison, J., Moora, M., Zobel, M., Coissac, E., Edwards, M.E., Lorenzen, E.D., Vestergard, M., Gussarova, G., Haile, J. et al. (2014) Fifty thousand years of Arctic vegetation and mega-faunal diet. *Nature* 506, 47–51.

Williams, J.W., Post, D.M., Cwynar, L.C. and Lotter, A.F. (2002) Rapid and widespread vegetation responses to past climate change in the North Atlantic region. *Geology* 30, 971–974.

Zimov, S.A., Zimov, N.S., Tikhonov, A.N. and Chapin, F.S. III (2012) Mammoth steppe: a high-productivity phenomenon. *Quaternary Science Reviews* 57, 26–45.

4 Evolution of Modern Landscapes

Keith J. Kirby[1]* and Charles Watkins[2]
[1]*Department of Plant Sciences, University of Oxford, Oxford, UK;*
[2]*School of Geography, University of Nottingham, Nottingham, UK*

4.1 Introduction

Wildwood, whatever it was like, now exists in Europe perhaps only in the most remote mountainous and boreal areas; even here, it can be affected by long-range pollution and climate change and be under threat from logging. Sites often described as 'primeval', such as the Białowieża Forest in Poland, or Fiby Urskog in Sweden, turn out to have had a more active management history than at first appears (Bradshaw and Hannon, 1992; Latałowa *et al.*, Chapter 17). The diverse ways in which the extent, structure and species composition of Europe's forest and woodland cover have been altered by humans over the last 6000 years (Table 4.1) has also had implications for a wide range of other (non-plant) species (Bengtsson *et al.*, 2000).

4.2 The Emergence of Woodland Management

Different tree species have different properties, even as firewood, and were selectively collected from the Neolithic period onward (Out, 2010). Our ancestors would have seen that cutting down trees does not necessarily kill them; multiple stems spring up from the stumps of most broadleaves, perhaps an evolutionary response to browsing by beavers, wild ox or even elephants (Koop, 1987; Buckley and Mills, Chapter 6). This habit could be turned to advantage because the regrowth was much easier to fell and handle than the original forest trees. The remains of a wooden well made of oak planks in the Rhine Valley (Kreuz, 1992) and the wooden planks and hurdles of Neolithic trackways through the peatlands of southern England testify to their woodworking skills (Rackham, 2003).

Wood-pasture systems could have developed from the general foraging of livestock across the landscape (Hartel *et al.*, Chapter 5; Plate 5). Lopping trees to provide leaf fodder would create pollards and allow more light to reach the ground, thus encouraging the field layer, making such areas more attractive to livestock and reducing the potential for tree regeneration. Early 'woodland management' could also have included selective preservation or encouragement of individual trees and species that were particularly useful, for example fruit trees.

From classical times onward, written accounts describe the variety of roles that trees and woods played in societies (Williams, 2006), the many ways in which they were treated and

---

*E-mail: keith.kirby@bnc.oxon.org

Table 4.1. Examples of ways in which humans have changed European forests.

Actions[a]	Possible consequences
Conversion to other land uses	Loss of woodland area, reduced patch size, increased edges, changes in balance of habitats across landscape
Woodland management	Generally loss of oldest age classes, reduced dead wood, more open space, shifts in tree species composition
Hunting	Reductions in top carnivores, changes in abundance/distribution of large herbivores; consequential effects on woodland composition and structure
Grazing by livestock	Change in nature, distribution and intensity of herbivore pressure with effects on understorey and regeneration in particular
Direct harvesting of litter, understorey plants and fungi	Effects on abundance of particular species, export of soil nutrients
Arable cultivation among the trees	Disruption to soils, damage to roots, alterations to nutrient cycles
Control of fire regime	Generally more fires (sometimes less) and changes in their intensity
Air pollution	Local and long-distance effects on lower plants, lichens and fungi in particular, but also on soil and water nutrient cycles
Climate change	Shifts in species distribution and abundance and in forest processes

[a]The order given broadly reflects the historical importance of these actions in influencing woods and forests.

the benefits derived from them. Wood, directly or as charcoal, was the primary fuel, and many items that are now made from metal and plastic would have been made in wood. Further, the vegetation under or between the trees could be grazed or cut for hay, and nuts, fruit and fungi provided food for humans or livestock; wild game might be taken, litter was collected for animal bedding and resinous trees were tapped for turpentine and tar.

The tending of woods grew more sophisticated and geared to particular markets, especially where the resource was limited or of high value (Plates 6 and 7). Stands, deliberately planted with the desired tree species rather than relying on what regenerated naturally, became more common from the 16th century onward, although orchard trees had been planted at least as far back as classical times. Later, with the development and spread of 'scientific' forestry, trees were seen and treated as crops (see Savill, Chapter 7; Bürgi, Chapter 8).

4.3 Changes in Forest Extent and Distribution

4.3.1 Reductions in forest cover

Clearing is not a technological problem. In the past, humans with stone or flint axes needed only boundless energy to fell trees; in contrast, fire and browsing animals can wreak havoc in forested areas with little effort.

(Williams, 2008)

The pre-Neolithic extent of European tree cover seems unlikely to have been less than 60–70%, compared with the current average of 30–40% (Williams, 2006; Kirby and Watkins, Chapter 1). The eastern Dutch landscape, for instance, switched from Neolithic/Bronze Age settlements being islands in a wooded landscape to relics of woodland surrounded by cultivated land in the Iron Age (Groenewoudt et al., 2007). Even with a stone axe, a tree of diameter 20–30 cm might only take an hour to fell (Mathieu and Meyer, 1997). Researchers trying to create a replica of the 'Dover Boat' (dated to c.3500 years BP) reported favourably on the efficiency of Bronze Age tools in felling and then working up oak trees of a metre in diameter (Clark, 2004). Fire and grazing would help to keep areas open once any large standing trees had been felled or killed.

The primary reason to clear trees was usually to increase the area of land available for farming, so fertile and easily worked soils were likely to have been cleared first. High population densities would also favour clearance, partly because of the need for more food, but also because more labour was available to carry out the work. With the development of long-distance trade, such as under the Roman Empire, a surplus of easily transported primary resources such as grain or hides would allow for exchange with manufactured or exotic goods from elsewhere.

The spread of settlements through the classical and medieval periods indicate an ongoing loss of woodland. Williams (2006) quotes a decline in woodland cover from about 30 to 13 million ha across France between 800 and 1300 AD, but what remained was still a quarter of the country, whereas in England cover was only about 15% by 1100 AD (Rackham, 2003). Within England, parts of the south-west and central regions had very little woodland, but in the south-east, woodland (probably as wood-pasture) appears to have been still extensive.

From 1500 AD onward, there are complaints in contemporary accounts across Europe about shortages, or the increasing price, of firewood, timbers for shipping, etc. (Williams, 2006). However, that there was increasing pressure on the woodland resource might not mean more clearance. If the woods were managed, a shortage or increase in price would be an incentive to keep or expand the woodland cover, and to improve its treatment (Warde, 2006). Nevertheless, further losses did occur (Williams, 2006), with, for example, woodland cover in England decreasing to 4–5% by 1900 AD (Rackham, 2003).

4.3.2 Increases as well as decreases

At different times and places, increases in woodland cover occurred. Some regeneration of woodland followed the withdrawal of the Roman influence in northern Britain (Dumayne-Peaty, 1999), in the Netherlands following the Black Death in the late 14th century (van Hoof *et al.*, 2006) and in Germany as a result of the disruptions caused by wars in the 17th century (Williams, 2006). Consequently, even 'ancient' woodland may have gone through a clearance phase in the distant past (Day, 1993; Barker, 1998) and overlie prehistoric, Roman, or Medieval settlements and fields, rather than being directly descended from the wildwood (Peterken, 1993; Dupouey *et al.*, 2002).

New woods were deliberately created in the past in response to market opportunities, as with the oak coppices planted following the opening of the Bonawe Ironworks in west Scotland (Lindsay, 1976). They were also planted to address environmental problems such as the spread of sand dunes, or for strategic reasons, as with the Forestry Commission's major afforestation programme in Britain (Tsouvalis, 2000; Foot, 2010). Expansion or contraction of woodland also depended on the success of farming. If climate or market changes led to farm abandonment, trees spread back naturally, as has happened in mountainous areas in recent decades (Debussche *et al.*, 1999; Olsson *et al.*, 2000; Prevosto *et al.*, 2004). Taking land out of farming has also been deliberately encouraged latterly, with grants to farmers in the European Union for the afforestation of farmland. This has led to much planting of Sitka spruce (*Picea sitchensis*) in Ireland, of *Eucalyptus* spp. in Spain and Portugal, and of pines in Greece (e.g. Weber, 2000; Arabatzis, 2005; Buscardo *et al.*, 2008).

4.3.3 Patterns of clearance and survival

Within particular landscapes, accessibility may have determined where forests were cleared and where they were left. The land closest to settlements tends to be used for intensive forms of farming and for valuable crops that need regular attention. In England, ancient woods are often associated with parish (local community) boundaries (Rackham, 1990; Peterken, 1993). Woodland may be left (or first regrow) on steep or difficult slopes. In mountainous regions, the woods are often concentrated on the mid-slopes between the more intensively farmed valley bottoms and free-range or seasonal grazing of the higher ground.

It can be inconvenient to move large quantities of wood around, so primary processing has often been done on site. Communities developed based around the woodland resource, such as the charcoal makers of medieval Bohemia (Woitsch, 2009) or the chair makers in the Chiltern beechwoods of southern England in the 19th century (Dallimore, 1911). Sawmills were located along rivers where they could take advantage both of water power (both directly and, later, via the generation of hydroelectricity) and of the river as a transport link (Kortelainen, 1999). This often leaves a legacy of extensive woodland in an area even after the industry has declined.

4.3.4 The ecological consequences of a patchy landscape

Millennia of forest clearance have led to an overall reduction in the extent of forest cover and in the size of individual patches. Clearance has also led to greater distances between patches, to much more forest edge and to more variety in the land cover between the patches. The ecological implications of such changes form the domain of landscape ecology (Forman, 1995).

First, the absolute loss of woodland cover increases the vulnerability of forest species to extinction because their potential population sizes are reduced. Particularly vulnerable are species affected by preferential loss of certain forest types and structures, e.g. flood plain forests (Wenger et al., 1990; Peterken and Hughes, 1995), or species that require large patches of forest, e.g. Lynx (Schadt et al., 2002; Herfindal et al., 2005). However, while some have argued that species richness should be greatest when a given area of forest survives in one block, in practice the same total area distributed across a number of patches (provided the patches are not too small) tends to give a higher species total because the patches cover a wider range of environmental conditions (Game and Peterken, 1984; Schwartz, 1999).

Secondly, the degree of connectivity between patches is important (Andrén and Delin, 1994; Bailey, 2007). Species survive in some woods, while they are lost from others. Empty patches may be recolonized from one of the adjacent populations, and the occasional exchange of genes contributes to the overall fitness of the group of populations. This concept of the metapopulation (a group of separated populations between which limited exchange can occur) has been developed particularly with respect to butterflies (Hanski et al., 1994), but applies also to other species groups, for example, woodland herbs and epiphytes (Valverde and Silvertown, 1997; Snäll et al., 2005).

Thirdly, there is the change in the relationship between the forest and its surroundings in that there is a major increase in the extent and diversity of edges (Herlin, 2001). Mixed landscapes may lead to increased pressure from generalist predators (Santos and Tellería, 1992; Storch et al., 2005), and grazing impacts may be higher from large herbivores that also use the adjacent open ground. In modern landscapes, adjacent roads can form a barrier to species movement, lead to direct wildlife casualties and affect the breeding success of woodland birds as a result of the associated traffic noise (Reijnen et al., 1995; Grilo et al., 2009; Kerth and Melber, 2009). In addition, forest edges tend to be drier because of higher transpiration and more air movement than occurs in woodland interiors (Herbst et al., 2007), and there is potential for pesticide drift and fertilizer influx from adjacent farmland (Gove et al., 2007).

The fragmentation of forests in time as well as space has proved to be a critical factor in whether various species have been able to survive or thrive. Some species are now found largely in sites that have been wooded for several hundred years – 'ancient woods' (Peterken, 1977; Hermy et al., 1999; Hermy, Chapter 13; Goldberg, Chapter 22). These species rarely occur in woods that have developed more recently, either because such woods do not provide the relevant habitat conditions or because the species cannot reach them. These 'ancient woodland indicator species' are most obvious in landscapes where the woods exist as isolated patches surrounded by farmland (Peterken, Chapter 18).

4.4 Changes in Structure and Composition Through Management

For many plants and animals associated with woodland, the age and size of the trees and the disposition of stands of different ages are as important as total woodland extent and patch size. By the Neolithic period, there are clear indications that trees were being harvested deliberately and repeatedly. Classical writings refer to different types of woodland management (Visser, 2010), and accounts of how woods should be treated increase rapidly thereafter (although as now, what was written may not reflect what people actually did on the ground). Wood in various forms was certainly a very useful product, but woodland had other values: the concept of multiple uses, multiple benefits from the same patch of trees, is new only in its terminology.

During the medieval period, many areas were designated for hunting deer and other large game by the monarch or nobles (Fletcher, Chapter 9) and, within these, tree regeneration might be restricted by browsing. The same could apply where livestock had free access across the land. These wood-pastures (Hartel *et al.*, Chapter 5) tended to have a relatively open canopy, and a wide range of woody species might be present, but particularly shrubby species and light-demanding trees. In Spain and Portugal, large areas of wood-pasture (*dehesa*, *montado*) survive associated with the production of cork from the bark of *Quercus suber* (Grove and Rackham, 2001; Plieninger and Wilbrand, 2001). Another variant of this open woodland type in the Baltic States and in Scandinavia is the wooded meadow, in which the herbage between the trees was cut for hay (Peterken, 2013).

The trees in wood-pastures have the potential to develop big crowns because they are open grown with little competition from younger growth, whereas in the past most were probably pollarded (the stem cut above browse height, typically 2–3 m) to provide leaf fodder or fuel (Plate 5). The ground flora tends to be a mixture of species now associated with grassland and with closed forest conditions; the species richness of other groups, such as birds and invertebrates, can be high (Hartel and Plieninger, 2014; Hartel *et al.*, Chapter 5). The old trees tend to be rich in epiphytic lichens (Rose, 1993) and saproxylic invertebrates (Siitonen and Ranius, Chapter 11).

Another common form of medieval woodland management was by coppicing (Plate 6). Where the woodland resource was limited, compared to the demand, there would be an incentive to ensure that it was well demarcated, for instance by the creation of boundary banks. Compared with high forests, coppices have a higher proportion of young growth and open stands, so favouring species associated with these structures (Buckley and Mills, Chapter 10); old tree and dead wood specialists tend to be scarce. Woodland plants that are tolerant of shaded conditions, but intolerant of grazing, are particularly evident (Rackham, 2003).

Most European conifers do not regrow well, if at all, if cut. However, under the right conditions, they regenerate freely in cut (or burnt) areas, producing a new stand of single-stemmed trees. From the 18th century onward, there was increasing interest in such high forest, of broadleaves as well as of conifers, to meet changing markets and new technologies (Savill, Chapter 7). Stands managed on an even-aged, clear-felling basis (Plate 7) provide large blocks of uniform condition, favouring species associated with young to middle-aged trees; the habitats associated with old trees and dead wood tend to be scarce except where specific provision has been made for them (see Quine, Chapter 15). Where forests are cut on a smaller scale, as under the selection system or other forms of what is now termed 'continuous cover forestry' (Bürgi, Chapter 8), the tree and understorey layers tend to be more varied at the stand scale, with higher levels of humidity and shade at ground level over the whole cycle. Species needing large gaps may be reduced compared with their occurrence in coppice or clear-fell systems.

The type of management practised, therefore, has had profound implications for the plants and animals associated with or found in a particular stand and wood. The variety of management regimes across the landscape then influences how species survive at a regional scale.

4.5 Deliberate Modification of the Tree and Shrub Composition of Forests

Different trees and shrubs survived better under some systems than others. Hazel, *Corylus avellana*, thrives under a coppice regime, but Scots pine, *Pinus sylvestris*, does not. Oak,

Quercus spp., is tolerant of grazing and forms long-lived trees which are often prevalent in wood-pastures. Limes, *Tilia* spp., are less tolerant of grazing and hence more common in high forest or coppice stands.

Useful or attractive trees and shrubs have been deliberately encouraged by, for example, planting seeds, moving young plants or layering stems, and this has become an increasingly common practice over the last 500 years. The species favoured might be locally or regionally native, as with much of the encouragement of Norway spruce, *Picea abies*, and beech, *Fagus sylvatica*, for timber or hazel for underwood. Where the species is already part of the system, or is at least of similar type, relatively little change in the associated plants and animals may occur.

More substantial changes follow where the favoured species differs substantially in character from those it replaces. There may be loss of obligate associated species, changes in associated species abundance because of changes in tree species traits (shade cast, litter composition and breakdown, bark characteristics, changes to the soil moisture or nutrient regimes). Change of tree species is often related to different forms of management as well, leading to further effects on associated species (Mitchell and Kirby, 1989; Quine, Chapter 15). This process is illustrated by studies of the potential impact of loss of ash, *Fraxinus excelsior*, through disease and its replacement by other trees (Mitchell *et al.*, 2014).

4.6 Other Species Gains and Losses

Directly or indirectly, humans have changed the distribution of many species, from tiny invertebrates to large mammals (Siitonen and Ranius, Chapter 11; Hearn, Chapter 14). Species seen as a direct threat to humans or to our interests have been actively persecuted, as the wolf has been across most of Europe. Range reductions have occurred in many smaller predatory birds and mammals where they were seen as interfering with game rearing (Yalden, 1999). Other species have been heavily overexploited to at least local extinction, notably the beaver, *Castor fiber*. Some of the 'lost' species are now spreading back, either naturally or through reintroduction programmes (Deinet *et al.*, 2013), and we are rediscovering their roles in woodland and other ecosystems, e.g. trophic cascades following the return of top predators such as the wolf or lynx (Ripple *et al.*, 2014).

Species have been deliberately encouraged outside their native range as sources of food and fur, for example, rabbits *Oryctolagus cuniculus*, and various species of deer, and now have significant impacts on woodland. Pet cats in urban areas exert a toll on small mammals and birds and interbreed with the wild cat, *Felis silvestris* (Hubbard *et al.*, 1992). Similar concerns have been raised about interbreeding between dogs and wolves in Italy (Boitani, 1992), although behavioural and physiological differences between the species may mean that mating is unlikely or that offspring rarely survive to reproduce in the wild (Vilà and Wayne, 1999). Introduced plants, such as *Rhododendron ponticum* and other deliberate ornamental introductions, as well as accidental escapes from gardens, can alter the composition of the ground and shrub layers and pose problems for woodland management (Godefroid and Koedam, 2003; Kelly, 2005; Chmura and Sierka, 2007).

Changes in the distribution of pests and diseases affecting trees, such as the emerald ash borer, *Agrilus planipennis*, and ash dieback, caused by *Hymenoscyphus fraxineus* (Paulasso *et al.*, 2013), have the potential to threaten the overall structure and composition of the woodland itself (Potter, Chapter 23). Other diseases may have more limited ecological impacts but can threaten iconic species, e.g. the role of squirrel pox in reducing red squirrel populations in Britain (Rushton *et al.*, 2006).

4.7 Changes to the Fire Regime

Closed broadleaved woodland, particularly in the wetter, western parts of the Continent, does not burn easily (Rackham, 2003). However, if the pre-Neolithic landscape were relatively open, natural fires might have ignited and spread in dry summers where there was tall grass, bracken, *Pteridium aquilinum*, or heather, *Calluna vulgaris*, as do human-set fires today, so helping to maintain that openness (Hannon et al., 2000). In the boreal and Mediterranean zones, the higher proportion of flammable trees and shrubs mean that fires are more frequent, even in closed woodland areas (Angelstam, 1998; Granström, 2001; Grove and Rackham, 2001; Angelstam and Kuuluvainen, 2004; Pausas et al., 2008; Thomas and McAlpine, 2010).

Humans have used fire since early times to modify their surroundings, and high levels of charcoal fragments in peat cores are often linked to anthropogenic changes in forest openness or composition (Bradshaw and Hannon, 1992; Tinner et al., 1999). In the last century, the planting of more flammable trees, such as conifers or eucalypts, in uniform stands, has increased the fire risk even in regions where natural fires might otherwise have been rare. Fire vulnerability may be further increased under current projections for climate change (Lindner et al., 2010).

4.8 Changes to the Forest Soil

Across Europe, the clearance and conversion of woodland to more open landscapes has been associated with soil erosion (e.g. Dearing et al., 1990; van der Knaap and van Leeuwen, 1995; Barker, 1998). Further changes to the structure and nutrient status of the soils may follow periods of agricultural use (mixing of topsoil, plough pans, removal of stones, nutrient depletion or nutrient enrichment, depending on the nature of the farming). These effects can persist, even if the land subsequently goes back under trees, and can contribute to the differences in plant composition between woodland stands with different histories (Dupouey et al., 2002; Hermy, Chapter 13).

Where the land remained under trees, the loss of the mega-herbivores would have altered the way that nutrients are redistributed across the landscape through dung (Doughty et al., 2013). Concentrations of livestock at night or around feeding and watering points, or closer to the settlement at particular seasons, would lead to the build-up of nutrients in some areas (often marked by patches of nettles, *Urtica dioica*) and their depletion elsewhere. Repeated harvesting and removal off-site of leaf fodder, young coppice growth, litter and ground vegetation (such as bracken) would remove nutrients and carbon from the soil (Dzwonko and Gawroński, 2002; Bürgi and Gimmi, 2007; Gimmi et al., 2013).

The woodland soil surface has been disturbed by the creation of banks, ditches, charcoal hearths and other features associated with the past management of the woods, and even by small scale arable cultivation (Rackham, 2003; Cervasco and Moreno, Chapter 16). There may be fewer pit-and-mound type features (Ulanova, 2000; Šamonil et al., 2010) because of reduced windthrow in managed woods. Localized compaction or disturbance of the soil may occur through the use of large-scale machinery within woods and the efforts to avoid this, through, for example, the creation of brash mats (Hutchings et al., 2002; Godefroid and Koedam, 2004; Frey et al., 2009). Wet areas, which are often hotspots of biodiversity, have often been drained (Prieditis, 1999; Heino et al., 2005).

The effects of intensification of soil preparation for forestry are particularly great on certain soil types, such as mobile sands and peatland, leading to major battles between the conservation and forestry sectors (e.g. Tsouvalis, 2000; Warren, 2000; Quine, Chapter 15). However, across Europe, the trend now is generally towards less intensive forms of preparation, particularly in second and subsequent rotations (Bürgi, Chapter 8).

4.9 Forests and Atmospheric Pollution

Nineteenth-century industrial pollution, particularly high levels of sulfur dioxide, severely reduced the abundance and diversity of epiphytic bryophytes and lichens over wide areas (Bates, 2002). A textbook example of natural selection (if often oversimplified) became the relative survival of different forms of the peppered moth on trees in unpolluted and soot-blackened environments (Cook, 2003). As air pollution from power generation has been reduced, some lichens are spreading back and this may be accompanied by increases in the abundance of some lichen-feeding moths (Conrad *et al.*, 2004).

During the 1980s, the condition of the forests themselves became of concern: ozone and compounds of nitrogen and sulfur were implicated in what became known as 'acid rain' impacts on trees (Plate 8), ground flora, soils, fresh waters, etc. (Muniz, 1990; Kubíková, 1991; Falkengren-Grerup, 1995). This led to the setting up of pan-European forest monitoring systems (Gorham, 1998; Fischer *et al.*, 2010).

In the acid rain debate, trees turned out to be culprits as well as victims. Trees scavenged particles and gases out of the atmosphere which were then washed into the soil, streams or lakes, leading to the acidification of soils and water, and reductions in freshwater invertebrates and their predators (Ormerod *et al.*, 1986, 1989). The worst predictions of forest die-off failed to materialize (Menz and Seip, 2004). Nevertheless, while there have been subsequent improvements in air quality, the recovery of some forests and streams has been limited (Karlsson *et al.*, 2011; Durford *et al.*, 2012).

Nitrogen emissions from power stations, road traffic and agricultural operations continue to pose a threat to semi-natural vegetation. Roelofs (1995) developed estimates of critical loads for a variety of vegetation types, including forests, which were later updated by Bobbink *et al.* (2002). These are currently exceeded in many parts of Europe (Erisman *et al.*, 2003). This excess nitrogen deposition appears to have had less effect on the general composition of the woodland vascular flora (Verheyen *et al.*, 2012) than it has in other vegetation types, e.g. heathland, perhaps because the more competitive, nitrophilous species remain limited by the shade from the tree layer. Still, as nitrogen levels build up in the soil, a tipping point may be reached, and then there could be a sudden change in the vegetation.

4.10 Climate Change

The climate has not remained constant since the pre-Neolithic period. There have been periods when parts of Europe were warmer than now, and other times when conditions were generally colder, and the vegetation has responded accordingly (Huntley, 1990). The nature and rate of change over the last 200 years have, however, been exacerbated by human activity, specifically by the increased output of 'greenhouse gases' following the industrial revolution. This has led to changes in species distributions, abundance and behaviour, but also to changes in where forests are grown and the threats to which they are exposed (Lindner *et al.*, 2008; Amano *et al.*, 2014).

The impacts of climate change are likely to differ both according to forest type and regionally across Europe (Lindner *et al.*, 2010). In northern and western Europe, the increasing atmospheric carbon dioxide content and warmer temperatures could increase forest growth and wood production in the short-to-medium term; in contrast, increasing drought and disturbance risks will have adverse effects. The negative impacts are more likely to outweigh the positive trends in southern and eastern Europe, so that in the Mediterranean regions productivity may decline as a result of strongly increased drought and fire risk. The threats from pests and diseases will alter in response to changed environmental conditions. Furthermore, the main problems may not be the changing climate, but its interaction with changed trade patterns as more potentially harmful organisms are transported around the world (Potter, Chapter 23).

Much work has been done on trying to project the likely impacts of climate change on biodiversity; this has used the climate records for where species currently occur, and looked at where such conditions might be found in future. For species living beneath the canopy, the microclimate in which they occur is, however, different to that measured in the open, and their response to overall climate change may, therefore, be delayed (De Frenne *et al.*, 2013). On the one hand, reduced woodland cover and isolation of forest patches in a farmed matrix may increase the vulnerability of species to climate-driven extinction; on the other, forest management may allow the maintenance of suitable local conditions for longer than might be expected.

Trees and forests also have a role to play in the response to climate change, both in terms of the adaptation of the landscape to changed environmental conditions and the adaptation of societies (Read *et al.*, 2009). Increased forest cover and productivity can contribute to carbon sequestration; trees and woods can contribute to shading of people, buildings and livestock; more wood could be used as a renewable energy source. Lindner *et al.* (2010) consider that the adaptive capacity in the European forest sector is relatively large in the boreal and the temperate oceanic regions, more constrained by socio-economic factors in the temperate continental region and most limited in the Mediterranean region.

4.11 Conclusion

The composition and structure of European woods and forests have not been constant, but have changed, sometimes slowly, sometimes rapidly, in response to human and natural pressures. The processes, though, have not been all one way; trees, woods and forests have influenced us as well, providing a source of spiritual and cultural inspiration, alongside many material benefits.

In the past, variation at a landscape scale maintained a rich diversity of species from the medieval period through to the 19th century, but greater homogenization of forests in the 20th century has led to overall species losses. In the future, our woods and forests will continue to change, irrespective of whether we intervene or not. Subsequent chapters describe the processes involved and variations on this theme in different parts of the continent.

References

Amano, T., Freckleton, R.P., Queenborough, S.A., Doxford, S.W., Smithers, R.J., Sparks, T.H. and Sutherland, W.J. (2014) Links between plant species' spatial and temporal responses to a warming climate. *Proceedings of the Royal Society B: Biological Sciences* 281(1779):20133017. Available at: http://rspb.royalsocietypublishing.org/content/281/1779/20133017 (accessed 19 November 2014).
Andrén, H. and Delin, A. (1994) Habitat selection in the Eurasian red squirrel, *Sciurus vulgaris*, in relation to forest fragmentation. *Oikos* 70, 43–48.
Angelstam, P.K. (1998) Maintaining and restoring biodiversity in European boreal forests by developing natural disturbance regimes. *Journal of Vegetation Science* 9, 593–602.
Angelstam, P. and Kuuluvainen, T. (2004) Boreal forest disturbance regimes, successional dynamics and landscape structures: a European perspective. *Ecological Bulletins* 51, 117–136.
Arabatzis, G. (2005) European Union, Common Agricultural Policy (CAP) and the afforestation of agricultural land in Greece. *New Mediterranean* 4, 48–54.
Bailey, S. (2007) Increasing connectivity in fragmented landscapes: an investigation of evidence for biodiversity gain in woodlands. *Forest Ecology and Management* 238, 7–23.
Barker, S. (1998) The history of the Coniston woodlands, Cumbria, UK. In: Kirby, K.J. and Watkins, C. (eds) *The Ecological History of European Forests*. CAB International, Wallingford, UK, pp. 167–183.
Bates, J. (2002) Effects on bryophytes and lichens. In: Bates, J.W. and Farmer, A.M. (eds) *Air Pollution and Plant Life*. Wiley, Chichester, UK, pp. 309–342.
Bengtsson, J., Nilsson, S.G., Franc, A. and Menozzi, P. (2000) Biodiversity, disturbances, ecosystem function and management of European forests. *Forest Ecology and Management* 132, 39–50.
Bobbink, R. and Roelofs, J.M. (1995) Nitrogen critical loads for natural and semi-natural ecosystems: the empirical approach. *Water, Air, and Soil Pollution* 85, 2413–2418.
Bobbink, R., Ashmore, M., Braun, S., Flückiger, W. and Van den Wyngaert, I.J.J. (2002) *Empirical Nitrogen Critical Loads for Natural and Semi-natural Ecosystems: 2002 Update*. Institute for Applied Plant Biology, Schönenbuch, Switzerland. Available at: http://www.iap.ch/publikationen/nworkshop-background.pdf (accessed 19 November 2014).
Boitani, L. 1992. Wolf research and conservation in Italy. *Biological Conservation* 61, 125–132.

Bradshaw, R. and Hannon, G. (1992) Climatic change, human influence and disturbance regime in the control of vegetation dynamics within Fiby Forest, Sweden. *Journal of Ecology* 80, 625–632.

Bürgi, M. and Gimmi, U. (2007) Three objectives of historical ecology: the case of litter collecting in Central European forests. *Landscape Ecology* 22, 77–87.

Buscardo, E., Smith, G., Kelly, D., Freitas, H., Iremonger, S., Mitchell, F.G., O'Donoghue, S. and Mckee, A.-M. (2008) The early effects of afforestation on biodiversity of grasslands in Ireland. *Biodiversity and Conservation* 17, 1057–1072.

Chmura, D. and Sierka, E. (2007) The invasibility of deciduous forest communities after disturbance: a case study of *Carex brizoides* and *Impatiens parviflora* invasion. *Forest Ecology and Management* 242, 487–495.

Clark, P. (2004) *The Dover Bronze Age Boat*. English Heritage, Swindon, UK.

Conrad, K., Woiwod, I., Parsons, M., Fox, R. and Warren, M. (2004) Long-term population trends in widespread British moths. *Journal of Insect Conservation* 8, 119–136.

Cook, L.C. (2003) The rise and fall of the *carbonaria* form of the peppered moth. *The Quarterly Review of Biology* 78, 399–417.

Dallimore, W. (1911) The beechwood industry of the Chilterns. *Bulletin of Miscellaneous Information (Royal Gardens, Kew)* 1911, 109–114.

Day, S.P. (1993) Woodland origin and 'ancient woodland indicators': a case-study from Sidlings Copse, Oxfordshire, UK. *The Holocene* 3, 45–53.

De Frenne, P., Rodríguez-Sánchez, F., Coomes, D.A., Baeten, L., Verstraeten, G., Vellend, M., Bernhardt-Römermann, M., Brown, C.D., Brunet, J., Cornelis, J., Decocq, G.M., Dierschke, H. *et al.* (2013) Microclimate moderates plant responses to macroclimate warming. *Proceedings of the National Academy of Sciences of the United States of America* 110, 18561–18565.

Dearing, J.A., Alström, K., Bergman, A., Regnell, J. and Sandgreen, P. (1990) Recent and long-term records of soil erosion from southern Sweden. In: Boardman, J., Foster, I.D.L and Dearing, J.A. (eds) *Soil Erosion on Agricultural Land: Proceedings of a Workshop Sponsored by the British Geomorphological Research Group, Coventry, UK, January 1989*. Wiley, Chichester, UK, pp. 173–191.

Debussche, M., Lepart, J. and Dervieux, A. (1999) Mediterranean landscape changes: evidence from old postcards. *Global Ecology and Biogeography* 8, 3–15.

Deinet, S., Ieronymidou, C., McRae, L., Burfield, I.J., Foppen, R.P., Collen, B. and Böhm, M. (2013) *Wildlife Comeback in Europe: The Recovery of Selected Mammal and Bird Species*. Final Report to Rewilding Europe by ZSL, Birdlife International and the European Bird Census Council. Zoological Society of London, London.

Doughty, C.E., Wolf, A. and Malhi, Y. (2013) The impact of large animal extinctions on nutrient fluxes in early river valley civilizations. *Ecosphere* 4, 1–17.

Dumayne-Peaty, L. (1999) Continuity or discontinuity? Vegetation change in the Hadrianic–Antonine frontier zone of northern Britain at the end of the Roman occupation. *Journal of Biogeography* 26, 643–665.

Dunford, R.W., Donoghue, D.N.M. and Burt, T.P. (2012) Forest land cover continues to exacerbate freshwater acidification despite decline in sulphate emissions. *Environmental Pollution* 167, 58–69.

Dupouey, J.L., Dambrine, E., Laffite, J.D. and Moares, C. (2002) Irreversible impact of past land use on forest soils and biodiversity. *Ecology* 83, 2978–2984.

Dzwonko, Z. and Gawroński, S. (2002) Effect of litter removal on species richness and acidification of a mixed oak-pine woodland. *Biological Conservation* 106, 389–398.

Erisman, J.W., Grennfelt, P. and Sutton, M. (2003) The European perspective on nitrogen emission and deposition. *Environment International* 29, 311–325.

Falkengren-Grerup, U. (1995) Long-term changes in flora and vegetation in deciduous forests of southern Sweden. *Ecological Bulletins* 44, 215–226.

Fischer, R., Lorenz, M., Köhl, M., Mues, V., Granke, O., Iost, S., van Dobben, H., Reinds, G.J. and de Vries, W. (2010) *The Condition of Forests in Europe, 2010 Executive Report*. ICP Forests (International Co-operative Programme on Assessment and Monitoring of Air Pollution Effects on Forests) and European Commission, Hamburg, Germany and Brussels.

Foot, D. (2010) *Woods and People: Putting Forestry on the Map*. The History Press, Stroud, UK.

Forman, R. (1995) *Land Mosaics. The Ecology of Landscapes and Regions*. Cambridge University Press, Cambridge, UK.

Frey, B., Kremer, J., Rüdt, A., Sciacca, S., Matthies, D. and Lüscher, P. (2009) Compaction of forest soils with heavy logging machinery affects soil bacterial community structure. *European Journal of Soil Biology* 45, 312–320.

Game, M. and Peterken, G.F. (1984) Nature reserve selection strategies in the woodlands of central Lincolnshire, England. *Biological Conservation* 29, 157–181.

Gimmi, U., Poulter, B., Wolf, A., Portner, H., Weber, P. and Bürgi, M. (2013) Soil carbon pools in Swiss forests show legacy effects from historic forest litter raking. *Landscape Ecology* 28, 835–846.

Godefroid, S. and Koedam, N. (2003) Identifying indicator plant species of habitat quality and invasibility as a guide for peri-urban forest management. *Biodiversity and Conservation* 12, 1699–1713.

Godefroid, S. and Koedam, N. (2004) Interspecific variation in soil compaction sensitivity among forest floor species. *Biological Conservation* 119, 207–217.

Gorham, E. (1998) Acid deposition and its ecological effects: a brief history of research. *Environmental Science and Policy* 1, 153–166.

Gove, B., Power, S.A., Buckley, G.P. and Ghazoul, J. (2007) Effects of herbicide spray drift and fertilizer overspread on selected species of woodland ground flora: comparison between short-term and long-term impact assessments and field surveys. *Journal of Applied Ecology* 44, 374–384.

Granström, A. (2001) Fire management for biodiversity in the European boreal forest. *Scandinavian Journal of Forest Research* 16, 62–69.

Grilo, C., Bissonette, J.A. and Santos-Reis, M. (2009) Spatial–temporal patterns in Mediterranean carnivore road casualties: consequences for mitigation. *Biological Conservation* 142, 301–313.

Groenewoudt, B., van Haaster, H., van Beek, R. and Brinkkemper, O. (2007) Towards a reverse image. Botanical research into the landscape history of the eastern Netherlands (1100 BC–AD 1500). *Landscape History* 29, 17–33.

Grove, A.T. and Rackham, O. (2001) *The Nature of Mediterranean Europe: An Ecological History*. Yale University Press, New Haven, Connecticut.

Hannon, G.E., Bradshaw, R. and Emborg, J. (2000) 6000 years of forest dynamics in Suserup Skov, a semi-natural Danish woodland. *Global Ecology and Biogeography* 9, 101–114.

Hanski, I., Kuussaari, M. and Nieminen, M. (1994) Metapopulation structure and migration in the butterfly *Melitaea cinxia*. *Ecology* 75, 747–762.

Hartel, T. and Plieninger, T. (eds) (2014) *European Wood-Pastures in Transition: A Social–Ecological Approach*. Earthscan from Routledge (imprint of Taylor & Francis), Abingdon, UK.

Heino, J., Virtanen, R., Vuori, K.-M., Saastamoinen, J., Ohtonen, A. and Muotka, T. (2005) Spring bryophytes in forested landscapes: land use effects on bryophyte species richness, community structure and persistence. *Biological Conservation* 124, 539–545.

Herbst, M., Roberts, J.M., Rosier, P.T.W., Taylor, M.E. and Gowing, D.J. (2007) Edge effects and forest water use: a field study in a mixed deciduous woodland. *Forest Ecology and Management* 250, 176–186.

Herfindal, I., Linnell, J.D.C., Odden, J., Nilsen, E.B. and Andersen, R. (2005) Prey density, environmental productivity and home-range size in the Eurasian lynx (*Lynx lynx*). *Journal of Zoology* 265, 63–71.

Herlin, I.S. (2001) Approaches to forest edges as dynamic structures and functional concepts. *Landscape Research* 26, 27–43.

Hermy, M., Honnay, O., Firbank, L., Grashof-Bokdam, C. and Lawesson, J.E. (1999) An ecological comparison between ancient and other forest plant species of Europe, and the implications for forest conservation. *Biological Conservation* 91, 9–22.

Hubbard, A.L., Mcoris, S., Jones, T.W., Boid, R., Scott, R. and Easterbee, N. (1992) Is survival of European wildcats *Felis silvestris* in Britain threatened by interbreeding with domestic cats? *Biological Conservation* 61, 203–208.

Huntley, B. (1990) European post-glacial forests: compositional changes in response to climatic change. *Journal of Vegetation Science* 1, 507–518.

Hutchings, T.R., Moffat, A.J. and French, C.J. (2002) Soil compaction under timber harvesting machinery: a preliminary report on the role of brash mats in its prevention. *Soil Use and Management* 18, 34–38.

Karlsson, P.G., Akselsson, C., Hellsten, S. and Karlsson, P.E. (2011) Reduced European emissions of S and N – effects on air concentrations, deposition and soil water chemistry in Swedish forests. *Environmental Pollution* 159, 3571–3582.

Kelly, D.L. (2005) Woodland on the western fringe: Irish oak wood diversity and the challenges of conservation. *Botanical Journal of Scotland* 57, 21–40.

Kerth, G. and Melber, M. (2009) Species-specific barrier effects of a motorway on the habitat use of two threatened forest-living bat species. *Biological Conservation* 142, 270–279.

Koop, H. (1987) Vegetative reproduction of trees in some European natural forests. *Vegetatio* 72, 103–110.

Kortelainen, J. (1999) The river as an actor-network: the Finnish forest industry utilization of lake and river systems. *Geoforum* 30, 235–247.

Kreuz, A. (1992) Charcoal from ten early Neolithic settlements in Central Europe and its interpretation in terms of woodland management and wildwood resources. *Bulletin de la Société Botanique de France: Actualités Botaniques* 139, 383–394.

Kubíková, J. (1991) Forest dieback in Czechoslovakia. *Vegetatio* 93, 101–108.

Lindner, M., Garcia-Gonzalo, J., Kolström, M., Green, T., Reguera, R., Maroschek, M., Seidl, R., Lexer, M.J., Netherer, S., Schopf, A. *et al.* (2008) *Impacts of Climate Change on European Forests and Options for Adaptation*. AGRI-2007-G4-06, Report to the European Commission Directorate-General for Agriculture and Rural Development, Brussels, Belgium. Available at: http://ec.europa.eu/agriculture/analysis/external/euro_forests/full_report_en.pdf (accessed 19 November 2014).

Lindner, M., Maroschek, M., Netherer, S., Kremer, A., Barbati, A., Garcia-Gonzalo, J., Seidl, R., Delzon, S., Corona, P., Kolström, M., Lexer, M.J. and Marchetti, M. (2010) Climate change impacts, adaptive capacity, and vulnerability of European forest ecosystems. *Forest Ecology and Management* 259, 698–709.

Lindsay, J.M. (1976) The commercial use of highland woodland, 1750–1870: a reconsideration. *Scottish Geographical Magazine* 92, 30–40.

Mathieu, J.R. and Meyer, D.A. (1997) Comparing axe heads of stone, bronze, and steel: studies in experimental archaeology. *Journal of Field Archaeology* 24, 333–351.

Menz, F.C. and Seip, H.M. (2004) Acid rain in Europe and the United States: an update. *Environmental Science and Policy* 7, 253–265.

Mitchell, P.L. and Kirby, K.J. (1989) *Ecological Effects of Forestry Practices in Long-established Woodland and their Implications for Nature Conservation*. Occasional Paper 39, Oxford Forestry Institute, Oxford, UK.

Mitchell, R.J., Beaton, J.K., Bellamy, P.E., Broome, A., Chetcuti, J., Eaton, S., Ellis, C.J., Gimona, A., Harmer, R., Hester, A.J. *et al.* (2014) Ash dieback in the UK: a review of the ecological and conservation implications and potential management options. *Biological Conservation* 175, 95–109.

Muniz, I.P. (1990) Freshwater acidification: its effects on species and communities of freshwater microbes, plants and animals. *Proceedings of the Royal Society of Edinburgh Section B: Biological Sciences* 97, 227–254.

Olsson, E.G., Austrheim, G. and Grenne, S. (2000) Landscape change patterns in mountains, land use and environmental diversity, mid-Norway 1960–1993. *Landscape Ecology* 15, 155–170.

Ormerod, S.J., Allenson, N., Hudson, D. and Tyler, S.J. (1986) The distribution of breeding dippers (*Cinclus* (L.); Aves) in relation to stream acidity in upland Wales. *Freshwater Biology* 16, 501–507.

Ormerod, S.J., Donald, A.P. and Brown, S.J. (1989) The influence of plantation forestry on the pH and aluminium concentration of upland Welsh streams: a re-examination. *Environmental Pollution* 62, 47–62.

Out, W.A. (2010) Firewood collection strategies at Dutch wetland sites in the process of Neolithisation. *The Holocene* 20, 191–204.

Pausas, J.G., Llovet, J., Rodrigo, A. and Vallejo, R. (2008) Are wildfires a disaster in the Mediterranean basin? – A review. *International Journal of Wildland Fire* 17, 713–723.

Pautasso, M., Aas, G., Queloz, V. and Holdenrieder, O. (2013) European ash (*Fraxinus excelsior*) dieback – a conservation biology challenge. *Biological Conservation* 158, 37–49.

Peterken, G.F. (1977) Habitat conservation priorities in British and European woodlands. *Biological Conservation* 11, 223–236.

Peterken, G.F. (1993) *Woodland Conservation and Management*, 2nd edn. Chapman and Hall, London.

Peterken, G.F. (2013) *Meadows*. British Wildlife Publishing, Gillingham, UK.

Peterken, G.F. and Hughes, F.M.R. (1995) Restoration of floodplain forests in Britain. *Forestry* 68, 187–202.

Plieninger, T. and Wilbrand, C. (2001) Land use, biodiversity conservation, and rural development in the dehesas of Cuatro Lugares, Spain. *Agroforestry Systems* 51, 23–34.

Prevosto, B., Curt, T., Dambrine, E. and Coquillard, P. (2004) Natural tree colonisation of former agricultural lands in the French Massif Central: impact of past land use on stand structure, soil characteristics and understorey vegetation. In: Honnay, O., Verheyen, K., Bossuyt, B. and Hermy, M. (eds) *Forest Biodiversity – Lessons from History for Nature Conservation*. CAB International, Wallingford, UK, pp. 41–53.

Prieditis, N. (1999) Status of wetland forests and their structural richness in Latvia. *Environmental Conservation* 26, 332–346.

Rackham, O. (1990) *Trees and Woodland in the British Landscape*, rev. edn. Dent, London.

Rackham, O. (2003) *Ancient Woodland: Its History, Vegetation and Uses in England*, rev. edn. Castlepoint Press, Dalbeattie, UK.

Read, D.J., Freer Smith, P., Morison, J., Hanley, N., West, C. and Snowdon, P. (eds) (2009) *Combating Climate Change: A Role for UK Forests*. The Stationery Office Limited, London.

Reijnen, R., Foppen, R., Braak, C.T. and Thissen, J. (1995) The effects of car traffic on breeding bird populations in woodland. III. Reduction of density in relation to the proximity of main roads. *Journal of Applied Ecology* 32, 187–202.

Ripple, W.J., Estes, J.A., Beschta, R.L., Wilmers, C.C., Ritchie, E.G., Hebblewhite, M., Berger, J., Elmhagen, B., Letnic, M., Nelson, M.P. *et al.* (2014) Status and ecological effects of the world's largest carnivores. *Science* 343(6167). Available at: https://www.sciencemag.org/content/343/6167/1241484?related-urls=yes&legid=sci;343/6167/1241484 (accessed 19 November 2014).

Rose, F. (1993) Ancient British woodlands and their epiphytes. *British Wildlife* 5, 83–94.
Rushton, S.P., Lurz, P.W.W., Gurnell, J., Nettleton, P., Bruemmer, C., Shirley, M.D.F. and Sainsbury, A.W. (2006) Disease threats posed by alien species: the role of a poxvirus in the decline of the native red squirrel in Britain. *Epidemiology and Infection* 134, 521–533.
Šamonil, P., Král, K. and Hort, L. (2010) The role of tree uprooting in soil formation: a critical literature review. *Geoderma* 157, 65–79.
Santos, T. and Tellería, J. (1992) Edge effects on nest predation in Mediterranean fragmented forests. *Biological Conservation* 60, 1–5.
Schadt, S., Knauer, F., Kaczensky, P., Revilla, E., Wiegand, T. and Trepl, L. (2002) Rule-based assessment of suitable habitat and patch connectivity for the Eurasian lynx. *Ecological Applications* 12, 1469–1483.
Schwartz, M.W. (1999) Choosing the appropriate scale of reserves for conservation. *Annual Review of Ecology and Systematics* 30, 83–108.
Snäll, T., Pennanen, J., Kivistö, L. and Hanski, I. (2005) Modelling epiphyte metapopulation dynamics in a dynamic forest landscape. *Oikos* 109, 209–222.
Storch, I., Woitke, E. and Krieger, S. (2005) Landscape-scale edge effect in predation risk in forest–farmland mosaics of Central Europe. *Landscape Ecology* 20, 927–940.
Thomas, P.A. and McAlpine, R. (2010) *Fire in the Forest*. Cambridge University Press, Cambridge, UK, 225 pp.
Tinner, W., Hubschmid, P., Wehrli, M., Ammann, B. and Conedera, M. (1999) Long-term forest fire ecology and dynamics in southern Switzerland. *Journal of Ecology* 87, 273–289.
Tsouvalis, J. (2000) *A Critical Geography of Britain's State Forests*. Oxford University Press, Oxford, UK.
Ulanova, N.G. (2000) The effects of windthrow on forests at different spatial scales: a review. *Forest Ecology and Management* 135, 155–167.
Valverde, T. and Silvertown, J. (1997) A metapopulation model for *Primula vulgaris*, a temperate forest understorey herb. *Journal of Ecology* 85, 193–210.
van der Knaap, W.O. and van Leeuwen, J.F.N. (1995) Holocene vegetation succession and degradation as responses to climatic change and human activity in the Serra de Estrela, Portugal. *Review of Palaeobotany and Palynology* 89, 153–211.
van Hoof, T.B., Bunnik, F.P.M., Waucomont, J.G.M., Kürschner, W.M. and Visscher, H. (2006) Forest re-growth on medieval farmland after the Black Death pandemic – implications for atmospheric CO_2 levels. *Palaeogeography, Palaeoclimatology, Palaeoecology* 237, 396–409.
Verheyen, K., Baeten, L., De Frenne, P., Bernhardt-Römermann, M., Brunet, J., Cornelis, J., Decoq, G., Dierschke, H., Eriksson, O., Hedl, R. et al. (2012) Driving factors behind the eutrophication signal in understorey plant communities of deciduous temperate forests. *Journal of Ecology* 100, 352–365.
Vilà, C. and Wayne, R.K. (1999) Hybridization between wolves and dogs. *Conservation Biology* 13, 195–198.
Visser, R.M. (2010) Growing and felling? Theory and evidence related to the application of silvicultural systems in the Roman period. In: Moore, M., Taylor, G., Harris, E., Girdwood, P. and Shipley, L. (eds) *TRAC 2009. Proceedings of the Nineteenth Annual Theoretical Roman Archaeology Conference, Michigan and Southampton 2009*. Oxbow Books, Oxford, UK, pp. 11–22.
Warde, P. (2006) Fear of wood shortage and the reality of the woodland in Europe, c.1450–1850. *History Workshop Journal* 62, 28–57.
Warren, C. (2000) 'Birds, bogs and forestry' revisited: the significance of the flow country controversy. *Scottish Geographical Journal* 116, 315–337.
Weber, N. (2000) *NEWFOR – New Forests for Europe: Afforestation at the Turn of the Century*. European Forest Institute, Joensuu, Finland.
Wenger, E.L., Zinke, A. and Gutzweiler, K.-A. (1990) Present situation of the European floodplain forests. *Forest Ecology and Management* 33–34, 5–12.
Williams, M. (2006) *Deforesting the Earth: From Prehistory to Global Crisis*. University of Chicago Press, Chicago, Illinois.
Williams, M. (2008) A new look at global forest histories of land clearing. *Annual Review of Environment and Resources* 33, 345–367.
Woitsch, J. (2009) Charcoal makers in Bohemia: from privileged craftsmen to strange forest dwellers. In: Saratsi, E., Bürgi, M., Johann, E., Kirby, K.J., Moreno, D. and Watkins, C. (eds) *Woodland Cultures in Time and Space – Tales from the Past, Messages for the Future. Proceedings of Conference in Thessaloniki, Greece in 2007, Organized by IUFRO Division 6.07.04*. Embryo Publications, Athens, pp. 80–88.
Yalden, D. (1999) *The History of British Mammals*. Poyser, London.

Part II

The Variety of Management Across European Woods and Forests

This part of the book looks at different ways in which woods and forests have been, and are being, managed. A particular feature of the last two decades has been the considerable increase in interest in and understanding of wood-pasture systems (Chapter 5). The traditional counterpart of such systems – coppice – has also not been neglected (Chapter 6). However, the tendency has been for woods and forests to be converted or to develop through neglect into high forest structures (Chapter 7). Concerns over the impact of large clear-felled areas have led to a revival of interest in smaller scale working, often grouped under the title of 'close-to-nature' forestry (Chapter 8). Hunting has often been as important as timber production in shaping where trees and woods occur in the landscape (Chapter 9) and has contributed to both species extirpations, such as the loss of the wolf from large parts of Europe, and to species introductions and translocations (pheasants, fallow deer, etc.).

5 Wood-pastures in Europe

Tibor Hartel,[1]* Tobias Plieninger[2] and Anna Varga[3]

[1]*Department of Environmental Studies, Sapientia Hungarian University of Transylvania, Cluj-Napoca, Romania; [2]Department of Geosciences and Natural Resource Management, University of Copenhagen, Frederiksberg, Denmark; [3]Institute of Ecology and Botany, Hungarian Academy of Sciences, Vácrátot, Hungary*

5.1 Introduction

Humans interact with their landscapes both physically, in how we extract resources, and culturally, through the different values that we place on landscapes and their components. These interactions have strong consequences for the structural and ecological properties of the landscape and its capacity to provide goods and services, as well as for the culture of the local human societies (Plieninger and Bieling, 2012). The values attached to cultural landscapes depend in part on their physical properties (their landscape structure and biodiversity content), but also reflect a particular set of human skills, knowledge and aesthetic judgements. Cultural landscapes are recognized for their potential for sustainable development and nature conservation (Barthel *et al.*, 2013a,b). However, to understand the limits to and possibilities for conserving the sociocultural heritage and the ecological properties of cultural landscapes, we need to consider their social and ecological dimensions, and their historical as well as their current interactions (Plieninger and Bieling, 2012; de Snoo *et al.*, 2013; Mikulcak *et al.*, 2013).

Woodland grazing with livestock has been part of the landscape since at least the Neolithic era (Luick, 2009) and wood-pastures represent an important part of the European cultural and ecological heritage (Bergmeier *et al.*, 2010; Hartel and Plieninger, 2014). Wood-pastures may have a wide diversity of forms and expressions in Europe, ranging from scattered trees in a pasture – a 'savannah' – to closed-canopy forests with high tree density grazed by livestock (Fig. 5.1 and Plate 5). The species composition of tree communities in wood-pastures and the age structure of their tree populations have also been shaped by human activity, most typically by logging, coppicing and pollarding (Jørgensen and Quelch, 2014) and, indirectly, through the impact of the grazing on regeneration.

The diverse and dynamic character of these systems and their uses are reflected in the various terms applied to them in Europe (Bergmeier *et al.*, 2010): 'ancient park', 'savannah', 'pasture woodland', 'semi-open pasture land', 'traditional orchard', 'woodland pasture', 'wooded pasture', 'wooded meadow' and 'pastoral woodland'. The term 'silvopastoral system' is a more technical and formal term often used to describe types of land use that integrate modern livestock grazing with trees and their associated goods and services (Mosquera Losada *et al.*, 2009).

---

*E-mail: hartel.tibor@gmail.com

© CAB International 2015. *Europe's Changing Woods and Forests: From Wildwood to Managed Landscapes* (eds K.J. Kirby and C. Watkins)

Fig. 5.1. Examples of wood-pastures in various locations of Europe: (a) a wood-pasture made by oaks scattered through pasture and arable fields in the western part of Lesvos Island, Greece; (b) *dehesa* wood-pasture in Monroy, Spain, scattered mostly with holm oaks; (c) traditionally grazed ancient oak wood-pasture in southern Transylvania, Romania; (d) wood-pasture with ancient willow pollards grazed by sheep in the lowlands of Transylvania, Romania; (e) oak wood-pasture grazed by pigs in Croatia; (f) 'Hatfield Forest' wood-pasture in Essex, UK. (From: (a–b) Tobias Plieninger; (c–d) Tibor Hartel; (e) Anna Varga; (f) Keith Kirby.)

In this chapter, we use the term 'wood-pasture' for treed landscapes in which livestock grazing co-occurs with woody vegetation (trees and shrubs). We emphasize the grazing by livestock in order to reflect the cultural nature of most European wood-pastures – pasturing is fundamentally a human activity (Hartel and Plieninger, 2014), although grazing by 'natural' large grazers such as deer can be important at some sites.

We provide a brief general historical overview of wood-pastures in Europe, and then look at how societies viewed and managed wood-pastures in different periods, and at the changing social, economic and cultural values attributed to wood-pastures. We highlight the recent fate of European wood-pastures and key messages for their conservation. Particular attention is paid to Eastern Europe, which is rich in wood-pastures that have been less well described and discussed than those in the central and western countries of Europe. We draw on an expert-based survey made in 2012 for 12 European countries (Spain, Portugal, Germany, Belgium, France, Switzerland, Italy, Hungary, the Czech Republic, Romania, Sweden and Greece) in order to produce a more complete picture of the extent of wood-pastures in these countries, the ecosystem services they provide and the threats that they face.

5.2 Wood-pasture: A Multi-purpose System

Trees and shrubs were considered important in pasture management because of their beneficial effects in protecting the soil from erosion, sheltering livestock against extreme weather conditions and providing favourable microclimatic conditions that might benefit grassland (Dorner, 1910; Gegesi, 1911; Manning et al., 2006). Trees in wood-pastures could provide various goods (wood, fruits) for local communities as well as being a regular or at least occasional source of fodder for livestock (Wessely, 1864; Vera, 2000; Oroszi, 2004; Jørgensen, 2013).

Oaks (*Quercus* spp.) were particularly important across Europe because of the diverse goods provided, including acorns, good-quality timber and cork. The pear (*Pyrus* sp.) was planted on many pastures in the 18th and 19th centuries in central and Eastern Europe because of its economic value and for protection against soil erosion. In lowland areas with high soil moisture content, the various willow species (*Salix* spp.) were maintained in grazed landscapes, their branches being used for making baskets, fences, binding, sticks or support for hay. These trees are common in many 'traditional' wood-pastures and hay meadows from the lowland and foothill regions of Europe, where the environmental conditions are suitable for them (Luick, 2009; Bergmeier et al., 2010; Oppermann, 2014).

Past and current land management often produces differences in the species composition and size distribution of trees between wood-pastures and closed forests. In the lowland foothills region of southern Transylvania (Romania) wood-pastures are still well represented (Hartel et al., 2013) and are generally dominated by oak (mostly *Q. robur* with some *Q. petraea*) and fruit trees, especially pear, with occasional dominance of hornbeam (*Carpinus betulus*) and beech (*Fagus sylvatica*) (Fig. 5.2). These tree species were selectively maintained by the local communities for their various services, at least in the 18th and 19th centuries (Dorner, 1910; Oroszi, 2004). The forests in this region (where livestock grazing is prohibited) are dominated by hornbeam, beech and oak; they lack pear and apple (*Malus* sp.) trees. The closed forests are mostly managed for timber production, trees being cut in 80–120 year cycles. Because the primary goal of the management of many wood-pastures was grazing, scattered trees in wood-pastures could be left to grow for several centuries (Hartel et al., 2013). Consequently, large, old trees are better represented in wood-pastures than in production forests. Similar management-induced differences are known in other parts of Europe (Rackham, 1989; Vera, 2000).

Regular grazing and other management interventions are needed in wood-pastures to maintain their often semi-open structure and the provision of ecosystem goods and services. The complete or even partial cessation of these activities will lead to structural changes in the wood-pastures, reflecting their intermediate state between open pastures and closed forests (Gillet, 2008). In the absence of management, shrubs and fast-growing trees tend to develop to the detriment of the grassland vegetation (Manning et al., 2006, 2009), but are also a serious threat to the old trees (Bergmeier et al., 2010). In many closed canopy forests in the traditional rural landscapes of central Romania, one can still commonly find overtopped large trees (hollowing, dying or dead oak, hornbeam

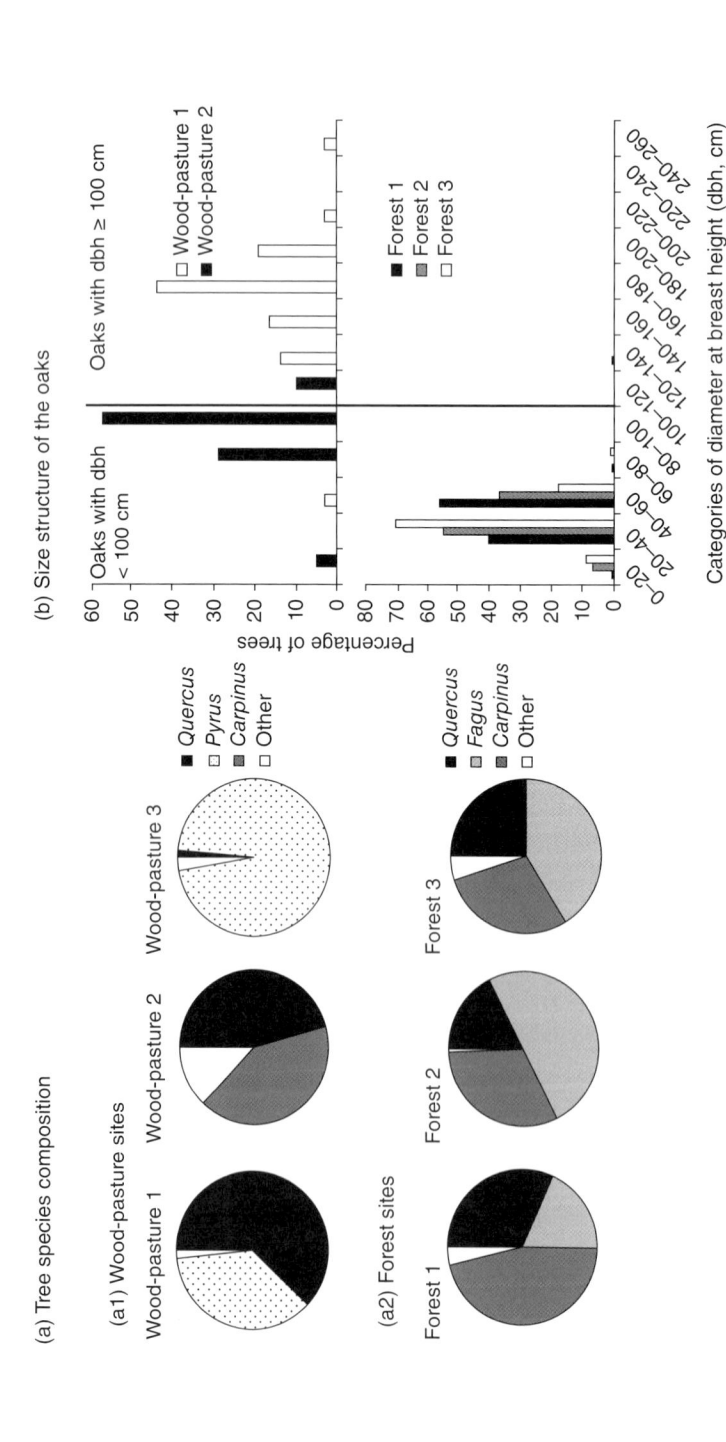

Fig. 5.2. (a) The tree species composition, and (b) the size distribution of oak trees (*Quercus robur* and *Q. petraea*) in three wood-pastures and three forest sites from the traditional rural landscapes of southern Transylvania (Romania) (from Hartel et al., 2013). In (a) the tree species include oaks (*Quercus* spp.), beech (*Fagus sylvatica*), hornbeam (*Carpinus betulus*), pear (*Pyrus*) and 'others', which are represented as shown in the keys; the four named species are always shown where they were identified. In 'wood-pasture 2' the high proportion of hornbeam is caused by the cessation of grazing in the past 10 years, as is probably the better representation of the smaller and presumably younger oaks in this wood-pasture compared with 'wood-pasture 1'. Oak size data for 'wood-pasture 3' are not shown because only one stem was identified in the sampling site.

and beech pollards), indicating that the area had a semi-open character in the past decades, most likely as wood-pasture. In contrast to regular management interventions in wood-pastures, intensive management of the grassland hampers the development of the woody vegetation. Both abandonment and intensification can thus lead to a shift from a multifunctional to a mono-functional land use, with a decrease in the richness and quality of the ecosystem goods and services provided (Bugalho et al., 2011).

Wood-pastures respond quickly to variations in the management regime and intensity which, in turn, are influenced by various historical and current shocks, demographic changes, industrialization and institutional development. The spatial and temporal dynamics of wood-pastures were and still are strongly linked to broad social and economic developments and to the different ways that societies have perceived and valued these landscapes and their components. To achieve sustainable use of the wood-pasture landscapes of Europe, we must understand how the nature of these links between societies and wood-pastures has changed over time (Hartel and Plieninger, 2014).

5.3 Historical Development of Wood-pastures in Europe

5.3.1 Forest grazing and pasturing in ancient times

Vera (2000) and Luick (2008), using a variety of written evidence, have reviewed the ways in which various societies from central and western Europe interpreted terms such as *forestis*, *silva*, *wald*, *silva glandiferae*, or *silva vulgaris ascuae*. These and similar terms used to name wood-pastures always included a 'non-tree' type of vegetation and land use as well, such as pastures, pasturing, cropped fields or even built areas. For example, in the English Domesday Book (1086 AD) *silva pastilis*, literally meaning wood-pasture, is distinct from *silva minuta*, which were coppices (see Buckley and Mills, Chapter 6).

Pollarding and coppicing were common management practices applied to trees in the Middle Ages and afterwards (Jørgensen and Quelch, 2014), with the former being most prevalent in wood-pastures. Pollarding was applied in cycles from less than 4 to 10 or more years (Rackham, 1989; Vera, 2000). Trees managed in this way may eventually become very old (Read, 2000; Lonsdale, 2013; Quelch, 2013), as pollarding slows the growth of the bole, and the relatively short height, with, for much of the time, relatively little canopy in relation to bole size, protects these trees from damage caused by storms.

Wood-pastures were at the heart of many local economies for centuries. For example 'pannage', the turning out of pigs into woods dominated by oak and beech so that they could eat the mast, was often combined with coppicing and pollarding, the pannage being critical to the fattening of the pigs. In many parts of lowland and mild-altitude regions of Europe, this practice continued up to the 20th century. The season usually started in September or November and lasted until December (Makkai, 2003; Jørgensen, 2013); during this time, the pigs were allowed to stay out in the woods the whole time (Hornyák, 1998). The value of the land might be measured not in terms of timber or firewood produced but in terms of the number of pigs that it could sustain in this way (Luick, 2009; Szabó, 2013).

Because of its importance, forest grazing came under strong regulation. For example, in the Saxon area of southern Transylvania, forest grazing was regulated by a document commonly referred to as 'Andreanum', The Golden Charter of Transylvanian Saxons of 1224. A document originating in the 16th century records the Transylvanian Saxons asking István Báthory (the Prince of Transylvania, King of Poland and Grand Duke of Lithuania) to empower local authorities to control grazing with sheep and pigs in the oak forests in their territory (Teşculă and Goţa, 2007). In 1808 grazing was apparently banned from nearly half of 20,000 ha of forest (Oroszi, 2004).

At about the same time, various local institutions such as the Church, the aristocrats or the community itself sought to restrict what could be done in their forests. There had long been cases where activities that interfered with the hunting interests of the nobility had been prohibited (Luick, 2009), but Oroszi (2004) notes moves to exclude livestock grazing and, especially, wood extraction. In one village of the Saxon region of southern Transylvania, ploughing was prohibited in the vicinity of forests in the 16th century, because 'the plough is not good neighbour with the forest'. Oroszi (2004) suggests that this represents the first steps towards the more formally regulated and institutionalized forest management that developed in the 18th and 19th centuries (Savill, Chapter 7).

5.3.2 Driving the livestock out of the forest (18th–19th centuries)

The demand for timber and agricultural products sharply increased in the 18th/19th century with the rapid growth in human population (the population of Europe doubled in just 200 years) and the development of trade, industry, infrastructure and urbanization. Production needed to be intensified and new models of agriculture and forestry were introduced (Dorner, 1910; Oroszi, 2004; Luick, 2009). Livestock came to be seen as an enemy of modern forestry because trampling and grazing damaged seeds, saplings and trees. This created tensions between those keen to maintain forest grazing and pasturing and those seeking to protect forests and increase wood production.

The Hungarian journal of forestry (*Erdészeti Lapok*, in Hungarian) of the 19th and early 20th century contains a lively debate about the topic of forest grazing which probably reflects the situation in western and central Europe more generally. Those interested in wood production and forestry were concerned that the expansion of forest grazing and an increase of livestock density would damage woodland; while forestry was often blamed for the crisis, it was really more a result of inefficiency in the agricultural sector. There were calls for more woodland to be converted into agricultural land, and military maps from the 18th and 19th centuries show that across Transylvania previously wooded landscapes became pastures with scattered trees. Similar changes happened elsewhere, for example in Hungary (Fig. 5.3) and Germany (Luick, 2009).

Foresters such as Emil Belházy (1888) set out the management interventions needed to achieve sustainable forest grazing, relating, for instance to tree densities, species composition and regeneration patterns. He considered that the primary aim of trees in grazed forests was to improve soil condition and grassland productivity.

Meanwhile, there were suggestions for agricultural changes, such as those made by Antal Lonkay (1903). He argued that increasing agricultural production using traditional (ancient) methods would degrade both woodland and farmland, and urged the intensification of farming so that there would be no need for the conversion of more forests in agricultural areas: 'The extensive agriculture is at its end; let's quickly help people and teach them how to do intensive agriculture in smaller land areas'. Lonkay called for the abandonment of the traditional production systems, which included crop rotation with a fallow period, and its replacement by cultivation with legumes to assure satisfactory production while allowing the land to recover. Such changes in the agricultural sector would allow forest grazing to be stopped and this would increase the potential for timber production. Hence, Lonkay also proposed the improvement of the forestry system and a departure from traditional woodland management. The efficiently managed forests would generate significant income for local communities to buy and plant fruit trees scattered through the pastures. Fruit trees would provide food for livestock, but also generate income from the fruit; this money could, in turn, be used to improve the maintenance of the pastures. He admitted that the strategy would require major institutional changes, but said 'we should not shrink from the difficulties at the beginning'.

The 18th and 19th centuries were thus marked by the widespread separation of the grazing from forests across Europe. Ancient wood banks and lynchets are still common in the historic landscapes of Southern Transylvania, delineating woodlands from other land uses (Szabó, 2010), separating ancient forestry from pasturing. Typical tree species planted along them were hornbeam (*Carpinus betulus*), black locust (*Robinia pseudoacacia*) and ash (*Fraxinus excelsior*) (Hartel, personal observations).

Ancient land management systems were abandoned and new forms of production

(a) 18th century

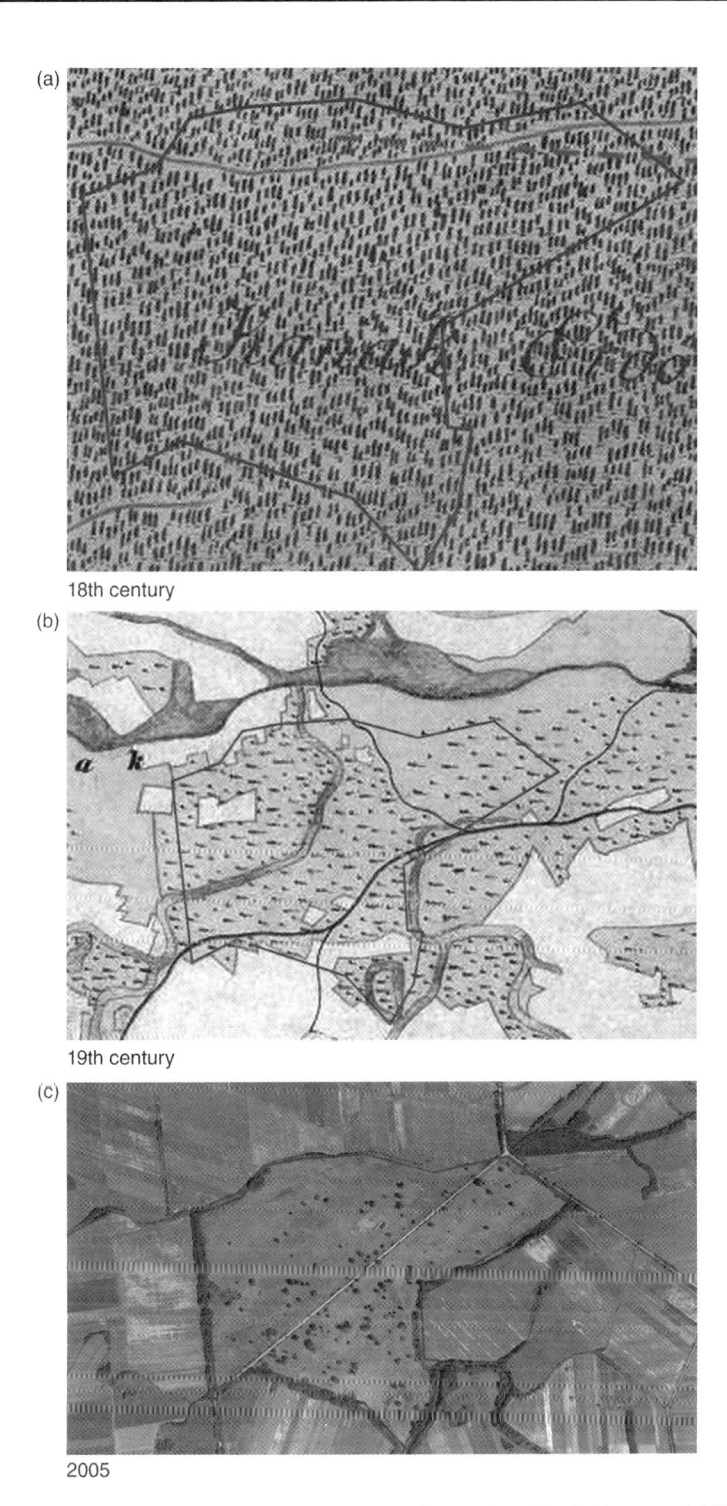

(b) 19th century

(c) 2005

Fig. 5.3. Rókás wood-pasture near Túristvándi in Hortobágyi National Park, Hungary: (a) First military survey, 1784; (b) Second military survey 1858, Kingdom of Hungary and Temes, 2005; (c) Aerial photo 2005. (From: (a–b) Arcanum Database Ltd, Budapest, ISBN: 963 7374 21 3; (c) Hortobágyi National Park.)

adopted, which led to a reorganization of the landscape. The separation of agricultural and forestry land uses was reinforced in the 20th century by the development in many countries of separate institutions dealing with forestry and farming. Forest grazing by cattle, horses, sheep, goats or pigs was frequently prohibited by law. The economic value of timber was higher than ever before, resulting in the formation of closed canopy forests managed for timber production. This was accompanied by the rise of formal forestry education and a forest management attitude that was strongly focused on economic production. Similar changes in other land-use sectors, such as agriculture and water management, resulted in a splitting of landscapes into monofunctional land-use units that were institutionally and ecologically isolated from the rest of the landscape.

During the second half of the 20th century, the above processes led to the destruction of many wood-pastures in Europe. These pastures were either improved to allow more intensive livestock production or converted to arable and other land-use types with the removal of trees in the process. Where livestock farming was unprofitable, the wood-pastures were abandoned and the gaps between trees became infilled by natural regeneration.

5.3.3 New recognition for wood-pastures?

Despite the destruction of many wood-pastures in Europe as outlined above, trees, nonetheless, continued to be maintained or planted in pastures where they were seen to be beneficial for the soil or grassland, or where they provided shelter against extreme weather conditions. In western Europe, wood-pastures have survived more often at higher elevations or on less productive soils in the lowlands where low intensity grazing is still maintained. They have also survived in areas that are protected for various reasons, such as land used for hunting or amenity parks (see Fletcher, Chapter 9). In eastern Europe, ancient wood-pastures are still common even in lowland areas (Hartel et al., 2013).

In recent decades, the value of wood-pastures has increasingly been recognized and this should translate into more sympathetic policies and management. Their aesthetic, heritage and biodiversity values contribute to local economies through cultural tourism (Bieling and Konold, 2014). Wood-pastures often contain large, old trees which can be highly valued by the public. In the UK, several thousands of old trees in wood-pastures and hedgerows have been recorded by citizens, which has improved our understanding of the distribution and condition of the trees, and also created public support for their conservation (Butler, 2014).

A revival of academic interest in wood-pastures and veteran trees, which started in Britain during the 1990s (Kirby et al., 1995), is reflected in studies on the biodiversity potential of wood-pastures and veteran trees in other countries: the Czech Republic (Spitzer et al., 2008; Horák and Rébl, 2012; Vojta and Drhovská, 2012; Sebek et al., 2013); Portugal (Gonçalves et al., 2011); Romania (Hartel et al., 2013, 2014); and Sweden (Paltto et al., 2011; Koch Widerberg et al., 2012). Vegetation dynamics and landscape changes related to management regimes have been studied in Switzerland (Gavazov et al., 2009), the Italian Alps (Garbarino et al., 2010), Belgium (Van Uytvanck et al., 2008), the Netherlands (Vandenberghe et al., 2009; Smit et al., 2010), Spain (Plieninger and Schaar, 2008), Romania (Öllerer, 2012, 2013) and Sweden (Brunet et al., 2011). Other research on vegetation structure and the conservation status of wood-pastures has been taking place in Turkey (Ugurlu et al., 2012), Hungary (Varga et al., 2015) and Greece (Chaideftou et al., 2011).

A common concern has been how scattered tree cover interacts with pasture productivity. Rivest et al. (2013) performed a global meta-analysis which showed that scattered trees (in densities ranging from $c.15–50$ trees ha^{-1}) do not compromise pasture yield. In the Swiss Jura mountains, extensively managed wood-pastures act as buffers against extreme climatic oscillations (Gavazov et al., 2009). The role of grazing regime and shrub composition and configuration in tree regeneration in wood-pastures has been another central topic (Pulido et al., 2001; Smit et al., 2006; Plieninger, 2007; Baraza et al., 2010; Cooper and McCann, 2010).

Some of these studies revive old knowledge, and Jorgensen (2013) highlights the need to consider the particular history of wood-pastures when planning management. The new and old evidence though seems to converge: having trees scattered across pastures can be good from both economic and ecological perspectives.

5.4 National Inventories of Wood-pastures

Most European countries lack a coherent nationwide assessment of wood-pastures. They are often not recognized as distinct landscape elements, partly because of the difficulty of identifying them consistently (Kirby and Perry, 2014). This applies at institutional and policy levels, but also with land-cover maps such as the CORINE database (a geographic land-cover/land-use database encompassing most of the countries of Europe and parts of the Maghreb). When a field assessment of wood-pastures in central Romania was compared with a subsequent CORINE land-cover classification (Hartel and Moga, 2010), the wood-pastures appeared as several categories: 'agroforestry areas', 'transitional elements with trees and shrubs', 'semi-natural grasslands', 'pastures', 'forests' or 'heterogeneous agricultural areas'. Similarly, a Greek study found wood-pastures in the CORINE land-cover categories for scrub and/or herbaceous associations under 'sclerophyllous vegetation', 'moors and heathland' and 'natural grasslands' (Kizos, 2014).

The difficulties of categorizing wood-pastures arise because they are determined not only by their structural features, such as the presence of trees at various densities, but also by the presence of grazing management with livestock. A wood with a closed canopy and grazed by livestock is a wood-pasture in land-use terms, even though it will also be classed in land-cover terms as 'forest'. What is certain is that the areas of wood-pasture involved are much larger than was previously thought.

The wood-pasture cover in Spain and Portugal is c.3.1 million ha (Costa et al., 2014). Sweden has about 50,000 ha of oak-dominated wood-pastures (Ihse and Lindahl, 2000; Ihse, 2011). Germany records 5500 ha of wood-pastures, although other estimates suggest there might be between 50,000 to 100,000 ha (Luick, 2009). Hungary has c.5500 ha of wood-pastures, mostly in protected areas (Varga and Bölöni, 2009), there is a growing interest also from farmers. The central part of Romania has over 5000 ha of oak-dominated wood-pastures (Hartel et al., 2013). In Belgium, there may be 3500–4000 ha in the Flanders region, largely as protected areas. Around 15% of the mountain forests in the Swiss Alps are grazed by livestock (Mayer et al., 2004), around 52,000 ha in the Swiss Jura (Herzog, 1998) and around 400,000 ha in the Austrian Alps (Herzog, 1998). In contrast, there are virtually no wood-pastures in the Czech Republic (J. Vojta and J. Horák, personal communication).

5.5 Wood-pastures as Multifunctional Landscape Elements: Past and Present

The overall tendency, particularly in the past century, has been for a reduction in the use of products from wood-pasture systems, although in the Mediterranean regions cork production is still significant. The harvesting of wood decreased or has stopped, while other products, such as medicinal plants and mushrooms, are now gathered much less often as traditional rural lifestyles have been abandoned. The grazing systems have also changed with shifts in the traditional livestock type and species.

However, new values connected to recreation and nature conservation have been recognized. Most of the wood-pastures from Hungary, for example, are maintained for their ecological values with traditional grazing. Table 5.1 presents data on the management and the use of a variety of goods by local communities, both historically and currently, from wood-pastures in 12 European countries. The table is based on the responses of 16 academics working in those countries

5.6 Threats to Wood-pastures

Even as the value of wood-pastures has become more widely recognized, so there has been a

Table 5.1. The management activities and the goods extracted from wood-pastures in the traditional past and currently. The extracted goods are presented for grassland (managed as pastures or hay meadows), woodland and where available for other resources.

Country	Traditional management	Current management	Trends in the diversity of goods extracted
Belgium	Hay meadow Pasturing (cattle) Wood-related goods: fodder (e.g. from pollarded *Fraxinus* sp. for cattle), scrub, firewood	Nature conservation Grazing management, sometimes haymaking	Reduction; extraction of wood-related goods stopped; new forms of valuation
Czech Republic	Pasturing (sheep, cattle, pigs) Wood-related goods: coppicing Game	Wood-pastures are forested	Virtually no wood-pastures in this country
France	Pasturing (cattle, horses) Fodder Wood-related goods: wood extraction, foliage for grazing animals Fruits, medicinal plants, game, e.g. berries, mushrooms, game	Pasturing (cattle, horses) Fodder Recreational activities	Reduction; extraction of wood-related goods stopped; new forms of valuation
Germany	Pasturing (cattle) Wood-related goods: firewood, fruit trees (from Streuobstwiesen – orchards)	Pasturing (cattle, goats) Wood-related goods: fuelwood, firewood Timber (from composite forests) Recreational activities Nature conservation	Reduction; new forms of valuation
Greece	Pasturing (sheep) Wood-related goods: cork, coppicing	Pasturing (cattle)	Reduction; change in livestock
Hungary	Pasturing (cattle, sheep, goats, pigs, horses) Fruits, medicinal plants, mushrooms Wood-related goods: pollarding, pear trees (fruit)	Haymaking for management purposes Grazing (cattle, sheep, goats, horses) for management purposes	Reduction; wood-related goods tend to be abandoned
Italy	Pasturing Wood-related goods: coppicing	Pasturing (on the way to being abandoned)	Reduction; extraction of wood-related goods stopped
Romania (central part)	Pasturing (cattle, pigs, horses, buffalo, sheep, goats) Wood-related goods: coppicing, acorns, apple and pear trees (wild or domestic) Fruits, medicinal plants and other, e.g. dog rose, mushrooms Recreation Skopationsfest (Saxons in Transylvania)	Pasturing (sheep) Wood-related goods Apple and pear trees (wild or domestic) Fruits, medicinal plants and other, e.g. dog rose, mushrooms Recreational activities Nature conservation	Reduction; change in livestock; new forms of valuation

Continued

Table 5.1. Continued.

Country	Traditional management	Current management	Trends in the diversity of goods extracted
Spain, Portugal	Pasturing (sheep, pigs) Wood-related goods: charcoal, firewood, cork, acorns	Pasturing (sheep, cattle) Wood-related goods Cork	Change in livestock; reduction of wood-related goods; new forms of valuation
	Other goods: honey, crop production	Recreational activities Tourism Nature conservation	
Sweden	Haymaking Fodder Pasturing (cattle, sheep, horses) Wood-related goods: pollarding, acorns (for pigs), scrub Fruits, medicinal plants, e.g. berries	Pasturing (cattle, sheep, horses)	Reduction; extraction of wood-related goods stopped
Switzerland	Pasturing (cattle) Fodder	Pasturing (cattle) Fodder	Reduction; extraction of wood-related goods stopped; new forms of valuation
	Wood-related goods	Recreational activities Nature conservation	

growing appreciation of the threats to their future persistence (Bergmeier et al., 2010; Beaufoy, 2014; Bergmeier and Roellig, 2014; Hartel and Plieninger, 2014).

5.6.1 Management changes

These can be summarized as land use intensification, land abandonment and the conversion into other land-cover forms. Land-use intensification typically involves the increase of grazing livestock density alongside the removal of woody vegetation and the use of chemical fertilizers. The result is a pasture with low biodiversity value and potentially low resilience, especially towards climatic variations. The decrease of grazing intensity, or its abandonment, results in shrub expansion and the reforestation of wood pastures. Land conversion may be to built-up areas or infrastructure such as roads, or conversion to arable land. Any woody vegetation is usually removed in the process. While the immediate driving force for these changes is local economics, often they are encouraged by national and regional policies.

5.6.2 Policy mismatch

Through much of the late 20th century, agricultural and rural development policies supported the destruction of many wood-pastures in Europe. Public infrastructure programmes converted thousands of hectares of Spanish and Portuguese cork oak (*Q. suber*) and holm oak (*Q. ilex*) wood-pastures into irrigated land, eucalyptus plantations, artificial water bodies or industrial units (Joffre et al., 1999). From the 1950s onward, public agricultural policies in southern Germany provided landowners with grants for clearing scattered fruit trees (Herzog, 1998). Even now, forest services seek to abolish the grazing rights in the mountain forests of the Alps (Mayer et al., 2004).

Public incentives for wood-pasture conservation may be available, for example through agri-environmental schemes, but land managers are reluctant to participate in them. An analysis of schemes for wood-pastures and other farm woodlands in the German state of Saxony identified high production costs and opportunity costs for land use, contractual uncertainties, land-tenure implications and variable societal preferences for the ecosystem services of wood-pastures as obstacles to scheme uptake (Schleyer and Plieninger, 2011).

The multifunctional nature of wood-pastures is difficult to manage under institutional structures that are organized as single land-use sectors. Under the Common Agricultural Policy of the European Union (EU), there are restrictions on agricultural support for wood-pastures, as some regulations consider them to be forests rather than pastures (Beaufoy, 2014) and forest policy is a matter for the member states, not the EU itself. However, the ecological values of wood-pastures make them an issue for environmental policy, and other societal values such as recreation, beauty, cultural history and sense of place relate to the spheres of public health and human well-being. Poor integration between these sectors poses major challenges to the design of effective mechanisms to safeguard wood-pasture.

5.6.3 Decline of old, hollowing or dying trees

The large number of old, hollow trees found in wood-pastures are important as keystone structures in ecosystems and the value of these has recently been recognized worldwide (Lindenmayer *et al.*, 2013; Siitonen and Ranius, Chapter 11). Their cultural value is also increasingly emphasized (Butler, 2014). These trees are in sharp decline due to human-related factors such as cutting, reforestation and uncontrolled pasture burning and land abandonment. High losses of these trees often go unnoticed, especially in the traditional rural landscapes of Europe. For example, in central Romania many ancient oaks collapsed as a result of the severe, uncontrolled burning of wood-pastures in 2012 (Hartel *et al.*, 2013).

5.6.4 Lack of regeneration

Intensive grazing, often linked with shrub removal, is detrimental to tree saplings. Lack of regeneration may also be an issue in traditionally managed wood-pastures (Plieninger, 2007), thus threatening the future continuity of the veteran tree populations. The regeneration of wood-pastures in Southern Transylvania is often linked with their overall abandonment following socio-political instability, such as the collapse of communism in 1989 (Hartel *et al.*, 2013). Individual trees might then be retained when the areas were restored to grazing.

5.7 Conclusion

European wood-pastures were created and managed by humans and, in the past, were intimately linked with farming systems. Their survival depends strongly on human intervention: wood-pastures have always undergone major structural changes as societies change. The most important factors threatening wood-pastures now are their structural simplification (typically by their transformation into either closed forest or open agricultural areas), the loss of scattered trees, in particular large and old trees, the lack of regeneration and inappropriate land-use policies. The change from multifunctional management into intensive mono-functional land use is the main driver of this change, reflecting the growing demands for timber and agricultural products.

We lack a coherent picture of the extent and status of wood-pastures in Europe. They are increasingly recognized for their historical, cultural, aesthetic and biodiversity values, but many gaps remain to be filled. A promising lead may be to encourage citizens to identify the scattered old trees in their neighbourhoods. Other local initiatives, with their successes and failures, should provide a valuable platform for knowledge sharing to improve our understanding and actions.

Recent research is focusing on understanding how agricultural production and trees can be better integrated alongside the other ecosystem services provided by wood-pastures. A new social and academic recognition of wood-pastures in Europe should, in turn, generate efficient policy measures to ensure their sustainable future.

Acknowledgements

We gratefully acknowledge support from the Alexander von Humboldt Foundation (TH) and from the European Community's Seventh Framework Programme under Grant Agreements No. 613520 (Project AGFORWARD; TH and TP) and No. 603447 (Project HERCULES; TP). We are grateful to the following researchers for sharing their knowledge of wood-pastures in their own countries: Jan Van Uytvanck, Jakub Horák, Jaroslav Vojta, Francois Gillet, Claudia Bieling, Matteo Garbarino, Augusta Costa, Dan Turtureanu, Zsolt Molnár, Jörg Brunet, Thomas Ranius, Margareta Ihse, Alexandre Buttler.

References

Baraza, E., Zamora, R. and Hódar, J.A. (2010) Species-specific responses of tree saplings to herbivory in contrasting light environments: an experimental approach. *Ecoscience* 17, 156–165.

Barthel, S., Crumley, C.L. and Svedin, U. (2013a) Biocultural refugia: combating the erosion of diversity in landscapes of food production. *Ecology and Society* 18(4): 71. Available at: http://www.ecologyandsociety.org/vol18/iss4/art71/ (accessed 20 November 2014).

Barthel, S., Crumley, C. and Svedin, U. (2013b) Bio-cultural refugia – safeguarding biodiversity for practices for food security and biodiversity. *Global Environmental Change* 23, 1142–1152.

Beaufoy, G. (2014) Wood-pastures and the Common Agricultural Policy: rhetoric and reality. In: Hartel, T. and Plieninger, T. (eds) *European Wood-Pastures in Transition: A Social–Ecological Approach*. Earthscan from Routledge (imprint of Taylor & Francis), Abingdon, UK, pp. 273–281.

Belházy, E. (1888) A legeltetésre szolgáló erdőkről [Regarding the forests used for grazing]. *Erdészeti Lapok* 27, 281–299. [In Hungarian.]

Bergmeier, E. and Roellig, M. (2014) Diversity, threats and conservation of European wood-pastures. In: Hartel, T. and Plieninger, T. (eds) *European Wood-Pastures in Transition: A Social–Ecological Approach*. Earthscan from Routledge (imprint of Taylor & Francis), Abingdon, UK, pp. 19–35.

Bergmeier, E., Petermann, J. and Schröder, E. (2010) Geobotanical survey of wood-pasture habitats in Europe: diversity, threats and conservation. *Biodiversity and Conservation* 19, 2995–3014.

Bieling, C. and Konold, W. (2014) Common management of wood-pastures and sustainable regional development in the southern Black Forest (Germany). In: Hartel, T. and Plieninger, T. (eds) *European Wood-Pastures in Transition: A Social–Ecological Approach*. Earthscan from Routledge (imprint of Taylor & Francis), Abingdon, UK, pp. 235–249.

Brunet, J., Valtinat, K., Mayr, M.L., Felton, A., Lindbladh, M. and Bruun, H.H. (2011) Understory succession in post-agricultural oak forests: habitat fragmentation affects forest specialists and generalists differently. *Forest Ecology and Management* 262, 1863–1871.

Bugalho, M.N., Caldeira, M.C., Pereira, J.S., Aronson, J. and Pausas, J.G. (2011) Mediterranean cork oak savannas require human use to sustain biodiversity and ecosystem services. *Frontiers in Ecology and the Environment* 9, 278–286.

Butler, J. (2014) Mapping ancient and other trees of special interest: UK citizens' contribution to world tree heritage. In: Hartel, T. and Plieninger, T. (eds) *European Wood-Pastures in Transition: A Social–Ecological Approach*. Earthscan from Routledge (imprint of Taylor & Francis), Abingdon, UK, pp. 203–215.

Chaideftou, E., Thanos, C.A., Bergmeier, E., Kallimanis, A.S. and Dimopoulos, P. (2011) The herb layer restoration potential of the soil seed bank in an overgrazed oak forest. *Journal of Biological Research* 15, 47–57.

Cooper, A. and McCann, T. (2010) Cattle exclosure and vegetation dynamics in an ancient, Irish wet oakwood. *Plant Ecology* 212, 79–90.

Costa, A., Madeira, M., Santos, J.L. and Plieninger, T. (2014) Recent dynamics of evergreen oak woodpastures in south-western Iberia. In: Hartel, T. and Plieninger, T. (eds) *European Wood-Pastures in Transition: A Social–Ecological Approach*. Earthscan from Routledge (imprint of Taylor & Francis), Abingdon, UK, pp. 70–83.

de Snoo, G.R., Herzon, I., Staats, H., Burton, R.J.F., Schindler, S., van Dijk, J., Lokhorst, A.M., Bullock, J.M., Lobley, M., Wrbka, T. et al. (2013) Toward effective conservation in farmland: making farmers matter. *Conservation Letters* 6, 66–72.

Dorner, E.B. (1910) *Az Erdélyi Szászok Mezőgazdasága* [The Agriculture of the Transylvanian Saxons]. Pannonia Könyvnyomda Nyomása, Győr, Hungary. [In Hungarian.]

Garbarino, M., Lingua, E., Martinez Subirà, M. and Motta, R. (2010) The larch wood pasture: structure and dynamics of a cultural landscape. *European Journal of Forest Research* 130, 491–502.

Gavazov, K.S., Peringer, A., Buttler, A., Gillet, F. and Spiegelberger, T. (2009) Dynamics of forage production in pasture-woodlands of the Swiss Jura Mountains under projected climate change scenarios. *Ecology and Society* 18(1): 38. Available at: http://www.ecologyandsociety.org/vol18/iss1/art38/ (accessed 20 November 2014).

Gegesi, K.E. (1911) Legelő erdő és fáslegelő [Pasture woodland and wood-pasture]. *Erdészeti Lapok* 526–534. [In Hungarian.]

Gillet, F. (2008) Modelling vegetation dynamics in heterogeneous pasture-woodland landscapes. *Ecological Modelling* 217, 1–18.

Gonçalves, P., Alcobia, S., Simões, L. and Santos-Reis, M. (2011) Effects of management options on mammal richness in a Mediterranean agro-silvo-pastoral system. *Agroforestry Systems* 85, 383–395.

Hartel, T. and Moga, C. (2010) *Manual de Bună Practică Pentru Managementul Pășunilor cu Arbori* [*Good Practice Management of Wood-pasture Habitats*]. Mihai Eminescu Trust, Sighișoara, Romania. [In Romanian.]

Hartel, T. and Plieninger, T. (eds) (2014) *European Wood-Pastures in Transition: A Social–Ecological Approach*. Earthscan from Routledge (imprint of Taylor & Francis), Abingdon, UK.

Hartel, T., Dorresteijn, I., Klein, C., Máthé, O., Moga, C.I., Öllerer, K., Roellig, M., von Wehrden, H. and Fischer, J. (2013) Wood-pastures in a traditional rural region of Eastern Europe: characteristics, management and status. *Biological Conservation* 166, 267–275.

Hartel, T., Hanspach, J., Abson, D.J., Máthé, O., Moga, C.I. and Fischer, J. (2014) Bird communities in the traditional wood-pastures with changing management in Eastern Europe. *Basic and Applied Ecology* 15, 385–395.

Herzog, F. (1998) Streuobst: a traditional agroforestry system as a model for agroforestry development in temperate Europe. *Agroforestry Systems* 42, 61–80.

Horák, J. and Rébl, K. (2012) The species richness of click beetles in ancient pasture woodland benefits from a high level of sun exposure. *Journal of Insect Conservation* 17, 307–318.

Hornyák, G. (1998) *Galvács (Fejezetek a falu történetéből)* [*Galvács (Episodes from the History of the Village)*]. Bibor Kiadó, Budapest. [In Hungarian.]

Ihse, M. (2011) *Fler träd i hagen – bättre för djuren och jorden. Utredning om Förutsättningar för en ny Definition av Ersättningsberättigade Svenska Betesmarker – Principer och Kriterier* [*More Trees in the Pasture – Better for Animals and Earth. Clarification of Requirements for a New Definition of Eligible Swedish Pastures – Principles and Criteria*]. Landsbygdsdepartementet [Swedish Ministry for Rural Affairs], Stockholm. [In Swedish.]

Ihse, M. and Lindahl, C. (2000) A holistic model for landscape ecology in practice: the Swedish survey and management of ancient meadows and pastures. *Landscape and Urban Planning* 50, 59–84.

Joffre, R., Rambal, S. and Ratte, J.P. (1999) The dehesa system of southern Spain and Portugal as a natural ecosystem mimic. *Agroforestry Systems* 45, 57–79.

Jørgensen, D. (2013) Pigs and pollards: medieval insights for UK wood pasture restoration. *Sustainability* 5, 387–399.

Jørgensen, D. and Quelch, P. (2014) The origins and history of medieval wood-pastures. In: Hartel, T. and Plieninger, T. (eds) *European Wood-Pastures in Transition: A Social–Ecological Approach*. Earthscan from Routledge (imprint of Taylor & Francis), Abingdon, UK, pp. 55–69.

Kirby, K.J. and Perry, S.C. (2014) Institutional arrangements of wood-pasture management: past and present (UK). In: Hartel, T. and Plieninger, T. (eds) *European Wood-Pastures in Transition: A Social–Ecological Approach*. Earthscan from Routledge (imprint of Taylor & Francis), Abingdon, UK, pp. 254–270.

Kirby, K.J., Thomas, R.C., Key, R.S., Mclean, I.F.G. and Hodgetts, N. (1995) Pasture woodland and its conservation in Britain. *Biological Journal of the Linnean Society* 56, 135–153.

Kizos, T. (2014) Social–cultural values of oak wood-pastures and transhumance in Greece. In: Hartel, T. and Plieninger, T. (eds) *European Wood-Pastures in Transition: A Social–Ecological Approach*. Earthscan from Routledge (imprint of Taylor & Francis), Abingdon, UK, pp. 171–184.

Koch Widerberg, M., Ranius, T., Drobyshev, I., Nilsson, U. and Lindbladh, M. (2012) Increased openness around retained oaks increases species richness of saproxylic beetles. *Biodiversity and Conservation* 21, 3035–3059.

Lindenmayer, D., Laurance, W.F., Franklin, J.F., Likens, G.E., Banks, S.C., Blanchard, W., Gibbons, P., Ikin, K., Blair, D., McBurney, L. *et al.* (2013) New policies for old trees: averting a global crisis in a keystone ecological structure. *Conservation Letters* 7, 61–69.

Lonkay, A. (1903) A legeltetés kérdése [The question of grazing]. *Erdészeti Lapok* 8, 687–697. [In Hungarian.]

Lonsdale, D. (ed.) (2013) *Ancient and Other Veteran Trees: Further Guidance on Management*. Tree Council, London.

Luick, R. (2008) *Transhumance in Germany. Report to European Forum on Nature Conservation and Pastoralism*. European Forum on Nature Conservation and Pastoralism (EFNCP) and University of Applied Sciences Rottenburg /Fachhochschule Rottenburg, Schadenweilerhof, Rottenburg, Germany. Available at: http://www.efncp.org/download/Swabian_Alb_F_F_Download.pdf (accessed 20 November 2014).

Luick, R. (2009) Wood pastures in Germany. In: Rigueiro-Rodríguez, A., McAdam, J. and Mosquera-Losada, M.R. (eds) *Agroforestry in Europe – Current Status and Future Prospects*. Springer, Dordrecht, Netherlands, pp. 359–376.

Makkai, G. (2003) *Az Erdélyi Mezőség Tájökológiája* [*The Landscape Ecology of the Transylvanian Plain*]. Mentor Kiadó, Marosvásárhely, Hungary. [In Hungarian.]

Manning, A.D., Fischer, J. and Lindenmayer, D.B. (2006) Scattered trees are keystone structures – implications for conservation. *Biological Conservation* 132, 311–321.

Manning, A.D., Gibbons, P. and Lindenmayer, D.B. (2009) Scattered trees: a complementary strategy for facilitating adaptive responses to climate change in modified landscapes? *Journal of Applied Ecology* 46, 915–919.

Mayer, A.C., Stöckli, V., Gotsch, N., Konold, W. and Kreuzer, M. (2004) Forest grazing in alpine regions: a re-evaluation of a multi-usage tradition. *Schweizerische Zeitschrift für Forstwesen* 155, 38–44.

Mikulcak, F., Newig, J., Milcu, A.I., Hartel, T. and Fischer, J. (2013) Integrating rural development and biodiversity conservation in central Romania. *Environmental Conservation* 40, 129–137.

Mosquera-Losada, M.R., Rodríguez-Barreira, S., López-Díaz, M.L., Fernández-Núñez, E. and Rigueiro-Rodríguez, A. (2009) Biodiversity and silvopastoral system use change in very acid soils. *Agriculture, Ecosystems and Environment* 131, 315–324.

Öllerer, K. (2012) The flora of the Breite wood-pasture (Sighişoara, Romania). *Brukenthal Acta Musei* 7, 589–604.

Öllerer, K. (2013). On the spatio-temporal approaches towards conservation of extensively managed rural landscapes in central-eastern Europe. *Journal of Landscape Ecology* 6, 32–46.

Oppermann, R. (2014) Wood-pastures as examples of European high nature value landscapes – functions and differentiations according to farming. In: Hartel, T. and Plieninger, T. (eds) *European Wood-Pastures in Transition: A Social–Ecological Approach*. Earthscan from Routledge (imprint of Taylor & Francis), Abingdon, UK, pp. 39–52.

Oroszi, S. (2004) *Az erdélyi szászok erdőgazdálkodása* [The forest management of Transylvanian Saxons]. *Erdészeti Egyesület, Erdészettörténeti közlemények* [Forestry Association Forest History Publications] 63, 1–153. [In Hungarian.].

Paltto, H., Nordberg, A., Nordén, B. and Snäll, T. (2011) Development of secondary woodland in oak wood pastures reduces the richness of rare epiphytic lichens. *PLoS ONE* 6(9): e24675.

Plieninger, T. (2007) Compatibility of livestock grazing with stand regeneration in Mediterranean holm oak parklands. *Journal of Nature Conservation* 15, 1–9.

Plieninger, T. and Bieling, C. (eds) (2012) *Resilience and the Cultural Landscape – Understanding and Managing Change in Human Shaped Environments*. Cambridge University Press, Cambridge, UK.

Plieninger, T. and Schaar, M. (2008) Modification of land cover in a traditional agroforestry system in Spain: processes of tree expansion and regression. *Ecology and Society* 13(2): 25. Available at: http://www.ecologyandsociety.org/vol13/iss2/art25/ (accessed 20 November 2014).

Pulido, F.J., Díaz, M. and Hidalgo de Trucios, S.J. (2001) Size structure and regeneration of Spanish holm oak *Quercus ilex* forests and dehesas: effects of agroforestry use on their long-term sustainability. *Forest Ecology and Management* 146, 1–13.

Quelch, P. (2013). Upland wood pastures. In: Rotherham, I.D. (ed.) *Cultural Severance and the Environment*. Springer, Dordrecht, Netherlands, pp. 419–430.

Rackham, O. (1989) *The Last Forest – The History of Hatfield Forest*. Dent, London.

Read, H. (2000) *A Veteran Tree Management Handbook*. English Nature, Peterborough, UK.

Rivest, D., Paquette, A., Moreno, G. and Messier, C. (2013) A meta-analysis reveals mostly neutral influence of scattered trees on pasture yield along with some contrasted effects depending on functional groups and rainfall conditions. *Agriculture, Ecosystems and Environment* 165, 74–79.

Schleyer, C. and Plieninger, T. (2011) Obstacles and options for the design and implementation of payment schemes for ecosystem services provided through farm trees in Saxony, Germany. *Environmental Conservation* 38, 454–463.

Sebek, P., Altman, J., Platek, M. and Cizek, L. (2013) Is active management the key to the conservation of saproxylic biodiversity? Pollarding promotes the formation of tree hollows. *PLoS ONE* 8(3): e60456.

Smit, C., Den Ouden, J. and Müller-Schärer, H. (2006) Unpalatable plants facilitate tree sapling survival in wooded pastures. *Journal of Applied Ecology* 43, 305–312.

Smit, C., Bakker, E.S., Apol, M.E.F. and Olff, H. (2010) Effects of cattle and rabbit grazing on clonal expansion of spiny shrubs in wood-pastures. *Basic and Applied Ecology* 11, 685–692.

Spitzer, L., Konvicka, M., Benes, J., Tropek, R., Tuf, I.H. and Tufova, J. (2008) Does closure of traditionally managed open woodlands threaten epigeic invertebrates? Effects of coppicing and high deer densities. *Biological Conservation* 141, 827–837.

Szabó, P. (2010) Ancient woodland boundaries in Europe. *Journal of Historical Geography* 36, 205–214.

Szabó, P. (2013) Rethinking pannage: historical interactions between oak and swine. In: Rotherham, I.D. (ed.) *Trees, Forested Landscapes and Grazing Animals: A European Perspective on Woodlands and Grazed Treescapes*. Earthscan from Routledge (imprint of Taylor & Francis), Abingdon, UK, pp. 51–62.

Teşculă, N. and Goța, A. (2007) Istoria ariei protejate [The history of the protected area]. In: Mihai Eminescu Trust (ed.) *Planul de Management al Reservației 'Stejari Seculari de la Breite'* [*The Management of the Breite Wood-Pasture Reserve*]. Mihai Eminescu Trust, Sighișoara, Romania, pp. 51–55. [In Romanian.]

Ugurlu, E., Rolecek, J. and Bergmeier, E. (2012) Oak woodland vegetation in Turkey – a first overview based on multivariate statistics. *Applied Vegetation Science* 15, 590–608.

Van Uytvanck, J., Maes, D., Vandenhaute, D. and Hoffmann, M. (2008) Restoration of woodpasture on former agricultural land: the importance of safe sites and time gaps before grazing for tree seedlings. *Biological Conservation* 141, 78–88.

Vandenberghe, C., Smit, C., Pohl, M., Buttler, A. and Freléchoux, F. (2009) Does the strength of facilitation by nurse shrubs depend on grazing resistance of tree saplings? *Basic and Applied Ecology* 10, 427–436.

Varga, A. and Bölöni, J. (2009) Erdei legeltetés, fás legelők, legelőerdők tájtörténete [The landscape history of forest grazing, treed pastures and wood-pastures]. *Természetvédelmi Közlemények* 15, 68–79. [In Hungarian.]

Varga, A., Ódor, P., Molnár, Zs., Bölöni, J. (2015) The history and natural regeneration of a secondary oak-beech woodland on a former wood-pasture in Hungary. *Acta Societatis Botanicorum Poloniae*. In press.

Vera, F.W.M. (2000) *Grazing Ecology and Forest History*. CAB International, Wallingford, UK.

Vojta, J. and Drhovská, L. (2012) Are abandoned wooded pastures suitable refugia for forest species? *Journal of Vegetation Science* 23, 880–891.

Wessely, J. (1864) Az erdő mint mentő a takarmány szűkében [The forest as refuge when the fodder is scarce]. *Erdészeti Lapok* 7, 209–221. [In Hungarian.]

6 Coppice Silviculture: From the Mesolithic to the 21st Century

Peter Buckley* and Jenny Mills
Peter Buckley Associates, Ashford, UK

6.1 Introduction

Coppice refers to the repeated cutting of stems regrowing from a stump or 'stool' at intervals, typically from 5 to 30 years (Plate 6). It is one of the oldest silvicultural systems known, with well-documented archaeological evidence dating from Mesolithic, Neolithic, Roman and Anglo-Saxon times in Europe. Such management would have maintained regular openness in woods, together with its many associated species. Rotation lengths are, however, generally too short to allow much development of conditions for late-successional species dependent on old growth, except in old coppice stools, or as occasional mature trees left on boundaries and as standards.

In modern Europe, the landscape now consists mainly of high forest patches, set in open landscapes dominated by agriculture. Coppices have either been abandoned or converted to high forests, thereby reducing the intensity and frequency of disturbances and thus discriminating against species of open woodlands. In Britain, for example, woodland census data indicate an 80% contraction in the area of coppice from the post-war period 1947–1949 (c.142,000 ha) to 1995–2000 (c.24,000 ha) (Forestry Commission, 1952, 2003).

Sustainable forest management policies adopted by several national forestry bodies frequently advocate harvesting regimes that mimic natural disturbances, implying shady woods and 'close-to-nature' forest structures. This view needs to be balanced against the long history of coppice management (Van Calster *et al.*, 2007) and awareness that the current forest flora and fauna may be adapted to its short-cycle felling regimes (Decocq *et al.*, 2004). This chapter explores the origins of coppicing, and its rise and fall. Its ecological legacies are dealt with in more detail in Buckley and Mills (Chapter 10).

6.2 The Physiological and Evolutionary Significance of Coppice

Most woody broadleaves coppice freely, at least when young. Among European species, oak (*Quercus* spp.), ash (*Fraxinus* spp.), hornbeam (*Carpinus betulus*) and lime (*Tilia* spp.) withstand repeated coppicing well, but birch (*Betula* spp.), sycamore (*Acer pseudoplatanus*) and beech (*Fagus sylvatica*) can be less responsive. Conifers lack basal buds and may sprout

---

*E-mail: peterbuckleyassociates@gmail.com

only weakly if damaged, with some exceptions such as the yew (*Taxus baccata*), Canary pine (*Pinus canariensis*) and sequoia (*Sequoiadendron sempervirens*).

The ability of trees to coppice or re-sprout is a natural response to wounding caused by the snapping and uprooting of stems by windstorms, fire, browsing by animals (including the lost mega-herbivores) and to damage caused by drought, flooding, pests and diseases. Feild *et al.* (2004) hypothesized that the earliest woody angiosperms had ecophysiological traits consistent with shady, disturbed and wet habitats, where clonal reproduction through sprouting would have been particularly advantageous.

Coppice originates from the activation of dormant buds present at the root collar and on stools formed from previous cutting. Shoots may also arise from adventitious buds that spontaneously form on callus tissue after injury at the cut surface (as in the European beech, *Fagus sylvatica*), or develop from roots, producing suckers as with aspen (*Populus tremula*). With repeated cutting, the stems that arise from coppice stools or from root suckers can produce clonal masses and a single genotype may come to dominate large parts of a wood. In an ancient wood in southern England, clonal trees accounted for 48 and 65% of all wild cherries (*Prunus avium*) present on managed and unmanaged sites, respectively (Vaughan *et al.*, 2007). Even larger 'individuals' occur elsewhere: in Utah, USA, the 'Pando' clonal colony of trembling aspen (*Populus tremuloides*), comprises 47,000 stems and covers 43 ha (DeWoody *et al.*, 2008).

Re-sprouting relies on the plant laying down multiple latent meristems and stored starch reserves. In canopy gaps, the re-sprouting mechanism allows a damaged individual to persist in the understorey and perhaps eventually to exploit any gap created; some shade-tolerant species are better coppicers than light demanders (Paciorek *et al.*, 2000; Tanentzap *et al.*, 2012). Coppice shoots grow more rapidly than seedlings, and reach the canopy sooner, but as they are usually cut before flowering takes place, there is a reduced capacity for dispersal by seed. Thus, sprouting species can hold ground in the face of competition from non-sprouters that exhibit rapid growth rates and a high reproductive turnover (Bellingham and Sparrow, 2000; Bond and Midgley, 2001).

Repeated vegetative reproduction from stump sprouts or root suckers and with reduced flowering, might eventually lead to genetic 'fixing' or 'freezing' of variation in coppices. The evidence for this is mixed (Aravanopoulos *et al.*, 2001; Cottrell *et al.*, 2003; Mattioni *et al.*, 2008; Valbuena-Carabaña *et al.*, 2008). However, if coppice stands have become 'fixed' in the past, they may have a reduced capacity to adapt to accelerating rates of climate change compared with sexually regenerating populations.

6.3 Historic Development of Coppice Silviculture

Early humans may have exploited coppice shoots regrowing naturally from trees that had been broken or browsed to make temporary bivouacs or huts. At Star Carr, a Mesolithic (9000 BC) site in Yorkshire, UK, aspen and willow (*Salix* spp.) timbers were split and smoothed, laid down near the shore and covered with brushwood, possibly as landing places for boats (Conneller *et al.*, 2012). A piece of coppiced wood, possibly a digging stick or a spear haft, was also found (Taylor *et al.*, 2010), as were other pieces that may have come from the base of a coppice stool (Hadley *et al.*, 2010). Palynological studies provide supporting evidence for early exploitation, for instance at Sarló-hát in north-east Hungary, where pollen profiles suggest that hazel (*Corylus avellana*) coppicing was practised between 6000 and 5500 BC (Magyari *et al.*, 2012). The abundance of charcoal from *Cornus* spp. analysed from occupation levels of another Neolithic site (6159–5630 BC) in south-west Bulgaria implies that the durable but supple twigs of *Cornus* were preferred for wattle and daub (Marinova and Thiebault, 2008).

At Smakkerup Huse, Zealand, Denmark (radiocarbon dated 5000–3900 cal BC), large numbers of hazel stakes were recovered, many more than 1 m in length and 2 to 6 cm in diameter, possibly the remains of a weir or fish trap. Their age, when cut, varied from 5 to 12 years, and many had a swollen base typical of coppiced material, indicating longer term management of the woodland (Price and Gebauer, 2005). Remains of weirs, mostly of hazel wickerwork, and basketry traps made of willow withies, have been found at many other sites in Denmark dating from 6450 to 2350 BC. At one Neolithic site, 6000–7000 hazel rods, 4 m long, and hundreds of larger 6 m long poles were used in the construction of a 250 m long, tightly meshed wattle weir (Fischer, 2007). Neolithic, Bronze and Iron Age wooden trackways incorporating coppiced material have been excavated at many wetland sites in the UK, Germany and the Netherlands (Hillam *et al.*, 1990; Rackham, 2003).

Coppice was also used in wattle-and-daub infilling for walls and for roofing. At Avgi, Kastoria, in north-west Greece (5650–4300 BC), fire-hardened daub shows impressions of branches or thin trunks, 0.06–0.10 m thick (Kloukinas, 2012). Reconstructions give an indication of the quantities used: for example, a 5.5 × 12 m building, based on the ground plan of one of the houses in the Neolithic village at Cuiry-les-Chaudardes (France), required 320 poles of 4–6 cm in diameter, 2.5 or 4 m long, for the roof and 6000 flexible sticks, 1.25 m long and 1–2 cm in diameter, for the walls (Chapelot and Fossier, 1985).

Formal, written prescriptions for coppice management appear much later in the historical period. Roman authors include coppice (*silva caedua*) in descriptions of estate management for firewood, fencing and building materials. Oak and sweet chestnut (*Castanea sativa*) were used to stake grapevines: Columella recommended a 5 year cycle for chestnut and 7 years for oak, while Pliny recommended 7 and 10 years respectively (Meiggs, 1982). In England, some Anglo-Saxon charters of the 8th and 9th centuries grant rights to cartloads of rods and fuelwood (Rackham, 1990; Hooke, 2010), and the Domesday Book, dating from 1086, distinguishes between *silva minuta* woods coppiced for underwood and pasture woods or *silva pastilis* (Rackham, 2003).

The 13th century Italian, Pierre Crescenzi, advised cutting hazel, ash and willow at least every 5 or 6 years, although longer periods are also mentioned (Huffel, 1927). Rotation lengths varied depending on the market demand for different coppice products and rates of growth, but cycles lengthened over time. Small bundles of faggots (fuelwood) gave way to firewood of larger diameters as flue chimneys were installed in houses; hop poles required longer rotations, and for oak tanbark coppices in England, rotations of 20–40 years were needed. At Hayley Wood in England, coppice cycles were every 7 years in the 14th century; 10–11 years in the 16th century and 15 years by the 18th century (Rackham, 1990). Děvín Wood in the Czech Republic was managed on cycles of 7 years in the late 14th/early 15th centuries, a 12 year cycle in the late 17th century and a 30 year cycle from the early 19th to the early 20th century, after which the cycle lengthened to 40 years (Szabó, 2010). In Spain, *Q. pyrenaica* coppice rotations for traditional charcoal production were 8–13 years, but increased from the mid-19th century to 20–30 years (Serrada *et al.*, 1992).

Scattered among the coppice might be larger maiden trees – the 'standards'. The proportion of standards to area of underwood varied and attempts were made through ordinances and laws to prescribe their number to safeguard future timber supplies (Warde, 2006). An English statute of 1544 stipulated 12 standards per acre (30 ha^{-1}). In medieval France, boundary trees (*arbres de laies*) were selected and left as standards to indicate the areas to be cut each year. A French forestry ordinance fixed 40 years as the minimum age for cutting standards (Huffel, 1927). For the royal forests another ordinance of 1376 stipulated that at least 8 or 10 per *arpent* (16 or 20 ha^{-1}) plus the boundary trees were to be reserved in high forest as well as coppice. A minimum rotation of 10 years for coppice was imposed in 1563 for all forests except those belonging to the state, where the age was usually higher (see also Rochel, Chapter 19).

From the Middle Ages, coppices began to be modified by regularizing the working of compartments (coupes), subdividing them using earthworks, tracks and rides, and permanently enclosing woodlands to prevent browsing. New and more productive species were encouraged, such as sweet chestnut in Britain (Peterken, Chapter 18), or black locust (*Robinia pseudoacacia*), and recently *Eucalyptus*, in other parts of Europe. Adjustments were also made to the proportions of other species, and the planting of replacement stools instead of relying on natural regeneration became more common.

Forest regulations were, though, often ignored by the rural population, who used their local coppice for firewood, fencing, grazing, fodder, game, mushrooms and medicinal plants. A large number of standards would have suppressed, through shading, the growth of underwood. The over-exploitation and poor condition of the French royal forests and a fall in revenue led to a reformation of forest management and a new *Code Forestier* in 1669 which, again, stipulated a coppicing cycle of not less than 10 years (Huffel, 1927). Many communal woods not covered by this Code were included in another of 1827 which sought to limit ancient firewood, grazing and other rights. This led to conflicts with communities whose very existence depended on them; for example, the so-called *guerre des demoiselles* in the Ariège region of the Pyrenees (Rochel, Chapter 19).

Conflicts over woodland management in part reflected fluctuating population levels and changing demands for fuel and other coppice products. The population of Europe is thought to have doubled between the 10th and 13th centuries, when woodland was increasingly cleared for crop cultivation but was also managed intensively as coppice. The 'Great Famine' in northern Europe at the beginning of the 14th century, and later the Black Death, reduced Europe's population and eased the pressure on woodland. However, when the population started to increase again in the 16th century, much greater volumes of wood were needed as fuel for the metal, glass, pottery and other industries, as well as larger timbers for construction.

6.4 The Rise and Fall of Coppice as an Industrial Resource

Bloomery smelting developed in the Iron Age. The Romans exploited forested areas with iron ore deposits, such as Catalonia, Spain (Tomàs, 1999) and the Weald of Kent and Sussex and the Forest of Dean in the UK. A constant supply of charcoal was needed to fuel a furnace, and the amount needed for the six Roman ironworks in Sussex between 120 and 240 AD has been estimated as the annual yield from an area of around 93 km^2 of coppice (Rackham, 2003). A large 16th century furnace in the Forest of Dean needed 53 km^2 of coppice to sustain long-term use; a smaller one in the Weald of Kent in the mid-17th century needed 16–20 km^2 (Crossley, 2013). In the Ariège region, there were 57 forges in 1840, each one needing up to 20 km^2 of beech coppice, depending on the rotation length (Cantelaube, 2008). Even with high-quality charcoal and increasingly efficient furnaces, supplies of wood and charcoal eventually had to be sourced from neighbouring regions. Coke-fired blast furnaces gradually took over from the 18th century, greatly reducing the need for charcoal.

Other extraction industries also needed large volumes of firewood. Glass was made in many parts of Europe, including Italy, England, France, Belgium, Germany and Bohemia (Jéhin, 2010). Furnaces had to reach high temperatures and were sited in woods to access the large quantities of fuel for which they were often in competition with from the iron industry. Between 1462 and 1823, an estimated 6.2 million m^3 of fuelwood were cut in the Tolfa mountains of Italy to process alunite mineral into alum (Jones, 2008). One 16th century furnace at Bagot's Park, Staffordshire, UK, would have used about 1560 t of wood a year, or 19 ha of 15-year-old coppice (Crossley, 1967). In England, glassmakers were forced to change to coal early in the 17th century after a Royal Proclamation of 1615 prohibited them from using wood owing to a fear of shortages.

In the late 18th and early 19th centuries, the price of tan bark increased as a result of the greater industrial use of leather and the availability of more hides. Oak coppice around 20 years old was preferred for this as the bark was richer in tannin, but other species such as chestnut were used as well. Harvesting was during the short period in spring when the sap was rising. The bark was dried and milled, then used to make the liquids in which the skins were soaked (Clarkson, 1974; Mead, 1989). The long tanning process slowed down production and faster chemical-based processes were introduced in the late 19th century; there was also substitution by bark from South America (Zarrilli, 2009). The demand for tan bark increased during the First World War, but the industry declined rapidly thereafter.

The decline in the value of charcoal and bark reduced the importance of coppice, except for domestic and agricultural use. During the First and Second World Wars, coppices were again heavily exploited, but population movement from rural areas to towns and cities, especially in the latter half of the 20th century, resulted in an abandonment of agricultural land, traditional farming and coppicing practices, particularly in mountainous and Mediterranean areas. The increasing availability of cheap coal, then the use of oil and gas rather than wood fuel, a decreasing market for small timber products, rising costs of felling (especially in more remote areas) and the small size of many holdings made most coppicing uneconomic.

Despite this general trend, in south-east Europe coppice still represents a significant percentage of total forest area. Much of it is in poor condition, because of intense exploitation in the 18th and 19th centuries for charcoal production, domestic and industrial fuel. However, coppice remains important, especially in poorer, rural areas where firewood is essential for domestic heating and cooking. Rising fuel prices have also revived the use of firewood in urban areas (Stajic et al., 2009).

6.5 Surviving and Neglected Coppice in Europe: The Extent of the Forest Estate

The vagaries of census statistics, compounded by variable reporting by different countries, make it difficult to gain an accurate picture of the forest area under coppice in Europe, but about 28 million ha, or 14.8% of the total forest area, is a reasonable reflection of the present situation (Fig. 6.1). This is a much larger figure than that of 2.8 million ha given in the

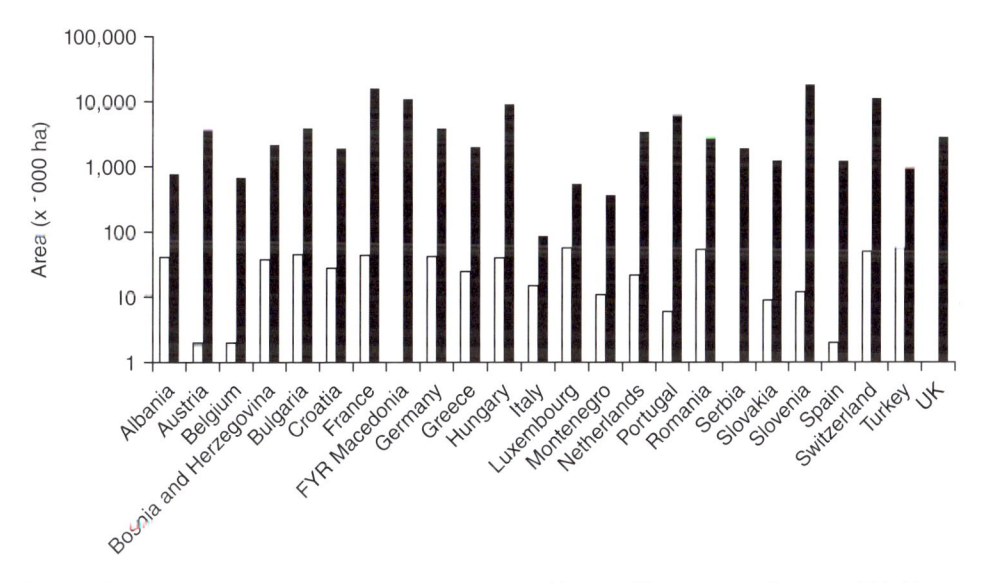

Fig. 6.1. Open columns show estimates of coppice areas (thousand ha on a logarithmic scale) in Europe for countries with 1% or more of their forest area recorded as coppice, including Turkey but excluding the Russian Federation. Black columns show total forest areas excluding 'other wooded land' (×1000 ha) taken from the *FAO Global Forest Resources Assessment 2010* (FAO, 2010). The coppice areas are taken from the following references: Albania: Pisanelli et al. (2010); Austria, Belgium, Germany, Montenegro, the Netherlands, Portugal, Slovakia, Switzerland, the UK: Forest Europe (2011); Bosnia and Herzegovina: FAO (2010); Bulgaria: Velichkov et al. (2009); Croatia: Croatia's submission to UNFCCC (2011); France: Forêt Privée Française (2008); FYR (Former Yugoslav Republic of) Macedonia, Romania, Serbia: Stajic et al. (2009); Greece: Meliadis et al. (2010); Hungary: Stajic et al. (2009); Italy: Bernetti and la Marca (2010); Luxembourg, Slovenia: UN-ECE/FAO Forest Resources (2000); Spain: Serrada et al. (1992); Turkey: Kiliç and Tüfekçioğlu (2010).

State of Europe's Forests 2011 (Forest Europe, 2011), but 11 of the 24 countries included in Fig. 6.1 submitted no data for coppice, including Greece, Serbia and Spain, which still have large areas of coppice. For Italy, most of its coppice was included in 'other naturally regenerated forest', whereas earlier *State of Europe's Forests* reports in 2003 and 2007 both indicated that Italy then had well over 40% coppice regeneration, some 3.5 million ha.

The overall figures include short-rotation coppices (plantations of fast-growing, closely planted clones of willows and poplars, *Populus* spp.) sited on ex-arable land that are mechanically harvested on rotations of 2 to 6 years. However, these are developing only slowly in Europe, with about 50,000 ha distributed mostly in Sweden, Italy, France, Poland and the UK (Leek, 2010), compared with more than 700,000 ha of other biomass systems (Don *et al.*, 2012). More significant, but still only forming a small part of the total estimate, are 1.2 million ha of *Eucalyptus* coppice, mainly *E. globulus* and *E. camaldulensis*, grown in Portugal and Spain on rotations of 10–15 years for pulp, paper and fibreboard production, and renewed every three rotations or so by replanting.

Coppicing has been of relatively little importance in the boreal zone, because of the poor re-sprouting ability of the dominant conifers there. In the temperate deciduous (nemoral) zone, coppice species include: hornbeam, hazel, oak, small-leaved lime (*T. cordata*), ash (*F. excelsior*), field maple (*Acer campestre*) and sycamore on mesophytic sites; sessile oak (*Q. petraea*) and birch on poorer soils; and alder (*Alnus glutinosa*) and wych elm (*Ulmus glabra*) along flood plains. Oaks are the usual species grown as standard trees in coppice-with-standards systems.

Beech is commonly grown as high forest because of its poor coppicing ability and is the dominant species through much of central Europe, extending into south-east Europe along mountain chains. In the Italian mountains, over half of these beech forests have been coppiced for firewood and charcoal since the Middle Ages and during intensive exploitation in the 18th century. The other naturally occurring species, Norway spruce (*Picea abies*) and silver fir (*Abies alba*), were shaded out by beech after clear-felling or deliberately eliminated by forest users (Nocentini, 2009).

In the thermophilous, deciduous and semi-deciduous forests of southern Europe, many of the above species co-occur with other coppicing species, such as Hungarian oak (*Q. frainetto*), Pyrenean oak (*Q. pyrenaica*), downy oak (*Q. pubescens*), Turkey oak (*Q. cerris*), hop hornbeam (*Ostrya carpinifolia*), manna ash (*F. ornus*), silver lime (*T. tomentosa*), eastern hornbeam (*C. orientalis*) and chestnut. Some of these species, in turn, extend further south into the Mediterranean region, where broadleaved, evergreen forests predominate with sclerophyllous species such as holm oak (*Q. ilex*), cork oak (*Q. suber*), kermes oak (*Q. coccifera*) and the strawberry tree (*Arbutus unedo*). In the south of France, holm oak and downy oak have long been the main coppice species. Palynological studies indicate that the latter may have been the natural dominant species, but its requirement for slightly better, moister soils probably led to its greater clearance for agriculture. In habitats where both species can coexist, coppicing seems to have favoured holm oak, which, while slow growing, is more resilient to the effects of fire, grazing and felling. This species has an enormous circum-Mediterranean distribution of over 9 million km^2, ranging from Portugal to Syria and Morocco to Algeria (Sánchez-Humanes and Espelta, 2011).

In southern Europe, the species formerly managed as coppice that have the potential to grow to large dimensions tend to be preferentially selected for conversion to high forest because of their ability to produce sawtimber. These include Pyrenean, Hungarian, downy and Portuguese (*Q. faginea*) oaks and sweet chestnut; coppice of less productive species with lower potential value such as holm and kermes oaks are more likely to be abandoned.

6.6 Coppice Silviculture

The basic form of coppice – simple coppice – produces an even-aged crop, although the stools themselves may vary in age. In coppice-with-standards, a more irregular structure is created. The density of the standards is a compromise between the productivity of the two components of the crop: too much shading from standard trees results in reduced underwood (coppice) production. A 50:50 cover of each component was considered normal stocking for working coppice in post-war England, with a minimum of 15 standards ha^{-1} (Forestry Commission, 1952), whereas in modern conservation coppicing a cover of not more than 30% of standards is often recommended. Felling of standards takes place concurrently with a coppice cut, but the length of the standard rotation may vary according to market requirements, with cohorts of new standards being identified at the same time.

A third, but rapidly disappearing, version of the coppice system is selection coppice, in which two or three age classes of stems are rotated on the same coppice stool. This existed mainly in the mountain beechwoods of the Italian Apennines (especially in Tuscany and Emilia-Romagna), as well as in parts of the Pyrenees (Coppini and Hermanin, 2007; Nocentini, 2009), often alongside conventional coppice-with-standards. At each harvest (c.6–12 years) the oldest, dominant stems are cut for firewood and the others are lightly thinned. The system has the advantage of retaining a light canopy cover after harvesting, protecting the soil surface against erosion on steep mountain slopes and also encouraging meristem activity in a species with otherwise poor coppicing ability.

6.6.1 Cutting methods

Traditionally, coppicing was carried out using an axe or billhook, although nowadays chainsaws or tree harvesters are used. Evidence for the effect of different cutting treatments is often contradictory (Harmer and Howe, 2003; Harmer, 2004), because it may be influenced by, *inter alia*, the quality and height of the cut, the season of cutting, the tree species, its biological age and site conditions. Most practical accounts suggest that low cuts on the coppice stool encourage new sprouts to develop at or below ground level, thus forming independent roots and creating more stable stems. Clean, sloping cuts may help to shed moisture and prevent rotting.

Ducrey and Turrel (1992) found no difference in sprouting recovery between axe and chainsaw cuts in holm oak. The traditional *saut du piquet* technique, involving cutting stems at 50 cm above stool height and splitting them off the stool, ostensibly to decrease sprout competition, actually increased stool mortality and reduced sprouting. Mechanized felling using a tree harvester, or high or low chainsaw cuts, made little difference to stool mortality or sprouting recovery in mature sessile oak (Pyttel *et al.*, 2013), but the low cuts (at c.11–12 cm above ground) caused more dominant shoots to develop from below the root collar. In contrast, Giudici and Zingg (2005) reported that sprouting after re-coppicing a 60-year-old chestnut crop was positively related both to a higher stool cutting position (although inevitably higher on larger stools) and to clean, regular and inclined cuts, in agreement with other studies on a range of species (e.g. Harmer and Howe, 2003; Croxton *et al.*, 2004).

At each harvest, a proportion of coppice stools die (commonly 5–10% per cut, but potentially more in over-mature stands) and need to be replaced, either by planting, natural regeneration or layering. Larger stools and those with fewer neighbours tend to show better recovery than small stools with weak sprouting. Shading, for example from the standard trees, may reduce re-sprouting. Competition between stems on the same stool can be important: Rydberg (2000) noted less sprouting from birch and aspen stools that had been selectively coppiced, leaving some stems uncut.

Coppice shoots rapidly self-thin: Gracia and Retana (2004) reported an average of 164 sprouts per stool in holm oak stands in the first year, reducing to 1–10 after 30 years. Again, the number of live stems in mature chestnut coppice had almost halved 4 years after harvesting (Giudici and Zingg, 2005), and by 70–90% after 5–10 years in *Q. petraea* (Šplíchalová *et al.*, 2012). Espelta *et al.* (2003) showed that actively thinning subordinate stems in Turkey oak and *Q. cerrioides* coppice substantially increased the height and diameter growth of the remainder, but this encouraged a new wave of basal sprouts.

6.6.2 Time of cutting

Coppice is usually cut during the dormant period (except for tan bark), between late autumn and early spring, as there will be less bark tearing and frost damage to developing shoots. Ducrey and Turrel (1992) found that cutting in March and November produced better sprout regrowth compared with harvesting during a cold period in January or cutting within the growing season (May and July), even though stool mortality was not affected. Shoot regrowth after cutting 8-year-old maiden saplings of ash, oak and sycamore at different periods in January, April and June showed only marginal differences after 6 years (Harmer and Howe, 2003). Clipping of established short-rotation osier willow (*S. viminalis*) coppice (to simulate deer browsing) tended to show the same pattern: the stools with late summer clipping treatments were less able to compensate than after winter or early summer clipping (Guillet and Bergström, 2006). However, over longer rotations of 10–20 years it is unlikely that summer cutting would have more than a marginal impact on yields.

6.7 Conversion to High Forest

Generally, across Europe, the conversion of coppices to high forest structures had begun by the 18th and 19th centuries and accelerated during the 20th century, encouraged by official forestry policies, but also happening by neglect. In the UK, in the early 1980s, two thirds of all coppice crops were no longer cut on rotation and were effectively being 'stored' (Evans, 1992). After six decades or more of neglect, many of these stands had reverted naturally to high forest and had begun to self-thin. In the Dolomiti Bellunesi National Park (north-eastern Italy), beech was heavily coppiced for firewood and charcoal for industrial use from the medieval period onward. From the 1960s, rotations were lengthened and subsequently abandoned, followed by conversion to high forest in the mid-1990s. Norway spruce, previously suppressed by the coppice canopy or eliminated by foresters, is now regenerating naturally (Andreatta, 2008).

Changes in species composition occur through deliberate selection or replacement with more productive species, such as sweet chestnut, and through natural succession. In Mediterranean coppices, the gradual abandonment of coppicing after the Second World War may favour downy oak colonization, as under a canopy of holm oak, downy oak seedlings may outcompete holm oak (Bonin and Romane, 1996). Increasing fire frequency and intensity may determine other species shifts; in central Catalonia, Spain, Bonfil *et al.* (2004) reported greater susceptibility in holm oak than in the deciduous *Q. cerrioides*, an unusual situation as deciduous oaks are normally considered more vulnerable to fire.

Disturbances may allow non-native species to invade, for instance black locust and the tree of heaven (*Ailanthus altissima*) in northern Italy (Radtke *et al.*, 2013). The tree of heaven, a prolific seeder, is also a problem in Bulgaria and Spain, while black locust, which coppices well and spreads by root suckers, is invasive in parts of Italy, Croatia and Slovenia (FAO, 2013). These species invade quickly after fire (Maringer *et al.*, 2012) and may therefore change the species composition of abandoned coppices, which are vulnerable to forest fires. Black cherry (*Prunus serotina*), another pioneer species, is able to invade degraded coppices in France and Italy (FAO, 2010; Forest Europe, 2011).

Some foresters have sought to maintain an economic output from ageing coppices by more active management, for example:

- lengthening the coppice rotation (e.g. from 20–50 years) to produce larger diameter poles or timber (Amorini *et al.*, 2001; Cutini, 2001; Grigoriadis and Zagas, 2005), often with a resulting increase in increment;
- implementing regular, low intensity 'improvement' cuts to establish a mature canopy (e.g. of oak, beech or ash), followed by 3–4 regeneration fellings, with the aim of establishing a new crop of seedlings on a uniform shelterwood principle (e.g. Van Calster *et al.*, 2007);
- selective crown thinning, including singling, around the best formed individual stems at regular but increasing intervals

from 4–10 years, or overall thinning treatments removing 15–20% of the volume around canopy closure (Dafis and Kakouros, 2006);
- pre-commercial thinning every 4 years, removing coppice, shrubs and other non-commercial species, followed by selective thinnings every 8 years to eventually establish a continuous cover silviculture (e.g. Decocq *et al.*, 2004), a process that can take more than a century;
- clearing undergrowth and thinning to select the best stems, creating an open, silvopastoral forest structure suitable for grazing by domestic animals (e.g. Camprodon and Brotons, 2006; Hernando *et al.*, 2010); and
- clearing and replacing the stands with commercial, even-aged plantations of conifers or broadleaves.

6.7.1 Coppice versus high forest yields

On good sites, some high-yielding chestnut coppices are capable of producing mean annual increments of 16 m^3 ha^{-1} on a rotation age of 25 years (Marziliano *et al.*, 2013), whereas under less favourable conditions mountain beech coppices above 1000 m in central Italy may still achieve 10 m^3 ha^{-1} (Coppini and Hermanin, 2007; Nocentini, 2009). However, on sites with pronounced summer drought, Mediterranean oaks such as the holm, Hungarian and Pyrenean oaks might yield annual increments of only 2–3 m^3 ha^{-1} over a similar period.

Comparing coppice and high forest stands may be misleading: the stem form factors developed for high forest mensuration are not necessarily applicable to coppice, and coppices are often situated on poorer quality sites (Suchomel *et al.*, 2012). The products from the two systems also differ in value. Yield tables for oak high forest and oak coppice on good quality sites in the Czech Republic showed that the gross value yields of coppice fell far short of those of high forest, such that prices for wood fuel would need to increase by 16.8% in order for the best coppice to outperform the worst high forest (Kneifl *et al.*, 2011). In mixed-species coppice forests in Bavaria, Mosandl *et al.* (2010) considered that oak standards of *Q. robur* and *Q. petraea* exhibited better growth than those grown in high forests, but were of poorer quality due to lower branching, variable epicormic growth and bole wounds caused during extraction.

Coppice yields are ultimately limited by the prevailing environmental conditions, particularly drought and soil fertility. Questions have been raised as to whether repeated harvesting of young stem wood (and sometimes even fine branch wood, foliage and litter) might gradually deplete soil resources. Rackham (2003) suggested that the gradual lengthening of coppice rotations recorded in Bradfield Woods, Suffolk, UK, might be partly explained by a decline in soil nutrients over 700 years of continuous wood removal. Rubio and Escudero (2003) found reductions in soil organic matter, total nitrogen percentage and colloidal potassium after coppicing in chrono-sequences of chestnut coppice on acid soils in Extremadura, Spain, although both nitrogen and potassium returned to pre-felling levels after 15–20 years. Higher turnover rates of nutrients in young coppices have also been recorded. These are attributed to increased soil moisture, insolation and soil temperatures following harvesting stimulating the decomposition of litter and redundant fine roots through microbial activity (Rubio *et al.*, 1999; Hölscher *et al.*, 2001; Tedeschi *et al.*, 2006). Ash deposition from the burning of waste material or wildfires, and the rapid decomposition of herbaceous litter of colonizing species after harvests, including nitrogen fixers, could also contribute to maintaining soil fertility. However, on xeric sites in some Mediterranean forests, this accelerated turnover of nutrients might be compromised by long periods of drought (Corcuera *et al.*, 2006).

6.8 Reinstating Coppice Management

Coppices can be rejuvenated by recutting long after abandonment, but the results may not be commercially viable if the degree of re-sprouting is reduced after prolonged periods of storage or neglect. In Mediterranean stands of over-mature holm oak coppice, senescence of the original stump led to the death of the whole tree (Panaïotis *et al.*, 1997). In the Czech Republic, the probability of re-sprouting in sessile oak aged up to 97 years fell to less than 50% when individual stem diameters exceeded 35 cm at breast height (Šplíchalová *et al.*, 2012). In south-west Germany, coppice stems of 80–100-year-old sessile oak suffered 16% stool mortality two growing seasons after felling (Pyttel *et al.*, 2013), with a slight increase in mortality with increasing stem sizes up to 25 cm diameter.

Matula *et al.* (2012) also found a decline in re-sprouting ability with increasing stump diameter, again in sessile oak, in a stand that had been converted to high forest almost 100 years previously; there was less than 40% probability of re-sprouting for stem diameters above 70 cm. Nevertheless, virtually all of the lime and most of the contemporary hornbeam stools in the same stand re-sprouted, the latter showing an increased probability of sprouting with increasing stem diameter. Re-coppicing a 60-year-old sweet chestnut stand in southern Switzerland resulted in only 4% stool mortality (Giudici and Zingg, 2005).

Variation in the ability of different species to recover from delayed re-coppicing may be due to differences in the allocation of resources between roots and shoots, or in the ability of some species to produce root suckers as well as coppice shoots. The axillary buds become progressively embedded as the stem increases in diameter, which may also mean that the ability of these buds to re-shoot gradually declines with time since last coppiced (Tredici, 2001).

6.9 Future Drivers of Change

A lack of markets lies behind the long-term decline of coppice silviculture. There is also the low yield and poor quality of wood and timber produced by coppice compared with high forest. The yield and quality of coppice is also affected by the browsing of young growth by increasing populations of deer and the difficult terrain on which coppices often grow, often with poor road access. Coppice is also often on widely dispersed and small parcels of forest in private ownership.

Potential future markets for coppice are biomass and wood fuel, grown and extracted on an industrial scale, and supported by the renewable energy targets of the European Union (EU) to cut emissions of carbon dioxide. Recently, there has been a small-scale revival of coppicing in Europe to supply wood-fuelled stoves and boilers, and for use in rustic garden products and engineered and 'glulam' wood for joinery, furniture and construction. There is some renewed interest in selection of coppices in mountain areas, as these can produce good quantities of firewood and at the same time protect against soil erosion (Coppini and Hermanin, 2007; Nocentini, 2009).

Short-rotation coppice (SRC), developed in Sweden after the oil crises of the 1970s, has succeeded there mainly due to the well-established use of wood fuel for communal heating systems (Christersson and Verma, 2006). However, the relatively small areas so far developed in Europe fall far short of the hundreds of thousands of hectares needed if SRC is to make a substantial contribution to renewable energy.

Some countries have developed wood-fuel strategies in order to target under-managed woods (e.g. Jansen and Kuiper, 2004; Forestry Commission England, 2007). Assuming that developing wood-fuel markets will require wood chips and wood pellets rather than logs, stems 7–18 cm in diameter will be at a premium, favouring the use of younger and recently worked stands. Although older and unmanaged stands can be brought back into management using specialized machinery, including whole-tree extraction, it is unlikely that this will return them and the species populations to their previous state. Modern economic and social requirements may mean that large areas are cut at one time, leading to unbalanced age structures, particularly in small woods. Mechanization of harvesting may mean more potential damage to ground vegetation and shrub layers, with unwanted small branches burnt or left in piles.

Another existing 'market' for coppices is their continued silvopastoral use, providing forage for domestic animals. In parts of the Mediterranean, coppices are traditionally grazed by sheep, cows and goats. At moderate stocking levels, silvopastoralism can maintain open conditions in woods, enhance species richness in the herb layer and reduce the fire hazard of accumulating litter. Even where coppicing has ceased, proposals to convert woodlands to a high forest structure may run in parallel with livestock management objectives: the animals control regrowth of coppice stools following operations of thinning and singling (Papachistou and Platis, 2011; Núñez et al., 2012).

The future value of coppices as carbon sinks to mitigate climate change is important, though their impact will be somewhat less than that of forests managed on longer rotations. Converting coppices to high forest stands by extending rotation lengths will increase above-ground carbon stocks as biomass accumulates. While coppice stands do accumulate carbon rapidly after harvesting, the average tree carbon stocks per unit area may be less than half that of close-to-nature forest systems (Mason et al., 2009). Below ground, soil carbon pools regularly exceed the average sequestered in woody biomass over a full forest rotation. The greater frequency of harvesting in coppice systems means that soil carbon levels will tend to be lower than under high forest, but still considerably more than in arable crops or pasture. SRC can, as a result, increase soil carbon stocks compared with the farmland it replaces.

Under all Intergovernmental Panel on Climate Change (IPCC) greenhouse gas emission scenarios, conditions for forest growth are projected to continue to deteriorate, particularly in southern Europe, with higher summer temperatures, longer periods of drought and increasing risks of wildfires. Worked coppices accumulate relatively little fuel load: if large areas are no longer cut, alternative strategies will be needed, such as prescribed burning (and possibly grazing), or conversion to closed high forest, so as to eliminate flammable ground and low scrub layers. The increasing likelihood of prolonged spring and summer drought may also affect the long-term viability of coppice, especially in the fully developed canopies of over-aged stands where a high water demand could lead to stagnation in growth. Several authors advocate thinning in order to improve plant water status and water-use efficiency (e.g. Gracia et al., 1999; Cañellas et al., 2004; Corcuera et al., 2006; Moreno et al., 2007; Gyenge et al., 2011). More radical thinning, clearance of undergrowth and wider spacing of coppice stools, perhaps combined with controlled grazing, may make more effective use of scarce moisture resources; at the same time, more drought-resistant species can be encouraged.

Coppice as a system is therefore likely to have a future, but for different products and managed in different ways from those in the past.

References

Amorini, E., Manetti, M.C., Turchetti, T., Sansotta, A. and Villani, F. (2001) Impact of silvicultural system on *Cryphonectria parasitica* incidence and on genetic variability in a chestnut coppice in Central Italy. *Forest Ecology and Management* 142, 19–31.

Andreatta, G. (2008) La diffusione dell'abete rosso negli ex-cedui di faggio del Parco delle Dolomiti Bellunesi. *Forest@* 5, 265–268. Available at: http://www.sisef.it/forest@/pdf/?id=efor0547-0005 (accessed 21 November 2014).

Aravanopoulos, F.A., Drouzas, A.D. and Alizoti, P.G. (2001) Electrophoretic and quantitative variation in chestnut (*Castanea sativa* Mill.) in Hellenic populations in old-growth natural and coppice stands. *Forest Snow and Landscape Research* 76, 429–434.

Bellingham, P.J. and Sparrow, A.D. (2000) Resprouting as a life history strategy in woody plant communities. *Oikos* 89, 409–416.

Bernetti, G. and la Marca, O. (2010) Il bosco ceduo nella realtà italiana. *Atti della Accademia dei Georgofili* 7, 542–585.

Bond, W.J. and Midgley, J.J. (2001) Ecology of sprouting in woody plants: the persistence niche. *Trends in Ecology and Evolution* 16, 45–51.

Bonfil, C., Cortés, P., Espelta, J.M. and Retana, J. (2004) The role of disturbance in the co-existence of the evergreen *Quercus ilex* and the deciduous *Quercus cerrioides*. *Journal of Vegetation Science* 15, 423–430.

Bonin, G. and Romane, F. (1996) Chêne vert et chêne pubescent. Histoire, principaux groupements, situation actuelle. *Forêt Méditerranéenne* 17, 119–128.

Camprodon, J. and Brotons, L. (2006) Effects of undergrowth clearing on the bird communities of the northwestern Mediterranean coppice holm oak forests. *Forest Ecology and Management* 221, 72–82.

Cañellas, I., del Rio, M., Roig, S. and Montero, G. (2004) Growth response to thinning in *Quercus pyrenaica* Willd. coppice stands in [the] Spanish central mountain[s]. *Annals of Forest Science* 61, 243–250.

Cantelaube, J. (2008) Le charbon de bois et la forge à la catalane (Pyrénées, XVIIe–XIXe siècles). In: Menozzi, M.-J., Flipo, F. and Pécaud, D. (eds) *Énergie et Société: Sciences, Gouvernances et Usages*. Editions Edisud, Aix en Provence, France, pp. 35–45. Available at: http://www.ecologie-humaine.eu/DOCUMENTS/SEH_Energie/Energie_07_Cantelaube.pdf (accessed 21 November 2014).

Chapelot, J. and Fossier, R. (trans. Cleere, H.) (1985) *The Village and House in the Middle Ages*. B.T. Batsford, London.

Christersson, L. and Verma, K. (2006) Short-rotation forestry – a complement to "conventional forestry". *Unasylva* 57, 34–39. Available at: http://www.fao.org/docrep/008/a0532e/a0532e07.htm (accessed 21 November 2014).

Clarkson, L.A. (1974) The English bark trade 1660–1830. *The Agricultural History Review* 22, 136–152.

Conneller, C., Milner, N., Taylor B. and Taylor, M. (2012) Substantial settlement in the European Early Mesolithic: new research at Star Carr. *Antiquity* 86, 1004–1020.

Coppini, M. and Hermanin, L. (2007) Restoration of selective beech coppices: a case study in the Apennines (Italy). *Forest Ecology and Management* 249, 18–27.

Corcuera, L., Camarero, J.J., Sisó, S. and Gil-Pelegrín, E. (2006) Radial-growth and wood-anatomical changes in overaged *Quercus pyrenaica* coppice stands: functional responses in a new Mediterranean landscape. *Trees* 20, 91–98.

Cottrell, J.E., Munro, R.C., Tabbener, H.E., Milner, A.D., Forrest, G.I. and Lowe, A.J. (2003) Comparison of fine-scale genetic structure using nuclear microsatellites within two British oakwoods differing in population history. *Forest Ecology and Management* 176, 287–303.

Crossley, D.W. (1967) Glassmaking in Bagot's Park, Staffordshire, in the sixteenth century. *Post-Medieval Archaeology* 1, 44–83.

Crossley, D.W. (2013) *The Supply of Fuel for Post-medieval Metal Industries*. Archaeology Datasheet 305, The Historical Metallurgy Society, Gateshead, UK. Available at: http://hist-met.org/images/pdf/HMSdatasheet305.pdf (accessed 21 November 2014).

Croxton, P.J., Franssen, W., Myhill, D.G. and Sparks, T.H. (2004) The restoration of neglected hedges: a comparison of management treatments. *Biological Conservation* 117, 19–23.

Cutini, A. (2001) New management options in chestnut coppices: an evaluation on ecological bases. *Forest Ecology and Management* 141, 165–174.

Dafis, S. and Kakouros, P. (eds) (2006) *Guidelines for the Rehabilitation of Degraded Oak Forests*. Greek Biotope/Wetland Centre, Thermi, Greece.

Decocq, G., Aubert, M., Dupont, F., Alard, D., Saguez, R., Wattez-Franger, A., de Foucault, B., Delelis-Dusollier, A. and Bardat, J. (2004) Plant diversity in a managed temperate deciduous forest: understorey response to two silvicultural systems. *Journal of Applied Ecology* 41, 1065–1079.

DeWoody, J., Rowe, C.A., Hipkins, V.D. and Mock, K.E. (2008) "Pando" lives: molecular genetic evidence of a giant aspen clone in Central Utah. *Western North American Naturalist* 68, 493–497.

Don, A., Osborne, B., Hastings, A., Skiba, U., Carter, M.S., Drewer, J., Flessa, H., Freibauer, A., Hyvonen, N., Jones, M.B. *et al.* (2012) Land-use change to bioenergy production in Europe: implications for the greenhouse gas balance and soil carbon. *GCB Bioenergy* 4, 372–391.

Ducrey, M. and Turrel, M. (1992) Influence of cutting methods and dates on stump sprouting in holm oak (*Quercus ilex* L.) coppice. *Annales des Sciences Forestières* 49, 449–464.

Espelta, J.M., Retana, J. and Habrouk, A. (2003) Resprouting patterns after fire and response to stool cleaning of two coexisting Mediterranean oaks with contrasting leaf habits on two different sites. *Forest Ecology and Management* 179, 401–414.

Evans, J. (1992) Coppice forestry – an overview. In: Buckley, G.P. (ed.) *Ecology and Management of Coppice Woodlands*. Chapman and Hall, London, pp. 233–245.

FAO (2010) *Global Forest Resources Assessment 2010*. FAO Forestry Paper 163, Food and Agriculture Organization of the United Nations, Rome. Available at: http://www.fao.org/forestry/fra/fra2010/en/ (accessed 21 November 2014).

FAO (2013) *State of Mediterranean Forests 2013*. Food and Agriculture Organization of the United Nations, Rome.

Feild, T.S., Arens, N.C., Doyle, J.A., Dawson, T.E. and Donoghue, M.J. (2004) Dark and disturbed: a new image of early angiosperm ecology. *Paleobiology* 30, 82–107.

Fischer, A. (2007) Coastal fishing in Stone Age Denmark – evidence from below and above the present sea level and from the bones of human beings. In: Milner, N., Craig, O.E. and Bailey G.N. (eds) *Shell Middens in Atlantic Europe*. Oxbow Books, Oxford, UK, pp. 54–69.

Forest Europe (2011) *State of Europe's Forests 2011 – Status and Trends in Sustainable Forest Management in Europe*. Forest Europe Ministerial Conference on the Protection of Forests in Europe, Oslo, pp. 70–71.

Forestry Commission (1952) *Census of Woodlands 1947–1949: Woodlands of Five Acres and Over*. Forestry Commission Census Report No. 1, Her Majesty's Stationery Office, London.

Forestry Commission (2003) *National Inventory of Woodland and Trees: Great Britain*. Forestry Commission, Edinburgh, UK.

Forestry Commission England (2007) *A Woodfuel Strategy for England*. Cambridge, UK.

Forêt Privée Française (2008) *Private Forest Property in France, 2008–2009*. Centre National de la Propriété Forestière (CNPF)/Institut pour le Développement Forestier (IDF)/Forestiers Privés de France/Union de la Coopération Forestière Française (UCFF), Paris. Available at: http://www.foretpriveefrancaise.com/data/info/141462-ChiffresCles_english_2008.pdf (accessed 21 November 2014).

Giudici, F. and Zingg, A. (2005) Sprouting ability and mortality of chestnut (*Castanea sativa* Mill.) after coppicing. A case study. *Annals of Forest Science* 62, 513–523.

Gracia, M. and Retana, J. (2004) Effect of site quality and shading on sprouting patterns of holm oak coppices. *Forest Ecology and Management* 188, 39–49.

Gracia, C.A., Sabate, S., Martinez, J.M. and Albeza, E. (1999) Functional responses to thinning. In: Roda, F., Retana, J., Gracia, C.A. and Bellot, J. (eds) *Ecology of the Mediterranean Evergreen Oak Forests*. Ecological Studies, Springer, Berlin, Heidelberg, pp. 329–337.

Grigoriadis, N. and Zagas T. (2005) Contribution of the extension of rotation to ecology and productivity in a Greek oak coppice forest. *Annali di Botanica* 5, 37–45.

Guillet, C. and Bergström, R. (2006) Compensatory growth of fast-growing willow (*Salix*) coppice in response to simulated large herbivore browsing. *Oikos* 113, 33–42.

Gyenge, J., Fernandez, M.E., Sarasola, M. and Schlichter, T. (2011) Stand density and drought interaction on water relations of *Nothofagus antarctica*: contribution of forest management to climate change adaptability. *Trees* 25, 1111–1120.

Hadley, P., Hall, A., Taylor, M., Needham, A., Taylor, B., Conneller C. and Milner N. (2010) To block lift or not to block lift? An experiment at the early Mesolithic site of Star Carr, North-East Yorkshire, UK. *Internet Archaeology* 28. Available at: http://intarch.ac.uk/journal/issue28/hadley_toc.html (accessed 21 November 2014).

Harmer, R. (2004) *Restoration of Neglected Coppice*. Forest Information Note 56, Forestry Commission, Edinburgh, UK.

Harmer, R. and Howe, J. (2003) *The Silviculture and Management of Coppice Woodlands*. Forestry Commission, Edinburgh, UK.

Hernando, A., Tejera, R., Velázquez, J. and Núñez, M.V. (2010) Quantitatively defining the conservation status of Natura 2000 forest habitats and improving management options for enhancing biodiversity. *Biodiversity Conservation* 19, 2221–2233.

Hillam, J., Groves, C.M., Brown, D.M., Baillie, M.G.L., Coles, J.M. and Coles, B.J. (1990) Dendrochronology of the English Neolithic. *Antiquity* 64, 210–220.

Hölscher, D., Schade, E. and Leuschner, C. (2001) Effects of coppicing in temperate deciduous forests on ecosystem nutrient pools and soil fertility. *Basic and Applied Ecology* 2, 155–164.

Hooke, D. (2010) *Trees in Anglo-Saxon England: Literature, Lore and Landscape*. The Boydell Press, Woodbridge, UK.

Huffel, G. (1927) Les méthods de l'aménagement forestier en France. *Annales de l'École Nationale des Eaux et Forêts et de la Station de Recherches et Expériences* 1(2), Berger-Levrault, Paris, France. Available at: http://documents.irevues.inist.fr/bitstream/handle/2042/33455/AEF_1927_1_2_001.pdf?sequence=1 (accessed 21 November 2014).

Jansen, P. and Kuiper, L. (2004) Double green energy from traditional coppice stands in the Netherlands. *Biomass and Bioenergy* 26, 401–402.

Jéhin, P. (2010) Verriers et forêts sous l'ancien regime en Alsace. *Les Actes de CRESAT* 7, 49–58.

Jones, A.L. (2008) Sustainable forest management in Late Renaissance Italy: an exploratory study of early forest management practices in the context of the Tolfa Mountains alum industry, 1462–1823. M.Sc. Dissertation, School of Oriental and African Studies, London.

Kiliç, M. and Tüfekçioğlu, U. (2010) Conversion of coppice forests obtained from shoots (coppice management). Abstracts. *Conference on the Oak – Ecology, History, Management and Planning II*. Suleyman Demirel University, Isparta, Turkey, pp. 30–31.

Kloukinas, D. (2012) The technology of house construction at the Neolithic settlement of Avgi. Centre for Documentation and Promotion of Excavations at Avgi – Kastoria Prefecture, IZ´ Ephorate of Prehistoric and Classical Antiquities, Ministry of Culture [and Sport], Athens. Available at: http://www.neolithicavgi.gr/?page_id=1120andlangswitch_lang=en (accessed 21 November 2014).

Kneifl, M., Kadavý, J. and Knott, R. (2011) Gross value yield potential of coppice, high forest and model conversion of high forest to coppice on best sites. *Journal of Forest Science* 57, 536–546.

Leek, N. (2010) Short rotation plantation. In: Mantau, U. (ed.) *EUwood – Real Potential for Changes in Growth and Use of EU Forests. Final Report*. Centre of Wood Science, University of Hamburg, Hamburg, Germany, pp. 89–92.

Magyari, E., Chapman, J., Fairbairn, A.S., Francis, M. and de Guzman, M. (2012) Neolithic human impact on the landscapes of north-east Hungary inferred from pollen and settlement records. *Vegetation History and Archaeobotany* 21, 279–302.

Maringer, J., Wohlgemuth, T., Neff, C., Pezzatti, G.B. and Conedera, M. (2012) Post-fire spread of alien plant species in a mixed broad-leaved forest of the Insubric region. *Flora* 207, 19–29.

Marinova, E. and Thiebault, S. (2008) Anthracological analysis from Kovacevo, southwest Bulgaria: woodland vegetation and its use during the earliest stages of the European Neolithic. *Vegetation History and Archaeobotany* 17, 223–231.

Marziliano, P.A., Iovino, F., Menguzzato, G., Scalise, C. and Nicolaci, A. (2013) Growth and yield models, assortment type and analysis of deadwood in chestnut coppice. *Forest@* 10, 14–25. Available at: http://www.sisef.it/forest@/contents/?id=efor0839-0010#efor839-ref37 (accessed 21 November 2014).

Mason, W.L., Nicoll, B.C. and Perks, M. (2009) Mitigation potential of sustainably managed forests. In: Read, D.J., Freer-Smith, P.H., Morison, J.I.L., Hanley, N., West, C.C. and Snowdon, P. (eds) *Combating Climate Change – A Role for UK Forests. An Assessment of the Potential of the UK's Trees and Woodlands to Mitigate and Adapt to Climate Change*. The Stationery Office, Edinburgh, UK, pp. 100–118.

Mattioni, C., Cherubini, M., Micheli, E., Villani, F. and Bucci, G. (2008) Role of domestication in shaping *Castanea sativa* genetic variation in Europe. *Tree Genetics and Genomes* 4, 563–574.

Matula, R., Svátek, M., Kůrová, J., Úradníček, L., Kadavý, J. and Kneifl, M. (2012) The sprouting ability of the main tree species in central European coppices: implications for coppice restoration. *European Journal of Forest Research* 131, 1501–1511.

Mead, G. (1989) The Sussex leather industry in the 19th century. *Journal of Sussex Industrial History* 19, 2–10.

Meiggs, R. (1982) *Trees and Timber in the Ancient Mediterranean World*. Clarendon Press, Oxford, UK.

Meliadis, I., Zagkas, T. and Tsitsoni, T. (2010) Chapter 15 Greece. In: Tomppo, E., Gschwantner, T., Lawrence, M. and McRoberts, R.E. (eds) *National Forest Inventories: Pathways for Common Reporting*. Springer, Dordrecht, The Netherlands, pp. 259–268.

Moreno, G., Obrador, J.J., Garciá, E., Cubera, E., Montero, M.J., Pulido, F.J. and Dupraz, C. (2007) Driving competitive and facilitative interactions in oak dehesas with management practices. *Agroforestry Systems* 70, 25–44.

Mosandl, R., Summa, J. and Stimm, B. (2010) Coppice-with-standards: management options for an ancient forest system. *Forestry Ideas* 16, 65–74.

Nocentini, S. (2009) Structure and management of beech (*Fagus sylvatica* L.) forests in Italy. *iForest* 2, 105–113. Available at: http://www.sisef.it/iforest/pdf/?id=ifor0499-002 (accessed 21 November 2014).

Núñez, V., Hernando, A., Velázquez, J. and Tejera, R. (2012) Livestock management in Natura 2000: a case study in a *Quercus pyrenaica* neglected coppice forest. *Journal for Nature Conservation* 20, 1–9.

Paciorek, C.J., Condit, R., Hubbell, S.P. and Foster, R.B. (2000) The demographics of resprouting in tree and shrub species of a moist tropical forest. *Journal of Ecology* 88, 765–777.

Panaïotis, C., Carcaillet, C. and M'hamedi, M. (1997) Determination of the natural mortality age of a holm oak (*Quercus ilex* L.) stand in Corsica (Mediterranean island). *Acta Oecologica* 18, 519–530.

Papachistou, T.G. and Platis, P.D. (2011) The impact of cattle and goats grazing on vegetation in oak stands of varying coppicing age. *Acta Oecologica* 37, 16–22.

Pisanelli, A., Bregasi, M., Jupe, A. and Toromani, E. (2010) Research priorities and opportunities within the forestry and agroforestry sectors in Albania. *iForest* 3, 113–117. Available at: http://www.sisef.it/iforest/pdf/?id=ifor0534-003 (accessed 21 November 2014).

Price, T.D. and Gebauer, A.B. (eds) (2005) *Smakkerup Huse: A Late Mesolithic Coastal Site in Northwest Zealand, Denmark*. Aarhus University Press, Aarhus, Denmark.

Pyttel, P.L, Fischer, U.F., Suchomel, C., Gärtner, S.M. and Bauhus, J. (2013) The effect of harvesting on stump mortality and re-sprouting in aged oak coppice forests. *Forest Ecology and Management* 289, 18–27.

Rackham, O. (1990) *Trees and Woodland in the British Landscape*. Dent, London.

Rackham, O. (2003) *Ancient Woodland: Its History, Vegetation and Uses in England*. Castlepoint Press, Dalbeattie, UK.

Radtke, A., Ambraß, S., Zerbe, S., Tonon, G., Fontana, V. and Ammer, C. (2013) Traditional coppice forest management drives the invasion of *Ailanthus altissima* and *Robinia pseudoacacia* into deciduous forests. *Forest Ecology and Management* 291, 308–317.

Rubio, A. and Escudero, A. (2003) Clear-cut effects on chestnut forest soils under stressful conditions: lengthening of time-rotation. *Forest Ecology and Management* 183, 195–204.

Rubio, A., Gavilán, R. and Escudero, A. (1999) Are soil characteristics and understorey composition controlled by forest management? *Forest Ecology and Management* 113, 191–200.

Rydberg, D. (2000) Initial sprouting, growth and mortality of European aspen and birch after selective coppicing in central Sweden. *Forest Ecology and Management* 130, 27–35.

Sánchez-Humanes, B. and Espelta, J.M. (2011) Increased drought reduces acorn production in *Quercus ilex* coppices: thinning mitigates this effect but only in the short term. *Forestry* 84, 73–82.

Serrada, R., Allué, M. and San Miguel, A. (1992) The coppice system in Spain. Current situation, state of art and major areas to be investigated. *Annali dell'Istituo Sperimentale per la Selvicoltura* 23, 266–275.

Šplíchalová, M., Adamec, Z., Kadavý, J. and Kneifl, M. (2012) Probability model of sessile oak (*Quercus petraea* (Matt.) Liebl.) stump sprouting in the Czech Republic. *European Journal of Forest Research* 131, 1611–1618.

Stajic, B., Zlatanov, T., Velichkov, I., Dubravac, T. and Trajkov, P. (2009) Past and recent coppice forest management in some regions of south-eastern Europe. *Silva Balcanica* 10, 9–19.

Suchomel, C., Pyttel, P., Becker, G. and Bauhus, J. (2012) Biomass equations for sessile oak (*Quercus petraea* (Matt.) Liebl.) and hornbeam (*Carpinus betulus* L.) in aged coppiced forests in southwest Germany. *Biomass and Bioenergy* 46, 722–730.

Szabó, P. (2010) Driving forces of stability and change in woodland structure: a case-study from the Czech lowlands. *Forest Ecology and Management* 259, 650–656.

Tanentzap, A.J., Mountford, E.P., Cooke, A.S. and Coomes, D.A. (2012) The more stems the merrier: advantages of multi-stemmed architecture for the demography of understorey trees in a temperate broadleaf woodland. *Journal of Ecology* 100, 171–183.

Taylor, B., Conneller, C. and Milner, N. (2010) Little house by the shore. *British Archaeology* No. 115 (Nov/Dec 2010). Available at: http://www.archaeologyuk.org/ba/ba115/feat6.shtml (accessed 21 November 2014).

Tedeschi, V., Rey, A., Manca, G., Valentini, R., Jarvis, P.G. and Borghetti, M. (2006) Soil respiration in a Mediterranean oak forest at different developmental stages after coppicing. *Global Change Biology* 12, 110–121.

Tomàs, E. (1999) The Catalan process for the direct production of malleable iron and its spread to Europe and the Americas. *Contributions to Science* 1, 225–232.

Tredici, P. (2001) Sprouting in temperate trees: a morphological and ecological review. *The Botanical Review* 67, 121–140.

UN ECE/FAO (2000) *Forest Resources of Europe, CIS, North America, Australia, Japan and New Zealand: Contribution to the Global Forest Resources Assessment (Industrialized Temperate/Boreal Countries) 2000. Main Report*. Geneva Timber and Forest Study Papers, No. 17, United Nations Economic Commission for Europe with Food and Agriculture Organization of the United Nations, New York and Geneva, Switzerland.

UNFCCC (2011) *Croatia's Submission of Information on Forest Management Reference Levels*. United Nations Framework Convention on Climate Change, Bonn, Germany Available at: http://unfccc.int/files/meetings/ad_hoc_working_groups/kp/application/pdf/awgkp_croatia_2011.pdf (accessed 21 November 2014).

Valbuena-Carabaña, M., González-Martínez, S.C. and Gil, L. (2008) Coppice forests and genetic diversity: a case study in *Quercus pyrenaica* Willd. from Central Spain. *Forest Ecology and Management* 254, 225–232.

Van Calster, H., Baeten, L., De Schrijver, A., De Keersmaeker, L., Rogister, J.E., Verheyen, K. and Hermy, M. (2007) Management driven changes (1967–2005) in soil acidity and the understorey plant community following conversion of a coppice-with-standards forest. *Forest Ecology and Management* 241, 258–271.

Vaughan, S.P., Cottrell, J.E., Moodley, D.J., Connolly, T. and Russell, K. (2007) Clonal structure and recruitment in British wild cherry (*Prunus avium* L.) *Forest Ecology and Management* 242, 419–430.

Velichkov, I., Zlatanov, T. and Hinkov, G. (2009) Stakeholder analysis for coppice forestry in Bulgaria. *Annals of Forest Research* 52, 183–190.

Warde, P. (2006) The fear of wood shortage and the reality of woodland in Europe *c.*1450–1850. *History Workshop Journal* 62, 29–57.

Zarrilli, A.G. (2009) 'Red gold': development and crisis of the forest exploitation in Gran Chaco, Argentina in the 20th century. In: Saratsi, E., Bürgi, M., Johann, E., Kirby, K., Moreno, D. and Watkins, C. (eds) *Woodland Cultures in Time and Space*. Embryo Publications, Athens, pp. 115–121.

7 High Forest Management and the Rise of Even-aged Stands

Peter Savill*
Department of Plant Sciences, University of Oxford, Oxford, UK

7.1 Introduction

In most of pre-industrial Europe, local supplies of wood were indispensable for everyday life, as necessary to a population as food. Forests provided the raw material for buildings, furniture, ploughs, carts and wheels. More importantly, gigantic quantities were also required for domestic heating and cooking (Williams, 2002). To meet these needs, broadleaved woodland was generally managed in the medieval period as coppice or forms of wood-pasture (Hartel *et al.*, Chapter 5; Buckley and Mills, Chapter 6). High forest was rare except in the boreal zone, whereas now it is the commonest structure for forests across the whole Continent. This chapter explores how and why this change happened and the systems of management developed to cope with a land use whose timescale exceeds human generations.

7.2 Changing from Coppice to High Forest Systems

Before canals, modern roads and railways, the movement of heavy loads, such as timber, was difficult, slow and expensive, so there was a strong incentive for districts to be more or less self-sufficient in wood and timber. Indeed, Brasnett (1953) considered that it was not until the 20th century that local self-sufficiency began to be seen as unnecessary, even in theory. Across Europe as a whole, there were ample supplies of wood for most purposes up to about the mid-14th century, and even until the mid-16th century supply and demand were more or less in balance. At local levels, however, demand exceeded supply, leading to timber being traded across regions and between countries: for example, deal boards from the Baltic States to Britain, firewood across the English Channel, e.g. Mollat (1947), Carus-Wilson (1962–1963), Lillehammer (1986, 1990). Yet of all the major industries, the timber trade has been one of the least recorded across Europe (Meiggs, 1982).

Environmental degradation also became problematic in western Europe from the late 16th century onward. Intensive grazing by sheep and cattle often prevented regeneration. Clearance of forest might lead to soil erosion on slopes or wind-blow of sandy soils (Darker, 1990, Dirlou, 1999). This pattern mirrored what had happened in the ancient

---

*E-mail: peter.savill@plants.ox.ac.uk

© CAB International 2015. *Europe's Changing Woods and Forests: From Wildwood to Managed Landscapes* (eds K.J. Kirby and C. Watkins)

Mediterranean world 2000–3000 years earlier (Meiggs, 1982; although see also Grove and Rackham, 2001).

Concerns about shortages and increased prices, particularly for fuel, occurred from the mid-17th century onward. These led to a phase of forest reconstruction which lasted until almost the end of the 18th century. Management systems were developed to provide sustainable yields of wood and timber, especially in France, Austria and some of the German states (Halkett, 1984; Williams, 2002; Warde, 2006).

Increasing fuel prices were the main concern for most people, but governments tended to be more interested in the effects on strategic supplies, such as naval timber. Williams (2002) proposed that this prompted both Evelyn's *Sylva* to be written in 1664 and the French Forest Ordinance of 1669 associated with Colbert. Glacken (1967) identifies these as the point when the need for forest conservation was officially recognized in Europe. Colbert's Ordinance was by far the more effective and enduring influence. It codified and updated the mass of ancient French forest laws which had become confused and unworkable. It regulated what could be done in forests, made forestry part of the state economy and ensured a supply of timber for the navy (Rochel, Chapter 19). This approach was widely copied across Europe by those who saw the need for systematic forest planning when shortages of timber developed and there was no opportunity to draw on extensive timber supplies from their colonies, as could the UK (until the 20th century). Thus, central European states became the leaders in planned forest management (Johann, 2006).

In 1713, Hans Carl von Carlowitz (1645–1714), a Saxon mining administrator, published *Silvicultura oeconomica*. This proposed and described the ideas behind sustainability in forestry, including sustainable yields from cutting and rotations. He recommended planting fast-growing conifers to replenish timber stocks, a policy that eventually spread through Silesia, in what is now Poland, and Moravia, in what is now the Czech Republic. The approach was also adopted from the plains in Germany to mountainous areas in Switzerland, Italy, Austria and Slovenia (Johann, 2006). In the more western states, deciduous trees were favoured but, overall, the broadleaved forest area in central Europe decreased dramatically as conifers were established.

By the second half of the 18th century, a spate of manuals had appeared and forestry was taught in several German universities. These provided the foundations of the reputation that Germany acquired throughout the world for the rational and scientific study of forests and forestry and the development of a sustained yield from even-aged woods. Heinrich Cotta (1763–1844), Director of the Forest Academy of Tharandt, Saxony, from 1816–1844, is particularly associated with this classical German plantation forestry.

During periods of re-establishing forests, economic and practical considerations nearly always favour the use of a few species and even-aged techniques. The Austrian Forest Ordinance of 1786 prescribed even-aged techniques; in Prussia, an ordinance of 1787 provided that even-aged stands should be grown; and the Darmstadt Ordinance of 1776 even prohibited the use of the uneven-aged selection system.

Even-aged stands, mostly of pine (*Pinus* spp.) or Norway spruce (*Picea abies*), promoted by 19th century German foresters, were seen as a way of imposing order upon nature. In this context, Pryor (1982) noted:

> the main advantages of planting arise from its artificiality and minimum reliance on unpredictable natural events. Enough plants can be ordered for the desired year and can then be evenly distributed across the whole area, in rows, to facilitate subsequent tending. This makes reliance on natural regeneration seem like a technique inherited from a primitive 'hunter–gatherer' technology, whereby the time of arrival and dissemination of seed, the genetic quality and even the species of the regeneration are largely outside the control of the manager.

7.3 The Need for New Administrative Tools

Even-aged monocultures can be far more profitable than other forms of forestry. Martin Faustmann, in 1849, developed what became known as the Faustmann formula, which deals with the principle of discounted cash flow; Max Robert Pressler then showed, in 1860, how to calculate the optimum rotation for a forest using discounting. The highest returns from investment in forestry occur, predictably, when:

- prices for the type of produce being grown are high;
- the costs of inputs (mostly labour) are low;
- there are good markets for small-sized material from thinnings or coppice;
- interest (or discount) rates are low; and
- tree growth is fast and hence rotations are short.

Today, short rotation (c.15 years) *Eucalyptus* coppice grown for pulpwood in Portugal and Spain gives a very good return on money invested. Oak (*Quercus* spp.) grown on rotations of 150–250 years is much less attractive to investors, even though the value of the timber at the end of an oak rotation can be extremely high, far more than the value of pulpwood.

High forest systems took a number of different forms. Felling and restocking a particular part of the forest might be carried out by one generation of foresters, tending the stands during the middle of their lives by a second generation, and tending the final mature stages by a third generation. Notwithstanding, any individual forester might be dealing with all three stages of stand development simultaneously in different parts of the forest. The formalization of a number of 'silvicultural systems' that described the different ways in which these high forest stands might be treated assisted in the management process.

Initially, there were still seen to be benefits from local sustainability, and foresters promoted the concept of a 'normal' age structure. At its simplest, this means that there is an even distribution of age classes by area, ranging from young stands to those of rotation age. So for oak stands grown on a 150 year rotation there would be an equal area of each, (say) 10-year-old, age class. In all-age selection forests normality is ensured through the distribution of trees of different stem diameters rather than ages, so there are many small, young trees (to allow for losses as they grow and are thinned) and few large, old ones per unit area.

'Normality' at the forest scale has less relevance today, as cheap and efficient transport is available, and no one industry or community depends upon a regular supply of material from just a single forest. There may though still be advantages to an owner or community in terms a steady income and giving continued employment to forest workers. There can also be ecological benefits. Normality guarantees that the proportion of the forest at risk from serious damage by wind, fire, pests and pathogens is always at a minimum over the period of a rotation, because stands tend to be vulnerable to any one form of damage only at particular stages in their lives. The heterogeneity conferred by normality may also increase the diversity of species that can survive continually in the forest and provide a measure of resilience that enables forests to persist and to continue to be productive, despite fluctuations in conditions (Malcolm, 1979).

7.1 Silvicultural Systems

A *silvicultural system* is the process by which the stands constituting the forest are tended, removed and replaced by new stands, resulting in the creation of woods of a distinctive form (Troup, 1928; Savill and Brown, 2010) (Box 7.1). However, in practice, each system tends to merge with the next along a continuum, and they are usually applied in a flexible manner.

Systems differ from one other in terms of age, size distribution of trees, rotation length and means and pattern of regeneration. An initial division is between coppice and coppice-with-standards (where most of the trees have vegetative origins) and high forest (where the origin is from seed). High forest may then be separated into selection, shelterwood and clear-felling systems based on the extent of the canopy felled at any one time. A further difference is whether the stand is restocked using natural regeneration or through planting, the latter being particularly prevalent with clear-felling systems.

Box 7.1. Silvicultural systems.

Coppice system

In the coppice system, the stand is replaced from shoots that arise primarily from concealed dormant buds in the stump of a tree following felling. The shoots can also develop as suckers from buds on roots in some species. Coppice is normally worked through clear-felling blocks of about 0.5 ha upwards every 3–30 years, the period depending on the produce required and the growth rate of the species (Buckley and Mills, Chapter 6). Traditional coppice has declined everywhere in Europe since at least the mid-1800s (Bürgi, 1998) as a result of industrialization, such that large areas of abandoned coppice and coppice-with-standards exist in Europe. In its modern form, *Eucalyptus* coppice is used for the production of pulpwood, and extensive areas have been planted in Portugal and Spain. New coppices of *Salix* are also being established in Sweden for short-rotation energy crops cut on a 3–5 year cutting cycle (Dimitriou and Aronsson, 2005). Poplars are also planted for energy, but so far only on a very small scale. They tend to be more suited to warmer, more southern climates.

High forest systems

Selection systems involve the manipulation of a forest to maintain a continuous cover, to provide for the regeneration of the desired species and for the controlled growth and development of trees through a range of diameter classes which are mixed singly (in single-tree selection systems) or in small groups (group selection systems). The system is particularly suited to shade-bearing species, and the best examples are in the silver fir (*Abies alba*) with beech (*Fagus sylvatica*) and Norway spruce (*Picea abies*) forests of central Europe (Troup, 1928; Knuchel, 1953). These forests are largely confined to mountainous regions where continuous protection of the soil against erosion and also often against avalanches is of great importance (see Fig. 7.1). Stands managed on a single-tree selection system are, at all times, an intimate mixture of trees of all age classes. The interval between successive selection fellings varies. Short periods (less than 5 years) allow better stand management, particularly of young trees. Longer intervals result in larger volumes of timber being removed at each visit, making them more economically viable. They also improve the success of regeneration of light-demanding species because the canopy is opened up more.

Shelterwood systems provide a method of establishing more or less even-aged stands, usually by natural regeneration, under a thinned overstorey that produces sufficient shade and a moderated environment for young trees to establish. The overstorey is removed as soon as establishment is complete. Treatments usually include a series of regeneration fellings, starting with a preparatory felling to encourage the development of the crowns of future seed bearers, and then a seeding felling at the time of a good seed crop when a third to a half of the stems are removed as well as the understorey and any regeneration already present. There are then typically two to four secondary fellings over a period of years, with timing and intensity regulated to allow seedlings to grow, but also to prevent serious weed growth. The final felling is the last secondary felling in which all of the remaining overstorey is removed. The whole series of operations normally takes 5–20 years. Infrequent mast years and frost-sensitive seedlings such as beech both necessitate long regeneration periods. The secondary fellings for a light-demanding species must be few and rapid, and the whole process may be completed in 5 years. Stands managed under a shelterwood system have many features in common with those established by planting under a clear-felling system. They can be pure, even aged and uniform in structure and density over large areas; see Troup (1928) for a description of the system in France and Köstler (1956) who focuses on Germany.

Clear-felling usually involves the cutting of a minimum of about a hectare, but blocks of between 5 and 20 ha are not uncommon. This is now probably the most widespread system of management in forests where profitable wood production is a primary objective. Its main advantages include simplicity, uniformity and, in particular, ease of felling and extraction. The use of clear-felling does not preclude the use of natural regeneration (see Fig. 7.1), but the system normally operates with establishment by planting. The stands produced tend to be uniform and even aged. Even-aged forests are also found naturally after widespread destruction by fire or windthrow followed by complete regeneration within a limited period of time (Jones, 1945; Peterken, 1996), and most successful clear-felling schemes use species that are adapted to such conditions. In contrast to the selection system, conditions at any one point may change dramatically over the course of a rotation. After felling the closed forest, the ground is exposed, more precipitation reaches the soil so that the water table rises, temperatures fluctuate more and nutrient

Continued

Box 7.1. Continued.

losses from leaching increase (Miller, 1979). A little later, there is a dense growth of ground vegetation. During the thicket stage, both precipitation and the light reaching the ground reduce so that ground vegetation is at a minimum. Relative humidity is high and relatively stable. Details about many of these effects are given by Geiger (1965), Helliwell and Harrison (1979), Horsley and Marquis (1983) and Lowman *et al.* (2012); see also Quine (Chapter 15).

Other high forest systems – group, strip, wedge and edge systems (Troup, 1928; Savill and Brown, 2010) can be considered as variants of the three basic high forest systems described above.

Woodland management can be thought of as grading from intensive through to extensive. Intensive management is normally associated with the coppice, clear-felling and shelterwood systems. The less intensive approach is more appropriate to selection and associated group systems, which need careful – but not capital-intensive – management to run well. The same distinctions apply to the strategy adopted for obtaining and using natural regeneration: one can either invest time and money in trying to get a full stocking from any one seed year, as with a shelterwood system with careful preparatory thinnings, cultivation and weed control; or operate a group, or selection, system with minimum preparation for seed, but accepting and using the steady trickle of saplings that establish themselves, largely unaided.

Tolerant (shade-bearing) species are best suited to irregular, uneven-aged silviculture, which is also most often practised on fragile sites, steep slopes, sites with high water tables

Fig. 7.1. Forests managed under two regimes at extreme ends of the range of possible silvicultural systems. (a) Strip clear-felling in Scots pine (*Pinus sylvestris*) in the Forest of Rouvray, France (Troup, 1928). This is a clear-cutting system that relies upon natural regeneration rather than planting. The width of cleared strips is about four times the height of the seed-bearing trees, which can be seen in lines.

Fig. 7.1. (b) Single-tree selection in the mountainous Couvet Forest in Switzerland (Troup, 1928). The main species are Norway spruce (*Picea abies*) and European silver fir (*Abies alba*).

and very dry sites that would be adversely affected by the complete removal of the forest cover, even for short periods. Even-aged systems are more appropriate for intolerant, colonizing (light-demanding) species and sites with few environmental limitations.

7.5 The Rise of Plantations

Plantations based on clear-felling have come to dominate large parts of European and other forests too. Globally, in the early 21st century, 70% of industrial forest products are sourced from the 7% of the world's forests that are predominantly planted or have a planted component (Evans, 2009).

The shortage of timber in the 17th and 18th centuries implied a need to increase the forest area. New woodland can and does develop spontaneously on abandoned land, as is very apparent in many mountainous regions of Europe at present. However, woodland could often be created more quickly through planting, and from about 1850, the loss of forest area in many countries was compensated for by planned reforestation with spruce and pine. When planting is carried out on a large scale, the economics tends to favour creating single-species, even-aged blocks. These tend to become ready for harvest at a similar time, and are then clear-felled and the sites replanted for the next generation.

The experience of planting that had developed in Germany in the 18th and 19th centuries was followed by similar operations in Britain and Ireland in the 20th century (Plate 7) – essentially, the establishment of even-aged coniferous monocultures (Tsouvalis, 2000). In Spain and Portugal, extensive stands of *Eucalyptus* have been created in response to moves to take land out of agricultural production across the European Union (EU). *Eucalyptus* now accounts for almost 633,000 ha in Spain (SECF, 2011) – some 3.5% of the total forest area, but 61% of the 1037 thousand ha of intensively managed plantations. In Portugal, where the climate is much better suited to its growth, *Eucalyptus* coppice has recently become the predominant species, at 26% of the forest area (ICNF, 2013), exceeding the area of native maritime pine (*P. pinaster*) (23%) and cork oak (*Quercus suber*) forest (also 23%). While *Eucalyptus* has mainly been established on surplus agricultural land, it is also replacing maritime pine plantations which suffer serious damage from pine wood nematodes and also from fires.

Concern about timber shortages also led to a drive to improve the production from the existing woods. Coppice was a very effective way of producing large amounts of small-sized wood, but it has limited scope for producing timbers of larger dimensions. Stems can be singled or 'stored' to grow into high forest but are likely to be rotten at the base where they join the stool, and they seldom grow to large sizes. Coppice depends on vegetative reproduction, and the species and even the genotypes differ little from one stand to its successor, so limiting opportunities to improve the quality of the stands. Hence, clear-felling and replanting allowed the creation of stands grown from seedlings that would eventually reach large dimensions as high forest; further, the trees used could be of a different, more desirable or productive species or genotype from that previously present.

7.6 Increased Use of Conifers and Introduced Species

Based on statistics from FAO (2010), it is possible to develop a rough index that shows the relative importance of introduced species (Fig. 7.2). In most places where plantations are established, native species predominate in the plantings: for example, the share of introduced species in Slovakia and Switzerland is 2%, in Germany 8%, in Italy 15% and in Norway and Sweden 18% of the planted forest area. However, in countries with poor or relatively unproductive native tree floras, there is a very heavy reliance on introduced species, for example in Hungary, but particularly in the six maritime countries in western Europe: Ireland, the UK, Belgium, Denmark, the Netherlands and Portugal.

The practice of introducing trees from elsewhere is as old as civilization itself, but it is only since the middle of the 19th century that their use in plantations has come to be widespread. Introductions frequently follow a pattern of initial use only in gardens and arboreta and then, if successful, perhaps half a rotation later, small-scale planting on estates and by government departments. Large plantations of the most promising species are established some 20–50 years after that (Streets, 1962).

Thus, Sitka spruce (*P. sitchensis*), the major plantation tree in Great Britain and Ireland, was introduced in 1831. It was established as one of the chief introduced species in the early 1920s, and by the mid-1950s, it became the most widely planted tree, occupying 30% of the forest area in Britain by about 2000 (Forestry Commission, 2003) and 52% in Ireland (NFI, 2007). In central Europe, black locust (*Robinia pseudoacacia*) was introduced into Hungary between 1710 and 1720. The first large plantations were established at the beginning of the 19th century on the Hungarian Plain to stabilize the wind-blown sandy soil. In 1885, the species occupied 37,000 ha, and by 2005, 4 million ha or 23% of the forest area of Hungary (Rédei *et al.*, 2008).

Individuals, companies and even governments that invest in growing trees normally wish to do so with as little risk as possible, and with a reasonable prospect of obtaining a return on their investment as soon as they can.

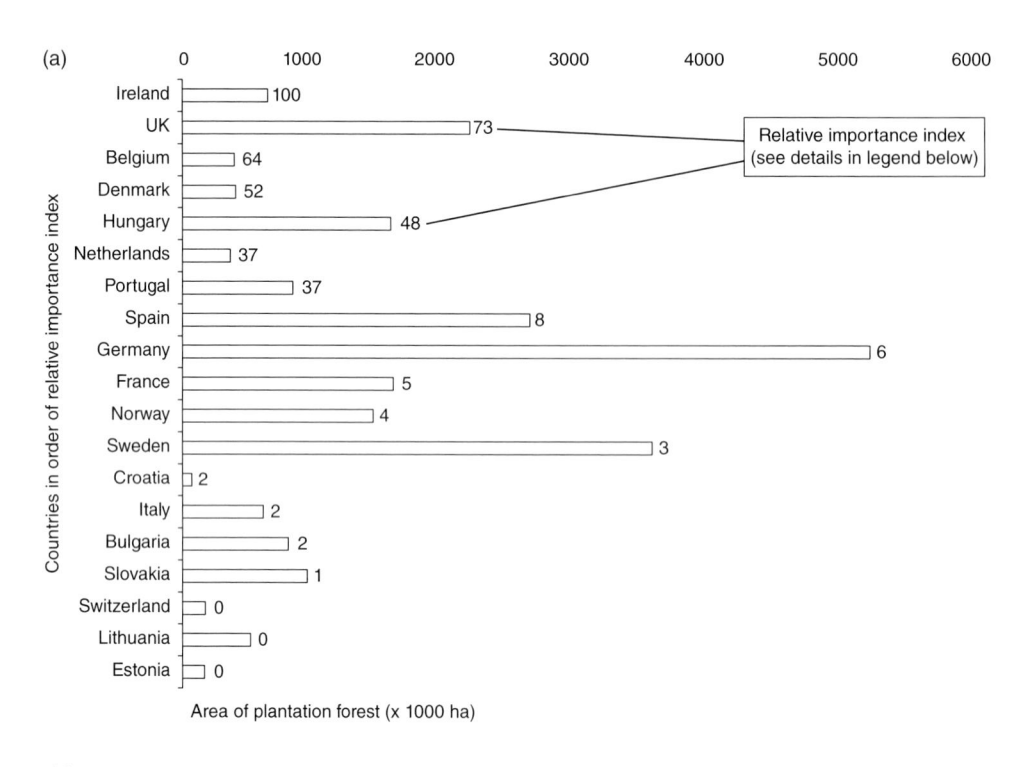

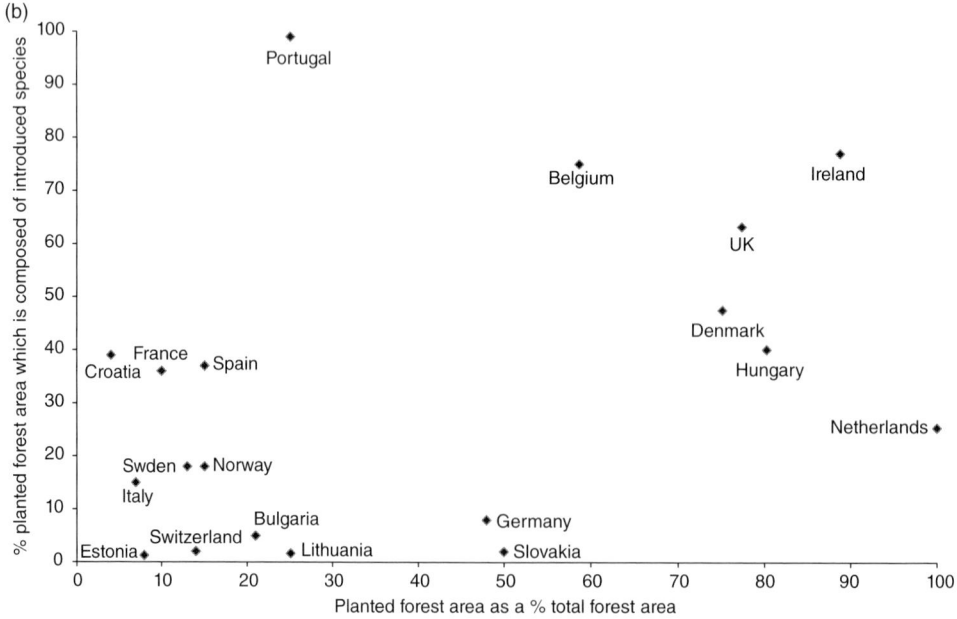

Fig. 7.2. (a) Planted forest areas in a selection of European countries and the importance of introduced species (from FAO, 2010), listed in descending order of relative importance. The relative importance index is derived from % total forest area planted × % total forest that is introduced species, expressed as a percentage of the value for Ireland, which is set at 100%. (b) Area of plantations as % total forest area versus % planted area made up of introduced species in different European countries.

Many species of conifers, but particularly species of pines and spruces in Europe, have offered both of these advantages although they are increasingly susceptible to diseases (Potter, Chapter 23). With the exception of species of *Eucalyptus* in some warm temperate regions, and to a much lesser extent poplars and *Robinia pseudoacacia*, broadleaves have proved to be considerably less successful plantation trees.

Conifers are much less 'site demanding' than broadleaves and will often grow well on many kinds of poor sites and in soils where there is little competition from agriculture. Such sites include exposed, often upland, situations or sites in climates that suffer from extremes of drought or high rainfall and waterlogging. Examples include the infertile, sandy soils of Lüneburg Heath in Lower Saxony, northern Germany, the Hungarian Plain and the Landes in Gascony, France (Clout, 1969), the mountainous Black Forest in Baden-Württemberg, south-west Germany, much of the land planted in the late 20th century in Greece, Turkey and Cyprus, and the peat-covered uplands of Great Britain and Ireland (Tsouvalis, 2000). All have been extensively afforested.

More of the timber produced by conifers is in the form of usable stem wood, rather than branches. They characteristically grow straight and without forks, thereby fulfilling the needs of modern sawmills and industry. Few broadleaves can do this, especially on exposed sites. For example, according to the British forest management tables (Hamilton and Christie, 1971), an average plantation-grown Sitka spruce of 40 cm diameter at breast height (dbh) would have a volume of about 2.0 m^3, and an oak with the same diameter a volume of 1.5 m^3. At 60 cm diameter, the volumes would be 3.3 and 2.4 m^3, respectively. As the annual production of wood per hectare for the median yield classes of Sitka spruce is 15 m^3 ha^{-1}, and that for oak is 6 m^3 ha^{-1}, a stand of spruce averaging 40 cm dbh would yield 1050 m^3 over a rotation of 75 years and an oak stand would produce 535 m^3 over 90 years. Conifer crops tend to be more uniform in size, form and quality than broadleaves, thus providing the standardized products preferred by the wood-processing industries.

In temperate countries, most conifers are much more productive than broadleaved trees. Christie and Lines (1979) showed that yield classes for Scots pine (*P. sylvestris*) in 29 regions of Europe range between 1 and 14 m^3 ha^{-1} a year, depending upon site quality and region where they are grown, with the highest average rates of growth in Eastern Europe. Annual production of Norway spruce varies between 2 and 20 m^3 ha^{-1}. However, the range of production for broadleaves such as oak and beech is usually between 1 and 6 m^3 ha^{-1} a year.

Rotations for plantation conifers (the interval between the establishment of the young seedlings and final felling) are commonly between 50 and 60 years. For broadleaves, they are typically longer: 70 years for ash (*Fraxinus excelsior*), alder (*Alnus glutinosa*) and sweet chestnut (*Castanea sativa*), 120 to 150 years for oak. Thus, conifers provide an earlier return on any investment and, though not immune, they are less prone to problems such as poorly formed stems (due to poor genetic stock or poor silviculture), shake, knots and damage by squirrels and deer.

The largest consumer of timber products in the world is the EU with its 730 million population, which used around 430 million m^3 of the roundwood equivalent of solid wood products in 2011 (EU Forestry Statistics, 2012). The bulk of that European production (71%) comes from coniferous trees, with only 29% from broadleaves. About 75% of the timber used in the EU goes for paper pulp, pallets and constructional work. The remaining 25% is used as fuel (EU Forestry Statistics, 2012).

Overall then, the dominance of conifers in modern forestry is market driven. Much of the EU's domestic production is concentrated in Sweden, Germany, France and Finland. Austria and Poland are also significant producers (EU Forestry Statistics, 2012).

7.7 How Forestry is Changing

Even as large-scale conifer stands were being created in the 19th century, concerns about their long-term sustainability were being raised. On some sites, growth began to decline after several rotations of monocultures, windthrow occurred and pathological disorders became evident. These problems were often ascribed to the use of even-aged stands, although more commonly they were caused by the incorrect matching of species to site (Jones, 1952; Evans, 1998). European foresters started to look again at uneven-aged systems. The Baden forest law of 1833 prohibited clear-felling. In the 1880s, Karl Gayer (1822–1907), Professor of Forest Science at Munich University, led a movement that called for natural regeneration with uneven-aged forest structures rather than even-aged stands and clear-felling. The Dauerwald system (Jones, 1952) – now called 'continuous cover' – was revived, and management in many parts of Germany and Prussia moved to single-tree selection (Bürgi, Chapter 8).

Since about 1970, there has also been an increasing interest in conserving natural and semi-natural ecosystems, as well as a wider appreciation of forests as part of the landscape, and for their recreation and protection values. What were at one time seen as purely 'production' systems must now deliver a wide range of additional ecosystem services (Picard *et al.*, 2012; Leslie, 2014). Certification systems have been introduced (Upton and Bass, 1995) that tend to discourage the use of introduced species in semi-natural woods and encourage uneven-aged forests. In some cases, this is in regions where such systems have not been traditional in the recent past and there may be considerable difficulties if production and profit are also important motives.

Over a century ago, William Schlich (1904) said of forestry in Great Britain:

> a sympathetic understanding of the business … will go a long way to prevent any oscillating policy, that otherwise might threaten to interfere with the progress of forest management. Continuity of action will then become the order of the day, without which no industry can flourish, the produce of which frequently requires a century and more to mature.

However, it is also worth noting that forceful advocacy by distinguished foresters has similarly led to oscillations in policy and practice (Jones, 1952).

Illustrations of how policy shifts can leave a legacy of what has become seen as inappropriate forest practices are provided by the changing approaches to even-aged monocultures in 19th century Germany (see above) and the first 100 years of the British Forestry Commission (founded in 1919). The initial priority of the Forestry Commission was to create a strategic reserve of timber, but this later developed into the ruthless pursuit of economic (i.e. profitable) forestry, almost to the exclusion of all else. Large areas of coniferous plantations were created on open semi-natural habitats – moors, heaths and bogs (Tsouvalis, 2000). Fifty years later, the rise in car ownership and resulting increased mobility led to some social and landscape objectives being given higher priority, so more mixed stands came to be used. More recently still, biodiversity concerns are leading to open habitats being restored and plantations being removed (Forestry Commission, 2010). A parallel set of changes has occurred with forestry in ancient woodland, where from the 1950s through to the 1980s broadleaved stands were replaced by conifers and much of their conservation value was lost. This policy changed in 1985 (Forestry Commission, 1985), and broadleaved stands are now being restored, often through the premature felling of the coniferous plantations established in the earlier policy era (Peterken, Chapter 18).

It has proved difficult to anticipate what society will need and want from its woods in 50 years' time, particularly as forests and society respond to changes in climate and the increased pressure from new pests and diseases. Forestry that satisfies, social, environmental *and* economic objectives provides an insurance against unknown changes in the economic and political climate of the future. Some countries, such as France and Germany, appear to have achieved stable forest policies designed to combine production with environmental and social values in a sensitive and sustainable manner (Westoby, 1987). In these countries, there is generally wide public support for timber harvesting and forest management because the practices applied are consistent with the accepted guidelines and standards that promote sustainable forest management.

7.8 Future High Forest and Natural Forest Structures

High forest systems of one sort or another are likely to continue to dominate future forests across Europe and so we need to understand how similar their ecology is likely to be – on the one hand to the original natural forests and on the other to cultural landscapes that have been created over the last 1000 years (Bürgi, Chapter 8; Quine, Chapter 15). The high forest structures that we have inherited can be organized along a gradient of artificiality between at one extreme, a stand planted on open ground with a species introduced from another continent, and at the other, naturally regenerated forest of locally native species (Savill et al., 1997). Five categories covering these extremes are described below.

1. Planting of bare land where there has been no forest for at least 50 years, and often for much longer. This includes all 20th century afforestation of grassland, such as: the beech plantations on chalk downland in southern England or the spruce planting in the uplands of Scotland; almost all 20th century forest in Ireland; and the great 18th century pine plantings of Les Landes in France (Osmaston, 1968). The lack of a woodland species legacy may mean that these stands will long remain impoverished compared with nearby ancient woodland. However, they can and do develop their own distinctive assemblages of species (Quine, Chapter 15).

2. Reforestation of land that has carried forest within the last 50 years and hence retains some of the features of the previous forest conditions, but where the previous stand is replaced by a different one, for example ancient oak woods replaced by *Robinia pseudoacacia* in Hungary (Molnár, 1998). In Britain, much broadleaved woodland was converted to coniferous plantations in the period 1950–1980, but this process is often now being reversed (Goldberg, 2003; Thompson et al., 2003). The former plant and animal assemblages may survive and be revived to varying degrees.

3. Replanting of land that has carried forest within the last 50 years by renewal of the same species as before, for example, most pine and spruce plantations in Scandinavia established in the last 70 years (Eggertsson et al., 2008). This is not as common as the first two categories because usually one reason for planting trees is to introduce a more productive species. Most past species assemblages should be able to survive under these conditions, subject to the maintenance of key habitats, such as wetlands, dead wood, etc.

4. Forests established by natural regeneration with deliberate silvicultural intervention and assistance, for example many of the beech and oak forests of northern France (Jones, 1952) and Germany, and most coniferous forests in Scandinavia and Finland. The species assemblages are likely to have much in common with those of more natural forests, except where they lack key structures such as large old trees.

5. Forests that have regenerated naturally without any artificial assistance, including all the remoter coniferous forests in north temperate and boreal regions and, much more rarely in Europe, broadleaved forests, such as Białowieża Forest in Poland and some regeneration on Mediterranean mountains. These provide the controls against which the impacts of past and present management interventions can be judged.

Clear-felling and even-aged stands have been important for wood production in the past and are likely to continue to be so in future, albeit as part of a much wider mix of systems designed to meet a much broader range of objectives. Depending on the balance of those objectives for a particular forest, so the nature of the clear-felling system can be modified.

References

Anon. (2013) *Jordbruksstatistik Årsbok 2013*. Sveriges officiella statistic Jordbruksverket Statistiska centralbyrån, p. 117. Available at: www.scb.se/statistik/_publikationer/ovo904_2013 (accessed March 2013).

Barker, S. (1998) The history of the Coniston woodlands, Cumbria, UK. In: Kirby, K.J. and Watkins, C. (eds) *The Ecological History of European Forests*. CAB International, Wallingford, UK, pp. 167–183.

Brasnett, N.V. (1953) *Planned Management of Forests*. George Allen and Unwin, London.
Bürgi, M. (1998) Habitat alterations caused by long-term changes in forest use in northeastern Switzerland. In: Kirby, K.J. and Watkins, C. (eds) *The Ecological History of European Forests*. CAB International, Wallingford, UK, pp. 203–211.
Carlowitz, H. von (1713) *Sylvicultura oeconomica, oder haußwirthliche Nachricht und Naturmäßige Anweisung zur wilden Baum-Zucht*. Johann Friedrich Braun, Leipzig, Germany. Reprinted 2012, Kessel, Remagen, Germany.
Carus-Wilson, E. (1962–1963) Medieval trade of the ports of the Wash. *Medieval Archaeology* 6–7, 182–201.
Christie, J.M. and Lines, R. (1979) A comparison of forest productivity in Britain and Europe in relation to climatic factors. *Forest Ecology and Management* 2, 75–102.
Clout, H.D. (1969) Country planning in Gascony. *Scottish Geographical Magazine* 85, 9–16.
Dimitriou, I. and Aronsson, P. (2005) Willows for energy and phytoremediation in Sweden. *Unasylva* 56, 46–20. Also available from FAO Corporate Document Repository at: ftp://ftp.fao.org/docrep/fao/008/a0026e/a0026e12.pdf (accessed 25 November 2014).
Dirkx, G.H.P. (1998) Wood-pasture in Dutch common woodlands and the deforestation of the Dutch landscape. In: Kirby, K.J. and Watkins, C. (eds) *The Ecological History of European Forests*. CAB International, Wallingford, UK, pp. 53–62.
Eggertsson, O., Nygaard, P.H. and Skovsgaard, J.P. (2008) *History of Afforestation in the Nordic Countries*. In: Halldorsson, G., Oddsdottir, E.S. and Sigurdsson, B.D. (eds) AFFORNORD: Effects of Afforestation on Ecosystems, Landscape and Rural Development. TemaNord, No. 2008:562, Nordic Council of Ministers, Copenhagen, pp. 16–27. Available at: http://www.skogoglandskap.no/filearchive/history_afforestation_nordic_countries.pdf (accessed June 2014).
EU Forestry Statistics (2012) Available at: http://epp.eurostat.ec.europa.eu/statistics_explained/index.php/Forestrystatistics http://epp.eurostat.ec.europa.eu/statistics_explained/index.php/Forestry_statistics#Primary_and_secondary_wood_products (accessed 25 November 2014).
Evans, J. (1998) The sustainability of wood production in plantation forestry. *Unasylva* 49, 47–52. Available at: http://www.fao.org/docrep/w7126e/w7126e07.htm#TopOfPage (accessed June 2014).
Evans, J. (ed.) (2009) *Planted Forests – Uses, Impacts and Sustainability*. Food and Agriculture Organization of the United Nations (FAO), Rome and CAB International, Wallingford, UK.
FAO (2010) *Global Forest Resources Assessment 2010. Main Report*. FAO Forestry Paper 163, Food and Agriculture Organization of the United Nations, Rome. Available at: http://www.fao.org/docrep/013/i1757e/i1757e.pdf (accessed 24 November 2014).
Forestry Commission (1985) *Broadleaves in Britain: A Discussion Paper*. Forestry Commission, Edinburgh, UK.
Forestry Commission (2003) *National Inventory of Woodland and Trees 1995–1998*. Forestry Commission, Edinburgh, UK.
Forestry Commission England (2010) *When to Convert Woods and Forests to Open Habitat in England: Government Policy*. Policy and Programmes Group, Forestry Commission, Bristol, UK. Available at: http://www.forestry.gov.uk/pdf/eng-oh-policy-march2010.pdf/$file/eng-oh-policy-march2010.pdf (accessed June 2013).
Geiger, R. (1965) *The Climate near the Ground*. Harvard University Press, Boston, Massachusetts.
Glacken, C. (1967) *Traces on the Rhodian Shore: Nature and Culture in Western Thought from Ancient Times to the End of the Eighteenth Century*. University of California Press, Berkeley, California.
Goldberg, E.A. (2003) Plantations on ancient woodland sites. *Quarterly Journal of Forestry* 97, 133–138.
Grove, A.T. and Rackham, O. (2001) *The Nature of Mediterranean Europe: An Ecological History*. Yale University Press, New Haven, Connecticut.
Halkett, J.C. (1984) The practice of uneven-aged silviculture. *New Zealand Journal of Forestry* 29, 108–118.
Hamilton, G.J. and Christie, J.M. (1971) *Forest Management Tables (Metric)*. Her Majesty's Stationery Office, London.
Helliwell, D.R. and Harrison, A.F. (1979) Effects of light and weed competition on the growth of seedlings of four tree species on a range of soils. *Quarterly Journal of Forestry* 73, 160–171.
Horsley, S.B. and Marquis, D.A. (1983) Interference by weeds and deer with Allegheny hardwood reproduction. *Canadian Journal of Forest Research* 13, 61–69.
ICNF (2013) *IFN6 – Áreas dos Usos do Solo e das Espécies Florestais de Portugal Continental. Resultados Preliminares*. Instituto da Conservação da Natureza e das Florestas, Lisbon.
Johann, E. (2006) Historical development of nature-based forestry in central Europe. In: Diaci, J. (ed.) *Nature-Based Forestry in Central Europe: Alternatives to Industrial Forestry and Strict Preservation*. Studia

Forestalia Slovenica No. 126, University of Ljubljana, Ljubljana, pp. 1–18. Available at: http://www.natura2000.gov.si/uploads/tx_library/Diaci_Nature_based_forestry.pdf (accessed January 2013).

Jones, E.W. (1945) The structure and reproduction of the virgin forest of the north temperate zone. *New Phytologist* 44, 130–148.

Jones, E.W. (ed.) (1952) *Silvicultural Systems by R.S. Troup*, 2nd edn. Clarendon Press, Oxford, UK.

Knuchel, H. (1953) *Planning and Control in the Managed Forest* (translated from the German by Anderson, M.L.). Oliver and Boyd, London.

Köstler, J. (1956) *Silviculture* (translated from the German by Anderson, M.L.). Oliver and Boyd, London.

Leslie, R. (2014) *Forest Vision: Transforming the Forestry Commission*. New Environment Books, Bristol, UK.

Lillehammer, A. (1986) The Scottish–Norwegian timber trade in the Stavanger area in the sixteenth and seventeenth centuries. In: Smout, T.C. (ed.) *Scotland and Europe 1200–1850*. John Donald, Edinburgh, UK, pp. 97–111.

Lillehammer, A. (1990) Boards, beams and barrel-hoops: contacts between Scotland and the Stavanger area in the seventeenth century. In: Simpson, G.G. (ed.) *Scotland and Scandinavia 800–1800*. John Donald, Edinburgh, UK, pp. 100–106.

Lowman, M.D., Schowalter, T.D. and Franklin, J.F. (2012) *Methods in Forest Canopy Research*. University of California Press, Berkeley and Los Angeles, California.

Malcolm, D.C. (1979) The future development of even-aged plantations: silvicultural implications. In: Ford, E.D., Malcolm, D.C. and Atterson, J. (eds) *Ecology of Even-Aged Forest Plantations*. Institute of Terrestrial Ecology, Cambridge, UK, pp. 481–504.

Meiggs, R (1982) *Trees and Timber in the Ancient Mediterranean World*. Clarendon Press, Oxford, UK.

Miller, H.G. (1979) The nutrient budgets of even-aged forests. In: Ford, E.D., Malcolm, D.C. and Atterson, J. (eds) *Ecology of Even-aged Forest Plantations*. Institute of Terrestrial Ecology, Cambridge, pp. 221–256.

Mollat, M. (1947) Anglo–Norman trade in the fifteenth century. *Economic History Review* 17, 143–150.

Molnár, Z. (1998) Interpreting present vegetation features by landscape historical data: an example from a woodland–grassland mosaic landscape. In: Kirby, K.J. and Watkins, C. (eds) *The Ecological History of European Forests*. CAB International, Wallingford, UK, pp. 241–263.

NFI (2007) *National Forest Inventory, Republic of Ireland – Results: Covering the National Forest Inventory, 2004 to 2006*. Forest Service, Department of Agriculture, Fisheries and Food, Dublin. Available at: https://www.agriculture.gov.ie/media/migration/forestry/nationalforestinventory/nationalforestinventorypublications/4330NFIResults.pdf (accessed 24 November 2014).

Osmaston, F.C. (1968) *The Management of Forests*. George Allen and Unwin, London.

Peterken, G.F. (1996) *Natural Woodland. Ecology and Conservation in Northern Temperate Regions*. Cambridge University Press, Cambridge, UK.

Picard, O., Dreyer, E. and Landmann, G. (2012) Synergy and contradictions between wood production and ecosystem services supplied to society – the case of private forests. *Revue Forestière Française* 64, 363–369.

Pryor, S.N. (1982) *An Economic Analysis of Silvicultural Options for Broadleaved Woodland*. Department of Forestry, University of Oxford, Oxford, UK.

Rédei, K., Osváth-Bujtás, Z. and Veperdi, I. (2008) Black locust (*Robinia pseudoacacia* L.) improvement in Hungary: a review. *Acta Silvatica et Lignaria Hungarica* 4, 127–132.

Savill, P. and Brown, N. (2010) Silvicultural systems. In: Bowes, B.G. (ed.) *Trees and Forests: A Colour Guide*. Manson Publishing, London, pp. 210–224.

Savill, P., Evans, J., Auclair, D. and Falck, J. (1997) *Plantation Silviculture in Europe*. Clarendon Press, Oxford, UK.

Schlich, W. (1904) *Forestry in the United Kingdom*. Bradbury, Agnew, and Co., London.

SECF (2011) *Situación de los Bosques y del Sector Forestal en España*. Based on MARM (Ministerio Agricultura Alimantación Medió Ambiente, Madrid), 2009, ICF3 (Tercer Inventario Forestal Nacional (1997–2007)) Informe 2010. La Sociedad Española de Ciencias Forestales, Palencia, Spain. Available at: http://www.secforestales.org/web/images/ISFE/inforestal2010.pdf (accessed 24 November 2014).

Streets, R.J. (1962) *Exotic Forest Trees in the British Commonwealth*. Clarendon Press, Oxford, UK.

Thompson, R., Humphrey, J., Harmer, R. and Ferris, R. (2003) *Restoration of Native Woodland on Ancient Woodland Sites*. Forestry Commission, Edinburgh, UK.

Troup, R.S. (1928) *Silvicultural Systems*. Clarendon Press, Oxford, UK.

Tsouvalis, J. (2000) *A Critical Geography of Britain's State Forests*. Oxford University Press, Oxford, UK.
Upton, C. and Bass, S. (1995) *The Forest Certification Handbook*. Earthscan Publications, London.
Warde, P. (2006) *Ecology, Economy and State Formation in Early Modern Germany*. Cambridge University Press, Cambridge, UK.
Westoby, J. (1987) *The Purpose of Forests: Follies of Development*. Basil Blackwell, Oxford, UK.
Williams, M. (2002) *Deforesting the Earth: From Prehistory to Global Crisis*. University of Chicago Press, Chicago, Illinois.

8 Close-to-nature Forestry

Matthias Bürgi*
Swiss Federal Research Institute WSL, Birmensdorf, Switzerland

8.1 Introduction

Throughout most of the 19th century, forestry in large parts of Europe was dominated by the German classical school of forestry advocated, for example, by Heinrich Cotta, Johann Christian Hundeshagen and Ludwig Hartig. This promoted the creation of pure, even-aged and often coniferous stands which were clear-cut and replanted in a rotation system. The approach was based on the principle of sustainability, traced back to Hans Carl von Carlowitz (1713), which could be fulfilled by making regular annual cuts controlled by area, followed by planting the next generation of trees, typically spruce (*Picea abies*) (Savill, Chapter 7).

The more homogenous a stand was in terms of growing conditions, species composition and age of the trees, the easier it was to make sure that every annual cut contained about the same amount of timber. However, these homogenous monocultures proved to be vulnerable to disturbance events such as insect outbreaks, windthrow and frost damage (Jacobsen, 2001). Moreover, on some soils, yields declined in the second or third generation of conifers (Heyder, 1986). Later, nature conservation organizations started to criticize the system, lamenting the loss of natural tree species composition, diversity of stand structures and the related biodiversity.

The perceived disadvantages of the even-aged stand approach encouraged innovative foresters to develop management schemes based on multi-species, uneven-aged stands. Early ideas for silvicultural systems designed according to ecological principles evolved in France (Adolphe Gurnaud), Germany (Karl Gayer), and Switzerland (Henry Biolley) (Pommering and Murphy, 2004). In Germany, the movement against clear-cutting and artificial regeneration can be traced back to the mid-19th century (Heyder, 1986). The attempts to manage forest in a more natural way also partly relate back to older selection systems used in parts of France, Germany, Austria and Slovenia, and the so-called 'Plenterwald' system practised in parts of Switzerland (Fig. 8.1; Balsiger 1914; Schütz, 2001). In the 20th century, the debate was enriched by the 'Dauerwald' Movement (Heyder, 1986) and, later, by discussions about mimicking natural disturbance regimes in forest management (Mitchell *et al.*, 2002).

The terms 'close-to-nature forestry', 'near-to-nature silviculture', 'naturalistic silviculture', 'natural forestry', or 'continuous cover forestry'

---

*E-mail: matthias.buergi@wsl.ch

Gruppenweise Mischung der Größeklassen im Plenterwald.
Groß-Doppwald bei Konolfingen.

Fig. 8.1. An early example of close-to-nature forest management was the Plenterwald system, promoted at the beginning of the 20th century in Switzerland: '*Gruppenweise Mischung der Grössenklassen im Plenterwald. Gross-Doppwald bei Konolfingen*', published in Balsiger (1914).

have been used to name various management practices that all share common traits of a focus on site conditions and no clear-cutting. However, the terms have been defined differently over time and by different authors, and their similarities and differences have been debated intensively over the years (Gamborg and Larsen, 2003). The situation is further complicated by the fact that the debate is either led by forest scientists who publish in scientific journals or by forest practitioners who discuss the issues in more practice-oriented publications; neither debate may fully reflect what is happening on the ground! Here, we outline some strands of the development of this approach without going too much into the details of the differences between followers of the various schools.

8.2 Roots and Prerequisites

Private forest owners managing their forests select the stems to be harvested based on their own needs and the demand for products such as firewood, wood for tools and furniture, and construction timber. This results in a single-stem management or, more appropriately, single-stem use, and was widespread before the introduction of more systematic, professional forestry. This traditional way of using the forest survived best in agricultural regions with small-scale private forest ownership, often individual farmers. It probably still goes on in many apparently 'unmanaged' woods. The absence of formal control over yield and harvest does mean though that this type of forest use did and does risk overexploiting the forests if the demand-driven harvest surpasses the growth. Other traditional uses, not related to timber extraction, but still performed in the same forests, such as grazing by livestock and the collection of leaves and needles, were also part of the intensive use of many forests up to the mid-20th century (Bürgi et al., 2013).

Schütz (2001) outlines how this unregulated single-stem management was often ignored or held in low regard by foresters. In German-speaking parts of Europe, the term 'Plenterwald' (selection forest) was sometimes even seen as being related to 'Plünderwald' (plunder forest), which nicely reflects the negative connotation of this use and management system. However, the original 'Plenterwald' system developed from being simply demand driven into a sophisticated form of forest management and in Switzerland, for example, Arnold Engler (1905) and Henry Biolley (1901) developed the theory of the single-tree selection system.

The development towards regulated single-stem management ('geregelte Plenterung') was inspired by the German forester Johann Karl Gayer, who, as early as the 1880s, was developing and promoting mixed forests ('Mischwald') and the shelterwood selection system, the so-called 'Femelschlag'. This latter system was applied in parts of Germany, Switzerland and the Vosges, a mountain region belonging at this time to Germany (Heyder, 1986).

These new developments at the beginning of the 20th century were inspired by and built on new insights from soil and vegetation sciences. The new science of ecology was being established and there was much debate around ideas of succession and the concept of climax woodland types adapted to local conditions of soil and climate. This helped to foster the rise of the new dogma of 'near-to-nature silviculture', in which the concept of natural vegetation communities was a guiding concept for species selection, and new ways to control the sustainability of timber yields were developed.

Silvicultural publications from the early 20th century often promoted management based on careful observation of soil conditions, tree growth and climatic constraints; see, for example; Jankowsky (1903) (*Die Begründung naturgemässer Hochwaldbestände*), Morosow (1928) (based on lectures from 1902/3), and Mayr (1909) (*Waldbau auf naturgesetzlicher Grundlage*). Instead of controlling the sustainability of timber production by the area, as in the clear-cutting system, the French forester Adolphe Gurnaud promoted the idea that silvicultural practice should aim at a state of equilibrium, where there was the optimal distribution of tree size classes in every compartment of the forest (Agnoletti *et al.*, 2009). Henry Biolley and Hermann Knuchel further developed a method that was based on full calipering of standing timber volume (Badré, 1983; Agnoletti, 2006). This new control method cleared the way for the use of spatially dispersed cuts, group felling or even single-stem management. In Switzerland, such developments led to a ban on clear-cutting in 1903 and, for a while, deciduous tree species were planted extensively to transform the existing coniferous monocultures back into stands with a more natural species composition (Bürgi and Schuler, 2003).

8.3 Developments in the 20th Century

The development of close-to-nature management systems at the beginning of the 20th century fitted well into the zeitgeist and the 'back-to-nature' movement (Johann, 2006). In countries such as Sweden (Lundell, 2011), continuous cover forestry also became popular at the beginning of the 20th century, illustrating the widespread acceptance of the basic ideas of close-to-nature management as a reaction against large-scale clear-felling in Europe.

In the 1920s, fostered by Alfred Möller and the example of pine management on the Bärenthoren estate, 'Dauerwald' (continuous forest) became a keyword in the forest debate in Germany. This was a debate with much definitional rumbling, as Dauerwald was often seen either as a reversion to Plenterwald or as vague to such a degree that any forest management considering the principles of sustainability would fulfil it anyway (Heyder, 1986). Neither interpretation is in line with the main elements of the Dauerwald idea, which focused on natural regeneration and single-stem management instead of large cuts, therefore fulfilling the core requirements of close-to-nature forestry. The idea was adopted by the Nazis in Germany in the 1930s, when the Dauerwald principles became part of forest legislation (Jacobsen, 2001), a development that did not make subsequent debate about close-to-nature forestry any the less emotional.

After the Second World War, close-to-nature forestry and the Dauerwald approach continued to be promoted by the 'Arbeitsgemeinschaft Naturgemässe Waldwirtschaft' (ANW, or 'Working Group on Close-to-nature Silviculture'), which was founded in 1950 (Schmidt, 2009). In a contribution to the journal *Der Forst- und Holzwirt*, Willy Wobst (1954) summarizes the rationale for following the 'naturgemässe Waldwirtschaft': continuous cover takes best care of the growing potential; preserves the interior forest climate; and leads to a more valuable timber assortment. The movement sought to discourage the system of clear-cutting. 'Plenterwald' was not the aim as such, but was seen as a very appropriate forest type under certain conditions.

Some of the reasons for this movement were spiritual arguments. Wobst writes: 'The forest is the last resort for the agitated and uprooted man to feel himself close to his creator and to regain the power and balance of his soul' (German original: 'Die Wald ist der letzte Ort, an dem der gehetzte und von völliger Entwurzelung bedrohte Mensch sich noch dem Schöpfer nahe fühlen und seelische Kräfte und seelisches Gleichgewicht wiederfinden kann'). In 1989, the organization ANW changed its name to Pro Silva (http://www.prosilvaeurope.org/). An updated list of principles of nature-based silviculture was compiled by Gamborg and Larsen (2003) based on information from Pro Silva, and these are presented with minor adaptations below:

- maintaining and protecting soil productivity through continuous forest cover and maintenance of forest biomass;
- maintaining natural forest vegetation, although the forest is used for production purposes;
- maintaining and protecting natural energy and mineral cycles, and improving carbon storage and forest climate;
- using natural dynamic forest processes, including using natural regeneration;
- propagating mixed forest with attention to rare species;
- minimizing the use of fertilizers, pesticides and drainage;
- restricting the use of non-native species to cases of economic necessity provided they can be mixed with native species, which are generally favoured;
- adding value and enhancing diversity in forest structure through forest regeneration,

stand tending and exploitation as a means of obtaining niches in time and space;
- using selective felling to avoid clear-cutting and other methods destroying forest conditions; and
- abolishing rotation age as an instrument for determining when a tree should be cut, and adopting methods based on a target-diameter felling.

Careful observation by the forestry practitioner and adapting the silvicultural practice to the local conditions are considered critically important. Close-to-nature forestry has also been called 'applied ecology' (Schütz, 1986), as it requires in-depth knowledge on the ecology and dynamics of forest ecosystems. However, while scientific insights from soil and vegetation sciences underlay the beginning of close-to-nature forestry, the diversity and complexity of growing conditions in the forest mean that the practitioner cannot simply follow too strict guidelines, but must adapt them to his or her particular situation. The resulting practice may still be worded in scientific terms as, for example, the so-called 'bio-rationalization' (biological optimization in Gamborg and Larsen, 2003), but the focus is on inspiration from natural processes (Schütz, 2011).

The ecological movement of the 1980s gave the idea of close-to-nature forestry additional support. Monocultures of conifers in large parts of central and Eastern Europe planted on sites naturally stocked by mixed forests and broadleaved species proved to be especially vulnerable to acid rain and affected by 'Waldsterben' (forest dieback). Concern was raised not just over the species composition of such stands as compared with natural conditions, but also over the difference in the structure created by modern forest management compared with previous disturbance regimes.

Ecological theory was used to propose that ecosystems should be more resilient to outside impacts if the management practices mimicked the characteristics of natural disturbance regimes (Hansen et al., 1991; Kuuluvainen, 2002; Mitchell et al., 2002; Seymour et al., 2002). In general, these tended to be on a smaller scale than the large clear-fells that had been practised. Whereas certain aspects of natural disturbance regimes of course can be copied, such as reducing the size and the average return interval of the cuts, other aspects inherently remain different: trees felled in the course of a harvesting operation are taken out of the forest, whereas trees felled by wind or burnt by forest fires remain in the system, but in very different states and offering very different habitats depending on the type of disturbance.

In France, from the 1980s onward, there has been the spread of an adapted form of close-to-nature forestry applied to broadleaves and exotic species such as Douglas fir (*Pseudotsuga menziesii*). Research results from networks such as the Association Futaie Irrégulière (AFI), which was formed in 1991 to promote the silviculture of irregular forest stands, demonstrate that overall stand structure and its composition can be adjusted to make the approach work in stands of light-demanding broadleaves. In France and the UK, the use of these techniques has largely been undertaken by private and other owners, often in the face of significant opposition from the state forest services. The adoption of irregular silviculture in the private sector has followed the economic failure of the even-aged plantation model in fertile lowland areas due to the high cost of planting, the relatively small scale and variability of the forest resource, the need to focus on the production of higher value material and the much higher risk profile associated with such 'conventional' approaches. While the wider benefits linked to biodiversity, sport and landscape are increasingly important, most examples of well thought out, sustained ir regular management established so far have been driven by economics. However, in parts of Scandinavia, the politics of close-to-nature forestry is such that its use is opposed by the forest industries, which desire large quantities of small roundwood (Susse et al., 2011; A. Pooro, Dorset, 2014, personal communication).

International developments have further fuelled interest in close-to-nature forestry, such as the forest principles agreed at the Rio Earth Summit in 1992 and the subsequent development of forest certification under the Forest Stewardship Council (FSC), and later the Programme for the Endorsement of Forest Certification (PEFC), with their call to consider all dimensions of sustainability in management decisions (Pommering and Murphy, 2004). In Europe, these have been taken forward through the MCPFE (*Ministerial Conference on the Protection of Forests in Europe*; see http://www.foresteurope.org/) process, which is based on the vision:

> To shape a future where all European forests are vital, productive and multifunctional; where forests contribute effectively to sustainable development, through ensuring human well-being, a healthy environment and economic development in Europe and across the globe; where the forests' unique potential to support a green economy, livelihoods, climate change mitigation, biodiversity conservation, enhancing water quality and combating desertification is realized to the benefit of society.

8.4 Ecological Implications

High forest management with clear-cutting and artificial regeneration favouring coniferous species led to homogenous stands and the expansion of conifers far beyond their natural distribution and abundance in Europe. This was seen as damaging both to nature conservation and to wider amenity considerations. The increase in public interest in biodiversity and its conservation, and in forests more generally, has consequently encouraged the spread of the principles of close-to-nature forestry (Dafis, 2001). The large-scale conversion of homogenous coniferous forests into more natural, site-adapted and often mixed stands has also been promoted (Spiecker, 2003). Concerns about the impacts of climate change and the resulting additional stress on forest ecosystems, compounded by the emergence of new pests and diseases, has further raised interest in management schemes increasing the resistance and resilience of forest ecosystems by emphasizing adaptation to local and regional prerequisites and needs (Spiecker, 2003).

Some of the principles of close-to-nature forestry correspond well with the sorts of principles developed independently by forest ecologists to promote biodiversity conservation within managed forests (Peterken, 1996; Colak *et al.*, 2003; Lindenmayer and Franklin, 2003). Advocates of 'close-to-nature' forestry are often thought to favour selection systems, but the all-aged forests that these produce are rare in natural conditions and do not necessarily mimic well the conditions under which many woodland species have survived or thrived over the last millennium. Jones (1945) argued that 'some degree of differentiation of discrete age classes seems to be the rule' in natural forests, suggesting that group systems of various kinds come closest to simulating the structure of natural stands in broadleaved and mixed temperate forests. In regions where fire and large-scale windstorms are common disturbance patterns, a large-scale stand may be appropriate, such as that produced by clear felling, and these are common and widespread in coniferous forest types and in mixed boreal forests. According to Peterken (1996), virtually all natural woodland is analogous to high forest, but not to coppice, which develops only in a fragmentary and temporary fashion.

However, there are still aspects of close-to-nature forestry that do not correspond well with pure nature conservation recommendations. There may still be a focus on production and non-native species, and fertilizers, pesticides and drainage are not ruled out. More fundamentally, questions remain as to whether all aspects of biodiversity will be protected and promoted to the necessary level if all forests are managed under close-to-nature principles. Not all species naturally occurring in unmanaged forests will necessarily benefit; nor will some of the species associated with traditional management practices such as coppice and wood-pasture that thrive in open and anthropogenically disturbed forest stands (Hartel *et al.*, Chapter 5; Buckley and Mills, Chapter 6).

In stands managed under close-to-nature principles, the amounts of standing and lying dead wood will be reduced in comparison with that in a natural stand, albeit they may be higher than under clear-fell systems. The average gap size in close-to-nature forestry is likely to be smaller than in natural forests, where natural disturbance events occasionally open up larger parts of a forest. Light-demanding species in the herb layer might therefore have trouble finding enough open stands and gaps.

Climate change adds an additional element to this discussion. In the future, local adaptation may require that not only species composition, but also harvesting techniques and infrastructural measures have to be planned in light of projected climatic developments (Spittlehouse and Steward, 2003; Brang *et al.*, 2014). If the species that are currently part of the natural forest vegetation of a site cannot cope with future climate conditions, the respective principles of close-to-nature forestry as promoted by Pro Silva will have to be adapted.

Close-to-nature forestry cannot by itself deliver the full range of conditions needed to conserve European forest conservation objectives. As Gamborg and Larsen (2003) note, 'Additional measures may be needed to protect biodiversity specifically connected with deadwood and both the late and early successional stages'. An increase in structural diversity on various scales will, in addition, support the whole range of ecological functions and processes of and within forest ecosystems.

8.5 Conclusion

The history of forestry and forest management schemes reads in parts like a sequence of ideas that turned into ideologies, of insights that grew into dogmas, of terms that were ill-defined and misunderstood, and of generalizations beyond the limits of the experience on which they were based. However, forest management schemes have to be interpreted in their historical and biogeographical contexts, taking account of the interrelationships between scientific development, insights from practical experience and topics of public interest.

The development of the German forest school stands in the context of the cameral sciences, which were developed in the Age of Enlightenment and aimed to improve fiscal administration and resource management by the state (Lowood, 1990). In contrast, close-to-nature forestry is based on advances in the natural sciences, such as soil sciences and vegetation science. Ecology, as a science and as a public concern, finally supported the spread of the Dauerwald Movement, as one form of close-to-nature forestry, and more recently the idea of mimicking natural disturbance regimes by forest management practices.

Today, concerns about global change, especially climatic shifts and changes in disturbance regimes, including the spread of pests and diseases, mean that more emphasis is put on resilient forests and adaptive management. Production remains important but, at the same time, many forests are intensively used for recreation. Nature conservation is also highly valued, although it is unclear whether there are still further declines in biodiversity yet to come due to the delayed effects of past forestry practice (Hermy, Chapter 13).

The historical change in demand for ecosystem goods and services has led to the rise and fall of various silvicultural practices, which all have their strengths and weaknesses. Close-to-nature forestry might not be the panacea that some of its promoters claim, but is definitely well suited to fulfil a wide range of societal demands. It has the strength that it does not consist of a dogmatic set of rules, but of adaptive measures around a set of principles. As long as its adherents continue to accept such adaptation in how the principles are applied, so its application may increasingly spread.

References

Agnoletti, M. (2006) Man, forestry, and forest landscapes. Trends and perspectives in the evolution of forestry and woodland history research. *Schweizerische Zeitschrift für Forstwesen* 157, 384–392.

Agnoletti, M., Dargavel, J. and Johann, E. (2009) History of forestry. In: Squires, V.R. (ed.) *The Role of Food, Agriculture, Forestry and Fisheries in Human Nutrition, Volume 1. Encyclopedia of Life Support Systems (EOLSS)*. UNESCO (United Nations Educational, Scientific and Cultural Organization)/EOLSS, Paris/Oxford, UK. Available at: http://www.eolss.net/sample-chapters/c10/E5-01A-02.pdf (accessed 25 November 2014).

Badré, L. (1983) *Histoire de la Forêt Française*. Arthaud, Paris.
Balsiger, R. (1914) *Der Plenterwald und seine Bedeutung für die Forstwirtschaft der Gegenwart*. Büchler, Bern, Switzerland.
Biolley, H. (1901) Le jardinage cultural. *Journal Forestier Suisse* 52, 97–104, 113–132.
Brang, P., Spathelf, P., Larsen, J.B., Bauhus, J., Bončĭna, A., Chauvin, C., Drössler, L., García-Güemes, C., Heiri, C., Kerr, G. et al. (2014) Suitability of close-to-nature silviculture for adapting temperate European forests to climate change. *Forestry* 87, 492–503.
Bürgi, M. and Schuler, A. (2003) Driving forces of changes in forest management – an analysis of regeneration practice in the forests of the Swiss lowlands during the 19th and 20th century. *Forest Ecology and Management* 176, 173–183.
Bürgi, M., Gimmi, U. and Stuber, M. (2013) Assessing traditional knowledge on forest uses to understand forest ecosystem dynamics. *Forest Ecology and Management* 289, 115–122.
Carlowitz, H. von (1713) *Sylvicultura oeconomica, oder haußwirthliche Nachricht und Naturmäßige Anweisung zur wilden Baum-Zucht*. Johann Friedrich Braun, Leipzig, Germany. Reprinted 2012, Kessel, Remagen, Germany.
Colak, A.H., Rotherham, I.D. and Calikoglu, M. (2003) Combining 'naturalness concepts' with close-to-nature silviculture. *Forstwissenschaftliches Centralblatt* 122, 421–431.
Dafis, S. (2001) Ecological impacts of close-to-nature forest on biodiversity and genetic diversity. In: Green, T. (ed.) *Ecological and Socio-Economic Impacts of Close-to Nature Forestry and Plantation Forestry: A Comparative Analysis. Proceedings of the Scientific Seminar of the 7th Annual EFI Conference, Instituto Superior de Agronomia – ISA, Lisbon, Portugal, 3 September 2000*. EFI-Proceedings No. 37, European Forest Institute, Joensuu, Sweden, pp. 21–25.
Engler, A. (1905) *Aus der Theorie und Praxis des Femelschlagbetriebes*. Francke, Bern, Switzerland.
Gamborg, C. and Larsen, J.B. (2003) 'Back to nature' – a sustainable future for forestry? *Forest Ecology and Management* 179, 559–571.
Hansen, A.J., Spies, T.A., Swanson, F.J. and Ohmann, J.L. (1991) Conserving biodiversity in managed forests: lessons from natural forests. *BioScience* 41, 292–312.
Heyder, J.C. (1986) *Waldbau im Wandel*. Sauerländer, Franfurt am Main, Germany.
Jacobsen, M.K. (2001) History and principles of close-to-nature forest management: a central European perspective. In: Read, H., Forfang, A.S., Marciau, R., Paltto, H., Andersson, L. and Tardy, B. (eds) *Textbook 2 – Tools for Preserving Woodland Biodiversity*. NACONEX (Nature Conservation Exchange Experience) 2001, Pro-Natura, Göteborg, Sweden, pp. 56–60. Available at: http://www.pro-natura.net/naconex/news5/E2_11.pdf (accessed 25 November 2014).
Jankowsky, R. (1903) *Die Begründung naturgemässer Hochwaldbestände*. Selbstverlag, Haslach, Germany.
Johann, E. (2006) Historical development of nature-based forestry in central Europe. In: Diaci, J. (ed.) *Nature-Based Forestry in Central Europe: Alternatives to Industrial Forestry and Strict Preservation*. Studia Forestalia Slovenica No. 126, University of Ljubljana, Ljubljana, pp. 1–18. Available at: http://www.natura2000.gov.si/uploads/tx_library/Diaci_Nature_based_forestry.pdf (accessed 25 November 2014).
Jones, E.W. (1945) The structure and reproduction of the virgin forest of the North temperate zone. *New Phytologist* 44, 130–148.
Kuuluvainen, T. (2002) Natural variability of forests as a reference for restoring and managing biological diversity in boreal Fennoscandia. *Silva Fennica* 36, 97–125.
Lindenmayer, D.B. and Franklin, J.F. (eds) (2003) *Towards Forest Sustainability*. Island Press, Washington, DC.
Lowood, H.E. (1990) The calculating forester: quantification, cameral science, and the emergence of scientific forestry management in Germany. In: Frängsmyr, T., Heilbron, J.L. and Rider, E.R. (eds) *The Quantifying Spirit in the 18th Century*. University of California Press, Berkeley, California, pp. 315–342.
Lundell, S. (2011) Family forestry in transition: times of freedom, responsibility and better knowledge. In: Antonson, H. and Jansson, U. (eds) *Agriculture and Forestry in Sweden Since 1900 – Geographical and Historical Studies*, 1st edn. Skogs- och Lantbrukshistoriska Meddelanden No. 54, Royal Swedish Academy of Agriculture and Forestry, Stockholm, pp. 406–422.
Mayr, H. (1909) *Waldbau auf Naturgesetzlicher Grundlage*. Paul Parey, Berlin.
Mitchell, R.J., Palik, B.J. and Hunter, M.L. (2002) Natural disturbance as a guide to silviculture. *Forest Ecology and Management* 155, 315–317.
Morosow, G.F. (1928) *Die Lehre vom Walde*. Aus dem Russischen übersetzt von Ruoff, S. und Rubner, K. [Translated from Russian by Ruoff, S. and Rubner, K.]. Neumann, Neudamm, Germany.

Peterken, G.F. (1996) *Natural Woodland. Ecology and Conservation in Northern Temperate Regions*. Cambridge University Press, Cambridge, UK.

Pommering, A. and Murphy, S.T. (2004) A review of the history, definitions and methods of continuous cover forestry with special attention to afforestation and restocking. *Forestry* 77, 27–44.

Schmidt, U.E. (2009) Wie erfolgreich war das Dauerwaldkonzept bislang: eine historische Analyse. *Schweizerischer Zeitschrift für Forstwesen* 160, 144–151.

Schütz, J.P. (1986) Charakterisierung des naturnahen Waldbaus und Bedarf an wissenschaftlichen Grundlagen. *Schweizerische Zeitschrift für Forstwesen* 137, 747–760.

Schütz, J.P. (2001) *Der Plenterwald und weitere Formen strukturierter und gemischter Wälder*. Parey Buchverlag, Berlin.

Schütz, J.P. (2011) Development of close-to-nature forestry and the role of ProSilva Europe. *Zbornik Gozdarstva in Lesarstva* 94, 39–42.

Seymour, R.S., White, A.S. and de Maynadier, P.G. (2002) Natural disturbance regimes in northeastern North America – evaluating silviculture systems using natural scales and frequencies. *Forest Ecology and Management* 155, 357–367.

Spiecker, H. (2003) Silvicultural management in maintaining biodiversity and resistance of forests in Europe – temperate zone. *Journal of Environmental Management* 67, 55–65.

Spittlehouse, D.L. and Steward, R.B. (2003) Adaptation to climate change in forest management. *BC Journal of Ecosystems and Management* 4, 1–11.

Susse, R., Allegrini, C., Bruciamacchie, M. and Burrus, R. (2011) *Management of Irregular Forests: Developing the Full Potential of the Forest*. English translation by Morgan, P. from SelectFor, Aberystwyth, UK. Association Futaie Irrégulière, Besançon, France.

Wobst, W. (1954) Zur Klarstellung über die Grundätze der naturgemässen Waldwirtschaft. *Der Forst- und Holzwirt* 13, 269–274.

9 The Impact of Hunting on European Woodland from Medieval to Modern Times

John Fletcher*

Reediehill Deer Farm, Auchtermuchty, UK

9.1 Introduction

The impact of early agriculture on the environment is constantly discussed and researched by ecologists, yet that of hunting, while of less significance than the more invasive effects of farming, is rarely considered. Nevertheless, throughout the historical era, hunting has influenced the natural environment, and especially woodland, to a greater extent than is often imagined; and its impact remains very significant today. About 33% of Europe is covered in woodlands and much of this is still used for hunting, even in reserves (Broekmeyer et al., 1993). To understand this potent influence, we need to consider the ways in which early hunting led to the development of the medieval *forestes* (royal or noble hunting grounds enshrined in forest law), and from there to the modern hunting reserve.

9.2 Early Impacts of Hunting

The concept of a wildwood evolving after the ice retreated from Europe, which remained pristine and unaffected by humans until the first farmers made their impact during the Neolithic era, is an oversimplification (Kirby and Watkins, Chapter 3). From the first arrival of hominids, about 700,000 years ago, woodland was affected by foraging and, to a greater extent, by hunting. The use of fire to move prey – where vegetation makes this possible – is common to most hunter–gatherer societies.

We know little of the scale of these early influences, but hunting probably contributed to the extinction of many of the Pleistocene megafauna. More specifically, hunting was a factor in the extinction of the aurochs (*Bos primigenius*) and the elk (*Alces alces*) from Britain during the Bronze Age (Legge, 2010). Historically, humans were responsible for hunting the wolf (*Canis lupus*) to extinction in England in the 14th or 15th centuries to protect the prized deer, cattle and sheep; the species survived in Scotland into at least the 17th century and in Ireland into the 18th century (Pluskowski, 2010). More significant in its effect on woodland, and on the landscape in general, was the extinction of the beaver (*Castor fiber*), probably in medieval times, although recent work suggests survival in northern England into the 1700s. Such losses paralleled a continuing reduction in these species across mainland Europe at least until the latter part of the 20th century (Fumagalli, 1994; Coles, 2010; Hearn, Chapter 14).

---

*E-mail: tjohn.fletcher@virgin.net

9.3 Meat or Merit?

As agro-pastoral systems spread and eroded hunting societies, the product of the hunt must have become less materially important. However, within human hunting societies, hierarchies of prestige seem inevitable and the association of the most prized quarry with an elite among hunting groups would have been forged at an early date. In Europe, such highly esteemed quarry probably included the red deer (*Cervus elaphus*), aurochs and the wild boar (*Sus scrofa*). The connection of these species with royalty certainly exists throughout history (Flannery and Marcus, 2012). Thus, a stag surmounts the 7th century sceptre found at Sutton Hoo in England (Enright, 2006). In the Anglo-Saxon poem *Beowulf*, dated between the 8th and 11th century, Beowulf's king of the Danes, Hrothgar, inhabits Heorot, the Hart Hall, which was probably bedecked with stag's antlers, while the hunting scenes in the late 14th century poem of *Sir Gawain and the Green Knight* are pivotal to that narrative.

Increasingly, royalty and the nobility were also able to control the landscape in order to maximize their hunting opportunities. For those living at a more menial level, access to game remained attractive despite the advent of agriculture. Poaching provided a rare but important source of meat for the many: the good poacher was often a local hero enshrined in the literature of many cultures, such as the English Robin Hood.

9.4 Medieval Hunting Reserves

In company with many later authors, Xenophon extolled hunting as 'the best training for war' (Anderson, 1985), and this link has remained into modern times. The English Duke of Wellington had packs of hounds despatched to his army in Spain during the Peninsular War (1807–1814) so that the cavalry could hone their horsemanship. During the two World Wars, Scottish deerstalkers proved their value as marksmen and snipers. It was in the interests of rulers to encourage a coterie of skilled hunters while also publicly demonstrating their own competence.

The peacetime rulers laid out their strategies for the hunt as carefully as generals in a war. Success brought public acclaim but failure ignominy, so that well-stocked royal hunting reserves were vital. The Romans enclosed substantial areas as game parks and Columella (*c*.65 AD) describes massive hunting reserves in Gaul (Anderson, 1985).

Early Chinese emperors burnt large tracts of ground to ensure a suitable kill. From eastern Asia, we have well-authenticated accounts of royal 'ring hunts' involving many tens of thousands of beaters gathering game for weeks to a point where, within an enclosure surrounded by mounted men, the ruler and his favourites could kill massive quantities of game (Allsen, 2006). These concepts spread west into central Europe and Germany, albeit in less favourable terrain and on a smaller scale, where they came to exert an impact on woodland.

The laws of European nations created hunting reserves from an early date to ensure aristocratic or royal access to game and to defuse disputes as hunters pursued their quarry into neighbouring landholdings (Giese, 2011b). Roman and Germanic law considered wild animals as belonging to no one until they were taken, *res nullius*, but the Franks imposed legislation to reserve hunting on their own extensive royal hunting reserves and appointed keepers to preserve the game. They coined the term *forestes* to denote land under forest law that might or might not be wooded (Vera, 2000). In some cases, the land was appropriated by the crown, but Charlemagne in the 9th century returned land to his subjects while retaining the right to 'vert and venison' and forbidding the creation of new forests by his vassals (Marvin, 2006; Ahrland, 2011).

The demand by Carolingian kings to preserve game probably saved woodland that would otherwise have been cleared for agriculture. Like monarchs before and after them, the East Frankish/German monarchs oversaw their *forestes* as the court processed from hunting lodge to hunting lodge, ruling 'from the saddle' (Giese, 2011a). Much of the *forestes* was wooded so that these hunting laws provided the protection of woodland from agriculture over several centuries.

Rulers made progresses in hunting as part of the process of stamping their authority over the natural resources of the environment – the process was ecological as well as social. Through the Middle Ages, feudal pressures had increased and the hunting rights of subjects had been progressively eroded. By the end of the 15th century, the Electors and Princes within the Holy Roman Empire claimed exclusive rights to hunt and, with few exceptions, everyone else was excluded. Eventually, by the end of the 18th century, even landowners were not permitted to hunt on their own land (Knoll, 2004).

The French Revolution of 1789 removed the privileges of the French aristocracy and dissolved ancient hunting laws, but in Germany changes came more slowly. Here, there was a gradual erosion of privileged hunting rights during the early 19th century, culminating in the short-lived revolution of 1848 which swept away the historic rights. Even after the collapse of the revolution, most territorial laws recognized the hunting rights of landowners, although those with small holdings often continued to be excluded from hunting. Others sold their rights to larger landowners, which allowed the system to move back to its previous domination by the elite (Blüchel, 1997). In England, interest in the Royal Forests waxed and waned, but some persisted into the 19th century (Langton and Jones, 2005).

9.5 Early Modern Hunting Parks in Europe

The most conspicuous impact of hunting on European woodland is perhaps seen today in the remaining deer parks which, at their best, remain enclaves of wood-pasture (Plate 5). In England, the 10th century laws of Cnut make it clear that hunting reserves had already been set aside, though every man was entitled to hunt on his own property. Anglo-Saxon *haga* are often described; these were enclosures or traps to permit the killing of deer. With the advent of Norman rule, *haia*, which seem to have been synonymous with *parcus* in the Domesday Book, are numerous and describe deer parks (Liddiard, 2003; Fletcher, 2011). Similar patterns existed elsewhere in Europe (Ahrland, 2011).

The parks provided the elite with an opportunity to demonstrate their mastery of wild nature and conspicuous control of wild animals, usually deer. The concept of the park reached Europe from Persia, where they had been described by Greek authors as *paradeisos*. They were indeed earthly symbols of paradise with their connotations of perfect enclaves in which the elite could lose themselves in recreation (Ahrland, 2011; Fletcher, 2011).

The surviving parks are most numerous in England and have contributed to the high number and density of veteran trees in Britain (Hartel *et al.*, Chapter 5; Siitonen and Ranius, Chapter 11). The English parks reached their apogee in the medieval period, and in the 13th century an estimated 2000–3000 deer parks came and went. These were largely constructed to contain the fallow deer (*Dama dama*) newly introduced by the Normans, probably from Sicily. However, the influence of the parks remained powerful into modern times, with a particular resurgence during the late 15th and 16th centuries: Queen Elizabeth I inherited around 200 deer parks on her ascendancy in 1558 (Ahrland, 2011; Fletcher, 2011).

Although it is often said that medieval English monarchs rarely hunted, especially within their deer parks, the Norman kings, and later Edward I, Edward III and Henry IV, and especially Henry VIII and Elizabeth I among the Tudor rulers in England, hunted regularly, in the case of Henry VIII almost daily for much of the year, with the hunting usually outside the parks but within their forests (Hore, 1895; Sabretache, 1948; Rackham, 1986; Cummins, 1988; Beaver, 2008). On mainland Europe, royal hunts seem to have been common throughout the medieval period (Rackham, 1986; Cummins, 1988). Hunting, especially of the supremely symbolic red deer, was of enormous importance and the management of the forest was directed to ensuring that the hunts were fruitful. The parks could provide hunting spectacles and they could always be called upon to supply gifts of venison as a material token of the privilege associated with the stag (Fletcher, 2011).

The parks had a significant impact on the wooded landscape throughout their existence. They incorporated many enterprises other than deer and hunting: mining, smelting, charcoal burning, tanning, rabbit warrens and, especially, the grazing of cattle and horses (Liddiard, 2007). Many were compartmentalized with areas of coppice from which the deer were excluded on a rotational basis; in other parts of the park, trees were pollarded to provide feed for deer, with the residue forming valuable firewood (Rackham, 1986). Within parks, trees were protected both administratively and physically, giving shade and shelter as well as acorns, leaves, beech mast and chestnuts for the animals; ultimately, in some parks, they also provided an invaluable source of building timber (Rackham, 1986). Britain probably had more parks than any other nation in Europe and some 300 still survive as working examples of wood-pasture (Fletcher, 2011).

Parks did not become so numerous in other parts of Europe, but were present nevertheless. Substantial game parks were described by Roman authors; Charlemagne had a hunting park at Aachen and probably elsewhere; and the Normans established parks in Sicily. Count Robert II of Artois created a massive park at Hesdin with a menagerie and gardens (Ahrland, 2011) Fallow deer reached Denmark by the 13th century and, with bear and rabbits, were contained on 50 islands listed as royal game reserves if not exactly parks (Radtke, 2011). In southern Sweden, several parks are known to have existed though fallow deer probably did not reach there until the 16th century. The noble families of Italy – the Visconti, the Este and Gonzaga – had hunting parks, while in France in 1519 Francois I commenced building the 20 mile long wall around Chambord, a deer park of 5500 ha (d'Anthenaise and Chatenet, 2007). Following this, many other parks were constructed along the Loire. Louis XIII built his hunting lodge at Versailles in 1624, later to become the basis for the Chateau de Versailles, where hunting remained the most popular activity after gambling and lovemaking (Mitford, 1966). The woodland was laid out with *allées* (roads and alleys) and *rond-points* (circular meeting points) to allow the progress of the hunt to be followed from coaches if necessary.

9.6 Hunting and the Wider Landscape

Outside the parks, human populations were sufficiently small until at least the 16th century as to permit hunting with relatively little impact on the woodland. Restrictions on grazing within forests during the 'Fence Month' when deer were calving may have exerted a small influence, and the game competed with domestic animals for the fruit of the trees, but we may imagine that these effects were small. However, as human populations grew there was a commensurate need for more agricultural land. The scale of royal hunts also had a significant ecological impact in many parts of Europe: in Germany rulers vied with each other to present ever more remarkable hunting spectacles and in so doing they exerted a growing influence on woodland.

Although the hunting of smaller landowners and poaching by peasants must have had some ecological impact, it was probably small compared with that of the royal holding hunts. In order to stage these events, game was gathered from large areas by the army and local peasantry over many days and then confined within enclosures created by sheets of canvas. There they were slaughtered by the aristocracy often after having been driven into water (Plate 4). The competition between German rulers to create ever more impressive water hunts had devastating effects on woodland in many areas. For example, in 1668 the Bavarian Elector created a fairly modest enclosure on the banks of Lake Starnberg south of Munich with a paling fence measuring only 2866 m, yet by 1677 the surrounding forests could not supply sufficient timber to maintain the paling, and in 1681 the park had to be abandoned (Knoll, 2004).

More scientific efforts to improve forestry and agriculture were obstructed by the old fashioned demands of the aristocracy for hunting. Where once the rulers had argued that woodland was too valuable a resource to allow access for peasants, now foresters criticized the damage caused to woodland by the nobility. The anonymous author of a memorandum discussed in the Bavarian Electoral administration in 1782 argued that:

> There can be no doubt, that both red deer and wild boar cause damage to the forests because they ruin the young seeds of timber plants and rip the trunks of trees in winter. There is no possibility in forests where red deer and wild boar are preserved that any seed can root and can be grown in time.
> In such regions the timber won't grow at all or only misgrown and damaged wood will appear.
> (Knoll, 2004)

Simon Rottmanner (1740–1813), a prominent Bavarian promoter of agricultural and forestry innovations, complained that 'The damage to the forests which is caused by major hunting events is beyond any description. Millions of trees could be where there are only small bushes, pathways, alleys' (Knoll, 2004).

In 1798, four villages situated in the big hunting park near Munich mentioned above succeeded in getting their agricultural acreage protected against wildlife by a new fence. The local Electoral forester was ordered to do the planning. His documents provide information about the mode of construction and the amount of wood that was needed for the 14.8 km fence (Knoll, 2004). The forester calculated 7275 posts and the same amount of 6 m horizontal rails. In addition, over 436,500 small trunks of young trees were used as palings. The local woodland situated near the town was part of the hunting preserve with its dense wildlife population and also had to supply the local peasants with wood.

Nor was this fence the only one. More than 18,000 posts were needed to build another fence at the outer border of the hunting park estimated at 37 km in length. Because of its durability, oak was used for these posts and each large oak yielded about 40 posts as the *Oberstjägermeister* (chief of the Electoral hunting authority) Sigmund von Preysing calculated. Thus, 480 oaks had to be cut for the 18,000 posts not including the horizontal rails! Finally, records show that the fence was frequently damaged by storm, rain, wild game, livestock and resisting peasants. In short, it had to be repaired constantly, which required further timber (Knoll, 2004). Simon Rottmanner criticized:

> I know well that they suggest fences as a method to prevent damage caused by wild game. But these fences have to be built very high, very strong and very narrow and they have to be maintained over years, so that they cause exhaustion of forests and other costs. ... One other forester ordered the construction of a fence made of planks. One can easily imagine how much timber and money was spent on this. As the woodlands of this forester are in the worst condition, one may doubt that in one century there will grow trees as have been cut into planks. ... Recently I have seen a coursing-garden being erected, which was fenced in with the best young timber. This will hurt the nearby forest for centuries.

The large numbers of game demanded by the German courts from the foresters in the 17th and 18th centuries devastated both agricultural and arboreal crops. The Bavarian Elector killed 1105 wild boar over 2 weeks in November 1735, and in 1763 some 5000 wild animals were killed by being driven into an artificial park (Knoll, 2004). In order to be able to gather such large numbers of game, the densities in the surrounding woodland must have been immense and damaging over huge areas. These set-piece staged hunts had a clear purpose in impressing the importance of royalty on the people. Large Baroque hunts in Germany were sometimes viewed by crowds of 10,000. In Germany, there was growing public protest and minor reforms, but it was not until the 1848 revolution that legal change linked hunting rights to land ownership.

In France, by contrast, elite hunting was generally *par force des chiens* (using dogs) with mounted followers. As pioneered at Versailles, the layout of the forests was designed to permit viewers to keep up with the action in carriages or at least view the chase down *allées*. Thus, the forest of Compiègne had around 1600 km of roads and alleys by 1763 specifically laid out for hunting. In France, the revolution reduced the population of game and made hunting the universal recreation that it remains. Nevertheless, the protection given to rabbits under Colbert at Compiègne in 1669 remained in place until the damage to forests became unsustainable, and the warrens were not finally destroyed until after the First World War (Reed, 1954).

Where hunting has historically been a popular pastime of the wealthy, then game numbers may be expected to increase to the point where they may damage the environment because they benefit from legal protection and gamekeeping. In contrast, where poaching is common, then game numbers may drop to the point of extinction. Although by its very nature the evidence for widespread clandestine deer killing in medieval England is not easy to find, analysis of the proceedings of 13th and 14th centuries forest courts by Jean Birrell has made it clear that poaching deer from the *forestes* took place very regularly. It was probably not normally punished by the full permitted weight of the law but rather by fines within the culprit's means (Birrell, 1996, 1999).

Poaching by peasants must have been on a considerable scale, at least in Britain. Today, we have a population of around 2 million free-ranging deer of six species. It is difficult to believe that medieval England would not have had the carrying capacity of at least half of that number, albeit of only the three deer species: roe (*Capreolus capreolus*), red and fallow. Now, in modern Britain as elsewhere in Europe, deer numbers continue to grow despite the best efforts of hunters with high powered rifles and the depredations of road traffic accidents (up to an estimated 74,000 deer are killed in vehicle collisions in Britain each year, and within Europe some 10–15 people die in consequence). Yet British history is not short of commentators bemoaning the shortage of deer from the 13th century onward, and by the end of the 18th century red deer had almost disappeared from England, and roe deer were probably extinct. Poaching must therefore have been carried out very effectively and on a grand scale and have contributed a significant quantity of meat and leather. The Robin Hood legends have close parallels in most of Europe and poaching was usually the reason for these heroic figures being outlawed, as well as a continuing important part of their romantic life in the greenwood.

Archaeological excavations of peasant middens occasionally yield significant numbers of deer bones, yet their absence in other sites may simply reflect a certain prudence in destroying evidence against discovery by the officers of the forest law, or the removal of bones in the woods (Sykes, 2007). However, poaching was not restricted to peasants with snares. Gangs of poachers are described from the 15th into the 18th and even 19th centuries, and these involved a disproportionate number of peers and gentry who were sufficiently brazen to break into parks and remove all the deer. Often, there were pitched battles between parties whose composition sometimes reflected competing factions at court (Manning, 1993).

Rabbits (*Oryctolagus cuniculus*) were another significant influence on European woodlands. They were initially introduced to Britain from Sicily by the Normans, not to provide sport but meat and fur, but in other parts of Europe, rabbits might be kept for hunting, especially by women, using small dogs and bows and arrows (Robinson, 1984). The rabbits were carefully garnered in enclosed warrens from the Roman era onward, but by the mid-14th century escapees were recorded as causing damage (Sykes and Curl, 2010). By the mid-16th century, feral rabbits were numerous and they further benefited from the improved agricultural practices of the 18th century and, later, a decline in predators that were being killed by gamekeepers. By the 19th century, they were ubiquitous and Britain's most serious agricultural pest (Yalden, 1999). In his book about animal communities, Elton (1966) makes many more comments about the impact of rabbits on the local woods than he does about that of deer, even though by this stage the numbers of rabbits had crashed because of the spread of myxomatosis in the 1950s.

9.7 Modern Hunting

As aristocratic monopolies of large wild game were eroded throughout Europe, deer and wild boar numbers fell and alternative sport had to be found. For example, in England red deer, the time-honoured quarry, had been hunted into a long decline from the 15th century and, apart from small numbers in deer parks, they were virtually extinct by the end of the 18th century. Some red deer were kept at the kennels, taken to the hunt in a cart and retrieved at the end of the day so that they could run again. Some 140 packs of hounds were hunting the carted stag in its heyday, and, although hunting carted stags remained popular into the 20th century, from about 1750 the hounds used to hunt stags were increasingly put to foxes (*Vulpes vulpes*). Hunting of the fox and to a lesser extent of the hare became so popular during the late eighteenth and early 19th century that landowners specifically planted fox coverts and spinneys which increased the number of small woods and later provided protection for pheasants. For example, in 1797, a fox covert of 10 acres in north Lincolnshire was planted with 49,000 oaks (*Quercus* spp.), 18,000 spruces (*Picea* spp.), 8200 ash (*Fraxinus excelsior*), 7400 privets (*Ligustrum vulgare*) and 5000 silver firs (*Abies alba*). It was later extended to 12 acres by planting blackthorn (*Prunus spinosa*) (Woodruffe-Peacock, 1918).

Foxes provided faster, and also more democratic, sport ideally suited to the faster horses then available and hounds were developed accordingly. The Enclosure Acts created obstructions, but foxes, hounds and followers could now jump hedges. Woodland was unpopular with the fox hunts and the development of the sport did not encourage large plantations. Instead, small coverts were required in which foxes' earths could be located and stopped on the day of the hunt, with open fields and hedgerows providing the main part of the country. However, while fox hunting increasingly demanded a good supply of foxes, elsewhere, gamekeepers struggled to keep them out of the park and woods where they would create havoc among the pheasants (*Phasianus colchicus*) being reared for the elite to shoot (de Belin, 2013).

9.7.1 The influence of driven pheasant shoots on British woodland

Many British woods remained surprisingly intact and unchanged from the medieval period until at least the late 19th century, when modern agriculture began to make a significant impact (Rackham, 1986). Since then, the planting and management of small woodlands for pheasants has also been a considerable factor influencing woodland structure (Coles, 1971; Fisher, 2000; Jones, 2009). Historically and ecologically, modern plantations, even carefully designed to provide a diversity of plantings and local strains of native trees, can never recreate ancient woodlands with their coppice stools, which may be a 1000 years old. Modern plantings may though enrich landscapes that at times seemed in danger of becoming an arable desert.

From as early as 1750 it had become a popular fashion to use increasingly effective guns to shoot birds flying. Parks had to accommodate this sport and the design of game coverts went hand in hand with the aesthetics of landscape (Williamson, 1995). Gentlemen assisted by servants with dogs walked through the woods and shot pheasants as they rose, but from about 1860 the servants were instead employed to beat the birds out of the woods and over stationary guns. Woodland plantations were laid out to create coverts from which high-flying pheasants could be flushed most effectively. By 1862, when Queen Victoria purchased Sandringham for the Prince of Wales, later King Edward VII, he had become a great enthusiast and this gave the sport a social cachet. The royal estate at Sandringham was rearing 12,000 pheasants a year by 1900 and nearby Elveden was rearing 20,000; between 1867 and 1923, one of the leading shots, the second Marquess of Ripon, shot 241,000 pheasants. In order to produce these huge numbers of birds, estates had to plan, plant and manage woodland with great care, and most of those woods still remain. At its height in the early 20th century, an estimated 25,000 gamekeepers were employed to manage the woodlands so as to maximize the numbers of pheasants put before the guns (Tapper, 1992).

The rearing and release of game birds for driven shoots in the 21st century dwarfs even the Edwardian grandeur in Britain. It is now a large industry involving and influencing 15 million ha which represents two thirds of the rural land. In 2004, some 35 million pheasants and 6.5 million partridge were reared and released, of which 15 million pheasants and 2.5 million partridge were shot (Olstead and Moore, 2006). There has been rapid growth since then, and it is probably reasonable to estimate that 50 million pheasants and 15 million partridge were being reared annually for shooting by 2013. Within the 15 million ha of land mentioned above, some 830,000 ha of woodland are managed for shooting. Unlike commercial coniferous plantations, this woodland usually exists as small strips of trees and shrubs often planted in areas dominated by arable farming and with little or no existing woodland.

After pest control, thinning and coppicing are the most time-consuming activities of the workers directly employed in managing shoots. The objective of the managed shoot is to create a suitable environment in which to release young birds and yet which also permits them to be easily driven out to the guns. The ideal woodland is mixed evergreen and deciduous with a wide fringe of shrubs, a canopy with plenty of gaps and open rides. Woods managed for pheasants are normally less than 2 ha in size to create an edge effect (Short et al., 1994). Many British arable farms plant 'game crop' – seed-rich cover that benefits many species other than the game birds, and many will leave beetle banks to provide invertebrate feed for partridge. In a survey of landowners, 61% of properties rearing pheasants also undertook new woodland planting (Tapper, 2005). If legislation to reduce the scale of this business were to be introduced, small woodlands throughout Britain would probably suffer dramatically.

9.7.2 The influence of modern hunting enclosures on Spanish woodland

Enthusiasm for big game hunting continues worldwide and many hunting enclosures remain in use throughout Europe, indeed, new deer enclosures are being constructed in the Baltic countries, western Russia, Romania, Bulgaria and other countries. Some of the many *fincas* (farms/country estates) in Spain designed for hunting by the *montería* system, in which deer and wild boar are driven by human and canine beaters to rifles, are now being abandoned. However, others continue with intensive management of indigenous vegetation to ensure good conditions for the deer and wild boar. These hunting reserves cater not only for Europeans but also for Chinese, American and other nationals.

During the last 150 years, the hunting reserves of the wealthy in the Iberian peninsula have exerted a growing influence in the *dehesa*, the management system that conserves a diverse Mediterranean savannah-like woodland dominated by oak species and extending to 3.5–4 million ha. This open woodland is a unique wood-pasture habitat in which there is a community of charcoal burners, cork oak (*Q. suber*) harvesters and livestock graziers herding sheep, cattle and pigs. Erosion of these traditional land uses tends to lead to their replacement by monocultures of vines, olive and citrus groves, or to cereal production. Elsewhere, cattle, sheep, goats and the Iberian black pigs which are grazed under the trees may be reaching stocking levels that might imperil the pasture woodland. The replacement of these domestic species by fenced game reserves in which roe, red and, occasionally, fallow deer are preserved along with mouflon (*Ovis musimon*), wild boar and partridge for sporting purposes provides an economic alternative which permits the growth of bushes and shrubs such as *Cistus* spp. This may increase biodiversity and reduce stocking density, but unless also grazed by livestock will not maintain the traditional landscape.

There are some 29,000 hunting estates in Spain occupying 36 million ha, which represents 72% of the Spanish land area. Of this area, approximately 2.7% of the hunting areas are enclosed, amounting to 1 million ha (Carlos Otero Muerza, 2013, Director APROCA, Spanish Landowners' Association for Hunting and Nature Conservancy, Madrid, personal communication). The biomass of game species in a hunting reserve is a small fraction of the biomass of domestic livestock under conventional *dehesa* grazing systems. Until the 1960s, such a change in land use was unusual, but the movement has gathered pace as hunting has become more highly valued. These commercial hunting enclosures demonstrate a significant increase in biodiversity. The game benefit from acorns, while the cork oaks provide bark for the cork industry, and charcoal burners and beekeepers can continue their crafts. Game management has encouraged the construction of dams and reservoirs, and the protection of oak trees yielding acorns.

9.8 Conclusion

In Britain, the remnants of traditional deer parks are now often carefully managed to preserve parkland and 'provide lungs' for urban populations. Elsewhere, woodland that often originally owed its preservation to hunting laws frequently remains intact today. In France and Germany, roe deer are shot by licensed recreational hunters to achieve a sustainable cull, but wild boar numbers are increasing faster than agriculturalists would like over much of their range. There is a growing shortage of hunters over much of Europe, with the danger that deer, like wild boar, will come to be seen as vermin.

As populations of deer grow, their impact on woodland becomes more and more serious. Plantations must now be laid out to take account of the need to cull deer, with high seats, rides and grazing lawns being planned at the planting stage (Putman *et al.*, 2011). Where roads are not fenced, expensive collisions of vehicles with deer leave a growing toll in maimed animals and injured motorists. Perhaps within these facts lie the seeds of a practical education of our increasingly urbanized populations.

A new angle to the hunting reserve has evolved in the form of nature reserves throughout Europe where animals can be stalked by camera and where management is directed to the new holy grail of biodiversity. Enterprises catering for those ecotourists who wish to see and photograph beavers, elk, wolves, bison, deer, wild boar and, of course, birds, are growing quickly. A study in 2013 showed that Finnish wildlife tourism enterprises realised €2.1 million in 2012. Ironically, wildlife tourism may reduce the number of people who are interested in hunting to kill at a time when the woodland environment is at risk from overpopulations of deer and wild boar.

There is a strange irony in reflecting that royalty and nobility throughout Europe, from the earliest recorded history, made strenuous efforts to create *forestes* so as to preserve their hunting rights and that in these, woodland was often preserved from agriculture. Yet that proved ultimately ineffective because wild game numbers fell, presumably in the face of extensive poaching. Now, the aristocratic hunting rights have been eroded and game numbers have risen to such a level as to threaten the woodlands and agriculture. Perhaps we can say that the greatest impact on woodland today may be the *absence* of hunters.

References

Ahrland, Å. (2011) Vert and Venison – high status hunting and parks in medieval Sweden. In: Grimm, O. and Schmölcke, U. (eds) *Hunting in Northern Europe until 1500 AD. Old Traditions and Regional Developments, Continental Sources and Continental Influences. Papers presented at a workshop organized by the Centre for Baltic and Scandinavian Archaeology (ZBSA) Schleswig, June 16th and 17th, 2011.* Wachholtz Verlag, Neumünster, Germany, pp. 439–464.
Allsen, T.T. (2006) *The Royal Hunt in Eurasian History.* University of Pennsylvania Press, Philadelphia, Pennsylvania.
Anderson, J.K. (1985) *Hunting in the Ancient World.* University of California, Berkeley, California.
Beaver, D.C. (2008) *Hunting and the Politics of Violence before the English Civil War.* Cambridge University Press, Cambridge, UK.
Birrell, J. (1996) Peasant deer poachers in the medieval forest. In: Britnell, R. and Hatcher J. (eds) *Progress and Problems in Medieval England. Essays in Honour of Edward Miller.* Cambridge University Press, Cambridge, UK, pp. 68–88.
Birrell, J. (ed.) (1999) *Forests of Cannock and Kinver: Select Documents 1235–1372.* Collections for a History of Staffordshire 4th Series, Volume XVIII. Staffordshire Record Society, Stafford, UK.

Blüchel, K.G. (1997) *Game and Hunting*. Könemann, Köln, Germany.
Broekmeyer, M.E.A., Vos, W. and Koop, H. (1993) Forest reserves in Europe – a review. In: Broekmeyer, M.E.A., Vos, W. and Koop, H. (eds) *European Forest Reserves. Proceedings of the European Forest Reserves Workshop, 6-8 May 1992, Wageningen, The Netherlands*. Pudoc, Wageningen, Netherlands, pp. 9–28.
Coles, B. (2010) The European beaver. In: O'Connor, T. and Sykes, N. (eds) *Extinctions and Invasions – A Social History of British Fauna*. Windgather, Oxford, UK, pp. 104–115.
Coles, C. (1971) *The Complete Book of Game Conservation*. Barrie and Jenkins, London.
Cummins, J. (1988) *The Hound and the Hawk*. Weidenfeld and Nicolson, London.
d'Anthenaise, C. and Chatenet, M. (2007) *Chasses Princières dans l'Europe de la Renaissance. Actes du Colloque de Chambord 1er et 2 Octobre 2004*. Actes Sud-Maison de la Chasse et de la Nature. Arles, Paris.
de Belin, M. (2013) *From the Deer to the Fox. The Hunting Transition and the Landscape, 1600–1850*. UH Press, Hatfield, UK.
Elton, C. (1966) *The Pattern of Animal Communities*. Methuen, London.
Enright, M.J. (2006) *The Sutton Hoo Sceptre and the Roots of Celtic Kingship Theory*. Four Courts Press, Dublin, Ireland.
Fisher, J. (2000) Property rights in pheasants: landlords, farmers and the game laws 1860–80. *Rural History* 11, 165–180.
Flannery, K. and Marcus, J. (2012) *The Creation of Inequality: How Our Prehistoric Ancestors Set the Stage for Monarchy, Slavery and Empire*. Harvard University Press, Cambridge, Massachusetts.
Fletcher, T.J. (2011) *Gardens of Earthly Delight – The History of Deer Parks*. Windgather, Oxford, UK.
Fumagalli, V. (1994) *Landscapes of Fear*. Polity, Cambridge, UK.
Giese, M. (2011a) Continental royal seats, royal hunting lodges and deer parks seen in the mirror of medieval written sources. In: Grimm, O. and Schmölcke, U. (eds) *Hunting in Northern Europe until 1500 AD. Old Traditions and Regional Developments, Continental Sources and Continental Influences. Papers presented at a workshop organized by the Centre for Baltic and Scandinavian Archaeology (ZBSA) Schleswig, June 16th and 17th, 2011*. Wachholtz Verlag, Neumünster, Germany, pp. 387–397.
Giese, M. (2011b) Legal regulations on hunting in the barbarian law codes of the Early Middle Ages. In: Grimm, O. and Schmölcke, U. (eds) *Hunting in Northern Europe until 1500 AD. Old Traditions and Regional Developments, Continental Sources and Continental Influences. Papers presented at a workshop organized by the Centre for Baltic and Scandinavian Archaeology (ZBSA) Schleswig, June 16th and 17th, 2011*. Wachholtz Verlag, Neumünster, Germany, pp. 485–505.
Hore, J.P. (1895) *The History of the Royal Buckhounds*. Privately published, Newmarket, UK.
Jones, E. (2009) The environmental effects of blood sports in lowland England since 1750. *Rural History* 20, 51–66.
Knoll, M. (2004) Hunting in the eighteenth century. An environmental history perspective. *Historical Social Research* 29, 9–36.
Langton, J. and Jones, G. (eds) (2005) *Forests and Chases of England and Wales c.1500–c.1850*. St John's College, Oxford, UK.
Legge, A.J. (2010) The aurochs and domestic cattle. In: O'Connor, T. and Sykes, N. (eds) *Extinctions and Invasions – A Social History of British Fauna*. Windgather, Oxford, UK, pp. 26–35.
Liddiard, R. (2003) The deer parks of Domesday Book. *Landscapes* 1, 4–23.
Liddiard, R. (ed.) (2007) *The Medieval Park – New Perspectives*. Windgather, Oxford, UK.
Manning, R.B. (1993) *Hunters and poachers: a social and cultural history of unlawful hunting in England 1485–1640*. Clarendon Press, Oxford, UK.
Marvin, W.P. (2006) *Hunting Law and Ritual in Medieval English Literature*. Brewer, Woodbridge, UK.
Mitford, N. (1966) *The Sun King*. Hamish Hamilton, London.
Olstead, J. and Moore, S. (eds) (2006) *The Economic and Environmental Impact of Sporting Shooting*. Report prepared by Public and Corporate Economic Consultants (PACEC), Cambridge, UK. British Association for Shooting and Conservation (BASC), Wrexham, UK.
Pluskowski, A.G. (2010) The wolf. In: O'Connor, T. and Sykes, N. (eds) *Extinctions and Invasions – A Social History of British Fauna*. Windgather, Oxford, UK, pp. 68–74.
Putman, R., Apollonia, M. and Andersen, R. (eds) (2011) *Ungulate Management in Europe: Problems and Practices*. Cambridge University Press, Cambridge, UK.
Rackham, O. (1986) *The History of the Countryside*. Dent, London.
Radtke, C. (2011) Lordship and hunting in Schleswig – a sketch. In: Grimm, O. and Schmölcke, U. (eds) *Hunting in Northern Europe until 1500 AD. Old Traditions and Regional Developments, Continental Sources and Continental Influences. Papers presented at a workshop organized by the Centre for

Baltic and Scandinavian Archaeology (ZBSA) Schleswig, June 16th and 17th, 2011. Wachholtz Verlag, Neumünster, Germany, pp. 419–439.
Reed, J.L. (1954) *The Forests of France*. Faber, London.
Robinson, R. (1984) Rabbit. In Mason, I.L. (ed.) *Evolution of Domesticated Animals*. Longman, London.
Sabretache [Albert Stewart Barrow] (1948) *The Monarchy and the Chase*. Eyre and Spottiswoode, London.
Short, C., Cox, G., Hallett, J., Watkins, C. and Winter, M. (1994) *Implications of Game Management for Woodland Management, Landscape Conservation and Public Recreation*. Countryside Commission, Cheltenham, UK.
Sykes, N.J. (2007) Animal bones and animal parks. In: Liddiard, R. (ed.) *The Medieval Park – New Perspectives*. Windgather, Oxford, pp. 49–62.
Sykes, N.J. and Curl, J. (2010) The rabbit. In: O'Connor, T. and Sykes, N. *Extinctions and Invasions – A Social History of British Fauna*. Windgather, Oxford, UK, pp. 116–126.
Tapper, S. (1992) *Game Heritage: An Ecological Review from Shooting and Gamekeeping Records*. Game Conservancy, Fordingbridge, UK.
Tapper, S. (2005) *Nature's Gain. How Gamebird Management Has Influenced Wildlife Conservation*. A Report from the Game Conservancy Trust, July 2005. Game Conservancy Trust, Fordingbridge, UK.
Vera, F.W.M. (2000) *Grazing Ecology and Forest History*. CAB International, Wallingford, UK.
Williamson, T. (1995) *Polite Landscapes – Gardens and Society in Eighteenth Century England*. Sutton, Stroud, UK.
Woodruffe-Peacock, E.A. (1918) A fox covert study. *Journal of Ecology* 6, 110–125.
Yalden, D.W. (1999) *History of British Mammals*. Poyser, London.

Part III

How Plants and Animals Have Responded to the Changing Woodland and Forest Cover

---

We focus in this section on patterns in dead-wood invertebrates, birds and vascular plants (Chapters 11, 12 and 13) because they operate at different scales and require different types of forest. As attitudes have changed, some of the previously 'lost' species, particularly large mammals, are making a comeback, whereas vigorous (but usually unsuccessful) attempts are made to stop the spread of other, introduced, species (Chapter 14). While much effort has gone into describing the wildlife of traditionally managed ancient woods by coppicing (Chapter 10), we are also starting to understand better the ecology of the new types of woodland that are being created, for example the Atlantic spruce forests of Britain and Ireland (Chapter 15).

10 The Flora and Fauna of Coppice Woods: Winners and Losers of Active Management or Neglect?

Peter Buckley* and Jenny Mills
Peter Buckley Associates, Ashford, UK

10.1 Introduction

Coppice management in various forms (Plate 6) was widespread through much of Europe alongside areas treated as wood-pastures (Hartel *et al.*, Chapter 5; Buckley and Mills, Chapter 6). In places, coppice systems have come to be seen as particularly rich in wildlife and this form of traditional management has been widely promoted for conservation purposes as the antithesis of even-aged plantation silviculture (Quine, Chapter 15). This chapter explores why particular groups of species are associated with worked coppice and how they fare when such stands cease to be cut or are converted to high forest.

10.2 The Diversity of Coppice

Actively managed coppices are particularly valued for the richness of their field layers, seed bank species, migrant warblers and abundant insect fauna. The lack of old trees means fewer niches for mature forest specialists, but the interspersing of small coupes, spanning early to mid-successional stages, sometimes complemented by standard trees, creates a varied environment with a distinctive character. This may be further diversified by variations in the length of the cutting cycle and activities such as reduction of the fuel load, grazing and litter raking. Rackham (1975) noted that at Hayley Wood in Cambridgeshire, UK, 'no two successive plots have behaved alike, and there are often big differences from one part of the plot to another'. The conversion of coppices to high forest (whether deliberately or by neglect) alters the intensity, frequency and extent of management interventions, and fundamentally changes the forest structure. Changes to the associated wildlife are therefore to be expected.

Studies of long term species changes resulting from the abandonment or conversion of coppice stands in Europe include those by Debussche *et al.* (2001), Decocq *et al.* (2004), Kirby *et al.* (2005), Van Calster *et al.* (2007, 2008a), Baeten *et al.* (2009), Amar *et al.* (2010), Itô *et al.* (2012) and Kopecký *et al.* (2013). Most of these have recorded a decline in richness of shrubs and trees and understorey layers under increasing shade, as well as a gradual tendency towards homogenization in the field layers with lowered beta diversity, but increases in shade-tolerant, vernal and eutrophic

---

*E-mail: peterbuckleyassociates@gmail.com

© CAB International 2015. *Europe's Changing Woods and Forests: From Wildwood to Managed Landscapes* (eds K.J. Kirby and C. Watkins)

species. The impacts are especially great on species with a high turnover that may have nowhere else to go in the immediate landscape, such as butterflies, reptiles, amphibians, carabid beetles and corticolous bryophytes.

Interpreting how change in the silvicultural system has affected the wildlife is complicated by other changes going on at the same time. In many temperate deciduous stands, increasing eutrophication and acidification, variously attributed to atmospheric deposition, biomass accumulation and increasing canopy cover (Verheyen *et al.*, 2012), have happened in parallel with coppice abandonment. The consequence of changes in tree species composition, for example deteriorating litter quality, especially on soils sensitive to acidification under pure plantations of beech or oak (Hölscher *et al.*, 2001; Strandberg *et al.*, 2005; Van Calster *et al.*, 2007, 2008a; Baeten *et al.*, 2009) may have as much an impact as the change in management system. Interactions between different processes make it difficult to predict how particular species will respond at any individual site, but general trends in coppice systems can be described.

10.2.1 Plants

The early stages of the coppice cycle provide many opportunities for herbaceous plants prior to canopy closure, uniquely maintaining a functionally diverse mix of open ground, intermediate and shade-tolerant plants (Hermy, Chapter 13). Immediately after cutting, the light levels in a coppice stand are very high, but they fall rapidly as the stand regrows, sometimes to only 1% of the ambient light level after 3–5 years: at this low level only a few shade-tolerant species can survive (Mitchell, 1992; Mason and MacDonald, 2002; Broome *et al.*, 2011). However, the pulses of light that occur under a regular cutting regime sustain many heliophilous plants, including ruderal species (annuals and biennials) and perennials that capitalize on the disturbance, either by immigrating through efficient seed dispersal, or by germinating from persistent seed banks *in situ*. Periodic gaps create seed regeneration niches for light-demanding trees and shrubs, whereas the temporarily open conditions benefit vernal and shade-tolerant species such as the bluebell (*Hyacinthoides non-scripta*), primrose (*Primula vulgaris*), wood anemone (*Anemone nemorosa*), dog's mercury (*Mercurialis perennis*) and wild daffodil (*Narcissus pseudonarcissus*), enabling them to flower profusely and set seed.

As the canopy closes, species richness in the field layer declines rapidly (Ash and Barkham, 1976; Ford and Newbould, 1977; Mason and MacDonald, 2002; Verdasca *et al.*, 2012) but shade-tolerant herbaceous species, shrubs and climbers such as bramble (*Rubus fruticosus*), ivy (*Hedera helix*) and honeysuckle (*Lonicera periclymenum*) persist later into the rotation. Standard trees in coppice-with-standards woods help to create a patchy environment that allows the coexistence of light-demanding and shade-tolerant species (Joys *et al.*, 2004). Standards also provide a more varied substrate for epiphytes; while the branches of young coppice may show only pioneer corticolous bryophytes, the bark and bases of old trees may be colonized by later successional species (Bardat and Aubert, 2007).

The conversion of coppices to high forest may not result in greater canopy dominance than in worked coppice. Decocq *et al.* (2004) found that regular disturbances and the lighter conditions created by selective cutting at 8 year intervals promoted an increase in understorey vegetation, particularly of bramble cover. However coppice-with-standards cut on 30 year rotations allowed greater vegetation recovery between harvests, and higher populations of vernal geophytes and shade-tolerant perennials.

Lengthened rotations under high forest may have long-term consequences for light-demanding plants that rely on persistent seed banks. These tend to decline in density and species richness in ageing or converted coppices after only 50 years – less than half the length of most high forest broadleaved rotations (Brown and Warr, 1992). These losses could accumulate over time (Fig. 10.1) without periodic reactivation of the seed bank through heavy thinning, ride management (Buckley *et al.*, 1997), fire or rare windthrow events. The species concerned may disappear completely, or persist only in small populations along edges and glades.

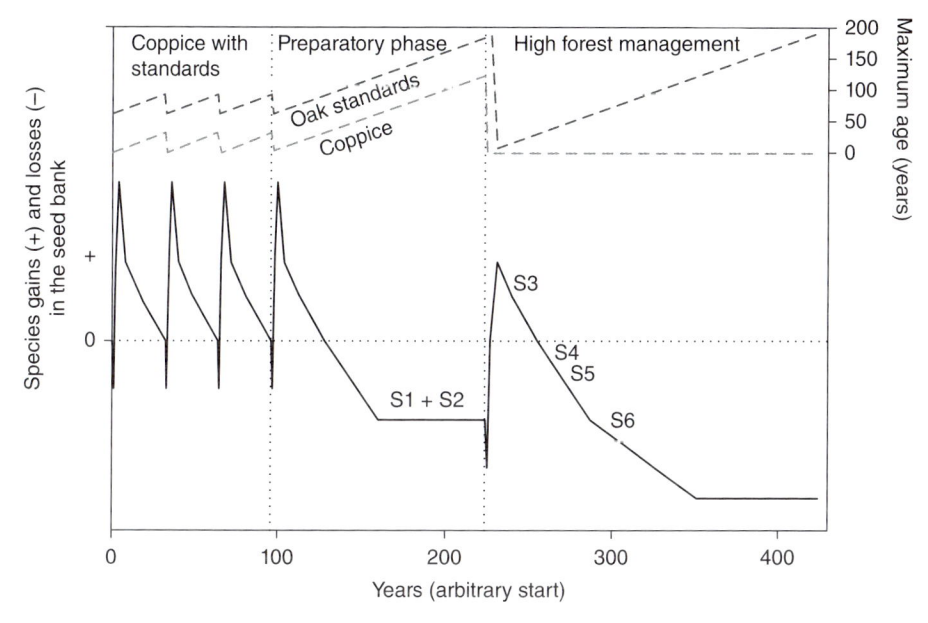

Fig. 10.1. Seed bank dynamics in terms of species gains and losses in the seed bank. Observed trends (S1–S6) were extended with expected trends derived from the literature to cover the full trajectory from coppice-with-standards to high forest management. Solid lines represent species gains (+) and losses (−). Dashed lines depict the maximum age of coppice shoots (light grey) and oak trees (dark grey) to show the timing of canopy disturbances. (From Van Calster et al., 2008b, reproduced with kind permission from John Wiley and Sons.)

10.2.2 Birds

Woodland structure has consistently been shown to be important as a factor determining the abundance of songbirds (Fuller, 2012; Hinsley et al., Chapter 12). Different species use different coppice stages (Fuller and Moreton, 1987). If the structure is altered by management or through neglect the assemblages change.

As a specific example, in coppiced holm oak forest in north-east Spain, a dense mixed shrub layer was critical in retaining subalpine (Sylvia cantillans), Sardinian (S. melanocephala) and garden (S. borin) warblers (Camprodon and Brotons, 2006). Clearance of the understorey still provided a habitat for the robin (Erithacus rubecula), wren (Troglodytes troglodytes) and blackbird (Turdus merula) if the main canopy was left intact; stands that had been both cleared and thinned attracted open-ground foragers such as the turtle dove (Streptopelia turtur), green woodpecker (Picus viridis), nightjar (Caprimulgus europaeus), mistle thrush (Turdus viscivorus) and cirl bunting (Emberiza cirlus).

Bird species numbers tend to increase with increasing variety of age structure in the coppice (Donald et al., 1998). Canopies developing into and beyond the pole stage are generally less attractive, but standard trees in coppice provide a niche for canopy-feeding and hole-nesting species. The shading out of low shrub cover resulting from the abandonment of coppice management and succession to high forest, or loss of the shrub layer through intensive animal browsing, directly affects woodland birds by reducing food availability (invertebrates and fruits), depleting nesting sites and exposing the birds to predators.

10.2.3 Invertebrates

The warm microclimates within early to mid-successional stages of coppice regrowth are literally hotspots for butterflies and many other insect species. The plant diversity of these coppice stages means that they also tend to be rich in nectar sources (Greatorex-Davies et al., 1993; Warren and Fuller, 1993; Sparks et al., 1996). In the Alsacian Hardt of the Upper Rhine Rift Valley, openness and nectar abundance were the best predictors of butterfly species richness and diversity, followed closely by host plant cover (Fartmann et al., 2013). Similarly, cyclical cutting of invading scrub to prevent wildfires in the *Quercus suber* wood-pastures (*montado*) of southern Portugal boosted butterfly populations in a period of up to 5 years, corresponding to increased cover of nectar-producing annual herbs. However, the populations returned to pre-disturbance levels after 10–20 years (Verdasca et al., 2012).

Most species that thrive in the early successional stages tend to be generalists, but some species of conservation concern may also use young coppice areas. In the UK, the heath fritillary (*Melitaea athalia*) and the pearl-bordered fritillary (*Boloria euphrosyne*) butterflies are both strongly associated with young coppice where their food plants thrive (the common cow-wheat, *Melampyrum pratense*, and *Viola* spp., respectively); these butterflies have declined as the area of worked coppice shrank (Hodgson et al., 2009).

Many nocturnal macro- and micro-moths depend less on the warm and sunlit conditions associated with young coppice and rides, and rely more on muscular energy for their thermoregulation. In southern England, Merckx et al. (2012) found more macro-moths in over-mature coppice and high forest than in younger coppice or wide rides. A number of formerly common, but now severely declining species, favoured more open conditions but, equally, Broome et al. (2011) identified scarce and threatened species that depended on old neglected coppice. A range of age classes in adjacent coupes, including over-stood areas, is thus desirable to cater for the widest range of species.

Other insect species, including sawflies, beetles, sap-sucking aphids and bugs feed on ground flora in cleared forest gaps, and on young shrub growth and tree seedlings in which the leaf chemistry may be more palatable than older foliage. In rides, Greatorex-Davies et al. (1994) found a greater species richness and abundance of leaf feeders than in shadier areas. The abundance of flowering plants in young coppice, such as bramble, blackthorn (*Prunus spinosa*), hawthorn (*Crataegus monogyna*), elder (*Sambucus nigra*), umbellifers and composites, supported insect pollinators. Sources of pollen and nectar are also important for the adults of many saproxylic species.

Increased plant cover in the early successional stages favours ground predators, such as carabid beetles (du Bus de Warnaffe and Lebrun, 2004) and web-spinning spiders. In the south-east of the Czech Republic, sparse stands of former ancient coppice in Milovicky Wood, singled and thinned, hosted more species of carabids associated with undisturbed habitats than dense, unthinned stands of over-mature (>80-year-old) coppice. Arachnids were less abundant in the dense stands and the species richness of isopods and myriapods was also lower (Spitzer et al., 2008). Many of the fauna sampled in these open woods were thermophilous steppe specialists of conservation concern. A tiered forest structure adds a further dimension to insect diversity; in northern Bavaria, Dolek et al. (2009) showed that standard oak trees in coppice woodland sustained greater numbers of arboreal ants than similar trees either in high-canopy forest or abandoned coppice.

Not all species, however, benefit from the recurrent disturbances and open conditions created by coppicing. Numbers of leaf-mining Lepidoptera and web-building spiders generally increased on old hazel (*Corylus avellana*) stools (Sterling and Hambler, 1988), and leaf-litter feeders may be subject to increased risk of desiccation after coppice cuts. In fragmented coppice forests in south-west France, collembolan communities were dominated by generalist forest species rather than by specialists associated with dense, larger forests. The fauna was more homogenous in coppice-with-standards woods than in clear-cut stands, presumably because the standard trees provided a wider variation in humidity (Lauga-Reyrel and Deconchat, 1999).

10.2.4 Dead wood and associated species

Dead wood is in short supply in regularly worked coppices, due both to the harvesting frequency and the relative immunity of low forest canopies to natural disturbances by windstorms. In southern Europe, the accumulated wood may also be consumed by wildfires. Estimates of fallen dead wood in managed British coppices range from 1 to 12 m^3 ha^{-1} compared with 10–60 m^3 ha^{-1} in neglected coppice, the latter overlapping the range of 50–150 m^3 ha^{-1} that is associated with temperate old-growth broadleaved forests (Hodge and Peterken, 1998; Kirby et al., 1998).

In coppice-with-standards (*Quercus* and *Carpinus*) forests in northern France, a lapse in the normal rotation of 20–25 years more than doubled the amount of woody debris in stands older than 60 years which, in turn, supported greater numbers of saproxylic beetle species (Lassauce et al., 2012a). These older coppices were characterized by self-thinning triggered by the natural senescence of pioneer species such as birch (*Betula* spp.) and aspen (*Populus tremula*), and the beginnings of windthrow disturbances.

For coppices managed on rotation (as also in intensively managed high forests) the lack of dead wood reduces the richness and diversity of fungi, bryophytes, saproxylic invertebrates and lichens, while also limiting the numbers of cavities for nesting birds and bats (Stokland et al., 2012). Some habitats are provided by mature standard trees, ancient pollards and old coppice stools, but these are likely to be less extensive than in areas managed as wood-pastures (Hartel et al., Chapter 5; Siitonen and Ranius, Chapter 11). Most of the fallen dead wood in worked coppice is also likely to be of small diameter, although fine logging debris, ranging from 2.5 to 7.5 cm in diameter, may still support some of the saproxylic beetles that use larger material. Lassauce et al. (2012b) found that 75% of the beetle species were shared with material 7.5–12.5 cm in diameter, and that tree species diversity and the degree of canopy closure were the main drivers of these assemblages.

10.2.5 Mammals

The short rotations mean that the proportion of coppice woods in temporarily open or partly shaded conditions can be 20% or more, compared with broadleaved high forest silviculture where the gap phases rarely amount to more than 5%. These open conditions, dispersed over the woodland as a whole, provide good habitat for colonizing voles, mice and shrews up to the point of canopy closure. In a Mediterranean coppice dominated by *Q. cerris* and other deciduous oaks, the numbers of wood mice (*Apodemus sylvaticus*) and shrews (*Crocidura* spp.) were inversely correlated with stand age (Capizzi and Luiselli, 1996). A similar abundance of small mammals has been found in the younger growth stages of temperate and boreal woodlands (Gurnell et al., 1992; Ecke et al., 2002). Short-rotation coppice (SRC) crops of willow and poplar also provide suitable conditions for small mammal generalists in agricultural surroundings, as long as the weedy herbaceous layers are retained at ground level (Riffell et al., 2011; Campbell et al., 2012).

The diversity of small mammals in coppice increases with greater densities of herbaceous and understorey cover, with the shelter and refuge points provided by coppice stools, fallen timber, a varied micro-topography and ecotones between stands of contrasting age. *Microtus* field voles and *Sorex* shrews favour unshaded grassy patches and herbaceous cover in rides, glades and recently cleared coupes; bank voles (*Myodes glareolus*) tend to prefer less open stages with bramble thickets, shrubs bearing fleshy fruits and tree seed shed from an overhead canopy (Flowerdew and Ellwood, 2001; Miklós and Žiak, 2002; Bush et al., 2012). Wood mice are adaptable in their habitat requirements and are present throughout the coppice cycle, but may be outcompeted under mature stands by the sympatric yellow-necked mouse (*A. flavicollis*) (Capizzi and Luiselli, 1996; Marsh and Harris, 2000; Miklós and Ziak, 2002).

The hazel dormouse (*Muscardinus avellanarius*) is strongly arboreal and favoured by coppice cycles, provided that these are long enough to allow good flowering, fruiting and seeding opportunities. Its habitat requirements include a relatively unshaded understorey, a high diversity of shrubs and climbers providing a continuity of food items throughout the season, and a scattering of canopy trees providing catkins, seeds, caterpillars and nesting places (Bright and Morris, 1990; Bright *et al.*, 2006; Juškaitis, 2007, 2008; Panchetti *et al.*, 2007; Mortelliti *et al.*, 2009, 2011; Trout *et al.*, 2012). Commercial coppice stands may be less suitable for this dormouse; for example, chestnut coppice tends to support few other shrubs and its vertical branching habit provides limited aerial routeways, whereas cutting in large coupes creates barriers to dispersing animals.

The edible or fat dormouse (*Glis glis*) is also a pulsed resource consumer, relying on mast fruiting, particularly of beech and oak, to successfully reproduce and to sustain it through winter hibernation. This is more a species of high forest, but it can maintain itself in suboptimal scrub and coppice habitats, sometimes skipping reproduction in poor seed years (Beiber and Ruf, 2009; Lebl *et al.*, 2010). The garden dormouse (*Eliomys quercinus*), is less arboreal than either of the other two species, but also tends to forage under shrubby layers and in young understorey thickets (Bertolino, 2007), such as are found in coppice.

The red and the (introduced) grey squirrels (*Sciurus vulgaris* and *S. carolinensis*) thrive in mature forests, but in coppice benefit from the canopies of moderate densities of standard trees and their seed crops. They also forage in understorey layers for nuts, berries and insects (Gurnell *et al.*, 1992).

The abundance of small mammals in young coppice stands attracts predators such as the red fox (*Vulpes vulpes*), weasel (*Mustela nivalis*), stoat (*M. erminea*), polecat (*M. putorius*) and pine martin (*Martes martes*). The open canopies following coppicing also enable owls (*Strix aluco* and *Tyto alba*), hawks and snakes to hunt for prey (Flowerdew and Ellwood, 2001).

Several bats, such as Bechstein's bat (*Myotis bechsteinii*) and the barbastelle bat (*Barbastella barbastellus*) – the latter a specialist moth predator – are capable of foraging within relatively dense understoreys, such as in coppice. Others, such as the greater and lesser horseshoe bats (*Rhinolophus ferrumequinum* and *R. hipposideros*), Leisler's bat (*Nyctalus leiseri*) and the noctule (*N. noctula*) patrol more open woodland margins and glades. Roosting sites are provided by cracks, holes and crevices in older trees, which may become more available in overstood coppice (Waters *et al.*, 1999; Schofield and Fitzsimmons 2004; Mackie and Racey, 2007).

10.3 Impacts of Deer Browsing on Flora and Fauna in Coppice

An increasing issue for those wishing to maintain or restore coppice is the increase in deer populations that has occurred across much of Europe. Deer populations may benefit from the mosaic of conditions provided by worked coppice stands, as the open areas provide good feeding conditions, while the adjacent dense pole stands provide cover for lying up. Increased browsing by wild deer damages the economic potential of the coppice, but affects also the ground flora, tree seedlings, shrubs and climbers. Frequently, increases in deer numbers have coincided with neglect of coppice and may encourage its conversion to less vulnerable high forest stands.

Young coppice shoots are particularly susceptible to damage from browsing, barking and rubbing. At moderate browsing levels, this damage may be more apparent than real, as the shoot densities developing on young coppice stools heavily self-thin, and thus provide a surplus for consumption. However, with increasing pressure, field layers tend to become grassier and shrub layers heavily reduced, delaying canopy closure. The browsing preferences of deer species can alter the balance of trees and shrubs within coppices and promote less palatable species in the field layer, such as the grass *Brachypodium sylvaticum* and bracken (*Pteridium aquilinum*). Browsing and changing light levels interact, making it difficult to predict the likely level of change at any one site (Joys *et al.*, 2004).

Intensive grazing can deplete soil seed banks as a result of losses of cover and the seed rain from understorey vegetation. Chaideftou *et al.* (2011) noted an almost threefold reduction in seed density in the upper 5 cm soil profiles of a heavily overgrazed game park compared with sporadically grazed mixed oak coppice (mainly *Q. frainetto*, *Q. pubescens* and *Q. cerris*) in north-west Greece. Typical forest herbs were significantly affected, and less palatable, ruderal species dominated the seed bank. The elimination of shrub layers by roe (*Capreolus capreolus*) and muntjac (*Muntiacus reevesi*) deer populations in young coppices has a direct impact on songbirds, and particularly migrant warblers that depend on low woody and herbaceous growth for foraging and nesting sites (Ballon and Hamard, 2007; Gill and Fuller, 2007; Holt *et al.*, 2010, 2011).

Also likely to be affected by high deer browsing in coppice are Lepidoptera, Heteroptera and other phytophagous invertebrates that utilize the vegetation in open woods and rides, although bare ground created at high grazing intensities may benefit solitary bees and wasps (Stewart, 2001). In Milovicky Wood, in an area heavily grazed by red (*Cervus elaphus*) and fallow (*Dama dama*) deer (0.5 animals ha^{-1}), several arachnids, including some of conservation concern, as well as isopods and diplopods, were less abundant among the poorly developed herbaceous and shrub layers than in stands with greater structural variety owing to depletion of the plant litter (Spitzer *et al.*, 2008).

Deer browsing and grazing directly influences small mammal populations by the removal of their food sources – herbaceous and woody vegetation, flowers, seeds and fruits, together with its insect biomass. There is also an indirect effect as the modified vegetation structure may lead to reduced cover and shelter, making the small mammals more vulnerable to predation (Flowerdew and Ellwood, 2001). At Wytham Woods in central England, following a severe deer culling programme, bramble cover increased and the vole population recovered after 7 years, showing that, with appropriate management, small mammal populations could be quickly restored (Bush *et al.*, 2012). Similarly in the Hoge Veluwe National Park in the Netherlands, excluding red deer, roe deer and mouflon (*Ovis musimon*) from previously heavily grazed pine woodland and heathland caused a rapid increase in small rodent densities (Smit *et al.*, 2001).

10.4 Conservation Strategies

Many species do thrive within woods managed under coppice regimes, but equally there are those that may benefit more from conversion to high forest in one of its many forms (Savill, Chapter 7; Bürgi, Chapter 8), or to wood-pasture (Hartel *et al.*, Chapter 5). Priorities need to be set for sites across a landscape according to the desired balance of species.

This also applies to decisions about the type of coppice treatment to adopt where this has been identified as the most appropriate management. The requirements of particular taxa differ; for example, the cutting of new coupes adjacent to old stands and the long (15–20 year) rotations recommended for dormice (Bright *et al.*, 2006) are less desirable for some sedentary fritillary butterflies, which benefit from adjacent coupes (Fuller and Warren, 1990), or for nightingales and other migrant warblers that prefer shorter rotations (Henderson and Bayes, 1989). Coupe size and distribution, and the density of standard trees, need to be chosen according to the priorities of the site, but also to take account of species sensitivities to habitat loss and fragmentation at the landscape scale.

Metapopulations of species that are of conservation concern might be more efficiently maintained by targeting the coppicing effort within their centralized zone of distribution, rather than at the periphery; but, at the same time, there is a risk in concentrating the effort in too few locations. Generalist species of young woodland might be encouraged by increasing connectivity between patches, including the strategic placement of SRC in the wider landscape.

10.5 Short-rotation Coppice

SRC crops can act as surrogates of early successional woodland and scrub habitats, they tend to be more species rich in plants, birds, butterflies and small mammals than arable crops or improved grassland, although less diverse than long-established deciduous woodland (Sage *et al.*, 2006; Dauber *et al.*, 2010; Campbell *et al.*, 2012). The benefits are greater, therefore, if they are sited on land previously intensively used for agriculture, as cultivation will be less frequent and there are fewer applications of chemicals (Dimitriou *et al.*, 2009).

Much depends on the regional context, such as whether the crops are introduced into primarily agricultural or forested landscapes (Baum *et al.*, 2012), the proportion of the area planted, and the size and juxtaposition of individual blocks. Optimizing headlands and edge effects, maintaining multiple age classes, using polyclonal mixtures and promoting herbaceous layers below the crop are all likely to increase diversity. However, there could be negative effects if the pressure to grow other biofuel crops and more cereals means the displacement of SRC crops on to land with a greater existing biodiversity value, including existing coppice woodland (Verkerk *et al.*, 2014).

10.6 Conclusion

Ultimately, the extent of future coppice will be driven by markets, including new ones such as for biofuels. The value of coppice wildlife – as part of an ecosystem service – can be incorporated in these calculations but will probably only make a marginal difference to the total area under coppice. If we wish these distinctive plant and animal assemblages to survive, we need to address three aspects simultaneously:

- identify the core areas where maintaining coppice, if necessary purely for conservation reasons, is essential;
- ensure that the new ways in which coppice is treated (including as biofuel) allow its associated wildlife to survive; and
- incorporate elements of the coppice system (open space, dense young growth, edge habitats) into other forms of woodland management.

The success of these approaches needs to be monitored and the results fed back into future management strategies at stand, wood and landscape scales.

References

Amar, A., Smith, K.W., Butler, S., Lindsell, J.A., Hewson, C.M., Fuller, R.J. and Charman, E.C. (2010) Recent patterns of change in vegetation structure and tree composition of British broadleaved woodland: evidence from large-scale surveys. *Forestry* 83, 345–356.

Ash, J.E. and Barkham, J.P. (1976) Changes and variability in the field layer of a coppiced woodland in Norfolk, England. *Journal of Ecology* 64, 697–712.

Baeten, L., Bauwens, B., De Schrijver, A., De Keersmaeker, L., Van Calster, H., Vandekerkhove, K., Roelandt, B., Beeckman, H. and Verheyen, K. (2009) Herb layer changes (1954–2000) related to the conversion of coppice-with-standards forest and soil acidification. *Applied Vegetation Science* 12, 187–197.

Ballon, P. and Hamard, J.-P. (2007) *Recherche de l'Équilibre Forêt – Cervidés dans le Massif du Cosson*. Unité de Recherche Ecosystèmes Forestiers, Centre National du Machinisme Agricole, du Genie Rural, des Eaux et des Forêts, Nogent-sur-Vernisson, France.

Bardat, J. and Aubert, M. (2007) Impact of forest management on the diversity of corticolous bryophyte assemblages in temperate forests. *Biological Conservation* 139, 47–66.

Baum, S., Bolte, A. and Weih, M. (2012) High value of short rotation coppice plantations for phytodiversity in rural landscapes. *GCB Bioenergy* 4, 728–738.

Beiber, C. and Ruf, T. (2009) Habitat differences affect life history tactics of a pulsed resource consumer, the edible dormouse (*Glis glis*). *Population Ecology* 51, 481–492.

Bertolino, S. (2007) Microhabitat use by garden dormice during nocturnal activity. *Journal of Zoology* 272, 176–182.

Bright, P.W. and Morris, P.A. (1990) Habitat requirements of the dormice *Muscardinus avellanarius* in relation to woodland management in southwest England. *Biological Conservation* 54, 307–326.

Bright, P., Morris, P. and Mitchell-Jones, T. (2006) *The Dormouse Conservation Handbook*, 2nd edn. English Nature, Peterborough, UK.

Broome, A., Clarke, S., Peace, A. and Parsons, M. (2011) The effect of coppice management on moth assemblages in an English woodland. *Biodiversity Conservation* 20, 729–749.

Brown, A.H.F. and Warr, S.J. (1992) The effects of changing management on seed banks in ancient coppices. In: Buckley, G.P. (ed.) *Ecology and Management of Coppice Woodlands*. Chapman and Hall, London, pp. 147–166.

Buckley, G.P., Howell, R. and Anderson, M.A. (1997) Vegetation succession following ride edge management in lowland plantations and woods. The seed bank resource. *Biological Conservation* 82, 305–316.

Bush, E.R., Buesching, C.D., Slade, E.M. and Macdonald, D.W. (2012) Woodland recovery after suppression of deer: cascade effects for small mammals, wood mice (*Apodemus sylvaticus*) and bank voles (*Myodes glareolus*). *PLoS ONE* 7(2): e31404.

Campbell, S.P., Frair, J.L., Gibbs, J.P. and Volk, T.A. (2012) Use of short-rotation coppice willow by birds and small mammals in central New York. *Biomass and Bioenergy* 47, 342–353.

Camprodon, J. and Brotons, L. (2006) Effects of undergrowth clearing on the bird communities of the north-western Mediterranean coppice holm oak forests. *Forest Ecology and Management* 221, 72–82.

Capizzi, D. and Luiselli, L. (1996) Ecological relationships between small mammals and age of coppice in an oak-mixed forest in central Italy. *Revue d'Écologie* 51, 277–291.

Chaideftou, E., Thanos, C.A., Bergmeier, E., Kallimanis, A.S. and Dimopoulos, P. (2011) The herb layer restoration potential of the soil seed bank in an overgrazed oak forest. *Journal of Biological Research – Thessaloniki* 15, 17 57.

Dauber, J., Jones, M.B. and Stout, J.C. (2010) The impact of biomass crop cultivation on temperate biodiversity. *GCB Bioenergy* 2, 289–309.

Debussche, M., Debussche, G. and Lepart, J. (2001) Changes in the vegetation of *Quercus pubescens* woodland after cessation of coppicing and grazing. *Journal of Vegetation Science* 12, 81–92.

Decocq, G., Aubert, M., Dupont, F., Alard, D., Saguez, R., Wattez-Franger, A., de Foucault, B., Delelis-Dusollier, A. and Bardat, J. (2004) Plant diversity in a managed temperate deciduous forest: understorey response to two silvicultural systems. *Journal of Applied Ecology* 41, 1065–1079.

Dimitriou, I., Baum, C., Baum, S., Busch, G., Schulz, U., Köhn, J., Lamersdorf, N., Leinweber, P., Aronsson, P., Weih, M. *et al.* (2009) The impact of short rotation coppice (SRC) cultivation on the environment. *Agriculture and Forestry Research* 3, 159–162.

Dolek, M., Freese-Hager, A., Bussler, H., Floren, A., Liegl, A. and Schmidl, J. (2009) Ants on oaks: effects of forest structure on species composition. *Journal of Insect Conservation* 13, 367–375.

Donald, P.F., Fuller, R.J., Evans, A.D. and Gough, S.J. (1998) Effects of forest management and grazing on breeding bird communities in plantations of broadleaved and coniferous trees in western England. *Biological Conservation* 85, 183–197.

du Bus de Warnaffe, G. and Lebrun, P. (2004) Effects of forest management on carabid beetles in Belgium: implications for biodiversity conservation. *Biological Conservation* 118, 219–234.

Ecke, F., Löfgren, O. and Sörlin, D. (2002) Population dynamics of small mammals in relation to forest age and structural habitat factors in northern Sweden. *Journal of Applied Ecology* 39, 781–792.

Fartmann, T., Muller, C. and Poniatowski, D. (2013) Effects of coppicing on butterfly communities of woodlands. *Biological Conservation* 159, 396–404.

Flowerdew, J.R. and Ellwood, S.A. (2001) Impacts of woodland deer on small mammal ecology. *Forestry* 74, 277–287.

Ford, E.D. and Newbould, P.J. (1977) The biomass and production of ground vegetation and its relation to tree cover through a deciduous woodland cycle. *Journal of Ecology* 65, 201–212.

Fuller, R.J. (ed.) (2012) *Birds and Habitat: Relationships in Changing Landscapes*. Cambridge University Press, Cambridge, UK.

Fuller, R.J. and Moreton, B.D. (1987) Breeding bird populations of Kentish sweet chestnut (*Castanea sativa*) coppice in relation to age and structure of the coppice. *Journal of Applied Ecology* 24, 13–27.

Fuller, R.J. and Warren, M.S. (1990) *Coppiced Woodlands: Their Management for Wildlife*. Nature Conservancy Council, Peterborough, UK.

Gill, R.M.A. and Fuller, R.J. (2007) The effects of deer browsing on woodland structure and songbirds in lowland Britain. *Ibis* 149, 119–127.

Greatorex-Davies, J.N., Sparks, T.H., Hall, M.L. and Marrs, R.H. (1993) The influence of shade on butterflies in rides of coniferized lowland woods in southern England and implications for conservation management. *Biological Conservation* 63, 31–41.

Greatorex-Davies, J.N., Sparks, T.H. and Hall, M.L. (1994) The response of Heteroptera and Coleoptera species to shade and aspect in rides of coniferized lowland woods in southern England. *Biological Conservation* 67, 255–273.

Gurnell, J., Hicks, M. and Whitbread, S. (1992) The effects of coppice management on small mammal populations. In: Buckley, G.P. (ed.) *Ecology and Management of Coppice Woodlands*. Chapman and Hall, London, pp. 213–232.

Henderson, A. and Bayes, K. (1989) *Conservation Advice: Nightingales and Coppice Woodland*. Royal Society for the Protection of Birds, Sandy, UK.

Hodge, S.J. and Peterken, G.F. (1998) Deadwood in British forests: priorities and a strategy. *Forestry* 71, 99–112.

Hodgson, J.A., Moilanen, A., Bourn, N.A.D., Bulman, C.R. and Thomas, C.D. (2009) Managing successional species: modelling the dependence of heath fritillary populations on the spatial distribution of woodland management. *Biological Conservation* 142, 2743–2751.

Hölscher, D., Schade, E. and Leuschner, C. (2001) Effects of coppicing in temperate deciduous forests on ecosystem nutrient pools and soil fertility. *Basic and Applied Ecology* 2, 155–164.

Holt, C.A., Fuller, R.J. and Dolan, P.M. (2010) Experimental evidence that deer browsing reduces habitat suitability for breeding common nightingales *Luscinia megarhynchos*. *Ibis* 152, 335–346.

Holt, C.A., Fuller, R.J. and Dolan, P.M. (2011) Breeding and post-breeding responses of woodland birds to modification of habitat structure by deer. *Biological Conservation* 144, 2151–2162.

Itô, H., Hino, T. and Sakuma, D. (2012) Species abundance in floor vegetation of managed coppice and abandoned forest. *Forest Ecology and Management* 269, 99–105.

Joys, A.C., Fuller, R.J. and Dolman, P.M. (2004) Influences of deer browsing, coppice history, and standard trees on the growth and development of vegetation structure in coppiced woods in lowland England. *Forest Ecology and Management* 202, 23–37.

Juškaitis, R. (2007) Habitat selection in the common dormouse *Muscardinus avellanarius* (L.) in Lithuania. *Baltic Forestry* 13, 89–95.

Juškaitis, R. (2008) Long-term common dormouse monitoring: effects of forest management. *Biodiversity Conservation* 17, 3559–3565.

Kirby, K.J., Reid, C.M., Thomas, R.C. and Goldsmith, F.B. (1998) Preliminary estimates of fallen dead wood and standing dead trees in managed and unmanaged forests in Britain. *Journal of Applied Ecology* 35, 148–155.

Kirby, K.J., Smart, S.M., Black, H.I.J., Bunce, R.G.H., Corney, P.M. and Smithers, R.J. (2005) *Long Term Ecological Change in British Woodlands (1971–2001)*. English Nature Research Reports Number 653, English Nature Peterborough, UK.

Kopecký, M., Hédl, R. and Szabó, P. (2013) Non-random extinctions dominate plant community changes in abandoned coppices. *Journal of Applied Ecology* 50, 79–87.

Lassauce, A., Anselle, P., Lieutier, F. and Bouget, C. (2012a) Coppice-with-standards with an overmature coppice component enhances saproxylic beetle biodiversity: a case study in French deciduous forests. *Forest Ecology and Management* 266, 273–285.

Lassauce, A., Lieutier, F. and Bouget, C. (2012b) Woodfuel harvesting and biodiversity conservation in temperate forests: effects of logging residue characteristics on saproxylic beetle assemblages. *Biological Conservation* 147, 204–212.

Lauga-Reyrel, F. and Deconchat, M. (1999) Diversity within the Collembola community in fragmented coppice forests in south-western France. *European Journal of Soil Biology* 35, 177–187.

Lebl, K., Kurbisch, K., Bieber, C. and Ruf, T. (2010) Energy or information? The role of seed availability for reproductive decisions in edible dormice. *Journal of Comparative Physiology B: Biochemical, Systems, and Environmental Physiology* 180, 447–456.

Mackie, I.J. and Racey, P.A. (2007) Habitat use varies with reproductive state in noctule bats (*Nyctalus noctula*): implications for conservation. *Biological Conservation* 140, 70–77.

Marsh, A.C.W. and Harris, S. (2000) Partitioning of woodland habitat resources by two sympatric species of *Apodemus*: lessons for the conservation of the yellow-necked mouse (*A. flavicollis*) in Britain. *Biological Conservation* 92, 275–283.

Mason, C.F. and MacDonald, S.M. (2002) Responses of ground flora to coppice management in an English woodland – a study using permanent quadrats. *Biodiversity and Conservation* 11, 1773–1789.

Merckx, T., Feber, R.E., Hoare, D.J., Parsons, M.S., Kelly, C.J., Bourn, N.A.D. and Macdonald, D.W. (2012) Conserving threatened Lepidoptera: towards an effective woodland management policy in landscapes under intense human land-use. *Biological Conservation* 149, 32–39.

Miklós, P. and Žiak, D. (2002) Microhabitat selection by three small mammal species in oak–elm forest. *Folia Zoologica* 51, 275–288.

Mitchell, P. (1992) Growth stages and microclimate in coppice and high forest. In: Buckley, G.P. (ed.) *Ecology and Management of Coppice Woodlands*. Chapman and Hall, London, pp. 31–51.

Mortelliti, A., Sanzo, G.S. and Boitani, L. (2009) Species surrogacy for conservation planning: three arboreal rodents' responses to habitat loss and fragmentation. *Biodiversity Conservation* 18, 1131–1145.

Mortelliti, A., Amori, G., Capizzi, D., Cervone, C., Fagiani, S., Pollini, B. and Boitani, L. (2011) Independent effects of habitat loss, habitat fragmentation and structural connectivity on the distribution of two arboreal rodents. *Journal of Applied Ecology* 48, 153–162.

Panchetti, F., Sorace, A., Amori, G. and Carpaneto, G.M. (2007) Nest site preference of common dormouse (*Muscardinus avellanarius*) in two different habitat types of central Italy. *Italian Journal of Zoology* 74, 363–369.

Rackham, O. (1975) *Hayley Wood, Its History and Ecology*. Cambridgeshire and Isle of Ely Naturalists' Trust, Cambridge, UK.

Riffell, S., Verschuyl, J., Miller, D. and Wigley, T.B. (2011) Meta-analysis of bird and mammal response to short-rotation woody crops. *Global Change Biology* 3, 313–321.

Sage, R., Cunningham, M. and Boatman, N. (2006) Birds in willow short-rotation coppice compared to other arable crops in central England and a review of bird census data from energy crops in the UK. *Ibis* 148, 184–197.

Schofield, H. and Fitzsimmons, P. (2004) The importance of woodlands for bats. In: Quine, C.P., Trout, R.C. and Shore, R.F. (eds) *Managing Woodlands and Their Mammals*. Forestry Commission, Edinburgh, UK, pp. 41–42.

Smit, R., Bokdam, J., den Ouden, J., Olff, H., Schot-Opschoor, H. and Schrijvers, M. (2001) Effects of introduction and exclusion of large herbivores on small rodent communities. *Plant Ecology* 155, 119–127.

Sparks, T.H., Greatorex-Davies, J.N., Mountford, J.O., Hall, M.L. and Marrs, R.H. (1996) The effects of shade on the plant communities of rides in plantation woodland and implications for butterfly conservation. *Forest Ecology and Management* 80, 197–207.

Spitzer, L., Konvicka, M., Benes, J., Tropek, R., Tuf, I.H. and Tufova, J. (2008) Does closure of traditionally managed open woodlands threaten epigeic invertebrates? Effects of coppicing and high deer densities. *Biological Conservation* 141, 827–837.

Sterling, P.H. and Hambler, C. (1988) Coppicing for conservation: do hazel communities benefit? In: Kirby K.J. and Wright, F.J. (eds) *Woodland Conservation in the Clay Vale of Oxfordshire and Buckinghamshire*. Nature Conservancy Council, Peterborough, UK, pp. 69–80.

Stewart, A.J.A. (2001) The impact of deer on lowland woodland invertebrates: a review of the evidence and priorities for future research. *Forestry* 74, 259–270.

Stokland, J.N., Siitonen, J. and Jonsson, B.G. (eds) (2012) *Biodiversity in Dead Wood*, 1st edn. Cambridge University Press, Cambridge, UK.

Strandberg, B., Kristiansen, S.M. and Tybirk, K. (2005) Dynamic oak scrub to forest succession: effects of management on understorey vegetation, humus forms and soils. *Forest Ecology and Management* 211, 318–328.

Trout, R.C., Brooks, S.E., Rudlin, P. and Neil, J. (2012) Effects of restoring a conifer plantation on an ancient woodland site (PAWS) in the UK on the habitat and local population of the hazel dormouse. *European Journal of Wildlife Research* 58, 635–643.

Van Calster, H., Baeten, L., De Schrijver, A., De Keersmaeker, L., Rogister, J.E., Verheyen, K. and Hermy, M. (2007) Management driven changes (1967–2005) in soil acidity and the understorey plant community following conversion of a coppice-with-standards forest. *Forest Ecology and Management* 241, 258–271.

Van Calster, H., Baeten, L., Verheyen, K., De Keersmaeker, L., Dekeyser, S., Rogister, J.E. and Hermy, M. (2008a) Diverging effects of overstorey conversion scenarios on the understorey vegetation in a former coppice-with-standards forest. *Forest Ecology and Management* 256, 519–528.

Van Calster, H., Chevalier, R., Van Wyngene, B., Archaux, F., Verheyen, K. and Hermy, M. (2008b) Long-term seed bank dynamics in a temperate forest under conversion from coppice-with-standards to high forest management. *Applied Vegetation Science* 11, 251–260.

Verdasca, M.J., Leitão, A.S., Santana, J., Porto, M., Dias, S. and Beja, P. (2012) Forest fuel management as a conservation tool for early successional species under agricultural abandonment: the case of Mediterranean butterflies. *Biological Conservation* 146, 14–23.

Verheyen, K., Baeten, L., De Frenne, P., Bernhardt-Römermann, M., Brunet, J., Cornelis, J., Decocq, G., Dierschke, H., Eriksson, O., Hedl, R. *et al.* (2012) Driving factors behind the eutrophication signal in understorey plant communities of deciduous temperate forests. *Journal of Ecology* 100, 352–365.

Verkerk, P., Zanchi, G. and Lindner, M. (2014) Trade-offs between forest protection and wood supply in Europe. *Environmental Management* 53, 1085–1094.

Warren, M.S. and Fuller, R.J. (1993) *Woodland Rides and Glades: Their Management for Wildlife*. Joint Nature Conservation Committee, Peterborough, UK/British Trust for Ornithology, Thetford, UK.

Waters D., Jones, G. and Furlong M. (1999) Foraging ecology of Leisler's bat (*Nyctalus leisleri*) at two sites in southern. *Britain Journal of Zoology* 249, 173–180.

11 The Importance of Veteran Trees for Saproxylic Insects

Juha Siitonen[1]* and Thomas Ranius[2]

[1]*Natural Resources Institute Finland, Vantaa, Finland;* [2]*Department of Ecology, Swedish University of Agricultural Sciences, Uppsala, Sweden*

11.1 Introduction

Old trees – often referred to as ancient or veteran – have always attracted attention, but recently there has been a revival of interest in them from an ecological and conservation perspective. Ancient trees are old individuals that have clearly passed beyond maturity and often show features such as cavities or hollow trunks, bark loss over sections of the trunk and a large quantity of dead wood in the canopy. The term 'veteran tree' includes younger individuals that have developed similar characteristics as a result of adverse growing conditions or injury (Woodland Trust, 2008; Lonsdale, 2013). Veteran trees are defined as being of interest biologically, culturally or aesthetically because of their age, size or condition (Read, 2000).

A large old tree has been described as an arboreal megalopolis for saproxylic species (Speight, 1989). We therefore address the following questions:

- What is it about these trees that is important – in terms of their structure and where they occur – for invertebrates?
- Which invertebrate taxa inhabit these trees?
- How are invertebrate species occurrence patterns and long-term persistence affected by the abundance, quality and spatial distribution of old trees?
- What practices should be encouraged to further the conservation of saproxylic species associated with old trees?

11.2 What Are Saproxylic Species?

Speight (1989) defined saproxylic species as those that depend, during some part of their life cycle, upon the dead or dying wood of moribund or dead trees (standing or fallen), upon wood-inhabiting fungi, or upon the presence of other saproxylic species. However, dead wood occurs regularly in mature living trees, and heartwood decay and hollowing is a normal part of the life cycle of almost all long-lived deciduous trees (Alexander, 2008). Hollow trees are not necessarily moribund. So the ecological definition of saproxylic species should also include species that depend, during some part of their life cycles, upon wounded or decaying woody material from living trees (Stokland *et al.*, 2012).

Studies of sub-fossil insect remains (Buckland and Dinnin, 1993; Whitehouse, 2006; Olsson and Lemdahl, 2009) suggest that many saproxylic species that are now rare or

*E-mail: juha.siitonen@metla.fi

regionally extinct were common across the landscape in the pre-Neolithic period. Some of the first species to disappear, such as *Rhysodes sulcatus* and *Prostomis mandibularis*, are inhabitants of old-growth forests with minimal human influence. Others have survived in the cultural habitats created by humans, in particular in wood-pastures (Hartel *et al.*, Chapter 5), but are now endangered or extinct regionally or across western Europe as a whole. In the *European Red List of Saproxylic Beetles* (Nieto and Alexander, 2010), 11% of about 450 species assessed were considered threatened in all of Europe, while at the European Union (EU) level, 14% were threatened, primarily through habitat loss in relation to logging and wood harvesting and the decline of veteran trees throughout the landscape.

11.3 Veteran Trees in Past and Present Landscapes

Veteran trees occur in different types of natural habitats including closed old-growth forests. Under the original natural conditions, the density of wide-crowned veteran trees would have been higher in more open woodland where there was less competition from young growth; for example, in wetland forests where only patches of higher ground are suitable for tree growth, rocky outcrops and mountain slopes, in flood-plain forests where regular disturbances keep the stands open, and in areas that are heavily grazed by large herbivores.

Agriculture and an increasing impact of humans spread into Europe starting from the Neolithic period some 8000 years ago. By about 3000 years ago, natural forests had been cleared from most cultivable lowland areas all over Europe (Kaplan *et al.*, 2009; Kirby and Watkins, Chapter 4). However, traditional land-use forms, such as grazing, and the coppicing and pollarding of trees, created new types of cultural woodland habitats that were to some degree analogous to natural habitats and allowed the survival of many saproxylic species (Siitonen, 2012b).

Of particular importance were the various forms of wood-pastures (Bergmeier *et al.*, 2010; Hartel *et al.*, Chapter 5; Plate 5). These range from forests grazed only occasionally by cattle to permanently grazed grasslands with scattered trees (Rackham, 1998), such as are often found in parks. From the medieval period onward, parks were commonly established around castles and manor houses. Many of the oldest parks were originally used for hunting, but later landscaped to create more attractive settings, sometimes incorporating veteran trees already on the site (Fletcher, Chapter 9). Post-medieval landscape parks that include older features host much richer saproxylic beetle faunas than parks lacking such continuity (Harding and Alexander, 1994; Alexander, 1998), but this richness also depends on whether the parks contain some semi-natural vegetation or are close to potential source areas. Hollow lime trees (*Tilia* spp.) in old manor parks in southern Sweden have been found to host as many specialist and red-listed saproxylic beetle species as similar trees in open wood-pastures or overgrown former wood-pastures (Jonsell, 2012). In Estonia, old manor parks have about three times as high a basal area of large (>40 cm) trees as mature forest patches in the surrounding landscape; the proportion of broadleaved trees (*Tilia* spp., *Quercus robur* and *Fraxinus excelsior*) is also about three times that found in the forests (Lõhmus and Liira, 2013).

Other cultural habitat types important for saproxylic species include orchards and avenues, the latter able to host saproxylic specialists such as the hermit beetle (*Osmoderma eremita*) (Plate 5) and the rose chafer species *Protaetia marmorata* (Oleksa *et al.*, 2006, 2013). Hedgerows form extensive networks in old agricultural landscapes (Baudry *et al.*, 2000) and the large pollarded trees in them are important habitats for saproxylic species such as the hermit beetle (Dubois *et al.*, 2009). Veteran trees may also be found along field margins, riverbanks and old roads.

11.4 Important Structures and Associated Species in Old Trees

11.4.1 Microhabitat diversity

The importance of old trees for saproxylic species lies in the large number and variety of special structures they have in comparison with younger trees. Microhabitats (distinct parts of a tree that host different species assemblages) include cavities with wood mould, water-filled rot holes, dead bark, exposed wood, sap flows, fruiting bodies and mycelia of fungi, dead branches and dead roots (Siitonen, 2012a) (Fig. 11.1).

On completely dead trees, these microhabitats gradually disappear as a result of decomposition. In temperate forests, the time required for a dead medium-sized tree to decompose until it is no longer a visible structure is only a few decades. Most saproxylic species are able to use dead trees only during a particular part of the decay succession (Stokland and Siitonen, 2012) and so the window of time during which a dead tree constitutes a suitable habitat may be only a few years to a few decades. Species feeding on fresh phloem utilize trees only during the first summer following tree death, after which they need to colonize new hosts. In contrast, dead-wood microhabitats in old living trees can last for much longer periods; cavities filled with wood mould, which start to develop in trees that have reached their maturity, may then persist for centuries within the live tree.

Tree species, growth rate, age and diameter of the tree, as well as the environment in which the tree is growing, all affect the number of microhabitats per tree (Winter and Möller, 2008; Hall and Bunce, 2011; Vuidot *et al.*, 2011; Regnery *et al.*, 2013). Wide-crowned trees growing in open conditions usually support a higher number of microhabitats than otherwise similar trees growing in closed forest.

11.4.2 Tree cavities and their invertebrates

Cavities are formed primarily by the action of fungi that decay the dead heartwood in mature living trees (heart rots). Heartwood

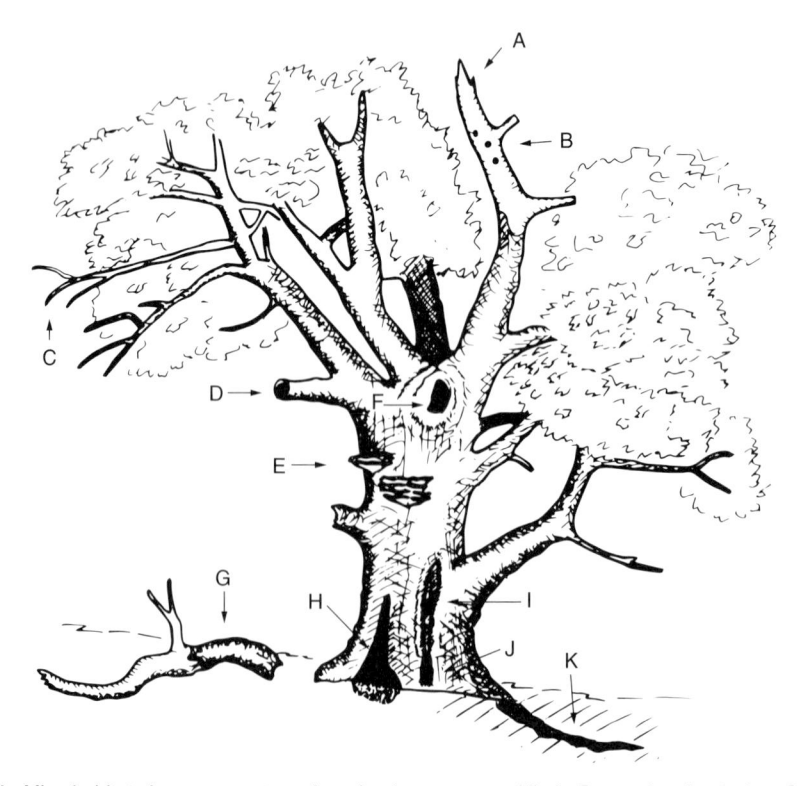

Fig. 11.1. Microhabitats in a veteran tree. A, a dead sun-exposed limb; B, woodpecker holes; C, dead attached branches in the canopy; D, a branch cavity; E, polypore fruiting bodies; F, a trunk cavity; G, a fallen branch on the ground; H, a basal cavity; I, an open wound surrounded by callus tissue; J, sap exudation; K, a dead root in the soil. (Drawing by Juha Siitonen.)

forms the inner core of the trunk and larger branches. Heart-rot fungi do not grow in the functional sapwood forming the outer wood layer, so only the inner trunk will become soft as a result of decay. Damage to, or breakage of, large branches provides saproxylic invertebrates with access to the exposed heartwood. Cavities then develop through the combined action of heart-rot fungi, invertebrates and the physical breakdown of the decaying wood.

In a study in Sweden, less than 1% of oak (*Q. robur*) trees up to 100 years old had cavities. By the age of 200–300 years, about 50% of the trees bore cavities, while trees older than 400 years were all hollow. Cavities form earlier in fast-growing trees than in slow-growing trees, probably because fast-growing trees have bigger branches that shed earlier (Ranius *et al.*, 2009). Some shorter lived broadleaved species, such as willows (*Salix* spp.), poplars (*Populus* spp.), maples (*Acer* spp.) and horse chestnut (*Aesculus hippocastanum*) start to become hollow at much younger ages. As a cavity develops and becomes larger, it becomes structurally more complex, and the diversity of the invertebrates it supports increases. Wood mould starts to accumulate in the bottom of the cavity and is the principal substrate of invertebrates living in cavities. Wood mould is a mixture of decaying wood, dead leaves and debris falling into the cavity, the residues of bird or mammal nests and droppings, and fungi and frass produced by the invertebrates themselves.

The cavity mould fauna is taxonomically very diverse, and across Europe includes several thousand species. Many are generalists that are also found in other dead-wood microhabitats, but others are specialists (Siitonen, 2012a). The most important groups are beetles (Coleoptera), gnats and flies (Diptera) and their dependent parasitic wasps (Hymenoptera). Ants (Hymenoptera: Formicidae), springtails (Collembola), mites (Acari), spiders (Araneae), harvestmen (Opiliones), pseudoscorpions (Pseudoscorpionida), woodlice (Isopoda), centipedes (Chilopoda) and millipedes (Diplopoda) are also frequently found. This species assemblage forms a complex food web of wood mould consumers and fungivores, their specialized predator and parasitoid species, and scavengers feeding on dead insects. Most beetle species associated with tree cavities seem to prefer the reddish brown, crumbly mould typical of, for example, oaks decayed by the fungus *Laetiporus sulphureus*. The moist, dark wood mould typical of several other broadleaved tree species, e.g. elms (*Ulmus* spp.), ashes (*Fraxinus* spp.) and horse chestnut, seems to be favoured by dipterans (Andersson, 1999; Alexander, 2002).

Large chafers (Scarabaeidae, Cetoniinae), such as the hermit beetle, *Protaetia* and *Gnorimus* spp., are among the most emblematic and functionally important species inhabiting cavities. Their larvae tunnel into the decaying walls of the cavity, thereby expanding it and creating large amounts of nitrogen-enriched frass (Jönsson *et al.*, 2004). Other typical beetle families found in cavities include comb-clawed beetles (Alleculidae), click beetles (Elateridae), and darkling beetles (Tenebrionidae) (Kelner-Pillault, 1974; Martin, 1989; Ranius and Jansson, 2000).

Among the specialist species, the hermit beetle is the most intensively studied and hence is commonly used here as an example. However, there are other cavity specialist species that are much rarer, which may therefore be more demanding in their habitat requirements. Even the hermit beetle may vary in its requirements; molecular genetic studies indicate that it is a complex of at least four distinct species (Audisio *et al.*, 2009): *O. eremita* occurs in western Europe up to southern Sweden in the north; *O. barnabita* is distributed through eastern Europe to central Russia and up to southern Finland; *O. cristinae* is confined to Sicily; and *O. lassallei* is found in Greece and the European parts of Turkey. The taxonomic rank of the southern Italian hermit beetles ('*O. italicum*') remains unresolved. This speciation may have occurred in temperate forest refugia in the Italian and Balkan peninsulas and Sicily before and during the Pleistocene (Kirby and Watkins, Chapter 3). Similar patterns of genetic diversity and cryptic species might be found in other saproxylic species, thus emphasizing the desirability of preserving species throughout their distribution.

11.4.3 Other microhabitats

Sap flows can also be long-lasting features of veteran trees. They may be caused either by a chronic infection of anaerobic bacteria, or by repeated damage to bark resulting from internal fractures. Many saproxylic species feed on them, but there are some that specialize on such flows, particularly among the Diptera. Dead branches host specialized fungi and the insects feeding on them. Individual branches decay and fall rapidly, but within the crown of an old tree new dead branches appear regularly. Similarly, the annual fruiting bodies of important heart-rot fungi of broadleaved trees, such as *L. sulphureus* in oaks and *Polyporus squamosus* in elms and ashes, appear and decay within a couple of months, but these fungi produce new fruiting bodies on their host tree each year for decades.

Many saproxylic species depend on the conditions created by other species. Galleries made by wood- and bark-feeding insects provide microhabitats for a wide array of associated species to the point that some, such as the great Capricorn beetle (*Cerambyx cerdo*), can be considered to be ecosystem engineers. The larval tunnels and adult exit holes of this beetle are used by other beetle species and, by weakening the tree, they open the way to other saproxylic species (Buse *et al.*, 2008). Emergence holes of wood-boring beetles, such as cerambycids and anobids, are used by solitary bees and wasps that are unable to dig out their own nesting holes. Different species select holes of different sizes, or in different positions and under different conditions, in a manner that is analogous to that of hole-nesting birds, but at a miniature scale. Social insects that build their nests in cavities provide habitats for a range of commensals, such as the large rove beetle species, *Velleius dilatatus*, which is found in hornet (*Vespa crabro*) nests.

11.5 Effects of Environmental Factors on the Invertebrate Fauna

11.5.1 Effects of tree characteristics on species assemblages

The habitat requirements of saproxylic insects have been described from field observations (Palm, 1959). Some are generalists and can occur as larvae in any deciduous tree species, but most favour one or a few tree species, and a limited number are very restricted in their host choice (Jonsell *et al.*, 1998; Stokland, 2012). All tree species host at least some specialist saproxylic species, but in northern Europe oaks (*Q. robur*, *Q. petraea*) are the most important, both in terms of total species richness, number of specialist species and number of red-listed species. Oaks tend to grow larger and older than most other tree species and therefore have more of the microhabitats associated with old trees for longer periods.

There are only a few quantitative studies on host-tree associations of saproxylic species. Milberg *et al.* (2014) compared saproxylic beetle faunas between four tree species (*Q. robur*, *Acer platanoides*, *F. excelsior* and *T. cordata*) and found that the great majority of species were not specialized to tree species. However, some beetle species clearly preferred a particular host-tree species, and more species were associated with *Q. robur* than with any of the other tree species studied.

Large trees generally harbour more saproxylic species than smaller ones. In a study of nine hollow-specialist invertebrates, six had a higher occurrence probability in a given volume of wood mould in large hollow trees (Ranius *et al.*, 2011). A large tree contains, on average, a larger number of more varied microhabitats, which may be larger and more complex because they have had a longer time to develop. Also, a large trunk has a more stable internal microclimate than a thin trunk, and a larger tree has existed for longer and provides more opportunity for species colonization. The relative importance of these factors in driving the positive tree size–species richness relationship is not known.

Exposure to sun is important for saproxylic fauna (Palm, 1959), and more saproxylic insect species are reported in trees that are exposed to sun (Ranius and Jansson, 2000; Vodka *et al.*, 2009; Koch Widerberg *et al.*, 2012; Horák and Rébl, 2013). Greater exposure to sun gives rise to higher temperatures, so less time is needed for larval development; conversely, exposure to sun makes the wood drier, which may be a limiting factor for the occurrence of species, especially in regions with a warm and dry climate. As a result, species may be more abundant in sun-exposed trees towards their northern distribution limit, but lack such a pattern further south (Chiari et al., 2012). There could be an interaction with how and where trees grow as well. For instance, beech (*Fagus sylvatica*) often occurs in more dense and shaded stands than oak, which might lead to the invertebrates associated with beech being better adapted to shaded conditions (Gärdenfors and Baranowski, 1992). Even in beech-dominated forests though, more species appear to be associated with warm than with cool sites (Lachat et al., 2012).

11.5.2 Effects of surrounding landscape on species assemblages

Trees die and so the recruitment of 'new' cohorts of veteran trees, and their colonization by saproxylic invertebrates, must take place. The probability of a species finding the right habitat in a tree and its long-term persistence at a landscape scale depend not only on the characteristics of individual trees, but also on the density of suitable trees in the surrounding landscape (Sverdrup-Thygeson et al., 2010; Götmark et al., 2011; Ranius et al., 2011; Bergman et al., 2012). At the landscape level, local extinctions from trees must be balanced by the colonization of new trees.

A metapopulation study was carried out in southern Sweden on hermit beetles living in hollow oaks in areas with oak wood-pastures that are a few kilometres apart (Ranius, 2007). The larval development of this species only occurs in the cavities of old oaks, so each tree was effectively a habitat patch that potentially harboured a local population. The probability of occurrence of a population increased with the number of oaks with cavities containing wood mould (potential dispersal sources) in the surroundings, with the strongest relationship obtained when including only trees within a radius of 200 m (Ranius et al., 2011). Dispersal was considerably more common within this distance than further away. The dispersal rate and range were assessed by capture–recapture and telemetry (Plate 5), which confirmed that most dispersal movements by the beetle took place to the nearest tree and that most individuals remained in their natal tree throughout their lifetime (Hedin et al., 2008).

In the Swedish study, the population size of hermit beetle per tree was relatively stable between years in comparison with that of many other insects. However, population size varied a lot between trees because the quality of the hollow trees differed widely, mainly as a result of differences in the volume of wood mould (Ranius, 2007). The extinction risk was higher in trees with small amounts of wood mould than in trees with high-quality cavities. Cavities usually develop in oaks when they are about 200–300 years old (Ranius et al., 2009), and after that the oaks may remain at least for another 100 years or more (Fig. 11.2); in the study area, the oldest oaks were up to 500 years old. Therefore, at many sites, species persistence relies heavily on a few hollow trees of very high quality, and the overall population is vulnerable to extinction if these trees are lost or deteriorate.

In Italy, a similar study of hermit beetles revealed dispersal distances about ten times longer than in Sweden and many more dispersal events (Chiari et al., 2013a). Thus, metapopulation dynamics may vary in different parts of the distribution of a species. This may depend on temperature differences, although a complicating factor is that the Italian hermit beetles may be a different species or subspecies: '*O. italicum*'.

Dispersal biology is a key factor for species persistence and occurrence patterns in landscapes with various levels of habitat fragmentation. Species adapted to naturally short-duration habitats are expected to have evolved a more opportunistic and wider ranging dispersal strategy than species adapted to more stable and long-duration habitats. For these, a better strategy may be to remain at the same habitat patch and spend fewer resources (or take risks) on dispersal (Travis and Dytham, 1999).

Many common dead-wood microhabitats (e.g. twigs and small dead branches) are ephemeral, and species adapted to such conditions are strong colonizers. Species developing in recently dead trees, such as bark beetles, tend also to have good dispersal ability. In contrast, species associated with longer lasting

Fig. 11.2. Storkeegen (the stork oak, *Quercus robur*) in Jägerspris Nordskov, Denmark: (a) a lithograph by Gurlitt from 1839 portraying the giant hollow oak standing in an open pasture in the early 1800s; (b) portrays the same tree almost 150 years later in a photograph taken by Ole Martin in 1974, which shows beech regrowth following the cessation of grazing. At the time the photo was taken the tree was about 800–900 years old and was still alive, and specialized saproxylic beetle species, such as the click beetles *Ampedus cardinalis* and *A. hjorti* and the darkling beetle *Tenebrio opacus*, were still living in the hollow (Ole Martin, personal communication). The tree is now dead.

microhabitats, such as cavities in old living trees, appear to be poorer colonizers (Nilsson and Baranowski, 1997; Hedin *et al.*, 2008). They are, consequently, likely to be particularly vulnerable to the habitat fragmentation that has been caused by the severe declines in the abundance of veteran trees across Europe over recent centuries. There may even have been recent selection against strong dispersal tendencies in species occupying rare and widely dispersed niches.

11.5.3 Catering for the needs of the adults as well as the larvae

Many species that are saproxylic in the larval stage depend on nectar as adults. Such species include, for example, many longhorn beetles (Cerambycidae), jewel beetles (Buprestidae) and hoverflies (Syrphidae). Therefore, the landscape needs to contain an adequate supply of flowers through shrubs such as hawthorn (*Crataegus* spp.) or herbs such as umbellifers, thistles or composites.

11.5.4 Survey methods

A problem with assessing the conservation needs of saproxylic species has been developing survey methods with known efficiencies. A traditional method of sampling invertebrates living in cavities is to spread wood mould on a white canvas, and to search for adult individuals, fragments and larvae. This works well for some larger species, such as the hermit beetle and the click beetle species *Crepidophorus mutilatus*, in which fragments remain identifiable for many years. Other conventional methods include flight interception traps placed inside hollow trunks or outside cavities, pitfall traps buried in the wood mould, and traps baited with mouldy grass and animal bones. All of these methods are selectively efficient for a particular set of species and thus complement one other (Ranius and Jansson, 2002).

More novel sampling techniques have been developed recently, partly prompted by research on the saproxylic species included in the Annexes of the 1992 European Union Habitat and Species Directive. A new type of emergence trap has made it easier to survey for the violet click beetle (*Limoniscus violaceus*) (Gouix and Brustel, 2012). Dispersal of the hermit beetle has been studied using marking–release–recapture studies (Ranius and Hedin, 2001) and radiotelemetry (Dubois and Vignon, 2008; Hedin *et al.*, 2008; Chiari *et al.*, 2013a).

Another new branch of research utilizes pheromones. The pheromone of the hermit beetle (γ-decalactone) can be synthetically produced (Larsson *et al.*, 2003; Svensson *et al.*, 2009) and traps baited with this pheromone can be more efficient in detecting the presence of the hermit beetle than conventional methods. Pheromone trapping methods have been used in studies of the occurrence patterns and abundance of the hermit beetle at the habitat patch scale, as well as in studies of its dispersal (Larsson and Svensson, 2011; Svensson et al., 2011; Chiari et al., 2013b). The pheromone can also act as an attractant for other rare beetle species associated with hollow trees, for example the large click beetle species *Elater ferrugineus*, which is a predator of the hermit beetle (Larsson and Svensson, 2011). This species is otherwise difficult to survey.

Some species are easier to detect from their galleries. The great Capricorn beetle produces unmistakeable, large exit holes on the oaks that it inhabits. These indicate the number of individuals that have emerged from a particular tree which makes it possible to assess local population sizes of the species, and the effects of host-tree characteristics and environmental variables on breeding success (Buse et al., 2007; Albert et al., 2012).

The red-list assessment of a species requires knowledge of the number of the occupied sites and total area inhabited by the species. The unsystematic nature of much past recording risks severely underestimating the number of occupied sites. Modelling the habitat requirements of species based on intensive sampling in a restricted area, then collecting data on habitat availability (frequency, aggregation and quality of potential host trees) over entire regions, may allow more consistent prediction of the area inhabited by a species (Ranius et al., 2011).

11.6 Current Situation in Europe

Large old trees are a globally declining habitat feature (Lindenmayer et al., 2012). At the current rate of loss, most of the wood-pasture systems that were analysed by Gibbons et al. (2008), including southern Spanish holm oak (*Q. ilex*) wood-pastures, would lose all of their veteran trees within the next 90–180 years. The mortality rate of existing trees far exceeds the recruitment of new veteran trees. Existing management recommendations focusing on recruitment will not reverse this trend. The more critical factor is the current mortality rate of veteran trees, which should be kept at around 0.5% per year (Gibbons et al., 2008). Veteran oak trees often have a mortality of about 1% a year, and if they are situated in high-density forests, the mortality becomes even higher (Drobyshev et al., 2008).

The threats to cultural habitats and their veteran trees fall into two broad categories: intensification of land use and the cessation of traditional management. The intensification of land use results in the felling of trees valuable for the saproxylic insect fauna or in increased tree mortality; the cessation of traditional management such as grazing leads to regrowth and closed-canopy thickets. Old trees that used to grow in open conditions cannot adapt to shading by younger trees and tend to die because of competition (Hartel et al., Chapter 5). These issues can be illustrated by the case study from southern Sweden reported in Box 11.1.

11.7 How to Preserve the Specialized Saproxylic Species?

11.7.1 Management for increasing habitat amount and quality

The priority in the short term is the survival of existing veteran trees. Where traditional agriculture maintained the old trees and this has now been abandoned, the maintenance of tree quality is likely to require the reinstatement of active management. Haymaking and grazing by cattle should be revived to keep the surroundings free from competing young growth, or bushes and young trees should be removed mechanically. This removal must be done regularly; if it is left for too long, the conditions for trees, as well as for their associated species, become unstable, which may increase tree mortality. Trees traditionally managed by pollarding should be re pollarded if possible, as pollarding prolongs the lifespan of trees (Read, 2000) and also increases the rate of cavity formation (Sebek et al., 2013) (Fig. 11.3).

Where there are mature trees or trees approaching maturity, these can be encouraged to develop into the next cohort of veteran

> **Box 11.1.** History of veteran oaks (*Quercus* spp.) in southern Sweden.
>
> Large-scale loss of veteran oaks in southern Sweden started more than 200 years ago (Eliasson and Nilsson, 2002). High-quality oak timber became an important strategic resource for building warships in the 16th century and oak was declared to be the property of the state by a royal decree in 1558. During the 17th century, large battle fleets were constructed, and the selective felling of high-quality oaks led to an alarming depletion of oak stocks regionally. An assessment of the timber resources was needed for long-term planning, so comprehensive inventories covering entire provinces were made in the early 1730s. In these inventories, mature oaks were classified as being of good quality, partly decayed, too decayed to be used, or misshapen with too many branches, thus providing a reference point for assessing the subsequent changes.
>
> Most of the large oaks across the quality range grew on meadows close to villages. As the rural population grew, these trees became an obstacle to the intensification of agriculture. By the end of the 18th century, direct resentment against oak was widespread. In 1789, a royal decree gave the peasants the same rights to their land as the nobility had had before, but with one exception – the decree excluded the right to fell oak and beech. In response, the peasantry called for a nationwide survey to determine which trees were useful for the navy and which could be released from state control. During the 1790s, provinces were surveyed and good trees marked with a crown stamp.
>
> Despite the ban on felling oaks, the cutting of branches, deliberate damaging of trees, illegal disposal of trees and also legal felling based on exemptions were widespread. When the nationwide survey of oaks was repeated in the early 1820s, the number of oaks fit for naval use had been reduced by over 80% between 1790 and 1825. During the early 19th century, an average of 70,000 trees a year were felled legally; between 1800 and 1835 at least one and a half million mature oaks were cut legally, and as many (if not more) cut illegally. The felling ban on oak was finally abolished in 1830.
>
> The dramatic decline of mature oaks undoubtedly caused an equal and immediate decline in the populations of oak-associated saproxylic species, and many regional extinctions. However, as the colonization–extinction processes of species associated with veteran trees are slow, the slow disappearance of species has continued and may still be happening in landscapes where the current density of potential host trees is too low to allow long-term persistence. A study on the occurrence patterns of lichens and wood-inhabiting fungi associated with old oaks in southern Sweden showed that, of the 21 species studied, the occurrence of 18 species was positively related to the current density of oaks but, in addition, the occurrence of 11 species was explained by the historical density of oaks in 1830 (Ranius *et al.*, 2008).

trees. The formation of important microhabitats on these trees might be speeded up by damaging them (Cavalli and Mason, 2003), for example by creating entrances for decay fungi by drilling holes, the pruning of branches or the removal of patches of bark. Fungi have also been inoculated into these injuries, but it will take several decades before it is possible to see the outcome. Also, not all hollows are of equal habitat quality (Ranius, 2007). As yet, we know little about how to manipulate trees so that they become 'good' hollow trees.

For some species, it is possible to create artificial substrates. In England, tree trunks have been filled with sawdust, leaves and dead animals, and used as a breeding substrate by the rare violet click beetle (Whitehead, 2003). Other species, such as the hermit beetle and the rose chafer *Protaetia marmorata*, have been bred in the laboratory in plastic containers with sawdust and leaves as substrate. Breeding species in the laboratory and the inoculation of sites with artificial substrates could be a means of producing individuals to be used for translocation to suitable sites where they do not currently occur. This sort of study has never been conducted in a systematic way, but as many species have a very fragmented distribution, it could be an efficient conservation measure for the preservation of individual species. The relevance of such a method could increase if we need to consider changes in the potential range of species resulting from climate change.

11.7.2 Management for securing spatio-temporal continuity

Conservation efforts are expensive, so which sites and landscapes should be given priority?

Fig. 11.3. A veteran beech (*Fagus sylvatica*) in Monte Gonnaro, Italy. The tree has obviously been pollarded in the past, and re-pollarding would probably prolong its lifespan. (Photo by Stefano Chiari.)

Invertebrate species that specialize in old trees occur with a higher frequency per tree at sites where the density of suitable trees is currently high (Ranius *et al.*, 2011; Bergman *et al.*, 2012). This is due to the limited colonization ability of many species. A tree is thus more useful (i.e. on average used by more species) if it is situated close to many other hollow trees. What counts as close is related to the dispersal biology of the species concerned; for beetles inhabiting hollow trees this can vary between 200 and 5000 m (Ranius *et al.*, 2011; Bergman *et al.*, 2012). These distances do not directly reflect the dispersal capacity of individual beetles, but they indicate the scale at which host-tree connectivity is important at the population level.

Sites that previously harboured a large number of old trees, but from which many have now gone, may still maintain more species than might be expected or that can be maintained (without action) in the long run. This surplus of species is often referred to as an 'extinction debt'. Positive correlations have, for instance, been found between the historical abundance of old trees and the current occurrences of lichens and fungi (Ranius *et al.*, 2008). Sites and landscapes with a high historical density of old trees, where there is still a particularly rich fauna, should be a high priority for restoration efforts to secure long-term species persistence.

11.8 Future Prospects

Increasing public awareness of the importance of veteran trees to biodiversity is a key issue for sympathetic management. There is an increasing interest in old trees, because of their aesthetic, cultural and historical values. Utilizing real examples of the often colourful and intriguing creatures that are associated with them adds another dimension to this and may, in the long run, lead to greater protection of these species and habitats.

As with other groups of organisms, there is the question of how the saproxylic species inhabiting old trees will respond to climate change. In principle, a warming climate might benefit species at the northern parts of their range. However, the effects of climate change are likely to be mainly deleterious. This is because the mortality of old trees can be expected to increase as a result of more frequent drought and storm events. A warmer climate could also enhance invasive pest and pathogen species that increase the mortality of veteran trees.

References

Albert, J., Platek, M. and Cizek, L. (2012) Vertical stratification and microhabitat selection by the great Capricorn beetle (*Cerambyx cerdo*) (Coleoptera: Cerambycidae) in open-grown, veteran oaks. *European Journal of Entomology* 109, 553–559.

Alexander, K.N.A. (1998) The links between forest history and biodiversity: the invertebrate fauna of ancient pasture woodlands in Britain and its conservation. In: Kirby, K.J. and Watkins, C. (eds) *The Ecological History of European Forests*. CAB International, Wallingford, UK, pp. 73–80.

Alexander, K.N.A. (2002) *The Invertebrates of Living and Decaying Timber in Britain and Ireland – a Provisional Annotated Checklist*. English Nature Research Reports 467, English Nature, Peterborough, UK.

Alexander, K.N.A. (2008) Tree biology and saproxylic Coleoptera: issues of definitions and conservation language. *Revue d'Écologie (La Terre et la Vie)* 63, 1–5.

Andersson, H. (1999) Red-listed or rare invertebrates associated with hollow, rotting or sapping trees or polypores in the town of Lund. *Entomologisk Tidskrift* 120, 169–183. [In Swedish with English summary.]

Audisio, P., Brustel, H., Carpaneto, G.M., Coletti, G., Mancini, E., Trizzino, M., Antonini, G. and Debiase, A. (2009) Data on molecular taxonomy and genetic diversification of the European hermit beetles, a species complex of endangered insects (Coleoptera: Scarabaeidae, Cetoniinae, *Osmoderma*). *Journal of Zoological Systematics and Evolutionary Research* 47, 88–95.

Baudry, J., Bunce, R.G.H. and Burel, F. (2000) Hedgerows: an international perspective on their origin, function and management. *Journal of Environmental Management* 60, 7–22.

Bergman, K.-O., Jansson, N., Claesson, K., Palmer, M.W. and Milberg, P. (2012) How much and at what scale? Multi-scale analyses as decision support for conservation of saproxylic oak beetles. *Forest Ecology and Management* 265, 133–141.

Bergmeier, E., Petermann, J. and Scröeder, E. (2010) Geobotanical survey of wood-pasture habitats in Europe: diversity, threats and conservation. *Biodiversity and Conservation* 19, 2995–3014.

Buckland, P.C. and Dinnin, M.H. (1993) Holocene woodlands, the fossil evidence. In: Kirby, K.J. and Drake, C.M. (eds) *Dead Wood Matters: The Ecology and Conservation of Saproxylic Invertebrates in Britain*. English Nature Science No. 7, Peterborough, UK, pp. 6–20.

Buse, J., Schröder, B. and Assmann, T. (2007) Modelling habitat and spatial distribution of an endangered longhorn beetle – a case study for saproxylic insect conservation. *Biological Conservation* 137, 372–381.

Buse, J., Ranius, T. and Assman, B. (2008) An endangered longhorn beetle associated with old oaks and its possible role as an ecosystem engineer. *Conservation Biology* 22, 329–337.

Cavalli, R. and Mason, F. (eds) (2003) *Techniques for Re-establishment of Dead Wood for Saproxylic Fauna Conservation*. LIFE Nature Project NAT/IT/99/6245 'Bosco della Fontana' (Mantova, Italy). Scientific Reports 2, Centro Nazionale per lo Studio e la Conservazione della Biodiversità Forestale di Verona – Bosco della Fontana. Gianluigi Arcari, Mantova, Italy. In Italian and English.] Available at: http://ec.europa.eu/environment/life/project/Projects/index.cfm?fuseaction=home.showFile&rep=file&fil=BOSCO_FONTANA_deadwood.pdf (accessed 27 November 2014).

Chiari, S., Carpaneto, G.M., Zauli, A., Marini, L., Audisio, P. and Ranius, T. (2012) Habitat of an endangered saproxylic beetle, *Osmoderma eremita*, in Mediterranean woodlands. *Écoscience* 19, 299–307.

Chiari, S., Carpaneto, G.M., Zauli, A., Zirpoli, G.M., Audisio, P. and Ranius, T. (2013a) Dispersal patterns of a saproxylic beetle, *Osmoderma eremita* in Mediterranean woodlands. *Insect Conservation and Diversity* 6, 309–318.

Chiari, S., Zauli, A., Mazziotta, A., Luselli, L., Audisio, P. and Carpaneto, G.M. (2013b) Surveying an endangered saproxylic beetle, *Osmoderma eremita*, in Mediterranean woodlands: a comparison between different capture methods. *Journal of Insect Conservation* 17, 171–181.

Drobyshev, I., Niklasson, M., Linderson, H., Sonesson, K., Karlsson, M., Nilsson, S.G. and Lannér, J. (2008) Lifespan and mortality of old oaks – combining empirical and modelling approaches to support their management in southern Sweden. *Annals of Forest Science* 65, 401.

Dubois, G.F. and Vignon, V. (2008) First results of radio-tracking of *Osmoderma eremita* (Coleoptera: Cetoniidae) in French chestnut orchards. *Revue d'Écologie (La Terre et la Vie)* 63, 123–130.

Dubois, G.F., Vignon, V., Delettre, Y.R., Rantier, Y., Vernon, P. and Burel, F. (2009) Factors affecting the occurrence of the endangered saproxylic beetle *Osmoderma eremita* (Scopoli, 1763) (Coleoptera: Cetoniidae) in an agricultural landscape. *Landscape and Urban Planning* 91, 152–159.

Eliasson, P. and Nilsson, S.G. (2002) 'You should hate young oaks and young noblemen.' The environmental history of oaks in eighteenth and nineteenth century Sweden. *Environmental History* 7, 659–677.

Gärdenfors, U. and Baranowski, R. (1992) Beetles living in open forests prefer different tree species than those in dense forests. *Entomologisk Tidskrift* 113, 1–11. [In Swedish with English abstract.]

Gibbons, P., Lindenmayer, D.B., Fischer, J., Manning, A.D., Weinberg, A., Seddon, J., Ryan, P. and Barrets, G. (2008) The future of scattered trees in agricultural landscapes. *Conservation Biology* 22, 1309–1319.

Götmark, F., Åsegard, E. and Franc, N. (2011) How we improved a landscape study of species richness of beetles in woodland key habitat, and how model output can be improved. *Forest Ecology and Management* 262, 2297–2305.

Gouix, N. and Brustel, H. (2012) Emergence trap, a new method to survey *Limoniscus violaceus* (Coleoptera: Elateridae) from hollow trees. *Biodiversity and Conservation* 21, 421–436.

Hall, S.J.G. and Bunce, R.G.H. (2011) Mature trees as keystone structures in Holarctic ecosystems – a quantitative species comparison in a northern English park. *Plant Ecology and Diversity* 4, 243–250.

Harding, P.T. and Alexander, K.N.A. (1994) The use of saproxylic invertebrates in the selection and evaluation of areas of relic forest in pasture woodlands. *British Journal of Entomology and Natural History* 7(supplement 1), 22–26.

Hedin, J., Ranius, T., Nilsson, S.G. and Smith, H.G. (2008) Restricted dispersal in a flying beetle assessed by telemetry. *Biodiversity and Conservation* 17, 675–684.

Horák, J. and Rébl, K. (2013) The richness of click beetles in ancient pasture woodland benefits from a high level of sun exposure. *Journal of Insect Conservation* 17, 307–318.

Jonsell, M. (2012) Old park trees as habitat for saproxylic beetle species. *Biodiversity and Conservation* 21, 619–642.

Jonsell, M., Weslien, J. and Ehnström, B. (1998) Substrate requirements of red-listed saproxylic invertebrates in Sweden. *Biodiversity and Conservation* 7, 749–764.

Jönsson, N., Méndez, M. and Ranius, T. (2004) Nutrient richness of wood mould in tree hollows with the scarabaeid beetle *Osmoderma eremita*. *Animal Biodiversity and Conservation* 27, 79–82.

Kaplan, J.O., Krumhardt, K.M. and Zimmermann, N. (2009) The prehistoric and preindustrial deforestation of Europe. *Quaternary Science Reviews* 28, 3016–3034.

Kelner-Pillault, S. (1974) Étude écologique du peuplement entomologique des terreaux d'arbres creux (châtaigners et saules). *Bulletin d'Ecologie* 5, 123–156.

Koch Widerberg, M., Ranius, T., Drobyshev, I., Nilsson, U. and Lindbladh, M. (2012) Increased openness around retained oaks increases species richness of saproxylic beetles. *Biodiversity and Conservation* 21, 3035–3059.

Lachat, T., Wermelinger, B., Gossner, M.M., Bussler, H., Isacson, G. and Müller, J. (2012) Saproxylic beetles as indicator species for dead-wood amount and temperature in European beech forests. *Ecological Indicators* 23, 323–331.

Larsson, M.C. and Svensson, G.P. (2011) Monitoring spatiotemporal variation in abundance and dispersal by a pheromone–kairomone system in the threatened saproxylic beetles *Osmoderma eremita* and *Elater ferrugineus*. *Journal of Insect Conservation* 15, 891–902.

Larsson, M.C., Hedin, J., Svensson, G.P., Tolasch, T. and Francke, W. (2003) Characteristic odor of *Osmoderma eremita* identified as male-released pheromone. *Journal of Chemical Ecology* 29, 575–587.

Lindenmayer, D.B., Laurance, W.F. and Franklin, J.F. (2012) Global decline in large old trees. *Science* 338, 1305–1306.

Lõhmus, K. and Liira, J. (2013) Old rural parks support higher biodiversity than forest remnants. *Basic and Applied Ecology* 14, 165–173.

Lonsdale, D. (ed.) (2013) *Ancient and Other Veteran Trees: Further Guidance on Management*. Tree Council, London.

Martin, O. (1989). Click beetles (Coleoptera, Elateridae) from old deciduous forests in Denmark. *Entomologiske Meddelelser* 57, 1–107. [In Danish with English summary.]

Milberg, P., Bergman, K.-O., Johansson, H. and Jansson, N. (2014) Low host-tree preference among saproxylic beetles: a comparison of four deciduous species. *Insect Conservation and Diversity* 7, 508–522.

Nieto, A. and Alexander, K.N.A. (2010) *European Red List of Saproxylic Beetles*. IUCN (International Union for Conservation of Nature), Gland, Switzerland with the European Union, Publications Office of the European Union, Luxembourg.

Nilsson, S.G. and Baranowski, R. (1997) Habitat predictability and the occurrence of wood beetles in old-growth beech forests. *Ecography* 20, 491–498.

Oleksa, A., Ulrich, W. and Gawroński, R. (2006) Occurrence of the marbled rose-chafer (*Protaetia lugubris* Herbst, Coleoptera, Cetoniidae) in rural avenues in northern Poland. *Journal of Insect Conservation* 10, 241–247.

Oleksa, A., Chybicki, I.J., Gawroński, R., Svensson, G.P. and Burczyk, J. (2013) Isolation by distance in saproxylic beetles may increase with niche specialization. *Journal of Insect Conservation* 17, 219–233.

Olsson, F. and Lemdahl, G. (2009) A continuous Holocene beetle record from the site Stavsåkra, southern Sweden: implications for the last 10,600 years of forest and land-use history. *Journal of Quaternary Science* 24, 612–626.

Palm, T. (1959) Die Holz- und Rindenkäfer der Süd- und Mittelschwedischen Laubbäume [The wood and bark living Coleoptera of deciduous trees in southern and central Sweden]. *Opuscula Entomologica*, Supplementum XVI.

Rackham, O. (1998) Savanna in Europe. In: Kirby, K.J. and Watkins, C. (eds) *The Ecological History of European Forests*. CAB International, Wallingford, UK, pp. 1–24.

Ranius, T. (2007) Extinction risks in meta-populations of a beetle inhabiting hollow trees predicted from time series. *Ecography* 30, 716–726.

Ranius, T. and Hedin, J. (2001) The dispersal rate of a beetle, *Osmoderma eremita*, living in tree hollows. *Oecologia* 126, 363–370.

Ranius, T. and Jansson, N. (2000) The influence of forest regrowth, original canopy cover and tree size on saproxylic beetles associated with old oaks. *Biological Conservation* 95, 85–94.

Ranius, T. and Jansson, N. (2002) A comparison of three methods to survey saproxylic beetles in hollow oaks. *Biodiversity and Conservation* 11, 1759–1771.

Ranius, T., Eliasson, P. and Johansson, P. (2008) Large-scale occurrence patterns of red-listed lichens and fungi on old oaks are influenced both by current and historical habitat density. *Biodiversity and Conservation* 17, 2371–2381.

Ranius, T., Niklasson, M. and Berg, N. (2009) Development of tree hollows in pedunculate oak (*Quercus robur*). *Forest Ecology and Management* 257, 303–310.

Ranius, T., Johansson, V. and Fahrig, L. (2011) Predicting spatial occurrence of beetles and pseudoscorpions in hollow oaks in southeastern Sweden. *Biodiversity and Conservation* 20, 2027–2040.

Read, H. (2000) *Veteran Trees: A Guide to Good Management*. English Nature, Peterborough, UK.

Regnery, B., Paillet, Y., Couvet, D. and Kerbiriou, C. (2013) Which factors influence the occurrence and density of tree microhabitats in Mediterranean oak forests? *Forest Ecology and Management* 295, 118–125.

Sebek, P., Altman, J., Platek, M. and Cizek, L. (2013) Is active management the key to the conservation of saproxylic biodiversity? Pollarding promotes the formation of tree hollows. *PLoS ONE* 8(3): e60456.

Siitonen, J. (2012a) Microhabitats. In: Stokland, J., Siitonen, J. and Jonsson, B.G. (eds) *Biodiversity in Dead Wood*, 1st edn. Cambridge University Press, Cambridge, UK, pp. 150–182.

Siitonen, J. (2012b) Dead wood in agricultural and urban habitats. In: Stokland, J.N., Siitonen, J. and Jonsson, B.G. (eds) *Biodiversity in Dead Wood*, 1st edn. Cambridge University Press, Cambridge, UK, pp. 380–401.

Speight, M.C.D. (1989) *Saproxylic Invertebrates and their Conservation*. Publications and Documents Division, Council of Europe, Strasbourg, France.

Stokland, J.N. (2012) Host-tree associations. In: Stokland, J., Siitonen, J. and Jonsson, B.G. (eds) *Biodiversity in Dead Wood*, 1st edn. Cambridge University Press, Cambridge, UK, pp. 82–109.

Stokland, J.N. and Siitonen, J. (2012) Mortality factors and decay succession. In: Stokland, J.N., Siitonen, J. and Jonsson, B.G. (eds) *Biodiversity in Dead Wood*, 1st edn. Cambridge University Press, Cambridge, UK, pp. 110–149.

Stokland, J.N., Siitonen, J. and Jonsson, B.G. (2012) Introduction. In: Stokland, J.N., Siitonen, J. and Jonsson, B.G. (eds) *Biodiversity in Dead Wood*, 1st edn. Cambridge University Press, Cambridge, UK, pp. 1–9.

Svensson, G.P., Oleksa, A., Gawroński, R., Lassance, J.-M. and Larsson, M.C. (2009) Enantiometric conservation of the male-produced sex pheromone facilitates monitoring of threatened European hermit beetles (*Osmoderma* spp). *Entomologia Experimentalis et Applicata* 133, 276–282.

Svensson, G.P., Sahlin, U., Brage, B. and Larsson, M.C. (2011) Should I stay or should I go? Modelling dispersal strategies in saproxylic insects based on pheromone capture and radio telemetry: a case study on the threatened hermit beetle *Osmoderma eremita*. *Biodiversity and Conservation* 20, 2883–2902.

Sverdrup Thygeson, A., Skarpaas, O. and Ødegaard, F. (2010) Hollow oaks and beetle conservation: the significance of the surroundings. *Biodiversity and Conservation* 19, 837–852.

Travis, J.M.J. and Dytham, C. (1999) Habitat persistence, habitat availability and the evolution of dispersal. *Proceedings of the Royal Society B: Biological Sciences* 266, 120–139.

Vodka, S., Konvicka, M. and Cizek, L. (2009) Habitat preferences of oak feeding xylophagous beetles in a temperate woodland: implications for forest history and management. *Journal of Insect Conservation* 13, 553–562.

Vuidot, A., Paillet, Y., Archaux, F. and Gosselin, F. (2011) Influence of tree characteristics and forest management on tree microhabitats. *Biological Conservation* 144, 441–450.

Whitehead, P.F. (2003) Current knowledge of the violet click beetle *Limoniscus violaceus* (P.W.J. Müller, 1821) (Col., Elateridae) in Britain. In: Poland Bowe, C. (ed.) *Proceedings of the Second Pan-European Conference on Saproxylic Beetles*. English Nature, Peterborough, UK, pp. 1–9.

Whitehouse, N.J. (2006) The Holocene British and Irish ancient forest fossil beetle fauna: implications for forest history, biodiversity and faunal colonization. *Quaternary Science Reviews* 25, 1755–1789.

Winter, S. and Möller, G.C. (2008) Microhabitats in lowland beech forests as monitoring tool for nature conservation. *Forest Ecology and Management* 255, 1251–1261.

Woodland Trust (2008) *Ancient Tree Guide 4. What Are Ancient, Veteran and Other Trees of Special Interest?* Ancient Tree Forum., Woodland Trust, Grantham, UK. Available at: http://www.woodlandtrust.org.uk/mediafile/100263313/pg-wt-2014-ancient-tree-guide-4-definitions.pdf?cb=435a97b193f048419903e8a34c51ea8a (accessed 27 November 2014).

12 The Changing Fortunes of Woodland Birds in Temperate Europe

Shelley A. Hinsley,[1]* Robert J. Fuller[2] and Peter N. Ferns[3]

[1]*Centre for Ecology and Hydrology, Wallingford, UK;* [2]*British Trust for Ornithology, Thetford, UK;* [3]*School of Biosciences, Cardiff University, Cardiff, UK*

12.1 Introduction

We explore what is known of the history of woodland birds in Europe and how they have responded to changes in woodland extent, composition and management. Beyond the simple availability of habitat, woodland structure is a critical factor in species survival and distribution. Despite the huge transformation of postglacial forests, no woodland bird species has actually become extinct, and with forest cover now increasing, as long as diverse habitat structures can be maintained across a range of scales, forest birds should not only survive, but also thrive.

12.2 The Birds of the Early Holocene

Our knowledge of the bird fauna of postglacial woodlands (from about 13,000 years ago) is based mainly on bone fragments left by predators and found in cave deposits. Additional information can be inferred from the present bird fauna of sites where the climate is similar today, and from molecular evidence indicating if, and when, species divergence occurred. At least 136 species have been identified from avian bones from over 30 caves in southern Britain, just south of the line of maximum ice advance (Harrison, 1987). These include arctic–alpine breeders such as the ptarmigan, shore lark, Lapland bunting and snow bunting, and boreal forest species such as the hazel grouse, pine grosbeak and a crossbill (see Table 12.1 for scientific names). The last was likely to be linked to the presence of conifers even though it was in a stratum with evidence of severe frost action thought to be indicative of tundra habitat. A previously undescribed *Alectoris* partridge species of small size was also found at this level. The hazel grouse and pine grosbeak occurred well to the west of their present range.

Birds adapted to broadleaved trees probably appeared at the subarctic dwarf birch stage, as redpolls exploit such habitats today, and their bones have been found in cave deposits. Molecular evidence has not yet established whether redpolls constitute more than one species, although the arctic–alpine birds are usually given specific status as arctic redpolls (Marthinsen *et al.*, 2008). As vegetation height increased, comparisons with similar habitats today suggest that species such as

*E-mail: sahi@ceh.ac.uk

the willow warbler would have increased and become dominant (i.e. present at the highest density in the greatest number of sites), with the redwing, bluethroat, spotted flycatcher, brambling and common redstart beginning to appear (Haapanen, 1965; Eriksson *et al.*, 1971), though the willow warbler, bluethroat and common redstart have not been identified in the British cave deposits.

Bones deposited largely by eagle owls in Polish caves since the last glaciation include a high proportion of game birds. These were initially species associated with subarctic habitats, i.e. the willow grouse and ptarmigan, but as boreal forest developed, they were replaced by the black grouse and capercaillie (Lorenc, 2006). Other cave deposits indicate that the ranges of conifer specialists such as the nutcracker and hawk owl extended further west than they do today (Yalden and Albarella, 2009). Studies of succession in Finland suggest that the chaffinch and siskin would have been dominant in many boreal forest types, including spruce, while tree pipits and siskins probably dominated drier and more open pine forests (Haapanen, 1965) Associated species would probably have included the coal tit, pied and spotted flycatcher, goldcrest, crossbills, brambling, crested and willow tit, common redstart and treecreeper.

As soon as temperate broadleaved woodlands developed in central/northern Europe (*c.*7000–8000 years ago), bird communities began to resemble modern ones, with the bones of currently common species, such as the wren, dunnock, robin, great tit, chaffinch, nuthatch, hawfinch, blackbird, song thrush, mistle thrush, blackcap, common redstart, nightingale, tawny owl and jay all being found (Yalden and Albarella, 2009). The frequency of occurrence of these bones cannot be used to estimate relative species abundance because of the selectivity of cave-dwelling predators and the difficulty of identifying smaller species, especially passerines. Hence, there are few records for many small migrants (although some, as noted above, have been identified), along with a few small residents or partial migrants such as the blue tit and goldcrest (Yalden and Albarella, 2009). Variations in tree composition across the landscape would have favoured particular species, e.g. redpolls and willow warblers in birch areas, nuthatches with hazel, jays with oak, tits and woodpigeons with beech, hawfinches with hornbeam, etc.

Further south, where Mediterranean sclerophyllous vegetation types such as maquis and garrigue are present today, cave deposits indicate that the bird community did not change much even at the height of the glaciation. For example, the chough was recorded in cave deposits from northern Spain during all of the glacial and interglacial interstadial stages represented, regardless of whether they were cold, warm, or wet temperate (Eastham, 1984). The crested lark was present in one cold and one warm phase. The bones of the snowy owl are the only clear sign of much colder conditions. These could be from wintering migrants (as is the case with the reindeer also recorded), but the presence of arctic fox bones points to a subarctic climate at times. More than one refugium with a temperate/Mediterranean climate probably existed in Iberia and other parts of southern Europe during glacial periods (Gómez and Lunt, 2006), and from these forests, birds such as the chaffinch presumably recolonized the rest of Europe (Griswold and Baker, 2002), alongside the spread of the tree species.

12.3 The Birds of the Wildwood: Alternative Models of Forest Dynamics

The nature of the pre-Neolithic landscape is a subject of debate (Kirby and Watkins, Chapter 3). Was it closed-canopy forest with gaps of various sizes, containing different stages of regeneration, created by natural processes such as soil conditions, flooding, fire, storms and disease (Peterken, 1996; Birks, 2005); or was it a much more open structure analogous to wood-pasture in which large herbivores drove cycles of gap generation followed by revegetation creating shifting, large-scale mosaics (Vera, 2000)? Here, we suggest how birds might have responded to these two alternative structures ('closed canopy' versus 'wood-pasture'). We have assumed

Table 12.1. Common and scientific names of bird species.

Group/Common name	Scientific name	Group/Common name	Scientific name	Group/Common name	Scientific name
Birds of prey		**Grouse**		**Warblers and allies**	
Hen harrier	*Circus cyaneus*	Willow grouse	*Lagopus lagopus*	Grasshopper warbler	*Locustella naevia*
Merlin	*Falco columbarius*	Ptarmigan	*Lagopus muta*	Icterine warbler	*Hippolais icterina*
		Black grouse	*Tetrao tetrix*	Barred warbler	*Sylvia nisoria*
Buntings		Capercaillie	*Tetrao urogallus*	Lesser whitethroat	*Sylvia curruca*
Lapland bunting	*Calcarius lapponicus*	Hazel grouse	*Bonasa bonasia*	Whitethroat	*Sylvia communis*
Snow bunting	*Plectrophenax nivalis*			Garden warbler	*Sylvia borin*
		Larks and pipits		Blackcap	*Sylvia atricapilla*
Chats and thrushes		Crested lark	*Galerida cristata*	Bonelli's warbler	*Phylloscopus bonelli*
Robin	*Erithacus rubecula*	Woodlark	*Lullula arborea*	Wood warbler	*Phylloscopus sibilatrix*
Nightingale	*Luscinia megarhynchos*	Shore lark	*Eremophla alpestris*	Chiffchaff	*Phylloscopus collybita*
Bluethroat	*Luscinia svecica*	Tree pipit	*Anthus trivialis*	Willow warbler	*Phylloscopus trochilus*
Common redstart	*Phoenicurus phoenicurus*			Goldcrest	*Regulus regulus*
Whinchat	*Saxicola rubetra*	**Hoopoes**		Firecrest	*Regulus ignicapillus*
Blackbird	*Turdus merula*	Hoopoe	*Upupa epops*		
Song thrush	*Turdus philomelos*				
Redwing	*Turdus iliacus*	**Owls and other nocturnal species**		**Waterbirds, herons, waders**	
Mistle thrush	*Turdus viscivorus*	Snowy owl	*Bubo scandiaca*	Red-throated diver	*Gavia stellata*
		Tawny owl	*Strix aluco*	Black-throated diver	*Gavia arctica*
Corvids		Hawk owl	*Surnia ulula*	Cattle egret	*Bubulcus ibis*
Jay	*Garrulus glandarius*	Long-eared owl	*Asio otus*	Black stork	*Ciconia nigra*
Nutcracker	*Nucifraga caryocatactes*	Short-eared owl	*Asio flammeus*	Dunlin	*Calidris alpina*
Chough	*Pyrrhocorax pyrrhocorax*	Eagle owl	*Bubo bubo*	Stone curlew	*Burhinus oedicnemus*
Jackdaw	*Corvus monedula*	Nightjar	*Caprimulgus europaeus*		

Finches		**Pigeons and doves**		**Wrens and dunnocks**
Chaffinch	*Fringilla coelebs*	Stock dove	*Columba oenas*	Wren	*Troglodytes troglodytes*
Brambling	*Fringilla montifringilla*	Woodpigeon	*Columba palumbus*	Dunnock	*Prunella modularis*
Siskin	*Carduelis spinus*				
Linnet	*Carduelis cannabina*	**Starlings**		**Woodpeckers**	
Lesser redpoll	*Carduelis cabaret*	Starling	*Sturnus vulgaris*	Wryneck	*Jynx torquilla*
Arctic redpoll	*Carduelis hornemanni*			Green woodpecker	*Picus viridis*
Pine grosbeak	*Pinicola enucleator*			Grey-headed woodpecker	*Picus canus*
Bullfinch	*Pyrrhula pyrrhula*	**Tits, nuthatches, treecreepers**		Great spotted woodpecker	*Dendrocopos major*
Hawfinch	*Coccothraustes coccothraustes*	Long-tailed tit	*Aegithalos caudatus*	Middle spotted woodpecker	*Dendrocopos medius*
		Marsh tit	*Poecile palustris*		
		Willow tit	*Poecile montanus*	Lesser spotted woodpecker	*Dendrocopos minor*
Flycatchers		Crested tit	*Lophophanes cristatus*	White-backed woodpecker	*Dendrocopos leucotos*
Spotted flycatcher	*Muscicapa striata*	Coal tit	*Periparus ater*		
Red-breasted flycatcher	*Ficedula parva*	Blue tit	*Cyanistes caeruleus*	Three-toed woodpecker	*Picoides tridactylus*
Collared flycatcher	*Ficedula albicolis*	Great tit	*Parus major*		
Pied flycatcher	*Ficedula hypoleuca*	Nuthatch	*Sitta europaea*	Black woodpecker	*Dryocopos martius*
		Tree creeper	*Certhia familiaris*		

that species would have occupied the same habitats and niches and shown the same area requirements as today, though there are obvious risks with these assumptions. In some situations and over time, bird behaviour, habitat preferences and niche occupation have changed (Hinsley and Gillings, 2012; Wesołowski and Fuller, 2012).

At a large scale, differences in bird species composition between the two landscape types would probably have been small, but differences in species relative abundances, in contrast to simple presence or absence, could have been substantial. Relative abundances would have been affected by the grain of the patchwork; large versus small and scattered patches of forest or open ground would tend to favour different suites of species. On smaller scales, bird species composition and abundance would have varied with local habitat diversity, much as they do in modern woodland, especially in relation to wetness, stand age structure, foliage complexity and the relative proportions of edge and openness (Fuller, 1995).

12.3.1 Largely closed forest – the 'closed canopy' scenario

The 'closed canopy' scenario, with a preponderance of mature, old-growth forest, would favour species with large habitat area requirements and those needing large trees and/or dead wood. In broadleaved old growth, such species could include black, lesser spotted and white-backed woodpeckers and the stock dove, and in conifers, the capercaillie and three-toed woodpecker. Hole-nesting passerines, including tits and flycatchers, e.g. collared, pied and red-breasted flycatchers, would also have been typical of deciduous old growth. The volume of dead wood would be high under both 'closed' and 'open' scenarios, but the microclimate, influencing decay rates and arthropod fauna and abundance, was likely to have been cooler and wetter under closed canopy conditions (Fayt, 2004), which would also affect bird species abundance.

The bird assemblages of the Białowieża National Park (BNP) in eastern Poland (Latałowa et al., Chapter 17) offer one of the best remaining insights into the likely characteristics of the bird communities of such closed canopy forests (Wesołowski and Tomiałojć, 1997; Tomiałojć and Wesołowski, 2005; Wesołowski et al., 2006; Wesołowski, 2007). Species diversity is high, as even though it is set within a larger tract of forest of about 1000 km^2, the area of the BNP, at 47 km^2, comprises <2% of the area of wooded habitat in Britain, but supports about 72 breeding species compared with about 90 species in the woodland and woodland edge of the whole of Britain (Simms, 1971; Fuller, 1995). Today, there is an east–west gradient in species richness of forest birds across temperate Europe, with higher richness in the east (Fuller et al., 2007), but whether or not this gradient existed in the pre-Neolithic period is unknown.

Many species in the BNP occur at low density, possibly because of high predation pressure (see below). The forest is characterized by large trees (e.g. *Picea abies* can exceed 50 m) and multiple canopy layers, but shrub and ground cover is relatively sparse except in wet areas and recent tree-fall gaps. Thus, species such as the blackcap, garden warbler, chiffchaff and dunnock, which require low cover and understorey, tend to be confined to gaps and the swamp forest. Tree-fall gaps contain concentrations of dead wood and associated invertebrates that provide food resource hotspots for species such as woodpeckers and the nuthatch. The large, up-ended root plates of fallen trees offer nest sites, roost sites and foraging opportunities (Tomiałojć and Wesołowski, 2005). The gaps are avoided, however, by other species, including the red-breasted flycatcher and wood warbler (Fuller, 2000; Brazaitis and Angelstam, 2004).

Various woodpecker – and other – species of mature forest require a mix of tree species, in both space and time. Some are adept at exploiting patchy resources (Enoksson et al., 1995; Mikusiński and Angelstam, 1998) while others, for instance the marsh tit and hazel grouse (Jansson et al., 2004; Broughton et al., 2010), are relatively sedentary with short dispersal distances and an apparent aversion to crossing gaps in tree cover. Grazing and browsing within the forest would not necessarily limit bird species requiring mature old growth as long as the scale of open areas did not exceed movement or dispersal capabilities. So open areas embedded in a largely wooded matrix could have had relatively minor impacts on both bird movement and resource availability. Even a relatively high degree of open grazing land, especially with scattered bushes and trees, might be perceived by a forest species as 'poor-quality habitat', rather than as 'non-habitat'. In contrast, for species primarily dependent on open habitat, mature forest might have little or no value, the closed structure potentially being actively avoided.

12.3.2 Open mosaic landscape – the 'wood-pasture' scenario

Under a 'wood-pasture' scenario, the bird assemblages would be expected to include a higher abundance (and at smaller scales, also a higher proportion) of grassland species such as larks, pipits, buntings and finches. Where grassland-dominated gaps were extensive, larger-bodied grassland species such as the stone curlew, hoopoe and bustards might have occurred, with storks, herons and waders such as the lapwing, snipe, redshank and curlew in wetter areas. A mosaic of grassland patches of a range of sizes would probably also be favourable for a wide range of predators, including shrikes, owls, kestrels and harriers. True edge species, i.e. those using the resources offered by the different vegetation structures that occur across edges (Fuller, 2012), and are likely to be favoured by a 'wood-pasture' structure, include the black grouse, starling, green woodpecker, wryneck and jackdaw. Species such as the cattle egret and swallows that are associated with large herbivores, and those adept at exploiting invertebrates in dung, especially corvids, could also have benefited. Corvids can also exploit edges with respect to nest predation, both within the edge and in adjacent grassland that is overlooked by large shrubs and trees.

Species requiring dense cover and those typical of scrub-type habitats, e.g. the dunnock, wren, nightingale, lesser whitethroat, barred warbler, blackcap and garden warbler, would have occurred in open areas with shrubs and would also benefit from the regeneration phase of Vera's cyclical model, which is dominated by unpalatable shrubs – although the extent and duration of this phase is not clear. In more wooded parts of the mosaic, moderate grazing pressure that reduced understorey complexity and cover could have been favourable for species such as the wood warbler, common redstart and pied flycatcher, but would have reduced habitat availability for other species needing a dense understorey (Fuller *et al.*, 2012).

12.3.3 Forest-dominated, but more varied – the 'closed but varied' scenario

The debate is still open, but much current thinking, based on a number of approaches including pollen and charcoal analysis, insect assemblages and fossil and historical records (Mitchell, 2005; Fyfe, 2007; Szabó, 2009; Sommer *et al.*, 2011), favours an overall closed-canopy scenario (Hodder and Bullock, 2005). Natural gaps, some of which may have been maintained by large herbivores, were probably more numerous, extensive and long lived (or essentially permanent) than previously envisaged (Soepboer and Lotter, 2009; Whitehouse and Smith, 2010). This third and essentially intermediate scenario, which we term 'closed but varied', coupled with topographical and climate effects, could have created a wildwood with a very varied composition and structure, and hence diverse conditions for birds, at both landscape and local scales (Svenning, 2002; Reif *et al.*, 2013). Community composition would be spatially and temporally dynamic, shifting in response to both successional changes and less frequent, but possibly catastrophic, events, such large storms, fires or disease affecting herbivore population dynamics and trees (Kirby and Watkins, Chapter 3). The hypothesized avifaunal characteristics of the three scenarios are summarized in Fig. 12.1.

12.4 Fragmentation of the Wildwood

The physical consequences of the clearance and fragmentation of the postglacial European wildwood can be compared with those

Scenarios	Grassland species	Scavengers	Scrub nesters	Edge species	Berry feeders	Hole nesters	Mature forest species
'Wood-pasture' (Vera model)	■■	■■	■■	■■	■■	▮	▮
'Closed but varied' (intermediate model)	▮	▮	■	■	■	■■	■■
'Closed canopy' (gap phase model)	▮	▮	▮	▮		■■	■■

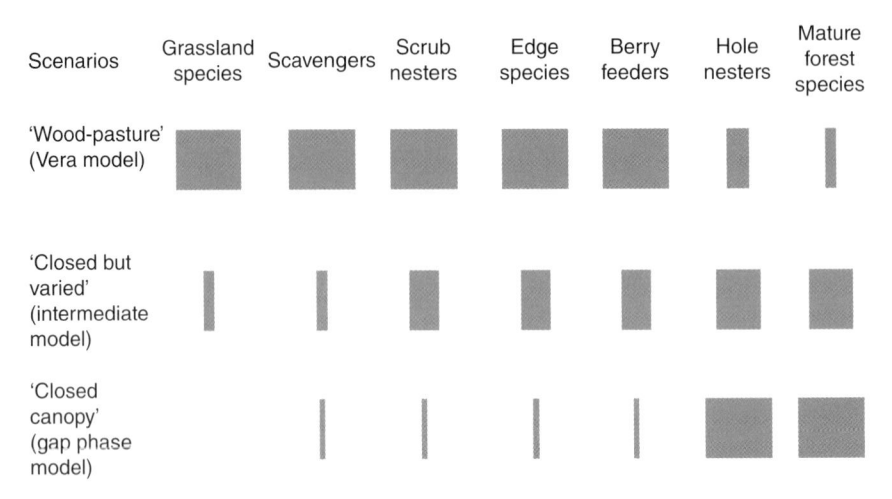

Fig. 12.1. Speculative avian guild composition under three forest landscape scenarios in the western European wildwood approximately 6000 BP. The sizes of the bars give an indication of the relative abundance of each group based on the likely availability of relevant resources. Comparisons should only be made within a group across the three scenarios, i.e. the sizes of bars do not indicate the abundance of groups relative to one another. Bird species of mature forest require extensive areas of large trees. Edge species are those that utilize a combination of forest and adjacent open habitat. In general, the 'wood-pasture' scenario favours species dependent on open or early successional vegetation and species needing complementary resources, whereas the resources for late successional species would be more limited. Overall, the 'closed but varied' scenario could have supported the most diverse avifauna.

seen today in, for example, large areas of tropical forest. However, whereas modern technology allows for rapid transformation within years or even months, the European timescale was of the order of decades or centuries, especially in the early stages when human populations were relatively small and localized (Kirby and Watkins, Chapter 4). Initial clearance of the wildwood by burning and temporary cultivation would have had greater impact in conifer-dominated landscapes (Granström and Niklasson, 2008), but being analogous to natural fire disturbance regimes, would probably have had little initial large-scale impact on birds. At smaller scales, clearance and subsequent abandonment might have added to the extent and variety of younger age classes in the landscape. This would have favoured bird species associated with natural clearings such as tree fall gaps, wet areas and poor soils, e.g. the tree pipit, wren, dunnock, blackcap, willow warbler and other warblers (Wesołowski, 1983; Fuller, 2000).

Clearance of trees by ring barking would have temporarily increased the supply of standing dead wood, and much cut wood may also have been left to decay (Rackham, 1990). This would have increased the supply of saproxylic invertebrates, to the benefit of woodpeckers (Angelstam and Mikusiński, 1994) and many other insectivorous species (Nilsson, 1979), as well as the supply of nest sites for both primary and secondary cavity nesters, including owls, treecreepers, nuthatches, flycatchers and tits. This was probably only a local effect though, as the natural forest was characterized by large volumes of dead wood anyway (Peterken, 1996; Angelstam et al., 2004).

Of more significance was the loss of forest per se and the increase in edge habitats as settlements became more permanent and larger, with clearings increasing in size and duration (Fagan et al., 1999). The change from a matrix of forest with gaps to a matrix of open habitats (crop and pasture land, fallows and wastes, settlements, roads, etc.) with embedded woodland/forest fragments of a range of sizes (Andrén, 1994; Turner et al., 2001) would have been largely detrimental for forest birds. In addition to habitat loss, the open matrix would hinder movement and dispersal, and the size of woodland patches would have consequences for minimum habitat area requirements and edge avoidance (Dolman, 2012). Loss of forest species may accelerate rapidly when woodland cover within a landscape decreases below 10–30% (e.g. Trzcinski *et al.*, 1999), but Radford *et al.* (2005) stress that decline may be substantial well above the threshold values at which species extinctions become detectable.

These newly developing landscapes would have favoured species typical of open country, edges, forest clearings and regenerating woodland. Mixed, small-scale extensive farming, along with the presence of livestock and the absence of pesticides, was probably very favourable for a wide range of bird species (Tubbs, 1997), notwithstanding the potential effects of hunting and persecution. The effects of fragmentation would have been most extreme where previously there had been largely closed-canopy forest, but there would still have been a loss of habitat for forest birds, even in areas that might have resembled wood-pasture.

Clearance was not, however, uniform across the continent, nor within single countries. In 11th century England, regional woodland cover varied from 1–3 to 40% or more in some localities (Rackham, 1990). Currently, forest cover varies from 0 to 73% by country across Europe (Kirby and Watkins, Chapter 1). So at some scales, bird assemblages might appear to have become more uniform, whereas at others, an increase in habitat heterogeneity would have increased diversity (Hinsley and Gillings, 2012). In general, human activity probably had more impact on large, specialized and/or carnivorous species than on smaller, generalist and/or omnivorous species (Mikusiński and Angelstam, 2004).

12.5 Effects of the Historical Emergence of Management

Initial 'management' of forest was probably largely incidental rather than active, for example, the grazing of re-sprouting stumps by livestock and the pasturage of cattle and pigs within the intact forest. While population levels remained relatively low, and most settlements were either shifting or small, human exploitation (for meat, eggs and feathers) of birds could still have had the greater impact on populations, especially of large, conspicuous or ground-nesting species (e.g. the black stork, eagle owl, forest grouse). Hunting and gathering activities may have targeted open areas like wetlands to take advantage of birds such as herons, egrets and wildfowl. The reduction of other forest fauna such as bison, deer, beaver, wolves and bears (Peterken, 1996; Svenning, 2002), the removal of dead wood for fuel, and an increasing proportion of young woodland, e.g. coppice and natural regeneration on abandoned land, would also have changed bird populations and ecological relationships.

The removal and alteration of the mammal fauna, especially predators, appears to have altered the dynamics of woodland bird populations in much secondary temperate woodland, increasing the significance of over-winter survival compared with that of predation. In the BNP, population densities of most birds are low compared with densities found in similar habitats across Europe (Tomiałojć and Wesołowski, 2005). Predation rates in the BNP are high (Walankiewicz, 2002; Broughton *et al.*, 2011). Wesołowski (1983) estimated that there were three times as many predator species in the BNP as in typical English secondary woodland where birds nest at higher densities and achieve higher breeding success. For example, wrens in the BNP occur at densities of about five territories per 10 ha, but in woodland in Britain (and elsewhere), densities of 15 to 32 territories are not unusual, and breeding success can be 50% greater than in the BNP.

Similar variations in nesting densities are also found for hole breeders such as the great tit and flycatchers, although other factors, e.g. climate or habitat structure, could also contribute to these differences. Fluctuations in the size of the migrant wood warbler breeding population, as well as their breeding success, in the BNP have also been related to predator pressure. As ground nesters, wood warblers are vulnerable to predators such as small mammals, and in years when rodent numbers are high, many birds apparently choose not to settle in the forest (Wesołowski et al., 2009). Thus, bird behaviour can change in response to human-mediated ecological changes (Naef-Daenzer, 2012).

The transition to much more open landscapes with greater proportions of early successional (including coppice) and young woodland would have been detrimental to bird species of mature forest, which tend to be mostly residents. Birds using younger growth stages, typically many migrant species, could, in contrast, have benefited enormously from greatly increased habitat availability. Increased densities of low vegetation would provide more nesting and foraging opportunities for species such as the nightingale, blackcap, barred warbler, garden warbler, chiffchaff and willow warbler (Fuller, 1995). The removal of large herbivores by humans would also have contributed to greater vegetation densities and to reduced disturbance of the forest floor by species such as wild boar, but this may have been offset by the activities of domestic animals and the collection of foliage for fodder and other human activities. Overall, human intervention in the wildwood and the subsequent forest was probably favourable for such young-growth species until as recently as the last 50–100 years (Fig. 12.2).

Some species of mixed and more open woodland, such as black and grey-headed woodpeckers and the wryneck, may also have benefited, perhaps not declining significantly

Fig. 12.2. The blackcap, Sylvia atricapilla (a), and the green woodpecker, Picus viridis (b), are examples of bird species requiring contrasting resources that are greatly affected by woodland management. The woodpecker requires mature trees for nesting and feeds mainly on the ground in places where the vegetation is short. Blackcaps, however, nest and feed in dense shrubby vegetation. Harvesting patterns and changes in canopy cover and grazing pressure can all potentially alter habitat quality for these species. Photo credits: BTO (British Trust for Ornithology) images – Ron Marshall (green woodpecker) and Liz Cutting (blackcap).

until large forests with old trees became uncommon. However, there is unlikely to have been a shortage of potential nest sites for small cavity nesters such as tits and nuthatches in mature secondary woodland (Broughton et al., 2011; Fuller et al., 2012). Human management in the form of the removal of dead wood, in tandem with the loss of the dead wood resource on large, old living trees, would have reduced invertebrate food supplies for many bird species, but especially for woodpeckers (Smith, 2007). Lack of dead deciduous wood could have extirpated the white-backed woodpecker from central and eastern Europe, and even Britain (Tomiałojć, 2000).

12.6 The Age of Managed Pasture Woods and Coppice

By 1000 years ago, much of the surviving woodland, at least in the temperate zone, was being exploited either as forms of wood-pasture or coppice (Plates 5 and 6); systems that persisted over much of lowland Europe into at least the late 19th century (Hartel et al., Chapter 5; Buckley and Mills, Chapter 6). We use examples from contemporary British woodland to explore how bird assemblages might have appeared in historic wood-pasture and coppice.

In both space and time, historic wood-pasture varied in terms of tree size and density, and in the quantity and extent of bushes and regenerating trees. Grazed woodland would have graded into open grassland and heath without trees. The locations of tree-covered and more open areas would shift over time, depending on the vicissitudes of grazing, and the bird communities would have tracked the spatial pattern of the vegetation. Wood-pasture on common land sometimes extended over substantial tracts and there may have been much carrion. This should have supported a rich fauna of scavengers and predatory birds, depending on local levels of human persecution. Overall, at large scales, wood-pasture probably held bird communities that today we would recognize as characteristic of: (i) closed canopy broadleaved woodland, with minimal woody understorey; (ii) woodlands with glades and open-grown trees; (iii) scrub; and (iv) open heath or grassland. Permanently wet areas would have provided additional habitat types.

The New Forest, in southern England, provides an insight into the bird assemblages that might have occurred in historic wood-pasture systems (Glue, 1973; Smith et al., 1992). During the breeding season, both the closed and more open-canopy New Forest woodlands are rich in hole and crevice-nesting species (six of the ten most abundant breeding species), including woodpeckers, nuthatches, treecreepers and tits, though the bird community is strongly dominated by blue tits and chaffinches. Wood warblers and common redstarts are also typical of pasture woods in the New Forest, apparently preferring the open woodland structures created by long-term grazing (Tubbs, 1986). Similar structures elsewhere in Europe support high densities of various flycatchers (Hartel et al., 2014). Although many of the New Forest pasture woods have an *Ilex aquifolium* shrub layer, this does not attract high densities of species that use shrub layer vegetation for feeding or roosting. With the exception of gorse (*Ulex* spp.), which occurs widely on heaths and forest fringes, scrub and regenerating woodland are currently scarce in the New Forest due to the high grazing pressure.

The cover and density of scrub, where it occurs, strongly affects the bird community, with the number of species and the density of birds generally increasing with canopy cover (Fuller, 2012). Characteristic species of relatively open, low scrub in the lowlands include the dunnock, linnet and whitethroat. The tree pipit may also occur in patchy scrub or woodland glades. As the scrub thickens, so the density of warblers, notably the garden warbler, willow warbler, lesser whitethroat and blackcap, tends to increase. These disappear as the scrub develops into woodland. As the availability of scrub or young regenerating woodland can vary over time depending on grazing pressure, so the balance of bird assemblages between those of early and late successional stages will shift.

Under coppice management, a relatively large proportion of the woodland area consists of rapidly growing young trees with high densities of stems, potentially offering very thick low vegetation structures. The rotational cutting that is performed creates a fine-grain patchwork of trees at different stages of regrowth. Studies of birds in 20th century coppiced woods all demonstrate rapid turnover of bird assemblages with growth of the coppice (Fuller, 1992). Three distinct phases of bird community development are typically evident: (i) 'establishment' when the underwood canopy is open and a vigorous ruderal non-woody field layer is often present; (ii) around canopy closure, when low woody vegetation, including bramble, is at its most impenetrable; and (iii) post canopy closure, when the low vegetation has been shaded out. Canopy closure can occur as early as 5 or 6 years after cutting depending on tree species and growth rates.

Most breeding bird species reach peak density during the canopy-closure period when shrub layer complexity is at its greatest (Fuller and Henderson, 1992). At this stage, densities of species such as the garden warbler, blackcap, willow warbler and nightingale can be very high. The diversity of bird species is lowest in maturing post canopy-closure coppice because the structural complexity is also at its lowest. Large standard trees and ancient stools provide some holes, but the densities of hole-nesting birds in coppiced woodland are generally lower than in wood-pastures.

Coppiced woods are highly variable (Peterken, 1993; Rackham, 2003), differing in rotation length, standard tree density, coupe size, stool density and tree species. Each of these variables affects the bird communities (Fuller, 1992). Rotation length influences the relative areas of young versus older coppice growth. Woods with longer rotations have relatively large areas of post canopy-closure coppice and, hence, less in the earlier canopy-closure phase which is associated with the highest diversity and overall density of bird species. High densities of standard trees typically result in heavily shaded and poor coppice regrowth, with insufficient vegetation density to attract many shrub-dependent birds, but may support more hole-nesting birds. Small coupes may be more heavily shaded than larger ones and, therefore, are also less attractive to shrub-dependent species. Similarly, a high density of stools may result in rapid canopy closure and shading of low vegetation. The tree species composition of coppice influences bird communities through effects on woodland structure and the availability of invertebrates. In the past, livestock or deer might be allowed to forage within the coppice once it had grown sufficiently to escape damage, but grazing and browsing could have rapidly reduced the habitat quality of the coppice for shrub-dependent birds, just as high deer pressure does today (Holt *et al.*, 2011).

12.7 The Shift Towards High Forest

From the 19th century onward, the rise of high forest management at the expense of coppice became increasingly widespread in Europe (Savill, Chapter 7). Relative to coppice, high forest provides: (i) more opportunities for bird species that depend on large trees, such as most hole/crevice nesters and canopy nesters including birds of prey and corvids; but (ii) fewer opportunities for birds dependent on the early establishment and canopy-closure phases of woodland development (Fuller, 1995). In a mixed coppice system cut on a 25 year rotation, one might expect bird species associated with dense understorey to occur in as much as 30% of the woodland area. In contrast, in shelterwood oak stands in central France, such species were mainly concentrated into just 10% of the woodland area, i.e. into stands of less than about 20 years growth (Ferry and Frochot, 1990). In continuous cover high forest, where trees are harvested in small groups or even individually, the extent of suitable habitat for bird species dependent on open space with low dense regeneration would be likely to be even less (Bürgi, Chapter 8).

12.8 Woodland Birds Today

12.8.1 Population trends

Woodland bird population trends in Europe across the last 10–20 years include the complete range of responses from increases (e.g. the black woodpecker, blackcap, chiffchaff) through stability (e.g. the blackbird, robin, great tit) to strong declines (BirdLife International, 2004; Gregory *et al.*, 2007). For example, using data for 52 species pooled across 18 European countries for 1980 to 2003, Wade et al. (2013) found that 47% of common forest species and 46% of specialists had declined, the average decline being 13% for common species and 18% for specialists. Gregory et al. (2007) cited the following species as declining across Europe: the wryneck, lesser spotted woodpecker, willow tit, marsh tit, nightingale, tree pipit, spotted flycatcher, icterine warbler, wood warbler, willow warbler and brambling.

Populations appeared to be more stable in eastern than in central and western Europe (Gregory et al., 2007), possibly linked to less intensive farm and forest management in the east and land abandonment following political change. Conditions on the wintering grounds and during migration were also implicated as factors by the greater declines associated with long-distance migrants. In addition, residents tended to show declines whereas short-distance migrants were more stable or increasing, suggesting a successful strategy of accessing resources across space and time.

Recent analysis of changes in British broadleaved woodland birds (Hewson et al., 2007; Baillie et al., 2013) have found similar patterns, with specialist species again faring less well than generalists, although increases were found for the great spotted woodpecker, nuthatch, great tit, blue tit, long-tailed tit, blackcap and chiffchaff. The importance of winter habitats in sub-Saharan Africa was again highlighted by declines in all long-distance migrants versus increases in short-distance migrants.

There are few, if any, other common ecological traits defining the species observed to be declining, stable or increasing. However, key drivers of change during the last century are thought to be the intensification of agriculture and the advent of modern forestry, as well as, more recently, the impact of increasing deer populations, tree diseases such as ramorum disease (caused by *Phytophthora ramorum*) and ash dieback (caused by *Hymenoscyphus fraxineus*), and climate change.

12.8.2 Influences of agriculture

Agricultural intensification, particularly in lowland Europe, has affected woodland birds both by causing direct habitat loss and by simplification of the landscape outside woodland. There has been replacement of mixed farming with simple arable rotations, intensive grassland management, increased field sizes and the general 'tidying up' of the countryside (Wilson et al., 2009). This reduces resources for generalist woodland species such as the blackbird, robin, wren and chaffinch, and increases the isolation of more specialist species such as the marsh tit. Certain agri-environment schemes have provided funding for the creation of farm woodlands, but these are generally small, a few hectares or less, and so are of more value to generalists and edge species than to specialists (Bellamy et al., 1996; Hinsley et al., 1996; Lampila et al., 2005). Small woods are also more vulnerable to the effects of drainage of agricultural land which has been implicated in the decline of ground feeding species such as the song thrush (Peach et al., 2004), and possibly also in the general decline of invertebrate food supplies in woodland (Fuller et al., 2005). The loss of trees from the countryside between woods may also decrease the connectivity of the landscape for forest birds.

In more marginal farming areas (e.g. uplands and mountainous areas), changing agricultural economics can result in land abandonment and reversion to scrub and forest as extensive 'traditional' practices become unprofitable. For example, the abandonment of wood-pasture systems across Europe (Bergmeier et al., 2010) can reduce bird diversity as grassland, scrub and veteran trees are lost to forest regeneration (Hartel et al., Chapter 5). Recent recognition of the value of such systems for both biodiversity and food quality, combined with the services that they provide, which include tourism, education, game hunting and cultural heritage, may help to reverse this trend through greater support within European Union rural development and land management policy (Pardini, 2009; Rigueiro-Rodríguez et al., 2009).

12.8.3 Forestry intensification

Forest management involving clear-cutting and replanting emerged in Europe in the middle of the 18th century and gathered pace at the beginning of the 19th century, though in the Nordic countries, the large-scale application of modern forestry only started in earnest in the 1950s (Thirgood, 1989; Angelstam, 1996; Savill, Chapter 7). The planting of conifers in deciduous woodland dates back to the mid-19th century, yet its ornithological consequences have been rather little studied. The local consequences for bird assemblages depend on the types of stands involved (tree species, foliage structure, age, management, etc.) and the resources they offer. Thus, although the bird communities of broadleaved and coniferous stands of similar growth stages in the same locations differ, the differences are not always consistent across studies (Donald *et al.*, 1998; Archaux and Bakkaus, 2007; du Bus de Warnaffe and Deconchat, 2008; Sweeney *et al.*, 2010).

Species especially favoured by the increasing proportion of conifers in European forests include crossbills, and the siskin, coal tit, goldcrest, firecrest and three-toed woodpecker. An expansion of the range of the black woodpecker into central and western Europe in the 20th century has also been linked to increased planting of conifers (Mikusiński, 1995). Siskins and crossbills have spread from Scotland across the whole of Britain and Ireland since the 1950s (Avery and Leslie, 1990; Balmer *et al.*, 2013). Several more generalist species have colonized the new conifer plantations, sometimes at high density, e.g. the song thrush, blackbird, robin and willow warbler. Species that appear to prefer broadleaved woodlands and have presumably been disadvantaged by their replacement with conifers include lesser spotted, middle and white-backed woodpeckers, pied and collared flycatchers, the nuthatch and marsh tit, wood and Bonelli's warblers and the hawfinch.

Other aspects of forest management also affect habitat suitability for many birds (Paillet *et al.*, 2010; Fuller *et al.*, 2012; Quine, Chapter 15). High planting densities, absent or impoverished shrub and ground flora, uniform thinning regimes, herbicide use and large areas of single-species even-aged stands are not generally conducive to avian diversity or abundance. Particularly affected are species (such as the capercaillie and hazel grouse, black, white-backed, middle and lesser spotted woodpeckers and the nuthatch) that variously require large old trees, mixed conifer and deciduous habitat, and more open structures with a well-developed ground flora (Enoksson *et al.*, 1995; Angelstam, 2004; Jansson *et al.*, 2004).

In contrast, recent clear-cuts and restocked areas can support a wide range of species that are absent from more mature forest, including the nightjar, wood lark, tree pipit, redpoll, whinchat and grasshopper warbler. The large-scale provision of such habitat, creating a mosaic of patches within grassland, heathland and moorland landscapes, can also benefit species such as the black grouse and hen harrier, and short and long-eared owls, though there may be large differences in habitat suitability between the first afforestation and restocks.

The balance of benefit versus detriment, and the consequences for population sizes and ranges, depend on the availability of species' preferred growth stages. The long timescale of woodland growth and development means that major shifts in policies and practices may leave legacies in bird populations for decades or even centuries. In Great Britain, the total area of broadleaved woodland increased by 34% between 1947 and 2002, but a decline in coppice management, combined with natural maturation, has shifted the age composition from 49% scrub and coppice to 97% high forest (Hopkins and Kirby, 2007). This increase in canopy cover and shading has been linked to recent declines in some woodland species associated with earlier growth stages or dense understorey, such as the nightingale, willow warbler and willow tit (Hewson *et al.*, 2007). However, woodland size and isolation (and other factors) can compromise the quality of mature woodland for some species typical of this growth stage. Despite the increase in mature woodland in Britain, the marsh tit, lesser spotted woodpecker and hawfinch have all declined for reasons that are not fully understood (Broughton, 2012; Charman *et al.*, 2012; Baillie *et al.*, 2013; Broughton *et al.*, 2013).

12.8.4 Birds and afforestation

Large-scale afforestation of previously open landscapes should be, by definition, good for forest birds, but has attracted criticism across Europe for a range of reasons, including aesthetic, economic and social factors (Dhubhain et al., 2009; Marey-Pérez and Rodríguez-Vincente, 2009), as well as ecological aspects. Commercial afforestation of the Flow Country in the north of Scotland, with its distinctive wet moorland avifauna (including red and black throated divers, and the dunlin, greenshank and merlin) was particularly contentious (Stroud et al., 1987). In Ireland, while forest cover increases resources for generalists such as the robin and blackbird, its extent within a given landscape should be limited to avoid detriment to farmland birds (Pithon et al., 2005). Although the composition of new Irish forests is dominated (c.80%) by exotic conifers, the value for bird diversity of broadleaves, mixed species planting and management to promote shrub cover and understorey vegetation has been recognized (Wilson et al., 2006; Sweeney et al., 2010). The large-scale planting of eucalypts in Portugal and Spain has also demonstrated the problems associated with monocultures of exotics. Such plantations support a lower diversity and abundance of both breeding and wintering birds compared with native oak forest and pine plantations (Pina, 1989; Calviño-Cancela et al., 2012). Moreover, they are of little or no value to the bird communities of the farmland that they displace (Telleria and Galarza, 1990).

12.9 Recent Trends

Interest in woodland for recreation, conservation and biodiversity has increased in the last 20–30 years and contributed to shifts in forest policy away from even-aged, coniferous and frequently exotic monoculture towards more diverse systems promoting the principles of sustainable forest management for multiple benefits (Zerbe, 2002; Forestry Commission, 2004; Mason, 2007; Savill, Chapter 7; Quine, Chapter 15). This is encouraging the planting and greater management of broadleaves and a general trend towards more mixed forest.

The potential increase in the market for wood fuel may promote increased stand thinning and production of small-volume wood, but could also increase disturbance and affect the availability of dead wood (Riffell et al., 2011). The shift away from clear-cutting in favour of continuous cover forestry and the promotion of natural regeneration may create mixed-aged stands with a more diverse composition and structure (Bürgi, Chapter 8). Reduction in clear-cutting could reduce habitat availability for species typical of early growth stages (Wilson et al., 2006), but the planting and management of new broadleaved woodlands (Zerbe, 2002) on a Europewide scale should help to offset this and increase the supply of early successional deciduous habitat for long distance migrants (Reif et al., 2013).

Greater diversity of structure and composition within and across woods should be generally positive for birds. An overall increase in deciduous forest in temperate and hemiboreal Europe could even shift the balance of composition back towards that of the earlier forest. In Ireland, afforestation has increased tree cover from about 5 to about 11% in the last 30 years, and may have contributed to the recent colonization by the great spotted woodpecker (McComb et al., 2010).

Increasing numbers of deer create management problems, especially for the establishment of new woodland. Evidence from Europe and elsewhere (Fuller and Gill, 2001; Martin et al., 2011) shows that large, uncontrolled numbers of deer can remove and damage the herbaceous and low woody cover required by many species including the willow warbler, bullfinch and dunnock. Excessive browsing can also prevent regeneration, especially in relation to coppice management and restocks, and has been implicated in the decline of nightingales (and other species) in eastern England (Gill and Fuller, 2007). Fences can be used to protect stands, but this can prove lethal for large-bodied species such as the capercaillie and other forest grouse (Baines and Andrew, 2003) due to collision mortality.

The provision of nest boxes, especially for small hole nesters such as tits and flycatchers, is now widespread and can greatly increase the size of breeding populations and, thereby, their potential competitive impact on other species (Maxwell, 2002; Hinsley et al., 2007). A large increase in collared flycatchers was recorded in forests in Baden-Württemberg, Germany, following the provision of nest boxes, but this was later followed by a population crash when edible dormice (*Glis glis*) moved into the boxes, predating both flycatcher eggs and young in the process (Gatter, 2007).

12.10 Conclusion

The timescale of woodland growth and development greatly exceeds that of the average human lifespan and hence planning for future forests requires vision, commitment and not a little luck. While some factors may be within the local control of forest managers and owners, others (such as pests and diseases) are not.

The greatest unknown impact on future forests and their birds is probably climate change (Leech and Crick, 2007; Milad *et al.*, 2011; Whitehouse *et al.*, 2013). Although effects on climate envelopes, migration strategies and potential range changes of both trees and birds can be modelled (Huntley *et al.*, 2007), the degree of uncertainty is huge and beset with complications. The planting by foresters of new mixtures of tree species thought to be relatively resilient to changing climate will have unknown consequences for bird populations. Social and political pressures, food security and mass migration of human populations may constitute even greater challenges.

Given our current knowledge of woodland ecosystems (Forest Europe, UNECE and FAO, 2011), and acceptance that they are dynamic, it should be possible to manage current threats and to integrate commercial and recreational interests (DeGraaf and Miller, 1996; Angelstam *et al.*, 2004; Wallenius *et al.*, 2010; Halme *et al.*, 2013). Despite the profound transformation of postglacial forests, no woodland bird species has yet become extinct, and with forest cover now increasing, and opportunities for strategic planning at international scales, forest birds should not only survive, but also thrive as long as diverse habitat structures and food resources can be maintained across a range of scales.

References

Andrén, H. (1994) Effects of habitat fragmentation on birds and mammals in landscapes with different proportions of suitable habitat: a review. *Oikos* 71, 355–366.

Angelstam, P. (1996) The ghost of forest past – natural disturbance regimes as a basis for reconstruction of biologically diverse forests in Europe. In: Degraaf, R.M. and Miller, R.I. (eds) *Conservation of Faunal Diversity in Forested Landscapes*. Chapman and Hall, London, pp. 287–337.

Angelstam, P. (2004) Habitat thresholds and effects of forest landscape change on the distribution and abundance of black grouse and capercaillie. *Ecological Bulletins* 51, 173–187.

Angelstam, P. and Mikusiński, G. (1994) Woodpecker assemblages in natural and managed boreal and hemiboreal forest – review. *Annales Zoologici Fennici* 31, 157–172.

Angelstam, P., Dönz-Breuss, M. and Roberge, J.-M. (eds) (2004) Targets and tools for the maintenance of forest biodiversity. *Ecological Bulletins* 51, 1–510.

Archaux, F.B. and Bakkaus, N. (2007) Relative impact of stand structure, tree composition and climate on mountain bird communities. *Forest Ecology and Management* 247, 72–79.

Avery, M. and Leslie, R. (1990) *Birds and Forestry*. T. and A.D. Poyser, London.

Baillie, S.R., Marchant, J.H., Leech, D.I., Massimino, D., Eglington, S.M., Johnston, A., Noble, D.G., Barimore, C., Kew, A.J., Downie, I.S. *et al.* (2013) *Bird Trends 2012: Trends in Numbers, Breeding Success and Survival for UK Breeding Birds*. BTO Research Report No. 644. British Trust for Ornithology, Thetford, UK. Available at: http://www.bto.org/birdtrends (accessed January 2014).

Baines, D. and Andrew, M. (2003) Marking of deer fences to reduce frequency of collisions by woodland grouse. *Biological Conservation* 110, 169–176.

Balmer, D., Gillings, S., Caffrey, B.J., Swann, R.L., Downie, I.S. and Fuller, R.J. (2013) *Bird Atlas 2007 – 11. The Breeding and Wintering Birds of Britain and Ireland*. British Trust for Ornithology, Thetford, UK.

Bellamy, P.E., Hinsley, S.A. and Newton, I. (1996) Factors influencing bird species numbers in small woods in south-east England. *Journal of Applied Ecology* 33, 249–262.

Bergmeier, E., Petermann, J. and Schröder, E. (2010) Geobotanical survey of wood-pasture habitats in Europe: diversity, threats and conservation. *Biodiversity and Conservation* 19, 2995–3014.

BirdLife International (2004) *Birds in Europe: Population Estimates, Trends and Conservation Status.* BirdLife Conservation Series No. 12, BirdLife International, Cambridge, UK.

Birks, H.J.B. (2005) Mind the gap: how open were primeval European forests? *Trends in Ecology and Evolution* 20, 154–156.

Brazaitis, G. and Angelstam, P. (2004) Influence of edges between old deciduous forest and clearcuts on the abundance of passerine hole-nesting birds in Lithuania. *Ecological Bulletins* 51, 209–217.

Broughton, R.K. (2012) Habitat modelling and the ecology of the marsh tit (*Poecile palustris*). PhD thesis, School of Applied Sciences, Bournemouth University, Bournemouth, UK.

Broughton, R.K., Hill, R.A., Bellamy, P.E. and Hinsley, S.A. (2010) Dispersal, ranging and settling behaviour of marsh tits *Poecile palustris* in a fragmented landscape in lowland England. *Bird Study* 57, 458–472.

Broughton, R.K., Hill, R.A., Bellamy, P.E. and Hinsley, S.A. (2011) Nest sites, breeding failure and causes of non-breeding in a population of British marsh tits *Poecile palustris*. *Bird Study* 58, 229–237.

Broughton, R.K., Hill, R.A. and Hinsley, S.A. (2013) Relationships between patterns of habitat cover and the historical distribution of the marsh tit, willow tit and lesser spotted woodpecker in Britain. *Ecological Informatics* 14, 25–30.

Calviño-Cancela, M., Rubido-Bará, M. and van Etten, E.J.B. (2012) Do eucalypt plantations provide habitat for native forest biodiversity? *Forest Ecology and Management* 270, 153–162.

Charman, E.C., Smith, K.W., Dillon, I.A., Dodd, S., Gruar, D.J., Cristinacce, A., Grice, P.V. and Gregory, R.D. (2012) Drivers of low breeding success in the lesser spotted woodpecker *Dendrocopos minor* in England: testing hypotheses for the decline. *Bird Study* 59, 255–265.

DeGraaf, R. and Miller, R.I. (eds) (1996) *Conservation of Faunal Diversity in Forested Landscapes.* Chapman and Hall, London.

Dhubhain, D.N., Flechard, M.C., Molony, R. and O'Connor, D. (2009) Assessing the value of forestry to the Irish economy – an input–output approach. *Forest Policy and Economics* 11, 50–55.

Dolman, P. (2012) Mechanisms and processes underlying landscape structure effects on bird populations. In: Fuller, R.J. (ed.) *Birds and Habitat: Relationships in Changing Landscapes.* Cambridge University Press, Cambridge, UK, pp. 93–124.

Donald, P.F., Fuller, R.J., Evans, A.D. and Gough, S.J. (1998) Effects of forest management and grazing on breeding bird communities in plantations of broadleaved and coniferous trees in western England. *Biological Conservation* 85, 183–197.

du Bus de Warnaffe, G. and Deconchat, M. (2008) Impact of four silvicultural systems on birds in the Belgian Ardenne: implications for biodiversity in plantation forests. *Biodiversity and Conservation* 17, 1041–1055.

Eastham, A. (1984) The bird bones in the Cave of Amalda. In: Altuna, J., Baldeon, A. and Mariezkurrena, K. (eds) *La Cueva de Amalda (Zestoa, Pais Vasco): Ocupaciones Paleoliticas y Postpaleoliticas.* Eusko Ikaskuntza, San Sebastian, Spain, pp. 239–254.

Enoksson, B., Angelstam, P. and Larsson, K. (1995) Deciduous forest and resident birds – the problem of fragmentation within a coniferous forest landscape. *Landscape Ecology* 10, 267–275.

Eriksson, C.G., Nilsson, L. and Svensson, B. (1971) The passerine bird fauna of some forest habitats in Stora Sjöfallet National Park, Swedish Lapland. *Bird Study* 18, 21–26.

Fagan, W.E., Cantrell, R.S. and Cosner, C. (1999) How habitat edges change species interactions. *American Naturalist* 153, 165–182.

Fayt, P. (2004) Old-growth boreal forests, three-toed woodpeckers and saproxylic beetles – the importance of landscape management history on local consumer-resource dynamics. *Ecological Bulletins* 51, 249–258.

Ferry, C. and Frochot, B. (1990) Bird communities of the forests of Burgundy and the Jura (Eastern France). In: Keast, A. (ed.) *Biogeography and Ecology of Forest Bird Communities.* SPB Academic Publishing, The Hague, The Netherlands, pp. 183–195.

Forest Europe, UNECE and FAO (2011) *State of Europe's Forests 2011. Status and Trends in Sustainable Forest Management in Europe.* Jointly prepared by Forest Europe, the United Nations Economic Commission for Europe and the United Nations Food and Agriculture Organization. Ministerial Conference on the Protection of Forests in Europe, Forest Europe Liaison Unit Oslo, Ås, Norway.

Forestry Commission (2004) *The UK Forestry Standard: The Government's Approach to Sustainable Forestry*, 2nd edn. Forestry Commission, Edinburgh, UK.

Fuller, R.J. (1992) Effects of coppice management on woodland breeding birds. In: Buckley, G.P. (ed.) *The Ecology and Management of Coppice Woodlands.* Chapman and Hall, London, pp. 169–192.

Fuller, R.J. (1995) *Bird Life of Woodland and Forest*. Cambridge University Press, Cambridge, UK.
Fuller, R.J. (2000) Influence of treefall gaps on distributions of breeding birds within interior old-growth stands in Białowieża Forest, Poland. *Condor* 102, 267–274.
Fuller, R.J. (2012) Avian responses to transitional habitat in temperate cultural landscapes: woodland edges and young-growth. In: Fuller, R.J. (ed.) *Birds and Habitat: Relationships in Changing Landscapes*. Cambridge University Press, Cambridge, UK, pp. 125–149.
Fuller, R.J. and Gill, R.M.A. (eds) (2001) Ecological impacts of deer in woodland. *Forestry* 74, 189–318.
Fuller, R.J. and Henderson, A.C.B. (1992) Distribution of breeding songbirds in Bradfield Woods, Suffolk, in relation to vegetation and coppice management. *Bird Study* 39, 73–88.
Fuller, R.J., Noble, D.G., Smith, K.W. and Vanhinsbergh, D. (2005) Recent declines in populations of woodland birds in Britain: a review of possible causes. *British Birds* 98, 116–143.
Fuller, R.J., Gaston, K.J. and Quine, C.P. (2007) Living on the edge: British and Irish birds in a European context. *Ibis* 149(supplement 2), 53–63.
Fuller, R.J., Smith, K.W. and Hinsley, S.A. (2012) Temperate western European woodland as a dynamic environment for birds: a resource-based view. In: Fuller, R.J. (ed.) *Birds and Habitat: Relationships in Changing Landscapes*. Cambridge University Press, Cambridge, UK, pp. 352–380.
Fyfe, R. (2007) The importance of local-scale openness within regions dominated by closed woodland. *Journal of Quaternary Science* 22, 571–578.
Gatter, W. (2007) Population dynamics, habitat choice, and breeding range of collared flycatcher *Ficedula albicollis* are determined by edible dormice *Glis glis*. *Limicola* 21, 3–47.
Gill, R.M.A. and Fuller, R.J. (2007) The effects of deer browsing on woodland structure and songbirds in lowland Britain. *Ibis* 149(supplement 2), 119–127.
Glue, D.E. (1973) The breeding birds of a New Forest valley. *British Birds* 66, 461–472.
Gómez, A. and Lunt, D. (2006) Refugia within refugia: patterns of phylogeographic concordance in the Iberian peninsula. In: Weiss, S. and Ferrand, N. (eds) *Phylogeography of Southern European Refugia*. Springer, Dordrecht, The Netherlands, pp. 155–188.
Granström, A. and Niklasson, M. (2008) Potentials and limitations for human control over historic fire regimes in the boreal forest. *Philosophical Transactions of the Royal Society B: Biological Sciences* 363, 2353–2358.
Gregory, R.D., Vorisek, P., Van Strien, A., Gmelig Meyling, A.W., Jiguet, F., Fornasari, L., Reif, J., Chylarecki, P. and Burfield, I.J. (2007) Population trends of widespread woodland birds in Europe. *Ibis* 149(supplement 2), 78–97.
Griswold, C.K. and Baker, A.J. (2002) Time to most recent common ancestor and divergence times of populations of common chaffinches (*Fringilla coelebs*) in Europe and North Africa: insights into Pleistocene refugia and current levels of migration. *Evolution* 56, 143–153.
Haapanen, A. (1965) Bird fauna of Finnish forests in relation to forest succession. *Annales Zoologici Fennici* 2, 153–196.
Halme, P., Allen, K.A., Auniņš, A., Bradshaw, R.H.W., Brūmelis, G., Čada, V., Clear, J.L., Eriksson, A.-M., Hannon, G., Hyvärinen, E. *et al.* (2013) Challenges of ecological restoration: lessons from forests in northern Europe. *Biological Conservation* 167, 248–256.
Harrison, C.J.O. (1987) Pleistocene and prehistoric birds of south-west Britain. *Proceedings of the University of Bristol Spelcological Society* 18, 81–104.
Hartel, T., Hanspach, J., Abson, D.J., Máthé, O., Moga, C.I. and Fischer, J. (2014) Bird communities in traditional wood-pastures with changing management in Eastern Europe. *Basic and Applied Ecology* 15, 385–395.
Hewson, C.M., Amar, A., Lindsell, J.A., Thewlis, R.M., Butler, S., Smith, K. and Fuller, R.J. (2007) Recent changes in bird populations in British broadleaved woodland. *Ibis* 149(supplement 2), 14–28.
Hinsley, S.A. and Gillings, S. (2012) Habitat associations of birds in complex changing cultural landscapes. In: Fuller, R.J. (ed.) *Birds and Habitat: Relationships in Changing Landscapes*. Cambridge University Press, Cambridge, UK, pp. 150–176.
Hinsley, S.A., Bellamy, P.E., Newton, I. and Sparks, T.H. (1996) Influences of population size and woodland area on bird species distributions in small woods. *Oecologia* 105, 100–106.
Hinsley, S.A., Carpenter, J.E., Broughton, R.K., Bellamy, P.E., Rothery, P., Amar, A., Hewson, C.A. and Gosler, A.G. (2007) Habitat selection by marsh tits *Poecile palustris* in the UK. *Ibis* 149(supplement 2), 224–233.
Hodder, K. and Bullock, J. (2005) The 'Vera model' of post-glacial landscapes in Europe: a summary of the debate. In: Hodder, K., Bullock, J., Buckland, P.C. and Kirby, K.J. *Large Herbivores in the Wildwood and in Modern Naturalistic Grazing Systems*. English Nature Research Reports No. 648, English Nature, Peterborough, UK, pp. 30–61.

Holt, C.A., Fuller, R.J. and Dolman, P.M. (2011) Breeding and post-breeding responses of woodland birds to modification of habitat structure by deer. *Biological Conservation* 144, 2151–2162.

Hopkins, J.J. and Kirby, K.J. (2007) Ecological change in British broadleaved woodland since 1947. *Ibis* 149(supplement 2), 29–40.

Huntley, B., Green, R., Collingham, Y. and Willis, S.G. (2007) *A Climate Change Atlas of Breeding Birds*. Lynx Edicions, Barcelona, Spain.

Jansson, G., Angelstam, P., Åberg, J. and Swenson, J.E. (2004) Management targets for the conservation of hazel grouse in boreal landscapes. *Ecological Bulletins* 51, 259–264.

Lampila, P., Monkkonen, M. and Desrochers, A. (2005) Demographic responses of birds to forest fragmentation. *Conservation Biology* 19, 1537–1546.

Leech, D.I. and Crick, H.Q.P. (2007) Influence of climate change on the abundance, distribution and phenology of woodland bird species in temperate regions. *Ibis* 149(supplement 2), 128–145.

Lorenc, M. (2006) On the taxonomic origins of Vistulian bird remains from cave deposits in Poland. *Acta Zoologica Cracoviensia* 49, 63–82.

Marey-Pérez, M.F. and Rodríguez-Vincente, V. (2009) Forest transition in northern Spain: local responses on large-scale programmes of field afforestation. *Land Use Policy* 26, 139–156.

Marthinsen, G., Wennerberg, L. and Lifjeld, J.T. (2008) Low support for separate species within the redpoll complex (*Carduelis flammea–hornemanni–cabaret*) from analyses of mtDNA and microsatellite markers. *Molecular Phylogenetics and Evolution* 47, 1005–1017.

Martin, T.G., Arcese, P. and Scheerder, N. (2011) Browsing down our natural heritage: deer impacts on vegetation structure and songbird populations across an island archipelago. *Biological Conservation* 144, 459–469.

Mason, W.L. (2007) Changes in the management of British forests between 1945 and 2000 and possible future trends. *Ibis* 149(supplement 2), 41–52.

Maxwell, J. (2002) Nest-site competition with blue tits and great tits as a possible cause of declines in willow tit numbers. *Glasgow Naturalist* 24, 47–50.

McComb, A.M.G., Kernohan, R., Mawhirt, P., Robinson, B., Weir, J. and Wells, B. (2010) Great spotted woodpecker (*Dendrocopos major*): proof of breeding in Tollymore Forest Park, Co. Down. *Irish Naturalists' Journal* 31, 66–67.

Mikusiński, G. (1995) Population trends in black woodpecker in relation to changes and characteristics of European forests. *Ecography* 18, 363–369.

Mikusiński, G. and Angelstam, P. (1998) Economic geography, forest distribution, and woodpecker diversity in central Europe. *Conservation Biology* 12, 200–208.

Mikusiński, G. and Angelstam, P. (2004) Occurrence of mammals and birds with different ecological characteristics in relation to forest cover in Europe: do macroecological data make sense? *Ecological Bulletins* 51, 265–275.

Milad, M., Schaich, H., Bürgi, M. and Konold, W. (2011) Climate change and nature conservation in central European forests: a review of consequences, concepts and challenges. *Forest Ecology and Management* 261, 829–843.

Mitchell, F.J.G. (2005) How open were European primaeval forests? Hypothesis testing using palaeoecological data. *Journal of Ecology* 93, 168–177.

Naef-Daenzer, B. (2012) Understanding individual life-histories and habitat choices: implications for explaining population patterns and processes. In: Fuller, R.J. (ed.) *Birds and Habitat: Relationships in Changing Landscapes*. Cambridge University Press, Cambridge, UK, pp. 408–431.

Nilsson, S.G. (1979) Density and species richness of some forest bird communities in south Sweden. *Oikos* 33, 392–401.

Paillet, Y., Bergès, L., Hjältén, J., Ódor, P., Avon, C., Bernhardt Römermann, M., Bijlsma, R.-J., De Bruyn L.U.C., Fuhr, M., Grandin, U.L.F. *et al.* (2010) Biodiversity differences between managed and unmanaged forests: meta-analysis of species richness in Europe. *Conservation Biology* 24, 101–112.

Pardini, A. (2009) Agroforestry systems in Italy: traditions towards modern management. In: Rigueiro-Rodríguez, A., McAdam, J. and Mosquera-Losada, M. (eds) *Agroforestry in Europe. Current Status and Future Prospects*. Springer, Dordrecht, The Netherlands, pp. 255–268.

Peach, W.J., Robinson, R.A. and Murray, K.A. (2004) Demographic and environmental causes of the decline of rural song thrushes *Turdus philomelos* in lowland Britain. *Ibis* 146(supplement 2), 50–59.

Peterken, G.F. (1993) *Woodland Conservation and Management*, 2nd edn. Chapman and Hall, London.

Peterken, G.F. (1996) *Natural Woodland*. Cambridge University Press, Cambridge, UK.

Pina, J.P. (1989) Breeding bird assemblages in eucalyptus plantations in Portugal. *Annales Zoologici Fennici* 26, 287–290.

Pithon, J.A., Moles, R. and O'Halloran, J. (2005) The influence of coniferous afforestation on lowland farmland bird communities in Ireland: different seasons and landscape contexts. *Landscape and Urban Planning* 71, 91–103.

Rackham, O. (1990) *Trees and Woodland in the British Landscape*, rev. edn. Dent, London.

Rackham, O. (2003) *Ancient Woodland: Its History, Vegetation and Uses in England*, rev. edn. Castlepoint Press, Dalbeattie, UK.

Radford, J.Q., Bennett, A.F. and Cheers, G.J. (2005) Landscape-level thresholds of habitat cover for woodland-dependent birds. *Biological Conservation* 124, 317–337.

Reif, J., Marhoul, P. and Koptik, J. (2013) Bird communities in habitats along a successional gradient: divergent patterns of species richness, specialization and threat. *Basic and Applied Ecology* 14, 423–431.

Riffell, S., Verschuyl, J., Miller, D. and Bently Wigley, T. (2011) Biofuel harvests, coarse woody debris, and biodiversity – a meta-analysis. *Forest Ecology and Management* 261, 878–887.

Rigueiro-Rodríguez, A., McAdam, J. and Mosquera-Losada, M. (eds.) (2009) *Agroforestry in Europe. Current Status and Future Prospects*. Springer, Dordrecht, The Netherlands.

Simms, E. (1971) *Woodland Birds*. Collins, London.

Smith, K.W. (2007) The utilization of dead wood resources by woodpeckers in Britain. *Ibis* 149(supplement 2), 183–192.

Smith, K.W., Burges, D.J. and Parks, R.A. (1992) Breeding bird communities of broadleaved plantation and ancient pasture woodlands of the New Forest. *Bird Study* 39, 132–141.

Soepboer, W. and Lotter, A.F. (2009) Estimating past vegetation openness using pollen–vegetation relationships: a modelling approach. *Review of Palaeobotany and Palynology* 153, 102–107.

Sommer, R.S., Benecke, N., Lougas, L., Nelle, O. and Schmölcke, U. (2011) Holocene survival of the wild horse in Europe: a matter of open landscape? *Journal of Quaternary Science* 26, 805–812.

Stroud, D.A., Reed, T.M., Pienkowski, M.W. and Lindsay, R.A. (1987) *Birds, Bogs and Forestry*. Nature Conservancy Council, Peterborough, UK.

Svenning, J.-C. (2002) A review of natural vegetation openness in north-western Europe. *Biological Conservation* 104, 133–148.

Sweeney, O., Wilson, M., Irwin, S., Kelly, T. and O'Halloran, J. (2010) Are bird density, species richness and community structure similar between native woodlands and non-native plantations in an area with a generalist bird fauna? *Biodiversity and Conservation* 19, 2329–2342.

Szabó, P. (2009) Open woodland in Europe in the Mesolithic and in the middle ages: can there be a connection? *Forest Ecology and Management* 257, 2327–2330.

Telleria, J.L. and Galarza, A. (1990) Avifauna y paisaje en el norte de España: efecto de las repoblaciones con arboles exoticos. *Ardeola* 37, 229–245.

Thirgood, J.V. (1989) Man's impact on the forests of Europe. *Journal of World Forest Resource Management* 4, 127–167.

Tomiałojć, L. (2000) Did white-backed woodpeckers ever breed in Britain? *British Birds* 93, 453–456.

Tomiałojć, L. and Wesołowski, T. (2005) The avifauna of Białowieża Forest: a window into the past. *British Birds* 98, 174–193.

Trzcinski, M.K., Fahrig, L. and Merriam, G. (1999) Independent effects of forest cover and fragmentation on the distribution of forest breeding birds. *Ecological Applications* 9, 586–593.

Tubbs, C.R. (1986) *The New Forest*. Collins, London.

Tubbs, C.R. (1997) A vision for rural Europe. *British Wildlife* 9, 79–85.

Turner, M.G., Gardner, R.H. and O'Neill, R.V. (2001) *Landscape Ecology in Theory and Practice*. Springer, New York.

Vera, F.W.M. (2000) *Grazing Ecology and Forest History*. CAB International, Wallingford, UK.

Wade, A.S.I., Barov, B., Burfield, I.J., Gregory, R.D. Norris, K. and Butler, S.J. (2013) Quantifying the detrimental impacts of land-use and management change on European forest bird populations. *PLoS ONE* 8(5): e64552.

Walankiewicz, W. (2002) Nest predation as a limiting factor to the breeding population size of the collared flycatcher *Ficedula albicollis* in the Białowieża National Park (NE Poland). *Acta Ornithologica* 37, 91–106.

Wallenius, T., Niskanen, L., Virtanen, T., Hottola, J., Brumelis, G. and Angervuori, A. (2010) Loss of habitats, naturalness and species diversity in Eurasian forest landscapes. *Ecological Indicators* 10, 1093–1101.

Wesołowski, T. (1983) The breeding ecology and behaviour of wrens *Troglodytes* under primaeval and secondary conditions. *Ibis* 125, 499–515.

Wesołowski, T. (2007) Primaeval conditions – what can we learn from them? *Ibis* 149(supplement 2), 64–77.

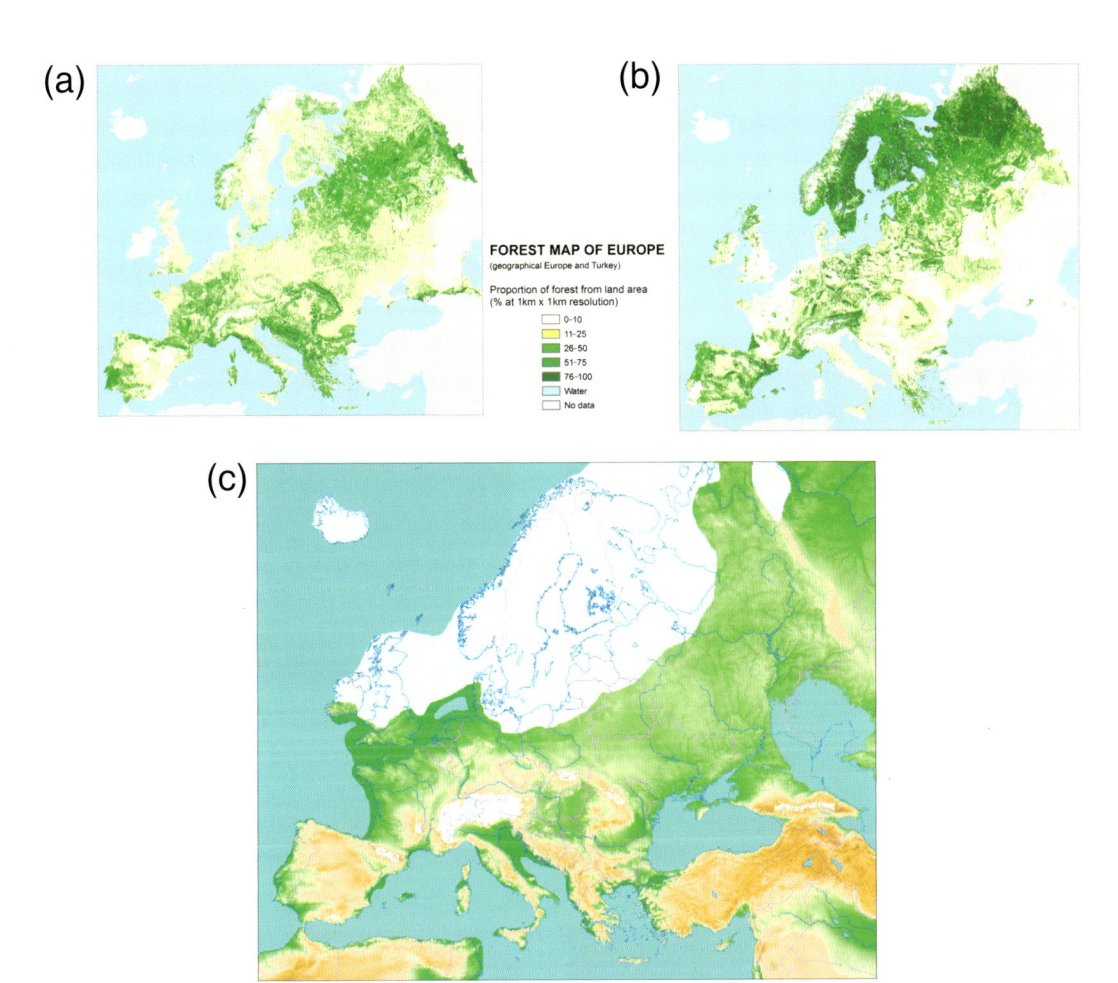

Plate 1 Landscapes present and past: (a) current distribution of broadleaved forest; (b) current distribution of coniferous forest; (c) extent of ice sheets in last glacial period; (d) reconstruction of Ice Age terrain and fauna in northern Spain. From: (a–b), European Forestry Institute, http://www.efi.int/portal/virtual_library/information_services/mapping_services/forest_map_of_europe (Päivinen et al., 2001; Schuck et al., 2002; Kempeneers et al., 2011, full details in Chapter 1); (c) http://en.wikipedia.org/wiki/Last_glacial_period#mediaviewer/File:Weichsel-W%C3%BCrm-Glaciation.png; (d) Public Library of Science, http://en.wikipedia.org/wiki/Mammoth_steppe#mediaviewer/File:Ice_age_fauna_of_northern_Spain_-_Mauricio_Ant%C3%B3n.jpg

Plate 2. Variations in European woodland: (a) open grazed woodland in Zaros, Crete (2012); (b) mature lime–hornbeam woodland Białowieża Forest, Poland (1988); (c) native pinewood at Creag Fhiaclach, Scotland (2007); (d) boreal forest landscape at Kuopio, Finland (1990). From: (a) Charles Watkins; (b–d) Keith Kirby.

Plate 3. Cultural features in woods: (a) a giant beech coppice stool at Monte Gottero, Liguria, Italy (2013); (b) a dead pine in Białowieża Forest, Poland (1985) with a beehive in a hollow; (c) wooded meadow in Laetatu, Estonia (2011); (d) outgrown pollard in the New Forest, England (2012). From: (a) Charles Watkins; (b–d) Keith Kirby.

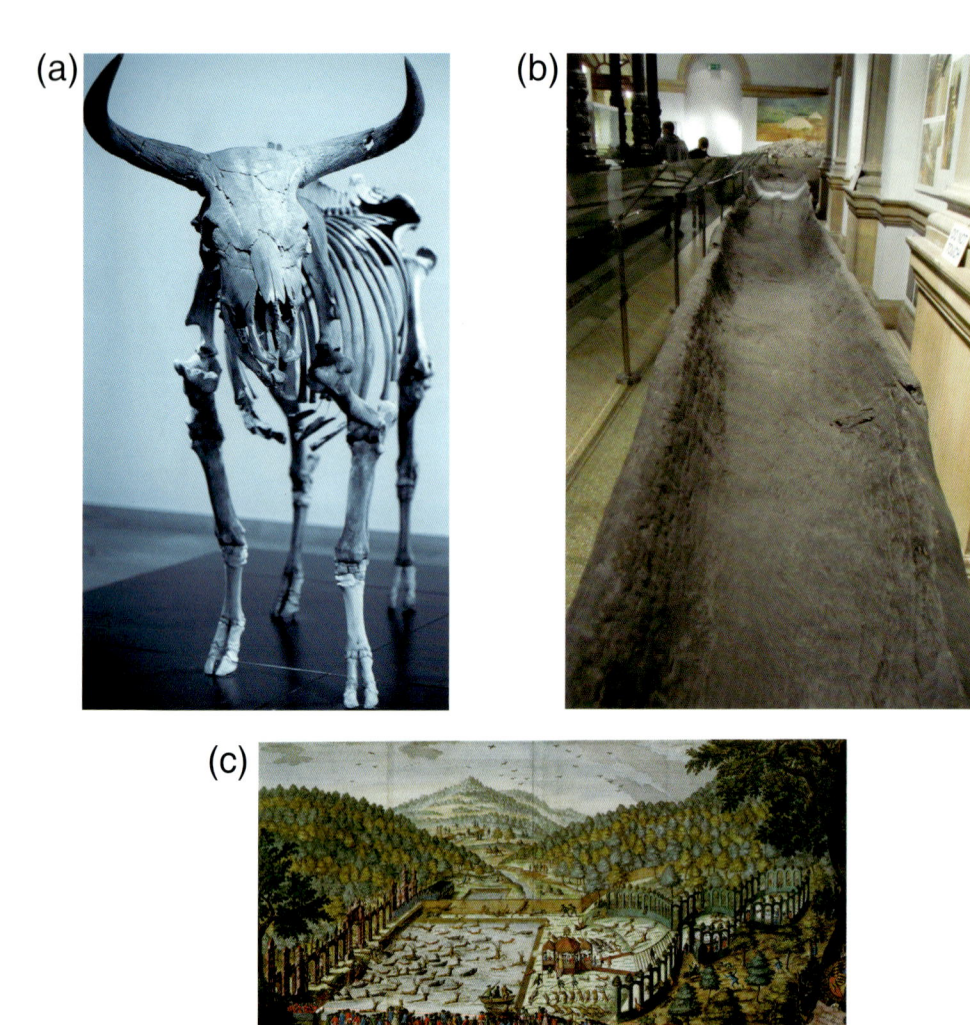

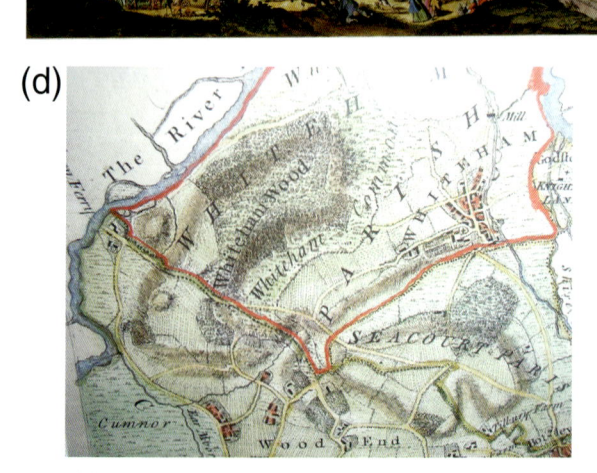

Plate 4. Evidence of the past: (a) skeleton of an aurochs; (b) a 15 m dugout canoe (c.2500 BC) in Dublin Museum; (c) coloured copperplate commemorating a hunt held near Stuttgart on 7 October 1748; (d) Extract from Rocque's map of Berkshire (1763) showing ancient woodland near Oxford. From: (a) image by Marcus Sümnick, National Museum, Copenhagen, http://en.wikipedia.org/wiki/File:Aurochse.jpg; (b) image by Keith Kirby; (c) courtesy of the Deutsches Jagd- und Fischereimuseum, Munich, Germany; (d) image by Keith Kirby.

Plate 5. Wood-pastures, veteran trees and their wildlife: (a) a wood-pasture in Cserépfalu, Hungary, with Hungarian grey cattle and their herder; (b) a hermit beetle, *Osmoderma eremita*, fitted for radio tracking; (c) a red deer stag, veteran trees and dead wood in Windsor Great Park, England; (d) veteran olive tree at Paliama, Crete (2012). From: (a) Anna Varga, 2012; (b) Roger Key, 2013; (c) Keith Kirby; (d) Charles Watkins.

Plate 6. Coppice working: (a) cut chestnut coppice poles in Valletti, Liguria, Italy (2013); (b) the spring flora, Swanton Novers, Norfolk (2009); (c) charcoal making using a kiln in Herefordshire, England (2012); (d) a traditional coppice worker in Blean Woods, Kent, England (1982). From: (a, c) Charles Watkins; (b, d) Keith Kirby.

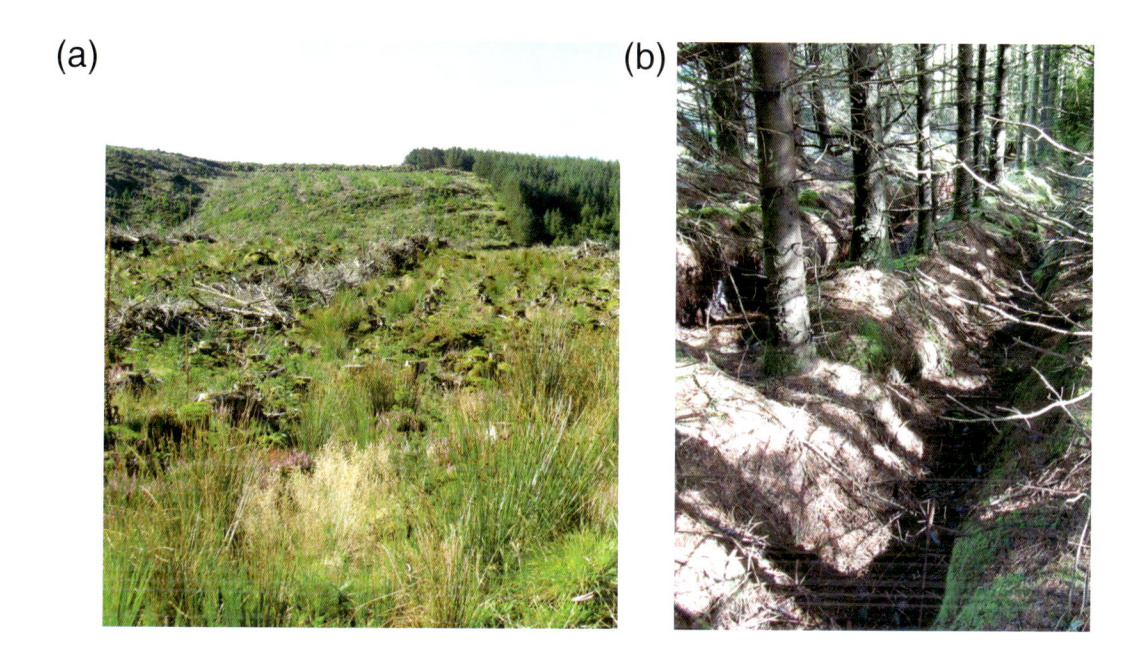

Plate 7. High forest: (a) large-scale clear-felled area in an afforestation site in Killarney, Ireland (2012); (b) deep ploughing to establish spruce on peatland in Killarney, Ireland (2012); (c) patch felling and regeneration in beech–conifer stands in the Black Forest, Germany (1991); (d) beech regeneration under oak at Fontainebleau, France (2013). From: (a–d) Keith Kirby.

Plate 8. Threats and opportunities: (a) Spruce affected by air pollution in the Black Forest, Germany (1991); (b) Dutch elm disease in Dukes Wood, Nottinghamshire, England (2013); (c) Heck cattle in Oostvaardersplassen, The Netherlands (2004); (d) Lady Slipper Orchid, *Cypripedium calceolus*, in Laetatu, Estonia (2011). From: (a, c, d) Keith Kirby; (b) Charles Watkins.

Wesołowski, T. and Fuller, R.J. (2012) Spatial variation and temporal shifts in habitat use by birds at the European scale. In: Fuller, R.J. (ed.) *Birds and Habitat: Relationships in Changing Landscapes.* Cambridge University Press, Cambridge, UK, pp. 63–92.

Wesołowski, T. and Tomiałojć, L. (1997) Breeding bird dynamics in a primaeval temperate forest: long-term trends in Białowieża National Park (Poland). *Ecography* 20, 432–453.

Wesołowski, T., Rowiński, P., Mitrus, C. and Czeszczewik, D. (2006) Breeding bird community of a primaeval temperate forest (Białowieża National Park, Poland) at the beginning of the 21st century. *Acta Ornithologica* 41, 55–70.

Wesołowski, T., Rowiński, P. and Maziarz, M. (2009) Wood warbler *Phylloscopus sibilatrix*: a nomadic insectivore in search of safe breeding grounds? *Bird Study* 56, 26–33.

Whitehouse, N.J. and Smith, D. (2010) How fragmented was the British Holocene wildwood? Perspectives on the 'Vera' grazing debate from the fossil beetle record. *Quaternary Science Reviews* 29, 539–553.

Whitehouse, M.J., Harrison, N.M., Mackenzie, J.A. and Hinsley, S.A. (2013) Preferred habitat of breeding birds may be compromised by climate change: unexpected effects of an exceptionally cold, wet spring. *PLoS ONE* 8(9): e75536.

Wilson, M.W., Pithon, J., Gittings, T., Kelly, T.C., Giller, P.S. and O'Halloran, J. (2006) Effects of growth stage and tree species composition on breeding bird assemblages of plantation forest. *Bird Study* 53, 225–236.

Wilson, J.D., Evans, A.D. and Grice, P.V. (2009) *Bird Conservation and Agriculture.* Cambridge University Press, Cambridge, UK.

Yalden, D.W. and Albarella, U. (2009) *The History of British Birds.* Oxford University Press, Oxford, UK.

Zerbe, S. (2002) Restoration of natural broad-leaved woodland in central Europe on sites with coniferous forest plantations. *Forest Ecology and Management* 167, 27–42.

13 Evolution and Changes in the Understorey of Deciduous Forests: Lagging Behind Drivers of Change

Martin Hermy*

Department of Earth and Environmental Science, University of Leuven, Leuven, Belgium

13.1 Introduction

The great changes in land cover that have occurred over the last few centuries are likely to continue over the coming decades (Goldewijk, 2001; Williams, 2006; Hansen *et al.*, 2013). In some places forests have been cleared, while elsewhere reforestation has taken place. This pattern of deforestation and reforestation is likely to recur, so a fragmented and changing forest cover either already is or is going to be the main characteristic of the world's future forests.

Some forests – such as ancient forests (Peterken, 1977; Hermy *et al.*, 1999) and old-growth forest (Nakashizuka, 1989; Mladenoff *et al.*, 1993) – have deep roots in the past. Others originated just a few centuries ago. Still others are mixtures of different types, often resulting in a complex mosaic of different origins (Verheyen *et al.*, 1999; Kirby and Watkins, Chapter 4).

In this chapter, we focus on the effects of former land use on the present-day composition and diversity of forest plant species in deciduous temperate forests, both in the above-ground vegetation and in the invisible but persistent seed banks. Most published work considers only land-use effects over the last two centuries, but recently, effects have been demonstrated of Gallo–Roman land use on the current species composition.

The imprint of the past is now widely recognized in what have been called 'ancient forest plants', often seen as flagship species, such as the bluebells (*Hyacinthoides non-scripta*) of lowland forests in the UK, the carpets of wood anemone (*Anemone nemorosa*) in many forests in other west European countries or the trilliums (e.g. *Trillium grandiflorum*) of the north-eastern USA. If we want to conserve or sustain these emblematic species we need to understand this phenomenon. Thus, we review the colonization rates found in ancient forest species and how these can be explained. Finally, we look at changes within forests and end with comments on the future of our forests.

13.2 Background

The history of the deciduous forests of Europe differs from that of North America where most of the large land clearance and land conversion to farmland goes back just a few centuries (Foster, 1992; Flinn and Vellend, 2005), but see also Burgess and Sharpe (1981). In contrast, particularly in western and southern Europe,

---

*E-mail: martin.hermy@ees.kuleuven.be

these changes go back to Gallo–Roman times and beyond (Dupouey *et al.*, 2002; Dambrine *et al.*, 2007; Etienne *et al.*, 2013). Further human influences arise from changes in forest management systems, for example from coppice to high forest (Decocq *et al.*, 2005; Van Calster *et al.*, 2008a; Hartel *et al.*, Chapter 5; Buckley and Mills, Chapter 6; Savill, Chapter 7), grazing effects (Kirby, 2001; Watkinson *et al.*, 2001; Heinken and Raudnitschka, 2002; Côté *et al.*, 2004; Boulanger *et al.*, 2009; Mudrak *et al.*, 2009) and fire management (Bergeron *et al.*, 2004; Certini, 2005).

Vast changes have also occurred in the landscape matrix around forest patches, such as the intensification of agricultural land (Firbank *et al.*, 2008) and urbanization (Godefroid and Koedam, 2003; Duguay *et al.*, 2007; McKinney, 2008; Vallet *et al.*, 2008). This can result in increased edge effects in forests (Fletcher *et al.*, 2007), such as nitrogen deposition (Gilliam, 2006; Wuyts *et al.*, 2008; Bobbink *et al.*, 2010) and the spread of introduced species (Martin *et al.*, 2009; Essl *et al.*, 2012).

The variation that we see in the forest flora reflects the interactions between the consequences of former land-use changes, management and environmental changes. There have been direct changes in the forest flora, such as loss of populations following reduction of forest cover, while other impacts may be delayed and are still to come – the so-called extinction debt (Vellend *et al.*, 2006; Kuussaari *et al.*, 2009). A long lifespan is a common feature of many forest plants (Bierzychudek, 1982; Ehrlen and Lehtilä, 2002), so such extinction debts may be more common than we yet realize.

In addition, we are faced with new impacts brought about by climate change (Sykes, 2009). The composition of forest ecosystems will alter as species adapt and migrate, increase and decline; some will become extinct. These effects will be felt over many decades which may, at least initially, mask the extent of the changes. It is extremely difficult to disentangle these interacting effects when looking at the composition of a particular forest or the distribution of a forest plant species. However, the past has left a vast imprint on the forests of today and understanding that relationship may help us to forecast or interpret future changes.

13.3 What Sorts of Plants Occur in Forests?

Most of the species in deciduous forests are arthropods; in temperate European forests only about 13% are plants (Fig. 13.1). Although trees are the dominant feature of forests, the number of tree species in temperate regions is

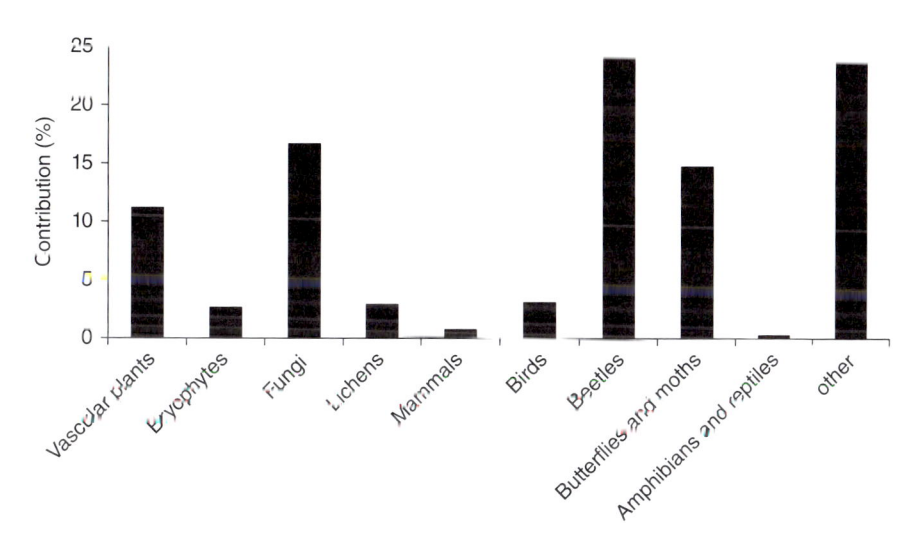

Fig. 13.1. Average contribution (%) of species groups in the total number of species (average number, 4228), as observed in six mixed deciduous forests in Europe. (Calculated from Hermy and Vandekerkhove, 2004.)

relatively small compared with the number of plant species in the understorey, especially those in the herbaceous layer. The ratio between the species richness of the herb layer and that of the overstorey tree layer varies between 2.0 and 10.0 (median, 5.1). In North American temperate forests, Gilliam (2007) found, similarly, that the herbaceous layer contained 84% of the total plant species of a forest.

The understorey (largely species less than 1 m high) of deciduous forest can be split into resident and transient species (Gilliam and Roberts, 2003). The resident species range from annuals to long-lived perennials (Ehrlen and Lehtilä, 2002). The transient species include seedlings and sprouts of shrubs and trees that may grow into the higher strata (Gilliam and Roberts, 2003), and those herbaceous species that emerge from the seed bank after disturbance events involving a temporary change of the light conditions (such as storm damage or tree cutting) and die out again above ground in later, more shady stages. Populations of transient species are, by definition, more dynamic than those of resident species, which may show only gradual changes through time.

In one sense, every species occurring in forests may be termed a 'forest species', but a more useful definition distinguishes species that depend to some extent on the physical existence or proximity of forests. Forest species in this strict sense may be subdivided into those in the forest interior (edge-avoiding or forest-core species) and those that are more common at forest edges (Matlack, 1994a; Ries *et al.*, 2004), which are also known as forest-peripheral species (Pellissier *et al.*, 2013). A further distinction is between species associated with interior edges such as clear-cuts or forest paths, and exterior edge species, which are more prone to the influence of surrounding land use (e.g. nitrogen deposition from farmland). The exterior edge species are often those of nutrient-rich environments, such as *Urtica dioica* and *Rubus fruticosus* agg.

In a study of 21 forests in Flanders, the author found 751 plant species (Fig. 13.2) or about 60% of the wild flora of Flanders. Only about 26% of the plant species found were forest-interior or forest-edge species; some 74% were more typical of open habitats, such as grasslands and heaths. The analysis of almost 20,000 plots from the French National Forest Inventory (Pellissier *et al.*, 2013) recorded 40 (18.7%) out of 214 species as forest-interior species and 38 (17.8%) as forest-edge species, a total of 78 forest species altogether. Forests in intensively managed cultural landscapes have become a last resort for a lot more than

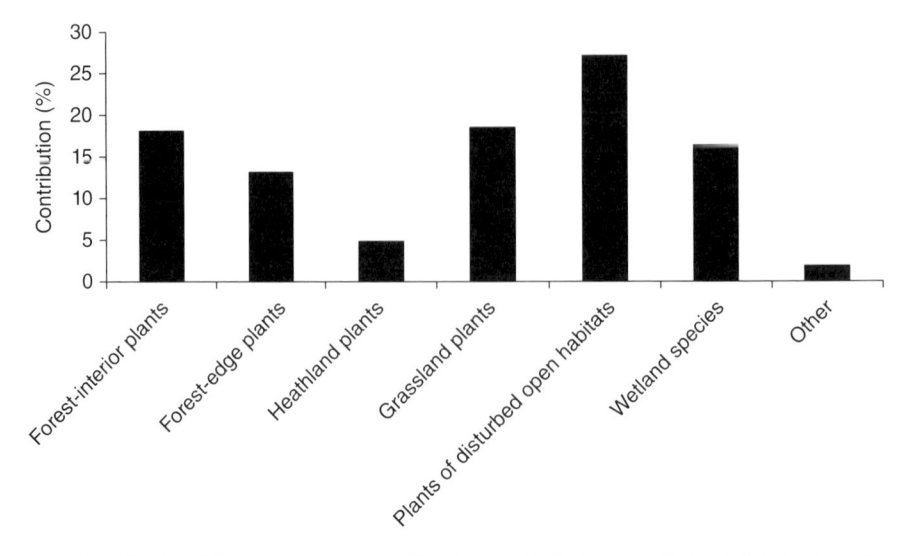

Fig. 13.2. Contribution (%) plant species classified into ecological groups in the total plant species richness in 21 forests in Flanders (751 species on a total area of 4060 ha). (Data from the author.)

just the forest species (Peterken and Francis, 1999).

The CSR (competitive, stress tolerant, ruderal) model of Grime (2001) considers stress and disturbance as the two main external factors determining the occurrence and abundance of species. Stress refers to the constraints limiting the growth of plant species, for example, shortages of light under the canopy or shortages of nutrients or water caused by competition from trees. Disturbance is linked with the partial or total destruction of plant biomass, perhaps as a consequence of tree felling, or the activities of pathogens or herbivores. The full spectrum of plant strategies is present in forests (Packham et al., 1992).

No forest plant can survive both high stress and high disturbance for long, but different species have developed different sets of traits in response to these challenges. Many forest species show greater tolerance of shade (Pellissier et al., 2013) and are more stress tolerant (cf. Hermy et al., 1999) than related species of open habitats. They also have heavy seeds (Thompson and Hodkinson, 1998; Verheyen et al., 2003), which means that their offspring are better prepared for the stressful (limited light, water, nutrients) but relatively stable environment that a forest normally offers (Crawley, 1997). Seed setting in a shaded environment may be irregular, so many forest plant species also reproduce vegetatively (Bierzychudek, 1982; Whigham, 2004). This is true both for forest herbaceous species, such as *Mercurialis perennis* and *Lamium galeobdolon*, and for some woody species, including *Ulmus procera*, *Sorbus torminalis*, *Populus tremula* and *Prunus spinosa*.

Disturbances in forests are, however, important. They usually result in more light reaching the forest floor and stimulate an increase in transient species, often species germinating from a persistent seed bank. These species 'wander' from opening to opening as biological nomads and, on average, have lower seed weights (cf. Salisbury 1942; Bierzychudek, 1982) than the typical forest-interior species. Thus, species display a variety of ecological 'strategies' to cope with life in forests.

13.4 Comparing Ancient and Recent Forests

In the fragmented and dynamic landscape that is typical of much of Europe, forests are a mosaic of stands with different histories. Patches of ancient forests (Hermy et al., 1999; Hermy and Verheyen, 2007) may be intermixed with recent forest that has grown up on open land in the last few 100 years. Although some plant species may colonize new forests quickly (e.g. *Arum maculatum*, *Geum urbanum*, *U. dioica*), others need centuries to do so (Peterken, 1981; Peterken and Game, 1984; Hermy et al., 1999; Verheyen et al., 2003; Hermy and Verheyen, 2007). These latter species are threatened by extinction where clearance rates for ancient forest are high.

Slow-colonizing species have been termed ancient forest species, as their presence suggests a long-continued existence of forest conditions (Hermy et al., 1999; Verheyen et al., 2003). Lists of such species in different European countries (and the USA) are available in the sources given in Table 13.1.

Hermy et al. (1999) found that about 30% of all European forest plant species have been described as ancient forest species. They tend to have a common 'ecological profile' (Hermy et al., 1999; Verheyen et al., 2003; Kimberley et al., 2014), being more shade tolerant than other forest plant species and avoiding very dry and wet sites. They also tend to be more stress tolerant than other forest plant species, which are more likely to belong to the competitive plant-strategy type. There is, though, considerable regional variation in ancient forest species; in some regions, a particular species may be typical of ancient forests, while in others it colonizes recent forests (Verheyen et al., 2003; De Frenne et al., 2011). Experience from the field suggests that all ancient forest species may be able to colonize to some extent and that on any particular site, each species responds in a unique way.

In most studies, the threshold date for defining ancient forest lies in the 17th to early 19th century (Goldberg, Chapter 22), so that ancient forests might have been cleared at some point before this, such as in early Roman times (Etienne et al., 2013). Even 2000 years after reforestation, forests may still exhibit

Table 13.1. Sources of lists of ancient forest species by country/region.

Country/region	Reference(s)
Belgium	Honnay et al., 1998
Denmark	Petersen, 1994; Graae, 2000; Graae et al., 2003
France	Sciama et al., 2009
Germany	Wulf, 1997
Great Britain	Rackham, 1980; Peterken and Game, 1984
Hungary	Kelemen et al., 2014
Ireland	Perrin and Daly, 2010
Italy	De Sanctis et al., 2010
Japan	Ito et al., 2004
Poland	Dzwonko and Loster, 1989; Jakubowska-Gabara and Mitka, 2007; Orczewska, 2010
Sweden	Brunet and Von Oheimb, 1998
USA	Matlack, 1994b
Europe and/or NE USA	Hermy et al., 1999; Verheyen et al., 2003; Hermy and Verheyen, 2007

clear floristic differences compared with forests that were never cleared in that period (Dupouey et al., 2002; Vanwalleghem et al., 2004; Dambrine et al., 2007; Plue et al., 2008). Thus, forests may not fully recover from large disturbance events such as deforestation and agricultural use, no matter how long ago this happened. Duffy and Meier (1992) reached a similar conclusion for the effects of clear-cutting of old-growth forest in the southern Appalachians of the USA.

Plue et al. (2008, 2009) showed that the effects of Roman land use persisted not only above ground, but also below ground in the seed banks. There were more occurrences of species such as *Verbascum thapsus*, *V. nigrum*, *Rorippa palustris*, *Cirsium arvense* and *Atropa bella-donna* that may have been introduced with Roman agriculture. Frequent recurring disturbances with regular light phases produced by coppicing every 15–30 years (Van Calster et al., 2008b) and natural events such as uprooting of trees (Jankowska-Blaszczuk and Grubb, 2006) would allow such seed-bank species to grow and flower, and enable the incorporation of fresh seeds into the soil. Ruderals found typically in arable fields, introduced via a period of agricultural use, often have highly persistent seed banks (Thompson and Hodkinson, 1998) and so would be able to bridge the periods between disturbances. Disturbance in forests may, therefore, allow persistence of both the resident and the transient species. However, the lengthening of forest management cycles, now common practice in Europe, may deplete the seed banks and ultimately change the composition of the forest (Van Calster et al., 2008b; Plue et al., 2010).

13.5 Colonization of New Forests

The overall forest cover in Europe (and in most European countries) has increased in past decades (FAO, 2010; Kirby and Watkins, Chapter 1). Nevertheless, these new forests will need time to develop to 'full-fledged' forest ecosystems, and the imprint of their (usually agricultural) past will be visible for centuries in their soils (Koerner et al., 1997; Verheyen et al., 1999; Flinn et al., 2005; Orczewska, 2010) and in their biota (Desender et al., 1999; Harmer et al., 2001; Verheyen et al., 2003). Soils under recent forest usually have higher pH and nutrient concentrations and lower organic matter than ancient forests soils, but the magnitude and persistence of these effects vary between regions (Koerner et al., 1997; Verheyen et al., 1999; Compton and Boone, 2000; Bellemare et al., 2002; Flinn et al., 2005) and between elements (for example, mobile basic cations versus persistent phosphate) (Honnay et al., 1999). That such differences persist in recent forests, and for total phosphate even after 2000 years (Provan, 1971; Dupouey et al., 2002), might contribute to the poor establishment of ancient forest species; but first, these species must be able to reach the new sites.

Where ancient forest is adjacent to recent forest, colonization rates of between c.20 to 100 m a century are typically found, depending on the plant species (NE USA, Matlack 1994b; Sweden, Brunet and Von Oheimb 1998; Belgium, Bossuyt et al., 1999), although in specific cases the rates may be higher (Orczewska, 2010). Rates are usually based on the furthest individual and the assumption that colonization started at the edge of the ancient forest (Hermy and Verheyen, 2007). The colonization success in isolated recent forest patches was found to be very low for 85% of the forest plant species and decreased strongly with increasing isolation (Honnay et al., 2002).

In a landscape with high connectivity between forests (42% forest cover), colonization success was clearly higher than in landscapes with low connectivity (7% forest cover) (Honnay et al., 2002). While colonization rates for most (ancient) forest plant species appeared to be less than a few metres a year, chance events may disperse plant seeds over longer distances (Cain et al., 1998; Vellend et al., 2003). Isolation might be less if species spread along hedgerows. The number of forest species in hedgerows does decline with increasing distance from the forest (Corbit et al., 1999; Sitzia, 2007; Roy and de Blois, 2008; Wehling and Diekmann, 2009), but a positive correlation of forest species number with the age of the hedgerow (Pollard et al., 1974; Deckers et al., 2005), suggests that the colonization of hedgerows takes a long time.

Recent forests thus have limited value for the conservation of forest plant species. Afforestation does increase the supply of some ecosystem services, such as biomass and carbon sequestration, and provide habitats for some plants and animals that are good dispersers, but the conservation value of these new forests will be low for centuries in contrast with the value of ancient forests (Peterken, 1977).

13.6 Dispersal and Recruitment Limitation

Dispersal limitation effects for forest species have been clearly shown (Honnay et al., 1999, 2002) (Fig. 13.3). Several studies have found that endo- and epi-zoochorous species are generally more successful at colonizing new forests than the other groups (Dzwonko, 1993; Matlack, 1994b; Brunet and Von Oheimb, 1998; Grashof-Bokdam and Geertsema, 1998; Bellemare et al., 2002; Honnay et al., 2002; Takahashi and Kamitani, 2004), although others have found no relation between dispersal mode and colonization ability (Mabry et al., 2000; Singleton et al., 2001; Wulf, 2004).

Dispersal categories often poorly represent realized dispersal distances (Vellend et al., 2003), and successful dispersal to a recent forest also depends on propagule pressure: the combination of the number of propagules (mostly seeds or fruits) produced by a species and the rate at which propagules arrive per unit time (Simberlof, 2009; Ricciardi et al., 2011). The reproductive output of forest species differs greatly (De Frenne et al., 2010) and so dispersal success might be expected to increase as the reproductive output increases. For seeds dispersed by animals, the colour of the fruit may influence the probability of dispersal (Galetti et al., 2003). Examples are the red berries of *A. maculatum*, a successful colonizer of recent forests versus the inconspicuous dark purple berries of *Paris quadrifolia*, an ancient forest plant species, and, in the USA, *T. undulatum* versus *Caulophyllum thalictroides* and *Allium tricoccum* (Flinn and Vellend, 2005).

Many typical forest species, and particularly ancient forest plant species, are usually assumed not to build up a persistent seed bank (Hermy et al., 1999; Verheyen et al., 2003), or where it exists, as for *Luzula pilosa*, it is small (Bossuyt and Hermy, 2001). If the sampling effort increases (both number of samples and sample plot area), then the probability of finding ancient forest plant species does increase (Plue et al., 2012), but the amount of viable seed of these species does remain low in comparison with that of species of edges and clear-cut areas. If land use for agriculture after the deforestation of ancient forest remains in place for a decade or more, then the survival probability of viable seeds of forest plant species is extremely low. Even if seeds did survive through to when the area becomes forest again they

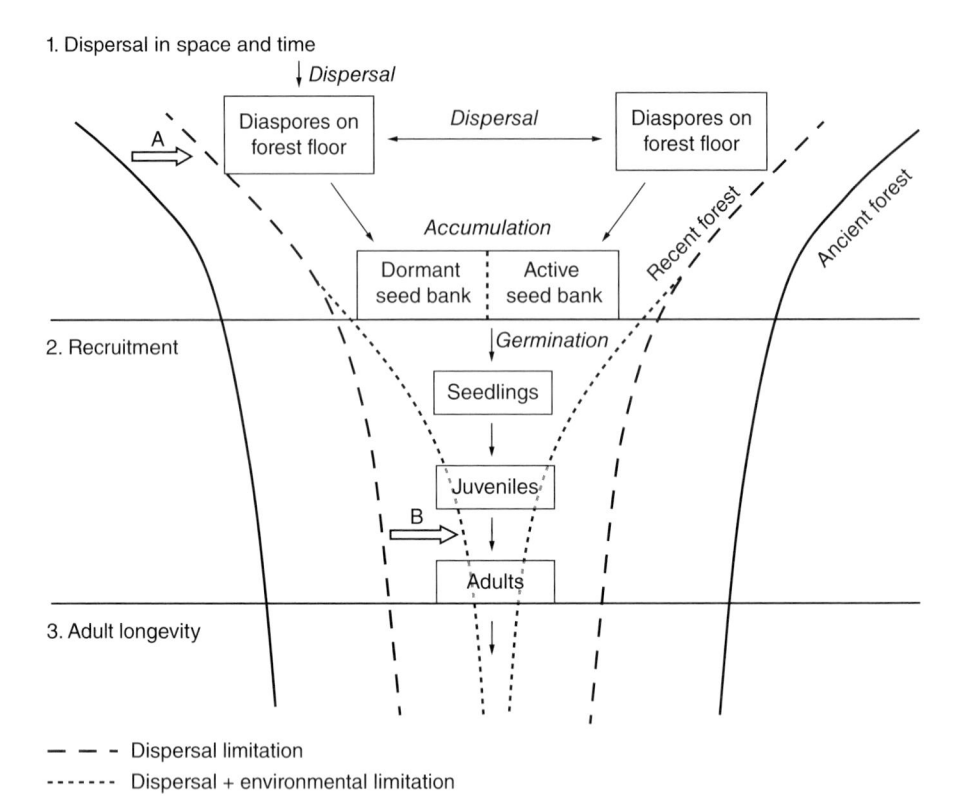

Fig. 13.3. Bottleneck model illustrating different successive processes of the colonization process, which can be impeded by a two stage-limitation (dispersal and environmental) in recent compared with ancient forest. The funnel-shaped lines represent the decreasing number of individuals through mortality and predation. Dispersal limitation causes lower seed occurrence on the forest floor in recent forest (arrow A). Environmental limitation then causes the funnel to narrow even more across different life stages (arrow B), depending on the species and the specific land-use history of the recent forest concerned (adapted from Baeten et al., 2009a,b,c). (Permission Wiley, © 2009 International Association for Vegetation Science.)

would recruit into a highly competitive environment in the new forest vegetation. This might mean that recruitment into new woods for ancient forest species is also a bottleneck (Flinn and Vellend, 2005) but, in fact, the performance of ancient forest plant species in recent forest is generally similar to or better than that in ancient forest (Donohue et al., 2000; Endels et al., 2004; but see Vellend, 2005). Introduction experiments using sowing and the transplanting of adults in recent forest have usually proved successful in the short term (Petersen and Philipp, 2001; Graae et al., 2004; Heinken, 2004; Verheyen and Hermy, 2004), but as the lifespans of many ancient forest plant species are relatively long (35 years and more; Ehrlen and Lehtilä, 2002), longer term experiments are needed to confirm this trend.

Baeten et al. (2009a) introduced a fast-colonizing (G. urbanum) and a slow-colonizing (P. elatior) species into both ancient and recent forest sites and monitored their performance for 8 years. Phosphorus availability was ten times greater in recent forest soils and this was also reflected in plant tissue samples. Species longevity was lower in recent forest sites, indicating faster turnover. The fast-colonizing G. urbanum counterbalanced this lower longevity by new establishment, whereas the number of slow-colonizing P. elatior fell. Thus, environmental constraints in recent forest may strengthen the differences in colonization capacity among forest herbs (Fig. 13.3).

Vellend (2005) and Flinn (2007) also found environmental barriers to the recolonization of forest herbs in recent forests established for more than 100 years on agricultural land.

Colonization can then be considered a three-stage process in which restricted seed availability, as a consequence of dispersal limitation, is followed by low recruitment and varying persistence (Fig. 13.3; Hipps *et al.*, 2005; Baeten *et al.*, 2009b). Van der Veken *et al.* (2007) reported on a cross-range edge transplant experiment established in 1960 with 27 experimental populations of *H. non-scripta*, 11 of which survived to 2005–2006. After 45 years, the populations remained small, covering from 0.23 to 26 m². Plant height, leaf length, number of flowers and seed size were generally lower in the introduced populations than in the original source populations, but the densities and proportions of non-flowering adults were higher, especially in the larger populations. This suggests that these populations are still expanding with more, but younger and so smaller, individuals. Observed migration rates from the original introduction points were very low, ranging from 0.006 to 0.06 m year^{-1}. Migration occurred by the establishment of isolated individuals, which were later absorbed by the advancing front of the main population. These results are probably typical for slow-dispersing forest plant species, for which occasional long-distance dispersal events cannot be excluded. However, even if these occasional events occur, it will take many decades to establish viable population sizes (Van der Veken *et al.*, 2007).

13.7 Changing Ancient Forests

In the study by Van der Veken *et al.* (2007) of *Hyacinthoides* populations, physical disturbance of the soil and major changes in the tree layer as a consequence of forest management proved to be the major determinants of population extinction over the 45 year period investigated. Therefore, we need to look at the consequences of relatively recent changes in management and environment alongside the effects of land-use history.

13.7.1 Management effects

Many ancient forests were formerly managed as coppice or coppice-with-standards. The regular occurrence of light phases and associated changes in environmental conditions, such as soil temperature, in these management systems may have favoured the abundance of many ancient forest plant species and, particularly, the vernal species. At each coppice cycle, the latter respond with abundant flowering and seed set (Ash and Barkham, 1976; Ford and Newbould, 1977; Rackham, 1980; Barkham, 1992). The carpets of species such as *A. nemorosa*, *H. non-scripta*, *P. elatior*, *P. vulgaris* and *Cardamine pratensis* in ancient forests are probably the result of centuries of coppicing (Buckley and Mills, Chapters 6 and 10). The recurrent disturbance also favours seed-banking species (Brown and Warr, 1992; Van Calster *et al.*, 2008b), and this may have a link back to Roman times (Plue *et al.*, 2008, 2009). In the 20th century, coppice and coppice-with-standards were gradually replaced by high forest management, which lengthens the cutting cycle to over 100 years for broadleaves and thereby severely reduces the frequency of occurrence of the light phase (Savill, Chapter 7).

At Montargis, south-east of Paris, Van Calster *et al.* (2008b) studied a chronosequence along the conversion pathway from coppice-with-standards to high forest. The seed-bank density and species richness both decreased with time since the last disturbance, which suggests that longer management cycles in forests tend to impoverish seed banks. The above-ground vegetation also changed following conversion to high forest. Low but significant overall increases were found for plot species richness and for the proportion of ruderal species. The differences in species composition between plots and the proportion of competitive and stress-tolerant species decreased significantly, suggesting an overall biotic homogenization (Vellend *et al.*, 2007).

Changes do not always result in a decline in species richness; losses in herbaceous diversity may be compensated for by an increase in tree seedling richness (Taverna *et al.*, 2005). In many cases, these changes are not driven by the invasion of non-native species, but by a reorganization of the native plant communities in response to eutrophication, especially nitrogen deposition, and to increasingly shaded conditions (Keith et al., 2009), something that is also seen in changes in regional floras (Smart et al., 2005).

Under coppice systems, heterogeneity was created at the forest scale by the short rotation cycles. Under high forest management, whether uneven or even aged, a higher proportion of the forest is in a relatively uniform young/mature closed canopy stage; there is also more systematic use of heavy machinery (Ampoorter et al., 2010). Even-aged high forest management systems may have a larger homogenization effect than uneven-aged high forest management systems (Van Calster et al., 2008a; Bürgi, Chapter 8).

13.7.2 Effects of environmental changes

Management change often occurs together with other environmental changes, especially acidification and eutrophication as a consequence of nitrogen deposition acting at both the forest and landscape scale (Thimonier et al., 1992; Lameire et al., 2000; Baeten et al., 2009c; Bobbink et al., 2010; Verheyen et al., 2012). In a study of the Meerdaal forest complex, which is located on loamy soils in central Belgium, Baeten et al. (2009c), using data from 1954 and 2000, found significant changes in plant species composition resulting from shifts in woody layer cover and increases in soil acidity. There was a strong decrease in the abundance of vernal species such as A. nemorosa, P. elatior and Paris quadrifolia, and an increase in shade-tolerant herbs and ferns such as Oxalis acetosella, Dryopteris dilatata and D. carthusiana.

Comparison of vegetation maps from 1954 and 2001 for the same forest (Hermy et al., 2009) showed spectacular changes in the area covered by various communities (Table 13.2): an almost total loss of non-forest communities; increases in oak–pine forest (with Deschampsia flexuosa and Molinia caerulea), in degraded oak–beech forest (with Rubus fruticosus agg., Prunus serotina, Quercus rubra, Holcus lanatus) and in oak–beech forest (with Milium effusum and Oxalis acetosella); and severe decreases in oak–beech forest (with Pteridium aquilinum, Convallaria majalis and Maianthemum bifolium), oak–ash forest (with Polygonatum multiflorum and Lonicera

Table 13.2. Changes in cover (in ha) of forest plant communities and associated open habitats between 1954 and 2001 in a mixed deciduous forest (Meerdaal Forest, central Belgium) (adapted from Hermy et al., 2009).

Forest plant community	Vegetation map 1954	% of total	Vegetation map 2001	% of total	% change in area from 1954 to 2001[a]
Ash–alder forest	16.6	1.3	19.1	1.5	15.1
Ash–oak forest with Polygonatum and Lonicera	425.7	34.5	234.3	19.0	−45.0
Ash–oak forest with Milium and Oxalis	30.1	2.4	11.1	0.9	−63.1
Elm–ash forest	7.4	0.6	7.3	0.6	−1.4
Oak–beech forest with Pteridum	551.5	44.7	265.3	21.5	−51.9
Oak–beech forest and Convallaria	10.1	0.8	123.6	10.0	1123.8
Oak–beech forest (degraded)	104.9	8.5	274.9	22.3	162.1
Oak–pine forest	8.3	0.7	297.6	24.1	3485.5
Communities from open habitats[b]	78.8	6.4	0	0.0	−100.0
Total with Milium and Oxalis area (as ha or %)	1233.6	100.0	1233.6	100.0	–

[a](2001 − 1954)/(1954) × 100
[b]Heathland with Calluna vulgaris, Genista anglica and Vaccinium myrtillus; fallow land with Agrostis capillaris; arable land with Arnoseris minima, Scleranthus annuus, Aphanes arvensis and Matricaria recutita; and grassland with Lolium perenne and Cynosurus cristatus.

periclymenum) and oak–ash forest (with *Milium effusum* and *Oxalis acetosella*). Overall, 444 ha of the forest complex kept the same forest type, whereas 789 ha (64%) changed in terms of the forest plant community. Most of this change can be considered as 'negative', i.e. acidification which lowers humus quality and an increase in common species such as *R. fruticosus* agg., *D. dilatata, Acer pseudoplatanus, Carex remota* (often along paths) and *Galeopsis tetrahit*. The species that increased included both common native species and invading species such as *A. pseudoplatanus*. Species that declined tended to be rarer or more specialized.

The increase in abundance of some native species has been referred to as 'over-dominance' and they have been called 'thugs' (Marrs *et al.*, 2010). Marrs *et al.* (2013) modelled the realized niche of potential over-dominant species in British forests and showed that four native species had a much larger cover than all other understorey species: *Hedera helix, M. perennis, Pteridium aquilinum* and *R. fruticosus* agg. Such over-dominance matches the field experience of many forest scientists and the competitive or stress-tolerant competitive strategy of these 'thugs' enables them to cope with modern conditions in our deciduous forests.

13.7.3 Effects of grazing

Eutrophication and management changes often interact with ungulate browsers (mainly deer) making it extremely difficult to disentangle individual influences. Most research on the impact of browsing on forest communities has revealed decreases in species richness or abundance of understorey plants (Rackham, 1980; Peterken and Jones, 1989; Kirby and Thomas, 2000; Kirby, 2001; Morecroft *et al.*, 2001; Rooney, 2001; Mudrak *et al.*, 2009). Trampling can also damage the herbaceous layer, particularly vernal species.

Yet at moderate levels of abundance, large herbivores can promote herbaceous diversity as some fast growing, often abundant and palatable species are browsed preferentially, notably in western Europe *Rubus fruticosus* agg., *H. helix, Vaccinium myrtillus, Vinca minor, M. perennis* (Kirby, 2001; Morecroft *et al.*, 2001; Van Uytvanck and Hoffmann 2009).

Other, often smaller and less palatable species or grazing-tolerant grasses that benefit from the created space include *A. nemorosa, P. elatior* and *Brachypodium sylvaticum* (Morecroft *et al.*, 2001; Van Uytvanck and Hoffmann 2009). A similar effect is seen in American forests (Potvin *et al.*, 2003; Royo *et al.*, 2010). Grazing management in forests is now a major tool for achieving conservation and restoration goals (Rodwell and Patterson, 1994; Wallis-DeVries *et al.*, 1998; Hartel *et al.*, Chapter 5), and large herbivores are accepted as a natural and essential part of forested ecosystems (Vera, 2000). They can also create shifting mosaics of grasslands, shrub thickets and trees, which are important for the conservation of both plants and fauna (Olff *et al.*, 1999).

13.7.4 Effects of invasive non-native species

Introduced understorey plant species have not generally colonized large areas of either ancient or recent forests, although locally there can be problems (Chabrerie *et al.*, 2010; Essl *et al.*, 2012). Many introduced and invasive plant species have early successional life-history traits (Martin *et al.*, 2009), which makes open places and forest edges more susceptible to invasion than the shady environment of the forest core. Only a few shade-tolerant species, often deliberately introduced trees and shrubs, have done well (Essl *et al.*, 2012). Invasive species are most likely to exploit gaps and other small-scale disturbances (Howard *et al.*, 2004), such as paths, skid trails and open places along rivers (e.g. *Impatiens glandulifera*), and often also conditions with high levels of nutrients. It may be that the small pool of exotic shade-tolerant invaders in western Europe, such as *A. platanoides* and *P. serotina*, will simply take longer to become a problem (Martin *et al.*, 2009).

13.8 Conserving and Expanding Forests: Does It Work?

In much of Europe, an active policy of forest protection is under way, including forest expansion as one element. However, both dispersal and recruitment limitations suggest that there will be only slow gains in forest species diversity from such expansion, and these gains may be offset if negative effects of management, grazing and environment on the flora of existing woodland are not countered, leading to the idea of a biodiversity balance (Jackson and Sax, 2009).

Baeten et al. (2010) resurveyed 20 plots in Flanders after a three-decade gap; half were in ancient forest and half were in recent forest or former agricultural land. The diversity in the ex-agricultural forests, contrary to expectation, did not increase, and the vegetation did not become more similar to that of ancient forest over time. Moreover, typical species in the ancient forest decreased, making the recovery of post-agricultural forests even more precarious (Fig. 13.4). These changes in the ancient forest may be interpreted as payment of an extinction debt.

In areas where fragmentation is a relatively 'recent' event (in the last few centuries), it might be that, given the longevity of many forest species, this extinction debt may be widespread (Vellend et al., 2006). Harmer et al. (2001) studied vegetation changes during 100 years of development of two ex-agricultural forests and similarly found a high turnover of species, but little colonization by typical ancient forest species, even after canopy closure. Lags in extinction and immigration, which lead extinction debt and immigration credit, respectively, are probably common features of forests. The longevity of most forest plant species postpones responses to important events such as land-use changes, climate change or even disturbances resulting in a delayed attainment of a diversity equilibrium (Jackson and Sax, 2009).

The changes in the flora within forests over time can be considerable, but they are often small compared with the changes in the flora of the surrounding landscape (Van Calster et al., 2008c). In the Thiérache region of northern France, which has had only small changes in land cover and urbanization between 1895 and 2005, forests lost about 13% (or 50 species) of their plant species, compared with over 20% for wetlands (39 species), arable (55 species) and other habitats (59 species). Across the major habitats, 31% of forest species decreased compared with 53% of arable land species (Fig. 13.5). Within forests, the

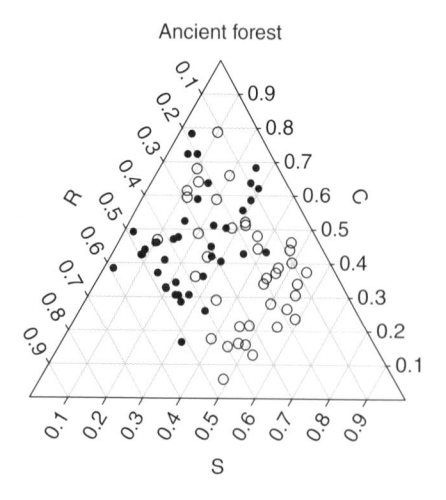

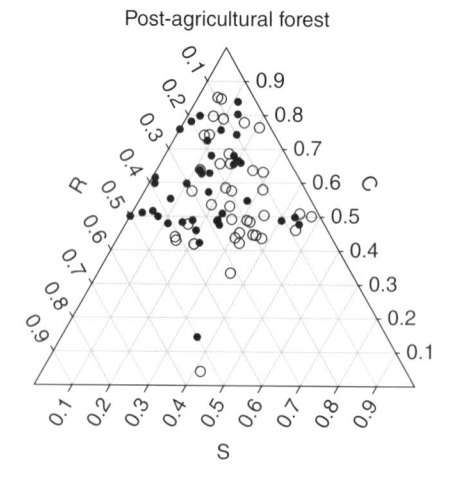

Fig. 13.4. Changes of the C–S–R signature of the vegetation in ancient and post-agricultural forest in Flanders between 1980 (open circles) and 2009 (filled circles). The plots from ancient forests (left) clearly show a significant shift towards the C component, particularly in contrast to the S component. For recent forest plots (right), no large shifts are visible. The C–S–R signature of a plot is a three-part numerical index (C, S and R coordinates) that represents the balance between the plant strategies (competitive–stress tolerant–ruderal, sensu Grime 2001) within the community based on the cover of individual species. The signature was calculated according to Hunt et al. (2004). (From Baeten et al., 2010.) (Permission Wiley, ©2010 The Authors. Journal compilation, ©2010 British Ecological Society.)

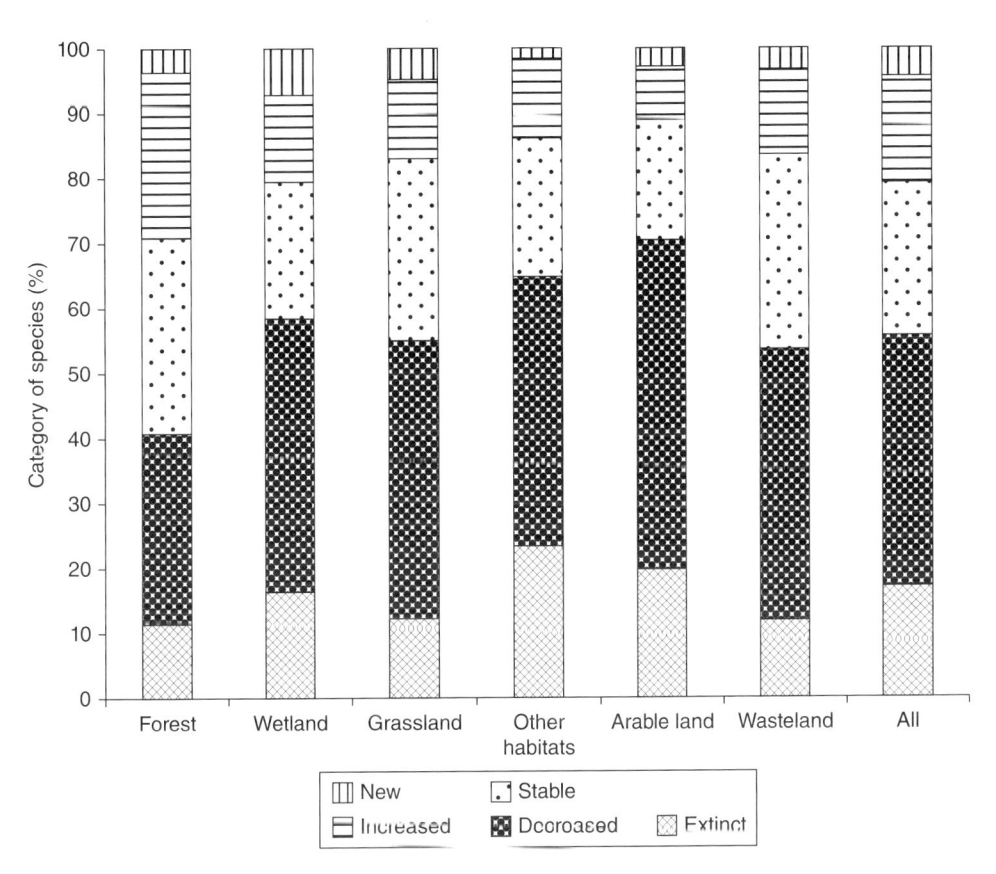

Fig. 13.5. Percentage of different categories of species – extinct, decreasing, stable, increasing and newly recorded – per habitat type between the end of the 19th century and 2000 in a rural area in the Thiérache region (N. France). For calculating the % stable and decreasing species, the number of newly recorded species was not taken into account. Total number of species per habitat type over both periods together were: forest, 413 species; wetland, 209 species; grassland, 248 species; arable land, 265 species; wasteland, 309 species; other habitats, 246 species. (Drawn from original data from Van Calster et al., 2008c.)

species that became extinct were those that were particularly localized and relatively rare species of open spaces and edges on both acid soils (e.g. *Genista pilosa, Erica cinerea, Euphrasia micrantha, Pedicularis sylvatica, Vaccinium myrtillus, V. vitis-idaea*) and basic soils (e.g. *Crepis praemorsa, Helleborus foetidus, Limodorum abortivum, Verbascum phlomoides*).

Many forests now contain sharp boundaries with the surrounding landscape, leaving almost no space for forest-edge species. Traditional forest uses, such as litter raking, forest grazing by livestock and the collection of bark for tannin were still widespread in the 19th century in the Thiérache (a region of France and Belgium). They have now disappeared completely, as have other traditional crafts (charcoal burners, clog makers, glassmakers) and the early iron industries and glass factories that depended on forest resources. Van Calster et al. (2008c) concluded that these practices had maintained forest ecosystems in a much more open state than nowadays by creating clearings in the forest interior and thus offered a habitat for a high diversity of plant species.

In forests, changes in community composition also lag behind climate change (Bertrand et al., 2011), as microclimate moderates the response of forest plant species to macro-climate warming (De Frenne et al., 2013). However, a thermophilization of the herbaceous forest plant species is already visible, and increasing the disturbance of forest ecosystems as a consequence of new economic perspectives (for example, harvesting biomass for bioenergy) may accelerate changes in biodiversity. So we are definitely facing new changes in forests, but when these will come about is much less clear. It is likely that shading by trees makes forest biodiversity lag behind major drivers of change, but for how long will this buffering effect hold?

References

Ampoorter, E., Van Nevel, L., De Vos, B., Hermy, M. and Verheyen, K. (2010) Assessing the effects of initial soil characteristics, machine mass and traffic intensity on forest soil compaction. *Forest Ecology and Management* 260, 1664–1676.

Ash, J.E. and Barkham, J.P. (1976) Changes and variability in the field layer of a coppiced woodland in Norfolk, England. *Journal of Ecology* 64, 697–712.

Baeten, L., Hermy, M. and Verheyen, K. (2009a) Environmental limitation contributes to the differential colonization capacity of two forest herbs. *Journal of Vegetation Science* 20, 209–223.

Baeten, L., Jacquemyn, H., Van Calster, H., Van Beek, E., Devlaeminck, R., Verheyen, K. and Hermy, M. (2009b) Low recruitment across life stages partly accounts for the slow colonization of forest herbs. *Journal of Ecology* 97, 109–117.

Baeten, L., Bauwens, B., De Schrijver, A., De Keersmaeker, L., Van Calster, H., Vandekerkhove, K., Roelandt, B., Beeckman, H. and Verheyen, K. (2009c) Herb layer changes (1954–2000) related to the conversion of coppice-with-standards forest and soil acidification. *Applied Vegetation Science* 12, 187–197.

Baeten, L., Hermy, M., Van Daele, S. and Verheyen, K. (2010) Unexpected understorey community development after 30 years in ancient and post-agricultural forests. *Journal of Ecology* 98, 1447–1453.

Barkham, J.P. (1992) The effects of coppicing and neglect on the performance of the perennial ground flora. In: Buckley, G.P. (ed.) *Ecology and Management of Coppice Woodlands*. Chapman and Hall, London, pp. 113–146.

Bellemare, J., Motzkin, G. and Foster, D.R. (2002) Legacies of the agricultural past in the forested present: an assessment of historical land-use effects on rich mesic forests. *Journal of Biogeography* 29, 1401–1420.

Bergeron, Y., Flannigan, M., Gauthier, S., Leduc, A. and Lefort, P. (2004) Past, current and future fire frequency in the Canadian boreal forest: implications for sustainable forest management. *Ambio* 33, 356–360.

Bertrand, R., Lenoir, J., Piedallu, C., Riofrio-Dillon, G., de Ruffray, P., Vidal, C., Pierrat, J.-C. and Gégout, J.-C. (2011) Changes in plant community composition lag behind climate warming in lowland forests. *Nature* 479, 517–520.

Bierzychudek, P. (1982) Life histories and demography of shade-tolerant temperate forest herbs: a review. *New Phytologist* 90, 757–776.

Bobbink, R., Hicks, K., Galloway, J., Spranger, T., Alkemade, R., Ashmore, M., Bustamante, M., Cinderby, S., Davidso, E., Dentener, F. *et al.* (2010) Global assessment of nitrogen deposition effects on terrestrial plant diversity: a synthesis. *Ecological Applications* 20, 30–59.

Bossuyt, B. and Hermy, M. (2001) Influence of land use history on seed banks in European temperate forest ecosystems: a review. *Ecography* 24, 225–238.

Bossuyt, B., Hermy, M. and Deckers, J. (1999) Migration of herbaceous plant species across ancient–recent forest ecotones in central Belgium. *Journal of Ecology* 87, 628–638.

Boulanger, V., Baltzinger, C., Saïd, S., Ballon, P., Picard, J.-F. and Dupouey, J.-L. (2009) Ranking temperate woody species along a gradient of browsing by deer. *Forest Ecology and Management* 258, 1397–1406.

Brown, A.H.F. and Warr, S.J. (1992) The effects of changing management on seed banks in ancient coppices. In: Buckley, G.P. (ed.) *Ecology and Management of Coppice Woodlands*. Chapman and Hall, London, pp. 147–166.

Brunet, J. and Von Oheimb, G. (1998) Migration of vascular plants to secondary woodlands in southern Sweden. *Journal of Ecology* 86, 429–438.

Burgess, R.L. and Sharpe, D.M. (eds) (1981) *Forest Island Dynamics in Man-Dominated Landscapes*. Springer, New York.

Cain, M.L., Damman, H. and Muir, A. (1998) Seed dispersal and the Holocene migration of woodland herbs. *Ecological Monographs* 68, 325–347.

Certini, G. (2005) Effects of fire on properties of forest soils: a review. *Oecologia* 143, 1–10.

Chabrerie, O., Loinard, J., Perrin, S., Saguez, R. and Decocq, G. (2010) Impact of *Prunus serotina* invasion on understory functional diversity in a European temperate forest. *Biological Invasions* 12, 1891–1907.

Compton, J.E. and Boone, R.D. (2000) Long-term impacts of agriculture on soil carbon and nitrogen in New England forests. *Ecology* 81, 2314–2330.

Corbit, M., Marks, P.L. and Gardescu, S. (1999) Hedgerows as habitat corridors for forest herbs in central New York, USA. *Journal of Ecology* 87, 220–232.

Côté, S.D., Rooney, T.P., Tremblay, J.-P., Dussault, C. and Waller, D.M. (2004) Ecological impacts of deer overabundance. *Annual Review of Ecology, Evolution and Systematics* 35, 113–147.

Crawley, M. (1997) *Plant Ecology*, 2nd edn. Blackwell Science, Oxford, UK.

Dambrine, E., Dupouey, J.-L., Laüt, L., Humbert, L., Thinon, M., Beaufils, T. and Richard, H. (2007) Present forest biodiversity patterns in France related to former Roman agriculture. *Ecology* 88, 1430–1439.

De Frenne, P., Graae, B.J., Kolb, A., Brunet, J., Chabrerie, O., Cousins, S.A.O., Decocq, G., Dhondt, R., Diekmann, M., Eriksson, O. et al. (2010) Significant effects of temperature on the reproduction output of the forest herb *Anemone nemorosa* L. *Forest Ecology and Management* 259, 809–817.

De Frenne, P., Baeten, L., Graae, B.J., Brunet, J., Wulf, M., Orczewska, A., Kolb, A., Jansen, I., Jamoneau, A., Jacquemyn, H. et al. (2011) Interregional variation in the floristic recovery of post-agricultural forests. *Journal of Ecology* 99, 600–609.

De Frenne, P., Rodríguez-Sánchez, F., Coomes, D.A., Baeten, L., Verstraeten, G., Vellend, M., Bernhardt-Romerman, M., Brown, C.D., Brunet, J., Cornelis, J. et al. (2013) Microclimate moderates plant responses to macroclimate warming. *Proceedings of the National Academy of Sciences USA* 110, 18561–18565.

De Sanctis, M., Alfo, M., Attorre, F., Francesconi, F. and Bruno, F. (2010) Effects of habitat configuration and quality on species richness and distribution in fragmented forest patches near Rome. *Journal of Vegetation Science* 21, 55–65.

Deckers, B., De Becker, P., Honnay, O., Hermy, M. and Muys, B. (2005) Sunken roads as habitats for forest plant species in a dynamic agricultural landscape: effects of age and isolation. *Journal of Biogeography* 32, 99–109.

Decocq, G., Aubert, M., Dupont, F., Bardat, J., Wattez-Franger, A., Saguez, R., De Foucault, B., Alard, D. and Delelis-Dusollier, A. (2005) Silviculture-driven vegetation changes in a European temperate deciduous forest. *Annals of Forest Science* 62, 313–323.

Desender, K., Ervynck, A. and Tack, G. (1999) Beetle diversity and historical ecology of woodlands in Flanders. *Belgian Journal of Zoology* 129, 139–156.

Donohue, K., Foster, D.R. and Motzkin, G. (2000) Effects of the past and the present on species distribution: land-use history and demography of wintergreen. *Journal of Ecology* 88, 303–316.

Duffy, D.C. and Meier, A.J. (1992) Do Appalachian herbaceous understories ever recover from clearcutting? *Conservation Biology* 6, 196–201.

Duguay, S., Eigenbrod, F. and Fahrig, L. (2007) Effects of surrounding urbanization on non-native flora in small forest patches. *Landscape Ecology* 22, 589–599.

Dupouey, J.-L., Dambrine, E., Laffite, J.D. and Moares, C. (2002) Irreversible impact of past land use on forest soils and biodiversity. *Ecology* 83, 2978–2984.

Dzwonko, Z. (1993) Relations between the floristic composition of isolated young woods and their proximity to ancient woodland. *Journal of Vegetation Science* 4, 693–698.

Dzwonko, Z. and Loster, S. (1989) Distribution of vascular plant species in small woodlands on the western Carpathian foothills. *Oikos* 56, 77–86.

Ehrlen, J. and Lehtilä, K. (2002) How perennial are perennial plants? *Oikos* 98, 308–322.

Endels, P., Adriaens, D., Verheyen, K. and Hermy, M. (2004) Population structure and adult performance of forest herbs in three contrasting habitats. *Ecography* 27, 225–241.

Essl, F., Mang, T. and Moser, D. (2012) Ancient and recent alien species in temperature forests: steady state and time lags. *Biological Invasions* 14, 1331–1342.

Etienne, D., Ruffaldi, P., Dupouey, J.-L., Georges-Leroy, M., Ritz, F. and Dambrine, E. (2013) Searching for ancient forests: a 2000 year history of land use in northeastern French forests deduced from the pollen compositions of closed depressions. *The Holocene* 23, 678–691.

FAO (2010) *Global Forest Resources Assessment 2010. Main Report.* FAO Forestry Paper 163, Food and Agriculture Organization of the United Nations, Rome. Available at: http://www.fao.org/docrep/013/i1757e/i1757e.pdf (accessed 24 November 2014).

Firbank, L.G., Petit, S., Smart, S., Blain, A. and Fuller, R. (2008) Assessing the impacts of agricultural intensification on biodiversity: a British perspective. *Philosophical Transactions of the Royal Society B: Biological Sciences* 363, 777–787.

Fletcher, R.J., Ries, L., Battin, J. and Chalfoun, A.D. (2007) The role of habitat area and edge in fragmented landscapes: definitively distinct or inevitably intertwined? *Canadian Journal of Zoology* 85, 1017–1030.

Flinn, K.M. (2007) Microsite-limited recruitment controls fern colonization of post-agricultural forests. *Ecology* 88, 3103–3114.

Flinn, K.M. and Vellend, M. (2005) Recovery of forest plant communities in post-agricultural landscapes. *Frontiers in Ecology and the Environment* 3, 243–250.

Flinn, K.M., Vellend, M. and Marks, P.L. (2005) Environmental causes and consequences of forest clearance and agricultural abandonment in central New York. *Journal of Biogeography* 32, 439–452.

Ford, E.D. and Newbould, P.J. (1977) The biomass and production of ground vegetation and its relation to tree cover through a deciduous woodland cycle. *Journal of Ecology* 65, 201–212.

Foster, D.R. (1992) Land-use history (1730–1990) and vegetation dynamics in central New England. *Journal of Ecology* 80, 753–772.

Galetti, M., Alves-Costa, C.P. and Cazetta, E. (2003) Effects of forest fragmentation, anthropogenic edges and fruit colour on the consumption of ornithocoric fruits. *Biological Conservation* 111, 269–273.

Gilliam, F.S. (2006) Response of the herbaceous layer of forest ecosystems to excess nitrogen deposition. *Journal of Ecology* 94, 1176–1191.

Gilliam, F.S. (2007) The ecological significance of the herbaceous layer in temperate forest ecosystems. *BioScience* 57, 845–858.

Gilliam, F.S. and Roberts, M.R. (2003) Conceptual framework for studies of the herbaceous layer. In: Gilliam, F.S. and Roberts, M.R. (eds) *The Herbaceous Layer in Forests of Eastern North America*. Oxford University Press, New York, pp. 3–14.

Godefroid, S. and Koedam, N. (2003) Distribution pattern of the flora in a peri-urban forest: an effect of the city–forest ecotone. *Landscape and Urban Planning* 65, 169–185.

Goldewijk, K.K. (2001) Estimating global land use change over the past 300 years: the HYDE database. *Global Biogeochemical Cycles* 15, 417–433.

Graae, B.J. (2000) The effect of landscape fragmentation and forest continuity on forest floor species in two regions of Denmark. *Journal of Vegetation Science* 11, 881–892.

Graae, B.J., Sunde, P. and Fritzboger, B. (2003) Vegetation and soil differences in ancient opposed to new forests. *Forest Ecology and Management* 177, 179–190.

Graae, B.J., Hansen, T. and Sunde, P.B. (2004) The importance of recruitment limitation in forest plant species colonization: a seed sowing experiment. *Flora* 199, 263–270.

Grashof-Bokdam, C.J. and Geertsema, W. (1998) The effect of isolation and history on colonization patterns of plant species in secondary woodland. *Journal of Biogeography* 25, 837–846.

Grime, J.P. (2001) *Plant Strategies, Vegetation Processes, and Ecosystem Properties*, 2nd edn. Wiley, Chichester, UK.

Hansen, M.C., Potapov, P.V., Moore, R., Hancher, M., Turubanova, S.A., Tyukavina, A., Thau, D., Stehman, S.V., Goetz, S.J., Loveland, T.R. *et al.* (2013) High-resolution global maps of 21st-century forest cover change. *Science* 342, 850–853.

Harmer, R., Peterken, G., Kerr, G. and Poulton, P. (2001) Vegetation changes during 100 years of development of two secondary woodlands on abandoned arable land. *Biological Conservation* 101, 291–304.

Heinken, T. (2004) Migration of an annual myrmecochore: a four year experiment with *Melampyrum pratense* L. *Plant Ecology* 170, 55–72.

Heinken, T. and Raudnitschka, D. (2002) Do wild ungulates contribute to the dispersal of vascular plants in central European forests by epizoochory? A case study in NE Germany. *Forstwissenschaftliches Centralblatt* 121, 179–194.

Hermy, M. and Vandekerkhove, K. (2004) Bosgebieden. In: Hermy, M., De Blust, G. and Slootmaekers, M. (eds) *Natuurbeheer*. Davidsfonds, Leuven, Belgium, pp. 307–359.

Hermy, M. and Verheyen, K. (2007) Legacies of the past in the present-day forest biodiversity: a review of past land-use effects on forest plant species composition and diversity. *Ecological Research* 22, 361–371.

Hermy, M., Honnay, O., Firbank, L., Grashof-Bokdam, C. and Lawesson, J. (1999) Ecological comparison between ancient forest plant species of Europe and the implications for forest conservation. *Biological Conservation* 91, 9–22.

Hermy, M., Baeten, L., Roelandt, B. and Plue, J. (2009) Flora en vegetatie: ook een spiegel van het verleden? In: Baeté, H., De Bie, M., Hermy, M. and Van den Bremt, P. (eds) *Miradal, Erfgoed in Meerdaalwoud en Heverleebos*. Davidsfonds, Leuven, Belgium, pp. 189–215.

Hipps, N.A., Davies, M.J., Dodds, P. and Buckley, G.P. (2005) The effects of phosphorus nutrition and soil pH on the growth of some ancient woodland indicator plants and their interaction with competitor species. *Plant and Soil* 271, 131–141.

Honnay, O., Hermy, M. and Degroote, B. (1998) Ancient forest plant species in Western Belgium. *Belgian Journal of Botany* 130, 139–154.

Honnay, O., Hermy, M. and Coppin, P. (1999) Impact of habitat quality on forest plant species colonization. *Forest Ecology and Management* 115, 157–170.

Honnay, O., Verheyen, K., Butaye, J., Jacquemyn, H., Bossuyt, B. and Hermy, M. (2002) Possible effects of climate change and habitat fragmentation on the range of forest plant species. *Ecology Letters* 5, 525–530.

Howard, T.G., Gurevitch, J., Hyatt, L., Carreiro, M. and Lerdau, M. (2004) Forest invasibility in communities in southeastern New York. *Biological Invasions* 6, 393–410.

Hunt, R., Hodgson, J.G., Thompson, K., Bungener, P., Dunnett, N.P. and Askew, A.P. (2004) A new practical tool for deriving a functional signature for herbaceous vegetation. *Applied Vegetation Science* 7, 163–170.

Ito, S., Nakayama, R. and Buckley, G.P. (2004) Effects of previous land-use on plant species diversity in semi-natural and plantation forests in a warm-temperate region in southeastern Kyushu, Japan. *Forest Ecology and Management* 196, 213–225.

Jackson, S.T. and Sax, D.F. (2009) Balancing biodiversity in a changing environment: extinction debt, immigration credit and species turnover. *Trends in Ecology and Evolution* 25, 153–160.

Jakubowska-Gabara, J. and Mitka, J. (2007) Ancient woodland plant species in a landscape park in central Poland. *Acta Societatis Botanicorum Poloniae* 76, 239–249.

Jankowska-Blaszczuk, M. and Grubb, P.J. (2006) Changing perspectives on the role of the soil seed bank in northern temperate deciduous forests and in tropical lowland rain forests: parallels and contrasts. *Perspectives in Plant Ecology, Evolution and Systematics* 8, 3–21.

Keith, S.A., Newton, A.C., Morecroft, M.D., Bealey, C.E. and Bullock, J.M. (2009) Taxonomic homogenization of woodland plant communities over 70 years. *Proceedings of the Royal Society B: Biological Sciences*, 276, 3539–3544.

Kelemen, K., Krivan, A. and Standovar, T. (2014) Effects of land-use history and current management on ancient woodland herbs in western Hungary. *Journal of Vegetation Science* 25, 172–183.

Kimberley, A.K., Smart, S., Blackburn, G., Whyatt, J. and Kirby, K.J. (2014) Identifying the trait syndromes of conservation indicator species: how distinct are British ancient woodland indicator plants from other woodland species? *Applied Vegetation Science* 16, 667–675.

Kirby, K.J. (2001) The impact of deer on the ground flora of British broadleaved woodland. *Forestry* 74, 219–229.

Kirby, K.J. and Thomas, R. (2000) Changes in the ground flora in Wytham Woods, southern England from 1974 to 1991 – implications for nature conservation. *Journal of Vegetation Science* 11, 871–880.

Koerner, W., Dupouey, J.L., Dambrine, E. and Benoît, M. (1997) Influence of past land use on the vegetation and soils of present day forest in the Vosges mountains, France. *Journal of Ecology* 85, 351–358.

Kuussaari, M., Bommarco, R., Heikkinen, R.K., Helm, A., Krauss, J., Lindborg, R., Öckinger, E., Pärtel, M., Pino, J., Rodà, F. et al. (2009) Extinction debt: a challenge for biodiversity conservation. *Trends in Ecology and Evolution* 24, 564–571.

Lameire, S., Hermy, M. and Honnay, O. (2000) Two decades of change in the ground vegetation of a mixed deciduous forest in an agricultural landscape. *Journal of Vegetation Science* 11, 695–704.

Mabry, C., Ackerly, D. and Gerhardt, F. (2000) Landscape and species level distribution of morphological and life history traits in a temperate woodland flora. *Journal of Vegetation Science* 11, 213–224.

Marrs, R.H., Le Duc, M.G., Smart, S.M., Kirby, K.J., Bunce, R.G.H. and Corney, P.M. (2010) Aliens or natives: who are the 'thugs' in British woods? *Kew Bulletin* 65, 583–594.

Marrs, R.H., Kirby, K.J., Le Duc, M.G., McAllister, H., Smart, S.M., Oksanen, J., Bunce, R.G.H. and Corney, P.M. (2013) Native dominants in British woodland – a potential cause of reduced species-richness? *New Journal of Botany* 3, 156–168.

Martin, P.H., Canham, C.D. and Marks, P.L. (2009) Why forests appear resistant to exotic plant invasions: intentional introductions, stand dynamics, and the role of shade tolerance. *Frontiers in Ecology and the Environment* 7, 142–149.

Matlack, G.R. (1994a) Vegetation dynamics of the forest edge – trends in space and successional time. *Journal of Ecology* 82, 113–123.

Matlack, G.R. (1994b) Plant species migration in a mixed-history forest landscape in eastern North America. *Ecology* 75, 1491–1502.

McKinney, M.L. (2008) Effects of urbanization on species richness: a review of plants and animals. *Urban Ecosystems* 11, 161–176.

Mladenoff, D.T., White, M.A., Pastor, J. and Crow, T.R. (1993) Comparing spatial pattern in unaltered old-growth and disturbed forest landscapes. *Ecological Applications* 3, 294–306.

Morecroft, M.D., Taylor, M.E., Ellwood, S.A. and Quinn, S.A. (2001) Impacts of deer herbivory on ground vegetation at Wytham woods, central England. *Forestry* 74, 252–257.

Mudrak, E.L, Johnson, S.E. and Waller, D.M. (2009) Forty-seven year changes in vegetation at the Apostle Islands: effects of deer on the forest understory. *Natural Area Journal* 29, 167–176.

Nakashizuka, T. (1989) Role of uprooting in composition and dynamics of an old-growth forest in Japan. *Ecology* 70, 1273–1278.

Olff, H., Vera, F.W.M., Bokdam, J., Bakker, E.S., Gleichman, J.M., de Maeyer, K. and Smit, R. (1999) Shifting mosaics in grazed woodlands driven by the alternation of plant facilitation and competition. *Plant Biology* 1, 127–137.

Orczewska, A. (2010) Colonization capacity of herb woodland species in fertile, recent alder woods adjacent to ancient forest sites. *Polish Journal of Ecology* 58, 297–310.

Packham, J.R., Harding, D.J.L., Hilton, G.M. and Stuttard, R.A. (1992) *Functional Ecology of Woodland and Forests*. Kluwer, Dordrecht, The Netherlands.

Pellissier, V., Bergès, L., Nedeltcheva, T., Schmitt, M.-C., Avon, C., Cluzeau, C. and Dupouey, J.-L. (2013) Understorey plant species show long-range spatial patterns in forest patches according to distance-to-edge. *Journal of Vegetation Science* 24, 9–24.

Perrin, P.M. and Daly, O.H. (2010) *A Provisional Inventory of Ancient and Long-established Woodland in Ireland*. Irish Wildlife Manuals No. 46, National Parks and Wildlife Service, Department of the Environment, Heritage and Local Government, Dublin.

Peterken, G.F. (1977) Habitat conservation priorities in British and European woodlands. *Biological Conservation* 11, 223–236.

Peterken, G.F. (1981) *Woodland Conservation and Management*. Chapman and Hall, London.

Peterken, G.F. and Francis, J.L. (1999) Open spaces as habitats for vascular ground flora species in the woods of central Lincolnshire, UK. *Biological Conservation* 91, 55–72.

Peterken, G.F. and Game, M. (1984) Historical factors affecting the number and distribution of vascular plant species in the woodlands of central Lincolnshire. *Journal of Ecology* 72, 155–182.

Peterken, G.F. and Jones, E.W. (1989) 40 Years of change in Lady-Park-Wood—the young-growth stands. *Journal of Ecology* 77, 401–429.

Petersen, P.M. (1994) Flora, vegetation, and soil in broadleaved ancient and planted woodland, and scrub on Røsnæs, Denmark. *Nordic Journal of Botany* 14, 693–709.

Petersen, P.M. and Philipp, M. (2001) Implantation of forest plants in a wood on former arable land: a ten year experiment. *Flora* 196, 286–291.

Plue, J., Hermy, M., Verheyen, K., Thuillier, P., Saguez, R. and Decocq, G. (2008) Persistent changes in forest vegetation and seed bank 1600 years after human occupation. *Landscape Ecology* 23, 673–688.

Plue, J., Dupouey, J.-L., Verheyen, K. and Hermy, M. (2009) Forest seed banks along an intensity gradient of ancient agriculture. *Seed Science Research* 19, 103–114.

Plue, J., Van Gils, B., Peppler-Lisbach, C., De Schrijver, A., Verheyen, K. and Hermy, M. (2010) Seed-bank convergence under tree species during forest development. *Perspective in Plant Ecology, Evolution and Systematics* 12, 211–218.

Plue, J., Thompson, K., Verheyen, K. and Hermy, M. (2012) Seed banking in ancient forest species: why total sampled area really matters. *Seed Science Research* 22, 123–133.

Pollard, E., Hooper, M.D. and Moore, N.W. (1974) *Hedges*. William Collins, London.

Potvin, F., Beaupré, P. and Laprise, G. (2003) The eradication of balsam fir stands by white-tailed deer on Anticosti Island. Quebec: a 150-year process. *Ecoscience* 10, 487–495.

Provan, D.M. (1971) Soil phosphate analysis as a tool in archaeology. *Norwegian Archive Review* 4, 37–50.

Rackham, O. (1980) *Ancient Woodland, Its History, Vegetation and Uses in England*. Arnold, London.

Ricciardi, A., Jones, L.A., Kestrup, A.M. and Ward, J.M. (2011) Expanding the propagule pressure concept to understand the impact of biological invasions. In: Richardson, D.M. (ed.) *Fifty Years of Invasion Ecology: The Legacy of Charles Elton*, 1st edn. Wiley-Blackwell, Chichester, UK, pp. 225–235.

Ries, L., Fletcher, R.J. Jr, Battin, J. and Sisk, T.D. (2004) Ecological responses to habitat edges: mechanisms, models and variability explained. *Annual Review of Ecology and Systematics* 35, 491–522.

Rodwell, J. and Patterson, G. (1994) *Creating New Native Woodlands*. Forestry Commission Bulletin 112, Her Majesty's Stationery Office, London.

Rooney, T.P. (2001) Deer impacts on forest ecosystems: a North American perspective. *Forestry* 74, 201–208.

Roy, V. and de Blois, S. (2008) Evaluating hedgerow corridors for the conservation of native forest herb diversity. *Biological Conservation* 141, 298–307.

Royo, A.A., Stout, S.S., deCalesta, D.S. and Pierson, T.G. (2010) Restoring forest herb communities through landscape-level deer herd reductions: is recovery limited by legacy effects? *Biological Conservation* 143, 2425–2434.

Salisbury, E.J. (1942) *The Reproductive Capacity of Plants*. G. Bell and Sons, London.

Sciama, D., Augusto, L., Dupouey, J.-L., Gonzalez, M. and Domínguez, C.M. (2009) Floristic and ecological differences between recent and ancient forests growing on non-acidic soils. *Forest Ecology and Management* 258, 600–608.

Simberlof, D. (2009) The role of propagule pressure in biological invasions. *Annual Review in Ecology and Systematics* 40, 81–102.

Singleton, R., Gardescu, S., Marks, P.L. and Geber, M.A. (2001) Forest herb colonization of post-agricultural forests in central New York, USA. *Journal of Ecology* 89, 325–338.

Sitzia, T. (2007) Hedgerows as corridors for woodland plants: a test on the Po Plain, northern Italy. *Plant Ecology* 188, 235–252.

Smart, S.M., Bunce, R.G.H., Marrs, R., Le Duc, M., Firbank, L.G., Maskell, L.C., Scott, W.A., Thompson, K. and Walker, K.J. (2005) Large-scale changes in the abundance of common higher plant species across Britain between 1978, 1990 and 1998 as a consequence of human activity: tests of hypothesised changes in trait representation. *Biological Conservation* 124, 355–371.

Sykes, M.T. (2009) Climate change impacts: vegetation. In: *Encyclopedia of Life Sciences (ELS)*. Wiley, Hoboken, New Jersey. Available at Wiley Online Library: http://dx.doi.org/10.1002/9780470015902.a0021227 (accessed 1 December 2014).

Takahashi, K. and Kamitani, T. (2004) Effect of dispersal capacity on forest plant migration at a landscape scale. *Journal of Ecology* 92, 778–785.

Taverna, K., Peet, R.K. and Phillips, L.C. (2005) Long-term change in ground layer vegetation of deciduous forests of the North Carolina Piedmont, USA. *Journal of Ecology* 93, 202–213.

Thimonier, A., Dupouey, J.L. and Timbal, J. (1992) Floristic changes in the herb-layer vegetation of a deciduous forest in the Lorraine plain under the influence of atmospheric deposition. *Forest Ecology and Management* 55, 149–167.

Thompson, K. and Hodkinson, D.J. (1998) Seed mass, habitat and life history: a re-analysis of Salisbury (1942, 1974). *New Phytologist* 138, 163–167.

Vallet, J., Daniel, H., Beaujouan, V. and Rozé, F. (2008) Plant species response to urbanization: comparison of isolated woodland patches in two cities of north-western France. *Landscape Ecology* 23, 1205–1217.

Van Calster, H., Baeten, L., Verheyen, K., De Keersmaeker, L., Dekeyser, S., Rogister, J.E. and Hermy, M. (2008a) Diverging effects of overstorey conversion scenarios on the understorey vegetation in a former coppice-with-standards forest. *Forest Ecology and Management* 256, 519–528.

Van Calster, H., Chevalier, R., Van Wyngene, B., Archaux, F., Verheyen, K. and Hermy, M. (2008b) Long-term seed bank dynamics in a temperate forest under conversion from coppice-with-standards to high forest management. *Applied Vegetation Science* 11, 251–260.

Van Calster, H., Vandenberghe, R., Ruysen, M., Verheyen, K., Hermy, M. and Decocq, G. (2008c) Unexpectedly high 20th century floristic losses in a rural landscape in northern France. *Journal of Ecology* 96, 927–936.

Van der Veken, S., Rogister, J., Verheyen, K., Hermy, M. and Nathan, R. (2007) Over the (range) edge: a 45-year transplant experiment with the perennial forest herb *Hyacinthoides non-scripta*. *Journal of Ecology* 95, 343–351.

Van Uytvanck, J. and Hoffmann, M. (2009) Impact of grazing management with large herbivores on forest ground flora and bramble understorey. *Acta Oecologica* 35, 523–532.

Vanwalleghem, T., Verheyen, K., Hermy, M., Poesen, J. and Deckers, S. (2004) Legacies of Roman land-use in the present-day vegetation in Meerdaal Forest (Belgium)? *Belgian Journal of Botany* 137, 181–187.

Vellend, M. (2005) Land-use history and plant performance in populations of *Trillium grandiflorum*. *Biological Conservation* 124, 217–224.

Vellend, M., Myers, J.A., Gardescu, S. and Marks, P.L. (2003) Dispersal of *Trillium* seeds by deer: implications for long distance migration of forest herbs. *Ecology* 84, 1067–1072.

Vellend, M., Verheyen, K., Jacquemyn, H., Kolb, A., Van Calster, H., Peterken, G. and Hermy, M. (2006) Extinction debt of forest plants persists for more than a century following habitat fragmentation. *Ecology* 87, 542–548.

Vellend, M., Verheyen, K., Flinn, K.M., Jacquemyn, H., Kolb, A., Van Calster, H., Peterken, G.F., Graae, B.J., Bellemare, J., Honnay, O. *et al.* (2007) Homogenization of forest plant communities and weakening of species–environment relationships via agricultural land use. *Journal of Ecology* 95, 565–573.

Vera, F.W.M. (2000). *Grazing Ecology and Forest History*. CAB International, Wallingford, UK.

Verheyen, K. and Hermy, M. (2004) Recruitment and growth of herb-layer species with different colonizing capacities in ancient and recent forest. *Journal of Vegetation Science* 15, 125–134.

Verheyen, K., Bossuyt, B., Hermy, M. and Tack, G. (1999) The land use history (1278–1990) of a mixed hardwood forest in western Belgium and its relationship with chemical soil characteristics. *Journal of Biogeography* 26, 1115–1128.

Verheyen, K., Honnay, O., Motzkin, G., Hermy, M. and Foster, D. (2003) Response of forest plant species to land-use change: a life-history based approach. *Journal of Ecology* 91, 563–577.

Verheyen, K., Baeten, L., De Frenne, P., Bernhardt-Römmerman, M., Brunet, J., Cornelis, J., Decocq, G., Dierschke, H., Eriksson, O., Hédl, R. *et al.* (2012) Driving factors behind the eutrophication signal in understorey plant communities of deciduous temperate forests. *Journal of Ecology* 100, 352–365.

WallisDeVries, M.F., Bakker, J.P. and Van Wieren, S.E. (1998) *Grazing and Conservation Management*. Kluwer, Dordrecht, Netherlands.

Watkinson, A.E., Riding, A.E. and Cowie, N.R. (2001) A community and population perspective of the possible role of grazing in determining the ground flora of ancient woodlands. *Forestry* 74, 231–239.

Wehling, S. and Diekmann, M. (2009) Importance of hedgerows as habitat corridors for forest plants in agricultural landscapes. *Biological Conservation* 142, 2522–2530.

Whigham, D.F. (2004) Ecology of woodland herbs in temperate deciduous forests. *Annual Review in Ecology and Systematics* 35, 583–621.

Williams, M. (2006) *Deforesting the Earth: From Prehistory to Global Crisis: An Abridgement*. University of Chicago Press, Chicago, Illinois.

Wulf, M. (1997) Plant species as indicators of ancient woodland in northwestern Germany. *Journal of Vegetation Science* 8, 635–642.

Wulf, M. (2004) Plant species richness of afforestation with different former use and habitat continuity. *Forest Ecology and Management* 195, 191–204.

Wuyts, K., Verheyen, K., De Schrijver, A., Cornelis, W.M. and Gabriels, D. (2008) The impact of forest edge structure on longitudinal patterns of deposition, wind speed, and turbulence. *Atmospheric Environment* 42, 8651–8660.

14 Gains and Losses in the European Mammal Fauna

Robert Hearn*
Laboratorio di Archeologia e Storia Ambientale, Università degli Studi di Genova, Genoa, Italy

14.1 Introduction

Since 1970, global vertebrate populations have declined by around 30% (McRae *et al.*, 2012), with mammals declining by 25% (Baillie *et al.*, 2010). In 2013, the IUCN (International Union for Conservation of Nature) Red List (http://www.iucnredlist.org/) categorized 25% of the assessed extant mammal species as threatened. Nevertheless, some species are reclaiming parts of their historic ranges across Europe (Deinet *et al.*, 2013).

We cannot track the gains and losses of most of the animals found in European woods and forests; for example, what changes in shrew distributions might there have been over the last 10,000 years? However, we have a considerable amount of information about such changes for the larger mammals. These larger mammals are important for the functioning of forests; and apart from the trees themselves, have been the species group most directly influenced by human activities, such as hunting.

In this chapter, I explore how and why selected species that have a particular significance in the character of the European forest, have declined or increased. My focus is on: the aurochs (the wild ox) (Plate 4), which would have been important in creating openness in the early Holocene landscape; large carnivores, such as wolves, that affect the abundance and behaviour of deer and other herbivores; beavers as important shapers of the riverine environments; and squirrels and deer that currently have major effects on woodland regeneration and management.

14.2 Aurochs

In October 2013, over 1000 delegates from 65 nations met in the medieval centre of Salamanca, Spain, for the 10th World Wilderness Congress. The image of an extinct species was omnipresent throughout proceedings: that of the aurochs (*Bos primigenius*).

The aurochs, the extinct progenitor of domestic cattle, was a 'ubiquitous species that was very successful in the late Pleistocene and early Holocene and was widespread over most of the northern hemisphere with the exception of North America' (Clutton-Brock, 1987). The image of the aurochs from cave paintings at Lascaux (France), combined with historical accounts and archaeological remains, have been used to reconstruct the physical characteristics of the beast. Aurochsen 'were

---

*E-mail: robert.hearn@edu.unige.it

relatively long-legged, especially when compared to modern cattle' with 'a plausible shoulder height of 160–180 cm in the bull, *c*.150 cm in the cow', the height of the aurochs 'at the withers nearly equalling the length of the trunk' (Van Vuure, 2002). Zeuner (1953, 1963; translated and summarized by Clutton-Brock, 1987) concluded that, in general, the aurochs bull was mostly black with white markings on the muzzle, forehead and running along the ridge of the back, with occasional light saddle patches, while the cow and calf were more or less red in colour.

Rare, written accounts add to our understanding. Julius Caesar's *The Gallic Wars* describes the 'aurochsen' that inhabited the Hercynian Forest as being 'extraordinary in speed and strength, sparing neither man nor wild beast which they have espied' (McDevitte and Bohn, 1869). The ferocity of the aurochs, and thereby the heroic connotations attached to its hunting, are a common component of later descriptions. Eginhard of Gall's *Early Lives of Charlemagne* details a violent confrontation between the Frankish king and a furious aurochs (Grant, 1922). These sources reveal the prized status attached to the aurochs' horns, conspicuous by their size, but the meat also garnered noble connotations. Ulrich Von Richental's *Chronique du Concile de Constance* (1417) contains both written and visual depictions of aurochs meat being sent as gifts from the Polish King Jogaila to the English King Henry V; King Sigismundus of Poland similarly sent meat to the Holy Roman Emperor Charles V in the 16th century (Baillie-Graham, 1913).

The symbolic, heraldic connotations imbued in the gifting of aurochs meat and horns would have been enhanced by the increasing rarity of the animals. The species is thought to have become extinct earlier in the south and west of Europe, but to have held on longer in the north and east (Van Vuure, 2002). On the Danish islands, the aurochs disappeared around 5500 BC, and from Britain in *c*.1500 BC (Yalden, 1999). There is no evidence of their presence in the Netherlands after 400 BC, or in Russia and Hungary after the 12th and 13th centuries, or in Germany after the late 15th century. Despite efforts to conserve the waning numbers through the suspension of hunting privileges and through winter feeding, the last aurochs died of natural causes in the Jaktorów Forest of east-central Poland in 1627 (Van Vuure, 2005).

The Swiss botanist Anton Schneeberger (1530–1581) described the Polish aurochsen and their habitat. The increasingly rare herds of aurochs sought refuge in dense, marshy woodlands, feeding on acorns, leaves and branches of shrubs and trees. In the summer months, the aurochsen left these areas in search of open pasture, often venturing into sown fields to feed on the grain, from which they had to be chased away by farmers and hunters (Van Vuure, 2005). Subsequent work tends to confirm this association with marshes and marshy forests, as does Hall (2008) for the species in Britain. Woodland clearance, problems with the animals trespassing on to agricultural land and competition with domestic stock could all have contributed to the species decline (Van Vuure, 2005).

In the 1920s, two German zoologists (the Heck brothers) tried to recreate the aurochs by cross-breeding primitive-type cattle and selecting for what they perceived to be aurochs-like characteristics. Some of the herd produced at Munich Zoo survived the war (Van Vuure, 2005). Their descendants form the current breed known as Heck cattle, or sometimes 'recreated aurochs', that have been introduced to various nature reserves, most notably The Oostvaardersplassen in the Netherlands (Lorimer and Driessen, 2013) (Plate 8). The Stichting Taurus project (www.taurosproject.com) is undertaking a breeding programme with the most promising 'primitive' breeds (i.e. those with many characteristics of the former aurochs) from Europe. Other researchers, such as the Polish Foundation for Recreating the Aurochs (PFOT), are trying to rebreed aurochsen using DNA samples derived from bone remains in European natural history museums to clone the species (Maier, 2012; Seddon *et al.*, 2014).

It is difficult to assess the extent of the role played by the aurochs in the landscape and its impact on tree and woodland cover. Over most of its range, it had ceased to be important ecologically long before it became extinct as a species, and the conditions where it was last recorded may not be typical of its habitat in its heyday. Vera (2000) proposes a key role for aurochs (along with bison and wild horse) in creating and maintaining half-open landscapes across much of Europe, but others question whether the animals would have had an impact on this scale (Van Vuure, 2005; Kirby and Watkins, Chapter 3).

Some of the ecological role of the aurochs may subsequently have been taken on by domestic stock grazing in wood-pastures, but domestic animals behave differently and their populations are more directly controlled by humans. Insights into the role of the species may come eventually from those areas where cattle (including Heck cattle) are being allowed to go feral in woodland systems as part of rewilding trials (Lorimer and Driessen, 2013, 2014). The aim is that this more naturalistic grazing will mimic the effects of extinct wild mega-herbivores on the floral composition and biodiversity of diverse habitats, landscapes and environments. These projects are still mostly too young to draw many conclusions and they raise the question that if we are reintroducing lost large herbivores, should the lost carnivores not also be added?

14.3 Carnivores

The Large Carnivore Initiative for Europe (www.lcie.org) seeks 'to maintain and restore, in coexistence with people, viable populations of large carnivores as an integral part of ecosystems and landscapes'. These include the wolverine (*Gulo gulo*), the Iberian lynx (*Lynx pardinus*) and the golden jackal (*Canis aureus*), but most attention has been paid to the recovery and return of the European populations of the brown bear (*Ursus arctos*), Eurasian lynx (*L. lynx*) and, particularly, the wolf (*Canis lupus*) (Breitenmoser, 1998).

14.3.1 The wolf

The wolf is one of the most adaptable and widely distributed of all land mammals. It tolerates a wide range of climatic and environmental conditions, and occupies a diverse range of habitats from arctic tundra and the taigas to temperate woodlands and prairies, Mediterranean maquis and sub-deserts, and even highly modified human environments (Mech and Boitani, 2003). In 2013, 7% (c.12,000 animals) of the global wolf population were found in ten European populations: from Finland to northern Ukraine; throughout the Balkan countries; around the Carpathians in eastern Europe; in central Scandinavia; along the Alps and the Italian peninsula; in northern Spain and Portugal, with small populations in south-central Spain; and across Germany and Poland (Deinet *et al.*, 2013). However, this distribution and abundance is only a shadow of what it once was, the species being once ubiquitous throughout the continent and present on many islands.

The decline and disappearance of wolves generally began in the late medieval to early modern period. By 1800, they could no longer be found in the British Isles, Ireland, the coastal lowlands of France, the low countries, Denmark, Germany and Poland (Boitani, 2000). Between 1800 and 1960, the wolf's European range contracted by nearly 50%. Their number and range reached their lowest levels between 1930 and 1960, by which time the species had disappeared from all central and northern European countries (Deinet *et al.*, 2013).

Various factors contributed to this range and population collapse, but hunting was a major factor. Wolf hunting is richly documented in archival repositories and in iconographical sources from the medieval and early modern period (Hearn *et al.*, 2015). Their depredations on livestock and wild game led to the hunting of wolves being encouraged through bounties offered by governments, and was further assisted by the development and incorporation of new technologies and techniques, particularly firearms (Bernard, 1981).

Rising human populations and the spread of farming contributed to the extermination of the wolf as it was brought into increasingly close proximity with people through competition for space and resources (as is seen with tigers and people on the Indian subcontinent today). The wolf, which had often occupied a position of cultural and totemic reverence among societies and communities was gradually transformed into a quasi-mythological beast, and a highly vilified animal (Ortalli, 1997). From the medieval period onward, the wolf was increasingly shrouded in a thick and elaborate cloak of allegory and metaphor, and the species became the subject of cultural demonization in a way that did not apply to other conspicuous, carnivorous mammals (Boitani, 1987). This would not seem an auspicious background for its revival.

In recent decades though, wolves have reappeared in a large number of habitats that they historically occupied (Deinet et al., 2013). The legal frameworks that once encouraged the elimination of the species have been repealed and replaced by those protecting it through much of Europe (Boitani, 2000): the Convention on International Trade of Endangered Species of Wild Fauna and Flora of 1975; the Bern Convention in 1979; and the EC Habitats Directive of 1992 (92/43/EEC). Signature parties are required to undertake measures geared towards the protection and promotion of populations of the species.

Local, national and international management action plans suggest that the increased public acceptance of the wolf has also been an important factor in the spread of the species in many parts of Europe (Musiani et al., 2009). Efforts have been made to raise its image through awareness and education campaigns to inform people of the complex ecological niche occupied by the wolf as a keystone species important in the regulation of ungulate populations, and to rehabilitate the image of the 'big, bad wolf' (Boitani, 2000).

The small, isolated, relic lupine populations found in southern and north-east Europe have played a crucial role in the natural recolonization by the species of many areas of the Continent. Rural depopulation and land abandonment, particularly in mountainous areas, has been accompanied by spread of woodland through natural regeneration and planting. Potential prey populations (such as deer and wild boar) have increased, while hunting pressure has often decreased. The range of the wolf quadrupled between 1970 and 2005 with consistently, albeit often small, positive changes occurring in each decade (Deinet et al., 2013).

Concerns and opposition to the reappearance of the species still occur, mainly from rural inhabitants (Bath, 2000; Buller, 2007; Mounet, 2007): wolves do pose threats to domestic livestock and quarry stocks, and, in doing so, to the traditions and practices of rural communities. Consequently, financial frameworks have been established to compensate individuals for wolf depredation and contribute to protective measures, such as fences and, most recently, new breeds of sheepdog. In some countries, such as Sweden, there is a selective cull of the wolf population in an attempt to reduce their impact. Conflicts between interest groups have not been helped by rumours of the clandestine reintroduction and translocation of wolves from across Europe by various conservation movements (Bath, 2000). Further reintroduction schemes are regularly debated and prove very lively topics of discussion, such as those that have taken place in Scotland (Nilson et al., 2007; Manning et al., 2009).

One of the arguments used to justify the reintroduction of wolves is that they might contribute to the regulation of deer populations. There is some indication that this can happen from observations in the Yellowstone National Park in the USA, although similar evidence from European areas is currently lacking. It is the turnaround in the image of the wolf, rather than hard evidence of woodland benefits, that has been critical.

The reinvention of the wolf is one of the most dramatic reappraisals of any species. Its image is changing from that of a bone-crunching mass of sinew and fur, waiting to snatch the lamb from the fold or babe from the cradle, to an animal sitting on its haunches at the top of the food chain, an iconic denizen of the wilderness and poster child for endangered species recovery. Many of the factors driving the recovery of wolves in Europe have also played significant roles in the recovery of other conspicuous mammalian predators, such as the brown bear and the lynx.

14.3.2 The brown bear

The brown bear is the largest terrestrial mammalian predator on the continent apart from the polar bear (*U. maritimus*). It was once widely distributed across mainland Europe, as well as being found on many of the larger islands, with the exceptions of Iceland, Gotland, Corsica and Sardinia; the former presence of the species in Ireland is still debated (Swenson *et al.*, 2000). By the 17th century, however, the brown bear could no longer be found in Denmark, Britain, the German lowlands and the Swiss plateau (Breitenmoser, 1998), and this decline continued until the early 20th century. In Switzerland, by 1900, the range of brown bears had contracted by 75% of its level in 1850, and by 1950 the species could only be found in 2% of its former range (Breitenmoser, 1998). The decline resulted from hunting, influenced by improved firearms, human population dynamics and habitat modification and loss, combined with the bear's relatively low reproductive rate. This pattern is generally representative of that in many other areas and, by 1955, the species occupied 37% of the range that it had had in the 1700s (Deinet *et al.*, 2013).

The survival of relic populations in the Cantabrian, Pyrenean, Apennine, Alpine, Eastern Balkan, Dinaric-Pindos, Carpathian, Scandinavian, Karelian and Baltic mountains enabled a recovery in the numbers of brown bears in these areas and subsequent spread into areas where they had become extinct. The most recent estimates (2010–2012) suggest that Europe is currently home to about 17,000 brown bears, which are found in inland forested and mountain areas where human activity is low (Deinet *et al.*, 2013).

14.3.3 The lynx

Europe is home to two species of lynx, the Eurasian lynx and the critically endangered Iberian lynx; these are the largest surviving European cats. The Eurasian species range covers boreal, deciduous and Mediterranean woodland throughout continental Europe, with the optimal habitat consisting of large forests that support stable populations of small ungulates as a prey basis (Breitenmoser *et al.*, 2000). During the last 500 years, populations of this species had declined in number and range, and by 1800 it had disappeared from the lowlands of western and central Europe, Italy and the plains of Hungary. Outside the large continuous boreal forests of Scandinavia and European Russia, forest clearance and hunting pressure during the 19th and 20th centuries meant that only relict populations survived. These were in the Pyrenees, the Massif Central (France), the Alps, the Bavarian and Bohemian forests, the Balkan mountains and the Carpathian mountains. The species range decreased by 48% between 1800 and 1960 (Deinet *et al.*, 2013).

The Eurasian lynx is now legally protected under the same legislation as that for the wolf and brown bear. Natural recolonization, supplemented by reintroductions and translocations, have led to a 37% range increase in the second half of the 20th century. The most significant recoveries have taken place in the alpine, Bohemian–Bavarian, Jura and Vosges-Palatinian regions of Austria, Germany, France and Italy, where translocation and population augmentation from captive breeding stock programmes have also been used (Deinet *et al.*, 2013). Similar schemes have been implemented in the Dinaric and Harz mountains (in the Balkans and Germany, respectively), and most recently (2012) from Estonia into Poland. In 2008, it was estimated that there were now between 9000 and 10,000 animals found in ten population groupings (Deinet *et al.*, 2013). There are also suggestions for trophic cascade effects on the population of the prey of the Eurasian lynx (Ripple *et al.*, 2014).

The maintenance and restoration of the 'European Big Three' (the wolf, bear and lynx) is becoming an increasingly prominent item on the agendas of national governments and international movements. The role of large carnivores as important ecological managers is frequently alluded to, but the evidence for such wider scale impacts is more obvious for some other reintroductions/recolonizations, in particular that of the beaver.

14.4 The Beaver

The European beaver (*Castor fiber*) was formerly distributed continuously across Eurasia, from the British Isles to eastern Siberia;

throughout the deciduous and coniferous forest zones, and extending into wooded river valleys far into the tundra of the north and the steppes of the south (Halley and Rosell, 2002). Its population started to decline principally because it was being hunted for its fur, meat and castoreum, a glandular secretion used in territorial marking that was highly valued as a base for medicine and perfumes. By the medieval period, the numbers and range of the beaver had become increasingly reduced across Europe; the advent of steel traps and firearms exacerbated the process from the 17th century onward.

During the 19th century, the beaver population underwent a further dramatic decline and the species disappeared from most European countries; by the beginning of the 20th century as few as 1200 animals survived in eight isolated and scattered European enclaves (Halley and Rosell, 2003). The key European remnant populations were located in the lower Rhone (France), the Elbe (Germany), Telemark (Norway), the Pripet marshes (Belarus, Ukraine, Russia) and Voronezh (Russia) (Deinet et al., 2013). The species teetered on the brink of extinction, but the legal protection of these remnant populations and targeted conservation measures, such as reintroduction and translocation programmes, combined with habitat restoration, began to turn the tide.

Initial efforts were motivated by the desire to safeguard and bolster the remnant European beaver population in the interest of fur harvesting prospects. The species was first legally protected in Norway in 1845, in France in 1909, in Germany in 1910 and in Russia and Ukraine in 1922. While also being protected in Finland, Sweden, Poland and Spain during the late 19th and early 20th century, such measures failed to prevent the extinction of the species in these countries (Halley and Rosell, 2003; Halley et al., 2012).

The protection and survival of the remnant populations of beavers preserved the genetic integrity of the European species and stimulated natural spread in these areas. It also provided source populations for reintroduction and translocation programmes in locations where the species had been completely extirpated. During this time, the primary motivation of those engaged with the programmes was still to replenish stocks for fur harvesting and, as a result, were predominately 'Hard Release' initiatives, largely devoid of prior habitat suitability assessments. The first beaver reintroduction programmes took place in 1922, when animals were translocated from Norway to Sweden. Subsequent intra-European schemes have been undertaken in at least 25 European countries (Deinet et al., 2013).

Based on the belief that the Eurasian and North American beaver were the same species, some early introduction attempts were made with what is now classed as *C. canadensis*. In most cases, such as in France, Poland and Austria, the North American animals failed to become established. However, following the joint introduction of *C. fiber* and *C. canadensis* in 1935–1937, the Finnish beaver population is predominately of the North American species (Halley and Rosell, 2002).

Since the 1970s, interest in the beaver as a fur bearer has been supplanted by the examination of the impact and possible role of the species in the creation and maintenance of habitats (Nolet and Rosell, 1998). Beavers strongly affect the ecology of the surrounding ponds, creeks and riparian forests (Baker and Hill, 2003), and can create and maintain open conditions. Their activities have attracted attention in relation to a number of ecological conceptual frameworks: as herbivores, disturbance agents, keystone species and ecosystem engineers and facilitators (Jones et al., 1994; Gurney and Lawton, 1996; Rosell et al., 2005; Nummi and Kuuluvainen, 2013).

Beavers, through the construction of dams, the excavation of canals and the building of lodges, convert previously terrestrial ecosystems into aquatic ecosystems. Their feeding affects tree stand structure and the composition of riparian woodlands. The activity of the animals therefore has implications at patch and at landscape level (Gurnell, 1998; Nummi and Kuuluvainen, 2013) which, in turn, has ramifications for other species, including people.

Opposition to the spread of beavers may come from anglers who consider (not necessarily on the basis of evidence!) that habitat modifications by beavers adversely affect salmon populations; similarly, they are often thought to have a serious impact on valuable timberland, leading to illegal hunting and trapping. Ecologists, conservation managers and environmental agencies have had to debate how far the biodiversity benefits might outweigh such considerations.

Where species have been reintroduced to cultural landscapes in which they have not been present, in some cases for centuries, there is perhaps a responsibility then to limit their impact (Sheail, 1999). This case is clearer when a species is introduced that has never been present, as with the grey squirrel (Smout, 2009).

14.5 A Species that Has Done Too Well: The American Grey Squirrel

From the mid-1800s to the early 1900s, it was fashionable for owners of country estates in Great Britain to populate their ground with specimens of exotic plants and animals (Yalden, 1999), some of which have become characteristic members of the modern faunal assemblage. The American grey squirrel (*Sciurus carolinensis*), a native of the deciduous forests of the eastern USA (Williamson, 1996, 109), is a conspicuous example.

It is thought that the grey squirrel was first introduced into Henbury Park in Cheshire, UK, by Mr Brocklehurst in 1876. Further releases took place in the UK at Woburn Park (Bedfordshire) in 1890 with eight secondary releases subsequently being drawn from this flourishing stock (Coates, 2011), including releases at various sites in Scotland 2 years later. Descendants of the population at Woburn were transported to Ireland, where the Earl of Granard introduced a dozen grey squirrels at Castle Forbes (County Longford) in 1911 (Boyd-Watt, 1923).

Between 1876 and 1929, grey squirrels were introduced to 30 sites in England and Wales, three in Scotland and one in Ireland (Middleton, 1931). Boyd-Watt (1923) confidently asserted that the grey squirrel could 'never become wide-spread and dominant like our other introduced animals'. However, in a manner consistent with patterns of growth predicted by Elton (1927), the grey squirrel range and abundance expanded rapidly during the early to mid-20th century. By 1950, the species had completely colonized central England and much of the south-east and north-east, with an estimated distribution range of 19,800 m^2 (Shorten, 1953). By 1995, it was estimated that there were about 2.5 million grey squirrels in Great Britain (Harris *et al.*, 1995), and they were found almost everywhere in central and southern England, Wales and the central lowlands of Scotland. The rapid spread of the grey squirrel is an excellent example of population 'explosion' (Elton, 1958).

The grey squirrel increasingly became recognized as a threat to the native red squirrel (*S. vulgaris*), with criticism of the 'American' squirrel emerging in tandem with its proliferation (Coates, 2011). There are three elements to the impact of the grey squirrel: the displacement and decline of the native red squirrel; the damage inflicted by the grey on young trees, which leads to loss of timber production and quality; and a less well-defined effect on other aspects of the woodland ecosystem, for example a perceived (not proven) impact on songbird populations.

The concomitant loss the native red squirrel has been widely observed following the arrival of the alien grey. Various mechanisms have been put forward to explain this, including interference with red squirrel mating behaviour, direct aggression and competitive exclusion (Skelcher, 1997; Gurnell and Pepper, 1998; Lurz *et al.*, 2001). Grey squirrels do affect the behaviour, diet choice, use of cached tree seeds, habitat use and body growth of the red. More recent research has shown that the presence of the grey affects the fecundity, residency and recruitment of the red, all of which have consequences at population level (Gurnell *et al.*, 2004). The role of disease in the ousting of the red by the grey has also been widely examined (Rushton *et al.*, 2000); the grey acts as a reservoir host of the squirrel poxvirus that is fatal to the red.

The extent to which disease has played a role in the decline of the red squirrel in Britain is debated (Tompkins *et al.*, 2002), and the exact mechanisms of the replacement of red by the grey squirrel are not fully understood and subject to evolving hypotheses (Gurnell *et al.*, 2004). Red squirrels have historically gone through periods of decline, followed by recovery, and one such decline period coincided with the arrival of the grey. There seems little doubt though that any subsequent recovery of the red population was hampered by its suppression in the presence of the grey.

The impact of the American squirrel on wood production and quality as a result of it stripping the bark from pole-sized trees is well documented. Grey squirrels strip bark in order to reach and eat the sweet, phloem tissue underneath. Antagonistic behaviour between squirrels also triggers attacks (Gurnell, 1987). Young stands with fast-growing beech (*Fagus sylvatica*) and sycamore (*Acer pseudoplatanus*) are especially vulnerable, but other broadleaves and conifers, including larch (*Larix decidua*), Norway spruce (*Picea abies*), Scots pine (*Pinus sylvestris*) and western red cedar (*Thuja plicata*), are susceptible to bark stripping as well (Mountford, 2006). Stands may suffer repeated attacks, particularly if the area hosts a large number of juvenile squirrels and there are mature seed-bearing trees. In wooded and forested areas exhibiting this combination of characteristics, the extent of damage can be considerable. For example, a recent study of the Chiltern forests in southern England found that 54% of beech and 44% of sycamore trees had evidence of squirrel damage (Rayden and Savill, 2004), as did 100% of sycamore, 66% of beech and 40% of oak (*Quercus petraea*) in broadleaved woodland in southern Britain (Huxley, 2003).

Bark-stripped trees are left prone to colonization by insects and fungi, and this can result in mortality or stunted growth. In turn, this may lead to changes in the composition and structure of woodland and forests (Mountford, 2006). Particularly susceptible species may no longer be planted, thus limiting the choices for the owner and reducing the opportunities to develop mixed stands.

The replacement of the native red squirrel and bark damage to trees by the grey squirrel have also been recorded in Italy. Grey squirrels transported from America and Canada were first introduced at Stupinigi near Turin (Piedmont) in 1948, with subsequent releases into the park of Villa Groppallo at Genoa Nervi in 1966 and at Trecate, Milan, in 1994. The animals released at Trecate were recaptured 2 years later and the potential range of the population at Nervi is limited by the sea, busy roads and residential districts. Following an initial period of establishment and stabilization, the Piedmont population has shown a rapid increase in its range in recent decades (Bertolino and Genovesi, 2003). An annual average range increase of 17.2 km^2 since 1948 means that the grey squirrel is now spread over an area of 880 km^2. The growth of this population has been particularly conspicuous in areas characterized by extensive woods with a large-seed producing deciduous trees (Lurz *et al.*, 2001).

Based on 100 year simulations beginning in 1996, the 'best' and 'worst' case scenarios have been predicted for the spread of the grey squirrel in continental Europe from these populations in northern Italy (Bertolino *et al.*, 2008). According to the best case scenario, it will take 30–40 years for the grey squirrel to colonize the Alps and 70–75 years to cross the border into France, with the first populations predicted in Switzerland in 2051–2056 (Bertolino *et al.*, 2008). The worst case scenario is dramatically quicker. Europe contains many ideal habitats for the grey squirrel, many of which are currently populated by the red. Therefore, a coordinated, pan-European programme for the eradication of the grey is desirable, and since 1997, has been sporadically conducted in Italy (Bertolino *et al.*, 2008). However, the lack of political commitment and an increasingly vocal animal rights movement has so far stunted efforts to eradicate the grey squirrel.

14.6 The Decline and Rise of Wild Boar and Deer

14.6.1 Wild boar

Europe is home to 20 species of wild ungulate, the majority of which are widespread and abundant in number and range, often reaching surprisingly high densities in many areas (Putman *et al.*, 2011). The range of the wild boar (*Sus scrofa*) contracted severely between the 17th and 19th centuries, but since the 1950s, their abundance and distribution has expanded enormously (Sáaez-Royuela and Tellerfía, 1986; Macdonald, 2001; Deinet *et al.*, 2013), a phenomenon mainly attributable to the species' high ecological plasticity. While this spread has been to the perceptible delight of hunting and environmentalist communities, the same perhaps cannot be said for agriculturalists (Hearn *et al.*, 2014).

Throughout their European range, wild boar cause significant damage to crops (Herrero *et al.*, 2006; Schley *et al.*, 2008). Its dependence on large quantities of energy-rich plants, worms and grubs, coupled with its physical size, large family groups and a propensity to root and trample as well as to consume plants, means that wild boar can be extremely damaging to crops (Schley and Roper, 2003). Wild boar are reservoirs for a number of viruses, bacteria and parasites that can be transmitted to domestic animals (e.g. classical swine fever, bovine TB, brucellosis, trichinellosis) and humans (e.g. hepatitis E, tuberculosis, leptospirosis and trichinellosis) (Meng *et al.*, 2009). In addition, hybridization through interbreeding with domestic swine affects the genetic integrity of both parties (Scandura *et al.*, 2011), an issue that can be particularly detrimental should the domestic pigs be a rare breed or a pedigree herd. The rooting behaviour of wild boar has also been shown to affect woodland composition and produce (Gómez and Hódar, 2008; Wirthner *et al.*, 2012).

While widespread across much of Europe, wild boar are a relatively new presence in the UK and the return of the species to southern England has provoked much debate (Goulding, 2011). The possibility of reintroducing it in Scotland is being investigated (Howells and Edwards-Jones, 1997; Leaper *et al.*, 1999) and research is under way on the potential role of wild boar as an ecosystem engineer due to the impact of its rooting on vegetation disturbance regimes (Sandom *et al.*, 2013).

Boar are increasingly being found in a number of urban areas such as Berlin (Kotulski and König, 2008), Barcelona (Cahill and Llimona, 2004), Gdansk (Mikos, 2002) and Geneva (Fischer *et al.*, 2002), thus increasing the potential for conflicts with people (Licoppe *et al.*, 2013). Various management strategies have been proposed and implemented across the continent, from government compensation schemes to a number of ingenious micromanagement techniques. However, hunting and selective culls remain the best strategies to reduce the numbers and the impact of the wild boar in Europe (Hearn *et al.*, 2014), although locally there may be objections to these.

14.6.2 Deer

Deer underwent an often dramatic decline in many countries in previous centuries, but during the 20th century, there has been a marked increase in their numbers across Europe. Deer are the largest wild herbivores in most lowland woodland ecosystems (Gill and Beardall, 2001), with an increasing amount of research highlighting their impact and the factors controlling this (Gerhardt *et al.*, 2013; Buckley and Mills, Chapter 6; Savill, Chapter 7; Fletcher Chapter 9; Hinsley *et al.*, Chapter 12; Hermy, Chapter 13).

For deer, decline began in the Middle Ages as a result of the combined effects of over-hunting, anthropogenic habitat modification and competition with domestic livestock. Both the roe deer (*Capreolus capreolus*) and the red deer (*Cervus elaphus*) experienced a significant reduction in their numerical abundance and spatial distribution from the early modern period, in some areas to the brink of local extinction. Both species recovered during the mid-to-late 20th century, benefiting from translocation and reintroduction programmes, natural recolonization facilitated by habitat creation and legislation changes (Deinet *et al.*, 2013).

Roe deer can now be found throughout continental Europe and Great Britain, with the exceptions of Ireland, Sardinia, Corsica, Sicily, Cyprus and Iceland (Burbaite and Csányi, 2009). The species occupies a wide variety of habitats, including forests, moorlands, pastures and arable and suburban areas, with densities generally being higher in woodland–field mosaic landscapes. The red deer has a similarly wide distribution throughout Europe and Great Britain, with the exception of northern Scandinavia, Finland and Iceland.

In most of its range, it inhabits broadleaved and coniferous forest and woodland margins (Deinet *et al.*, 2013), the red deer of open moorland in Scotland being something of an anomaly.

Deer play a major role in forest ecosystems and can change the composition and structure of the vegetation significantly (Putman *et al.*, 2011); they contribute to nutrient cycling and the dispersal of seeds through their guts or on their coats (Gill and Beardall, 2001). They generally tend to reduce stem densities, limit height growth and reduce foliage density (Gill and Beardall, 2001) through browsing on seedlings, leading shoots, side shoots and climbers. Browsing and bark stripping by deer affect forest regeneration, which may be crucial for forest functions such as erosion control and high-value timber production (Putman *et al.*, 2011).

Management of deer through shooting is well understood, but can be difficult to put into practice because of potentially conflicting interests. Hunters may wish to keep deer numbers at higher levels than the forester would like; landowners may not allow shooting, such that it compromises efforts made on the neighbouring land; some people object to the killing of animals as a matter of principle and campaign against deer management. As a result, deer numbers are still tending to increase.

14.7 Conclusion

The wild mammalian faunal assemblages of Europe underwent often dramatic declines from the medieval and early modern period. Many species, such as the aurochs, disappeared entirely; large carnivores were driven either past or to the brink of extinction, mainly as a result of their incompatibility with rural economies. Herbivorous species also declined, because of either exploitation or the rituals surrounding their chase. Some species declined following the arrival of foreign competitors.

However, since the 1970s, some of these trends have been reversed and both the ranges and populations of wild mammals are expanding. A novel element of recent decades has been a process of rewilding, both from deliberate attempts to reconstruct the more natural plant and animal assemblages (Latham, Chapter 21), and simply as an indirect consequence of a reduction in farming in marginal areas.

What is clear is that, particularly for the larger mammals, managing the interactions between people and animals remains a critical issue. This is as much about people's perceptions of an animal, its impacts, and how it is treated, as it is about evidence of its actual behaviour and effects. Sociological factors as well as biological and ecological factors will determine the future mammal fauna of Europe's forests.

References

Baillie, J.E.M., Griffiths, J., Turvey, S.T., Loh, J. and Collen, B. (2010) *Evolution Lost: Status and Trends of the World's Vertebrates*. Zoological Society of London, London.

Baillie-Graham, W.A. (1913) *Sport in Art*. Ballantyne, London.

Baker, B.W. and Hill, E.P. (2003) Beaver (*Castor canadensis*). In: Feldhamer, G.A., Thompson, B.C. and Chapman, J.A. (eds) *Wild Mammals in North America: Biology, Management and Conservation*, 2nd edn. The Johns Hopkins University Press, Baltimore, Maryland, pp. 288–310.

Bath, A.J. (2000) *Human Dimensions in Wolf Management in Savoie and Des Alpes Maritimes, France*. France LIFE-Nature Project Le Retour du Loup dans Les Alpes Françaises, European Commission, Brussels/IUCN-Large Carnivore Initiative for Europe, Gland, Switzerland.

Bernard, D. (1981) *L'homme et le Loup*. Berger-Levrault, Paris.

Bertolino, S. and Genovesi, P. (2003) Spread and attempted eradication of the grey squirrel (*Sciurus carolinensis*) in Italy, and consequences for the red squirrel (*Sciurus vulgaris*) in Eurasia. *Biological Conservation* 109, 351–358.

Bertolino, S., Lurz, P.W.W., Sanderson, R. and Rushton, S.P. (2008) Predicting the spread of the American grey squirrel (*Sciurus carolinensis*) in Europe: a call for a co-ordinated European approach. *Biological Conservation* 141, 2564–2575.

Boitani, L. (1987) *Dalle Parte del Lupo: La Riscoperta Scientifica e Culturale del Mitico Predatore*. Mondadori, Milan, Italy.

Boitani, L. (2000) *Action Plan for the Conservation of Wolves* (Canis lupus) *in Europe*. Council of Europe, Strasbourg, France.

Boyd-Watt, H.B. (1923) On the American grey squirrel (*Sciurus carolinensis*) in the British Isles. *The Essex Naturalist* 20, 189–205.

Breitenmoser, U. (1998) Large predators in the Alps: the fall and rise of man's competitors. *Biological Conservation* 83, 279–289.

Breitenmoser, U., Breitenmoser-Würsten, C., Okarma, H., Kaphegyi, T., Kaphegy-Wallmann, U. and Müller, U.M. (2000) *Action Plan for the Conservation of the Eurasian lynx* (Lynx lynx) *in Europe*. Council of Europe, Strasbourg, France.

Buller, H. (2007) Safe from the wolf: biosecurity, biodiversity and competing philosophies of nature. *Environment and Planning A* 40, 1583–1597.

Burbaite, L. and Csányi, S. (2009) Roe deer population and harvest changes in Europe. *Estonian Journal of Ecology* 58, 169–180.

Cahill, S. and Llimona, F. (2004) Demographics of a wild boar *Sus scrofa* Linnaeus, 1758 population in a metropolitan park in Barcelona. In: Fonseca, C., Herrero, J., Luís, A. and Soares, A.M.V.M. (eds) *Wild Boar Research 2002. A Selection and Edited Papers from the Fourth International Wild Boar Symposium, Lousã, Portugal, September 19–20, 2002. Galemys* 16(Nº especial), 37–52. Available at: http://www.secem.es/wp-content/uploads/2013/03/Galemys-16-NE-003-Cahill-37-52.pdf (accessed 2 December 2014).

Clutton-Brock, J. (1987) *A Natural History of Domesticated Animals*. Cambridge University Press, Cambridge, UK.

Coates, P. (2011) Over here: American animals in Britain. In: Lambert, R.A. and Rotherham, I.D (eds) *Invasive and Introduced Plants and Animals*. Earthscan, London, pp. 39–54.

Deinet, S., Ieronymidou, C., McRae, L., Burfield, I.J., Foppen, R.P., Collen, B. and Böhm, M. (eds) (2013) *Wildlife Comeback in Europe: The Recovery of Selected Mammal and Bird Species*. Final report to Rewilding Europe by Zoological Society of London (ZSL), BirdLife International and European Bird Census Council. ZSL, London.

Elton, C. (1927) *Animal Ecology*, 1st edn. Sidgwick and Jackson, London.

Elton, C. (1958) *The Ecology of Invasions by Animals and Plants*. Methuen, London.

Fischer, C., Gourdin, H. and Obermann, M. (2002) Spatial behaviour of wild boar in Geneva (Switzerland). Testing the methods and first results. In: Fonseca, C., Herrero, J., Luís, A. and Soares, A.M.V.M. (eds) *Wild Boar Research 2002. A Selection and Edited Papers from the Fourth International Wild Boar Symposium, Lousã, Portugal, September 19–20, 2002. Galemys* 16(Nº especial), 149–155. Available at: http://www.secem.es/wp-content/uploads/2013/03/Galemys-16-NE-012-Fischer-149-155.pdf (accessed 2 December 2014)

Gerhardt, P., Arnold, J.M., Hackländer, K. and Hochbichler, E. (2013) Determinants of deer impact in European forests – a systematic literature analysis. *Forest Ecology and Management* 310, 173–186.

Gill, R.M.A. and Beardall, V. (2001) The impact of deer on woodlands: the effects of browsing and seed dispersal on vegetation structure and composition. *Forestry* 74, 209–218.

Gómez, J.M. and Hódar, J.A. (2008) Wild boar (*Sus scrofa*) affect the recruitment rate and spatial distribution of holm oak (*Quercus ilex*). *Forest Ecology and Management* 256, 1384–1389.

Goulding, M. (2011) Native or alien? The case of the wild boar in Britain. In: Rotherham, I.D. and Lambert, R. (eds) *Invasive and Introduced Plants and Animals: Human Perceptions, Attitudes and Approaches to Management*. Earthscan, London, pp. 289–300.

Grant, A.J. (1922) *Early Lives of Charlemagne by Eginhard and the Monk of St Gall*. Chatto and Windus, London.

Gurnell, A.M. (1998) The hydrogeomorphological effects of beaver dam-building activity. *Progress in Physical Geography* 22, 167–189.

Gurnell, J. (1987) *The Natural History of Squirrels*. Christopher Helm, London.

Gurnell, J. and Pepper, H. (1998) Grey squirrel damage to broadleaf woodland in the New Forest: a study on the effects of control. *Quarterly Journal of Forestry* 92, 117–124.

Gurnell, J., Wauters, L.A., Lurz, P.W.W. and Tosi, G. (2004) Alien species and interspecific competition: effects of introduced eastern grey squirrels on red squirrel population dynamics. *Journal of Animal Ecology* 73, 26–35.

Gurney, W.S.C. and Lawton, J.H. (1996) The population dynamics of ecosystem engineers. *Oikos* 76, 273–283.

Hall, S.J.G. (2008) A comparative analysis of the habitat of the extinct aurochs and other prehistoric mammals in Britain. *Ecography: Patterns and Diversity in Ecology* 31, 187–190.
Halley, D.J. and Rosell, F. (2002) The beaver's re-conquest of Eurasia: status, population development and management of a conservation success. *Mammal Review* 32, 153–178.
Halley, D.J. and Rosell, F. (2003) Population and distribution of European beavers (*Castor fiber*). *Lutra* 46, 91–101.
Halley, D., Rosell, F. and Saveljev, A. (2012) Population and distribution of Eurasian beaver (*Castor fiber*). *Baltic Forestry* 18, 168–175.
Harris, S., Morris, P., Wray, S. and Yalden, D. (1995) *A Review of British Mammals: Population Estimates and Conservation Status of British Mammals Other than Cetaceans*. Joint Nature Conservation Committee, Peterborough, UK.
Hearn, R., Watkins, C. and Balzaretti, R. (2014) The cultural and land use implications of the reappearance of wild boar in north west Italy: a case study of the Val di Vara. *Journal of Rural Studies* 36, 52–63.
Hearn, R., Balzaretti, R. and Watkins, C. (2015) The wolf in the landscape: Antonio Cesena and attitudes to wolves in Sixteenth-Century Liguria. *Rural History* 26, 1–16.
Herrero, J., García-Serrano, A., Couto, S., Ortuño, V.M. and García-González, R. (2006) Diet of wild boar *Sus scrofa* L. and crop damage in an intensive agroecosystem. *European Journal of Wildlife Research* 52, 245–250.
Howells, O. and Edwards-Jones, G. (1997) A feasibility study of reintroducing wild boar to Scotland: are existing woodlands large enough to support minimum viable populations? *Biological Conservation* 81, 77–89.
Huxley, L. (2003) *The Grey Squirrel Review: Profile of an Invasive Alien Species*. European Squirrel Initiative, Dorset, UK [now Woodbridge, UK]. Available at: http://www.europeansquirrelinitiative.org/RevComplete.pdf (accessed 2 December 2014).
Jones, C.G., Lawton, J.H. and Shachak, M. (1994) Organisms as ecosystem engineers. *Oikos* 69, 373–386.
Kotulski, Y. and König, A. (2008) Conflict, crises and challenges: wild boar in the Berlin City: a social, empirical and statistical study. *Natura Croatica* 17, 233–246. Available at: http://hrcak.srce.hr/file/53983 (accessed 2 December 2014).
Leaper, R., Massei, G., Gorman, M.L. and Aspinall, R. (1999) The feasibility of reintroducing wild boar to Scotland. *Mammal Review* 29, 239–259.
Licoppe, A., Prévot, C., Heymans, M., Bovy, C., Casaer, J. and Cahill, S. (2013) Wild boar/feral pigs in (peri-)urban areas. International survey report as an introduction to the workshop: *Managing Wild Boar in Human-dominated Landscapes. International Union of Game Biologists, Congress IUGB 2013, Brussels, Belgium, August 28*. Available at: http://www.iugb2013.org/docs/Urban%20wild%20boar%20survey.pdf (accessed 2 December 2014).
Lorimer, J. and Driessen, C. (2013) Bovine biopolitics and the promise of monsters in the rewilding of Heck cattle. *Geoforum* 48, 249–259.
Lorimer, J. and Driessen, C. (2014) Wild experiments at the Oostvardersplassen: rethinking environmentalism for the Anthropocene. *Transactions of the Institute of British Geographers* 39, 169–181.
Lurz, P.W.W., Rushton, S.P., Wauters, L.A., Bertolino, S., Currado, I., Mazzoglio, P. and Shirley, M.D.F. (2001) Predicting grey squirrel expansion in north Italy: a spatially explicit modelling approach. *Landscape Ecology* 16, 407–420.
Macdonald, D. (2001) *The New Encyclopaedia of Mammals*. Oxford University Press, Oxford, UK.
Maier, S. (2012) *What's So Good About Biodiversity: A Call for Better Reasoning About Nature's Values*. Springer, Dordrecht, The Netherlands.
Manning, A.D., Gordon, I.J. and Ripple, W.J. (2009) Restoring landscapes of fear with wolves in the Scottish Highlands. *Biological Conservation* 142, 2314–2321.
McDevitte, W.A. and Bohn, W.S. (1869) Julius Caesar. *Caesar's Gallic War*. Harper's New Classical Library, New York.
McRae, L., Collen, B., Deinet, S., Hill, P., Loh, J., Baille, J.E.M. and Price, V. (2012) The state of the planet. In: Grooten, M., Almond, R., McLellan, R., Dudley, N., Duncan, E., Oerlemans and Stolten, S. (eds) *The Living Planet Report 2012: Biodiversity, Biocapacity and Better Choices*. WWF International, Gland, Switzerland, pp. 14–62.
Mech, L.D. and Boitani, L. (eds) (2003) *Wolves: Behavior, Ecology, and Conservation*. University of Chicago Press, Chicago, Illinois.
Meng, X.J., Lindsay, D.S. and Sriranganathan, N. (2009) Wild boars as sources for infectious diseases in livestock and humans. *Philosophical Transactions of the Royal Society of London, Series B: Biological Sciences* 364, 2697–2707.
Middleton, A.D. (1931) *The Grey Squirrel*. Sidgwick and Jackson, London.

Mikos, J. (2002) Management strategy of wild boar in Wejhero-Gdansk suburban area, north-central Poland. In: *Abstracts of the Fourth International Wild Boar Symposium, Lousã, Portugal, September 19–20, 2002*, p. 28.

Mounet, C. (2007) Living with 'problem' animals: the case of the wolf and the boar in the French Alps. *Journal of Alpine Research* 93, 65–76.

Mountford, E. (2006) Long-term patterns and impact of grey squirrel debarking in Lady Park Wood young-growth stands. *Forest Ecology and Management* 232, 100–113.

Musiani, M., Boitani, L. and Paquet, P.C. (eds) (2009) *A New Era for Wolves and People: Wolf Recovery, Human Attitudes, and Policy*. University of Calgary Press, Calgary, Alberta, Canada.

Nilson, E.R., Milner-Gulland, E.J., Schofield, L., Mysterud, A. Stenseth, N.C. and Coulson, T. (2007) Wolf reintroduction to Scotland: public attitudes and consequences for red deer management. *Proceedings of the Royal Society B: Biological Sciences* 274, 995–1003.

Nolet, B.A. and Rosell, F. (1998) Comeback of the beaver *Castor fiber*: an overview of old and new conservation problems. *Biological Conservation* 83, 165–173.

Nummi, P. and Kuuluvainen, T. (2013) Forest disturbance by an ecological engineer: beaver in boreal forest landscapes. *Boreal Environmental Research* 18A, 13–24.

Ortalli, G. (1997) *Lupi, Genti, Culture: Uomo e Ambiente nel Medioevo*. Einaudi, Turin, Italy.

Putman, R., Apollonio, M. and Andersen, R. (eds) (2011) *Ungulate Management in Europe: Problems and Practices*. Cambridge University Press, Cambridge, UK.

Rayden, T. and Savill, P. (2004) Damage to broadleaved woodland in the Chilterns by the grey squirrel. *Forestry* 77, 249–253.

Ripple, W.J., Estes, J.A., Beschta, R.L., Wilmers, C.C., Ritchie, E.G., Hebblewhite, M., Berger, J., Elmhagen, B., Letnic, M., Nelson, M.P. et al. (2014) Status and ecological effects of the world's largest carnivores. *Science* 343, 1241484.

Rosell, F., Bozsér, O., Collen, P. and Parker, H. (2005) Ecological impact of beavers *Castor fiber* and *Castor canadensis* and their ability to modify ecosystems. *Mammal Review* 35, 248–276.

Rushton, S.P., Lurz, P.W.W., Gurnell, J. and Fuller, R. (2000) Modelling the spatial dynamics of parapoxvirus disease in red and grey squirrels: a possible cause of the decline in the red squirrel in the United Kingdom? *Journal of Applied Ecology* 37, 1–18.

Sáaez-Royuela, C. and Tellería, J.L. (1986) The increased population of the wild boar (*Sus scrofa*, L.) in Europe. *Mammal Review* 16, 97–101.

Sandom, C.J., Hughes, J. and Macdonald, D.W. (2013) Rewilding the Scottish Highlands: do wild boar (*Sus scrofa*) use a suitable foraging strategy to be an effective ecosystem engineer? *Restoration Ecology* 21, 336–343.

Scandura, M., Iacolina, L. and Apollonio, M. (2011) Genetic diversity in the European wild boar (*Sus scrofa*): phylogeography, population structure and wild x domestic hybridization. *Mammal Review* 41, 125–137.

Schley, L. and Roper, T.J. (2003) Diet of wild boar (*Sus scrofa*) in Western Europe, with particular reference to consumption of agricultural crops. *Mammal Review* 22, 43–56.

Schley, L., Dufrêne, M., Krier, A. and Frantz, A.C. (2008) Patterns of crop damage by wild boar (*Sus scrofa*) in Luxembourg over a 10-year period. *European Journal of Wildlife Research* 54, 589–599.

Seddon, P.J., Moehrenschlager, A. and Ewen, J. (2014) Reintroducing resurrected species: selecting DeExtinction candidates. *Trends in Ecology and Evolution* 29, 140–147.

Sheail, J. (1999) The grey squirrel (*Scirius carolinensis*): a UK perspective on a vertebrate pest species. *Journal of Environmental Management* 55, 145–156.

Shorten, M. (1953) Notes on the distribution of the grey squirrel (*Sciurus carolinensis*) and the red squirrel (*S. vulgaris leucourus*) in England and Wales from 1945 to 1952. *Journal of Animal Ecology* 2, 134–140.

Skelcher, G. (1997) The ecological replacement of red by grey squirrels. In: Gurnell, J. and Lurz, P.W.W. (eds) *The Conservation of Red Squirrels, Sciurus vulgaris L*. People's Trust for Endangered Species, London, pp. 67–78.

Smout, T.C. (2000) The alien species in twentieth-century Britain: inventing a new vermin. In: Smout, T.C. (ed.) *Exploring Environmental History*. Edinburgh University Press, Edinburgh, UK, pp. 169–181.

Swenson, J.E., Gerstl, N., Dahle, B. and Zedrosser, A. (2000) *Action Plan for the Conservation of the Brown Bear (Ursus arctos) in Europe*. Council of Europe, Strasbourg, France.

Tompkins, D., Nettleton, P., Buxton, D. and Gurnell, J. (2002) Parapoxvirus causes a deleterious disease of red squirrels associated with UK population declines. *Proceedings of the Royal Society of London B: Biological Sciences* 269, 529–533.

Van Vuure, T. (2002) History, morphology and ecology of the aurochs (*Bos taurus primigenius*). *Lutra* 45, 1–45.
Van Vuure, T. (2005) *Retracing the Aurochs: History, Morphology and Ecology of an Extinct Wild Ox*. Pensoft Publishers, Sofia, Bulgaria.
Vera, F.W.M. (2000) *Grazing Ecology and Forest History*. CAB International, Wallingford, UK.
Williamson, M. (1996) *Biological Invasions*. Chapman and Hall, London.
Wirthner, S., Schütz, M., Page-Dumroese, D.S., Busse, M.D., Kirchner, J.W. and Risch, A.C. (2012) Do changes in soil properties after rooting by wild boar (*Sus scrofa*) affect understory vegetation in Swiss hardwood forests? *Canadian Journal of Forest Research* 42, 585–592.
Yalden, D. (1999) *The History of British Mammals*. Poyser, London.

15 The Curious Case of the Even-aged Plantation: Wretched, Funereal or Misunderstood?

Chris P. Quine*
Forest Research, Northern Research Station, Roslin, UK

15.1 Introduction

Plantations have proved to be an effective way of delivering the wood and wood products that we consume at alarming rates, paralleling to some degree the intensification and specialization that has been seen in farming (Brockerhoff et al., 2008). Indeed, much of the wood and wood products that we consume depends upon the production from plantations, with some estimates suggesting up to 35% currently and more in the future (Carle and Holmgren, 2008; Sutton, 2014); such production may limit the further loss of natural and semi-natural forests.

However, we also look for a wider range of 'ecosystem services' from our forests, including from plantations (Quine et al., 2011, 2013). Hence, the policy and practice developments of the latter half of the 20th century have often been around moderating the pursuit of timber products, for example by introducing more structural complexity and extending rotations, so that other objectives can be met (Fig. 15.1).

This means that even-aged plantations should not be dismissed as ecologically irrelevant or 'wretched and deprived of individuality' (Tsouvalis, 2000). They can provide multiple goods and services including, with thought and care in land-use allocation and forest design, significant levels of biodiversity through the provision of woodland and woodland edge habitats. Management choices influence the extent to which plantations stay like agricultural crops or develop features more like those of natural forests (Fig. 15.1).

15.2 What is an Even-aged Plantation?

Are even-aged plantations that different from other forest types? Many ecologists instinctively do not like them: the label comes with a raft of emotion related to land-use change and loss of other habitats. Such views have been prominent in north-west Europe where plantations have been used extensively to reforest land devoid of trees for many centuries. The drivers for such enterprises might be economic, but this afforestation was also seen as a means of land improvement, protecting habitations and farmland by stabilizing sand dunes, for instance (Savill, Chapter 7). Notable examples include: the afforestation of the Danish heathlands and sand dunes

---

*E-mail: chris.quine@forestry.gsi.gov.uk

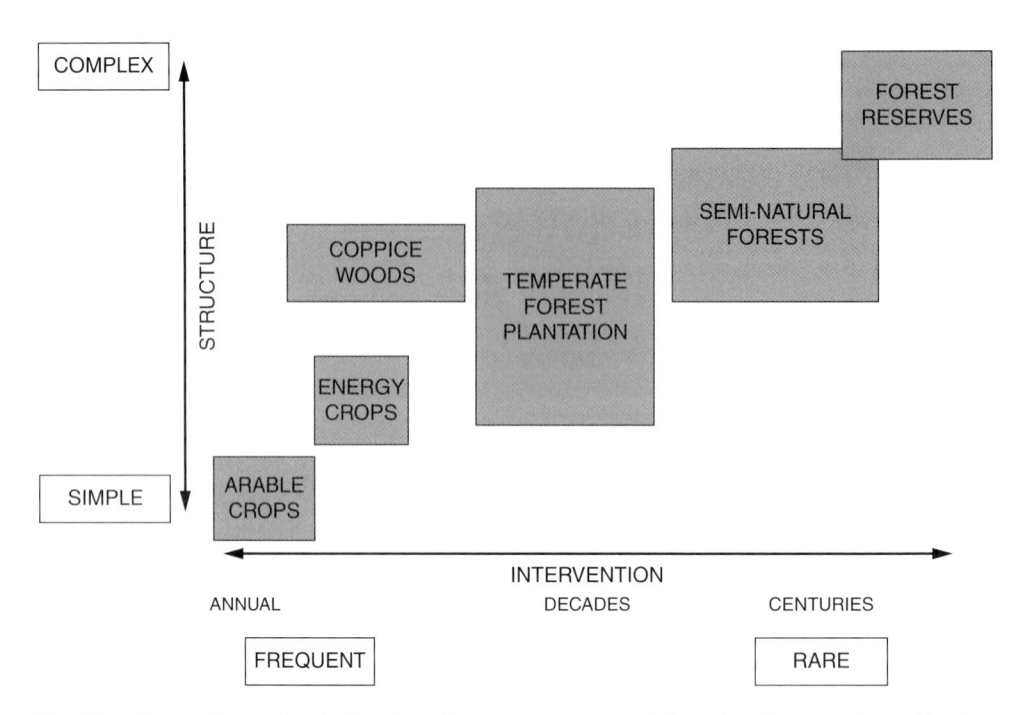

Fig. 15.1. Schematic showing the location of temperate, even-aged plantations in comparison with other forest types with respect to structural complexity and frequency of intervention. Choices over rotation length and stand management influence the positioning of plantations on these two axes.

(Olwig, 1984; Wilkie, 2001); the great Maritime pine forests of Les Landes (Maizeret, 2005); the pine forests of northern Germany (Ozenda, 1994) and east England (Ratcliffe and Claridge, 1996); and the recent *Eucalyptus* forests of the Iberian peninsula (Calvino-Cancela *et al.*, 2012).

Even-aged plantations may be caricatured as large areas of a single species of tree (often non-native), planted over a period of a few years across a previously largely treeless landscape. The legacy of this uniform establishment (afforestation) then follows through for many decades (Plate 7). The limited interventions aim to ensure a consistency of growth rates across the plantation and maintenance of canopy closure; in due course, clear-felling of large areas to harvest the forest products is followed by replanting, which starts the cycle all over again.

This is almost what has happened over hundreds of thousands of hectares of western Europe, but the real-life variant is somewhat fuzzier. Soil variations, climate effects, extreme weather events such as windstorms, the impacts of browsing animals, fire, pests, poor practice and genetic variation all create varying degrees of heterogeneity, even in first-rotation stands. The second and subsequent rotations start from a different point, which changes the potential occurrence of some species and features compared with the first planting. Self-seeding of both native and main crop tree species has occurred, sometimes encouraged; other changes reflect the changing environmental conditions and reductions of grazing pressure.

Diversification has increasingly been built into forest management and planning, whether in the form of tree species, age class, stand size and the patterns of roads and open spaces. In the UK, for instance, this diversity was initially driven by aesthetic standards for forest design, then for biodiversity benefits and, more recently, for risk management (Gardiner and Quine, 2000) to provide the basis for improving resilience and adaptation to climate change (Kirby *et al.*, 2009) or the impacts of tree disease.

Many 'even-aged plantations' are thus morphing into uneven-aged forests managed under the clear-fell and replant, or clear-fell and regenerate, management systems. The even-aged afforestation legacy will diminish over subsequent rotations. In parts of Europe, there are already managed forests that originated as something like an even-aged stand, but where the label 'plantation' would be strenuously denied; a case in point is the pine forests of northern Germany, which were created during the 1700s and 1800s, but are now classed as 'nemoral Scots pine forest' in the European forest typology (EEA, 2006).

Some of the most dramatic programmes of afforestation in the last century have been the planting of introduced spruce over large areas of marginal agricultural land along the Atlantic seaboard, particularly in the UK and Ireland. For example, more than 50% of the forests in the Republic of Ireland are of Sitka spruce (*Picea sitchensis*) (Horgan *et al.*, 2003; Coote, *et al.*, 2008), with 68% of the woods in Northern Ireland (Christie *et al.*, 2011), 35% of Scotland's forests and 29% of all woodland in Great Britain (Forestry Commission, 2003); for coniferous forest alone, Sitka spruce comprises 25.4% of English, 57.7% of Scottish and 59.1% of Welsh stocked forest area (Forestry Commission, 2012). Tittensor (2009) describes in detail the creation of one such new forest, Whitelee, in south-west Scotland. These programmes were also highly contentious and conflicts arose over loss of cherished cultural landscapes such as valleys in the English Lake District (Symonds, 1936) and areas important for nature conservation in the peatlands of north Scotland (Stroud *et al.*, 1987).

15.3 A Brief Historical Overview of Atlantic Spruce Forests

Centuries of exploitation and neglect of existing woodland, coupled with competition for land from agriculture, had meant there was little woodland (or potential for timber production remaining) in Britain and Ireland at the end of the 19th century. Britain had been reduced to about 4–5% woodland cover (Peterken, Chapter 18) and in Ireland perhaps only 1.5% of the land area was covered in woodland (Rackham, 2006). In both countries, growing stock was further reduced by the ravages of subsequent wartime felling necessary to replace sources of timber from overseas made unavailable as a result of naval blockade.

The desire to secure the supplies necessary to maintain industrial production for any further war created a demand for a strategic reserve and support for afforestation (though in World War II, Foot (2010) suggests that 46% of British growing stock still had to be cut down). Since then, there has been a consistent encouragement of woodland expansion although the motivation and specification has evolved. At different times, the objectives have included strategic reserve, rural development, industrial support, private investment and agricultural diversification – supported by various domestic and, latterly, European Union (EU) grants and some favourable tax regimes.

There has been an evolving and complex mix of contributions from state forestry and private afforestation – with the highest planting rates being achieved when both were active. In Britain, the annual planting peaked at approximately 30,000 ha in 1988/89; in the Republic of Ireland, the peak was achieved in 1995 at 23,710 ha a year, of which 17,353 ha were through private planting with EU financial assistance (Forest Service, 2008). By 2006, woodland cover in the Republic of Ireland was 10% of the land area, of which approximately 75% was coniferous in nature, in Britain in 2013, the figures were 13% of land area with 50% of this coniferous.

In the last two decades, the rate of coniferous forest expansion has slowed dramatically; in the Republic of Ireland, for example, afforestation had declined to 6947 ha in 2007, partly as a result of a shift to replanting of first rotation sites and more attractive finances for agriculture (including regular returns rather than long-term income) (Forest Service, 2008). In Britain, the reduction in afforestation was largely in response to conflicts over change of land use and the loss of open-ground species. Coniferous afforestation was also perceived as being fuelled by tax benefits for the rich and famous, and declined dramatically with tax changes in 1988 (Tsouvalis, 2000; Foot, 2010).

This emphasis on conifers partly reflected the nature of the demands for timber, but also the fact that the sites available (or in the case of Ireland, the only sites permitted) for afforestation were poor, typically very low in nutrients and often very poorly drained. The range of native trees available, particularly conifers, across the Atlantic seaboard, was restricted because of limited recolonization following the last glaciation. However, there had been a long history of plant collecting from other temperate regions, which increased the potential palette of suitable tree species. In particular, trees from the Pacific North-West of America grew well under Atlantic conditions and often better than European conifers such as Norway spruce (*Picea abies*), European larch (*Larix decidua*) and Corsican pine (*Pinus nigra* var. *maritima*). These new species included Douglas fir (*Pseudotsuga menziesii*), lodgepole pine (*P. contorta*), western hemlock (*Tsuga heterophylla*) and, particularly, Sitka spruce (Zehetmayr, 1954, 1960; Lines, 1987; Aldhous, 1997). Even Symonds (1936) noted: 'True, there are hundreds of acres on the fells where the ground is too wet for anything but conifer, and it is on such land that the funereal Sitka spruce shows its commercial value'.

15.3.1 The dominance of Sitka spruce

Sitka spruce has been the key species in the formation of the Atlantic spruce forests. In Britain, it was the principal species planted in every decade from 1941 to 1980, when it increased in proportion from 16% to 60% of plantings per decade (Locke, 1987). By 2006, in the Republic of Ireland, it comprised 52.3% of the forest estate (compared with 4.1% Norway spruce and 1.2% Scots pine) (Government of Ireland, 2007).

In the Pacific North-West, Sitka spruce occupies a narrow coastal strip as part of the temperate rainforests, but stretches over a very wide latitudinal range in this maritime climate from south-east Alaska to Oregon and Washington (Day, 1957; Roche and Haddock, 1987). Archibald Menzies, in 1793, is credited with the first records (and herbarium specimens) by a European, but the tree is more strongly associated with the plant explorer David Douglas who first sourced seed for planting in the UK in 1831. He reported 'It has the great advantage that it thrives on poor soils, and could become a large and useful tree in Great Britain'.

Sitka spruce showed a remarkable ability to tolerate both the poor wet soils and the degree of exposure, particularly the strong winds of a maritime climate (Malcolm, 1987; Pyatt *et al.*, 2001). Only on sites with low nitrogen, such as heather moorland and deep peat soils, was it surpassed by species such as lodgepole pine and the native Scots pine (*P. sylvestris*), and also in frost hollows by Norway spruce. Mark Anderson (1950) noted these site differences in his guide to relating choice of tree species to vegetation types and this shaped the composition of the early plantation forests. Subsequently, the superior growth rate, resilience and market preference for Sitka spruce led to increasing efforts to transform site suitability with intensive site preparation and nutrient inputs (Taylor, 1970, 1991; Hendrick, 1979; Paterson and Mason, 1999).

15.3.2 Breaking up the conifer blanket

The plantations were meant to be regularly thinned through the life of the crop before their eventual clear-felling, at perhaps age 45–60 years old. In practice, the thinning ambitions were thwarted by the cost of the operations, the insecurity of markets and the realization that, on exposed sites, thinning often precipitated windthrow (Booth, 1977).

Criticism of the scale of landscape change led to consideration of visual quality, with the pioneering work in the 1960s of Sylvia Crowe and, later, that a cadre of landscape specialists (Bell, 2012). The growing interest in diversification for landscape design was supplemented by the notion of restructuring, whereby fellings were staged such that there were several years between the cutting of adjacent compartments. This diversification of age classes provided a number of benefits, including limiting the rate of change in the landscape, creating a more uniform supply of timber, spreading the risk of extensive windthrow and producing a mosaic landscape in forests such as Kielder Forest in Northumberland, UK (Hibberd, 1985; McIntosh, 1995).

The requirements to deliver more than just timber led to the development of guidelines to safeguard watercourses, open habitats, special species, heritage features, etc. which, in the UK, were brought together in the UK Forestry Standard (Forestry Commission, 2011). The biodiversity guidelines have emphasized the desirability for the diversification of tree species (including the use of native species where possible) and of age structure (including retaining trees for a very long time), and the retention of standing and fallen dead wood. Most recently, the diversification of structure and tree species has also been recommended as a form of climate change adaptation (Read *et al.*, 2009; Ray *et al.*, 2010), and may contribute to reducing the vulnerability of our forests to pests and diseases.

15.4 Species Composition of Spruce Plantations

The Atlantic spruce forests are novel habitats. Arguably, they have been neglected by ecologists, many of whom had a great antipathy towards them because of their origin and the damage wrought in the process to valued semi-natural habitats. Indeed, there were some who considered that they were going to be ecological deserts of no interest.

Others suggested that these forests had potential, and evidence that this might be so is beginning to become available. There have been two substantial campaigns to capture and document the biodiversity of the Atlantic spruce forests: the Biodiversity Research Programme of the British Forestry Commission (Humphrey *et al.*, 2003a); and studies in the Republic of Ireland funded by COFORD (the Council for Forest Research and Development) and others (see Iremonger *et al.*, 2007; O'Halloran *et al.*, 2011) (Table 15.1).

Some of the predictions about the low biodiversity of large-scale spruce forests were overly pessimistic. In the British biodiversity assessment, ten Red Data listed macro-fungi were found in Sitka spruce stands – out of total of 29 Red Data listed species found in the entire plantation study (Humphrey *et al.*, 2000). In the Irish surveys, plantation forests supported species of conservation concern, including five vascular plant species, three bryophytes, nine spiders, one beetle and five bird species (Irwin *et al.*, 2013). Across the

Table 15.1. Summary of the sampling methods used for the different taxonomic groups in the British and Irish biodiversity assessments. (Adapted from Quine and Humphrey, 2010.) Detailed descriptions of the sampling methods and results are given in the publications quoted.

Taxonomic groups	Sampling method	Details for British studies	Details for Irish studies
Overview		Humphrey *et al.*, 2003a; Quine and Humphrey, 2010	O'Halloran *et al.*, 2011; Irwin *et al.*, 2013, 2014
Canopy invertebrates	Insecticide fogging of a single tree	Jukes *et al.*, 2002	Pedley *et al.*, 2014
Sub-canopy invertebrates	Malaise trapping	Jukes and Peace, 2003	
Ground invertebrates	Pitfall trapping	Jukes *et al.*, 2001	Oxbrough *et al.*, 2010; Mullen *et al.*, 2008
Dead wood and associated lichens, bryophytes and invertebrates	Forest plots Emergence traps over a period of months	Jukes and Peace, 2003 Humphrey *et al.*, 2002	Sweeney *et al.*, 2010a
Lepidoptera	Light traps		O'Halloran *et al.*, 2011
Fungi	Frequency and abundance of fruiting bodies	Humphrey *et al.*, 2000	O'Hanlon and Harrington, 2011
Epiphytes in the trunk and canopy	Forest plots		Coote *et al.*, 2008
Vascular plants	Vegetation plots	Ferris *et al.*, 2000	Coote *et al.*, 2012
Songbirds	Point counts and territory mapping	Fuller and Browne, 2003	Sweeney *et al.*, 2010b; O'Connell *et al.*, 2012

British data, species richness was often similar for a wide range of species groups between the spruce plantations and native pine woodland (Quine and Humphrey, 2010) (Fig. 15.2). In the Irish study, Irwin *et al.* (2014) suggest that, while there are differences in community composition, 'The species richness of non-native spruce-dominated plantations can be as high as that found in semi-natural woodlands, which suggests that temperate plantation forests, with appropriate management, can provide habitat for plant and animal species'.

The ecological contribution of the spruce forests is complicated by their mixed origin (e.g. preceding land use) and dynamic nature; many are still developing and there are few historical precedents to suggest what the end point will be.

15.5 Ecological Implications of Stand Dynamics

Models of natural stand dynamics following disturbance (Oliver and Larson, 1990; Mason and Quine, 1995) typically identify four stages of subsequent stand development: stand initiation; canopy closure; pole stage; multistorey. To these, we should add the stand-replacing disturbance itself, which resets the cycle; and in the case of afforestation, we should distinguish the single occurrence/act of change of land use. Most commercial woodland practice prevents the achievement of the multistorey stage 4 as the stands are felled at the pole stage.

The British studies that have been carried out captured all four stages of spruce forest (with the oldest trees being 69 years old), while the Irish studies largely covered the first three stages and tended to record lower volumes of dead wood (especially of large dimension) (Sweeney *et al.*, 2010a). The Irish studies also more formally addressed comparisons with open-ground habitats (Oxbrough *et al.*, 2006) and between afforestation and reforestation (Oxbrough *et al.*, 2010; Pedley *et al.*, 2014).

The length of time from initiation of the stand tends to be associated with the loss of open ground and some generalist species, but with the acquisition of forest species,

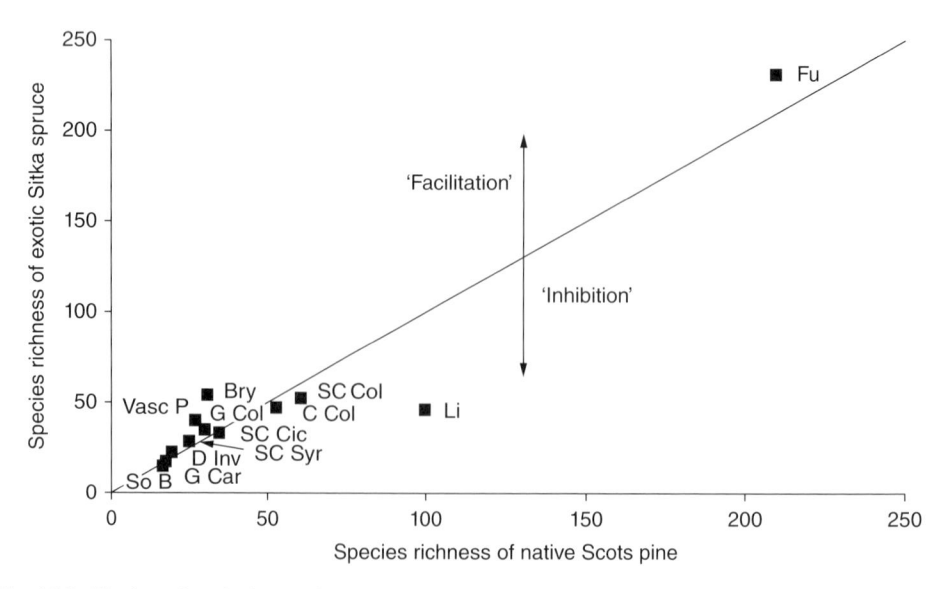

Fig. 15.2. The broad equivalency of species richness by group in exotic Sitka spruce (*Picea sitchensis*) and native Scots pine (*Pinus sylvestris*) from the British Biodiversity Assessment Programme. (From Quine and Humphrey (2010) reproduced under RightsLink licence 3438230916023.) Key: Bry, bryophytes; C Col, canopy coleopteran; D Inv, dead wood invertebrates; Fu, fungi; G Car, ground Carabidae; G Col, ground Coleoptera (excluding Carabidae); Li, lichens; SC Cic, sub-canopy Cicadamorpha; SC Col, sub-canopy coleopteran; SC Syr, sub-canopy Syrphidae; So B, songbirds; Vasc P, vascular plants.

particularly in the oldest stands with multiple canopy strata and a greater diversity of microhabitats. The levels of dead wood also builds up rapidly in the oldest stands. These changes are now summarized by broad stand stage (see also Fig. 15.3).

15.5.1 Precursors – the creation of woodland through afforestation (Stage 0)

The initial establishment of forests through afforestation necessarily begins a dramatic change in habitat. The character of the preceding land

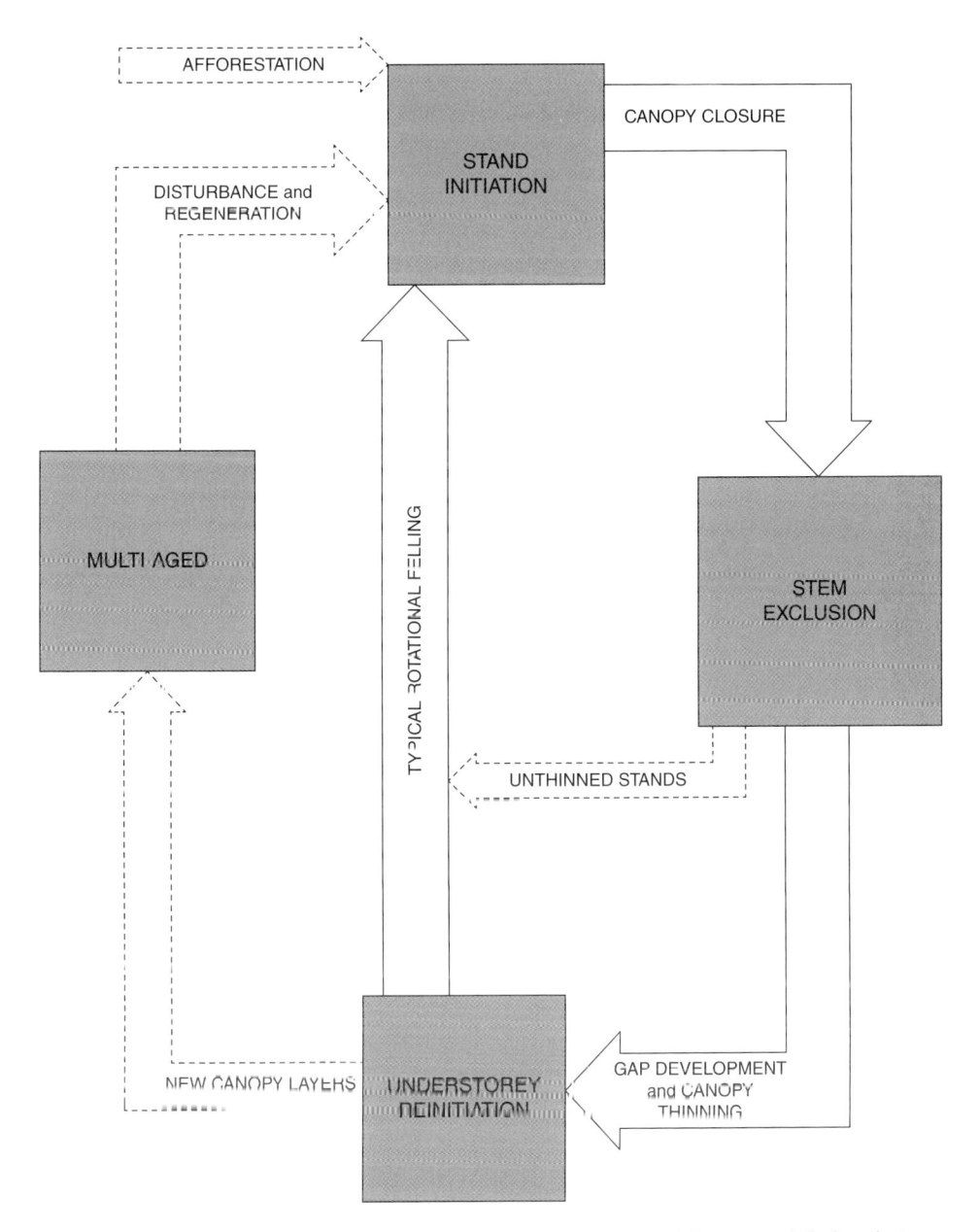

Fig. 15.3. The stand dynamics cycle for Atlantic spruce (*Picea sitchensis*) forests modelled on that initiated by natural disturbance.

use and community composition (grassland, heathland or bog) affects forest development. Many upland sites in Britain were previously managed (at least in part) for game shooting, and the reduction in intensity of wildlife control can be beneficial for many mammalian and avian predators.

More generally, in both Britain and Ireland, the land has been grazed and often fertilized/limed. Fencing, by reducing the impact of large herbivores, can lead to an increase in the vegetation biomass (Hill, 1979). While domestic grazing animals are generally removed, wild herbivores may still be present and subsequently increase. Site preparation measures such as ploughing and drainage alter the hydrology and expose bare soil for revegetation (Buscardo *et al.*, 2008). In due course, there is the contribution of the shade and needle fall from the growing trees (Wallace and Good, 1995).

The years immediately after afforestation can see rapid increase in vegetation, booms of small mammals, for example, field voles, *Microtus agrestis*, and associated predators such as mustelids or raptors (Petty *et al.*, 2000). Notable species that may benefit from the change to forests include hen harriers (*Circus cyaneus*) and black grouse (*Tetrao tetrix*) (Wilson *et al.*, 2009). Other species, such as waders of open habitats have been negatively affected (Stroud *et al.*, 1987; Avery and Leslie, 1992; Pearce-Higgins *et al.*, 2009) through loss of habitat due to site preparation, by apparent edge effects reflecting behavioural responses to the presence of trees (Wilson *et al.*, 2014) or, over time, due to increased predation from animals sheltering in the established woodland (Douglas *et al.*, 2013).

For some species, initial fears of displacement have only partially been confirmed. Species such as merlin (*Falco columbarius*) have displayed some plasticity in behaviour such that they have begun to nest in trees at the forest edge (Little *et al.*, 1995), while in Ireland hen harriers use restock areas (Irwin *et al.*, 2008). Oxbrough *et al.* (2006) showed loss of peatland specialist spiders on afforestation, but an increase in species richness when the preceding land use was improved grassland.

15.5.2 Stand initiation (Stage 1)

Once the trees are on the site, their influence over the stand develops during the next perhaps 15 years, the rate depending on factors such as stocking density and site quality, and the extent of vegetation left from the preceding rotation. At a typical planting spacing of <2 m (necessary to obtain adequate timber quality), Sitka spruce is particularly efficient at intercepting light compared with most broadleaves and other conifers such as Scots pine (Cannell, 1987).

The interaction of a changing stand environment with the consequences of colonization from nearby wooded or open habitats leads to the formation of distinct communities which are more diverse than in subsequent stages (Humphrey *et al.*, 2003b). There may be a substantial increase in ruderal weed species (Abdy and Mayhead, 1992), and the flora may initially be more closely matched with heathland and upland grassland semi-natural vegetation communities than with those of woodland (Good *et al.*, 1990). The remoteness of woodland seed sources may limit development of a woodland ground flora (Wallace and Good, 1995).

Organisms that benefit from this phase include light-requiring plants, small mammals and birds/invertebrates favouring shrub layers and flowering plants. Some of these may be of conservation concern – in Ireland for example, the garden warbler (*Sylvia borin*), grasshopper warblers (*Locustella naevia*) and linnets (*Carduelis cannabina*) were all recorded (Irwin *et al.*, 2013). On former woodland sites, there may be overlap with species of early coppice stages (Buckley and Mills, Chapter 10). There may also be invertebrates and lower plants on any dead wood left from a previous crop.

15.5.3 The impact of stand development – canopy closure and mortality (Stages 2 and 3)

Following canopy closure, there is a protracted period in which the trees grow in height (and stem diameter), but the canopy (leaf area index) remains largely static; in effect, the live canopy is elevated further and further above the ground. The closure of the canopy and the development of shaded conditions results in a loss of ground flora (Hill, 1986) and truncates community development (Wallace and Good, 1995), particularly in unthinned stands. In the absence of thinning, competition starts to bring some vertical structure to the stand and a hierarchy of stems. In moderately shade-tolerant species, such as Sitka spruce, subdominants begin to lose position in the canopy, even though they stay alive; only in the latter stages (Stage 3) of stand development are snags and large-diameter dead wood logs likely to be present (Humphrey *et al.*, 2003a).

Canopy cover is high and can result in the almost complete eradication of vascular plants, but near to the ground, the high moisture levels and shade can suit bryophytes and some invertebrates. This mid-rotation stage may have the greatest soil microbial community (Humphrey *et al.*, 2003a).

Conditions in the canopy may differ little from that just before canopy closure, but there have been relatively few studies of the assemblages of canopy invertebrates (Ozanne, 1996). Studies in Ireland have shown a high abundance of aphids and midges, which seem to benefit birds such as the gold crest (*Regulus regulus*) and coal tit (*Periparus ater*), but species richness is low compared with broadleaved woodland (Pedley *et al.*, 2014). Conditions are less favourable for birds favouring open habitats, such as black grouse (Pearce-Higgins *et al.*, 2007), but other woodland grouse can benefit (Picozzi *et al.*, 1996). Cone production is generally unlikely at this stage and so cone-feeding birds are not attracted.

In due course, canopy gaps appear, whether deliberately where thinning is possible (although such gaps are transient as the aim is for relatively rapid closure), or unplanned through crop mortality from abiotic factors such as wind (Quine, 2003) and biotic factors (pests and diseases). These gaps start to bring light to the forest floor once more, albeit still at low levels. This can be sufficient to allow a few flowering plants to grow, particularly on upturned stumps (Humphrey, 2005). There may also be the beginnings of tree regeneration (Quine, 2001).

At this stage, the economics of timber production usually dictate that the stands are felled – starting the cycle over again. Such a reset clearly constrains the structural development that is possible – but does keep a substantial proportion of the forest in the early seral stages; this is advantageous to species that benefit from such transient habitats – but it is less favourable to mature forest species that might benefit from Stage 4 development.

15.5.4 Prolonging the rotation and developing multiple storeys (Stage 4)

To date, most spruce forests have contained combinations of the first three stages of stand development and there is little experience of stands of Sitka spruce beyond economic rotation and achieving a multi-aged structure; for instance, of the 282,000 ha of spruce forest in south-west Scotland, only 2.7% was greater than 60 years old (Forestry Commission, 2012). There are aspirations to transform many forests currently managed as even-aged systems to more diverse species and structural mixtures, including through continuous cover type approaches (Mason, 2007). For example, the Woodlands for Wales strategy (Forestry Commission Wales, 2009) wanted 'appropriate diversification of...particularly non-native woodlands, at a range of scales and using mechanisms suited to the site and the woodland management objectives.... In practice, diversity will be achieved on a site by site basis and will include characteristics such as species, genetic diversity, age, structure and extent of woodland cover. A range of management techniques and silvicultural systems will all play a part in creating diversity and will be matched to site conditions and management objectives'.

Successful transformation of even-aged stands to uneven-aged stands is most likely on sheltered sites and freely draining soils that are suitable for a number of tree species and for natural regeneration. Biodiversity benefits are often claimed for such woodland, but there is remarkably little evidence to underpin such claims, particularly in countries with a long history of low forest cover (Quine *et al.*, 2007; Bürgi, Chapter 8).

Peterken *et al.* (1992), Humphrey *et al.* (2003a) and Humphrey (2005) concluded that there could be benefits from extending rotations, even if they remain largely even aged. This stage tends to have the greatest structural complexity (both vertical stratification and horizontal patchiness), and a more diverse flora with a greater complement of woodland herbs and bryophytes. The flora may also show the greatest similarity to semi-natural woodland vegetation communities (usually forms of native oak woodland) (Ferris *et al.*, 2000; Wallace, 2003; French *et al.*, 2008; Smith *et al.*, 2008).

Dead wood volumes may be very high; Humphrey *et al.* (2003a) quote >300 m^3 ha^{-1} in unthinned and partially windthrown stands, with a substantial proportion in large diameter (>20 cm) and well-decayed pieces. Lichens benefit from the increased light and dead wood substrate and are found in greater numbers in this than in other stages.

In many upland sites, there is likely to be attrition of old stands over time due to windthrow, but if the stand is left unharvested, this may lead to complex stand structures including a partial wave of tree regeneration within the gaps (Quine and Malcolm, 2007). Market demands also militate against retention stands, as small volumes of large-diameter stems are more difficult to handle. However, there is increasing evidence of the survival of large old conifers in sheltered valleys such as at Coed y Brenin in Wales and Glentress in Scotland. Modelling suggests that there could be the potential to manage some stands for the very long term across a substantial proportion of some landscapes (Quine *et al.*, 1999).

15.5.5 Resetting the woodland through disturbance

An event at the end of Stages 2, 3 or 4, usually clear-felling, but sometimes fire or storm, resets the cycle. Site quality will remain important in determining what happens but, increasingly, even-aged plantations will commence not from open land but from land that has recently borne trees. There has been less research on such second-rotation stands, but what there is suggests that it does not necessarily follow the same path as afforestation, so it is important to distinguish these two origins (Oxbrough *et al.*, 2010). The vegetation of restock sites may differ from that of the first rotation, with an initial phase unlike typical open ground vegetation units changing over time towards a woodland assemblage (Wallace and Good, 1995). The nature of the preceding tree crop, and the manner of its removal, will influence the availability of seed sources. There may be biological legacies such as dead wood, though these may harbour pests and diseases, and the size and surroundings of the gap are likely to be different second time around. Pedley *et al.* (2014) found little difference between second and subsequent rotations for canopy invertebrates, but Sweeney *et al.* (2010b) showed that the second rotation was better for migrant birds. The suitability of the subsequent restock sites is inconclusive for some animals, but some species, such as the nightjar, benefit.

15.6 Forest Design

The levels of diversity found in afforested sites depend on the scale at which the study is carried out, and how the stands studied and the spaces between them are arranged. In general, we know more about the ecology of individual stands and the influence of change in structure than we do about the influence of landscape scale and pattern. Yet it is often the comparison with other choices/options at these wider spatial scales that would be most useful and most contentious.

No one stand stage (Fig. 15.3) is 'best' for biodiversity as different taxonomic groups respond differently to the changing woodland conditions (Humphrey *et al.*, 2002, 2003a,b; Smith *et al.*, 2008). Bird communities may be particularly distinct in the pre-thicket stage (Humphrey *et al.*, 2003a,b), but other groups such as bryophytes benefit from the more closed stages. Forest-associated species generally increase with stand age, but the open and generalist species decline (Mullen *et al.*, 2008), although they may be able to survive along roads and rides (Wallace and Good, 1995; Fuller *et al.*, 2013).

The recommendations, for example in the UK Forestry Standard, the associated biodiversity guidelines (Forestry Commission, 2011) and voluntary certification schemes such as UKWAS (2012), are that forests should contain a range of age classes and open ground to increase species diversity. However there is relatively little evidence as to the most appropriate proportions for the different stages. In practice, the age structure and landscape pattern of forests will be shaped by design planning that incorporates visual landscape concerns, management practicalities, production forecasting and risk management as much as by biodiversity.

15.7 The Landscape Setting

The landscape setting of even-aged plantation forests is important, in particular the proximity to other long-established woodland (Humphrey et al., 2003b; Smith et al., 2008). Distance to forest edge (Smith et al., 2008) can also influence biodiversity in these human-made forests, as in more natural systems. There are, therefore, opportunities to improve habitat connectivity by the appropriate placement of new plantations, and by modifying the management of the intervening matrix to be less hostile to movement of organisms (Quine and Watts, 2009; Watts et al., 2010). For some species, plantations of non-native trees need to be treated as matrix, whereas for others the plantations may be suitable additional habitat. Managers should thus try to create a landscape that improves conditions for both sets of species.

An elegant description of this was proposed by Hunter (1999). He envisaged a landscape triad consisting of ecological reserves (high-quality habitat perhaps with minimum intervention), a broad sweep of ecological forestry (productive forests managed to support biodiversity as well as deliver timber) and some highly intensive forests (managed to maximize production and arguably substitute for the loss of production in making the wider matrix more diverse). These are variants of the land-sharing/land-sparing paradigms (Fischer et al., 2014).

Landscape planning at an even larger scale is needed to help to resolve the problem of obvious losses for high-profile species in regions with large-scale and rapid afforestation, notably large raptors, such as golden eagles (*Aquila chrysaetos*) (McGrady et al., 2003), and open-ground waders, for instance the dunlin (*Calidris alpina*), greenshank (*Tringa nebularia*) and golden plover (*Pluvialis apricaria*) in the Flow Country of Scotland (Wilson et al., 2014). The direct footprint of the forest is not the only effect – some hydrological effects may extend up to 100 m beyond the forest edge, and the behavioural impacts on birds (for whatever reason) may extend to a number of kilometres.

Growing interest in ecosystem services may govern how the proportion of broad forest types are determined across a landscape, but also the balance between high forest and low forest type landscapes.

15.8 Where Next?

Even forests largely made up of even-aged plantations will in future have greater provision for open habitats, the protection of riparian and other scarce habitats, and the diversification of tree species, age and structure than first time around. Recent concerns over tree health are adding to the interest in and demand for this. Over time, these plantations might be encouraged to acquire more of the semi-natural characteristics found to be beneficial to many species (Irwin et al., 2013).

However, the evidence base upon which to build these requirements and refine the specifications is surprisingly scant, given their extent. Conservation ecologists should perhaps give more attention to the potential that even minor modifications to standard production practices might bring in improving the biodiversity values of these new and extensive cultural landscapes.

Other options that are emerging include:

- Restoration to open habitats. This seeks to redress the worst excesses (ambitions) of the formation of the strategic reserve and the forest as solely a financial investment. Particular examples are of the peatlands (Patterson and Anderson, 2000) and lowland heaths (Barwick, 2009).
- Industrialization and intensification, including short-rotation forestry and biomass

production (including with yet more non-native species – such as *Eucalyptus*).
- Deforestation for wind turbines and other developments. The exposed, poor agricultural quality ground that has been occupied by spruce plantations does attract this latest of the upland land uses – namely renewable energy generation. More than 1200 ha of largely spruce forest has been felled near Moffat, in southern Scotland, to site 100 wind turbines, with restored heathland in between!
- A further option, naturalization, letting natural processes rip and seeing what happens seems less likely than perhaps a decade ago (Quine *et al.*, 1999), though there is sustained interest in variants of rewilding in locations such as Ennerdale in the English Lake District (Convery and Dutson, 2008).

15.9 Conclusion

Atlantic spruce forests can deliver multiple ecosystem services; they have contributed to an increase in domestic production of timber in the UK from 4 to 20% since World War II. Some have become popular sites for recreation: Kielder Forest attracts more than 300,000 visitors per year (Quine *et al.*, 2011, 2013). The forests are increasing in biodiversity value, albeit of a distinct and different nature to that of native woodland or the open habitats that they replaced; whether the overall balance is judged positive or negative depends on the values placed on particular species or assemblages.

The Atlantic spruce forests are in many ways unique, reflecting a particular response to the long-term loss of woodlands in Britain and Ireland, but they can offer useful insights, even for countries with a substantial forest area and a long tradition of forest stewardship.

- A land-use change focused on one particular end point can bring about unexpected benefits, as well as the anticipated losses. The loss of open ground habitat has clearly had an impact upon open-ground communities, but there has been the creation of novel habitats which contain some remnants of the open ground plus increased provision for forest specialists.
- Even if such habitats appear to be unpromising, these preconceptions deserve to be tested and evidence developed. Without it, ecologists are ill equipped to provide advice on technical specifications which indicate the potential trade-offs and gains that might be achieved with relatively little cost.

Tsouvalis (2000) suggested that: 'With the expected functions of forests becoming ever more complex and multiple, there is hope that plantation forests will cease to be treated as mere timber factories'. Irwin *et al.* (2013) noted that their 'intensive survey of plantation forests throughout Ireland has revealed that these sites offer habitats to some nationally rare and threatened species of plants, invertebrates and birds. The potential of plantation forests to enhance national biodiversity is thus evident and so the planning and management of these forests should incorporate this goal'. Christopher Smout (2003) talked about a nation rediscovering a woodland culture. Can we imagine the transformation of the even-aged plantation from villain to (minor) hero?

References

Abdy, E.M. and Mayhead, G.J. (1992) The ground vegetation in upland spruce plantations in north Wales. *Aspects of Applied Biology* 29, 73–81.
Aldhous, J.R. (1997) British forestry: 70 years of achievement. *Forestry* 70, 283–292.
Anderson, M.L. (1950) *The Selection of Tree Species*. Oliver and Boyd, Edinburgh, UK.
Avery, M. and Leslie, R. (1992) *Birds and Forestry*. Poyser, London.
Barwick, P. (2009) Making a silk purse out of a sow's ear – heathland restoration from forestry plantations. *Journal for Practical Ecology and Conservation Special Series No. 5*, 11–15.
Bell, S. (2012) *Landscape: Pattern, Perception and Process*, 2nd edn. Routledge, London.

Booth, T.C. (1977) *Windthrow Hazard Classification*. Forestry Commission Research Information Note 22, Forestry Commission, Edinburgh, UK.

Brockerhoff, E.G., Jactel, H., Parrotta, J.A., Quine, C.P. and Sayer, J. (2008) Plantation forests and biodiversity: oxymoron or opportunity? *Biodiversity and Conservation* 17, 925–951.

Buscardo, E., Smith, G.F., Kelly, D.L., Freitas, H., Iremonger, S., Mitchell, F.J.G., O'Donoghue, S. and Mckee, A.-M. (2008) The early effects of afforestation on biodiversity of grasslands in Ireland. *Biodiversity and Conservation* 17, 1057–1072.

Calvino-Cancela, M., Rubido-Bara, M. and van Etten, E.J.B. (2012) Do eucalypt plantations provide habitat for native forest biodiversity? *Forest Ecology and Management* 270, 153–162.

Cannell, M.G.R. (1987) Photosynthesis, foliage development and productivity of Sitka spruce. *Proceedings of the Royal Society of Edinburgh Section B: Biological Sciences* 93, 61–74.

Carle, J. and Holmgren, P. (2008) Wood from planted forest – a global outlook 2005–2030. *Forest Products Journal* 58, 5–18.

Christie, S., McCann, D., Annett, J., Bankhead, J., Burgess, D., Casement, P., Christie, P., Cooper, A., Griffin, J., Halliday, N. et al. (2011) Chapter 18: Status and changes in the UK ecosystems and their services to society: Northern Ireland. In: Watson, R. and Albon, S. (eds) *UK National Ecosystem Assessment: Understanding Nature's Value to Society*. United Nations Environment Programme World Conservation Monitoring Centre (UNEP-WCMC), Cambridge, UK. Available at: http://uknea.unep-wcmc.org/LinkClick.aspx?fileticket=MCZWs8Rz3fg%3D&tabid=82 (accessed 3 December 2014).

Convery, I. and Dutson, T. (2008) Rural communities and landscape change: a case study of wild Ennerdale. *Journal of Rural and Community Development* 3, 104–118.

Coote, L., Smith, G.F., Kelly, D.L., O'Donoghue, S., Dowding, P., Iremonger, S. and Mitchell, F.J.G. (2008) Epiphytes of Sitka spruce (*Picea sitchensis*) plantations in Ireland and the effects of open spaces. *Biodiversity and Conservation* 17, 953–968.

Coote, L., French, L.J., Moore, K.M., Mitchell, F.J.G. and Kelly, D. (2012) Can plantation forests support plant species and communities of semi-natural woodland? *Forest Ecology and Management* 283, 86–95.

Day, W.R. (1957) *Sitka Spruce in British Columbia*. Forestry Commission Bulletin 28. Her Majesty's Stationery Office, London.

Douglas, D.J.T., Bellamy, P.E., Stephen, L.S., Pearce-Higgins, J.W., Wilson, J.D. and Grant, M.C. (2013) Upland land use predicts population decline in a globally near-threatened wader. *Journal of Applied Ecology* 51, 194–203.

EEA (2006) *European Forest Types: Categories and Types for Sustainable Forest Management Reporting and Policy*. European Environment Agency, Copenhagen.

Ferris R., Peace, A.J., Humphrey, J.W. and Broome, A.C. (2000) Relationships between vegetation, site type and stand structure in coniferous plantations in Britain. *Forest Ecology and Management* 136, 35–51.

Fischer, J., Abson, D.J., Butsic, V., Chappell, M.J., Ekroos, J., Hanspach, J., Kuemmerle, T., Smith, H.G. and Von Wehrden, H. (2014) Land sparing versus land sharing: moving forward. *Conservation Letters* 7, 149–157.

Foot, D. (2010) *Woods and People: Putting Forests on the Map*. The History Press, Stroud, UK.

Forest Service (2008) *Irish Forests – A Brief History*. Forest Service, Wexford, Irish Republic.

Forestry Commission (2003) *National Inventory of Woodland and Trees – Great Britain*. Forestry Commission, Edinburgh, UK.

Forestry Commission (2011) *The UK Forestry Standard*. Forestry Commission, Edinburgh, UK.

Forestry Commission (2012) *Standing Timber Volume for Coniferous Trees in Britain*. Forestry Commission, Edinburgh, UK.

Forestry Commission Wales (2009) *Woodlands for Wales: The Welsh Assembly Government's Strategy for Woodlands and Trees*. Aberystwyth, UK.

French, L.J., Smith, G.F., Kelly, D.L., Mitchell, F.J.G., O'Donoghue, S., Iremonger, S.E. and McKee, A.-M. (2008) Ground flora communities in temperate oceanic plantation forests and the influence of silvicultural, geographic and edaphic factors. *Forest Ecology and Management* 255, 476–494.

Fuller, L., Oxbrough, A., Irwin, S., Kelly, T.C. and O'Halloran, J. (2013) The importance of young plantation forest habitat and forest road-verges for ground-dwelling spider diversity. *Biology and Environment: Proceedings of the Royal Irish Academy* 113B, 259–271.

Fuller, R. and Browne, S. (2003) Effects of plantation structure and management on birds. In: Humphrey, J.W., Quine, C.P. and Ferris, R. (eds) *Biodiversity in Britain's Planted Forests. Results from the Forestry Commission's Biodiversity Assessment Project*. Forestry Commission, Edinburgh, UK, pp. 93–99.

Gardiner, B.A. and Quine, C.P. (2000) Management of forests to reduce the risk of abiotic damage – a review with particular reference to the effects of strong winds. *Forest Ecology and Management* 135, 261–277.

Good, J.E.G., Williams, T.G., Wallace, H.L., Buse, A. and Norris, D.A. (1990) *Nature Conservation in Upland Conifer Forest. Report to the Forestry Commission and Nature Conservancy Council.* Institute of Terrestrial Ecology, Bangor, UK.

Government of Ireland (2007) *National Forest Inventory of Republic of Ireland – Results.* Forest Service, Wexford, Irish Republic.

Hendrick, E. (1979) Site amelioration for reforestation. *Irish Forestry,* 36, 89–98.

Hibberd, B.G. (1985) Restructuring of plantations in Kielder Forest District. *Forestry* 58, 119–130.

Hill, M.O. (1979) The development of flora in even-aged plantations. In: Ford, E.D., Malcolm, D.C. and Atterson, J. (eds) *The Ecology of Even Aged Plantations.* Institute of Terrestrial Ecology, Cambridge, UK, pp. 175–192.

Hill, M.O. (1986) Ground flora and succession in commercial forests. In: Jenkins, D. (ed.) *Trees and Wildlife in the Scottish Uplands.* Institute of Terrestrial Ecology, Edinburgh, UK, pp. 71–78.

Horgan T., Keane, M., McCarthy, R., Lally, M. and Thompson, D. (2003) *A Guide to Forest Tree Species Selection and Silviculture in Ireland.* Council for Forest Research and Development (COFORD), Dublin.

Humphrey, J.W. (2005) Benefits to biodiversity from developing old-growth conditions in British upland spruce plantations: a review and recommendations. *Forestry* 78, 33–53.

Humphrey, J.W., Newton, A.C., Peace, A.J. and Holden, E. (2000) The importance of conifer plantations in northern Britain as a habitat for native fungi. *Biological Conservation* 96, 241–252.

Humphrey, J.W., Davey, S., Peace, A.J., Ferris, R. and Harding, K. (2002) Lichens and bryophyte communities of planted and semi-natural forests in Britain: the influence of site type, stand structure and deadwood. *Biological Conservation* 107, 165–180.

Humphrey, J.W., Ferris, R. and Quine, C.P. (eds) (2003a) *Biodiversity in Britain's Planted Forests: Results from the Forestry Commission's Biodiversity Assessment Project.* Forestry Commission, Edinburgh, UK.

Humphrey, J.W., Ferris, R. and Peace, A.J. (2003b) Relationship between site type, stand structure and plant communities. In: Humphrey, J.W., Quine, C.P. and Ferris, R. (eds) *Biodiversity in Britain's Planted Forests. Results from the Forestry Commission's Biodiversity Assessment Project.* Forestry Commission, Edinburgh, UK, pp. 23–30.

Hunter, M.L. Jr (ed.) (1999) *Maintaining Biodiversity in Forested Ecosystems.* Cambridge University Press, Cambridge, UK.

Iremonger, S.J., O'Halloran, J., Kelly, D.L., Wilson, M.W., Smith, G.F., Gittings, T., Giller, P.S., Mitchell, F.J.G., Oxbrough, A., Coote, L. et al. (2007) *Biodiversity in Irish Plantation Forests: Final Report.* ERTDI Report Series No. 51, Environmental Protection Agency (EPA) and Council for Forest Research and Development (COFORD), Dublin. Available at: http://www.epa.ie/pubs/reports/research/biodiversity/ERTDI%20Report%2051.pdf (accessed 3 December 2014).

Irwin, S., Wilson, M.W., Kelly, T.C., O'Donoghue, B., O'Mahony, B., Oliver, G., Cullen, C., O'Donoghue, T. and O'Halloran, J. (2008) Aspects of the breeding biology of hen harriers *Circus cyaneus* in Ireland. *Irish Birds* 8, 331–334.

Irwin, S., Kelly, D.L., Kelly, T.C., Mitchell, F.J.G., Coote, L., Oxbrough, A., Wilson, M.W., Martin, R.D., Moore, K.A., Sweeney, O.F. et al. (2013) Do Irish forests provide habitat for species of conservation concern? *Biology and Environment: Proceedings of the Royal Irish Academy* 113B, 273–279.

Irwin, S., Pedley, S.M., Coote, L., Dietzsch, A.C., Wilson, M.W., Oxbrough, A., Sweeney, O., Moore, K.M., Martin, R., Kelly, D.L. et al. (2014) The value of plantation forests for plant, invertebrate and bird diversity and the potential for cross-taxon surrogacy. *Biodiversity and Conservation* 23, 697–714.

Jukes, M.R. and Peace, A.J. (2003) Invertebrate communities in plantation forests. In: Humphrey, J.W., Quine, C.P. and Ferris, R. (eds) *Biodiversity in Britain's Planted Forests. Results from the Forestry Commission's Biodiversity Assessment Project.* Forestry Commission, Edinburgh, UK, pp. 75–91.

Jukes, M.R., Peace, A.J. and Ferris, R. (2001) Carabid beetle communities associated with coniferous plantations in Britain: the influence of site, ground vegetation and stand structure. *Forest Ecology and Management* 148, 271–286.

Jukes, M.R., Ferris, R. and Peace, A.J. (2002) The influence of stand structure and composition on diversity of canopy Coleoptera in coniferous plantations in Britain. *Forest Ecology and Management* 163, 27–41.

Kirby, K., Quine, C. and Brown, N. (2009) The adaptation of UK forests and woodlands to climate change. In: Read, D.J., Freer-Smith, P.H., Morison, J.I.L., Hanley, N., West, C.C. and Snowdon, P. (eds) *Combating Climate Change – A Role for UK Forests. An Assessment of the Potential of the UK's Trees and Woodlands to Mitigate and Adapt to Climate Change.* The Stationery Office, Edinburgh, UK, pp. 164–179.

Lines, R. (1987) Seed origin variation in Sitka spruce. *Proceedings of the Royal Society of Edinburgh Section B: Biological Sciences* 93, 25–40.
Little, B., Davison, M. and Jardine, D. (1995) Merlins *Falco columbarius* in Kielder Forest: influences of habitat on breeding performance. *Forest Ecology and Management* 79, 147–152.
Locke, G.M.L. (1987) *Census of Woodlands and Trees 1979–82*. Forestry Commission Bulletin 63, Her Majesty's Stationery Office, London.
Maizeret, C. (2005) *Les Landes de Gascogne*. Delachaux et Niestlé, Paris.
Malcolm, D.C. (1987) Some ecological aspects of Sitka spruce. *Proceedings of the Royal Society of Edinburgh Section B: Biological Sciences* 93, 85–92.
Mason, W.L. (2007) Changes in the management of British forests between 1945 and 2000 and possible future trends. *Ibis* 149(Supplement 2), 41–52.
Mason, W.L. and Quine, C.P. (1995) Silvicultural possibilities for increasing structural diversity in British spruce forests: the case of Kielder Forest. *Forest Ecology and Management* 79, 13–28.
McGrady, M.J., Petty, S.J., McLeod, D.R.A., Mudge, G. and Bainbridge, I.P. (2003) Potential impacts of native woodland expansion on golden eagles (*Aquila chrysaetos*) in Scotland. In: Thompson, D.B.A., Redpath, S.M., Fielding, A.H., Marquiss, M. and Galbraith, C.A. (eds) *Birds of Prey in a Changing Environment*. [Proceedings of the Conference held by the Scottish Natural Heritage/British Ornithologists' Union/Joint Nature Conservation Committee.] The Natural Heritage of Scotland Series No. 12, The Stationery Office, Edinburgh, UK, pp. 341–350.
McIntosh, R.M. (1995) The history and multi-purpose management of Kielder Forest. *Forest Ecology and Management* 79, 1–11.
Mullen, K., O'Halloran, J., Breen, J., Giller, P., Pithon, J. and Kelly, T. (2008) Distribution and composition of carabid beetle (Coleoptera, Carabidae) communities across the plantation forest cycle – implications for management. *Forest Ecology and Management* 256, 624–632.
O'Connell, S., Irwin, S., Wilson, M.W., Sweeney, O., Kelly, T.C. and O'Halloran, J. (2012) How can forest management benefit bird communities? Evidence from eight years of research in Ireland. *Irish Forestry* 69, 44–57.
O'Halloran, J., Irwin, S., Kelly, D.L., Kelly, T.C., Mitchell, F.J.G., Coote, L., Oxbrough, A., Wilson, M.W., Martin, R.D., Moore, K. *et al.* (2011) *FORESTBIO Final Report. Management of Biodiversity in a Range of Irish Forest Types*. Report prepared for the Department of Agriculture, Fisheries and Food, Dublin. Available at: http://www.ucc.ie/en/media/research/planforbio/pdfs/FORESTBIOFinalProjectReport.pdf (accessed 3 December 2014).
O'Hanlon, R. and Harrington, T.C. (2011) The macrofungal component of biodiversity in Irish Sitka spruce forests. *Irish Forestry* 68, 40–53.
Oliver, C.D. and Larson, B.C. (1990) *Forest Stand Dynamics*. McGraw-Hill, New York.
Olwig, K. (1984) *Nature's Ideological Landscape*. Unwin Hyman, London.
Oxbrough, A., Gittings, T., O'Halloran, J., Giller, P.S. and Kelly, T.C. (2006) The influence of open space on ground-dwelling spider assemblages within plantation forests. *Forest Ecology and Management* 237, 404–417.
Oxbrough, A., Irwin, S., Kelly, T.C. and O'Halloran, J. (2010) Ground-dwelling invertebrates in reforested conifer plantations. *Forest Ecology and Management* 259, 2111–2121.
Ozanne, C.M.P. (1996) The arthropod communities of coniferous forest trees. *Selbyana* 17, 43–49.
Ozenda, P. (1994) *Végétation du Continent Européen*. Delachaux et Niestlé, Lausanne, France.
Paterson, D.B. and Mason, W.L. (1999) *Cultivation of Soils for Forestry*. Forestry Commission Bulletin 119, Edinburgh, UK.
Patterson, G.S. and Anderson, A.R. (2000) *Forests and Peatland Habitats*. Forestry Commission Guideline Note 1, Forestry Commission, Edinburgh, UK.
Pearce-Higgins, J.W., Grant, M.C., Robinson, M.C. and Haysom, S.L. (2007) The role of forest maturation in causing the decline of black grouse *Tetrao tetrix*. *Ibis* 149, 143–159.
Pearce-Higgins, J.W., Grant, M.C., Beale, C.M., Buchanan, G.M. and Sim, I.M.W. (2009) International importance and drivers of change of upland bird populations. In: Bonn, A., Allott, T., Hubaceck, K. and Stewart, J. (eds) *Drivers of Environmental Change in Uplands*. Routledge (imprint of Taylor & Francis), Abingdon, UK, pp. 209–227.
Pedley, S.M., Martin, R.D., Oxbrough, A., Irwin, S. Kelly, T.C. and O'Halloran, J. (2014) Commercial spruce plantations support a limited canopy fauna: evidence from a multi taxa comparison of native and plantations forests. *Forest Ecology and Management* 314, 172–182.
Peterken G.P., Ausherman D., Buchanan M. and Forman R.T.T. (1992) Old growth conservation within British upland conifer plantations. *Forestry* 65, 127–144.

Petty, S.J., Lambin, X., Sherratt, N.T., Thomas, C.J., Mackinnon, J.L., Coles, C.F., Davison, M. and Little, B. (2000) Spatial synchrony in field vole *Microtus agrestis* abundance in a coniferous forest in northern England: the role of vole-eating raptors. *Journal of Applied Ecology* 37, 136–147.

Picozzi, N., Moss, R. and Catt, D.C. (1996) Capercaillie habitat, diet, and management in a Sitka spruce plantation in central Scotland. *Forestry* 69, 373–388.

Pyatt, D.G., Ray, D. and Fletcher, J. (2001) *An Ecological Site Classification for Forestry in Great Britain*. Bulletin 124, Forestry Commission, Edinburgh, UK.

Quine, C.P. (2001) A preliminary survey of regeneration of Sitka spruce in wind-formed gaps in British planted forests. *Forest Ecology and Management* 151, 37–42.

Quine, C.P. (2003) Wind-driven gap formation and expansion in spruce forests of upland Britain. In: Ruck, B., Kottmeier, C., Mattheck, C., Quine, C.P. and Wilhelm, G. (eds) *Wind Effects on Trees: International Conference*. University of Karlsruhe, Karlsruhe, Germany, pp. 101–108.

Quine, C.P. and Humphrey, J.W. (2010) Plantations of exotic tree species in Britain: irrelevant for biodiversity or novel habitat for native species? *Biodiversity Conservation* 19, 1503–1512.

Quine, C.P. and Malcolm, D.C. (2007) Wind-driven gap development in Birkley Wood, a long-term retention of planted Sitka spruce in upland Britain. *Canadian Journal of Forest Research* 37, 1787–1796.

Quine, C.P. and Watts, K. (2009) Successful de-fragmentation of woodland by planting in an agricultural landscape? An assessment based on landscape indicators. *Journal of Environmental Management* 90, 251–259.

Quine, C.P., Humphrey, J.W. and Ferris, R. (1999) Should the wind disturbance patterns observed in natural forests be mimicked in planted forests in the British uplands? *Forestry* 72, 337–358.

Quine, C.P., Fuller, R.J., Smith, K.W. and Grice, P.V. (2007) Stand management: a threat or opportunity for birds in British woodland? *Ibis* 149, 161–174.

Quine, C.P., Cahalan, C., Hester, A., Humphrey, J., Kirby, K., Moffat, A. and Valatin, G. (2011) Chapter 8: Woodlands. In: Watson, R. and Albon, S. (eds) *UK National Ecosystem Assessment: Understanding Nature's Value to Society*. United Nations Environment Programme World Conservation Monitoring Centre (UNEP-WCMC), Cambridge, UK. Available at: http://uknea.unep-wcmc.org/LinkClick.aspx?fileticket=EuaMBUTBZIU%3D&tabid=82 (accessed 3 December 2014).

Quine, C.P., Bailey, S.E. and Watts, K. (2013) Sustainable forest management in a time of ecosystem services frameworks: common ground and consequences. *Journal of Applied Ecology* 50, 863–867.

Rackham, O. (2006) *Woodlands*. Collins New Naturalist, London.

Ratcliffe, P. and Claridge, J.E. (1996) *Thetford Forest Park: The Ecology of a Pine Forest*. Technical Paper 13, Forestry Commission, Edinburgh, UK.

Ray, D., Morison, J. and Broadmeadow, M. (2010) *Climate Change: Impacts and Adaptation in England's Woods*. Forestry Commission Research Note 201, Forestry Commission, Edinburgh, UK.

Read, D.J., Freer-Smith, P.H., Morison, J.I.L., Hanley, N., West, C.C. and Snowdon, P.R. (eds) (2009) *Combating Climate Change – A Role for UK Forests*. The Stationery Office, Edinburgh, UK.

Roche, L. and Haddock, P.G. (1987) Sitka spruce (*Picea sitchensis*) in North America with special reference to its role in British forestry. *Proceedings of the Royal Society of Edinburgh Section B: Biological Sciences* 93, 1–12.

Smith, G.F., Gittings, T., Wilson, M., French, L., Oxbrough, A., O'Donoghue, S., O'Halloran, J., Kelly, D.L., Mitchell, F.J.G., Kelly, T.C. et al. (2008) Identifying practical indicators of biodiversity for stand-level management of plantation forests. *Biodiversity and Conservation* 17, 991–1015.

Smout, T.C. (ed.) (2003) *People and Woods in Scotland*. Edinburgh University Press, Edinburgh, UK.

Stroud, D.A., Reed, T.M., Pienkowski, M.W. and Lindsay, R.A. (1987) *Birds, Bogs and Forestry. The Peatlands of Caithness and Sutherland*. Nature Conservancy Council, Peterborough, UK.

Sutton, W.R.J. (2014) Save the forests: use more wood. In: Fenning, T. (ed.) *Challenges and Opportunities for the World's Forests in the 21st Century*. Forestry Sciences, Vol. 81, Springer, Dordrecht, The Netherlands, pp. 213–227.

Sweeney, O.F.M., Martin, R.D., Irwin, S., Kelly, T.C., O'Halloran, J., Wilson, M.W. and McEvoy, P.M. (2010a) A lack of large-diameter logs and snags characterises dead wood patterns in Irish forests. *Forest Ecology and Management* 259, 2056–2064.

Sweeney, O.F.M., Wilson, M.W., Irwin, S., Kelly, T.C. and O'Halloran, J. (2010b) Breeding bird communities of second rotation plantations at different stages of the forest cycle. *Bird Study* 57, 301–314.

Symonds, H.H. (1936) *Afforestation in the Lake District: A Reply to the Forestry Commission's White Paper of 26th August 1936*. Dent, London.

Taylor, C.M.A. (1991) *Forest Fertilisation in Britain*. Her Majesty's Stationery Office, London.

Taylor, G.G.M. (1970) *Ploughing Practice in the Forestry Commission*. Her Majesty's Stationery Office, London.
Tittensor, R. (2009) *From Peat Bog to Conifer Forest: An Oral History of Whitelee, Its Community and Landscape*. Packard Publishing, Chichester, UK.
Tsouvalis, J. (2000) *A Critical Geography of Britain's State Forests*. Oxford University Press, Oxford, UK.
UKWAS (2012) *The UK Woodland Assurance Standard*. Version 3.0. United Kingdom Woodland Assurance Standard, Edinburgh, UK. Available at: http://ukwas.org.uk/ (accessed February 2014).
Wallace, H. (2003) Vegetation of plantation Sitka spruce – development of new 'forest noda'. In: Goldberg, E.A. (ed.) *National Vegetation Classification – Ten Years' Experience of Using the Woodland Section*. Joint Nature Conservation Committee (Report 335), Peterborough, UK, pp. 36–50.
Wallace, H.L. and Good, J.E.G. (1995) Effects of afforestation on upland plant communities and implications for vegetation management. *Forest Ecology and Management* 79, 29–46.
Watts, K., Eycott, A.E., Handley, P., Ray, D., Humphrey, J.W. and Quine, C. (2010) Targeting and evaluating biodiversity conservation action within fragmented landscapes: an approach based on generic focal species and least-cost networks. *Landscape Ecology* 25, 1305–1318.
Wilkie, M.L. (2001) From dune to forest: biological diversity in plantations established to control drifting sand. *Unasylva* 53, 64–69.
Wilson, J.D., Anderson, R., Bailey, S., Chetcuti, J., Cowie, N.R., Hancock, M.H., Quine, C.P., Russell, N., Stephen, L. and Thompson, D.B.A. (2014) Modelling edge effects of mature forest plantations on peatland waders informs landscape-scale conservation. *Journal of Applied Ecology* 51, 204–213.
Wilson, M.W., Irwin, S., Norriss, D.W., Newton, S.F., Collins, K., Kelly, T.C. and O'Halloran, J. (2009) The importance of pre-thicket conifer plantations for nesting Hen Harriers *Circus cyaneus* in Ireland. *Ibis* 151, 332–343.
Zehetmayr, J.W.L. (1954) *Experiments on Tree Planting on Peat*. Forestry Commission Bulletin 22, Her Majesty's Stationery Office, London.
Zehetmayr, J.W.L. (1960) *Afforestation of Upland Heaths*. Forestry Commission Bulletin 32, Her Majesty's Stationery Office, London.

Part IV

A Variety of Woodland Histories

There is not one woodland history for European woods and forests but many, and the fourth section of the book illustrates this diversity. What we see is a product of human as much as natural processes and that simply abandoning land can lead to the loss of valued features and landscapes (Italy, Chapter 16). Even what is often described as a near-natural site – the Białowieza Forest in Poland – proves to have a much more complicated management history (Chapter 17) than might be expected. Chapters 18 and 19 show the importance of regional variation within Britain and France, respectively, in describing woodland histories. In trying to understand the past and use it to chart the future, we need to bring together social and natural science studies, as in the case of Bergslagen in Sweden (Chapter 20).

16 Historical Ecology in Modern Conservation in Italy

Roberta Cevasco[1]* and Diego Moreno[2]

[1]*Centro per l'Analisi Storica de Territorio, Università del Piemonte Orientale, Alessandria, Italy;* [2]*Laboratorio di Archeologia e Storia Ambientale, Università delgi Studi di Genova, Genoa, Italy*

16.1 Introduction

Alongside technological and production standardization, a sort of nature standardization has made the re-naturalization of the territory the cornerstone of the actions of landscape importance in many rural areas.

(Agnoletti, 2013)

In this chapter, we consider 'woodland' from the perspective of historical ecology, the 'classical' historical approach that evolved in Britain and northern Europe during the 1960s. We suggest that this approach could be beneficial in conservation policies in southern Europe, where it has been relatively infrequently applied.

The southern European mountain landscape represents a rich but fragile environmental heritage that still exists, albeit at the mercy of marginalized farming and fragile rural communities (Saratsi, 2005; Montiel-Molina, 2007; Kizos *et al.*, 2013). Re-naturalization – land abandonment – is a symptom of the threats to this heritage. Specific examples of these processes are discussed that show the need for a profound change in European conservation politics and strategies.

16.2 Background

The leading international journal *Landscape Ecology* has, in the last 10 years, published over 1300 papers, but less than 70 refer to history or the use of historical data, and less than 15 come from southern European countries. Nevertheless, there appears to be a growing interest in historical ecology, and in a recent issue of the journal, Gimmi and Bugmann (2013) explicitly combine historical ecology with landscape ecology modelling problems at different spatial and temporal scales.

This remarkable general lack of interest in the adoption of a historical approach is also reflected in programmes and texts produced by official international institutions such as IUCN (International Union for Conservation of Nature) and UNESCO (United Nations Educational, Scientific and Cultural Organization), national institutions and, above all, the community of European conservation planners. The phrase 'historical ecology' appears rarely in UNESCO documentation and made its debut as late as 2008; it is related to 'cultural landscape', 'biocultural heritage' and 'biodiversity', a relationship symptomatic of its American origins (Persic and Martin, 2008).

---

*E-mail: robcev@gmail.com

It is also noticeable that there is no mention of any historical approach in the recent document IUCN *Red List of Threatened Ecosystems* (Rodriguez *et al.*, 2010).

Yet the last decade has witnessed a significant shift in the direction of ecosystem conservation. Proactive management, looking to new 'functional ecological equivalents', has been proposed as a viable alternative to the long tradition within planning and management, which has been dominated by natural sciences geared towards 'protecting' imagined 'unimpeded natural processes' (Editorial in *Nature*, 2008). In a broader scientific context, this is linked to the concept of 'Historical Range of Variation', although this cannot be understood without some understanding of the interactions of past 'anthropogenic disturbance' and the hypothetical but largely conventional 'potential natural vegetation types' (e.g. Southwest Forest Project, 2006).

These types of approaches have provoked scientific and political disagreements in Italy, for example in relation to the politics of Ligurian Regional Natural Parks (Lagomarsini, 2004), the European Landscape Convention (Moreno and Montanari, 2008) and the conservation of rural landscapes (Agnoletti, 2010, 2013). In Italy, the conservation planning community – made up of naturalists, architects, geographers and, not least, foresters – are operating in a countryside scarred by the increasingly visible consequences of more than 30 years of misled 're-naturalization' politics in the conservation of both sites of nature conservation importance and rural landscapes (Fig. 16.1). Many specific sites and broader landscapes, especially in the uplands, have been transformed by more than half a century of rural depopulation and the consequent regeneration of woodland on abandoned arable land and pastures. This process, called since the 1980s 're-naturalization', has influenced the conservation politics of almost all of the Italian regional governments and institutions such as Regional Natural Parks.

Fig. 16.1. Bandito in the Aveto Valley, Italy. *Quercus cerris* wood-pasture in 1955 (a) has by 2003 become overgrown (b). The soil slope (c) has been made unstable by the overgrown hazel bushes (*Corylus avellana*), which have not been managed for 3–4 decades (Cevasco and Moreno, 2014). Liguria is the Italian region with the maximum number of dead trees per hectare of woodland (Bertani, R., personal communication).

Fig. 10.1. Continued.

The political idiom of 're-naturalization' could be said to be derived from 'think global' models of ecosystem assessment or management: 'core' areas of biosphere reserves, the promotion of wilderness, laments for the loss of pristine nature. An example from Greece would be the designation of the North Pindos National Park mainly for the protection of the brown bear (Merzanis *et al.*, 2009; Saratsi *et al.*, 2009). However, an interesting consequence of the recent shift in conservation thinking towards 'acting local' means that conservation management is beginning to take account of the practices associated with particular places. At this scale, woodland dynamics are seen to be immediately and intimately related to the history of local practices, as exemplified by Rackham's (1967) discussion of the importance of knowledge of woodmanship and its history in understanding British woodland structure and composition.

16.3 The Spread of a Historical Ecological Approach in European Conservation Thinking

The historical approach to the study of environmental systems has spread in a strange and almost immeasurable way since its definition and application to the study of hedges (Sheail, 1980). The origin of this approach in Britain can be attributed to the existence of a relatively small interdisciplinary research team which could conceive a historical approach, a backdrop of a unique tradition of naturalist field studies and the unusual (from a continental perspective) practice of carrying out field studies as part of local or topographical history (Beckett and Watkins, 2011). The 'historical–topographical' approach and the use of multiple sources is not a minor point in understanding the present interest in (and potential of) the application of historical ecology in modern conservation. Conservation policies and actions need to take account of precise locations and precise environmental effects and impacts, including the implications for local people. If developed carefully, and applied critically, historical ecology studies at the site scale can contribute to the required agility for new European, and consequently national, conservation policies.

Historical ecology (even before it was formally recognized as such) has been part of some developments of forestry and legal sciences since the 19th century (Moreno, 1986, 1990a; Tack and Hermy, 1998). A major role in the worldwide promotion of the historical approach was played in the 1980s by the IUFRO (International Union of Forest Research Organizations) Forest History Research Group. Nevertheless, there was a difference between the previous continental scientific heritage of forest history (historiography) and more recent developments that take into account environmental history and archaeology, landscape history and archaeology, rural history and archaeology, vegetation history, etc. Interest in the latter cultural approach has flourished in Italy with the rise of environmental history studies among Italian foresters, historians and geographers (Moreno *et al.*, 1982; Caracciolo, 1988; Ingold, 2011). By the end of 1990s, this change in approach became clear as indicated by two points raised by Watkins and Kirby (1998): forestry versus 'woodmanship'; and 'woodland' as opposed to 'land bearing trees'. These are addressed in the following sections (16.3.1 and 16.3.2).

16.3.1 Forestry versus woodmanship

The first issue raised by Watkins and Kirby (1998) is the relationship between forestry (scientific knowledge) as opposed to woodmanship (place-based or local knowledge). Conservation management of woodland and trees requires more than just conventional forestry training (or pure biological/nature conservation training). As a result, there has been an increase in historical ecology studies that stress the importance of understanding local and forgotten practices of management and production, and local environmental knowledge. The activity of the IUFRO Research Group on Forest and Woodland History offers an example of the rapid shift that is occurring in the way in which a more historical–cultural approach is taken in interpreting the environmental effects of local practices. From this have been developed 'guidelines for the implementation of social and cultural values in sustainable forest management' (Agnoletti *et al.*, 2008). These acknowledge historical structures (artefacts) of historical, archaeological importance, although it is still difficult to find references to the historical ecology of trees and woodland.

In a substantial book about the role of 'traditional knowledge' (Parrotta *et al.*, 2006), there are just a few pages with explicit reference to historical ecology. Also, in a recent overview of the role of traditional forest-related knowledge in shaping European forests (a 46 page document, written by 16 authors) the concept of 'historical ecology' only comes up once in a reference (Johann *et al.*, 2011). Could it be that the word 'ecology' embraces a certain scientific association that distances it from the idea of history and the accumulative effects of human action through the centuries? Or is this lack of connection an effect of the concurrence of 'environmental meta-history models' (Sörlin and Warde, 2007), in which the history of the environment is derived from 'mental representations' rather than from the material processes of the woodland itself?

16.3.2 Woodland or land bearing trees

The second issue raised by Watkins and Kirby (1998) was the problematic definition of 'woodland' as opposed to 'land bearing trees'. This divide does not make sense in terms of the innovative contributions made by historical ecology. A woodland is a precise site of vegetation cover and soils, but it is also a historically stratified artefact; using the Italian micro-historical geographical terminology, it has a 'green memory' which should be seen as part of the historical characterization of a place (Cevasco, 2007).

We must avoid simple ecological (or legal) categorization and classification of wooded ecosystems, habitats and landscapes. In the south of Europe, such categorization is confronted with an extremely varied environmental heritage of multiple agro-silvo-pastoral local systems that were active at the end of the 19th century. From a historical perspective, this mosaic of micro-ecosystems would be better described, in field and documentary level observations, as 'land/sites bearing trees' (Moreno, 1985; Grove and Rackham, 2001). Research needs to be carried out on the topographical and ecological connections of this mosaic with other components of the environmental system, such as grasslands, wetlands and heaths (Balzaretti et al., 2004).

16.3.3 The need for an interdisciplinary approach

The inclusion of woodland heritage in the conservation studies of the 'Historical Rural Landscape' has received recent boosts in Italy and France but, in practice, very little reference is made to historical ecology in these examples (Agnoletti, 2010; Guillerme et al., 2013).

Recognition of the need for more interdisciplinary approaches can perhaps be found in the call from the Intergovernmental Platform on Biodiversity and Ecosystem Services (IPBES) in 2012 for 'diverse representation in activities and decisions'. Turnhout et al. (2012) suggest that an 'expert panel should include natural scientists, social scientists, humanities researchers, biodiversity practitioners and indigenous-knowledge networks'. The composition of the required expert panel recalls the multidisciplinary foundation group that worked on terrestrial ecology in the late 1960s in the UK (Rackham, 1967, 1980; Sheail, 1980) and also corresponds to the approach of American historical ecologists working on ecosystems and landscape restoration (Egan and Howell, 2005).

Looking at the IPBES statement in the context of southern European mountain biodiversity and environmental heritage conservation, 'biodiversity practitioners and indigenous knowledge' appear to be new, important actors. There is, however, a danger that 'indigenous' could be interpreted as 'local/localized knowledge'. The cliché of 'traditional practices' and the timeless existence of 'traditional knowledge' imagined to be 'conserved' in 'peasant' (or rural or mountain) 'civilization' or 'local community' could be used rather than a critical micro-historical approach, i.e. an approach that actually analyses the reconstruction of interpersonal social relations and tries to understand the social mechanisms of change and conflicts (Raggio, 2004). Such a replacement would be the equivalent of returning to the imagined unimpeded natural processes linking southern European woodlands and forests to primeval, 'primary' or 'undisturbed virgin' vegetation cover and ignoring our knowledge of the dynamics of the present mosaic of historically stratified micro-ecosystems (Clark, 1996). The IPBES text can be seen as evidence that the potential of historical ecology analysis is officially recognized, but we have yet to see how well it will be assimilated into official conservation practice.

A robust topographical approach which makes use of a combination of historical, archaeological and ethnographical evidence helps us to understand how to conserve woodland habitat and ecosystems. The broad cultural approach in history frequently emphasizes a use of sources (mainly textual and iconographical) that base the 'extraction of meaning' on a subjective, symbolic or perceptive understanding. These sources are contrasted with the materiality of historical environmental evidence and processes and the necessarily connected fine analysis of social local structures and processes (Torre, 2011).

For example, the general problem of common land conservation has often been posed (Rotherham *et al.*, 2010), but a great deal of work remains to be done on how the ecology of woodland resources worked (or in some cases still works) locally in multi-use common access and management systems in south European mountains (Saratsi, 2003; Arvanitis, 2011; Beltrametti *et al.*, 2014). Such past systems produced – until the end of 19th century – the most remarkable sites of natural interest (known as SIC – Siti di Interesse Comunitario, or Sites of Community Interest, SCI) that we have inherited as a 'positive environmental externality' of past economies (Moreno and Poggi, 1996). These are direct experiments in the environmental sustainability of a site which, when compared and validated on a micro-scale, and acknowledging the historical dynamics of local social and economic changes, may provide lessons for modern conservation management.

16.3.4 The role of historical ecology

Historical ecology offers a way of analysing the effect of actions, strategies and practices on environmental resources in their particular local contexts. In Italy, the combination of historical ecology with theoretical developments in the social history of material culture and topographical approaches to micro-history was influenced by the direct connections made by the social and economic historian Edoardo Grendi (1932–1999) with British local history (Grendi, 1996; Raggio, 2004; Balzaretti, 2013; Moreno, 2013).

Problems of scale, both spatial and temporal, are of current interest in conservation studies. In palaeoecological and environmental archaeological studies, high definition analysis (short time periods, small spatial scales) is increasingly used. Encouraging the same approach to the 'economic–social dimension or processes' would bring historical ecology into line with the basic assumption of social micro-history (Grendi, 1996). Both may be seen as a counter to the spread of the rhetorical use of history, widely adopted during recent decades in cultural studies, to characterize geographical spaces, landscapes and also environmental resources (Torre, 2008; Ingold, 2011).

16.4 Integrating Historical and Local Knowledge into Management Strategies

We now apply this last described concept of historical ecology to the integration of historical and local knowledge into land management strategies through four case studies. The recently developed ideas concerning 'Rural Landscapes of Historical Interest' (Agnoletti, 2010, 2013) have introduced into the administrative language a new approach to woodland and forestry resources in Italian conservation policies. In line with these ideas, woods are no longer seen as isolated elements but as an integral part of the landscape system.

Proposals made by the Italian *National Catalogue of Italian Historical Rural Landscapes* (or LRHIs) (Agnoletti, 2010) have been implemented in specific legislation for both rural landscapes and woods (Presidential Decree DPR 41 12-2-2012). In the catalogue, only 10 of the total of 123 historical rural landscapes are specifically described or named as woodlands:

1. Wood of Sorti della Partecipanza di Trino (Vercelli).
2. Fir and Spruce Woods of Val Cadino (Trentino).
3. Ampezzo Forest (Friuli Venezia Giulia).
4. Forest of Cansiglio (Veneto).
5. San Vitale Pinewoods (Emilia Romagna).
6. Fir Forest of the Monastery of Vallombrosa (Toscana).
7. Sant'Antonio Woods (Abruzzo).

8. Historical afforestation in the Sele basin (Campania).
9. Monumental Turkey oak woods of Valle Ragusa (Puglia).
10. Ficuzza Woods (Sicilia).

There are also six landscapes classified as 'wood-pastures', including broad areas with a mosaic of other landscapes, as well as specific wood-pasture sites: Roccaverano, Aveto Valley, Salten, Moscheta, Monte Minerva and Gallura.

More wide and complex are the choice of examples of 'land bearing trees' because individual landscapes are included in a wider context. In north-west Italy (Liguria, Piedmont and Val d'Aosta), for example, only the wood of Sorti della Partecipanza di Trino (Vercelli) can be considered an ancient coppice wood (http://eunis.eea.europa.eu/sites/IT1120002), the other landscapes being mosaics with fragments of wood-pastures (Roccaverano-Asti and High Aveto Valley-Genova) or 'land bearing trees' deeply integrated with agriculture and pasture: the *baraggia* (uncultivated and scarcely fertile) land (Adami, 2012), sweet chestnut groves (High

Table 16.1. Piedmont and Liguria 'wood', 'wood-pastures' and 'land bearing trees' in Italy; extracted from the catalogue of *Historical Rural Landscapes*. (From Agnoletti, 2013.)

Landscape type	Province	Region	Historical and present land use systems
Wood			
Wood of Sorti della Partecipanza di Trino[a]	Vercelli	Piedmont	13th century coppiced common wood with oaks, hop hornbeam, poplars and alder
Wood-pasture			
The wooded pastures of Roccaverano	Asti, Alessandria	Piedmont	Terraced pastures and meadows, oak wood-pastures (disappearing since the last half of the 19th century), wood meadows, etc.
Wooded meadows and pastures in the Santo Stefano d'Aveto Cheese Area	Genova	Liguria	17th century Turkey oak and beech wood-pastures, 19th century terraced wood-pasture meadows, wet mountain pastures, etc.
Land bearing trees			
The Plateau of the Vauda[a]	Torino	Piedmont	Medieval moorland with broadleaved trees (oak, alder, beech, etc.), once winter pastures
The Baraggia Land in the Vercelli and Biella Area[a]	Vercelli, Biella	Piedmont	Medieval moorland with broadleaved trees (oaks, beech, alder, etc.), once winter pastures
Historical Polyculture of Valle Uzzone (Alta Langa)	Cuneo	Piedmont	Small agricultural areas with wooded areas and meadows, 'gerbidi' (moors), etc.
The Galarei Vineyard	Cuneo (Serralunga d'Alba, Diano d'Alba)	Piedmont	Specialized vineyard, once 'wedded' vine ('vite maritata'), wood-pasture, hazelnut orchard, etc.
Chestnut Groves in the High Bormida Valley	Savona	Liguria	Post-medieval sweet chestnut orchards
Wooded Olive Groves of Lucinasco	Imperia	Liguria	Post-medieval terraced olive orchards once grazed in winter
Terraced and Irrigated Chestnut Groves and Vegetable Gardens in Upper Sturla Valley	Genova	Liguria	10th century irrigated and terraced chestnut orchards and vegetable gardens
Terraced Hazelnut Groves of Tigullio	Genova	Liguria	15th century terraced hazelnut groves with terraced olive groves

[a]Named in the catalogue (Agnoletti, 2010).

Bormida Valley-Savona), terraced olive orchards (Lucinasco-Imperia), irrigated terraced chestnut orchards (High Sturla Valley-Genova), terraced hazelnut groves (Tigullio-Genova), etc. (Table 16.1).

The *National Catalogue of Italian Historical Rural Landscapes* responded to an urgent need for the conservation and management of the rural heritage that took account of the historical dimension of the landscape. It was informed by the pioneering studies of Emilio Sereni (1907–1977) on Italian agricultural and landscape history (Quaini, 2011). The catalogue of rural landscapes initiative came, however, from the Ministry of Agriculture, Food and Forestry, not from the ministries of the Environment or Cultural Heritage, because it prioritized rural landscapes and their local production rather than being based on more aesthetic or natural landscape values.

Currently in Italy, new woodland is increasing at the rate of $c.70,000$ ha year^{-1} and this is now contributing to a reduction in the richness of formerly complex rural landscape mosaics (Agnoletti, 2013). The 20th century history of the south European mountains is predominately a history of large-scale depopulation, land abandonment, cessation of grazing by domestic animals and the rapid spread of secondary woodland over agricultural land. A major problem for the next few decades will be how to manage – and recover for productive and environmental purposes – what is left of the heritage of historical rural landscapes. In this context, historical ecology can help with communicating to politicians, administrators and planners the conservation importance of the local products and practices that shaped these real places.

16.4.1 An introduction to the case studies

In the following case studies from protected areas (Regional Parks, Natura 2000 sites) in north-west Italy, we consider the difference in environmental performance between restoring forms (aesthetic principles) or restoring systems (ecological principles). The models of woodland restoration examined are derived from locally documented historical patterns, and take into account both the general interests of conservation and the work practices of local users (farmers, hunters, etc.). The aim was to perpetuate the effects (e.g. biodiversity) of those historical activities and practices within the present local environmental–social–economic context.

Liguria is the most wooded region in Italy, with 70% of the surface area being covered with trees, largely due of the abandonment of previous agrarian/pastoral land use, although embedded in this developing new tree cover are a small number of ancient woods. Very recently, conservation policy has recognized that 're-naturalization' processes are one of the most direct causes of the loss of mountain biodiversity.

Re-naturalization has also exacerbated the effect of dramatic floods in recent years because, with the abandonment of management, as shown by pollen diagrams, the ecology of shrub and herbaceous layers changes, as does their capacity to react to rainfall events. The new woodlands are growing on former agricultural soils and former pastures which are extremely vulnerable to abandonment. For instance, in the terraced area of Cinque Terre, after the rainstorms of 25 October 2011, 88% of the landslides occurred on abandoned terraced areas covered with secondary woodlands (Agnoletti, 2013). Bringing this secondary woodland into more active management informed by historical local practices might thus improve the delivery of environmental services from these areas.

16.4.2 Trees and woodlands producing leaf fodder

Until the 20th century, historical transhumance systems in south European mountains were linked to the distribution and management of trees producing leaf fodder for sheep and goats. The effects of shredding and pollarding practices on trees – especially oaks – are still evident in some regions of Greece (Sidiropoulou and Ispikoudis, 2009; Sioliou and Ispikoudis, 2009), southern France, Spain, Portugal and Italy (Grove and Rackham, 2001).

Examples of wooded meadow systems can be found in the *National Catalogue of Italian*

Historical Rural Landscapes, but there is much variation in the level of awareness of the role of local practices in conservation policies.

- The larch (*Larix decidua*) wooded pastures of Salten (San Genesio, Bolzano) have been maintained by grazing and preserved with assistance from the Province of Bolzano since 1981 (through the *Piano paesaggistico* – or Landscape Plan), and the Rural Development Plan 2007–2013 provided incentives to continue 'traditional' forms of agriculture and farming (Biasi, 2013).
- The wooded meadow and pastures in the Santo Stefano d'Aveto area (Genova) and Roccaverano (Asti) are historically connected with important cheese production but persist thanks to marginal and small-scale agricultural practices rather than to any public initiatives. They are not recognized officially as of importance for environmental conservation.
- Bosco di Sant'Antonio in the Majella National Park (Abruzzo) conserves monumental pollard beech trees which represent a significant example of 'difesa' woods (Hermanin, 2010). However, these are included in a 'core' reserve where crucial management practices – such as pollarding – are forbidden (Agnoletti, 2013); this represents conflicting policies in agriculture, environment and cultural conservation.

In the Eastern Ligurian Apennines, woodland of beech (*Fagus sylvatica*) and Turkey oak (*Quercus cerris*) conserves relict parcels of non-woodland ancient trees used for fodder production in relationship with transhumance systems from the *Riviera Ligure* and the Po Plain (Moreno and Raggio, 1990; Cevasco, 2014). The importance of the Trebbia/Aveto watershed for transit and transhumance is documented in classical sources (Petracco, 1965), and evidence of prehistoric pastoral use has been recorded (Maggi and De Pascale, 2011; Branch, 2012).

At Montarlone (the High Trebbia Valley, northern Apennines, Italy), a comparison of ground flora with soil pollen analysis revealed pollen evidence of persistent ancient grasslands, with important changes in the herb composition dated to 1000 AD and 1800 AD. The grass and fodder resources of the ancient wood-pasture started to decline in the 19th century with the reduction in sheep transhumance systems and local goat pasture, and increasing shade from the beech trees (Cevasco and Molinari, 2009; Cevasco and Moreno, 2013). The old trees, currently being studied for saproxylic invertebrates, have not been pollarded since at least the 1930s, and now need urgent care and management to reduce the risk of collapse from the increasing weight of their branches. The reintroduction of pollarding in some experimental sites would help to conserve these veteran trees and their living associated fauna (invertebrates), fungi, lichens, etc. However this is difficult because of the age of the stems – will they respond having been so long out of the pollarding cycle? – and the very small number of local inhabitants. Although economically unsustainable, the action would have a high environmental and cultural interest in a Natura 2000 Site and hunting reserve (*Azienda faunistica venatoria Montarlone*) and help conservationists to recognize sites like this as rare, living archives for mountain environmental history.

Despite their conservation importance, the fragments of ancient wood-pasture in the Apennines are not being conserved as systems (Hartel *et al.*, Chapter 5). Where protected by parks, nature reserves and SIC, such protection only covers the shape of the landscape and the monumental trees, and not the environmental content of the system. Unless specific laws for historical rural landscape are developed and instituted, these systems are likely to disappear in a few decades.

If appropriate policies could be adopted for sites like Montarlone, it would be easier to reconnect present inhabitants (older and young locals, immigrants, the neo-rural population, others) to woodland practices and management with added positive environmental effects. For instance, pollarding, or less expensive coppicing, would assist *Vaccinium* and the herb populations of the ancient wood-pasture. This would be useful for the development of a sustainable stock economy with high environmental value, as long as game management succeeds in limiting the devastating impact of wild boars. The conservation of many wood-pastures in the south European region could also be helped by the growing interest in the history of commons management (Rotherham *et al.*, 2010; Torre and Tigrino, 2013).

16.4.3 Trees, woodland and soil fertility

Alnocoltura, a reconstructed recent name, refers to a cyclic agro-silvo-pastoral system producing cereals for human consumption documented for the post-medieval Ligurian Apennines through archive and field evidence (Moreno *et al.*, 1998). This 'cultural landscape' of alder woods of the upper Aveto valley was formerly widespread in the whole of this and neighbouring valleys (Cevasco, 2010) in the 19th century. The *alnocoltura* system in Liguria was also developed into a second local agroforestry system – documented from the 18th century – in which black alder (*Alnus glutinosa*) was co-planted with sweet chestnut (*Castanea sativa*) along terraces in an area near the coast (Vaccarezza, 2013).

Pollen evidence dates the adoption of the *alnocoltura* system in the Aveto valley back to at least the 11th century (Cevasco and Moreno, 2007; Guido *et al.*, 2013). *A. incana* trees fixed nitrogen in the soil and improved its fertility (Balzaretti, 2013). The alder woods were important for the woodcock (*Scolopax rusticola*) and snipe (*Gallinago gallinago*) because the nitrogen-rich soils sustained high populations of earthworms (*Lumbricus terrestris*), especially in combination with abundant cow dung. The mosaic of woodland with many clearings (of mowed and grazed land) provided excellent habitats for game.

The local environmental knowledge of the fertilizing capacity of alder had mostly disappeared among current inhabitants of the Aveto valley, and the system itself had been forgotten, with only indicative place names, traces in the soil and the ecological continuity of alder stands remaining. Nevertheless, at the end of the 1990s, restoration of the system was proposed in the Aveto valley by the Province of Genova's mountain natural environment administration as one of a number of actions to increase the potential resource for game (such as hares and partridge).

Restoration of the *alnocoltura* system (Fig. 16.2), which remained in use in the Aveto

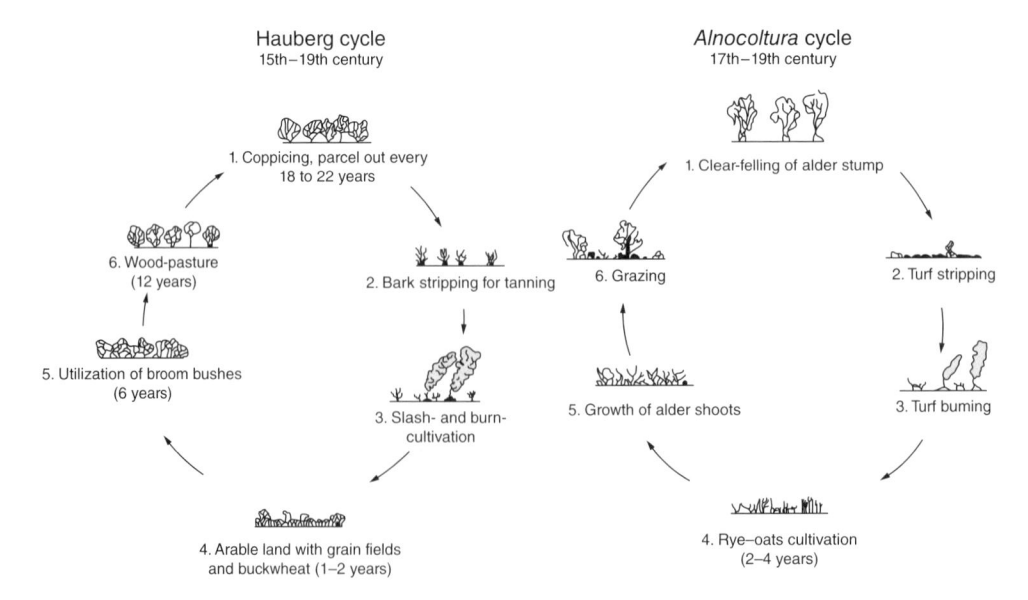

Fig. 16.2. The Hauberg cycle for the Siegerland area (NW Germany) from 1452 to 1850 (reconstructed from Pott, 1988, and Pott *et al.*, 1992) and the *alnocoltura* cycle (a cyclic agro-silvo-pastoral system) for the Aveto valley in NW Italy from the 17th to 19th centuries. These are both cyclic agro-silvo-pastoral systems; see text for details.

valley until the early 20th century, involves conserving relict alder woods through coppicing, grazing and the temporary production of grain crops (*Secale cereale*, *Avena sativa*, *Fagopyrum esculentum*, *Hordeum* sp., etc.). The increase of fodder resources (herbs and leaves), an expected positive effect of these practices, would fit with the local *Cabannina* cattle breed, which is historically well adapted to the woodland and shrub mountain pastures of the valley (Bertolotto and Cevasco, 2000). However, the proposed integration between the two environmental actions of alder wood restoration and cattle rearing has, in practice, been difficult to apply owing to the conflicting requirements of forestry and environmental regional and national legislation.

By contrast the similar Hauberg system (Fig. 16.2) has been successfully restored in the Siegerland (North Rhine-Westphalia, Germany). In this system, in use from the 15th to 19th centuries, oak–birch coppices replaced former beech forests. This was a cyclic agro-silvo-pastoral system based on a cooperative coppice system in which birch and oak were felled with a rotation of 18–22 years and made into charcoal for ore smelting. The bark of oak was peeled off and used in industry as a tanning agent. The remaining wood was burnt and the ash used to manure the soil; rye (*Secale*), buckwheat (*Fagopyrum*) and oats (*Avena*) were then sown for 1–2 years. For the following 5–7 years, the harvested area was left fallow to allow the tree stumps to produce fresh shoots. Fire promoted the germination of broom (*Cytisus scoparius*), which can be used for many purposes. The area was then used as a pasture with *Vaccinium* and *Calluna* or *Juniperus* (Pott, 1988; Pott *et al.*, 1992; Hoppe, 2009).

Restoring alder stools and managing their fertility remains an objective both for game management, site biodiversity, research and 'living rural heritage' issues. The management and conservation policy of the Regional Natural Park, Ente Parco Aveto, and of protected areas (SIC and ZRC – Zona di Ripopolamento e Cattura, an Area of Repopulating and Capture protected and managed for local game management by the Province of Genova) could provide assistance (Cevasco, 2009).

16.4.4 The collection of litter

Local nutrient transfers within the landscape are brought about by litter collection from the forests (Stuber and Bürgi, 2002). Impacts on the flora and soil chemistry have been experimentally analysed in the Ligurian Apennines and the Swiss lowlands (Bürgi and Gimmi, 2007). In the Apennines (Borbera Valley, Alessandria), raking in the spring clears the turf of debris produced by trimming or pruning chestnut trees and controls moss growth; late autumn raking was intended for gathering dead leaves. Simulating the effect of these raking practices showed an increase in hay production and an equivalent of nutrients for the input of human energy has been measured (Moreno, 1990b). This is an example of 'positive environmental externality' of a production practice that helped the reconstitution of the resource. Bürgi and Gimmi (2007) suggest that litter collecting might be an additional tool for nature conservation to foster species that suffer from the abandonment of traditional forest use.

In the Vara valley of Liguria, the effects of the abandonment of litter collecting on the herb layer in chestnut woods have been documented (Cevasco *et al.*, 1999). Winter gathering and burning of fallen chestnut leaves allowed grassland species to survive and be suitable for sheep grazing. Individual chestnut trees (or small groups of trees) were managed by shredding (in late summer) for fodder leaves and pruning (in spring–late winter) for chestnut fruits. Both practices increase the light available to the herb layer. Winter burning is a recent adaptation of the previous practice of raking leaves in the autumn for cattle bedding. To store these leaves, huts were built on the chestnut terraces, mainly between 1850 and 1950, in a phase of intensification of cattle farming. Observational data links the decline of biodiversity with the abandonment of these practices. The employment of fire is now illegal in regional forestry law, so this practice too has been lost for conservation.

16.4.5 Trees invading bogs: an experiment in applied historical ecology

An experiment in the SIC Roccabruna has recently been carried out to examine the extent to which biodiversity was affected by the local practices of grazing, mowing, burning, and spring and surface water regulation (Cevasco, 2013). Roccabruna, on the watershed of the Trebbia-Aveto, is a Natura 2000 site that is particularly rich in bogs and ancient grassland, and heathland interspersed with ancient woodlands and secondary woodland. Habitat restoration was implemented to assist the biodiversity of locally managed common lands. The project was conducted together with game managers from the Province of Genova and was based on previous historical ecological research (Maggi et al., 2002) using pollen and indicator plants (Fig. 16.3).

The integration of locally based management and knowledge was promoted in this project. Restoration management was seen as an opportunity for reconnecting local groups to the management of their environmental resources. Place-based knowledge and practices were identified on the ground: the management of water in the wetland 'perimeter' of alders and *Phragmites* populations; the use of fire; and the dynamics of herbs and shrubs threatened by the expansion of alder tree cover. This local knowledge is shared by many older people, but it is not clear how it will be passed on to younger generations and the new rural inhabitants. The restoration experiment has also shown positive effects at the 'Moglia di Casanova' site for the newt populations (*Triturus alpestris* and *T. vulgaris*), which are endangered species that contributed to the designation of the SIC in the 1990s. In other nearby sites, fire and mowing practices have promoted, respectively, *Arnica montana* and *Polygala chamaebuxus* populations, and hare and other small animals of value for the local game economy. However, the work has also shown the fragility of such conservation actions given that the north-west Italian mountains, due to large-scale abandonment, are currently experiencing devastation by increasing deer and wild boar populations which will also have an impact on the soil, vegetation structure and composition (Hearn et al., 2014).

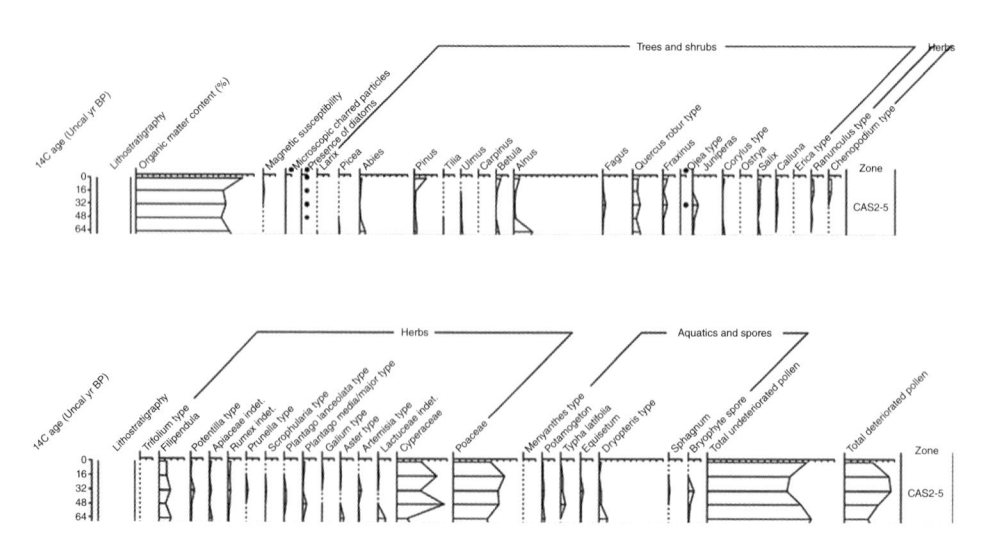

Fig. 16.3. The gradual abandonment of local management practices at the Moglia di Casanova site, part of a Natura 2000 site in the Province of Genova, Italy, as documented in the last 16 cm of the pollen diagram (Branch unpublished) with a rise in *Alnus* and *Pinus* and a decrease of *Poaceae* and *Cyperaceae* (modified by Cevasco, 2007).

16.5 Conclusion

Historical ecologists seek to extract precise past local practices for particular portions of the present environment: their 'green memory'. Past management and local environmental responses can be tested and evaluated for future conservation schemes and employed in diverse political and economic contexts. Generalized models in conservation need to be tested on the ground and, at the site scale, historical ecology provides the tools for such a test. This rethinking in conservation suggests a new policy towards placed practices, knowledge, regulations and rights, and will support a new strategy for the conservation of resources in local forestry, agricultural and pastoral management.

References

Adami, I. (2012) *Terre di Baraggia. Pascoli, Acque, Boschi e Risaie: per una Storia del Paesaggio Vercellese*. Dell'Orso, Alessandria, Italy.

Agnoletti, M. (ed.) (2010) *Paesaggi Rurali Storici. Per un Catalogo Nazionale (Historical Rural Landscapes. For a National Register)*, 1st edn. Laterza, Bari, Italy.

Agnoletti, M. (ed.) (2013) *Italian Historical Rural Landscapes. Cultural Values for the Environment and Rural Development*. Springer, Dordrecht, The Netherlands.

Agnoletti, M., Anderson, S., Johann, E., Kulvik, M., Kushlin, A.V., Mayer, P., Montiel Molina, C., Parrotta, J., Rotherham, I.D. and Saratsi, E. (2008) The introduction of cultural values in the sustainable management of European forests. *Global Environment* 2, 172–199.

Arvanitis, P. (2011) Traditional forest management in Psiloritis, Crete (c.1850–2011): integrating archives, oral history and GIS. PhD thesis, University of Nottingham, UK.

Balzaretti, R. (2013) *Dark Age Liguria. Regional Identity and Local Power, c.400–1020*. Bloomsbury, London.

Balzaretti, R., Pearce, M. and Watkins, C. (eds) (2004) *Ligurian Landscapes: Studies in Archaeology, Geography and History*. Accordia Research Institute, University of London, London.

Beckett, J. and Watkins, C. (2011) Natural history and local history in late Victorian and Edwardian England: the contribution of the Victoria County History. *Rural History* 22, 59–87.

Beltrametti, G., Cevasco, R., Moreno, D. and Stagno, A.M. (2014) Les cultures temporaires entre longue durée et chronologie fine (montagne ligure, Italie). In: Viader, R. and Rendu, C. (eds) *34ᵉ Journées Internationales d'Histoire. Cultures Temporaires et Féodalité. Les Cycles Culturaux et l'Appropriation du Sol dans l'Europe Médiévale et Moderne*. Colloque 12 et 13 octobre 2012, Calenda, Abbaye de Flaran, Valence-sur-Baïse, France. (In press.)

Bertolotto, S. and Cevasco, R. (2000) The 'alnoculture' system in the Ligurian eastern Apennines: archive evidence. In: Agnoletti, M. and Anderson, S. (eds) *Methods and Approaches in Forest History*. CAB International, Wallingford, UK, pp. 169–182.

Biasi, R. (2013) The meadows and wooded pastures of Salten. In: Agnoletti, M. (ed.) *Italian Historical Rural Landscapes. Cultural Values for the Environment and Rural Development*. Springer, Dordrecht, The Netherlands, pp. 255–257.

Branch, N.P. (2012) Early–Middle Holocene vegetation history, climate change and human activities at Lago Riane (Ligurian Apennines, NW Italy). *Vegetation History and Archaeobotany* 22, 315–334.

Bürgi, M. and Gimmi, U. (2007) Three objectives of historical ecology: the case of litter collecting in central European forests. *Landscape Ecology* 22, 77–87.

Caracciolo, A. (1988) *L'Ambiente come Storia*. il Mulino, Bologna, Italy.

Cevasco, R. (2007) *Memoria Verde. Nuovi Spazi per la Geografia*. Edizioni Diabasis, Reggio Emilia, Italy.

Cevasco, R. (2009) Alnocoltura: a traditional farming system in the northern Apennines, Italy. In: Krzywinski, K., O'Connell, M. and Kuster, H.J. (eds) *Europäische Kulturlandschaften. Wo Demeter ihre Felder hat und Pan zu Hause ist*. Bremen, Aschenbeck, Germany, pp. 110–111.

Cevasco, R. (2010) The environmental heritage of a past cultural landscape: the alderwoods (*Alnus incana* Moench) in the upper Aveto Valley (NW Apennines). In: Armiero, M. and Hall, M. (eds) *Nature and History in Modern Italy*. Ohio University Press, Athens, Ohio, pp. 126–140.

Cevasco, R. (2013) Storie per la gestione di una zona umida: le 'specie indicatrici'. In: Cevasco, R. (ed.) *La Natura della Montagna. Scritti in Ricordo di Giuseppina Poggi*. Oltre Edizioni, Sestri Levante, Italy, pp. 156–171.

Cevasco, R. (2014) Biodiversification processes: the cultural systems of Turkey oak in the northern Apennines (16th–20th c. A.D.). *1st European Conference for the Implementation of the UNESCO–SCBD Joint Programme on Biological and Cultural Diversity: Linking Biological and Cultural Diversity in Europe, 8–11 April 2014, Florence, Italy. Book of Abstracts.* United Nations Educational, Scientific and Cultural Organization, New York and Secretariat of the Convention on Biological Diversity, Montreal, Canada, p. 23. Available at: http://www.iufro.org/download/file/10655/5477/90300-florence14-abstracts_pdf/ (accessed 4 December 2014).

Cevasco, R. and Molinari, C. (2009) Microanalysis in woodland historical ecology. Evidences of past leaf fodder production in NW Apennines (Italy). In: Saratsi, E., Bürgi, M., Johann, E., Kirby, K.J., Moreno, D. and Watkins, C. (eds) *Woodland Cultures in Time and Space.* Embryo Publications, Athens, pp. 147–154.

Cevasco, R. and Moreno, D. (2007) Microanalisi geo-storica o geografia culturale della copertura vegetale? Sull'eredità ambientale dei 'paesaggi culturali'. *Trame dello Spazio,* no. 3, Dipartimento di Storia, Università degli Studi di Siena, Firenze, Italy, pp. 83–101.

Cevasco, R. and Moreno, D. (2013) Rural landscapes: the historical roots of biodiversity. In: Agnoletti, M. (ed.) *Italian Historical Rural Landscapes. Cultural Values for the Environment and Rural Development.* Springer, Dordrecht, The Netherlands, pp. 141–152.

Cevasco, R. and Moreno, D. (2014) Pendici liguri: riscoprire le relazioni tra suoli e copertura vegetale. In: Cesaretti, P. and Ferlinghetti, R. (eds) *Uomini e Ambienti. Dalla Storia al Future.* Bolis Edizioni per UBI Banca, Bergamo, Italy, pp. 46–67.

Cevasco, R., Moreno, D., Poggi, G. and Rackham, O. (1999) Archeologia e storia della copertura vegetale: esempi dall'Alta Val di Vara. *Memorie della Accademia Lunigianese di Scienze 'Giovanni Capellini',* LXVII–LXVIII–LXIX (1997–1998–1999), pp. 244–256.

Clark, D.B. (1996) Abolishing virginity. *Journal of Tropical Ecology* 12, 735–739.

Egan, D. and Howell, E.A. (eds) (2005) *The Historical Ecology Handbook. A Restorationist's Guide to Reference Ecosystems.* Island Press, Washington, DC.

Gimmi, U. and Bugmann, H. (2013) Preface: integrating historical ecology and ecological modelling. *Landscape Ecology* 28, 785–787.

Grendi, E. (1996) *Storia di una Storia Locale. L'Esperienza Ligure 1792–1992.* Marsilio, Venice, Italy.

Grove, A.T. and Rackham, O. (2001) *The Nature of Mediterranean Europe. An Ecological History.* Yale University Press, New Haven, Connecticut.

Guido, M.A., Menozzi, B.I., Bellini, C., Placereani, S. and Montanari, C. (2013) A palynological contribution to the environmental archaeology of a Mediterranean mountain wetland (North West Apennines, Italy). *The Holocene* 23, 1517–1527.

Guillerme, S., Jimenez, Y. and Moreno, D. (2013) Les paysages d'arbres hors forêt, des paysages porteurs des enjeux du développement durable. In: Luginbühl, Y. and Terrasson, D. (eds) *Paysage et Développement Durable.* Éditions Quæ, Paris.

Hearn, R., Watkins, C. and Balzaretti, R. (2014) The cultural and land use implications of the reappearance of the wild boar in North West Italy: a case study of the Val di Vara. *Journal of Rural Studies* 36, 52–63.

Hermanin, L. (2010) Bosco di Sant'Antonio. In: Agnoletti, M. (ed.) *Paesaggi Rurali Storici. Per un Catalogo Nazionale (Historical Rural Landscapes. For a National Register),* 1st edn. Laterza, Bari, Italy, pp. 391–393.

Hoppe, A. (2009) Hauberg: a cyclical agro-silvo-pastoral economy in the Siegerland, western Germany. In: Krzywinski, K., O'Connell, M. and Küster, H.J. (eds) *Europäische Kulturlandschaften. Wo Demeter ihre Felder hat und Pan zu Hause ist.* Aschenbeck, Bremen, Germany, pp. 138–141.

Ingold, A. (2011) Écrire la nature. De l'histoire sociale à la question environnementale? *Annales Histoire, Sciences Sociales* 66, 11–29.

Johann, E., Agnoletti, M., van Benthem, M., Bölöni, J., Holl, K., Kusmin, J., van Laar, J., Latorre, J.G., Latorre, J.G., Molnár, Z. et al. (2011) Europe. In: Parrotta, J.A. and Trosper, R. (eds) *Traditional Forest-Related Knowledge. Sustaining Communities, Ecosystems and Biocultural Diversity.* Springer, Dordrecht, The Netherlands, pp. 203–249.

Kizos, T., Plieninger, T. and Schaich, H. (2013) Instead of 40 sheep there are 400: traditional grazing practices and landscape change in western Lesvos, Greece. *Landscape Research* 38, 476–498.

Lagomarsini, S. (2004) Urban exploitation of common rights: two models of land use in the Val di Vara. In: Balzaretti, R., Pearce, M. and Watkins, C. (eds) *Ligurian Landscapes: Studies in Archaeology, Geography and History.* Accordia Research Institute, University of London, London, pp. 179–188.

Maggi, R. and De Pascale, A. (2011) Fire making water on the Ligurian Apennines. In: Van Leusen, M., Pizziolo, G. and Sarti, L. (eds) *Hidden Landscapes of Mediterranean Europe. Cultural and Methodological*

Biases in Pre- and Protohistoric Landscape Studies. Proceedings of the International Meeting Siena, Italy, May 25–27, 2007. BAR International Series 2320. Archaeopress, Oxford, UK, pp. 105–112.

Maggi, R., Montanari, C. and Moreno, D. (eds) (2002) *L'Approccio Storico-ambientale al Patrimonio Rurale delle Aree Protette. Materiali di Studio dal '2nd Workshop on Environmental History and Archaeology'. Archeologia Postmedievale* 6. All'Insegna del Giglio, Firenze, Italy.

Merzanis, G., Korakis, G., Tsiakanos, K. and Aravidis, E. (2009) Expansion of brown bear in the course of rural abandonment during the 20th century – a case study from the Pindos Mountain Range (Greece). In: Saratsi, E., Bürgi, M., Johann, E., Kirby, K., Moreno, D. and Watkins, C. (eds) *Woodland Cultures in Time and Space*. Embryo Publications, Athens, pp. 330–337.

Montiel-Molina, C. (2007) Cultural heritage, sustainable forest management and property in inland Spain. *Forest Ecology and Management* 249, 80–90.

Moreno, D. (1985) The agricultural uses of tree-land in the north-western Appennines since the Middle Ages. In: *History of Forest Utilization and Forestry in Mountain Regions. Symposium an der ETH Zurich, 3–7 1984*. ETH Zurich, Zurich, Switzerland, pp. 77–88.

Moreno, D. (1986) Boschi, storia e archeologia. Riprese, continuità, attese. *Quaderni Storici* 62, 435–444.

Moreno, D. (1990a) *Dal Documento al Terreno. Storia e Archeologia dei Sistemi Agro-silvo-pastorali*. il Mulino-Ricerche, Bologna, Italy.

Moreno, D. (1990b) Past multiple use of tree-land in the Mediterranean mountains: experiments on the sweet chestnut culture. *Environmental History Newsletter* 2, 37–49.

Moreno, D. (2013) L'altro lato della via Balbi. Ricerche di terreno in Liguria (1990–2010). In: Cevasco, R. (ed.) *La Natura della Montagna. Studi in Ricordo di Giuseppina Poggi*. Oltre Edizioni, Sestri Levante, Italy, pp. 32–42.

Moreno, D. and Montanari, C. (2008) Màs allà de la percepción: hacia una ecología histórica del paisaje rural en Italia. *Cuadernos Geográficos de la Universidad de Granada* 43, 29–49.

Moreno, D. and Poggi, G. (1996) Ecologìa historica, caracterizaciòn etnobotànica y valorisaciòn de los productos de la tierra. *Agricultura y Sociedad* 80–81, 169–180.

Moreno, D. and Raggio, O. (1990) The making and fall of an intensive pastoral land-use system. Eastern Liguria, 16–19th centuries. *Rivista di Studi Liguri* 56, 193–217.

Moreno, D., Piussi, P. and Rackham, O. (eds) (1982) *Boschi Storia e Archeologia. Quaderni Storici* 49.

Moreno, D., Cevasco R., Bertolotto, S. and Poggi, G. (1998) Historical ecology and post-medieval management practices in alder woods (*Alnus incana* (L.) Moench) in the northern Apennines, Italy. In: Kirby, K. and Watkins, C. (eds) *The Ecological History of European Forests*. CAB International, Wallingford, UK, pp. 185–201.

Nature (2008) Editorial. Ecologists must research how best to intervene in and preserve ecosystems. *Nature* 455, 263–264.

Parrotta, J., Agnoletti, M. and Johann, E. (eds) (2006) *Cultural Heritage and Sustainable Forest Management: The Role of Traditional Knowledge. Proceedings of the IUFRO Conference Held in Florence, Italy, 8–11 June 2006*. MCPFE (Ministerial Conference on the Protection of Forests in Europe) Liaison Unit, Warsaw, Poland.

Persic, A. and Martin, G. (2008) *Links between Biological and Cultural Diversity – Concepts, Methods and Experiences. Report of an International Workshop*. UNESCO (United Nations Educational, Scientific and Cultural Organization), Paris.

Petracco Sicardi, G. (1965) Toponimi Veleiati II. In: *Bollettino Ligustico* XVII, Liguria, Italy.

Pott, R. (1988) Impact of human influences by extensive woodland management and former land-use in north-western Europe. In: Salbitano, F. (ed.) *Human Influences on Forest Ecosystems Development in Europe*. ESF FERN-CNR (European Science Foundation, European Forest Ecosystem Research Network–Consiglio Nazionale delle Ricerche). Pitagora Editrice, Bologna, Italy, pp. 263–270.

Pott, R., Freund, H. and Speier, M. (1992) Anthropogenic changes of landscape by extensive woodland management and charcoal production tree-in Siegerland (North Rhine-Westphalia, Germany). In: Métaillé, J.-P. (ed.) *Protoindustries et Histoire des Forets: Actes du Colloque Tenu à la Maison de la Forêt, Loubières, Ariège, les 10-13 Octobre 1990*. Les Cahiers de l'ISARD 3, GDR ISARD-CNRS, Toulouse, France, pp. 163–184.

Quaini, M. (ed.) (2011) *Paesaggi Agrari. L'Irrinunciabile Eredità Scientifica di Emilio Sereni*. Silvana Editoriale, Milan, Italy.

Rackham, O. (1967) The history and effects of coppicing as a woodland practice. In: Duffey, E. (ed.) *The Biotic Effects of Public Pressures on the Environment: Proceedings of a Symposium Held at Monks Wood Experimental Station, March 20–21, 1967. Monks Wood Experimental Station Symposium 3*, Nature Conservancy/Natural Environment Research Council, London, pp. 82–93.

Rackham, O. (1980) *Ancient Woodland. Its History, Vegetation and Uses in England.* Edward Arnold, London.

Raggio, O. (2004) Microhistorical approaches to the history of Liguria: from microanalysis to local history. Edoardo Grendi's achievements. In: Balzaretti R., Pearce, M. and Watkins C. (eds) *Ligurian Landscapes: Studies in Archaeology, Geography and History.* Accordia Research Institute, University of London, London, pp. 97–104.

Rodríguez, J.P., Rodríguez-Clark, K.M., Baillie, J.E.M., Ash, N., Benson, J., Boucher, T., Brown, C., Burgess, N.D., Collen, B., Jennings, M. *et al.* (2010) Establishing IUCN Red List Criteria for threatened ecosystems. *Conservation Biology* 25, 21–29.

Rotherham, I.D., Agnoletti, M. and Handley, C. (2010) *End of Tradition? Part 1 History of Commons and Commons Management (Cultural Severance and Commons Past). Landscape Archaeology and Ecology* 8(1); *End of Tradition? Part 2 Commons: Current Management and Problems (Cultural Severance and Commons Present). Landscape Archaeology and Ecology* 8(2). Wildtrack Publishing, Sheffield, UK.

Saratsi, E. (2003) Landscape history and traditional management practices in the Pindos Mountain, northwest Greece *c.*1850–2000. PhD thesis, University of Nottingham, Nottingham, UK.

Saratsi, E. (2005) The cultural history of 'kladera' in Zagori area of Pindos Mountain, NW Greece. *News of Forest History* III, 107–117.

Saratsi, E., Bürgi, M., Johann, E., Kirby, K.J., Moreno, D. and Watkins, C. (eds) (2009) *Woodland Cultures in Time and Space.* Embryo Publications, Athens.

Sheail, J. (1980) *Historical Ecology. The Documentary Evidence.* National Environment Research Council, Huntingdon, UK.

Sidiropoulou, A. and Ispikoudis, I. (2009) Ancient trees as indicators of agroforestry systems – three case studies from northern Greece. In: Saratsi, E., Bürgi, M., Johann, E., Kirby, K.J., Moreno, D. and Watkins, C. (eds) *Woodland Cultures in Time and Space.* Embryo Publications, Athens, pp. 138–146.

Sioliou, M.K. and Ispikoudis, I. (2009) The cultural landscape of Pomaks in South Eastern Rodopi, Thrace, Greece. In: Saratsi, E., Bürgi, M., Johann, E., Kirby, K.J., Moreno, D. and Watkins, C. (eds) *Woodland Cultures in Time and Space.* Embryo Publications, Athens, pp. 291–298.

Sörlin, S. and Warde, P. (2007) The problem of the problems of environmental history: a re-reading of the field. *Environmental History* 12, 107–130.

Southwest Forest Project (2006) The Nature Conservancy's Center for Science and Public Policy, Phoenix, Arizona. Available at: http://azconservation.org/projects/southwest_forest_assessment (accessed July 2014).

Stuber, M. and Bürgi, M. (2002) Agricultural use of forests in Switzerland 1800–1950. Needles and leaves for litter harvesting. *Schweizerische Zeitschrift für Forstwesen* 153, 397–410.

Tack, G. and Hermy, M. (1998) Historical ecology of woodlands in Flanders. In: Kirby, K.J. and Watkins C. (eds) *The Ecological History of European Forests.* CAB International, Wallingford, UK, pp. 283–292.

Torre, A. (2008) Un tournant spatial en histoire? Paysages, regards, ressources pour une historiographie de l'espace. *Annales Histoire Sciences Sociales* 63, 1127–1144.

Torre, A. (2011) *Luoghi: La Produzione di Località in età Moderna e Contemporanea.* Pomezia Donzelli Editore, Rome.

Torre, A. and Tigrino, V. (2013) Beni comuni e località: una prospettiva storica. *Ragion Pratica* 41, 333–346.

Turnhout, E., Bloomfield, B., Hulme, M. and Vogel, J. (2012) Conservation policy: listen to the voices of experience. *Nature* 488, 454–455.

Vaccarezza, C. (2013) Castagni e ontani nel Tigullio. In: Cevasco, R. (ed.) *La Natura della Montagna. Studi in Ricordo di Giuseppina Poggi.* Oltre Edizioni, Sestri Levante, Italy, pp. 493–499.

Watkins, C. and Kirby, K. (1998) Introduction – historical ecology and European woodland. In: Kirby, K.J. and Watkins, C. (eds) *The Ecological History of European Forests.* CAB International, Wallingford, UK, pp. ix–xv.

17 Białowieża Primeval Forest: A 2000-year Interplay of Environmental and Cultural Forces in Europe's Best Preserved Temperate Woodland

Małgorzata Latałowa,[1]* Marcelina Zimny,[1] Bogumiła Jędrzejewska[2] and Tomasz Samojlik[2]

[1]Department of Plant Ecology, University of Gdańsk, Poland;
[2]Mammal Research Institute, Polish Academy of Sciences, Białowieża, Poland

17.1 Introduction

Białowieża Forest covers about 1500 km² along the border between Poland and Belarus in central eastern Europe (Fig. 17.1) (Faliński, 1986). The unique preservation of the forest ecosystem and rich historical documents describing use of the forest resources during the last several 100 years, make Białowieża Primeval Forest (BPF) of special value as a subject for long-term ecological studies. The history of this forest can be used to explore ideas arising from the ongoing discussion on the natural openness of European primeval forests and the role of game in shaping landscape structure (Vera, 2000; Birks, 2005; Mitchell, 2005; Holl and Smith, 2007; Whitehouse and Smith, 2010). Comparison of the data on historical forest management from documents and maps with the results from palaeoecological records provides a kind of 'historical analogue', which can be useful in the interpretation of the role of past human impact in pollen diagrams elsewhere (Nielsen and Ødgaard, 2004).

This chapter presents an overview of the environmental history of BPF in the last two millennia based on palaeoecological, archaeological and historical investigations carried out between 2003 and 2013.

17.2 Previous Studies

Until recently, the Holocene history of BPF was little known. Two old pollen profiles (Dąbrowski, 1959; Borowik-Dąbrowska and Dąbrowski, 1972) do not meet the modern standards in pollen analysis. There is heavy contamination of pollen spectra due to the low-quality coring device used, low pollen counts, low stratigraphic and taxonomic resolution and only three ^{14}C dates. Hence, these pollen spectra represent little more than a very general account of the Late Glacial and Holocene forest succession. Two disturbed profiles without ^{14}C dating, analysed with low taxonomic and stratigraphic resolution (Milecka et al., 2009) add little more to the forest history. More interesting information comes from Mitchell and Cole (1998) who analysed two short profiles (covering the last 1300 years) from small hollow

*E-mail: m.latalowa@ug.edu.pl

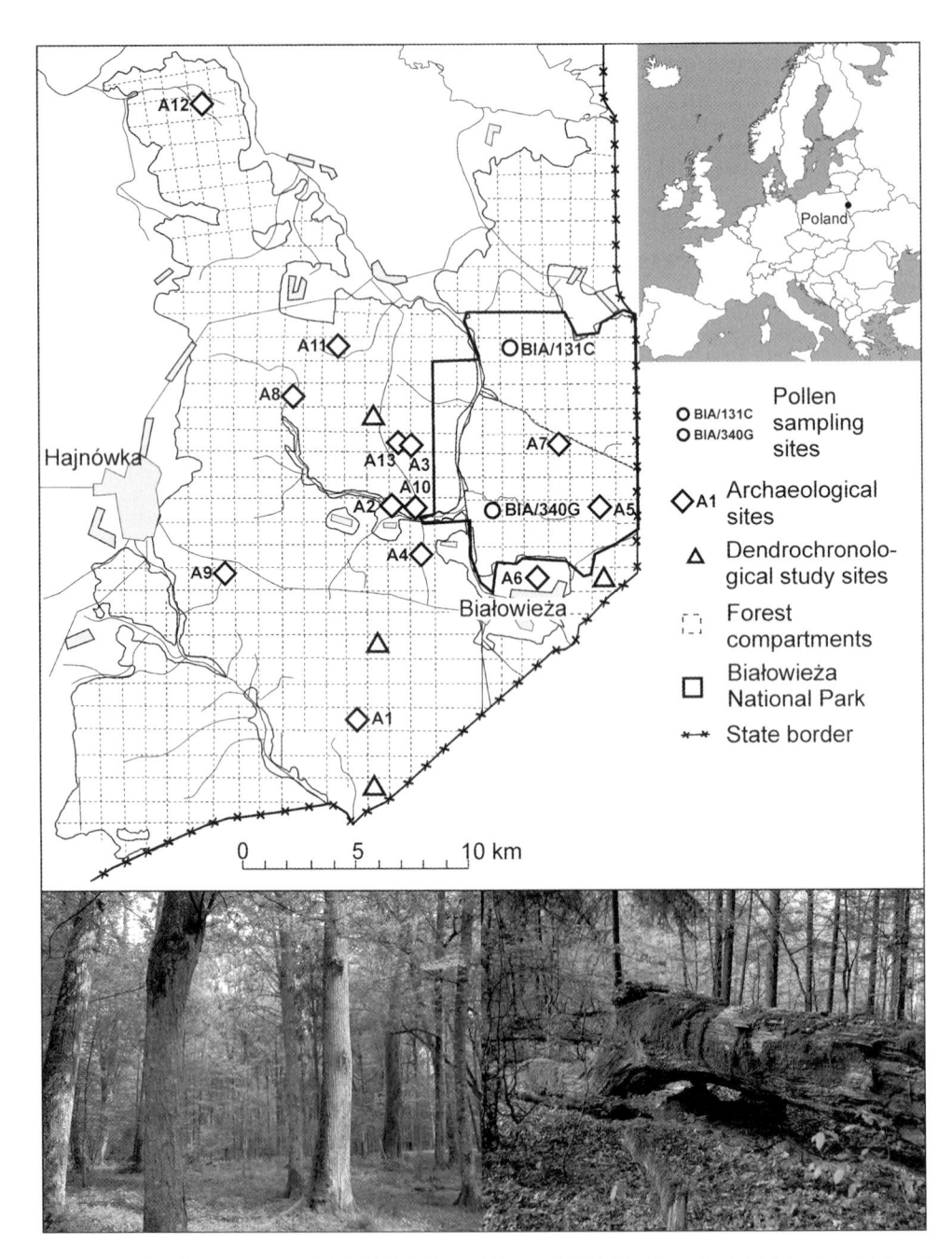

Fig. 17.1. Map of contemporary Białowieża Primeval Forest (BPF) (Polish part) with Białowieża National Park (BNP), and pollen sampling, archaeological and dendrochronological study sites marked. Below the map are two examples of forest landscapes in BNP. (Photographs by T. Samojlik.)

sites for pollen and charcoal content, focusing on long-term succession patterns in the *Tilio-Carpinetum* and *Pino-Quercetum* forest communities.

Similarly, the forest has been outside the scope of detailed archaeological investigations. Earlier works (Götze, 1929; Walicka, 1958; Dzierżykray-Rogalski and Jaskanis, 1961;

Żurowski, 1963; Górska, 1973, 1976) were based on occasional finds or short-term excavations, and lacked modern datings. The history of BPF in the modern period (15th to 20th centuries) has been more extensively described (e.g. Hedemann, 1939; Więcko, 1984; Daszkiewicz *et al.*, 2004, 2012), but few works have analysed the historical data in the context of human impact on the forest environment (Samojlik and Jędrzejewska, 2004; Samojlik *et al.*, 2013a,b,c).

17.3 A New Palaeoecological Record for Białowieża Primeval Forest

17.3.1 Methods

New, high-resolution pollen records from two sites provide evidence for the last *c*.2000 years of changes in the forest composition and structure. The BIA/131C site (see Fig. 17.1) is an irregularly shaped, shallow bog *c*.500 m long and *c*.250 m wide in the north-western part of the Białowieża National Park (BNP). It borders coniferous forests to the north-west and mixed deciduous forest to the south-east. The pollen profile was taken *c*.50 m from the eastern bog margin. The second profile (BIA/340G) (see Fig. 17.1) was collected in the western part of the Park, in a small (*c*.100 × 150 m), shallow bog surrounded by mixed coniferous forest with much spruce (*Picea*). There were also patches of mixed deciduous forest with oak (*Quercus*), hornbeam (*Carpinus*) and lime (*Tilia*) a short distance from this site. The relevant source area of pollen (RSAP) (Sugita, 1994) for both sites should not exceed 2–3 km radius (Sugita, 1994; Nielsen and Sugita, 2005; Gaillard *et al.*, 2008; Hellman *et al.*, 2009).

The upper parts of both profiles, deposited during the last *c*.2000 years, were sampled with 1 cm resolution and processed in the laboratory according to standard methods (Faegri and Iversen, 1989). At least 1000 pollen grains were counted in each sample. Micro charcoal particles were measured and counted along with the pollen. The percentage values were calculated in relation to the tree pollen sum. We present cumulative curves for the particles smaller than 40 µm and equal or larger than 40 µm, respectively (Figs 17.2 and 17.3). Macro-charcoal particles were counted in the BIA/131C profile under 16× magnification using a stereoscopic microscope; the results are expressed as the number of particles in 20 cm^3 of sediments, classified to three classes (<1 mm, 1–5 mm and >5 mm). Charred remains of herbs were counted separately (Fig. 17.3).

The micro- and macro-charcoal results (Fig. 17.3) do not fully correspond with each other, but the peaks in micro-charcoal particles are positively correlated with occurrences of macroscopic charcoal fragments. This reflects the different dispersal abilities of airborne particles according to their size and weight (Tolonen, 1986; Blackford, 2000). Charcoal of size <40 µm is likely to have come from long-distance transport, thus illustrating not only local, but also regional, fire activity (Olsson *et al.*, 2010). This micro-charcoal may derive not only from large-scale forest fires, but also from the burning of wood associated with the production of charcoal, potash, wood tar and iron, and even beekeeping in tree cavities, as well as from domestic fires in the nearby settlements. The threshold of >40 µm used here for macro-fragments is close to a value of 50 µm proposed by Tinner *et al.* (1998) as reflecting local events; we assume that the source area of this charcoal particle class is similar to that of the pollen source area.

The chronostratigraphy was established according to composite age/depth models (Zimny, 2014) based on calibrated ^{14}C (radiocarbon) accelerator mass spectrometry (AMS) (Reimer *et al.*, 2013) and ^{210}Pb (lead) dating (Binford, 1990) with the OxCal software (version 4.2; Bronk Ramsey and Lee, 2013). The models reveal high variations in peat accumulation rates and suggest the presence of short-term hiatuses that affect the precision of the palaeoecological reconstruction at some points. However, because these profiles are complementary to one another, this is not a serious limitation. Table 17.1 presents all of the radiocarbon dates used in this chapter and the results of their calibration.

The reconstruction of settlement and economic development in the area and their impact on forest changes used alterations in the abundance and taxonomic composition of pollen representing cultivated plants, weeds, meadow vegetation and plants spreading through disturbed forest habitats, as well as

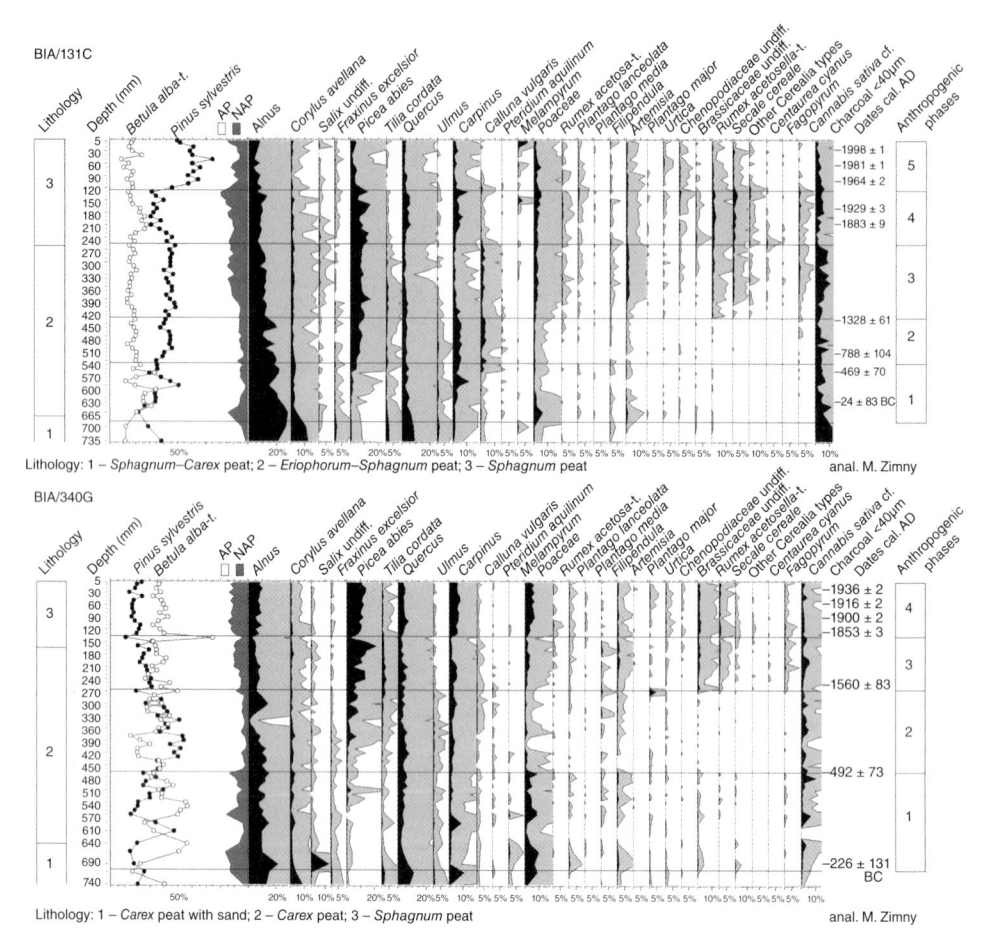

Fig. 17.2. Percentage pollen diagrams from the BIA/131C and BIA/340G sites in Białowieża National Park (BNP), Poland (selected data). AP, arboreal pollen grains; NAP, non-arboreal pollen grains; -t, type.

fluctuations in the tree pollen curves and the micro-charcoal particle content in the sediments. The two pollen diagrams (Fig. 17.2) display their own individual features depending on the local ecological conditions and the extent of past human interference, but four common zones representing different anthropogenic phases are defined. A fifth phase is reflected solely in the BIA/131C diagram, because the most recent deposit in the BIA/340G profile was missing.

17.3.2 Results

Phase 1, recorded in the bottom parts of both diagrams, is from the Iron Age, as indicated by the dates 226 ± 131 BC and AD 492 ± 73 in the BIA/340G profile (Fig. 17.2) and 24 ± 83 BC and AD 469 ± 70 in the BIA/131C profile (Figs 17.2 and 17.3). This long-lasting occupation phase was characterized by distinct variation in settlement activity. The initial development period appears to be due to the pre-Roman Iron Age settlement. In this stage, anthropogenic disturbances affected the whole range of forest habitats. An abrupt decline of oak, hornbeam and lime indicates the exploitation of mixed deciduous forest, while a similar drop in the alder (*Alnus*) pollen curve and short-lived willows (*Salix*) peak most probably reflects deforestation of the river valleys. A high proportion of charcoal particles in all classes, and an increased frequency of the palynological indices of forest fires such as *Melampyrum* pollen and *Gelasinospora* spores,

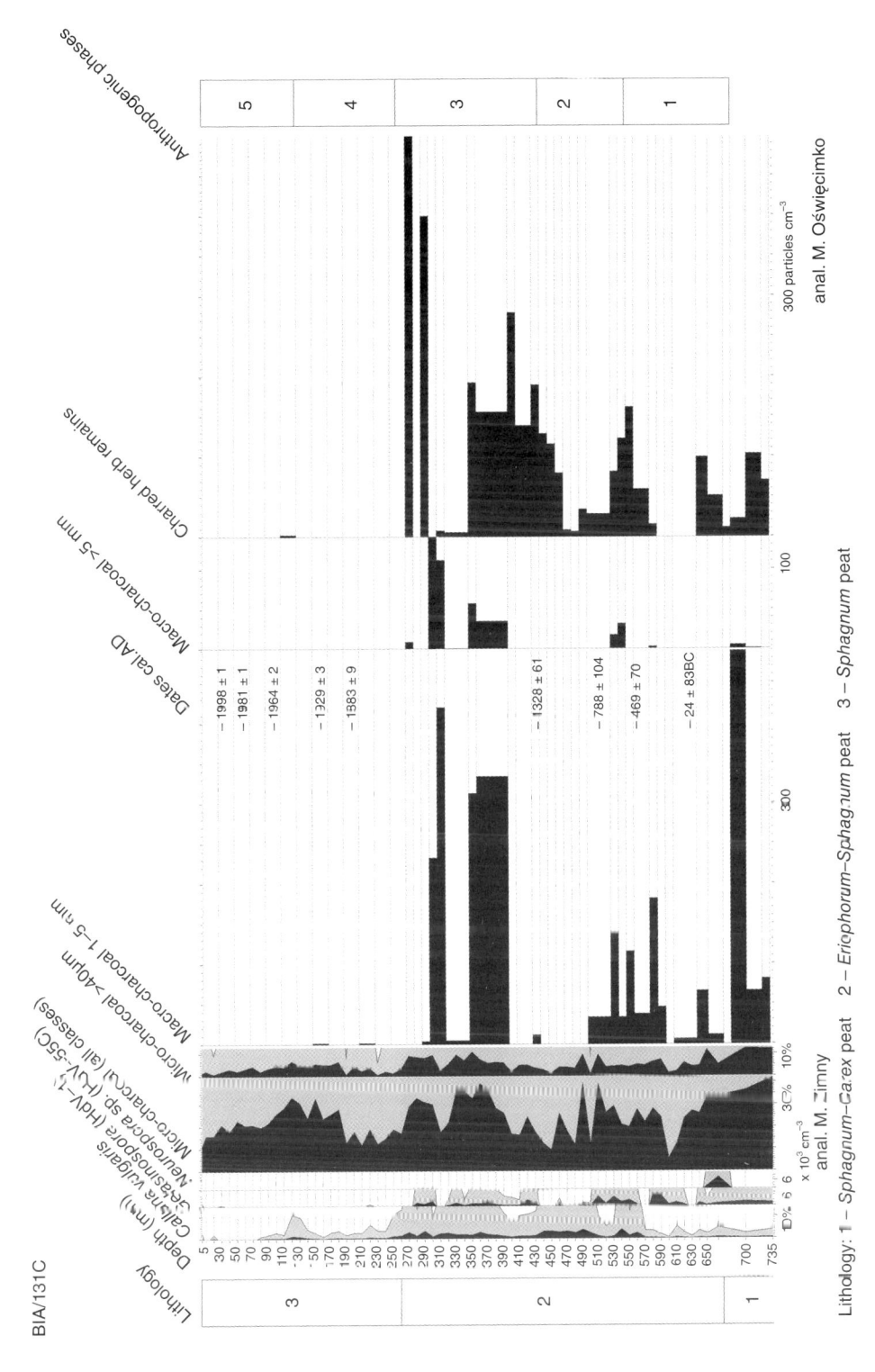

Fig. 17.3. Proxy data on forest fires in Białowieża Primeval Forest (BPF) based on the BIA/131C peat profile content. Macro-charcoal data from Oświęcimko (2012).

Table 17.1. The results of radiocarbon dating using accelerator mass spectrometry (AMS).

Site type	Laboratory code	Age (^{14}C BP)	Calibrated date (2σ)[a]	Material dated
Palaeoecological sites (site code – depth range, cm)				
BIA/340G – 69–68.5	Poz-28534	2155 ± 30	226 ± 131 BC	Peat
BIA/131C – 63.5–63	Poz-39679	2020 ± 30	24 ± 83 BC	*Andromeda polifolia* seeds (2), *Pinus* bark, Ericaceae stems and bark, Bryophyta stems
BIA/131C – 56.5–56	Poz-35663	1600 ± 30	AD 469 ± 70	Peat
BIA/340G – 47–46.5	Poz-52248	1560 ± 30	AD 492 ± 73	*Carex* fruits (c.20), fragment of leaf
BIA/131C – 51.5–51	Poz-39672	1220 ± 40	AD 788 ± 104	*A. polifolia* seeds (3) and stem, Bryophyta stems and leaves
BIA/131C – 43.5–43	Poz-35662	685 ± 30	AD 1328 ± 61	Peat
BIA/131C – 34.5–34	Poz-39671	180 ± 30	Modern	*Sphagnum* stems and heads, *Rubus idaeus* fruit stone (1)
BIA/340G – 26–25.5	Poz-52247	330 ± 30	AD 1560 ± 83	*Betula* periderm and leaf scars
BIA/340G – 14–13.5	Poz-28533	165 ± 25	Modern	*Carex lasiocarpa* fruits (9)
Archaeological sites (site code)				
A1	Poz-13154	2280 ± 35	306 ± 98 BC	Charcoal
A1	Poz-13156	1970 ± 30	AD 20 ± 65	Charcoal
A5	Poz-4506	1840 ± 30	AD 164 ± 78	Charcoal
A5	Poz-4508	1610 ± 30	AD 465 ± 73	Charcoal
A2	Poz-9872	1205 ± 30	AD 816 ± 121	Charcoal
A13	Poz-5870	924 ± 30	AD 1105 ± 79	Human bone fragments
A13	Poz-4407	880 ± 30	AD 1132 ± 90	Human bone fragments
A10	Poz-27170	335 ± 30	AD 1558 ± 83	Charcoal
Forest compartment no. 183, ground mound	Poz-9870	280 ± 30	Modern	Charcoal
Forest compartment no. 189, ground mound	Poz-10471	140 ± 30	Modern	Charcoal

[a]Standard deviation at 2σ gives confidence interval 95.4%.

suggest the use of fire either for forest clearance or in charcoal production. High proportions of birch (*Betula*) pollen in the subsequent regeneration phase might indicate the spread of birch on the earlier burnt land.

The land cleared from forest appears to have been used for cultivation (pollen of Cerealia-t. (type) and *Triticum*-t.). Meadows and pastures (suggesting animal husbandry) were located on a range of soil types as indicated by grassland taxa with different moisture requirements, e.g. *Plantago lanceolata, P. media, Centaurea jacea*-t., *Trifolium pretense*-t., *Filipendula* (*ulmaria*). There is also a relatively high frequency of other taxa typical of anthropogenic habitats, such as *Artemisia, P. major, Solanum nigrum* and *Chenopodiaceae*.

A further episode of new colonization appears to have taken place in the Roman Iron Age and to have continued, with fluctuating intensity, up to the 5th century AD. The palynological record suggests long-lasting and multiple impacts on the forest ecosystem. Human-made fires (suggested by the high charcoal frequency) would have deeply affected woodland in all kinds of habitats, but especially patches of deciduous forests (deep declines in *Carpinus*) where slash-and-burn cultivation was probably applied (high *Betula* pollen values).

Phase 2 spans the end of the 5th through to the 16th century, including the Migration Period, medieval times (until 15th century) and probably the beginnings of the modern times. There is a distinct decline in all kinds of anthropogenic indicators, especially between the 6th and 10th centuries; pollen of cultivated plants disappears and that typical of meadows, pastures and weeds declines to a minimum. There is only a slight increase (especially of indicators of wet meadows) in the later period, up to the 16th century. Fluctuations of tree pollen curves suggest forest disturbances, perhaps through the use of fire.

There is a continuous micro-charcoal presence in the sediments and charcoal frequencies are correlated with the *Calluna vulgaris* pollen curve; heather might have spread under a regime of regular burning of the forest floor and a more open tree canopy. This could also favour pine (*Pinus*) expansion and establishment, particularly on sandy substrates or other less fertile soils (BIA/131C). Forest disturbances could also allow the expansion of spruce on formerly deciduous forest areas.

The slight decreasing trend in the alder pollen curve may indicate human activity in the river valleys. Generally though, the whole phase seems to represent a period of forest regeneration after the decline of the Roman era occupation. Hornbeam, oak and lime increased their share and there was also a period of spruce expansion. We interpret this phase as a period of moderate economic exploitation of forest resources, and none or very weak agricultural activity in the area under consideration.

Phase 3 covers the 17th and most of the 18th century. There is a distinct increase in indicators of human activity. Disturbance to forest habitats by fire is suggested by high proportions of *Calluna* and charcoal particles (both profiles), and strong fluctuation of *Betula* and *Pinus* pollen (BIA/340G). Distinct declines of *Carpinus* (BIA/131C) and then of *Quercus* and *Tilia* (both profiles) point to the exploitation of mixed deciduous forest, whereas spruce increases. Cereal pollen (with rye as the most frequent), buckwheat (*Fagopyrum*) and hemp (cf. *Cannabis*) indicate cultivated fields close by, as does the frequent occurrence of cornflower (*Centaurea cyanus*) – the pollen of this weed of winter-sown maize fields does not spread very far. The low levels of *Poaceae* and *P. lanceolata* suggest that open meadows and pastures were limited in extent. A systematic decline of alder and relatively high frequencies of *Filipendula* pollen could indicate that animal husbandry was allowed in the river valleys.

Phase 4 (the end of the 18th century up to the mid-20th century) shows higher proportions of the pollen taxa of cultivated plants, but a lower frequency of taxa indicative of forest disturbances (e.g. *Calluna*) and some of the ruderal plants (*Artemisia*), and distinctly lower micro charcoal content. Hornbeam and oak increased, pine declined and spruce continued to be an important forest constituent. In the later part of this phase (the end of the 19th century to mid-20th century), the frequency of charcoal particles increases again, peaks of charcoal dust are accompanied by small peaks of *Melampyrum* and *Calluna* pollen.

Overall, Phase 4 appears to be a period of generally lower exploitation of woods in the area. However, our data also indicate contemporary development of local settlements based on agriculture, with cereals, hemp, flax and buckwheat being grown. The high frequency of *Chenopodiaceae* and *Brassicaceae* pollen may represent both common weeds and some popular root crops, such as beetroot, sugarbeet, mangolds, cabbage and turnips.

The most recent period (Phase 5) is only distinguishable in the upper part of the BIA/131C diagram; it starts around the mid-20th century and continues up to recent times. The drop in anthropogenic indicators reflects a decrease of settlement and economic activity within the forest. The pollen diagram illustrates vigorous expansion of pine and suggests the restriction of deciduous trees and spruce stands in the area surrounding the site, although this may reflect in part the spread of pine across the bog sample site itself. The decline in the frequency of micro-charcoal particles in the last few decades and the absence of macro-charcoal are a consequence of recent fire protection policies across the forest.

17.4 Archaeological Evidence

Between 2003 and 2011, the Mammal Research Institute, in cooperation with the Institute of Archaeology and Ethnology, carried out a field survey to look for potential archaeological sites in the Polish part of BPF, eight of which were excavated. Their chronology was determined by ^{14}C AMS dating of charcoal and bone fragments which are presented as calibrated ages (Reimer *et al.*, 2013) based on OxCal software (version 4.2) (Table 17.1). The cultural identification of sites was based on artefact analysis. Our review of the archaeological evidence is based on these and five other sites from studies A1–A13 in Fig. 17.1. These sites are mostly within a few kilometres of the pollen sampling sites.

Excavations identified two Iron Age cultures. Traces of Hatched Pottery Culture (HPC) (pre-Roman Iron Age; also called Stroked Ware Culture; see Barford *et al.*, 1991) and the Wielbark Culture (WC) (Fig. 17.4) were discovered at three sites (A1–A3, Fig. 17.1) (Krasnodębski and Olczak, 2006b). Charcoal samples from a settlement (site A1) were dated to 306 ± 98 BC and AD 20 ± 65. The HPC is identified as belonging to the Eastern Baltic tribes that occupied mainly central and western Belarus and eastern Lithuania between the 7th to 6th centuries BC and 4th to 5th centuries AD (Medvedev, 1996; Rusin, 1998; Beljavec, 1999–2001). Their sites in north-eastern Poland (including BPF) (Rusin, 1998) are scattered open settlements located peripherally to the main area occupied by this culture.

Traces of WC (Roman Iron Age) settlements and cemeteries were discovered at four sites (A4–A7 in Fig. 17.1) (Dzierżykray-Rogalski and Jaskanis, 1961; Samojlik, 2007; Krasnodębski *et al.*, 2008). Other WC sites are also known from the Belarussian part of BPF (Kavalenia *et al.*, 2009). Charcoal fragments from a cemetery in BNP (A5) were dated to AD 164 ± 78 and AD 465 ± 73; the artefact analysis dated the presence of WC in BPF to the 3rd to 5th centuries AD (Dzierżykray-Rogalski and Jaskanis, 1961; Krasnodębski *et al.*, 2008). The WC has been identified with German tribes migrating from Scandinavia towards the south-east: from Pomerania to Masovia, Podlasie (Poland), Polesie (Belarus), Western Wołyń and Podole to the Black Sea shores (Ukraine) (Barford *et al.*, 1991; Kokowski, 2005; Jaskanis, 2012). WC sites in BPF are peripheral to the main centres of this culture located to the north-west of our study area, but the presence of artefacts imported from the Roman Empire among finds in WC sites in BPF is evidence for the exchange of goods with south-west Europe (Kokowski, 2005; Krasnodębski *et al.*, 2008).

The decline of the WC was connected with the beginning of the Migration Period. A notable hiatus in archaeological evidence for human presence in BPF overlaps with the period of mass migration of people between AD 400 and 800, when vast areas of central Europe were abandoned until the new waves of settlers – Slavonic tribes – claimed those lands in the Early Medieval Period (Buko, 2005; Kobyliński and Szymański, 2005; Kokowski, 2005).

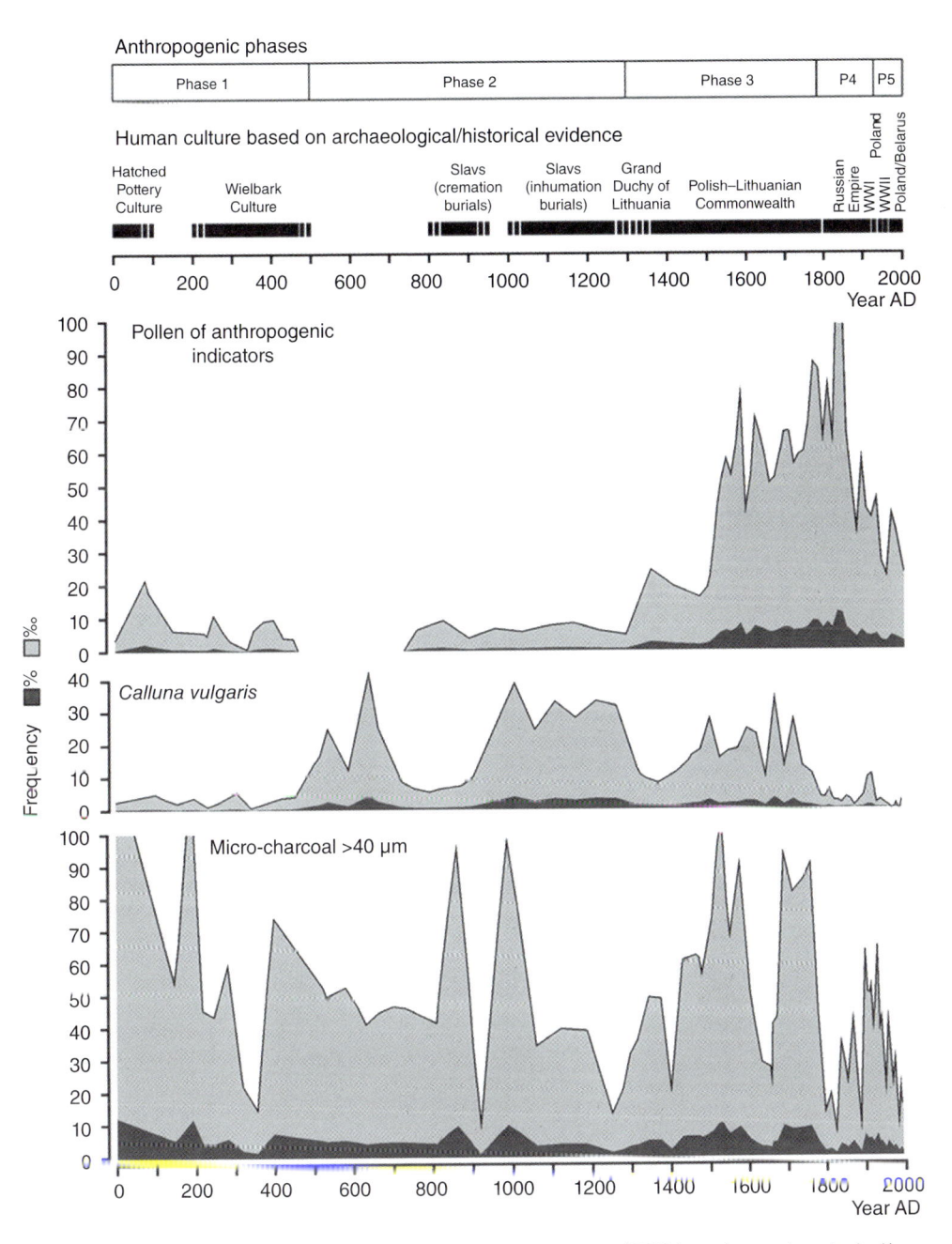

Fig. 17.4. Phases of human presence in Białowieża Primeval Forest (BPF) based on archaeological/historical evidence, pollen of anthropogenic indicators, *Calluna vulgaris* and micro-charcoal >40 µm from peat profile BIA/131C in the period 0–2000 AD.

Traces of Slavonic presence in BPF were discovered at a total of six sites and covered two periods. The earlier one (before the 9th to 10th centuries) includes three sites: two cemeteries (A8, A9, Fig. 17.1) and a complex of a cemetery and a nearby settlement (A2, A10) (Götze, 1929; Krasnodębski and Olczak, 2006a; Krasnodębski *et al.*, 2011). Charcoal samples from a barrow at site A2 were radiocarbon dated to AD 816 ± 121; the pottery came from between just before the 9th century (Götze, 1929) and the first half of the 10th century (Krasnodębski and Olczak, 2006a).

A second period of Slavonic presence in BPF spans the 10th to 12th centuries, the time of emergence and consolidation of the neighbouring states – Poland and Kievian Rus. BPF and the whole Podlasie region were colonized by both Eastern Slavs (the Dregoviche tribe) and Western Slavs (Mazovians) (Barford et al., 1991; Zoll-Adamikowa, 1996; Krasnodębski et al., 2005, 2011). In BPF, archaeological evidence from this period included four sites: three cemeteries (A8, A11–A12) and a cemetery with a nearby settlement (A3, A13) (Götze, 1929; Walicka, 1958; Krasnodębski et al., 2005, 2011; Samojlik, 2007). Only one of the sites showed continuous use over the two periods. Radiocarbon dating of human bone fragments from site A13 gave dates of AD 1105 ± 79 and AD 1132 ± 90; the artefacts were dated to the 11th to 12th centuries (Krasnodębski et al., 2005).

In the early 14th century, most of Podlasie region, including BPF, was taken over by the Grand Duchy of Lithuania (Sahanowicz, 2002). In 1385, the Kingdom of Poland and the Grand Duchy of Lithuania were joined, and this union lasted until 1795 (Davies, 1986). BPF became the private property and hunting ground of the grand dukes, a specially protected royal forest with the main purpose of preserving large game (European bison, *Bison bonasus*; brown bear, *Ursus arctos*; red deer, *Cervus elaphus*; and moose, *Alces alces*) (Samojlik et al., 2013c). Large game hunting was reserved for monarchs and usually took place in 'hunting gardens' (*kletna*), incorporating an area of 3–5 km². Eighteenth century maps document the existence of two hunting gardens (Samojlik et al., 2013c). At least five royal hunts in BPF are recorded between 1400 and 1546, and a further nine royal visits between 1581 and 1764 (Samojlik, 2006a, 2010).

Two sites from the 13th to 16th centuries have been excavated: the remnants of a hunting manor in Stara Białowieża (A10); and a settlement (A3, 2.5 km from the manor). Based on pottery and stove tiles, Stara Białowieża was in use in the first half of the 16th century (Samojlik, 2007; Jędrzejewska, 2011), and this is supported by a calibrated radiocarbon date for charcoal of AD 1558 ± 83. Pottery from the settlement appeared to be from the second half of 13th/14th century, but higher quality pieces were manufactured in the 14th to 15th centuries. After the hunting manor in Stara Białowieża was abandoned, a new manor was erected 5 km to the east, in the centre of the nearby Białowieża village (site A6). This new manor persisted until the second half of the 17th century, and had a fish pond, fields and gardens covering c.30 ha. The adjacent area of 15 ha belonged to Białowieża village itself (Samojlik et al., 2014).

17.5 Archival Studies

17.5.1 The royal forest of Polish kings

More detailed information on the management and conservation of BPF from the 15th to 20th centuries was extracted from historical archives in Poland and abroad, and from published literature (Samojlik, 2007; Samojlik et al., 2013a). Before 1760, apart from the royal manor, no villages were erected inside the forest, and the royal servants protecting the forest were settled in villages on its borders (Hedemann, 1939). Thereafter, new permanent villages and small ephemeral hamlets were established inside the forest (Samojlik, 2007), one in the immediate vicinity of the BIA/340G pollen site.

In the 14th to 16th centuries, BPF was subject to traditional utilization based on kings' permissions, so called access rights. Access rights for 44 localities, all outside the central part of BPF, have been identified (Samojlik and Jędrzejewska, 2004). These related particularly to: (i) the making of hay along river valleys, which might then be transported to villages outside the forest or left in haystacks in the meadows; (ii) bee-keeping, with the right to carve beehives in trees within mixed or coniferous forests (mainly in Scots pine, *P. sylvestris*) (Plate 3); and (iii) constructing small dams on forest rivers to create ponds for fishing and fish farming

(Samojlik and Jędrzejewska, 2004). The first two activities (in particular beekeeping) introduced fire to the forest. In 1589, the forest became a Royal Economy, with all income from the forest allocated to the royal court only (Samojlik, 2005a, 2009; Samojlik et al., 2013a).

In the second half of the 17th century, new forms of utilization appeared: bog iron ore exploitation in river valleys, potash manufacturing, and wood tar and charcoal production; after 1765 there was commercial timber extraction (Hedemann, 1939; Samojlik, 2009). The historical sources document four bog iron ore extraction sites on the border of the forest (at least 10 km away from the BIA/131C pollen site), that were active 1639–1747 AD (Samojlik, 2009). Potash production was introduced in BPF in 1675 (Jędrzejewska and Samojlik, 2004). The highest yielding tree species for potash burning were maple (*Acer*), elm (*Ulmus*) and ash (*Fraxinus*) (Hedemann, 1934). Over 100 small mounds – remnants of potash production – were mapped in BPF, one being less than 100 m from pollen site BIA/340G. Charcoal from two of those mounds in compartments 183 and 189 of BPF was radiocarbon dated (Samojlik, 2007); however, their calibrated age is beyond the recent calibration curve (Table 17.1).

Wood tar production in BPF is first mentioned in documents in 1696. Charcoal burning similarly probably started in the late-17th century, but was most intensive in the second half of the 18th century. Wood tar kilns and charcoal hearths were mapped in BPF. Based on the estimated density of charcoal hearths (2–4 sites per 100 km^2), and wood tar kilns (2–6 sites per 100 km^2; Samojlik et al., 2013b) there were probably some within a kilometre of the BIA/340G pollen site. Charcoal remains show that Scots pine was almost exclusively used for wood tar production (98.7% of samples), whereas charcoal was produced mainly from hornbeam (52.3%), birch (17.5%) and small-leaved lime (14.0%). Birch tar was made exclusively from birch (Samojlik et al., 2013b).

Records for commercial timber extraction in the 1770s and 1780s list the products for which trees were felled and the species exploited. Oak and Scots pine were valued the most, whereas maple, ash, elm, hornbeam, lime, Norway spruce, aspen (*Populus*), alder and birch had much lower prices (Anon., 1780). The average yearly extraction of wood in BPF in the second half of the 18th century can be estimated to 0.05–0.3 m^3 ha^{-1} (Samojlik, 2007).

Such activity implies a high population that must be supplied with food. In 1792, in 31 forest villages and small hamlets settled by royal foresters and guards, 5638 livestock were recorded (48% cows, 31% sheep, 16% pigs and 5% horses (Lithuanian State Historic Archives, 1792)). Crops cultivated in the fields around the forest included rye, wheat, barley, buckwheat and hemp (confirmed by both pollen and historical sources) with peas, broad beans and flax also mentioned in historical documents (Lithuanian State Historic Archives, 1786). High proportions of *Artemisia* in the pollen diagrams may be a general indicator of the spread of ruderal weeds from around villages to forest interiors along communication tracts and, especially, on and around production areas such as charcoal hearths or potash-making sites.

17.5.2 Under Russian rule

In 1795, Poland was seized by three neighbouring countries: Habsburg Austria, the Kingdom of Prussia and the Russian Empire. The entire forest fell under the rule of Russia, so the status of BPF changed from the royal to the imperial forest for the next 120 years (Samojlik, 2005a).

After 1795, potash and charcoal burning, fishing and creating fish ponds, and bog iron ore extraction ceased completely. Traditional beekeeping was reduced in extent to local use only and was prohibited in 1888. Haymaking, the utilization of forest meadows and livestock pasturing increased in extent and intensity along with the rise of local population. Livestock grazing mainly took place on land near to settlements, but in 1875 it posed such a threat to the bison population that wooden fences were built to separate the outer parts of the forest used for livestock from the inner part populated by bison (Genko, 1902–1903; Karcov, 1903). By the end of the 19th century, the total number of cattle pastured in clearings around settlements reached 8300 (Jędrzejewska et al., 1997).

The tradition of royal hunts, after being abandoned for over 60 years, was revived with the first tsar's hunt in BPF in 1860, in the area of old royal hunting enclosure. This and the subsequent imperial hunts were much larger in scale than those of earlier centuries (Daszkiewicz et al., 2012).

Exploitation of the timber resources became more important for the new administration, yet attempts at introducing large-scale logging were not successful (Daszkiewicz et al., 2012). Brincken's (1826) management plans for BPF as the reservoir of game for tsars' hunts included the removal of villages inside the forest, changing BPF into an even-aged forest plantation, the planting of coniferous trees and removing beehives from the forest. Despite the plans, the administration in the first 30 years of the 19th century remained almost the same as it had been in the 18th century.

Only after the fall of the Polish national uprising in 1831 did the implementation of the modern 'scientific' forestry model begin. This incorporated the following: the introduction of fire protection measures; prohibition of or a limit imposed on almost all traditional forms of utilization that introduced fire to the forest; forest taxation, including the division of the forest into sections/compartments and the cutting of section lines in 1843–1846; and the periodic exploitation of timber in 1796–1802, 1838–1841 (oaks), 1845–1863 (Scots pine), 1871–1884 (Scots pine, oak and ash) and 1885–1897 (all species) (Więcko, 1984; Daszkiewicz et al., 2012) (Fig. 17.3). Ten new hamlets for forest guards were erected inside the forest (Karcov, 1903).

From 1796 to 1888, the Russian administration had no clear goal in forest management, other than attempting to reconcile timber exploitation with the conservation of European bison and the preservation of its habitats. In 1888 though, BPF became the tsars' private property. The new priority was to maximize the game numbers in BPF through: the creation of hundreds of feeding sites and supplementary feeding applied for up to 6 months of the year; the introduction of native but extinct (red deer, *Cervus elaphus*) and alien species (fallow deer, *Dama dama*) to the forest; the creation of a breeding reserve in the area of the old royal hunting garden; and the extermination of all predators (Karcov, 1903; Jędrzejewska et al., 1997; Samojlik et al., 2013c).

17.5.3 World War I to the present

During the German occupation in World War I, a massive, and unprecedented, exploitation of timber began. Annual timber extraction increased to an estimated 7.7 m^3 ha^{-1} (Więcko, 1984), and only the heart of the forest was excluded from logging. Uncontrolled killing of game during and after World War I and in the subsequent Polish–Soviet War (1919–1921) led to extinction of the European bison (in 1919) and near extinction of other ungulate populations (Jędrzejewska et al., 1997).

In 1921, the core area of the BPF was proclaimed a Nature Reserve (later the Białowieża National Park) by the Polish government, which administered the entire forest in the period 1921–1939. From 1929 until 1939, BPF served as a hunting ground of the Polish president and foreign diplomats (Więcko, 1984). In addition, under the Polish State Forest Service, timber extraction continued outside the reserve (approx. 5–7 m^3 ha^{-1} annually; Więcko, 1984; Jędrzejewski and Jędrzejewska, 1995), along with large-scale planting of pine and spruce. Wars and political disturbances in the period 1915–1921 led to the depopulation of forest villages. However, the rapid development of the wood industry after 1921 attracted a wave of settlers repopulating towns, villages and hamlets (Bajko, 2001).

During World War II, BPF was first taken over by the Soviet Union (1939–1941), then by Germany (1941–1944), with both periods marked by persecution of the local inhabitants, local population decline, and the abandonment of fields. In the latter period, the forest was turned into a 'Reichjagdgebiet' – a hunting reserve of the Nazi occupiers (Jędrzejewska and Jędrzejewski, 1998).

In 1944, BPF was divided between the People's Republic of Poland and the Belarusian Soviet Socialist Republic of the USSR, a division strengthened by a wire fence built along the border in 1981 (Jędrzejewska and Jędrzejewski, 1998). In 1945–2010, the Polish side of the forest (595 km²) was managed by the State Forestry Service. At its core was the BNP (47 km², enlarged to 100 km² in 1996). The Belarussian part (880 km²) was given partial protection in 1945, and in 1991 was made a State National Park (Kavalenia *et al.*, 2009).

In the Polish part of the forest (outside the BNP area) commercial forestry dominated up to 1990. On average, 3–5 m³ ha⁻¹ of wood were extracted every year (Jędrzejewski and Jędrzejewska, 1995). Plantations, mainly of pine and spruce, were strongly promoted, with the aim of balancing the area felled and the area planted. The policies changed after 1990. New nature reserves were created throughout the previously managed forest. The timber extraction limits were lowered and the use of plantations almost completely abandoned; the pollen site BIA/131C is located in the area that was under regular forest management until 1996, when the BNP was enlarged. In the Belarussian part of the forest, selective cutting (dead and dying trees, no clear-cuts) amounting to 0.5–1 m³ ha⁻¹ annually continues.

Cattle grazing has been gradually removed from the Polish part of the forest. In 1957, some 3136 cattle grazed over 10,000 ha of forest; 10 years later, there were 814 cattle on half that area; in 1969, only 405 cattle were allowed; and cattle grazing was finally forbidden in 1973 (Więcko, 1984, Jędrzejewska *et al.*, 1997). Recent decades have also witnessed a significant shift in forest transportation; in the managed part of the forest, horse-drawn carts have been replaced by trucks and heavy machinery, although in the BNP, horses are kept for the transportation of tourists.

17.5.4 Changes in land-use extent and character

An analysis of historical maps of BPF shows that forest covered 81.3% of the BPF area in 1793, 74.4% in 1830, 75.6% in 1936 and 80.1% in 1999. Meadows and shrubs increased from 8.1% in 1793 to 10.2% in 1830 and 17.2% in 1900, before decreasing to 14.4% in 1936 and 7.4% in 1999. The cultivated and built-up area covered 10.6% of BPF in 1793 and 15.4% in 1830, but declined to 9.8% in 1900 and 10% in 1936, before increasing again to 12.5% in 1999 (Mikusińska *et al.*, 2013). Since the 1970s, the utilization of riverside meadows has been gradually abandoned, mainly as a result of the depopulation of villages inside and on the border of the forest, the abandonment of small hamlets of forestry workers and a general decline in farming in the villages in the Polish part of the forest. Since 2000, there has been almost no cultivation in the forest villages.

17.6 Dendrochronological Analyses of Fire Dynamics

Dendrochronological data provide an additional means of looking at forest fire history from the late 16th century onward. Four sites were sampled (Fig. 17.1) all between 5 to 15 km of each other, in different forest types of BPF: moist mixed coniferous forest (*c*.8 km from the BIA/340G pollen site), wet coniferous forest (*c*.12 km away), moist and wet mixed deciduous forest (*c*.5 km away) and moist and wet deciduous forest (*c*.5 km away). Fire scars were counted in samples from dead and living Scots pine.

We dated 76 fire events across the sites, mostly small-scale ground fires that, in 55% of cases, had left scars only on a single tree. The oldest sample was dated to 1592. The majority of fires (67%) occurred from late summer to early spring (Niklasson *et al.*, 2010; T. Samojlik and co-workers, unpublished). The long-term pattern of changes was similar across all sites.

The mean fire interval was 17 years in the first half of the 17th century, declined to 8 years in the second half of the 18th century, and then rose to 30 years in the second half of the 19th century and 76 years in the 20th century (Fig. 17.5). These changes correspond well with the number and intensity of various human activities that introduce fire risk to the forest: haymaking, beekeeping, cattle pasturing, potash and charcoal burning, and wood- and birch-tar production (Harnak,

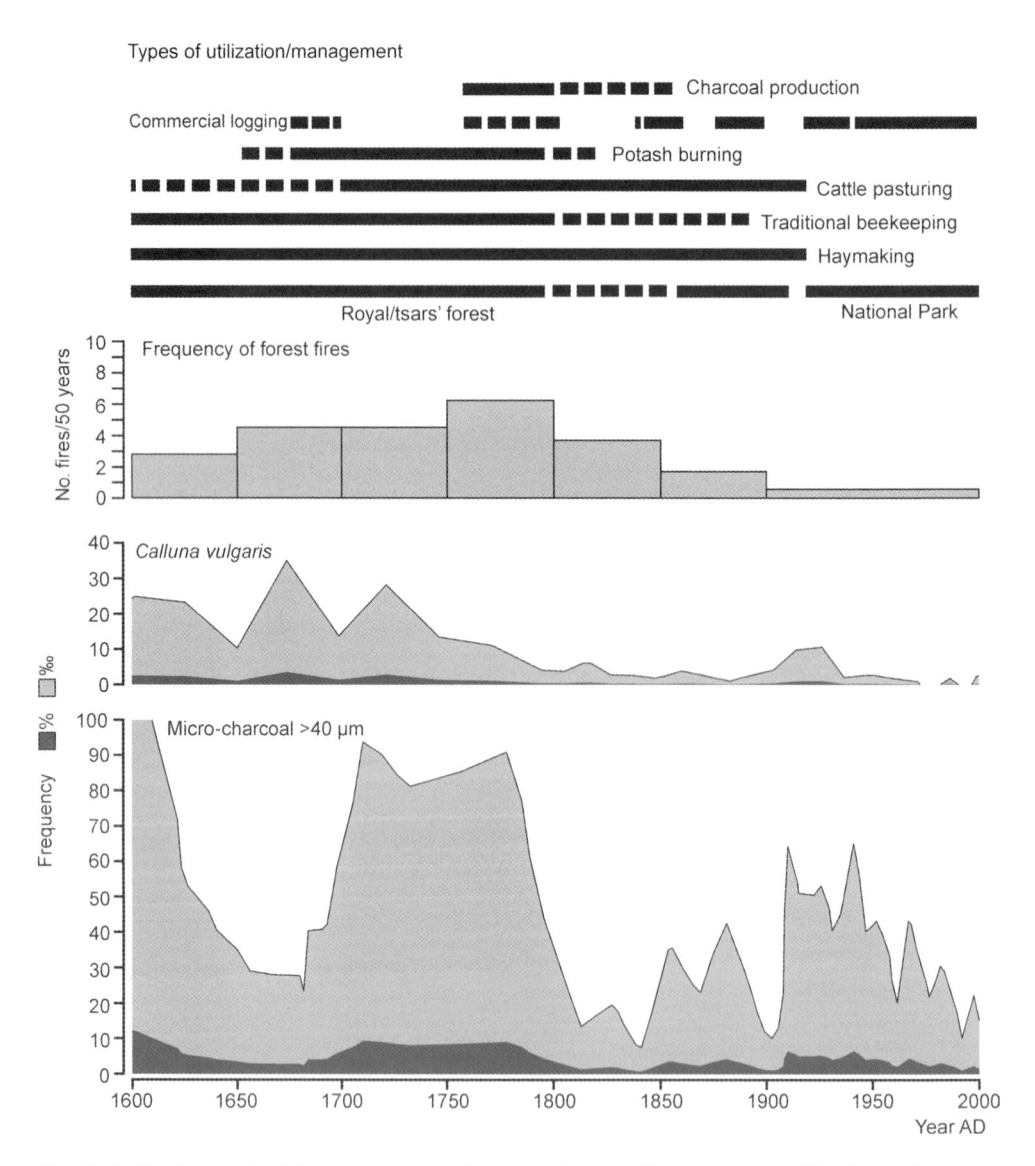

Fig. 17.5. The four parts of the diagram show: the chronology of different types of utilization and management of Białowieża Primeval Forest (BPF) based on historical sources; the frequency of forest fires based on dendrochronological studies at four sites within the BPF; and the frequency of pollen of *Calluna vulgaris* and micro-charcoal at >40 μm from the peat profile of BPF site BIA/131C in the period 1600–2000 AD.

1764; Brincken, 1826) that have been established by other techniques. The results are also consistent with the strengthening of fire protection policies from the early 19th century onward.

The reduction in fires has had ecological consequences. Small but frequent forest floor fires removed young deciduous trees and Norway spruce, and promoted relatively fire-resistant Scots pine. This led to the creation of a specific forest type: pure pine tree stands, called locally *bór lado* (from the Polish *lado*, meaning cleared area) (Samojlik and Jędrzejewska, 2004; Niklasson *et al.*, 2010). These covered 39% of the forest area in 1889 (Genko, 1902–1903). In the late 19th century, *lado* forest started to disappear from forest landscapes and Norway spruce began its expansion in coniferous stands (Niklasson *et al.*, 2010).

17.7 Interplay of Natural and Cultural Forces

17.7.1 The Iron Age

The archaeological data indicate that the BPF area was settled and exploited during the Iron Age by populations of the Hatched Pottery and then the Wielbark Cultures (Fig. 17.4). There is only limited knowledge of the distribution and extension of settlements of these cultures in the region because forests generally tend to have had only limited archaeological exploration, while in the strictly protected reserve of the BNP, only a few accidental discoveries have been investigated under rigorous restrictions (Samojlik *et al.*, 2013a).

The settlers grew cereals using land cleared from deciduous forests, as is shown by strong depressions in broadleaf tree pollen curves and distinct fluctuations of birch and pine pollen values characteristic of forest regeneration phases. The high proportions of birch pollen may reflect slash-and-burn cultivation – a technique widely used in that period (Poska *et al.*, 2008; Wacnik *et al.*, 2012). Farming the rather fertile soils found under oak–hornbeam forests during this period has been documented elsewhere in central and northern Poland (Ralska-Jasiewiczowa, 1964; Ralska-Jasiewiczowa and van Geel, 1998; Makohonienko, 2000).

Animal husbandry was important, and fertile meadows, probably grazed by cattle, replaced the alder woodland of the river valleys. However, changes in the forest cover were not necessarily all caused by demands for food production. Rich deposits of bog iron ore (Ratajczak and Skoczylas, 1999) could have been an important reason for the settlement of the area during that period. Iron production has been confirmed by the discovery of remains of iron smelting furnaces and abundant pieces of slag on archaeological sites (Jażdżewski, 1939; Samojlik, 2009). The extraction of bog iron ore could involve the destruction of extensive riverine habitats and would need a large amount of wood charcoal (Bauer, 1962, quoted in Faliński, 1986), which may explain the relatively large scale of forest disturbances indicated by the pollen diagrams, but with only a weak occurrence of indicators of agriculture (pollen of cereal types and *P. lanceolata*). Timber for wood charcoal could also have been taken from patches of mixed and deciduous forests being cleared for pastures and crop fields. Thus, human activity during the Iron Age could have had a considerable impact on the BPF ecosystems and have led to large-scale forest disturbance.

17.7.2 The Migration Period, medieval and early modern times

During this period, the climate seems to have been quite variable (Büntgen *et al.*, 2011; Steinhilber and Beer, 2011; PAGES 2k Consortium, 2013). The hydro-climatic reconstruction for central Europe indicates a drier climate for AD 600–800, for c.AD 1000 1200 and again in the 15th century (Büntgen *et al.*, 2011). If this created a negative water balance on the bog surface, it would have slowed down peat/sediment growth or might even have stopped it. Prolonged periods of bog surface dryness also usually lead to gradual decay of the existing peat layers, which are subsequently identified as hiatuses in the profile. Our study sites, as with those studied by Mitchell and Cole (1998), are sensitive to hydrological change because they are located in very shallow local depressions, with only a few tens of centimetres of organic deposits. Disturbances in the accumulation rates in the investigated profiles prevent more precise reconstruction of environmental changes century by century.

After a collapse of the late Roman/Iron Age settlement around the 5th century, human activity in the BPF area decreased and may even, for a short time, have ceased. Ongoing work on several new sites should help to determine the duration of this settlement decline. Signs of low-intensity human-induced forest disturbances resume around the 8th century, and then further increase between the 14th and 16th centuries. None the less, the whole period seems to be characterized by forest regeneration.

The main species that gained ground during this process were hornbeam, spruce and pine; alder displayed strong fluctuations, but with a decreasing trend. The persistent evidence of fire (micro- and macro-charcoal), and indicators of disturbances in forest habitats (mainly *Calluna* pollen), are accompanied by only scattered traces of agriculture (Fig. 17.4). The archaeological and historical data seem to confirm low permanent settlement activity in the whole medieval period, with only two known sites. However, charcoal burners, tar makers and people occupied with iron production could live in small huts in the forest, similar to those still present in this area in the beginning of the 20th century (Faliński, 1986), and found in more recent times in other regions. Such dwellings do not usually leave distinct archaeological traces. These small-scale forest uses were commonly practised until around the 15th century, when restrictive regulations were brought in by the royal administration.

17.7.3 The 17th and 18th centuries

This period was an important turning point. More liberal royal regulations and the growing need for forest products made this the most destructive period for the forest complex now protected within the BNP. The palaeoecological data indicate settlement development, a high frequency of fire and distinct changes in both the forest cover and composition. The gradual decline of alder forests probably reflects continuing extraction of iron ore and the subsequent development of meadows on the riverine habitats. Steep declines of deciduous trees – and especially of hornbeam – in the pollen diagrams, may reflect charcoal and potash production, which increased considerably in this period. Lime was among the biggest losers; while the species was of special value for the large number of beekeepers, it was also heavily damaged as a result of large-scale collection of bast – the inner bark of lime, which is used for cordage production and other purposes (Daszkiewicz *et al.*, 2004; Samojlik, 2005b).

The production of charcoal and wood tar, which involved burning, as well as beekeeping and cattle pasturing, facilitated the spread of fires (Fig. 17.5). These maintained a high proportion of birch, especially on deciduous forest habitats, and encouraged the spread of pine and *Calluna* on less fertile soil. High, stable proportions of *Pinus* pollen in the BIA/131C profile may also reflect the presence of *bór lado* – an open pine forest maintained by regular burning of the forest floor.

17.7.4 The 19th to mid-20th centuries

The pollen diagrams show an increase in indicators both for agriculture and for the regeneration of deciduous forest. The historical sources help to resolve this paradox. At the end of the 18th century and in the 19th century, settlement and agricultural activity increased, leading to gradual shrinkage of the forest cover (Faliński, 1986; Mikusińska *et al.*, 2013). The large-scale extraction of timber documented for the whole 19th century certainly resulted in forest damage (Więcko, 1984), but other forms of forest production, such as charcoal burning, bog iron ore extraction, potash making and wood tar and pitch smelting declined (Samojlik, 2007), enabling regeneration of the most devastated patches of the *Carpinus–Quercus–Tilia* forests.

The negative impact of the economic exploitation of the forest resources led to more restrictive policies on the use of BPF (Samojlik, 2007). Restrictions on access to the forest and on all activities using fire reduced forest fires, as confirmed by both tree-ring data and charcoal particle content in the peat deposits (also found by Mitchell and Cole, 1998).

Hornbeam and oak expanded on to the previously disturbed habitats, attaining similar or even higher abundance than in the early medieval period, but lime remained at a proportionally very low level. The regeneration of deciduous and mixed forests has reduced the occurrence of pine and other light-demanding plants of the forest floor, such as *Calluna*. The reductions in burning and grazing since the 19th century have meant that the *bór lado* forest community has declined, being overgrown by spruce and then by deciduous trees (Samojlik, 2006b).

Temporary peaks in the incidence of fire at the end of the 19th century and in the first half of the 20th century (Figs 17.4 and 17.5) may reflect the activity of the army during the insurrection of 1863 and in World Wars I and II (Bajko, 2001). Nevertheless, in contrast to the earlier, regular fires caused by the specific types of production within the forest, these incidental fires seem not to have had long-lasting or large-scale impacts on forest succession.

17.7.5 The recent decades

The creation of the Nature Reserve in 1921, and then that of the BNP in 1947, as well as strong depopulation of forest villages during World War II, reduced agricultural activity in the area bordering the strictly protected part of the BPF (Faliński, 1986). This is reflected in the topmost part of profile BIA/131C, where the pollen of cultured plants and weeds decline to a minimum (see also Mitchell and Cole, 1998).

In the pollen profile BIA/131C, the recent decades are also characterized by an unprecedented expansion of Scots pine. However, this does not reflect what is happening on the dry ground, but rather the localized recent spread of pine (and also spruce and birch) across peat bogs (Keczyński, 2005; Czerepko, 2008), possibly exacerbated by the cutting of a narrow-gauge railway through the area during World War I, which disturbed the hydrology.

17.8 The Role of Large Herbivores in Shaping Białowieża National Park

There is a particular interest in the effects of large herbivores on the primeval forest of BNP, because their role in the natural landscape of Europe has recently been much debated (Vera, 2000; Kirby and Watkins, Chapter 3).

Białowieża is one of the few areas of Europe still to have the complete extant ungulate assemblage – the European bison, moose, red deer, roe deer (*Capreolus capreolus*) and wild boar (*Sus scrofa*), as well as major carnivores – the wolf (*Canis lupus*) and lynx (*Lynx lynx*) (Jędrzejewska and Jędrzejewski, 1998). In addition, at various times, the forest has been used for domestic livestock grazing (Jędrzejewska *et al.*, 1997).

Currently, in the presence of wild ungulates but not livestock, there are very few gaps in the forests of the BNP, and only less than 5% of the area is open grasslands, and sedge and reed marshes in the flooded river valleys. Tree regeneration occurs continuously under the canopy of old stands (Kuijper *et al.*, 2010a) and in canopy gaps (Bobiec, 2007). The current (and historic) suite of browsing ungulates could slow down the growth of young cohorts of trees and affect their species composition (Kuijper *et al.*, 2010a) but, ultimately, the ungulates were not able to maintain the type of 'half-open' landscape suggested by Vera (2000). Nor has the past presence of aurochs (*Bos primigenius*) or the wild horse (*Equus ferus*), which would have added other large grazers to the system, ever been confirmed (Daszkiewicz *et al.*, 2004).

The long-term (70 years) study in the natural forests of the BNP has shown that changes in ungulate density played an important role in young tree recruitment (Kuijper *et al.*, 2010b). While some species suffered from high browsing pressure, species that are less preferred by ungulates or browsing tolerant benefited in periods of a high population density of ungulates. Moreover, the cascading effect of wolves on tree regeneration – via creating a landscape of fear for deer – was found to enhance the recruitment success of young trees in the forests of the BNP (Kuijper *et al.*, 2013). We therefore consider that predominantly closed, albeit browsed, high forest is the natural state here.

17.9 Conclusion

Białowieża Forest is special because it appears never to have been substantially cleared for farming as have many other landscapes.

Palaeoecological, archaeological and historical data do all indicate though that during the last 2000 years, BPF has been subject to almost continuous human intervention, albeit that the type and intensity of exploitation has varied. Two questions follow: is the present state of the BNP comparable to any earlier phase in this forest history; and has the scale of past human interventions here been significantly different (lower) from that determined in other regions?

The Roman Iron Age was a period of heavy anthropogenic change to the BPF ecosystems, probably due to the exploitation of forest resources for iron production. The overall result of vegetation changes in that period is comparable to that of many other sites in northern and central Poland. In contrast, published data from other parts of Poland and temperate Europe suggest substantial differences from the BPF history for the period from early medieval to modern times. Permanent settlement activity in BPF was rather low despite evidence of continuous disturbances in forest habitats through the use of fire.

In modern times, both settlement activity and forest disturbances increased, but the scale of these changes was much lower than in other areas. If the proportion of non-arboreal pollen in consecutive pollen zones illustrates changes in the relative openness of the landscape, then there has never been a time during the modern period when forest disturbances exceeded the scale of those that occurred in the Iron Age. The tree density in the forests is probably higher than around 2000 years ago, but similar to that in the Middle Ages.

The forest resources were exploited, and clearings were used for settlement and agriculture. However, even the artificially increased populations of ungulates at the end of the 19th and beginning of the 20th centuries did not break the ability of the forest communities to regenerate through natural succession. This underlines the importance of the BNP as a flagship area for investigating natural ecological processes at various timescales.

The most recent history of the BPF plant communities is well known through numerous ecological studies, including long-term observations on permanent plots (Faliński, 1986, 1988; Keczyński, 2005; Kuijper et al., 2010b; Bobiec et al., 2011). Our data extend this information up to two millennia backwards, thus enabling a better understanding of the interactions between forest dynamics and anthropogenic and natural factors.

Acknowledgements

The authors wish to thank Marta Oświęcimko who conducted the macro-charcoal analysis in the BIA/131C profile as a part of her MSc thesis, and Joanna Święta-Musznicka, who helped in supervising this thesis. The research reported in this study was funded by the Polish Ministry of Science and Higher Education/Polish National Research Center under: MNiSW/NCN: N N305 167839 – 'Natural history of Białowieża Primeval Forest in the light of palaeoecological research' and MNiSW: N309 013 31/1718 – 'The history of fires and their role in shaping treestands of Białowieża Primeval Forest'; and the research project 'Historical and current human impact on biodiversity' at the Mammal Research Institute PAS.

References

Anon. (1780) *Taxa, za Którey Opłatą Drzewo Wszelakie, i Różne Materyały, Tudzież Inne Rzeczy z Produktu Puszczy do Użycia i Wywozu z Lasów Ekonomii Brzeskiey Mają Bydź Dozwolone, od Dnia Pierwszego Nowembra 1780 Roku*. Central State Historical Archive of Ukraine in Lviv, Record Group 181, Series 2, File 228. [In Polish.]

Bajko, P. (2001) *Białowieża. Zarys Dziejów do 1950 Roku*. Białowieski Ośrodek Kultury (Białowieża Cultural Centre), Białowieża, Poland.

Barford, P., Kobyliński, Z. and Krasnodębski, D. (1991) Between the Slavs, Balts and Germans: ethnic problems in the archaeology and history of Podlasie. *Archeologia Polona* 29, 123–160.

Bauer, E. (1962) *Der Soonwald im Hunsrück; Forstgeschichte eines deutschen Waldgebeites*. Mitteilungen des Forstgeschichtlichen Instituts der Albert-Ludwigs-Universität Freiburg im Breisgau, Germany.

Beljavec, V. (1999–2001) A middle La Tene period grave from a flat cemetery at Radość Kamieniecka, western Belarus. *Wiadomości Archeologiczne* 55, 47–54. [In Polish with an English summary.]

Binford, M.W. (1990) Calculation and uncertainty analysis of ^{210}Pb dates for PIRLA project lake sediment cores. *Journal of Paleolimnology* 3, 253–267.

Birks, H.J.B. (2005) Mind the gap: how open were European primeval forests? *Trends in Ecology and Evolution* 20, 154–156.

Blackford, J.J. (2000) Charcoal fragments in surface samples following a fire and the implications for interpretation of sub-fossil charcoal data. *Palaeogeography, Palaeoclimatology, Palaeoecology* 164, 33–42.

Bobiec, A. (2007) The influence of gaps on tree regeneration: a case study of the mixed lime–hornbeam (*Tilio-Carpinetum* Tracz, 1962) communities in the Białowieża Primeval Forest. *Polish Journal of Ecology* 55, 441–455.

Bobiec, A., Kuijper, D.P.J., Niklasson, M., Romankiewicz, A. and Solecka, K. (2011) Oak (*Quercus robur* L.) regeneration in early successional woodlands grazed by wild ungulates in the absence of livestock. *Forest Ecology and Management* 262, 780–790.

Borowik-Dąbrowska, M. and Dąbrowski, M.J. (1972) Natural and anthropogenic vegetation changes in Białowieża National Park. *Archeologia Polska* 18, 181–200. [In Polish.]

Brincken, J. (1826) *Mémoire Descriptif sur la Forêt Impériale de Białowieża, en Lithuanie*. Glücksberg, Warsaw.

Bronk Ramsey, C. and Lee, S. (2013) Recent and planned developments of the program OxCal. *Radiocarbon* 55, 720–730.

Buko, A. (2005) *Archeologia Polski Wczesnośredniowiecznej. Odkrycia – Hipotezy – Interpretacje*. Wydawnictwo Trio, Warsaw.

Büntgen, U., Tegel, W., Nicolussi, K., McCormick, M., Frank, D., Troue, V., Kaplan, J.O., Herzig, F., Heussner, K.-U., Wanner, H. *et al.* (2011) 2500 years of European climate variability and human susceptibility. *Science* 331, 578–582.

Czerepko, J. (2008) A long-term study of successional dynamics in the forest wetlands. *Forest Ecology and Management* 255, 630–642.

Dąbrowski, M.J. (1959) Late Glacial and Holocene vegetation history of Białowieża Primeval Forest. *Acta Societatis Botanicorum Poloniae* 28, 197–248. [In Polish.]

Daszkiewicz, P., Jędrzejewska, B. and Samojlik, T. (2004) *Puszcza Białowieska w Pracach Przyrodników 1721–1831*. Wydawnictwo Naukowe Semper, Warsaw.

Daszkiewicz, P., Samojlik, T. and Jędrzejewska, B. (2012) *Puszcza Białowieska w Pracach Przyrodników i Podróżników 1831–1863*. Wydawnictwo Naukowe Semper, Warsaw.

Davies, N. (1986) *Heart of Europe. A Short History of Poland*. Oxford University Press, Oxford, UK/New York.

Dzierżykray-Rogalski, T. and Jaskanis, J. (1961) Grób szkieletowy dziecka z późnego okresu rzymskiego, odkryty w 1959 roku w Białowieży, pow. Hajnówka. *Rocznik Białostocki* 1, 283–291.

Faegri, K. and Iversen, J. (1989) *Textbook of Pollen Analysis*. Wiley, Chichester, UK.

Faliński, J.B. (1986) *Vegetation Dynamics in Temperate Lowland Primeval Forests*. Junk, Dordrecht, The Netherlands.

Faliński, J.B. (1988) Succession, regeneration and fluctuation in the Białowieża Forest (NE Poland). *Vegetatio* 77, 115–128.

Gaillard, M.-J., Sugita, S., Bunting, M.J., Middleton, R., Broström, A., Caseldine, C., Giesecke, T., Hellman, S.E.V., Hicks, S., Hjelle, K. *et al.* (2008) The use of modelling and simulation approach in reconstructing past landscapes from fossil pollen data: a review and results from the POLLANDCAL network. *Vegetation History and Archaeobotany* 17, 419–443.

Genko, N. (1902–1903) Kharakteristika Belovezhskoi Pushchi i istoricheskiya o nei dannyya. *Lesnoi Zhurnal* 21, 1014–1056; 22, 1269–1302; 23, 22–98.

Górska, I. (1973) Najdawniejsze Slady człowieka w Puszczy Białowieskiej. *Z Otchłani Wieków* 39, 270–273.

Górska, I. (1976) Badania archeologiczne w Puszczy Białowieskiej. *Archeologia Polski* 21, 109–134.

Götze, A. (1929) Archäologische Untersuchungen im Urwalde von Bialowies. *Beiträge zur Natur- und Kulturgeschichte Lithauens und angrenzenden Gebiete. Bayerische Akademie der Wissenschaften.; Mathematisch-Naturwissenschaftliche Abteilung: Abhandlungen der Mathematisch-Naturwissenschaftlichen Abteilung der Bayerischen Akademie der Wissenschaften*. Supplement 11–14, 511–550.

Harnak, G.H. (1764) *Summariusz z Podatków Łowieckich 1764*. National Historic Archive, Vilnius, inventory no. SA 11575.

Hedemann, O. (1934) *Dawne Puszcze i Wody*. Księgarnia Św. Wojciecha, Vilnius.

Hedemann, O. (1939) *L'Histoire de la Forêt de Białowieża (Jusqu'a 1798)*. Instytut Badawczy Lasów Państwowych, Rozprawy i Sprawozdania Seria A, Nr 1, Warsaw. [In Polish with a French summary.]

Hellman, S., Gaillard, M.-J., Bunting, J.M. and Mazier, F. (2009) Estimating the relevant source area of pollen in the past cultural landscapes of southern Sweden – a forward modeling approach. *Review of Palaeobotany and Palynology* 153, 259–271.

Holl, K. and Smith, M. (2007) Scottish upland forests: history lessons for the future. *Forest Ecology and Management* 249, 45–53.

Jaskanis, J. (2012) *Elite Barrows of the Wielbark Culture in Podlasie*. Muzeum Podlaskie w Białymstoku, Białystok, Poland. [In Polish with an English summary.]

Jażdżewski, K. (1939) O kurhanach nad górną Narwią i o hutnikach sprzed 17 wieków. *Z Otchłani Wieków* 14, 2–22.

Jędrzejewska, B. and Jędrzejewski, W. (1998) *Predation in Vertebrate Communities. The Białowieża Primeval Forest as a Case Study*. Springer, Berlin.

Jędrzejewska, B. and Samojlik, T. (2004) Kontrakty Jana III Sobieskiego z lat 1675–1686 w sprawie dzierżawy i użytkowania Leśnictwa Białowieskiego. *Kwartalnik Historii Kultury Materialnej* 52, 321–330.

Jędrzejewska, B., Jędrzejewski, W., Bunevich, A.N., Miłkowski, L. and Krasiński, Z.A. (1997) Factors shaping population densities and increase rates of ungulates in Białowieża Primeval Forest (Poland and Belarus) in the 19th and 20th century. *Acta Theriologica* 42, 399–451.

Jędrzejewska, H. (2011) Stanowisko Stara Białowieża w Świetle najnowszych badań archeologicznych (2006–2008). MSc. thesis, Instytut Archeologii, Uniwersytet Warszawski, Warsaw.

Jędrzejewski, W. and Jędrzejewska, B. (1995) Proposed National Park of Białowieża Primeval Forest. *Chrońmy Przyrodę Ojczystą* 51, 16–36. [In Polish with an English summary.]

Karcov, G. (1903) *Belovezhskaya Pushcha. Eya istoricheskii ocherk, sovremennoe okhotniche khozaistvo i Vysochaishie okhoty v Puchche*. A. Marks, St Petersburg, Russia.

Kavalenia, A.A., Danilovich, V.V., Dounar, A.B., Zhylinski, M.G., Kalechits, A.G., Litvin, A.M., Lysenka, P.F., Liauko, V.M., Nichyparovich, S.A., Stashkevich, A.I. and Ianouskaia, V.V. (eds) (2009) *Belavezhskaia Pushcha. Vytoki zapavednastsi. Gistoria i suchasnasts*. Belaruskaia Navuka, Minsk, Belarus.

Keczyński, A. (2005) Changes in stand structure on hydrogenic soils on the example of selected study sites in the Białowieża National Park. *Leśne Prace Badawcze* 4, 87–102.

Kobyliński, Z. and Szymański, W. (2005) Pradziejowe i wczesnośredniowieczne osadnictwo w zespole kemów w Haćkach. In: Faliński, J.B., Ber, A., Kobyliński, Z., Szymański, W. and Kwiatkowska-Falińska, A.J. (eds) *Haćki. Zespół Przyrodniczo-archeologiczny na Równinie Bielskiej*. Białowieska Stacja Geobotaniczna Uniwersytetu Warszawskiego, Białowieża-Warszawa, pp. 43–74.

Kokowski, A. (2005) *Starożytna Polska. Od Trzeciego Tysiąclecia Przed Narodzeniem Chrystusa do schyłku starożytności*. Wydawnictwo Trio, Warsaw.

Krasnodębski, D. and Olczak, H. (2006a) Badania archeologiczne przeprowadzone na uroczysku Stara Białowieża w oddz. 367A Puszczy Białowieskiej (AZP 45–92). *Podlaskie Zeszyty Archeologiczne* 2, 74–79.

Krasnodębski, D. and Olczak, H. (2006b) Badania archeologiczne w Puszczy Białowieskiej na stanowisku Teremiski-Dąbrowa, oddz. 338A i B (AZP 45–92). *Podlaskie Zeszyty Archeologiczne* 2, 80–83.

Krasnodębski, D., Samojlik, T., Olczak, H. and Jędrzejewska, B. (2005) Early mediaeval cemetery in the Zamczysko range, Białowieża Primeval Forest. *Sprawozdania Archeologiczne* 57, 555–583.

Krasnodębski, D., Dulinicz, M., Samojlik, T., Olczak, H. and Jędrzejewska, B. (2008) A cremation cemetery of the Wielbark culture in Kletna range (Białowieża National Park, Podlasie Province). *Wiadomości Archeologiczne* 60, 361–376. [In Polish with an English summary.]

Krasnodębski, D., Olczak, H. and Samojlik, T. (2011) Early mediaeval cemeteries in Białowieża Forest. In: Cygan, S., Glinianowicz, M. and Kotowicz, P. (eds) *In Silvis, Campis…et Urbe Średniowieczny Obrządek Pogrzebowy na Pograniczu Polsko-ruskim*. Instytut Archeologii Uniwersytetu Rzeszowskiego, Rzeszów-Sanok, Poland, pp. 144–174. [In Polish with an English summary.]

Kuijper, D.P.J., Cromsigt, J.P.G.M., Jędrzejewska, B., Miścicki, S., Churski, M., Jędrzejewski, W. and Kweczlich, I. (2010a) Bottom-up versus top-down control of tree regeneration in the Białowieża Primeval Forest, Poland. *Journal of Ecology* 98, 888–899.

Kuijper, D.P.J., Jędrzejewska, B., Brzeziecki, B., Churski, M., Jędrzejewski, W. and Żybura, H. (2010b) Fluctuating ungulate density shapes tree recruitment in natural stands of the Białowieża Primeval Forest, Poland. *Journal of Vegetation Science* 21, 1082–1098.

Kuijper, D.P.J., de Kleine, M., Churski, M., van Hooft, P., Bubnicki, J. and Jędrzejewska, B. (2013) Landscape of fear in Europe: wolves affect spatial patterns of ungulate browsing in Białowieża Primeval Forest. *Ecography* 36, 1263–1275.

Lithuanian State Historic Archives (1786) *Inwentarz Ekonomii Brzeskiey y Kobrynskiey*. Lithuanian State Historic Archives in Vilnius, SA 11518. [In Polish.]

Lithuanian State Historic Archives (1792) *Inwentarz Czterech Kwater Leśnictwa JKM Białowieży*. Lithuanian State Historic Archives in Vilnius, SA 11526. [In Polish.]

Makohonienko, M. (2000) *Przyrodnicza Historia Gniezna*. Prace Zakładu Biogeografii i Paleoekologii Uniwersytet im. Adama Mickiewicza. Homini, Bydgoszcz–Poznań, Poland.

Medvedev, A.M. (1996) Belarusskoie ponemanie v rannem zheleznom veke (1 tisachletie do n.e. – 5 v. n.e.). Institut Istorii Belarussi, Minsk, Belarus.

Mikusińska, A., Zawadzka, B., Samojlik, T., Jędrzejewska, B. and Mikusiński, G. (2013) Quantifying landscape change during the last two centuries in Białowieża Primeval Forest. *Applied Vegetation Science* 16, 217–226.

Milecka, K., Noryśkiewicz, A.M. and Kowalewski, G. (2009) History of the Białowieża Primeval Forest, NE Poland. *Studia Quaternaria* 26, 25–39.

Mitchell, F.J.G. (2005) How open were European forests? Hypothesis testing using palaeoecological data. *Journal of Ecology* 93, 168–177.

Mitchell, F.J.G. and Cole, E. (1998) Reconstruction of long-term successional dynamics of temperate woodland in Białowieża Forest, Poland. *Journal of Ecology* 86, 1042–1059.

Nielsen, A.B. and Odgaard, B.V. (2004) The use of historical analogues for interpreting fossil pollen records. *Vegetation History and Archaeobotany* 13, 33–43.

Nielsen, A.B. and Sugita, S. (2005) Estimating relevant source area of pollen for small Danish lakes around AD 1800. *The Holocene* 15, 1006–1020.

Niklasson, M., Zin, E., Zielonka, T., Feijen, M., Korczyk, A.F., Churski, M., Samojlik, T., Jędrzejewska, B., Gutowski, J.M. and Brzeziecki, B. (2010) A 350-year tree-ring fire record from Białowieża Primeval Forest, Poland: implications for Central European lowland fire history. *Journal of Ecology* 98, 1319–1329.

Olsson, F., Gaillard, M.J., Lemdahl, G., Greisman, A., Lanos, P., Marguerie, D., Marcoux, N., Skoglund, P. and Wäglind, J. (2010) A continuous record of fire covering the last 10,500 calendar years from southern Sweden – the role of climate and human activities. *Palaeogeography, Palaeoclimatology, Palaeoecology* 291, 128–141.

Oświęcimko, M. (2012) Sukcesja lokalnych zbiorowisk torfowiskowych na stanowiskach BIA/131C i BIA/318C w Białowieskim Parku Narodowym. MSc. thesis, University of Gdańsk, Gdańsk, Poland.

PAGES 2k Consortium (2013) Continental-scale temperature variability during the past two millennia. *Nature Geoscience* 6, 339–346.

Poska, A., Sepp, E., Veski, S. and Koppel, K. (2008) Using quantitative pollen-based land-cover estimations and a spatial CA-Markov model to reconstruct the development of cultural landscape at Rõuge, South Estonia. *Vegetation History and Archaeobotany* 17, 527–541.

Ralska-Jasiewiczowa, M. (1964) Correlation between the Holocene history of *Carpinus betulus* and the prehistoric settlement in North Poland. *Acta Societatis Botanicorum Poloniae* 33, 463–468.

Ralska-Jasiewiczowa, M. and van Geel, B. (1998) Human impact on the vegetation of the Lake Gościąż surroundings in prehistoric and early-historic times. In: Ralska-Jasiewiczowa, M., Goslar, T., Madeyska, T. and Starkel, L. (eds) *Lake Gościąż, Central Poland. A Monographic Study Part 1*. W. Szafer Institute of Botany, Polish Academy of Sciences, Kraków, Poland, pp. 267–293.

Ratajczak, T. and Skoczylas, J. (1999) *Polskie Darniowe Rudy żelaza*. Wydawnictwo Instytutu Gospodarki Surowcami Mineralnymi i Energią Polskiej Akademii Nauk, Kraków, Poland.

Reimer, P.J., Bard, E., Bayliss, A., Beck, J.W., Blackwell, P.G., Bronk Ramsey, C., Brown, D.M., Buck, C.E., Edwards, R.L., Friedrich, M. *et al.* (2013) Selection and treatment of data for radiocarbon calibration: an update to the International Calibration (INTCAL) Criteria. *Radiocarbon* 55, 1–23.

Rusin, K. (1998) Wstępne wyniki badań dwóch kurhanów z późnego okresu rzymskiego w Grochach Starych, gm. Poświętne, woj. białostockie. In: Ilkjaer, J. and Kokowski, A. (eds) *20 Lat Archeologii w Masłomęczu 1*. Maria Curie Skłodowska University, Lublin, Poland, pp. 189–209.

Sahanowicz, H. (2002) *Historia Białorusi od Czasów Najdawniejszych do Końca XVIII Wieku*. Instytut Europy Środkowo Wschodniej, Lublin, Poland.

Samojlik, T. (ed.) (2005a) *Conservation and Hunting. Białowieża Forest in the Time of Kings*. Mammal Research Institute, Polish Academy of Sciences, Białowieża, Poland.

Samojlik, T. (2005b) A tree of many uses – the history of small-leaved lime (*Tilia cordata*) in Białowieża Primeval Forest. *Rocznik Dendrologiczny* 53, 55–64. [In Polish with an English summary.]

Samojlik, T. (2006a) Hunts and stays of Polish kings and grand dukes of Lithuania in Białowieża Primeval Forest in the 15–16th century. *Kwartalnik Historii Kultury Materialnej* 54, 293–305. [In Polish with an English summary.]

Samojlik, T. (2006b) The grandest tree – a history of Scots pine (*Pinus sylvestris* L.) in Białowieża Primeval Forest until the end of the 18th century. *Rocznik Dendrologiczny* 54, 7–27. [In Polish with an English summary.]

Samojlik, T. (2007) *Antropogenne przemiany środowiska Puszczy Białowieskiej do końca XVIII wieku*. PhD thesis, Mammal Research Institute, Polish Academy of Sciences, Białowieża, Poland.

Samojlik, T. (2009) Bog iron ore extraction sites in the Białowieża Primeval Forest in the 17th–18th centuries. *Kwartalnik Historii Kultury Materialnej* 3–4, 399–411. [In Polish with an English summary.]

Samojlik, T. (2010) Traditional utilisation of Białowieża Primeval Forest (Poland) in the 15th to 18th centuries. *Landscape Archaeology and Ecology* 8, 150–164.

Samojlik, T. and Jędrzejewska, B. (2004) Utilization of Białowieża Forest in the times of Jagiellonian dynasty and its traces in the contemporary forest environment. *Sylwan* 148, 37–50. [In Polish with an English summary.]

Samojlik, T., Rotherham, I.D. and Jędrzejewska, B. (2013a) Quantifying historic human impacts on forest environments: a case study in Białowieża Forest, Poland. *Environmental History* 18, 576–602.

Samojlik, T., Jędrzejewska, B., Michniewicz, M., Krasnodębski, D., Dulinicz, M., Olczak, H., Karczewski, A. and Rotherham, I. (2013b) Tree species used for low-intensity production of charcoal and wood-tar in the 18th-century Białowieża Primeval Forest, Poland. *Phytocoenologia* 43, 1–12.

Samojlik, T., Rotherham, I.D. and Jędrzejewska, B. (2013c) The cultural landscape of royal hunting gardens from the fifteenth to the eighteenth century in Białowieża Primeval Forest. In: Rotherham, I.D. (ed.) *Cultural Severance and the Environment*. Springer, Dordrecht, The Netherlands, pp. 191–204.

Samojlik, T., Jędrzejewska, B., Krasnodębski, D. and Olczak, H. (2014) Vasa dynasty's hunting manor in Białowieża in the light of archival documents and archaeological excavations. *Kwartalnik Historii Kultury Materialnej* 62, 73–90. [In Polish with an English summary.]

Steinhilber, F. and Beer, J. (2011) Solar activity – the past 1200 years. *PAGES News* 19, 5–6.

Sugita, S. (1994) Pollen representation of vegetation in Quaternary sediments: theory and method in patchy vegetation. *Journal of Ecology* 82, 881–897.

Tolonen, K. (1986) Charred particle analysis. In: Berglund, B.E. (ed.) *Handbook of Holocene Palaeoecology and Palaeohydrology*. Wiley, New York, pp. 485–496.

Tinner, W., Condera, M., Ammann, B., Gäggler, H.W., Gedye, S., Jones, R. and Sägesser, B. (1998) Pollen and charcoal in lake sediments compared with historically documented forest fires in southern Switzerland since AD 1920. *The Holocene* 8, 31–42.

Vera, F.W.M. (2000) *Grazing Ecology and Forest History*. CAB International, Wallingford, UK.

Wacnik, A. Goslar, T. and Czernik, J. (2012) Vegetation changes caused by agricultural societies in the Great Mazurian Lake District. *Acta Palaeobotanica* 52, 59–104.

Walicka, E. (1958) Wczesnośredniowieczne kurhany w Puszczy Ladzkiej, pow. Bielsk Podlaski. *Wiadomości Archeologiczne* 25, 157–158.

Whitehouse, N. and Smith, D. (2010) How fragmented was the British Holocene wildwood? Perspectives on the "Vera" grazing debate from the fossil beetle record. *Quaternary Science Reviews* 29, 539–553.

Więcko, E. (1984) *Puszcza Białowieska*. Państwowe Wydawnictwo Naukowe, Warsaw.

Zimny, M. (2014) *Późnoholoceńska historia roślinności Puszczy Białowieskiej*. PhD Thesis, University of Gdańsk, Gdańsk, Poland.

Zoll-Adamikowa, H. (1996) Wczesnośredniowieczny obrządek pogrzebowy a zróżnicowanie etniczne na pograniczu polsko-ruskim. In: Parczewski, M. and Czopek, S. (eds) *Początki Sąsiedztwa. Pogranicze Etniczne Polsko-Rusko-Słowackie w Średniowieczu. Materiały z Konferencji – Rzeszów 9–11 V 1995*. Mitel, Rzeszów, Poland, pp. 81–90.

Żurowski, T. (1963) Cmentarzysko kurhanowe w Białowieży. *Biuletyn Informacyjny Zarządu Muzeów i Ochrony Zabytków* 50, 5.

18 Woodland History in the British Isles – An Interaction of Environmental and Cultural Forces

George F. Peterken*
St Briavels Common, Lydney, UK

18.1 Introduction

Most of the British Isles has a moist, cool climate, but parts of East Anglia are classified as semi-arid; whereas the western seaboard remains largely free of frost and snow in winter, the central Highlands of Scotland have sufficient snow cover to support a skiing industry. Furthermore, a complex geology gives rise to a wide variety of soil types. Hence, despite having among the lowest levels of woodland cover in Europe, Britain retains a surprisingly large range of woodland types. Beech, the dominant tree of European temperate deciduous forests, is prominent as a native tree only in the south and east of England and Wales, while in parts of Highland Scotland enclaves of birch–Scots pine woodland have much in common with the boreal forests of Scandinavia.

Several thousand years of human intervention have further affected this variation, as different regions within Britain each have distinctive patterns of woodland history. The broad pattern and regional differences in the transformation of natural forest to the current woodland cover are described below.

18.2 Outline of British Woodland History

An interest in the history of medieval royal forests and recent timber growing (see, for example, James, 1981), has now been eclipsed by a surge of interest in the ecological history of Britain's woodland. This was initiated by Steven and Carlisle (1959) and Tubbs (1968), comprehensively elaborated by Rackham (1976, 1980, 1986, 2006), Linnard (2000) and Smout *et al.* (2005), and has been incorporated into nature conservation practice by Peterken (1981).

Much of the natural forest was cleared in the Neolithic and Bronze Ages, and by 1086 AD only 15% of England was wooded (Rackham, 1980). Clearance continued until the late 19th century, when just 4% of England remained wooded; most of Ireland, Scotland and Wales were even more denuded. This decline in woodland cover was reversed during the 20th century by planting for timber and natural succession on land abandoned by agriculture and, to a lesser extent, by industries such as mining. Forests of all kinds now occupy 13% of Britain.

Foreign tree species have been introduced to Britain from Roman times onward and were widely planted in the countryside

---

*E-mail: gfpeterken@tiscali.co.uk

from the late 18th century. Conifers, often larch (*Larix* spp.), were planted in the 19th century, but in trivial amounts compared with the vast 20th century plantations of Sitka spruce (*Picea sitchensis*), Douglas fir (*Pseudotsuga menziesii*) and other conifers on upland grazing areas and lowland heaths (Quine, Chapter 15). In the 1960s, Sitka spruce replaced oak (*Quercus* spp.) as the commonest tree species in Britain and, in the 1970s, over 99% of all state planting was coniferous. Foresters redistributed Britain's native trees, notably by planting beech (*Fagus sylvatica*) in the north and west and Scots pine (*Pinus sylvestris*) in the south, both of these in areas outside their accepted native ranges. Some non-native broadleaved species have spread extensively, notably sycamore (*Acer pseudoplatanus*).

Extant woodland in Britain is thus a mosaic of 'ancient' woodland (originally defined as woods that have been present since at least 1600 AD for England and Wales, and 1750 for Scotland) and more recent woodland, formed either naturally or by planting. Ancient woods, which include any remnants of the original forest, survived in tens of thousands of discrete patches, mostly less than 20 ha, each surrounded by cultivated farmland or grazing land (Roberts *et al.*, 1992; Spencer and Kirby, 1992; Goldberg, Chapter 22). Most were located in remote parts of parishes. Some districts had very little woodland (the Fens of eastern England and chalk escarpments of the south-east), while a few districts retained extensive woodland, including the Weald, New Forest, Lower Wye Valley/Forest of Dean and some Highland valleys. Nevertheless, small amounts of woodland persisted almost everywhere, because everyone needed wood.

Most ancient woods were managed in the past, usually as coppice or wood-pasture (Hartel *et al.*, Chapter 5; Buckley and Mills, Chapter 6), and this remained the case until the early 20th century. However, by the 1970s, these systems had largely died out – neglected coppices had grown into high forest, many ancient woods had been clear-felled and planted with non-native conifers, and wood-pastures had ceased to be grazed.

Planting trees to create woods on farmland started in a small way before 1600. During the 18th and 19th centuries, numerous small woods and belts were planted in the lowlands for landscape and sport; only in the 20th century did large-scale woodland creation through planting become common. At times, there was also extensive natural development of woodland on open ground, for example in post-Roman times on open ground on the southern Chalk and limestone lands (e.g. Dewar, c.1926). In the 20th century, as agriculture became more mechanized, secondary woodland again developed naturally on steep slopes and unused common pastures.

Since 1985, when a significant shift in forestry policy occurred (Foot, 2010) there has been a small retreat from this polarized state of affairs. Forms of traditional management have been revived, mainly in nature reserves and public open spaces, and plantation forestry has been diversified by a degree of reversion to native trees, mixed age structures and more restocking by natural regeneration. Landscape-scale approaches designed to maintain ecologically resilient, productive and visually harmonious woodland and plantations have been increasingly encouraged (see www.forestry.gov.uk/fr/habitatnetworks).

18.3 Historical Stages and Processes of Change

This broad history has often been presented as a three-stage process whereby pre-Neolithic original and natural forests were transformed by traditional management such as coppicing and changed again by modern forestry. This oversimplifies the processes and masks important regional differences in the character and timing of changes, the consequences of which are still seen in today's woods. Here, I recognize six stages (see Table 18.1), by adding: (i) a stage of exploited wildwood between natural woodland and traditionally managed woodland; (ii) a late, improvement stage in traditionally managed woodland; and (iii) a 'post-modern' forestry, based on restoring and recreating native and relatively natural woodland (Tsouvalis, 2000).

Forest history can thus be represented as a series of transitions (Rackham, 1980, 1986, 2006). The transformation of natural to

Table 18.1. Stages in the treatment of ancient woodland in Britain.

	State or stage	Brief description	Period
1	Original natural forest	'Wildwood' with only limited human influences, dominated by natural processes	Pre-Neolithic, lingering locally
2	Exploited wildwood	'Wildwood' informally exploited variously as mosaics of coppice, wood-pasture, wood-meadow	Mostly Neolithic to medieval
3	Traditionally managed woodland	Managed woodlands and forests treated as either coppice or wood-pasture	Pre-Roman to 20th century
4	Improved, traditionally managed woodland	Coppices and wood-pastures improved for particular markets, usually by planting locally native species	17th–19th centuries
5	High forest plantations	High forest managed for timber and amenity, usually even-aged and established by planting; latterly dominated by introduced conifer species	18th century onward
6	New native woodland	Woodland restored to native stands by natural succession or planting after clearance or treatment as plantation high forest	Late 20th century onward

exploited wildwood (stages 1 to 2 in Table 18.1) involved settlement, cultivation, extensive pasturage and a great deal of forest clearance. Exploited wildwood (stage 2), in which there was little differentiation between wood-pasture and coppice, prevailed well into historical times, but was increasingly partitioned into clearly bounded coppices and wood-pastures (stages 2 to 3). Some traditional wood-pastures and coppices (stage 3) survived as such into recent times, but others were transformed in the 18th and 19th centuries (stages 3 to 4) in response to changing markets, either by conversion into high forest (usually of beech or oak), or by improving the coppice by weeding out the less useful species and planting the most useful. Meanwhile, many wood-pastures were rendered virtually treeless by timber exploitation and sustained grazing.

Planting on open ground accelerated in the 18th and 19th centuries and became the dominant process in the 20th century. Native trees and shrubs in ancient woods were increasingly replaced by plantations for timber production (stages 3 and 4 to 5), initially mainly mixtures of broadleaves and conifers, but eventually almost wholly of conifers. Latterly, some of these plantations have been removed in order to restore native mixtures and more natural structures (stages 5 to 6), especially where coppices had never passed through an improved stage (stages 3 directly to 5). Starting somewhat earlier in the 20th century, new native woodland also developed on land left unused by modern agriculture (stage 6).

Today, Britain has woodland at all stages from 3 to 6 and even small remnants of stage 2, but no original and natural woodland (stage 1) survives. Some ancient woods have passed successively through each stage, though many remain at stage 5 or neglected versions of stages 3 or 4. The transformation from stages 5 to 6 has been accompanied by an increased willingness to accept naturalized trees as de facto natives, especially sweet chestnut (*Castanea sativa*) and sycamore, as well as native species outside their native ranges, particularly beech and Scots pine. So the new native woodland (stage 6) is significantly different in composition from the original natural woodland (stage 1).

18.4 Regions

The histories of adjacent woods may differ as a result of the different circumstances, ideas and actions of their individual owners, although the existence of large land ownerships tends to generate clusters of woods with similar histories. At a larger scale, there are also regional differences of emphasis in the timing of transitions and the character of the stages they produced.

The traditional division of British vegetation is between: (i) lowland types concentrated in the south and east; (ii) upland types located mostly in the Highland Zone of Palaeozoic rocks in the north and west; and (iii) boreal types located mainly in the eastern and northern Highlands of Scotland; distinctions that are broadly reflected in the National Vegetation Classification (Rodwell, 1991). Here, I also insert a 'Borderland' Zone comprising the less oceanic east of the Highland Zone and some enclaves in the Lowland Zone (Peterken, 2008) and distinguish a South-east Lowland Zone which has always been well wooded by British standards (Fig. 18.1). The characteristic geology, land form, climate, farming and ancient woodland cover of these regions/zones are summarized in Table 18.2. The zones do not have sharp boundaries; rather, they merge and mix in response to climate, site characteristics and landscape history.

18.5 Pre-Neolithic Wildwood

The precise character of the original natural forest has been hotly debated. There was a widespread assumption that closed forests dominated the land, relieved only by the open water of rivers and lakes, the highest mountain tops and open ground around the coast, in valley mires and as local openings associated with inland cliffs and thin soils on steep slopes. The structure of this forest would have varied in response to several kinds of natural disturbance, particularly windthrow, and at any moment a proportion of the forest would have been temporarily open (Peterken, 1996); but even greater openness, maintained by large herbivores, has been implied by the pattern of

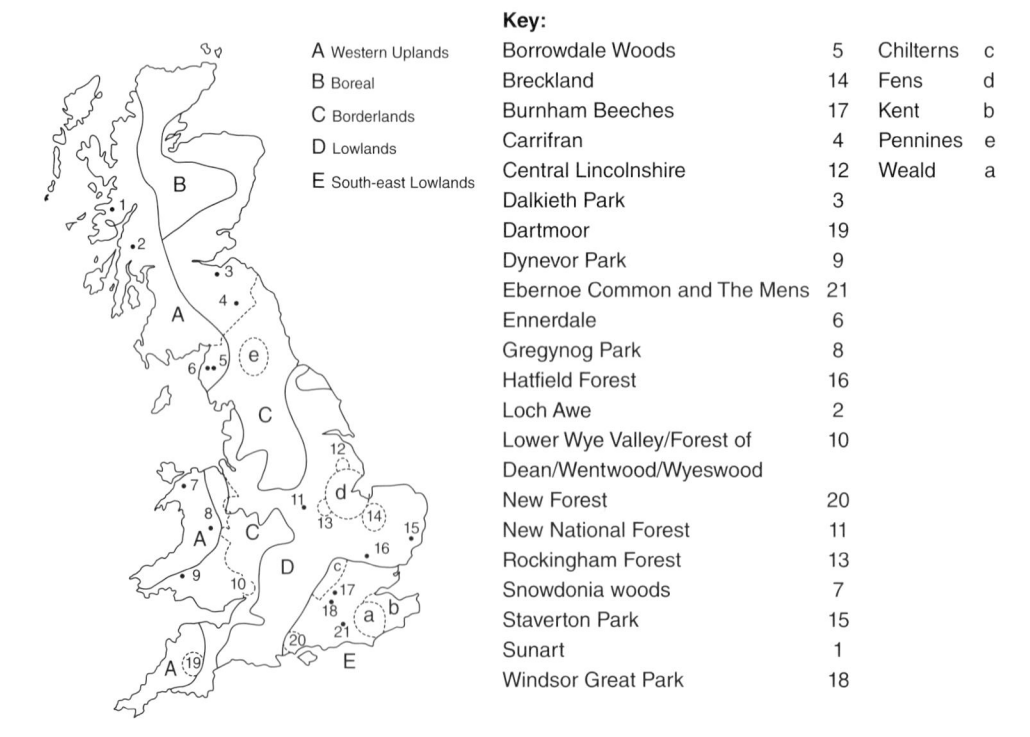

A Western Uplands	Borrowdale Woods	5	Chilterns c
B Boreal	Breckland	14	Fens d
C Borderlands	Burnham Beeches	17	Kent b
D Lowlands	Carrifran	4	Pennines e
E South-east Lowlands	Central Lincolnshire	12	Weald a
	Dalkieth Park	3	
	Dartmoor	19	
	Dynevor Park	9	
	Ebernoe Common and The Mens	21	
	Ennerdale	6	
	Gregynog Park	8	
	Hatfield Forest	16	
	Loch Awe	2	
	Lower Wye Valley/Forest of Dean/Wentwood/Wyeswood	10	
	New Forest	20	
	New National Forest	11	
	Rockingham Forest	13	
	Snowdonia woods	7	
	Staverton Park	15	
	Sunart	1	
	Windsor Great Park	18	

Fig. 18.1. Map of regions/zones of British vegetation (A–E); see Table 18.2 for summary descriptions of each region/zone. As indicated by the key, places and areas referred to in the text are shown on the map by the numbers 1–20 and the letters a–e, respectively.

Table 18.2. Summary descriptions of regions/zones.

Region	Description
Borderland	Irregular belt to east of Oceanic uplands. Mainly Palaeozoic rocks, but also younger hard limestones; fairly warm, fairly moist; mixture of rolling hills with ravines and gently undulating land; pastoral farming becoming increasingly arable. Ancient woods mostly associated with ravines, valley sides and higher ground.
Boreal	Eastern Highlands to northern Scotland. Palaeozoic rocks; cool, severe winters; mountains, hills and valleys; pastoral farming and moorlands used for sport. Ancient woods extensive within some valleys, but sparse to north.
Lowlands	English lowlands. Younger strata; relatively warm, dry climate with winter frosts; flat, gently undulating land; arable farming. Ancient woods thinly scattered, but with clusters.
South-east Lowlands	South-east England south and east of the Chalk escarpments. Relatively warm, dry climate; rolling hills and flat, gently undulating land; densely developed mixed with largely arable farming. Relatively high density of ancient woods, except on the Chalk and near the coast.
Western Uplands	Western seaboard. Palaeozoic rocks; mild, wet, oceanic climate; mountains, hills and valleys; pastoral farming and forestry. Ancient woods mainly on valley sides.

epiphytic lichens (Rose and James, 1974). More recently, Vera (2000) promoted the idea that large herbivores maintained a dynamic mosaic of scrub, glades and open-canopied stands of mature trees, leaving only limited parts as closed groves, perhaps mainly on waterlogged ground, marshland and infertile, thin soils (see also Svenning, 2002; Kirby and Watkins, Chapter 3).

Whatever its structure, strong regional differences developed in the composition of the wildwood. Birks et al. (1975) divided Britain into provinces based on tree pollen evidence. Their Lime province corresponds with the Lowland and South-east Lowland Zones combined, and their Birch and Pine provinces correspond with the Boreal Zone. The combined Oak–Hazel and Hazel–Elm Provinces broadly correspond with my Borderland and Western Upland Zones. Thus, the underlying patterns of individual tree species established 5000 years ago largely survive to the present time (Table 18.3).

18.6 Exploited Wildwood

Neolithic agriculture and settlement expanded the area of open ground for cultivation and grazing and modified the remaining woodland by various forms and intensities of use. The degree of clearance and the timing of the transition from predominantly wooded to predominantly open varied regionally (Rackham, 2003, 2006)

Breckland and the chalk downlands of southern England were cleared early, to the degree that secondary woodland was forming in those areas in the Neolithic (Rackham, 2006). Southern parts of the uplands were largely cleared in the Bronze and Iron Ages (Moore, 1968; Simmons, 2003), along with many flood plains, but the northern uplands remained well forested into Roman times (Turner, 1965). Prehistoric clearance thus left some well-wooded districts in the Lowlands and Borderland, and extensive well-wooded districts within the South-east Lowlands (the Weald, Chilterns).

The remnant woodland was probably de facto wood-pasture, i.e. subject to extensive grazing through into early historic times, in which trees were lopped for fodder and the woodland was functionally a form of meadow. The Somerset trackways show that pole regrowth from stumps was being exploited, which implies that coppicing took place (Rackham, 1980), and the analysis of pollen deposits implies that coppicing was well established in Neolithic East Anglia (Waller et al., 2012).

18.7 Traditional Woodland Management

By the end of the Roman period, the lowland clay lands had been extensively settled and

Table 18.3. Current occurrence of forest dominant tree species compared with pre-Neolithic provinces.

	Boreal	Western Uplands	Borderland	Lowlands	South-east Lowlands
Wildwood province[a]	Birch; Pine	Oak–Hazel; Hazel–Elm	Oak–Hazel; Hazel–Elm	Lime	Lime
Alnus glutinosa	Frequent	Common	Common	Frequent	Common
Betula pendula, B. pubescens	Abundant	Common	Common	Frequent	Frequent
Carpinus betulus	–	–	Very localized	Localized	Abundant
Corylus avellana	Rare	Local as dominant	Underwood	Underwood	Underwood
Fagus sylvatica	–	Almost all introduced	Localized	Very localized	Abundant
Fraxinus excelsior	Localized	Abundant	Abundant	Abundant	Abundant
Pinus sylvestris	Abundant	Localized	Localized	Probably introduced	Probably introduced
Quercus petraea	Occasional	Abundant	Common	Localized	Localized
Quercus robur	Rare	Frequent	Common	Abundant	Abundant
Tilia cordata	–	Rare	Frequent	Locally abundant	Very localized
Tilia platyphyllos	–	–	Localized	Very rare	Rare
Ulmus glabra	Rare	Common	Common	Occasional	Frequent

[a]Corresponding province in the wildwood, as summarized by Rackham (2003).

the distribution of woodland was very similar to that in the 11th century (Rackham, 1980; Hooke, 2010). The transition from exploited wildwood to 'traditional' management involved the segregation of coppice woods from the wood-pastures, thereby preventing new growth in the coppices from being damaged by grazing. The change was driven by the need to increase productivity as populations increased and woodland of all kinds was progressively cleared to pasture and cultivation. The pasture woodlands were themselves segregated into forests, chases, parks and the remaining wooded commons.

In the Lowlands, coppicing appears to have remained subordinate until post-Norman times (Rackham, 2006). Anglo-Saxon perambulations only occasionally mention coppice, and in the Domesday Book (1086 AD) wood-pastures (*silva pastilis*) remained dominant in Nottinghamshire and Derbyshire, although 50% of all woodland in Lincolnshire was coppice (*silva minuta*). However, by 1250, most woods were coppices (Rackham, 1980) and were clearly distinguished from the remaining wood-pastures. Many of those that survived various bouts of clearance were coppiced into the 20th century (Box 18.1).

In the South-east Lowlands, wood-pastures were still extensive in 1086, when the Domesday Book recorded woodland only in terms of swine numbers. In the Weald, some woodland must have been coppiced to fuel the Roman iron industry (Rackham, 1986). In the well-wooded Chilterns (Roden, 1968), the pressure to partition coppice from wood-pasture was limited by the high density of woodland, but coppices were recorded by the 13th century and underwood increased in the enclosed woods of the north-east Chilterns, where the woodland was sparser than further south. In contrast, in the New Forest some medieval coppices were allowed to revert to wood-pasture.

Woodland pasturage remained dominant until much later in the Uplands. Pasturage was an important function of woods in Wales (Linnard, 2000) and in Scotland (Smout *et al.*, 2005). Cattle and other stock retained access to woods until the 19th century, with the woods providing shelter in winter and even hay in summer. Summer regrowth of trees and shrubs

> **Box 18.1.** Central Lincolnshire, part of the Lowland Zone.
>
> Central Lincolnshire is a low-lying district of clays, sands, alluvium and peats on the eastern seaboard, through which the rounded hills of the Chalk Wolds and Limestone Heath run north to south. The land is intensively cultivated. Most hedges and hedge trees have been removed, and most streams and rivers ditched and wetlands drained. Woodland and plantations occupy just 5% of the land. Most are small, conspicuous and sharply defined islands in a matrix of arable land.
>
> Roughly half the woodland is ancient, managed in the past as coppices dominated by small-leaved lime (*Tilia cordata*), ash (*Fraxinus excelsior*), birches (*Betula* spp.) and hazel (*Corylus avellana*), with alder (*Alnus glutinosa*) on the few remaining wet peats, all with pedunculate oak (*Q. robur*) standards. Coppicing lapsed by the mid-20th century, so many woods were either converted to plantations, cleared to arable farmland or left unmanaged. Much of the more recent woodland was planted on farmland in the 18th and 19th centuries as small fox coverts and as narrow and elongated belts within designed landscapes around the great country houses. This became possible with the advance of enclosure, which broke up the common fields and the extensive sheep walks of the Wolds and Heath. Most of these new woods were treated as high forest and only a little as coppice. Into this small-scale patchwork, large plantations were inserted during the mid-20th century, mainly on unproductive sandy heaths and poorly drained clays.
>
> Apart from a few parklands around the great houses, wood-pasture is now almost unknown. However, in the Medieval period, the Domesday Book records both *silva minuta* (coppice wood) and *silva pastilis* (wood-pasture) in roughly equal proportions. The records left by the numerous monasteries and other landowners make it clear that the woods were not only well stocked with a variety of timber – the same species as now, but that many, if not most, were coppiced, although they were also used as pasture. Woodland regeneration required that such grazing be carefully controlled and penalties were imposed when the controls broke down.
>
> In 1259 (Foster, 1920), the Abbot of Kirkstead granted Robert of Tateshale access to Bracken Wood to hunt, but he had to call at the monastery for the key. If this was not forthcoming, he was allowed to break the lock on the gate and enter. He was expressly forbidden to enter anywhere except by one of the gates, and would have to make good any damage to the wood itself. Moreover, if his animals ran illegally into neighbouring woods, Robert had to call them back 'by mouth and horn' while waiting outside the wood.
>
> A year later, another tenant had to come to the abbey gates and 'swear upon the most holy relics' that he would not let his cattle enter the woods. In 1249, the same abbot had granted Philip Margun the right to enclose with a hedge a third part of the wood situated in the pasture of Buckenale and Burrethe whenever 'they wish to cut it, and to keep it enclosed for six years, but after six years, the said third of the wood shall be common to the vills of Buckenale and Burrethe, so that two-thirds of the wood shall always remain unenclosed and commonable'. In 1431, Robert Slyngesby spent five days remaking hedges round the lord of the manor's wood in Coningsby and blocking gaps (Cragg, 1912).
>
> Hints of the enduring character of woodland pasturage emerge in the assessment of Thomas, Earl of Kent's possessions in 1401: there was no profit of pasturage or pannage in his woods of Minting Park because 'the tenants of the manor have common of pasture there with their beasts throughout the year and always have had' (Calendar of Inquisitions Miscellaneous (Chancery) 7 (1399–1422), No.53, Public Record Office. HMSO, London).

remained an important source of fodder. The main surge of coppicing came in the 18th and 19th centuries as part of the improvement phase.

In the Borderland, coppicing was recorded in medieval woods owned by Cistercian Monasteries in South Wales (Linnard, 2000) and lowland Scotland (Lindsay, 1980), while medieval wood-pastures in the Dean encompassed coppice as well as groves of ancient trees (Hart, 1966). Woods around the Wye Gorge were partitioned in the 16th and 17th centuries between coppice and wood-pasture (Box 18.2).

Coppice management changed during the centuries of traditional management. Rotations were generally 4–8 years before 1400, but lengthened to 10–20 years by the 18th century, possibly due to declining fertility arising from repeated cropping and a change from faggots to fuel domestic hearths (Rackham, 1980). Cycles were initially irregular, but latterly some coppice owners aspired to cut fixed proportions of their woods each year.

> **Box 18.2.** Lower Wye Valley, part of the Borderland Zone.
>
> The Lower Wye valley and its surroundings form one of the few districts in Britain that is and always has been well wooded (Peterken, 2008). Growing on a range of limestones, sandstones, conglomerates and other hard rocks, the ancient semi-natural woods comprise mixtures of ash (*Fraxinus excelsior*), beech (*Fagus sylvatica*), small-leaved lime (*Tilia cordata*) and sessile oak (*Quercus petraea*), with an underwood of hazel (*Corylus avellana*), holly (*Ilex aquifolium*), yew (*Taxus baccata*), service (*Sorbus torminalis*) and rowan (*S. aucuparia*), and an admixture of birches (*Betula* spp.) and gean (*Prunus avium*), supplemented on the limestones with large-leaved lime (*T. platyphyllos*), wych elm (*Ulmus glabra*), field maple (*Acer campestre*) and whitebeam (*S. aria*). Pedunculate oak (*Q. robur*) is largely confined to heavy flood plain soils, while alder (*Alnus glutinosa*) dominates around seepages and flood plain swamps. On the most base-poor conglomerates, birch–sessile oak woodland develops.
>
> By the 17th century, the valley was lined with coppice woods, felled regularly to provide, inter alia, charcoal for the local metal industries, but to both east and west stretched very large wood-pastures within which only a small proportion of the woodland was enclosed and coppiced – Wyeswood and Wentwood to the west and the Forest of Dean to the east. The post-medieval distinction between the coppice woods of the Wye Valley and the wood-pastures on either side apparently developed in the 16th and 17th centuries.
>
> Medieval records of the Forest of Dean indicate mosaics of coppice, high forest and wood-pastures stocked with ancient and decaying trees, with only limited stock-proof boundaries to regulate pasturage and regeneration. By 1608, Hadnock Wood, where the Dean woods reached the Wye valley, was a coppice and, likewise, parts of the great Wyeswood in Monmouthshire, both of these having been wood-pastures earlier.
>
> Coppet Hill Wood and Doward were both medieval woods subjected to common grazing, but whereas the former continued to be grazed until woodland remained only on the inaccessible crags, an agreement on the Doward in 1718 enabled what is now Lords Wood to be treated as mixed coppice.

Rotations may have been shorter in the Lowlands and South-east Lowlands, simply because growth was faster in warmer summers and on more fertile soils. Certainly, rotations in later years were longer in the north around 1800 (Jones, 1961), but by then rotations were influenced by industrial markets.

Grazing, which had once been permitted in all woodland, was first controlled by excluding stock and/or deer for 5–7 years after felling in order to let fresh shoots grow out of reach, then excluded altogether. Grazing lingered longer in the South-east Lowlands, where grazing or grazing rights continued in some instances into the 20th century. Grazing also remained important in the west and north until the era of coppice improvement.

18.8 Parks, Forests and Wooded Commons

Many wood-pastures were incorporated into parks and forests, which were expanded by the Normans and reached their heyday in the 14th century. Only part of the emparked and afforested ground was wood-pasture – less than half in the case of forests – and much was either open heath and grassland or woodland treated as coppice. The wooded commons were the residue, subject to various rights to use trees, bushes and ground vegetation for fuel, building, bark, pasture, fodder and pannage, and again with varying degrees of tree cover.

Parks were commonest in well-wooded districts. Nationally, they were scarce in the Western Uplands, and scattered through the northern Borderlands and poorly wooded parts of the Lowlands, but common in the English Borderlands and the South-east Lowlands (Cantor and Hatherley, 1979; Rackham, 1986).

Forests were commonest in the Borderlands, including the Scottish lowlands, sparse throughout the Western Uplands and thinly scattered through the Lowlands and South-east Lowlands. Most of the Lowland forests were compartmented and used as coppice, with just a small wood-pasture component. Wooded commons survived best in well-wooded districts of the South-east Lowlands and southern Borderlands.

Trees were commonly pollarded, which enabled the crowns to be cropped and to regenerate. Nevertheless, the trees and shrubs of the wood-pastures tended to become sparse, especially on commons and wood-pasture forests subject, as in the 17th century Forest of Dean for example, to massive timber sales (Hart, 1966). Changing grazing regimes might permit episodes of regeneration (Peterken and Tubbs, 1965).

18.9 Improved Traditional Management

Improvement to the silvicultural treatment and composition of traditionally managed woods became common in the 18th and 19th centuries in order to generate greater productivity from the most useful native species and respond to changing demands for wood and timber (Jones, 1961). At the same time, common fields and pastures were being enclosed, which afforded opportunities to create new woodland by planting for timber, shelter, amenity and sport, thereby evening out the distribution of woodland in the Lowlands. Tree species were introduced from abroad, often in conifer broadleaf mixtures; and native trees, notably beech, pine and pedunculate oak (*Q. robur*), were redistributed. While in many woods coppice improvement was limited to little more than filling gaps with ash (*Fraxinus excelsior*) and other useful species, the composition of other woods was changed radically.

Regional differences developed in: (i) the amount of conversion of coppice to high forest; (ii) modifications of parkland wood-pastures; and (iii) the distribution and amounts of new woodland; but (iv) particularly in the incidence and choice of species for improving coppices. In the South-east Lowlands, the planted species were mostly sweet chestnut, hazel (*Corylus avellana*), ash and alder (*Alnus glutinosa*), i.e. much the same species as were there already, but concentrated to conform with local markets (Box 18.3). In the Western Uplands, the principal species were sessile oak (*Q. petraea*) and, to a lesser degree, pedunculate oak, which yielded charcoal for the metal industries and bark for tanning leather. In both regions, coppices were also improved by removing unwanted species from mixtures, in particular hazel from the oak woods of the Uplands and maple (*Acer campestre*) and ash from the hazel coppices in parts of the South-east Lowlands. In the Western Uplands, new oak coppices were also planted (Box 18.4).

The conversion of mixed coppices and wood-pastures to high forest of native species was determined more by local than regional factors, in some cases happening by neglect, in others through deliberate action. Oak in particular was widely planted in places as different as valley sides in west Wales, and lowland forests such as Salcey in Northamptonshire. Mixed coppices in the Chilterns were promoted to beech-dominated high forest

---

Box 18.3. Kent and the Chilterns, South-east Lowlands.

Coppice improvement in Kent was recorded by Boys (1794). Twelve of 26 woods mentioned had been improved, principally by planting chestnut (*Castanea sativa*), ash (*Fraxinus excelsior*) and willow (*Salix* spp.). The 'produce seemingly natural' (i.e. the species being replaced) comprised mixtures of oak (*Quercus* spp.), ash, hazel (*Corylus avellana*), birch (*Betula* spp.), beech (*Fagus sylvatica*), hornbeam (*Carpinus betulus*) and occasionally aspen (*Populus tremula*). The vast Bridge Woods were 'beginning to improve by ash and chestnut plants' and no doubt many more woods were improved in the 19th century. Ash was planted in six woods where it already grew naturally, but this was in fact a widespread measure, for gaps in coppices throughout the lowlands were often made good by planting ash and other species.

At the same time, equivalent improvements were under way in the woods of the Chilterns. Medieval woods were a mixture of oak, ash and beech (Roden, 1968), and many of the common woods were wood-pastures, but with the decline of the London firewood market and the rise of furniture manufacturing in High Wycombe, many of the woods were converted to beech-dominated high forest and managed on a form of selection system (Mansfield, 1952).

> **Box 18.4.** West Wales, part of the Western Uplands Zone.
>
> The oceanic western half of Wales is now characterized by sessile oak (*Quercus petraea*) woods mostly on steep valley sides on base-poor and often thin soils; ash–elm (*Fraxinus excelsior–Ulmus glabra*) and alder (*Alnus glutinosa*) woods on the deeper, more fertile soils of lower slopes and along watercourses; and alder-dominated woods on flood plains and flushed slopes, often with an admixture of ash and pedunculate oak (*Q. robur*). From the 18th to the early 20th centuries they had been managed as coppices with stock-proof walls, hedges and banks on their boundaries, some as simple coppice, others as coppice with oak standards, and some as even-aged high forest. However, for most of the 20th century, these woods served only as shelter and supplementary grazing for sheep.
>
> By the late 20th century, the oak woods all looked mature and seemingly natural, and were easily regarded as remnants of the original forest cover, but pollen analysis indicates that most have changed a good deal during recent centuries and that each wood has its own individual history. For example, Edwards (1986) concluded that Coed Cymerau was open woodland in the 17th century, but became closed woodland at various dates in the 19th century. Coed Llenyrych had been well wooded until the 17th century, was largely cleared in about 1630 and only returned to coppice between 1880 and 1930. Coed Ganllwyd was well wooded in the medieval period, but the oaks were felled in about 1660 and 1840. They recovered via birch (*Betula* spp.)-dominated phases on both occasions, the first time by natural succession, the second through planting. Likewise, woodland cover was sharply reduced at Coed y Rhygen in the 17th century and the late medieval period, the change being associated with farming. This wood has been an open wood-pasture for at least 1600 years, and it still contains 300-year-old open-grown oaks.
>
> More generally, Linnard (2000) reported that the forests and woods of upland Wales provided pasturage for a variety of livestock in medieval times, as well as pannage for pigs, while trees were also pollarded for fodder. Land held by the Cistercian monasteries at this time included woodland, some at least of which was maintained systematically as coppice or high forest. Elsewhere, woodland was in decline, especially in the west, leaving, for example, almost no woodland in the vicinity of Strata Florida abbey by the 16th century. The legacy of prolonged woodland pasturage is the scatter of trees on the hills – most of which are in protected locations on outcrops or by streams, a limited number of large trees within the higher level farmland, and a few wooded parks at, for example, Gregynog and Dynevor.
>
> The west Wales history of prolonged and extensive wood-pasturage giving way to coppice and high forest in the 18th century is a general feature of native woodland in the Western Upland zone. In the western Highlands around Loch Sunart, oak woods were extensively planted from the late 18th century onward, and many were coppiced before being allowed to grow to high forest. This planting was often inserted into a matrix of open wood-pasture of oak, ash, alder, birch and hawthorn (*Crataegus* spp.), some of which survives to this day. Around Loch Awe, the woodland is now dense and dominated by oak, but for most of the last millennium it was an open mixture of alder, birch, oak and hawthorn (Sansum, 2005).
>
> In Cumbria, the oak-dominated Johnny's Wood and Borrowdale Great Wood were planted, having previously been mixed stands with oak, ash and holly (*Ilex* spp.), which survived unchanged at Seatoller Wood. Likewise, oak coppices were widely planted in Scotland (Lindsay, 1980). In the south-west, the many valley-side oak woods of Somerset, Devon and Cornwall are all remarkably pure, seemingly planted to yield timber, charcoal (for smelting) and bark (for tanning). Even the famous high-altitude pedunculate oak woods of Dartmoor appear to be recent, though in the case of Wistman's Wood, the accumulation of depauperate (stunted, contorted) oaks was probably a natural process (Mountford *et al.*, 2001).

and treated by a system that would now be called 'continuous cover', yielding timber of all sizes for furniture manufacturers. Beech, notoriously weak to coppice in Britain, was also promoted to, or planted as, high forest elsewhere in the South-east Lowlands, parts of the Cotswolds (Lowlands) and the southern Borderlands, and more locally through the Western Uplands (Boxes 18.3 and 18.4). Oak planting as high forest was more evenly spread.

Wood-pastures were also modified. Trees on wooded commons were felled with no provision for regeneration. Parklands became less 'meat on the hoof' and more embellishments to country estates to be improved and expanded by tasteful plantings, often of exotic species. New parks were created by design. The remnants of the wooded Royal Forests were disafforested, leaving the wooded parts as ordinary coppices, while the open ground became farmland.

Those wood-pastures that survived – notably the New Forest and the Forest of Dean – were substantially enclosed and planted with oak. The surviving traditional wood-pastures were concentrated in the Borderland and South-east Lowlands. However, until the mid-20th century, trees within fields and on field boundaries remained common and must have increased with enclosure, so farmland looked much like parkland. This resemblance would be reinforced by the continued pollarding and shredding of farmland trees. Even in the Western Uplands, boundary trees could be found up to the limits of enclosure and sometimes beyond.

18.10 Plantations

There was some planting for timber production from the 17th century onwards (Edlin, 1956), but it was only after 1919, with the establishment of the Forestry Commission, that extensive planting became national policy. A massive expansion of plantations followed. Until the 1950s, oak, beech and other broadleaves were commonly included in the plantings, but eventually conifers dominated so completely that 99% of plantations established in the 1970s were coniferous (see Quine, Chapter 15).

Plantations in ancient woodland were concentrated in the larger individual woods and clusters of woods. The small, scattered woods in the Lowlands largely escaped. In the Borderlands and Western Uplands, some ancient woods were engulfed in afforestation schemes, but were spared complete change themselves. Only in the southern Borderlands and some of the Boreal pinewoods were large tracts of ancient woodland felled and replanted.

Afforestation of open ground, primarily with conifers initially, took in the heaths and low-grade farmland in the Lowlands and South-east Lowlands, but after 1945 it was concentrated in the Western Uplands, the northern parts of the Borderlands and the Boreal region. Extensive plantations were created on hill pastures, notably in the northern Pennines, Galloway and the Highlands.

These changes were contemporary with the modernization of agriculture, which led to countless hedges being removed and farmland trees being felled and not replaced (including those of many former parks) (Peterken and Allison, 1989). Many ancient woods were destroyed to expand farmland and the remaining coppices were largely neglected. None the less, some land was left unused, notably pastures on commons, steep slopes and other kinds of difficult ground, and here woodland colonized naturally. Agricultural mechanization thus generated a high-definition landscape as diffuse boundaries were replaced by sharp boundaries. The impacts were strongest in the Lowlands, but they were felt everywhere.

18.11 Revival and Restoration of Native Woodland

Many coppices and some wood-pastures survived into modern times in an unimproved condition: Lowland and southern Borderland coppices were still being cut regularly into the 20th century. A few are still cut today, albeit mostly self-consciously as nature reserves and usually following a period of neglect. Wood-pastures have mostly been abandoned or improved: those that survive are often infilled with younger trees and are no longer pollarded, as in the New Forest, Ebernoe Common and The Mens. In the lower Wye valley, several small-leaved lime–beech–oak common wood-pastures were cut and grazed well into the 20th century. Exceptionally, a few wood-pastures survive today, neither improved nor infilled, e.g. Windsor Great Park, Dalkeith Park and Staverton Park.

Public reaction against planted conifers, combined with lower prices for timber, led in the 1980s to programmes to diversify the new plantations and to reverse the 'coniferization' of ancient woods. There is still strong support for an increase in forest cover, recognizing that Britain is still poorly wooded in comparison with mainland Europe, but with a wider range of objectives than just wood production (e.g. Forestry Commission Wales, 2009). Additional woodland can improve the appearance of the landscape, benefit wildlife by restoring networks of forest habitats and rendering ecological services, such as flood control, reduced erosion and carbon sequestration (see the UK National Ecosystem Assessment, at http://uknea.unep-wcmc.org/Resources/tabid/82/Default.aspx). While this interest has encouraged more use of native woodland, the concept of 'native species' has broadened subtly with greater acceptance of naturalized species, such as sycamore and sweet chestnut, and natives beyond their natural range, notably beech (Peterken, 2001).

The ongoing reversal of the conversion of ancient woods to high forest has probably been fairly evenly spread across regions. The changes were driven by national-scale economics and nationally available incentives, and were implemented by the Forestry Commission under national policies. Geographical differences in the changes arise from the proportion of state-owned forests, the size and amounts of woodland (and thus the incentive for forest management) and the particular choices of individual large forestry estates.

Likewise, the creation of new native woodland on farmland has not been related to regions. Rather, national-level decisions to create a New National Forest (see www.nationalforest.org) and new urban-fringe forests have been complemented by large private initiatives and coordinated projects designed to reconnect particular clusters of woods. Some of this new woodland forms part of 'rewilding' projects, such as the notable enterprises in Ennerdale, which also involve naturalistic grazing (Hodder and Bullock, 2009) and at Carrifran (Ashmole and Ashmole, 2009). Such projects seem likely to be concentrated in the Western Uplands, with smaller projects in the Borderlands and South-east Lowlands.

Many former wood-pastures and surviving wooded commons have become dense woodland, in effect developing into the new wildwoods. At other sites, existing pollards have been lopped again and new ones have been started, e.g. Hatfield Forest and Burnham Beeches. More attention has also been paid to veteran trees elsewhere in the landscape (Read, 2000). Coppice restoration in nature reserves has, however, been hampered by the burgeoning deer populations that damage much of the regrowth.

Some regional distinctions can still be seen in the above. The South-east Lowlands region has a high density of ancient woodland and an urban-based population strongly influenced towards amenity forestry, and retained a few markets for coppice management until well after 1945. Here, there was relatively little coniferization and therefore correspondingly less need for restoration. In the rest of the Lowlands, restoration has been concentrated in the densely wooded districts, e.g. Rockingham Forest.

In the Borderland, the dense ancient woodland of the southern Marches contrasts with the immense conifer plantations of the northern Pennines and eastern Highlands. The Forestry Commission has been strongly involved throughout, which has led to a high incidence of both coniferization and restoration. In the Western Uplands, the substantial use of ancient woods as sheltered pasture meant that they were not so subject to conversion to plantations.

18.12 Some Consequences of Differences in Regional History

Broadly, the composition of semi-natural woods in the Lowlands and Borderlands is more natural than that of the Western Uplands and South-east Lowlands. In the former regions, stands are mostly mixtures of several tree and shrub species whose precise combinations respond to variations in site conditions, whereas woods in the latter regions tend more to be dominated by the one species that was most useful 150–200 years ago – beech in the Chilterns, hazel in the southern chalklands, sweet chestnut in the south-east and sessile oak in the Western Uplands.

The impact of pasturage in the woods likewise varies. In the Western Uplands, historic and recent grazing has turned the ground vegetation of many woods into pasture and heath, and conspicuously limited many woodland herbs and some trees to outcrops, ravines and other protected locations. Similar impacts can be seen, but less often, in southern Borderland woods, but here the soils tend to be more base rich, many woods had extensive protected locations on bouldery ground and more time has generally elapsed since grazing ceased.

The impact of grazing in the Lowlands and South-east Lowlands is difficult to assess; it may once have been considerable, but it must have been alleviated by the need for periodic tree regeneration. Moreover, Lowland and Borderland woods may have had a closer association with meadows, which would also have provided refuges for some woodland species. Nevertheless, the widespread elimination of lime from much of the Lowlands and South-east Lowlands must be due partly to historic and prehistoric grazing (Grant et al., 2011). The modern spread of deer throughout the Lowlands and into the southern Borderlands, combined with the enduring presence of red deer in the Highlands, is again reducing many woods to pastures (Fuller and Gill, 2001).

The species associated with mature and decaying trees in the original natural woodlands survived in wood-pastures and in the pollards associated with boundaries in coppices and farmland. Air pollution has reduced epiphyte populations on old trees across most of the country, while the loss of traditional tree management and of the veteran trees themselves has removed the habitat for many of the saproxylic invertebrates. The quality of lowland sites has been impaired.

A new element in regional differentiation (or in some cases homogenization) has been the expansion of forestry in the 20th century. The majority of British woodland came into being within the last 100 years, and it remains far from biological maturity (Quine, Chapter 15). Many woodland plants and animals have not colonized, or have colonized only the new woodland close to, ancient woodland (Peterken, 1981). Even on base-rich lowland soils, the developing assemblages are incomplete forms of communities found in nearby ancient woods, and in upland plantation forests on base-poor soils the vegetation remains closer to moorland vegetation than woodland. The differential impacts of climate change and tree diseases will presumably add another layer to the regional distinctiveness of our woodland, but we cannot yet know what form this will take.

References

Ashmole, P. and Ashmole, M. (2009) *The Carrifran Wildwood Story*. Borders Forest Trust, Jedburgh, UK.
Birks, H.J.B., Deacon, J. and Peglar, S. (1975) Pollen maps for the British Isles 5000 years ago. *Proceedings of the Royal Society of London B: Biological Sciences* 189, 87–105.
Boys, J. (1794) *General View of the Agriculture of the County of Kent*. Board of Agriculture, London
Cantor, L.M. and Hatherley, J. (1979) The medieval parks of England. *Geography* 64, 71–85.
Cragg, W.A. (1912) Conyngesby Compotus Roll A.D.1431. *Lincolnshire Notes and Queries* 12, 70–76.
Dewar, H.S.L. (c.1926) The field archaeology of Doles. *Papers and Proceedings of the Hampshire Field Club and Archaeological Society* 10, 118–126.
Edlin, H.L. (1956) *Trees, Woods and Man*. The New Naturalist 32, Collins, London.
Edwards, M.E. (1986) Disturbance histories of four Snowdonia woodlands and their relations to Atlantic bryophyte distribution. *Biological Conservation* 37, 301–320.
Foot, D. (2010) *Woods and People: Putting Forests on the Map*. The History Press, Stroud, UK.
Forestry Commission Wales (2009) *The Welsh Assembly Government's Strategy for Woodlands and Trees*. Available at: http://wales.gov.uk/docs/drah/publications/090324woodlandsforwalesstrategyen.pdf (accessed 8 December 2014).
Foster, C.W. (1920) *Final Concords of the County of Lincoln, A.D.1242–1272 with Additions A.D. 1176–1250*, Vol. 2. Lincoln Record Society, Lincoln, UK.
Fuller, R.J. and Gill, R.M.A. (2001) Ecological impacts of increasing numbers of deer in British woodland. *Forestry* 74, 193–199.
Grant, M.J., Waller, M.P. and Groves, J.A. (2011) The *Tilia* decline: vegetation change in lowland Britain during the mid and late Holocene. *Quaternary Science Reviews* 30, 394–408.
Hart, C.E. (1966) *Royal Forest. A History of Dean's Woods as Producers of Timber*. Clarendon Press, Oxford, UK.

Hodder, K.H. and Bullock, J.M. (2009) Really wild? Naturalistic grazing in modern landscapes. *British Wildlife* 20(Supplement), 37–43.
Hooke, D. (2010) *Trees in Anglo-Saxon England: Literature, Lore and Landscape*. Boydell Press, Woodbridge, UK.
James, N.D.G. (1981) *A History of English Forestry*. Blackwell, Oxford, UK.
Jones, E.W. (1961) British forestry in 1790–1813. *Quarterly Journal of Forestry* 55, 36–40, 131–138.
Lindsay, J.M. (1980) The commercial use of woodland and coppice management. In: Parry, M.L. and Slater, T.R (eds) *The Making of the Scottish Countryside*. Croom Helm, London, pp. 271–289.
Linnard, W. (2000) *Welsh Woods and Forests. A History*. Gomer, Llandysul, UK.
Mansfield, A. (1952) The historical geography of the woodland of the southern Chilterns, 1600–1947. MSc thesis, University of London, London.
Moore, P.D. (1968) Human influence upon vegetational history in north Cardiganshire. *Nature* 217, 1006–1009.
Mountford, E.P., Backmeroff, C.E. and Peterken, G.F. (2001) Long-term patterns of growth, mortality, regeneration and natural disturbance in Wistman's Wood, a high altitude oakwood on Dartmoor. *Report and Transactions of the Devon Association for the Advancement of Science* 133, 227–262.
Peterken, G.F. (1981) *Woodland Conservation and Management*. Chapman and Hall, London.
Peterken, G.F. (1996) *Natural Woodland*. Cambridge University Press, Cambridge, UK.
Peterken, G.F. (2001) Ecological effects of introduced tree species in Britain. *Forest Ecology and Management* 141, 31–42.
Peterken, G. (2008) *Wye Valley*. The New Naturalist 105, HarperCollins, London.
Peterken, G.F. and Allison, H. (1989). *Habitat Change: Woodlands, Hedgerows and Non-woodland Trees*. Focus on Nature Conservation No. 22, Nature Conservancy Council, Peterborough, UK.
Peterken, G.F. and Tubbs, C.R. (1965) Woodland regeneration in the New Forest, Hampshire, since 1650. *Journal of Applied Ecology* 2, 159–170.
Rackham, O. (1976) *Trees and Woodland in the British Landscape*. Dent, London.
Rackham, O. (1980) *Ancient Woodland*. Arnold, London.
Rackham, O. (1986) *The History of the Countryside*. Dent, London.
Rackham, O. (2003) *Ancient Woodland*, New edn. Castlepoint Press, Dalbeattie, UK.
Rackham, O. (2006) *Woodlands*. The New Naturalist 100, HarperCollins, London.
Read, H. (2000) *Veteran Tree Management Handbook*. English Nature, Peterborough, UK.
Roberts, A.J., Russell, C., Walker, G.J. and Kirby, K.J. (1992) Regional variation in the origin, extent and composition of Scottish woodland. *Botanical Journal of Scotland* 46, 167–189.
Roden, D. (1968) Woodland and its management in the medieval Chilterns. *Forestry* 41, 59–71.
Rodwell, J.S. (1991) *British Plant Communities. 1. Woodlands and Scrub*. Cambridge University Press, Cambridge, UK.
Rose, F. and James P.W. (1974) Regional studies on the British lichen flora I. The corticolous and lignicolous species of the New Forest, Hampshire. *Lichenologist*, 6, 1–72.
Sansum, P. (2005) Argyll oakwoods: use and ecological change, 1000–2000 AD – a palynological–historical investigation. *Botanical Journal of Scotland* 57, 83–97.
Simmons, I.G. (2003) *Moorlands of England and Wales – An Environmental History 8000 BC–AD 2000*. Edinburgh University Press, Edinburgh, UK.
Smout, T.C., MacDonald, A.R. and Watson, F. (2005) *A History of the Native Woodlands of Scotland, 1500–2000*. Edinburgh University Press, Edinburgh, UK.
Spencer, J.W. and Kirby, K.J. (1992) An inventory of ancient woodland for England and Wales. *Biological Conservation* 62, 77–93.
Steven, H.M. and Carlisle, A. (1959) *The Native Pinewoods of Scotland*. Oliver and Boyd, Edinburgh, UK and London.
Svenning, J.C. (2002) A review of natural vegetation openness in north-west Europe. *Biological Conservation* 104, 133–148.
Tsouvalis, J. (2000) *A Critical Geography of Britain's State Forests*. Oxford University Press, Oxford, UK.
Tubbs, C.R. (1968) *The New Forest: An Ecological History*. David and Charles, Newton Abbot, UK.
Turner, J. (1965) A contribution to the history of forest clearance. *Proceedings of the Royal Society B: Biological Sciences* 161, 343–354.
Vera, F.W.M. (2000) *Grazing Ecology and Forest History*. CAB International, Wallingford, UK.
Waller, M., Grant, M.J. and Bunting, M.J. (2012) Modern pollen studies from coppiced woodlands and their implications for the detection of woodland management in Holocene pollen records. *Review of Palaeobotany and Palynology* 187, 11–28.

19 Forest Management and Species Composition: A Historical Approach in Lorraine, France

Xavier Rochel*
Département de Géographie, Université de Lorraine, Nancy, France

19.1 Introduction

Almost all of France's forest ecosystems bear the mark of centuries of human activities. Some human 'disturbances' are obvious in the landscape, such as many archaeological remains; others, including some kinds of forest stands, are recognizable only to forest specialists. In some very natural-looking woodland, many of the marks left by human activities are so discreet as to be almost unnoticed, and history, geography and other disciplines rely on several types of archives to understand them.

This contribution focuses on the early modern period to the present, and aims to show how the implementation of different silvicultural systems led to important, either intentional or unintentional, changes in forest composition. We concentrate on the better documented regions of north-east France, and specifically on the Lorraine region. Here, historical research has been carried out for more than a century and a half thanks to the establishment of the School of Forestry at Nancy in Lorraine in 1824.

19.2 The Study of Forest History in France

Until the 1960s, forest history in France lay in the domain of professional foresters, especially the renowned professors of the Forestry School in Nancy. The works of Gustave Huffel (1926), Lucien Turc and the like centred on the successes of their predecessors, and left social and ecological aspects aside. Michel Devèze (1961) was the first prominent historian to focus his work on the relations between woodland and society, and during the 1960s–1970s, forest history ceased to focus on foresters and began to embrace larger segments of the relations between wooded ecosystems and human societies.

An important quantitative turn took place during the 1980s through the work of Andrée Corvol (1984), who studied the implementation of Colbertian forestry using a wide range of data sources. At the same time, geographers such as Gérard Houzard (1982) and Jean-Jacques Dubois (1989) began to translate forest-related historical data into maps. The union of history, geography, archaeology and natural sciences brought the focus on to the ecosystems themselves, rather

*E-mail: xavier.rochel@univ-lorraine.fr

than focusing only on those humans who inhabited them, worked in them and transformed them. The use of palynology (pollen records), anthracology (charcoal records) and other palaeoenvironmental techniques in relation to history has been particularly successful in the Pyrénées mountains owing to the Toulouse School of Physical Geography following the very early works of Maximilien ('Max') Sorre (1913).

As elsewhere in Europe, the relatively recent expansion of historical ecology has led to a much better understanding of the relations between humans and their forest environment. Dupouey et al. (2002) thus showed that the present-day soils and flora of the Haye Forest, near Nancy, a supposedly 'ancient woodland', was noticeably influenced by Gallo–Roman pastoral and agricultural activities that took place almost 2000 years ago.

19.3 Historical Forest Uses and Their Consequences for Forest Management

There is little difference between the traditional functions of woodlands in France and those elsewhere in Europe: wooded spaces were useful for their fuelwood, charcoal, timber, pasture and some other agricultural practices. Especial importance was attached to fuelwood and charcoal. In France, as opposed to Great Britain for example, mineral coal was not commonly used until late in the 19th century (Woronoff, 1994). Other fuels, such as peat, were confined to a few places. Forests were, therefore, the main provider of energy during what has been called the Little Ice Age. They had to be managed so as to provide small-diameter wood, fit for the easy making of charcoal – one could not spend too much time splitting logs. Thus, one of the main trends of forest history in France during the *Ancien Régime* (the times prior to the 1789 Revolution) was the transformation of woodland into coppices, or coppice-with-standards (Corvol, 1987).

The emphasis on producing fuel, particularly for the domestic market, had many consequences. French industries were forced out of a very large area around Paris, as historian Jean Boissière has shown: fuelwood prices were so high, as a result of the importance of urban demand, that wood- or charcoal-using industries could not compete (Arnould, 1978; Boissière, 1990, 1993). Nevertheless, a need for cheap energy did not mean that furnaces of all kinds, for enterprises from tile works to glassworks, resulted in the devastation of the forests. This has been one of the great myths of forest history, as rather, in many places, the importance of woodland for providing charcoal led to its protection.

Industrial establishments were set in the vicinity of forests, and their owners would not countenance losing, or seriously endangering, their main source of energy. The best illustration of this very close link between industry and forestry is the royal salt works at Arcet-Senans, Franche-Comté, in the east of the country. In this already quite continental region, salt was obtained from salt springs, and the water boiled off over wood fires. This consumed an extraordinary quantity of wood. Many forests of the Jura Mountains and the neighbouring plains were reduced to poor quality coppices, felled at very frequent intervals – for instance, every 5 years or so. However, the economic and strategic importance of salt making was such that in order to protect it, forest laws were passed, and forest management was enhanced to a very high technical level in Franche-Comté as early as the 16th century. The Arc-et-Senans salt works were established in 1775 on the edge of the large Chaux forest, and not directly in the vicinity of the salt springs, whose waters were transported through a 21 km long aqueduct to the works. This did not lead to the disappearance or the reduction of the Chaux forest, which is still nowadays one of the largest and better national forests in the region (Vion-Delphin, 1995).

Timber production was not necessarily a key function of forests until the 20th century, but it used to be quite important for building houses or ships. Traditionally managed coppice produces small-diameter stems that are fit for providing fuelwood or charcoal, but not so useful for construction. So traditional coppice would be turned into coppice-with-standards whenever it was possible and whenever public authorities were in a position to impose their ideas. The bent and sometimes gnarled trees grown in such woods were exceptionally useful for the making of ships (Ballu, 2008).

Figure 19.1a shows a classical coppice-with-standards landscape. The coppice (*taillis*) is composed of coppice stools that produce new shoots (*cépées*) after every felling. In order to provide timber and allow natural regeneration, foresters may protect young trees of seed origin (*baliveaux*) and leave them to grow until they reach a mature stage. These standard trees are called – according to their age from youngest to oldest – *baliveaux, modernes, anciens, vieilles écorces*. These names may

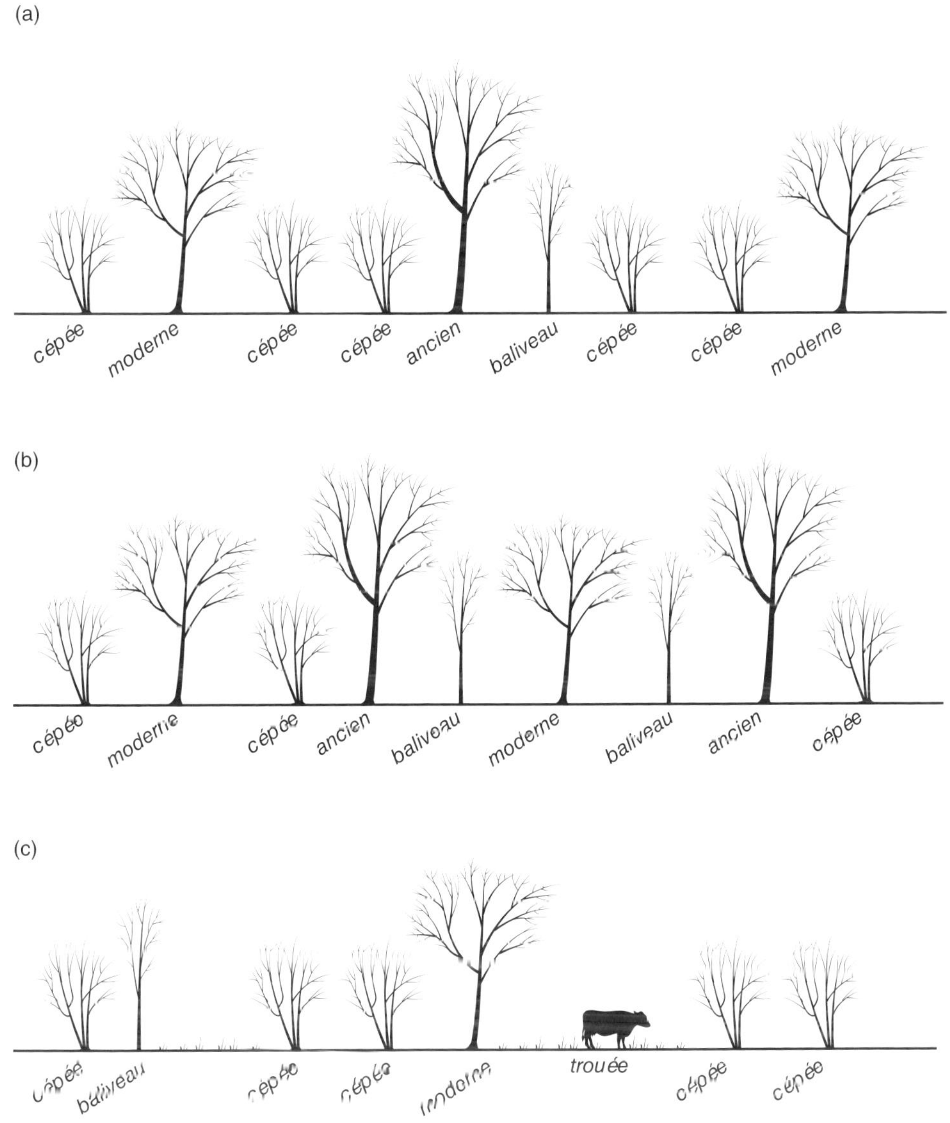

Fig. 19.1. The coppice-with-standards forestry system in Lorraine, France, as seen from a historical point of view: (a) classical coppice-with-standards landscape; (b–c) variants of the scenario shown in (a). Key: *cépée*, shoot regrowth from the coppice stool; *baliveau, moderne* and *ancien* refer to standard trees held over for increasing numbers of rotations; *trouée* are gaps or cleared areas. See text for further details. (From X. Rochel 2013–2014.)

change according to the region, and country to country. Parts (b) and (c) of Fig. 19.1 show variants on the above scheme. In Fig. 19.1b, foresters have protected too many standard trees. The coppice dwindles, so commoners complain that grazing is impossible, and that they lack fuelwood; shade-tolerant species like beech (*Fagus sylvatica*) are favoured. In Fig. 19.1c, there are fewer standard trees and heavy grazing pressure may hinder regeneration. Shade-tolerant species disappear, and pedunculate oak (*Quercus robur*) is favoured at the expense of sessile oak (*Q. petraea*); high-value species may give way to low-value, heliophile species such as birch (*Betula pendula*).

For a 21st century person, there is nothing surprising in the fact that the production of fuelwood and timber were among the most important functions of wooded spaces. What may seem more striking or exotic is the extent to which forests were used for different agricultural practices. The boundary between woodland and agricultural lands used to be quite permeable; indeed, in some times and places, it did not exist at all. Historians have come to coin the term '*forêt nourricière*' (sustaining or, more literally, 'nourishing' forest) in order to reflect the way that forests were directly useful for the human diet, through hunting certainly, but also through grazing and temporary cultivation. This led to the development of complex management systems which had very important ecological consequences.

Woodland grazing used to be very important until late in the 19th century (see also Hartel *et al.*, Chapter 5). Many pre-1900 forests were quite open, with a thin canopy and much light coming in on to the underwood, thus allowing rich grazing resources to exist among the trees. Many parts of forests were reduced to wastelands by peasants or shepherds through burning or girdling of the trees, and the animals themselves maintained – by browsing and grazing – this relatively open and probably species-rich semi-forested landscape.

In 1869, more than a third of the vast Orléans forest was occupied by heathlands, although state foresters were already doing their best to expel pastoral activities and to plant vast areas with pines. Temporary cultivation was also something quite frequent until the 19th century, at least in some regions. Cultivation in woodlands, in clearings or in a long (forest) fallow rotation system, was especially important in mountainous regions, or more generally where the soil was not very good for conventional agriculture.

Not all activities could coexist easily. Woodland grazing implies a rich herbaceous plant or shrub layer, but too much tree cover means poor grazing and heavy grazing means a degraded tree cover. These different functions could lead to tensions and conflicts among commoners, woodcutters, colliers, foresters and other parts of the rural society. Their activities had to be controlled, regulated and organized. In some cases, this led to different functions being assigned to different parts of the woodlands; in other cases, different functions had to coexist at the same places.

In the Basque mountains for instance, abundant charcoal had to be produced for the regional iron industries; but conventional coppicing could hardly be used because of the importance of pastoral activities. So the trees were not cut at ground level, but fuelwood and charcoal were obtained by pollarding (Fig. 19.2). This left the lower strata of the

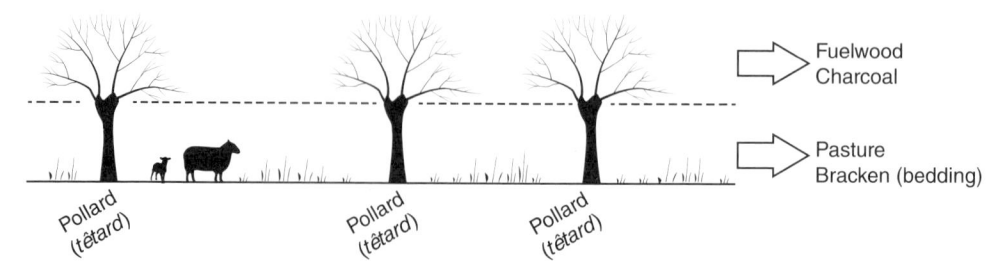

Fig. 19.2. Pollarded woodlands of the Basque country of France (*têtard* = pollard). (From X. Rochel 2013–2014.)

forest for grazing, but also for cutting bracken which was (and still is) used for animal bedding, as Basque agriculture depended heavily on maize, which was unsuitable for use as livestock bedding. These practices are at the origin of vast woods composed of pollards (*zuhaitz motzak* in Euskara – the Basque language; *arbres têtards* in French), which were replaced whenever necessary by young beech or oak saplings grown in forest nurseries. These woods were also sometimes called *taillis perchés* ('off-ground coppices') and the system has much in common with the Iberian dehesa system.

In the Ardennes mountains, the management system born from the interaction of agriculture and wood production was much more complex – a forest–fallow or *sartage*, system (Fig. 19.3). The coppice was harvested every 18–20 years. After felling, the area was temporarily cleared for cultivation through slash-and-burn. Rye was planted and harvested in the first year, and then broom (*Cytisus scoparius*), which was used as a fuel for heating bakers' ovens. The coppice stools were then left to regrow until, a few years later, the young oak shoots were strong enough for livestock to be admitted for grazing. After 18 years or so, the coppice stems were barked for the tannery industries, then felled and often turned into charcoal.

These different management systems, and the way they were more or less appropriately supervised (or not) by foresters, led to important ecological consequences. Some practices were particularly destructive for forest soils. In many regions of France, as elsewhere in Europe, litter raking was used in order to provide large quantities of material for animal bedding; written archives from 100 or 200 years ago give us many accounts of people swarming into the most accessible woodlands to rake until the last leaf was gone: no litter, no more organic matter and no more soil renewal.

19.4 The Making of the Technical and Legislative Framework

In order to protect the forests, to organize the coexistence of different forest functions and to impose the most satisfactory forestry techniques, the royal authority and its fiefs began in the Middle Ages to build up a legislative framework, as well as justice and police forces for law enforcement. The first truly notable steps were in the 13th and 14th centuries, with the Ordinance of Givors (1219) and, especially, the Ordinance of Brunoy (1346). This latter is regarded by French forest historians as a pioneer text in the field of sustainable

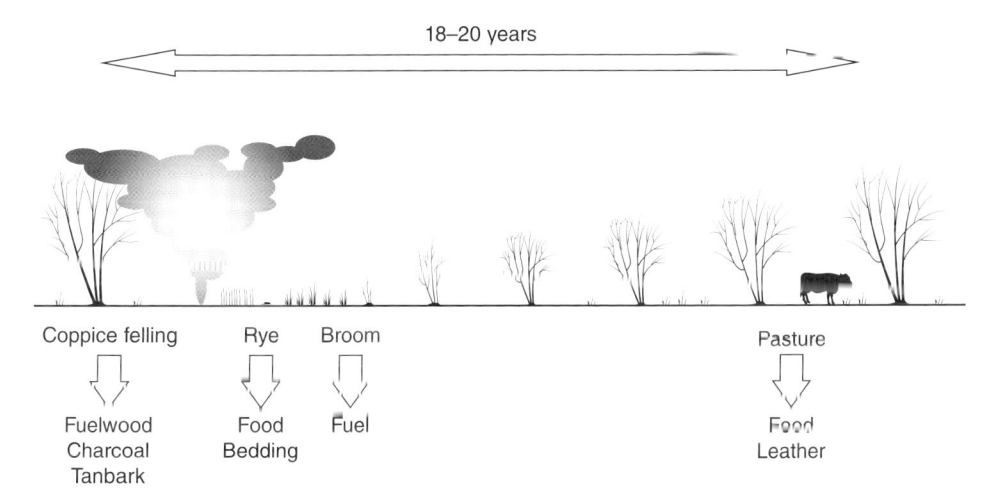

Fig. 19.3. The multiple-use *sartage* (forest–fallow) system in the Ardennes highlands, 18th–19th centuries: coppice felling, slash-and burn, followed by rye harvesting in the first year and then broom harvesting. Coppice stools were then left to grow, with grazing later on. (From X. Rochel 2013–2014.)

development, inasmuch as it was intended to ensure that 'forests and wood would perpetually sustain themselves in good condition' (*'que lesdits forêts et bois se puissent perpétuellement soutenir en bon état'*) (Devèze, 1965).

This legislative production accumulated during the following centuries, as concerns grew about the extent of forests, their condition and, specifically, the problem of wood supply in rural and urban households. During the 16th century, forestry issues were of particular concern for royal authorities, and in 1516 there was enacted what can be considered as a kind of forest code, as the accumulated forest laws had become practically unreadable. However, by the mid-1600s, the fuelwood supply of Paris and the charcoal supply of industries were under stress and, moreover, there was also a need for a supply of timber to the royal shipyards. The Royal Navy had to be powerful, shipyards had to be well supplied, and forests had to be preserved and managed. These considerations explain the upheavals of the 1660s.

When Louis XIV came to the throne and succeeded in establishing his power after a long period of turmoil, he decided to put forest issues in the hands of a trustworthy man. That man was Jean-Baptiste Colbert (1619–1683), who launched what is commonly called the Colbertian Reformation or '*Réformation colbertienne*' in 1661. In the few next years, the monarch's power over state forests was restored, and commissioners were sent to different parts of France to ensure that forests were well preserved and managed. In 1669, a legislative code was set out to clarify forest law: the Forest Ordinance of 1669, which would later be used as a model for several other European states.

The various legal texts enacted from the Middle Ages to 1669 aimed at standardizing forest management. Coppice-with-standards and, secondarily, high forest management were regularized; farming and grazing practices in woodland were meant to cease or at least to be drastically restricted. Old regional forestry techniques were to disappear, so that forest management would proceed and yield according to a generalized model (Gallon, 1752). The *sartage* technique in the Ardennes, for instance, or the pollarded *taillis perchés* of the Basque country, could not survive if foresters and peasants had to abide by the new rules. None the less, passing laws is one thing, enforcing them is another (Corvol, 1984), so, for instance, *taillis perchés* can still be observed in south-western France.

Dealing with coniferous forests or mixed forests that combined, for instance, beech and silver fir (*Abies alba*), or beech, fir and Norway spruce (*Picea abies*), proved exceptionally difficult. These woodlands were in the mountains, far from the political centres and the forests where the official forestry model was built. Some forest officers tried to impose the coppice-with-standards model in high altitude, mixed or coniferous woodlands and this ended in a complete failure. Here and there, traditional techniques were retained, such as *jardinage*, a traditional forestry system similar to our present-day selection system.

Furthermore, the French law could not be imposed everywhere. The conquests that had increased the territory of the French kingdom did not necessarily erase local rights and customs. The case of the Duchy of Lorraine (see below) shows how the implementation of forestry desired by the royal power could not be achieved without many concessions because of the political, technical, cultural and ecological peculiarities of the regions conquered by France.

The centuries-long accumulation of law texts ended with the political turmoil around 1800. Then, in the 1820s, two events of major importance for French forestry took place. In 1824, a school specializing in higher forestry training was established at Nancy, Lorraine, following the example of the neighbouring German states and, in 1827, a Forest Code was established that succeeded and replaced the obsolete 1669 Ordinance. Subsequently, silvicultural techniques were heavily transformed, and the coppice-with-standards system was very gradually abandoned in favour of the regular high forest system, which was deemed more appropriate for satisfying the needs of the booming industrial economy. A new French model was built around the idea of natural regeneration and, more precisely, the shelterwood system (Fig. 19.4), which had the advantage of being applicable to both deciduous and coniferous forests.

Figure 19.5 summarizes the historical background to French forestry both up to the shelterwood system mentioned above and beyond to the new legislation that was established in 2001.

19.5 Consequences of Forestry Policies for Forest Composition in the Woodlands of Lorraine

Interpreting written archives, and finding, verifying and correctly interpreting ancient

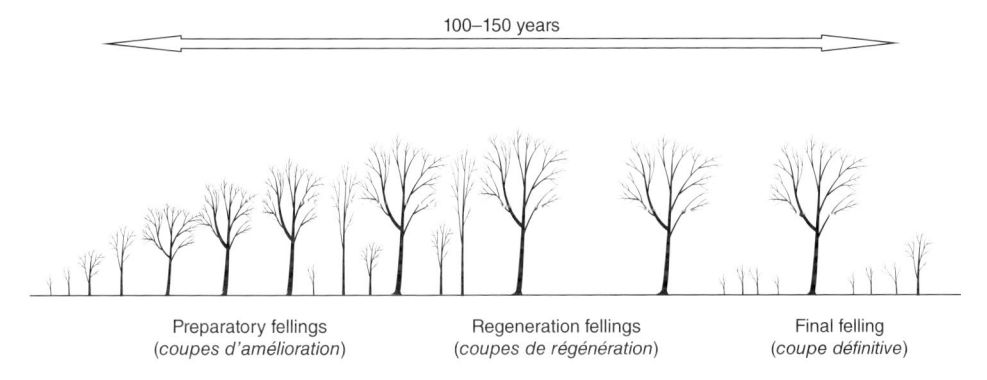

Fig. 19.4. Shelterwood, the dominant forestry model in France after the 1820s. (From X. Rochel 2013.)

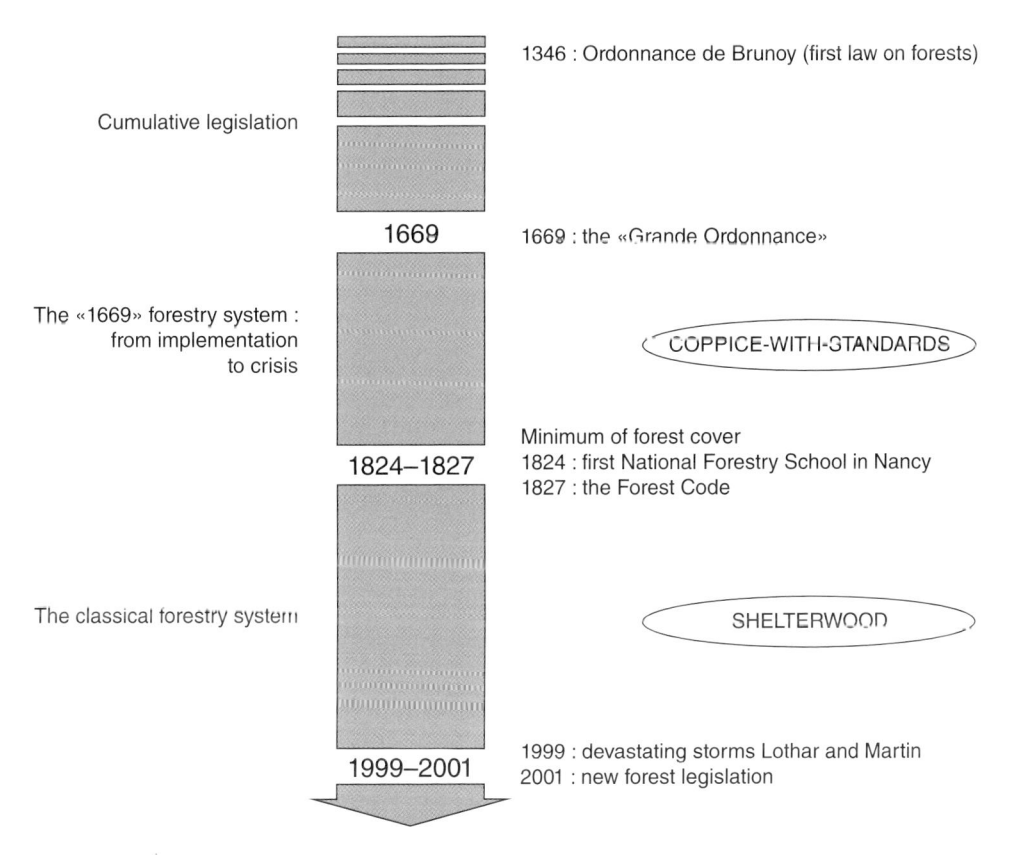

Fig. 19.5. A summary of French forestry history, 1346–2001. (From X. Rochel 2013.)

texts can be a difficult task in various fields of history; with forestry material there are questions around the use of vernacular and regional vocabulary, or translation of the old measurement units. However, the task is eased by the fact that the protection, surveillance, management and exploitation of forests were among the main priorities of royal and feudal authorities. Thus, there are large volumes of forest archives well before the critical changes of the 1820s, and these allow us to understand how foresters envisaged their work and what their objectives and methods were. Some of the data allow us to measure their actions and the results that they brought about.

We can, for example, look at the records left by the foresters of the Duke of Lorraine, in which were recorded all operations, felling by felling, and sometimes tree by tree (Rochel, 2004). The twin dukedoms of Lorraine and Bar were political entities of medieval origins. United politically in the 15th century, they were maintained as an independent, or semi-independent, state until 1766, precariously situated between two political giants: the Kingdom of France and the Holy Roman Empire. When the French progressively conquered the region between 1552 and 1766, local laws remained in place; the 1669 Ordinance, for instance, was not implemented, but a similar text, better adapted to local conditions, was issued by the Duke of Lorraine in 1701 and remained effective until the 1789 Revolution (Guyot, 1886; Husson, 1991).

The impact of this latter ordinance is revealed in 18th century documents which were analysed through a subset of 1859 felling areas and more than 498,775 felled or marked trees (Rochel, 2004). The fellings were located in the south-east part of the Duchy, on the border of the Vosges Mountains. They cover quite a small period of time – less than a century, and the information they contain is limited to forest composition without any reference to other flora and fauna; but the records do show that choices made by foresters had a very significant impact on the shaping of wooded ecosystems.

The first identified impact concerns the deciduous forests of the lowlands surrounding the Vosges Mountains. The high forest selection system was the traditional way of managing these forests. Implementation of modern (that is, coppice-with-standards dominated) forestry, should, according to the strictures of the Duchy's foresters (under the Ordinance of 1701; Moncherel, 1778), have favoured heliophile, coppice-supporting species such as oak (mostly *Q. petraea* or *Q. pedunculata*) or hornbeam (*Carpinus betulus*), but these forests are now often dominated by a single species, beech. This was already more or less the case in the 19th century, and so was it before the forests of the area were turned into shelterwood – *futaie régulière*. Although only 250 of our felling areas (1021 ha) concern these particular forests (Rochel, 2013), the analysis of foresters' archives explains this paradox.

In the mid-1700s, when ducal foresters were in a position to impose their actions and implement coppice-with-standards forestry, they were dealing with oak- and beech-dominated woodlands. Single-species forests appear nowhere in our data, except in one single wood. Beech represents 59% of marked or felled trees, oak 31% and hornbeam 6%. These woodlands were composed, surprisingly, of many old trees, an occurrence that was becoming rare in the region.

The 1701 Ordinance asked for the maintenance of 58 *baliveaux* (the youngest category of standard trees) per hectare, but the foresters found few stems that could be marked for such a purpose, and marked only 39. Instead, they left more older trees than was necessary: 32 *modernes* (rather than the 19 such trees mentioned by the ducal ordinances); there were also 22 *anciens* (rather than the 19 in the ducal ordinance), and eight *vieilles écorces* per hectare (rather than ten). This makes for a very dense forest cover. Many of these trees should have been felled in order to maintain the proper equilibrium for a classical coppice-with-standards; but foresters could not bring themselves to fell these aged trees, which were rare in other parts of Lorraine. The result matches with the situation depicted in Fig. 19.1b. Too many standard trees led to the decline of coppice, with shade-tolerant species such as beech favoured by this outcome, even though oak was the preferred crop species at the time.

The second impact of the implementation of ducal forestry in the region concerns the vegetation of the lower slopes of the Vosges Mountains. Like many central European highlands, these mountains are today predominantly covered with monospecific conifer forests, and managers complain that this situation threatens biodiversity and forest health. Coniferous stands amount to 56.3% of the forest cover, while mixed stands amount to 22.5% and deciduous forests 17.4% according to the *Inventaire Forestier National*. The proportion of pure conifer stands was strongly increased by spruce plantations, but even in old forests, the species composition was significantly altered by the activities of foresters over centuries. The archives show that the forests of the Vosges Mountains were quite different in the 17th and 18th centuries. They were mostly mixed forests (fir, beech, with some spruce), and almost no monospecific fir or spruce stands; typically, they may have been one quarter to one third deciduous species (mostly beech).

Archives dealing with highland forests show policies favouring different species, and this affected forest composition in the Vosges Mountains from the 16th century onward. Oak (*Q. petraea*) seems to have been particularly favoured in the 16th and early 17th centuries (Rochel, 2007). After the mid-17th century, foresters tended to favour conifers, and especially *Abies alba*, as being the most lucrative species. Beech was mostly used for fuelwood and oak for construction timber (but in the mountains, the poor quality of most oaks led to them also being used as fuelwood), while silver fir and spruce, united under the term *sapin* in most archives, are used locally for construction timber. The *sapin* could be used in sawmills where logs were sawn into planks.

This was the main source of income for forest owners, in particular for the Duke of Lorraine, who owned most of the largest forests in the Vosges. From the 14th century onward, the mountains provided useful forest resources to the regional vicinity, the most important of which were softwood logs and planks. These were tied into rafts and floated down the main rivers through the dukedom of Lorraine, sometimes as far as the Netherlands. The lowland regions had by this time cleared too much forest or overused the rest for fuelwood, and thus lacked timber.

For forest owners in the Vosges Mountains, revenue depended strongly on this commercial network, above all on coniferous species, which could be sawn into planks. Much of the income was generated by sawmill concessions and the selling of trees to sawmills; for instance, up to 90% of the income from the Mortagne forest to its owners in the 18th century (Rochel, 2004). The price of a mature fir tree was usually much higher than that of an oak, or beech, so at the beginning of the 19th century, the financial value of a conifer stand was much higher than that of a hardwood stand.

Landowners and foresters were (and still are!) prone to promoting species that bring in a large income. Our documents show this tendency in an unambiguous way: for example, '*tous les propriétaires de sapinières cherchent à les augmenter par tous les moïens qu'ils croient les plus convenables à la multiplication du sapin et à l'acceleration de sa croissance*' (all owners of conifer woods endeavour to extend them by all means that they deem favourable for the increasing of the number of fir trees, and the increasing of their growth). An appetite for money was not always the only reason for this activity however, as many foresters feared that beech could overcome fir and spruce: 'practice led us to learn that in forests rich in fir trees, one cannot preserve beeches without peril of smothering the birth of fir trees, given that beech seeds drop into fir stands, grow there, sink into the soil, lift the spreading roots of the fir trees and smother by growing those that it cannot kill'.

On one occasion only did a senior forester, Florent Bazelaire de Lesseux, acknowledge the important role of deciduous trees in mixed forests; he was convinced that beech trees helped to shelter young fir saplings, and help the whole forest to resist against winds. He noted in the margin of a report to his superiors that: 'the fir tree grows its roots on the surface of the soil...violent mountain winds uproot it easily if other, better rooted species are not there to resist and support it. But these species in too great a number are harmful to the fir tree, because under their cover its regrowth is too sheltered and chokes. Therefore, in order to preserve softwoods (which are the valuable species in the mountains) one must cut down beeches and preserve only an appropriate proportion' (Archives Départementales de Vosges, B 525).

Many foresters, having witnessed in some forests a development of beech over fir stands, worked to remove broadleaved species from all mixed stands. This operation was called *nettoiement*, or 'cleaning'. Sometimes they were felled one by one, but most of the time the broadleaved trees were cut in massive fellings. Woodcutters had to harvest all beech, oak, birch or other deciduous trees, thereby providing fuelwood in large quantities and allowing conifers to replace them. Conifers were then managed according to the selection system, as it was acknowledged, mostly thanks to unhappy experiences in the neighbouring Franche-Comté and Alsace provinces, that other forestry systems were unlikely to succeed. In mixed forests dominated by deciduous species, foresters chose a different way; those woods were managed as coppice-with-standards, with all conifers, whether small, crooked, or otherwise, being reserved as standards.

19.6 The Modern Forest – Conclusion

The impact of forestry management on forest composition, which seems quite strong in the Age of Enlightenment in Lorraine, was further intensified in the following two centuries. A new age for forestry started in the 1820s when (i) forest laws were standardized in the whole of the country and (ii) new forest techniques were implemented as a result of the new Forestry School established in Nancy in 1824, and the new Forest Code of 1827 (Badré, 1983).

During the 19th and 20th centuries, the planting and sowing of softwood species (especially Norway spruce) were attempted in several forests of the Vosges Mountains c.1820–1830, with massive success after 1830 in certain areas, such as dry, south-facing slopes on sandstone substrates (*Pinus sylvestris*, and more recently, the Douglas fir (*Pseudotsuga menziesii*). 'Cleaning' operations continued until the end of the 19th century. These considerations show how 18th century foresters initiated an enduring evolution which led from an initial mixed forest composed of *Abies alba* and *Fagus sylvatica* ('*hêtraie-sapinière*'), with some scattered oak here and there, to wholly coniferous stands in many places.

Outside the Vosges Mountains, but still in Lorraine, the change from coppice-with-standards dominated forestry to shelterwood dominated forestry was very often made in favour of shade-tolerant species such as beech, or in favour of sessile oak at the expense of pedunculate oak, for example. Today's woodlands are inherited from these several centuries of intense silvicultural practices. Very precise archive records help us to understand the impact of those particular activities. However, in the quest for a thorough understanding of the human-driven evolution of French woodlands through the last centuries, we still lack, and will always lack, written information on other activities, such as hunting, silvopastoral activities or litter gathering, which have left us very few written records.

References

Arnould, E. (1978) Métallurgie au bois et utilisation de la forêt. *Revue Forestière Française* 30(No. 6), 459–477.
Badré, L. (1983) *Histoire de la Forêt Française*. Arthaud, Paris.
Ballu, J.-M. (2008) *Bois de Marine – les Bateaux Naissent en Forêt*. Gerfaut, Paris.
Boissière, J. (1990) La consommation parisienne de bois et les sidérurgies périphériques: essai de mise en parallèle (milieu XVe-milieu XIXe siècles). In: Woronoff, D. (ed.) *Forges et Forêts: Recherches sur la Consommation Proto-industrielle de Bois*. École des Hautes Études en Sciences Sociales (EHESS), Paris, pp. 29–56.
Boissière, J. (1993) Populations et économies du bois dans la France moderne. Contribution à l'étude des milieux forestiers entre Paris et el Morvan au dernier siècle de l'Ancien Régime. Thesis, University of Paris I, Paris.

Corvol, A. (1984) *L'Homme et l'Arbre sous l'Ancien Régime*. Economica, Paris.
Corvol, A. (1987) *L'Homme aux Bois. Histoire des Relations de l'Homme et de la Forêt, XVIIe-XXe Siècles*. Fayard, Paris.
Devèze, M. (1961) *La Vie de la Forêt Française au XVIe Siècle*. SEVPEN, Paris.
Devèze, M. (1965) *Histoire des Forêts*. Presses Universitaires de France (PUF), Paris.
Dubois, J.-J. (1989) Espaces et milieux forestiers dans le Nord de la France: étude de biogéographie historique. Thesis, Université de Lille, Lille, France.
Dupouey, J.-L., Dambrine, E., Laffite, J.-D. and Moares, C. (2002) Irreversible impact of past land use on forest soils and biodiversity. *Ecology* 83, 2978–2984.
Gallon (1752) *Conférence de l'Ordonnance de Louis XIV du Mois d'Août 1669 sur le Fait des Eaux et Forêts*. Jacques Rollin, Paris.
Guyot, Ch. (1886) *Les Forêts Lorraines jusqu'en 1789*. Crépin-Leblond, Nancy, France.
Houzard, G. (1982) Poids de l'histoire et importance de la durée en biogéographie. *Annales de Normandie* 1, 269–288.
Huffel, G. (1926) *Les Méthodes de l'Aménagement Forestier en France: Étude Historique*. Berger-Levrault, Nancy, France.
Husson, J.-P. (1991) *Les Hommes et la Forêt en Lorraine*. Bonneton, Paris.
Moncherel, M. (1778) *Commentaire sur les Ordonnances de Lorraine, Civile, Criminelle, et Concernant les Eaux et Forêts, Combinées avec celles de France*. Société Typographique, Bouillon, Belgium.
Rochel, X. (2004) Gestion forestière et paysages dans les Vosges d'après les registres de martelages du XVIIIe siècle. Essai de biogéographie historique. Thesis, Université de Nancy 2, Nancy, France.
Rochel, X. (2007) Un faciès forestier relictuel: les chênaies montagnardes des Vosges lorraines. *Revue Géographique de l'Est* 4, 209–218.
Rochel, X. (2013) Sources d'archives et histoire de l'environnement. L'origine des hêtraies vosgiennes d'après les registres de martelages du XVIIIe siècle. In: Galop, D. (ed.) *Paysages et Environnement: de la Reconstitution du Passé Aux Modèles Prospectifs*. Presses Universitaires de Franche-Comté, Besançon, France, pp. 15–24.
Sorre, M. (1913) *Les Pyrénées Méditerranéennes: Étude de Géographie Biologique*. Armand Colin, Paris.
Vion-Delphin, F. (1995) La forêt comtoise de la conquête française à la Révolution (1674-fin du XVIIIème siècle). Thesis, Université de Franche-Comté, Besançon, France.
Woronoff D. (1994). *Histoire de l'Industrie en France du XVIe Siècle à nos Jours*. Seuil, Paris.

20 Barriers and Bridges for Sustainable Forest Management: The Role of Landscape History in Swedish Bergslagen

Per Angelstam,[1]* Kjell Andersson,[1] Robert Axelsson,[1] Erik Degerman,[2] Marine Elbakidze,[1] Per Sjölander[3] and Johan Törnblom[1]

[1]*School for Forest Management, Swedish University of Agricultural Sciences, Skinnskatteberg, Sweden;* [2]*Institute of Freshwater Research, Swedish University of Agricultural Sciences (SLU), Örebro, Sweden;* [3]*Academy North Development Unit, Storuman, Sweden*

20.1 Introduction

In the first part of the chapter we discuss the Pan-European context for sustainable forest management (SFM) policy, and how this translates into regional and local contexts. Next, we introduce Bergslagen in southern Sweden and review its forest landscape history of over the past 2000 years. We then summarize the present barriers resulting from this long forest landscape history for different dimensions of the sustainability of landscapes as social–ecological systems.

We also examine how local and regional actors and stakeholders in Bergslagen support the implementation of Pan-European, European Union (EU) and Swedish policies directed towards the different dimensions of sustainable forest management. Finally, we discuss the development of integrated landscape approaches as a bridge for policy implementation in terms of place- and evidence-based collaborative learning within and among European forest and woodland landscapes as social–ecological systems with different histories.

20.2 The European Scale

Forestry is not covered by EU processes in the way that agriculture is, but there is a Pan-European forest policy process (see Forest Europe, UNECE and FAO, 2011), which states that sustainable forest management is 'the stewardship and use of forests and forest lands in a way, and at a rate, that maintains their biodiversity, productivity, regeneration capacity, vitality and their potential to fulfil, now and in the future, relevant ecological, economic and social functions, at local, national, and global levels, and that does not cause damage to other ecosystems'. The recent forest action plan for the EU (European Commission, 2013a,b) highlighted several initiatives across the EU to support, implement and assess sustainable forest management.

*E-mail: per.angelstam@slu.se

Criteria have been developed by Forest Europe (The Ministerial Conference on the Protection of Forests in Europe, MCPFE) for countries to report on the implementation of sustainable forest management. Indicators were initially adopted by the Ministerial Conference in Lisbon in 1998, and later simplified and further improved for endorsement at the Vienna MCPFE in 2003. These indicators are used to assess progress towards sustainable forest management in the Pan-European region at the national level using, for example, national forest inventories (Forest Europe, UN-ECE and FAO, 2011). Similarly, indicators and performance targets have been developed within countries, such as the Swedish Environmental Quality Objectives, which were established as short-term targets relevant for securing the ecological sustainability of forests (MOE, 2013). For the social and cultural sustainability criteria (Axelsson et al., 2013a), there are ongoing attempts to define the appropriate indicators and determine how they can be assessed at multiple levels.

Forest management plans are then an important tool for the implementation of sustainable forest management on the ground by forest owners at the strategic, tactical and operational levels of management. Such plans provide information in terms of maps and databases. Large forest companies carry out their own forest inventories, and operations are planned for individual stands to reach the appropriate management goals. However, forest management plans are generally not communicated among landowners at the aggregated forest management unit level, such as watersheds, municipalities or other wider units (Angelstam et al., 2011a). This becomes a particular problem where the land is divided among several forest owners. Moreover, forest management plans often focus on growing trees for economic purposes and do not necessarily plan for other dimensions of sustainability.

Criticisms of the nature of much forest management worldwide, even where there were at least nominal forest plans, led to the development of forest certification as another tool to encourage sustainable forest management (Bass, 2004). The area of certified land and the number of products carrying a certification logo have increased rapidly, but there are large differences between countries in the popularity and standards of certification.

Elbakidze et al. (2011), for example, compared biodiversity conservation indicators at different spatial scales in Sweden and northwest Russia in terms of the areas of formally and voluntarily set aside forests for biodiversity conservation. They evaluated structural habitat connectivity by applying morphological spatial pattern analysis, and potential functional connectivity by using habitat suitability index modelling. While the Russian standard included indicators for all spatial scales of biodiversity conservation – from tree and stand to landscape and ecoregions – the Swedish standard focused mainly on stand and tree scales. The area of voluntary set-asides through the Forest Stewardship Council (FSC) was similar in Sweden and Russia, but formal protection in the Russian case study was three times higher than in Sweden. Swedish set-aside core areas were two orders of magnitude smaller, had much lower structural and potential functional connectivity and were located in a fragmented forestland holding. Additionally, certification standards in Sweden did not mirror evidence-based knowledge about ecological sustainability (Angelstam et al., 2013d). The potential of FSC certification for biodiversity conservation can thus only be assessed if both the standard content and its implementation on the ground are considered.

20.3 The Regional Scale

Within countries on the Fennoscandian peninsula, there are considerable differences among different regions, which have very diverse forest landscape histories (Bradshaw and Zackrisson, 1990; Bradshaw and Hannon, 1992; Lehtonen and Huttunen, 1997; Östlund et al., 1997; Hæggström, 2000; Storaunet et al., 2000).

The natural potential forest vegetation of Fennoscandia ranges from temperate to boreal and mountain forests, but the vast majority of the temperate forests with soils and climate favourable for agriculture was cleared for agriculture thousands of years ago. Later on, the temperate forests were also cleared for transport infrastructural and urban development. In contrast, most boreal forests remained relatively intact until the end of the 19th century (e.g. Björn, 2000; Angelstam *et al.*, 2004b), although in the past they had been influenced by the activities of hunter–gatherers, and grazing–herding would have created more open forests (Lagerås, 2007).

The gradients among landscapes with different trajectories of natural forest and cultural woodland histories, and between temperate and boreal forest ecoregions, are particularly sharp in Sweden's historic Bergslagen region (Fransson, 1965). Given a much shorter history of intensive forest and woodland use than in southern Europe, as well as in western and central Europe, large boreal forest resources were available for use by the iron and forest industries in the 18th to 19th centuries when the wood resources of other regions had already been depleted.

20.4 Bergslagen – An Introduction

Northern Fennoscandia contains the westernmost part of the Eurasian boreal forest biome, while southern Fennoscandia has the northernmost temperate forests in Europe (Fig. 20.1). There is a clear trend in forest cover from heavily cleared in the south to largely retained in the north, which is due to successive frontiers of agricultural development and forest use. These gradients in forest landscape history are particularly sharp in Sweden's Bergslagen region (Geijerstam and Nisser, 2011; Angelstam *et al.*, 2013c), which has long been recognized as a borderland (Fransson, 1965).

Bergslagen is an informal region of south-central Sweden (*c*.59–61°N latitude, 13–15°E longitude), which has a long history of mining and the consequent use of forests and streams. The bedrock is rich in minerals (Stephens *et al.*, 2009), and metals were first produced there more than 2000 years ago (Nelson, 1913). The name Bergslagen comes from the Swedish words for mine (berg) and law – or people working together (lag). The biophysical landscapes of Bergslagen form a gradient between urban areas in the Mälardalen valley with temperate forest remnants in agricultural settings,

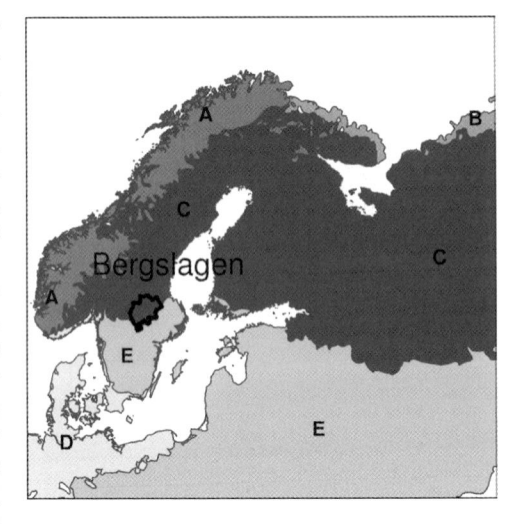

Fig. 20.1. Location of Bergslagen (southern Sweden) in relation to vegetation zones in north-west Europe (derived from Ahti *et al.*, 1968): (A) boreal mountain system; (B) boreal tundra woodland; (C) boreal coniferous forest; (D) temperate oceanic forest; and (E) temperate continental forest.

and rural remote boreal forest regions. Bergslagen includes parts of nine counties, and all municipalities and communities in its core area are peripheral to regional centres of economic growth (Andersson *et al.*, 2013a,b).

20.5 Forests, Forest Ownership and Land-use Dynamics

The forests in Bergslagen range from temperate broadleaved forests to the conifer-dominated boreal areas. The history and the current use of the forest landscapes are linked to the four main types of land ownership: non-industrial private (57.4% of the area), large forest companies (28.3%), the state forest company Sveaskog Co. (9.8%) and public bodies such as municipalities, the National Property Board and forest commons (4.6%). The non-industrial private forest tends to be at lower altitudes, the industrial forest companies at mid altitudes and the public forests close to the treeline (Fig. 20.2).

Good soils for agriculture are confined to areas below the marine limit during the latest glaciation (below 200 m altitude); since then

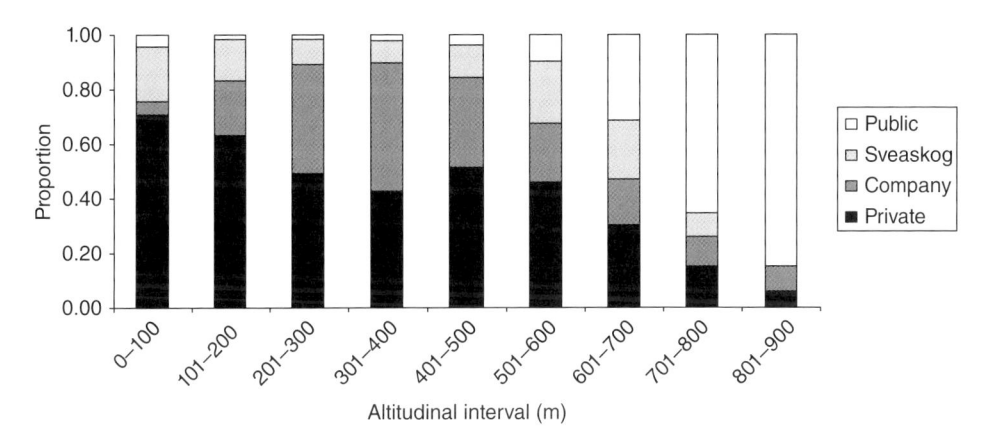

Fig. 20.2. The total area (top of each bar) and the proportions (portions of each bar) of forest owned by the four forest owner categories indicated in the key at different altitudes in the nine counties where the Bergslagen region is located in southern Sweden. The treeline is located in the highest altitudinal interval. Sveaskog is the state forest company.

there has been a land upheaval rate of about 1 m a century. The temperate forests along the shores of the large lakes of Mälaren and Hjälmaren began to be transformed by agricultural development in 8000 BP (Sporrong, 2008). Today, the traditional cultural landscape, with large areas of wooded grasslands, has been converted to agriculture or forest, and has lost some of its ecological and cultural value (Mikusinski *et al.*, 2003). The areas at the lowest altitudes have always been used for agriculture. Thus, below 100 m altitude, 30.6% of the land base has been cleared from forest. In the next 100 m interval (101–200 m) below the marine limit, 13.1% of the land has been cleared.

In contrast, areas immediately to the north were permanently colonized by agriculture only during the late 16th century when Finns settled to practice slash-and-burn farming (Montelius, 1953; Emanuelsson and Segerström, 2002). These areas are located above the marine limit, and have a lower proportion of soils suitable for agriculture. Hence, the proportion of cleared forests is only 2.2% at 201–300 m altitude, and less than 1.0% at altitudes higher than 300 m. However, this does not mean that the forests were left untouched. They provided grazing for cattle, fuel and materials for buildings, and fences, tools and household items for a long time. Semi-natural forest remnants remained as late as the 18th and 19th centuries, and large carnivore species such as wolves, the lynx, wolverine and bears (often used as indicators of ecological integrity) were present until the latter half of the 19th century.

The mines, forests and streams of Bergslagen were a very important part of the natural capital base underpinning the economic development of Sweden (Heckscher, 1935–1949). Small-scale production of iron began more than 2000 years ago (Geijerstam and Nisser, 2011) and mining for copper started in the 8th century. An important historical description of charcoal production for the mine in Falun, the main mining city in the region, survives from 1347 (Söderberg, 1932). Industrial iron mining commenced during the medieval period (Bindler *et al.*, 2011). From the 17th until the early 20th century, charcoal was the key industrial wood product. Increased demands for charcoal, and later for roundwood, triggered the development of sustained yield forestry from the 1840s (Brynte, 2002).

At present, the main forest products of Bergslagen are sawlogs and pulpwood from Scots pine (*Pinus sylvestris*) and Norway spruce (*Picea abies*), and biomass for biofuel production, including that from logging residues (branches and tops) and stumps. The main goal of current industrial forest management is to sustain and gradually increase the output of these products from the forests. At the same time, biodiversity conservation, social values and postmodern use of forest landscapes for recreation and tourism are expected under Pan-European, EU, national and regional policies.

The role of forest landscapes has, then, evolved from providing game, pasture for animal husbandry, fuelwood and charcoal for the iron industry, as well as timber and pulp for the forest industry, to also accommodate recreation and tourism. This long history of demands for forest goods, services and values in Bergslagen (Andersson *et al.*, 2013a,b) has resulted in considerable changes in land cover (Fig. 20.3).

Bergslagen supplied the main source of income for the Swedish state for several centuries, but today it suffers from economic vulnerability (Tillväxtverket, 2011) and human depopulation (Pettersson, 1999). Economic development (which can be regarded as a means) and social and cultural sustainability (the goal) depend on ecological sustainability and the carrying capacity of ecosystems. We therefore review the consequences of the landscape history of Bergslagen for ecological and other aspects of sustainable forest management.

20.6 Barriers to Sustainability

20.6.1 Ecological sustainability

The long history of forest use in Bergslagen has resulted in the survival of very few semi-natural forest remnants and undisturbed streams (Angelstam *et al.*, 2013c). The extent of protected land, a common indicator of biodiversity conservation, increases from the core to the periphery of the Bergslagen region (Fig. 20.4a). Ecological sustainability includes the presence of naturally occurring species, habitats and ecosystem processes (Brumelis *et al.*, 2011). Hence, sufficient habitat should be present as functionally connected habitat networks such that species and processes can operate across the landscape as a whole in both terrestrial and aquatic systems.

Angelstam *et al.* (2011a) modelled connectivity for specialized focal species such as resident bird species and insects in dead wood in four representative forest habitats in Bergslagen – natural old spruce, old pine and old deciduous forests, as well as forest–field edges of the cultured landscape. On average, 15% of the total forest belonging to the four different forest habitats was part of functional habitat networks that satisfied the requirements of the selected focal species. Overall, habitat network functionality for coniferous forests was better than for deciduous forest, but there were significant differences between the four forest habitats in the different boreal ecoregions. In general, the proportion of functional old spruce forest area was highest (15–42%) among the four forest types, especially in the mid and south boreal ecoregion. The functionality

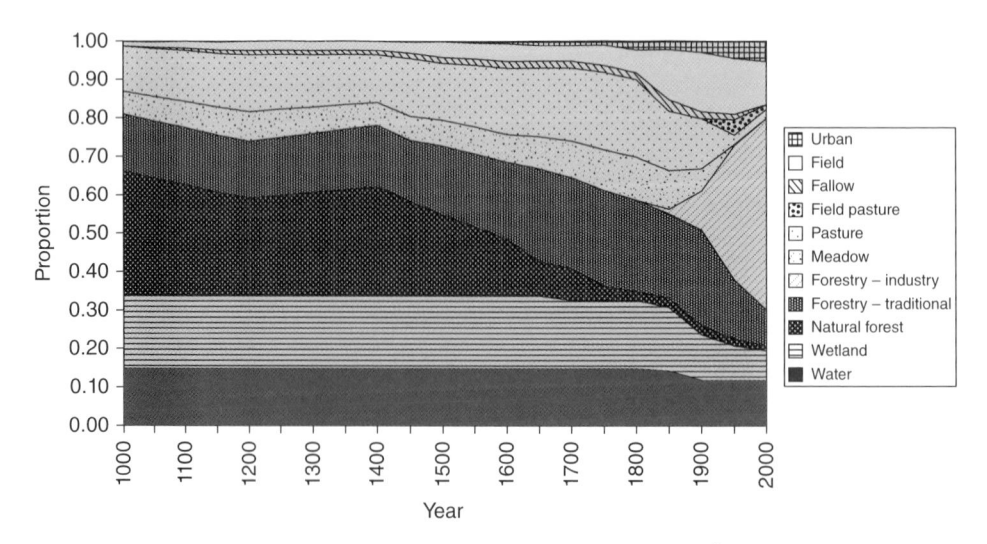

Fig. 20.3. Land cover dynamics during the past 1000 years in Bergslagen's Örebro County in southern Sweden. (Data from Angelstam, 1997, the Swedish Forest Agency and Statistics Sweden.)

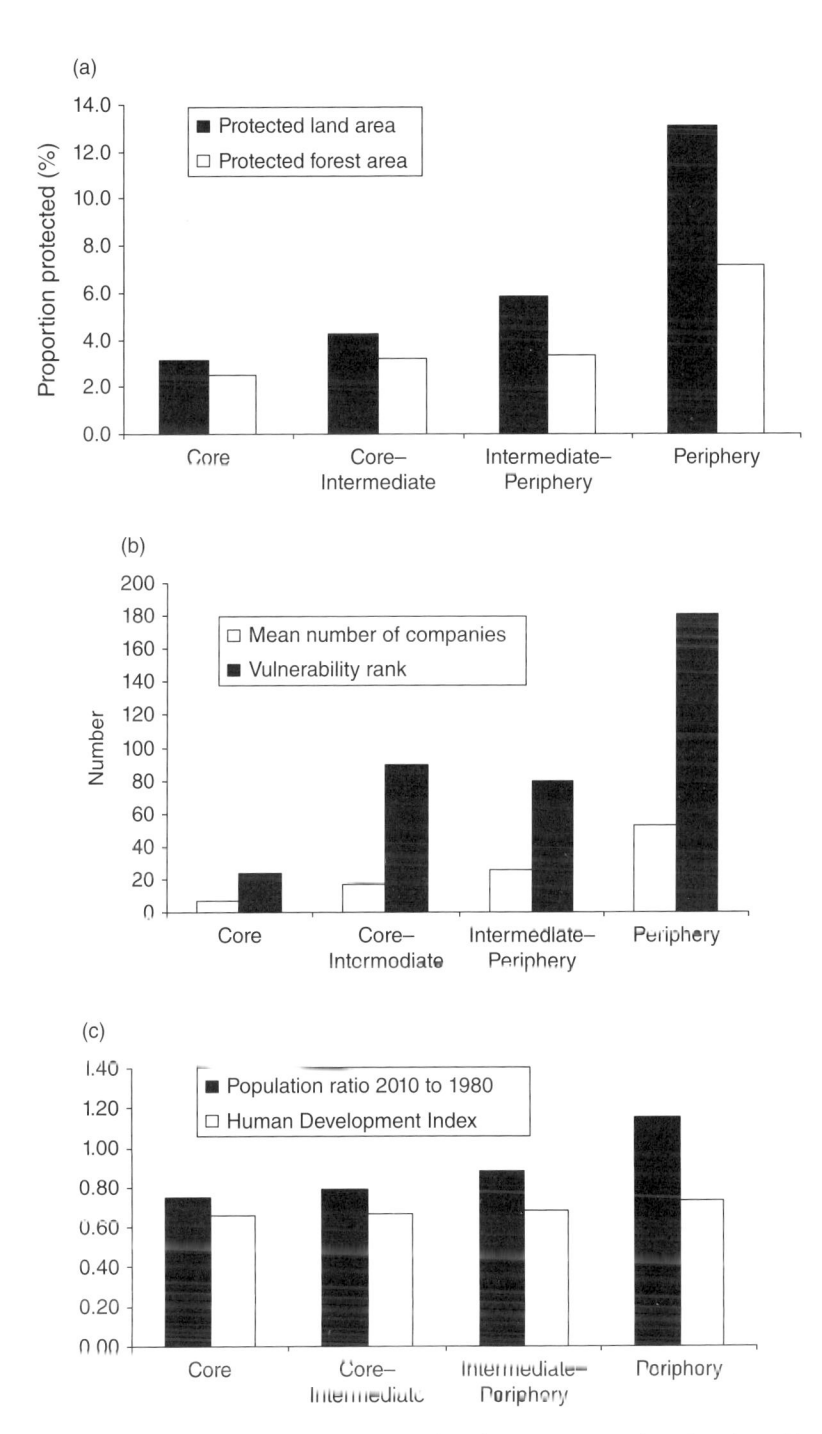

Fig. 20.4. Trends for the core to the periphery of the region of Bergslagen, southern Sweden, with respect to (a) ecological (Angelstam et al., 2013c), (b) economic (Angelstam et al., 2013c), (c) social (Axelsson et al., 2013a) and (d) cultural (Axelsson et al., 2013a) dimensions of sustainability.

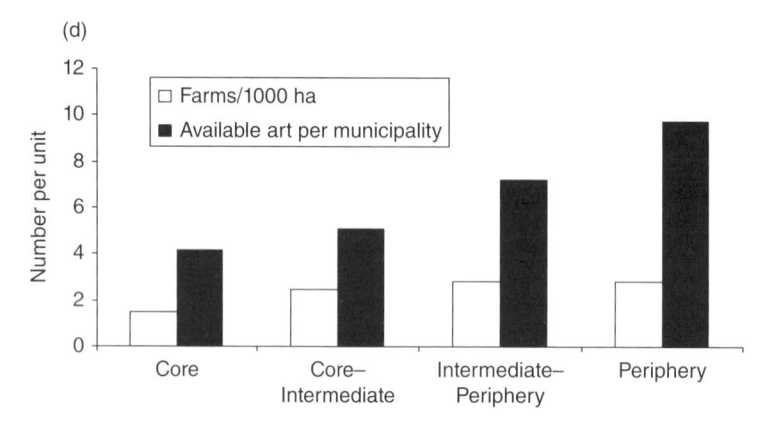

Fig. 20.4. Continued

of old pine forest was highest (42%) in the north boreal ecoregion and considerably lower (5–14%) in the three other ecoregions. Old deciduous forests had the highest percentage of functional networks in the hemiboreal forest ecoregion (21%), where the amounts were two to four times as high as in the other ecoregions. Finally, the proportion of functional forest–field edge habitat was generally very low (0–11%). Thus, only a small proportion of the four forest habitats of high conservation value were functional for demanding focal species, and the figures for the functionality of the protected area as functional habitat networks presented in Fig. 20.4a are seriously overestimated.

Key terrestrial ecosystem processes, such as fire or browsing by large herbivores, have been substantially altered by humans. Fire was an important disturbance regime in past boreal forests (Zackrisson, 1977), and was also a management tool for swidden (or slash-and-burn) agriculture (Montelius, 1953; Sarmela, 1987). An understanding of the historical range of variation in forest landscape dynamics can be achieved by using historical documents and maps and ecological records (such as pollen, charcoal and macrofossils in sediments, and fire scars), and by studying remnants of natural or culturally intact areas, as well as by modelling (see also Latałowa et al., Chapter 17).

Axelsson et al. (2007) compared the estimated natural and anthropogenic cultural forest disturbance regimes in Sweden, and found that the diversity had declined dramatically compared with reference landscapes where natural and cultural disturbance regimes were still dominant. Deciduous forest habitat networks are particularly important for forest biodiversity conservation in both boreal and temperate forests (Mikusinski et al., 2003). However, due to the long history of intensive forest management in Fennoscandia, deciduous forest patches resulting from stand-replacing forest disturbances such as fire do not form a functional green infrastructure (Manton et al., 2005) compared with regions with a short history of intensive forestry (e.g. Edman et al., 2011). Under the current forest industrial regime (*sensu* Lehtinen, 2006), which focuses on maximum sustained-yield forestry based on conifers (Elbakidze et al., 2013a), natural forest components such as dead wood, large and old trees, and old coniferous and deciduous stands have declined in the long term.

Altered trophic interactions between large carnivores, large herbivores such as moose (*Alces alces*) and forest vegetation are an additional factor. The extirpation of viable populations of large carnivores contributed to high densities of large herbivores, and consequent negative biodiversity impacts (Suominen, 1999; Mathisen and Skarpe, 2011). A key aspect of this is limited recruitment of deciduous trees of aspen (*Populus tremula*) and sallow (*Salix caprea*), which provide habitat for specialized species (Angelstam et al., 2000). The recent reappearance of large carnivore populations may help to remedy this situation.

The sustainability of aquatic ecosystems has also been affected by pressures at multiple scales. The most obvious changes concern streams and rivers (Ramberg, 1976), which were modified by water regulation and used for floating timber (Törnlund and Östlund, 2002; Bladh, 2008). The construction of dams has impaired connectivity for some species (Angelstam *et al.*, 2013c), but so has the clearance of obstacles such as boulders, rocks and large woody debris from headwater streams to improve timber floating (Jutila, 1985; Törnlund and Östlund, 2002). Streams have been canalized, straightened and flumed for logging (Nilsson *et al.*, 2005). Land drainage activities were originally driven by agricultural expansion during the 19th century, but during the 20th century became more common in forestry, as part of road construction or to improve tree growth in what were permanently moist forests and other wetlands (Hånell, 2006). The complete loss of forest affects streams and lakes in other ways as well (Törnblom *et al.*, 2011).

Riparian forests that produce and maintain coarse woody debris in streams are now scarce, and the amount of woody debris in streams is much reduced (Dahlström *et al.*, 2005), which has a negative impact on fish (Degerman *et al.*, 2004). Woody debris also affects water current, erosion and the retention of organic material. In addition, it affects the distribution of bottom substrates, all of which are crucial for ecological integrity and biodiversity within streams in the forest landscape (Dahlström and Nilsson, 2005).

Overall, the extensive alteration of streams, riparian forests and entire catchments has caused:

- Reduced diversity of habitats and species (Brookes, 1988; Dahlström and Nilsson, 2005; Nilsson *et al.*, 2005).
- Increased flow variations due to faster runoff and less capacity for water storage in forest soils (Ramberg, 1976). This variability results in disruption and stress that may adversely affect organisms. Reduced flow can cause problems with the clogging of spawning habitats for fish (Degerman and Nyberg, 2002).
- Increased erosion in some forest streams (Degerman and Nyberg, 2002). The increased sediment transport and deposition in slow-flowing sections means that streams become shallower and the range of associated habitats is reduced.
- Deterioration in water quality in streams because faster runoff means less opportunity for water purification (Ramberg, 1976; Foster *et al.*, 2005).

Landscape restoration (Mansourian *et al.*, 2005) is needed in both terrestrial and aquatic ecosystems, but there is a gap between evidence-based knowledge and policy on the one hand, and reality on the other (Angelstam *et al.*, 2011a, 2013a). National-level rhetoric tends to stress either the continued pressure on valuable natural systems or the responses made in terms of set-asides and new management practices.

The net effect on forest landscapes of pressures due to intensive management, and responses aimed at the mitigation of negative effects on the state of the ecosystem, are poorly understood by practitioners, policy makers and the public. Many in the forestry sector believe that employment is threatened by forest protection, but in fact the impact of internal reorganization and mechanization on employment is much greater (Keskitalo and Lundmark, 2009). Natural science is thus insufficient to guide the management of valuable natural and cultural landscapes. Understanding of the prevailing economic, social and cultural conditions is also needed in order to be able to plan at both strategic and tactical levels to guide local management.

20.6.2 Economic sustainability

Swedish forestry has long contributed to the national economy through exports with increasing added value. Britain's removal of import tariffs triggered a dramatic increase in wood exports from the boreal forests north of Bergslagen in the mid-19th century (Törnlund and Östlund, 2002) (Fig. 20.5). However, in Bergslagen, local wood use was already intense during the 18th century (Wieslander, 1936), and about 80% of the wood harvest was made into charcoal for the iron industry (Angelstam *et al.*, 2013c). Sustained yield forestry based on clear-felling was introduced into Bergslagen in the mid-19th century from

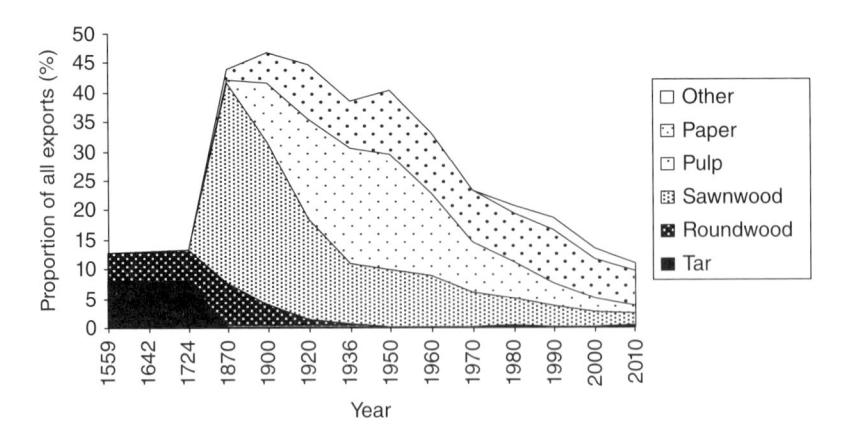

Fig. 20.5. The proportion of all contemporary Swedish monetary export value that was made up by different kinds of forest goods from 1559 to 2010 (data from Heckscher 1935–1949; Statistics Sweden).

Germany, where it was the first in Fennoscandia (Brynte, 2002), and it is still practised effectively (Axelsson et al., 2007; Elbakidze et al., 2013a). The knowledge and experience of management and silvicultural systems gained in Bergslagen were important in north Sweden where sustained yield wood production was introduced later (Streyffert, 1957); it was also important to other countries in similar phases of sustained yield development (Andrews, 1872; Nordberg et al., 2013).

Farmers owned most of the forests in the temperate forest zone, and also in river valleys up into the upland areas, where agriculture was the main land use. The farming and the mining economies were linked by food exports to the uplands, and by the transport of iron and charcoal to ports in the lowlands. To secure further wood resources for charcoal production, in the 17th century the government allowed the use of Crown land. The location of mineral deposits had a major influence on forest ownership and also for farmers who already owned forests, although extensive forest areas were increasingly becoming owned, or used, by large estates.

Increased globalization of the mining sector during the late 19th and 20th centuries led to a gradual decline in iron production. A transition from producing wood for charcoal to wood for sawmilling, paper and pulp, as well as, later, for bioenergy, followed in the late 20th century. Mechanization and rationalization of the forestry sector, especially since the 1970s, severely reduced job opportunities. These transitions have resulted in an increased role for public sector jobs linked to health, care and education, and to economic vulnerability (Tillväxtverket, 2011; Andersson et al., 2013a), which is greatest (i.e. vulnerability ranks around 25, few companies providing jobs) in the core of Bergslagen (Fig. 20.4b).

Traditionally, forest owners have focused on the sale of wood and timber products. However, forests also provide carbon sequestration, have an influence on water quantity and quality, can conserve biodiversity and can provide opportunities for rural development by maintaining natural and cultural values (Länsstyrelsen i Västmanlands Län, 2014). This has led to a debate about the purpose of forestry (Duncker et al., 2012). The General Director of the Swedish Forest Agency recently highlighted this by stressing the need for new forest products based on wood, fibre and biorefinery, as well as recreation and tourism (Stridsman, 2013), which might also provide future opportunities for increased employment.

20.6.3 Social and cultural sustainability

Magis and Shinn (2009) identify four constituents of social sustainability: human well-being, equity, democratic governance and democratic civil society. There is increasing interest in the non-material aspects linked to social capital, in addition to basic social needs and material cultural heritage (Axelsson et al., 2013a).

Social sustainability has historically focused on the availability of jobs in local communities and Reed (1999) and Rannikko (1999) argued for the need for production methods and innovations that sustain employment. However, social values are also related to quality of life, including such things as equity, participation in democratic life, security and health (Rosenström et al., 2006). Yet in Swedish forest policy, social values are often reduced just to recreational issues (Royal Swedish Academy of Agriculture and Forestry, 2009; Zaremba, 2012; Birkne et al., 2013).

The long history of the industrial use of natural resources once provided many local jobs, but resulted in relatively low levels of entrepreneurship and education among local people compared with that in regions with more diversified livelihoods (Bergdahl et al., 1997). Mechanization reduced employment severely between 1970 and 1990, and contributed to urbanization and thus the need for novel approaches to rural development. Figure 20.4c shows that two indicators of social sustainability, population change over the past 30 years and the Human Development Index (HDI), are much lower in the core than in the periphery of Bergslagen.

The declining population has led to the continued reduction of active agriculture, and the core of Bergslagen has the lowest density of active farms per unit area, as well as the lowest availability of art per municipality (Fig. 20.4d). None the less, the long history and associated cultural heritage of the area are attractive to tourists and there is potential for expansion (Vail and Hultkrantz, 2000), though this requires that people and local communities are willing to learn and be part of this change (Reed, 1999). The long history of stable land ownership by non-industrialized private landowners means that they frequently have a strong tie to their land. The associated social capital could enhance the resilience and ability of people in a local area to cope with uncertainty and change (Putnam, 2001). We confirmed the prediction that the proportion of non-industrial private forest land is correlated with population change in municipalities, using data from 114 municipalities in the nine counties where the Bergslagen region is located (Fig. 20.6).

20.7 Bridges Towards Sustainable Forest Management

Criteria and indicator systems for sustainable forest management, forest management plans and forest certification systems have contributed to increasing awareness about what sustainable forest management means. Nevertheless, there are gaps that need to be bridged for the implementation of sustainable forest management policy (Angelstam et al., 2004a; McDonald and Lane, 2004). The zoning of forest management intensity, traditional agroforestry systems and the application of integrated landscape approaches at local and regional levels of governance are three emerging tools.

The incompatibility of achieving different forest functions in the same forest stand has triggered discussions about forest zoning (Ranius and Roberge, 2011; Andersson et al., 2013b). For example, the Swedish state forest company Sveaskog introduced several categories of forest management intensity to focus on either biodiversity conservation or wood production, respectively (Angelstam and Bergman, 2004). However, the spatial complexity of forest ownerships hampers a zoning approach more generally across landscapes and regions.

Europe has a heritage of traditional agroforestry systems in which trees, crops and livestock are managed on the same area of land, resulting in a high environmental and cultural value (Mikusinski et al., 2003; Agnoletti, 2013). These systems are also important in Bergslagen, which is located at the interface between regions dominated by agriculture and forestry. Trees can stand inside parcels or on the boundaries as parts of hedges. Once widespread in traditional village systems, this integrated form of land use, which combined agriculture, animal husbandry and forestry (e.g. Elbakidze and Angelstam, 2007), can provide a range of benefits for production, landscape and biodiversity. Knowledge of agroforestry systems is often limited, and so its past and potential contribution to rural development is frequently undervalued, but it is starting to be recognized through support via agro-environmental schemes (Boonstra et al., 2011).

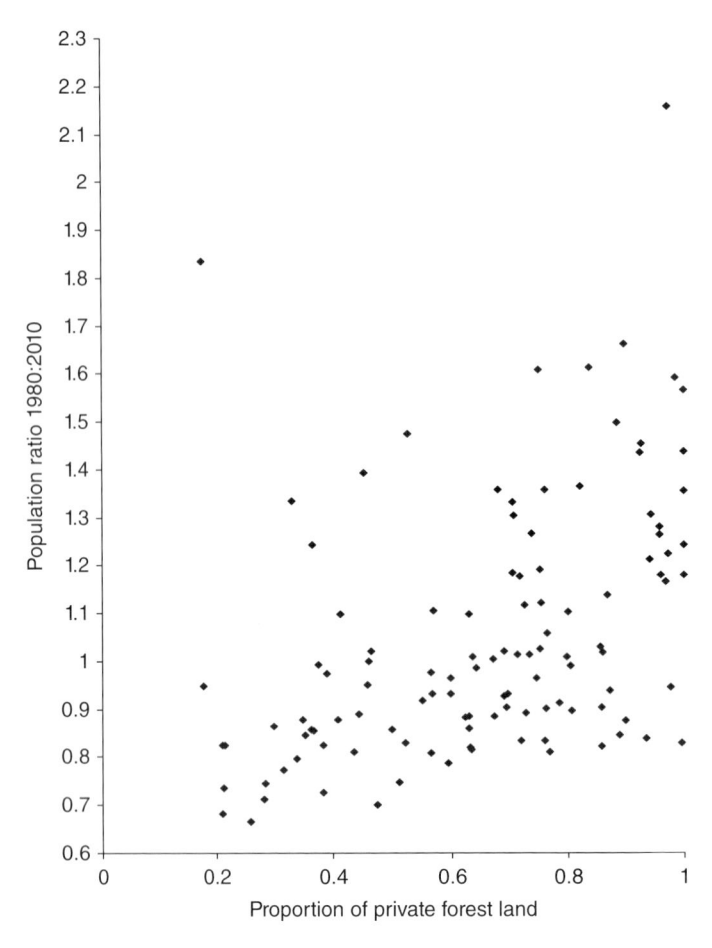

Fig. 20.6. Relationship between the proportion of forest land owned by non-industrial forest owners and the ratio of the human population in 114 municipalities in 1980 to that in 2010 ($n = 114$; $r = 0.44$; $r^2 = 0.19$; $F(1,112)$; $P = 26.9$, $P = 9.58\text{E-}07$).

The complexity of implementing sustainable forest management on the ground can be approached through place-based platforms or initiatives. Using these, analyses of states and trends and broad stakeholder engagement can be combined to develop and test local-scale innovation and adaptation strategies, and to monitor such efforts over the long term. Often summarized under the term 'landscape approach' (Axelsson et al., 2011), examples include the EU Leader ('Links between the rural economy and development actions') scheme (European Commission, 2006), Model Forests (IMFN, 2008), Biosphere Reserves (Elbakidze et al., 2013b), Ecomuseums (Davis, 2011) and long-term socio-ecological research (LTSER; Haberl et al., 2006; Singh et al., 2013). All five approaches are being used in the Bergslagen region (Andersson et al., 2013a,b; Axelsson et al., 2013b).

20.8 Discussion

20.8.1 From forest history to history of forest landscapes

Historical lessons can help us to address today's problems. To support implementation of sustainable forest management policy on the ground we need to study the changing interaction of humans and the natural environment (Marsh, 1864; Lowenthal, 2000). Landscapes are social–ecological systems: understanding their history requires collaboration across the disciplines found within the 'two cultures' of human and natural sciences *sensu* Snow (1959), as well as between scholars and practitioners (Axelsson *et al.*, 2013b). Joint learning and communication can then provide the appropriate knowledge for decision makers, and thus the basis for action. Landscape restoration should be 'a planned process that aims to retain ecological integrity and enhance human well-being' (Mansourian *et al.*, 2005).

However, improved territorial planning is needed to apply historical knowledge in an effective way. Integration within and across sectors is required to handle conservation, management and restoration in order to deliver an increased range of goods, services and landscape values (Angelstam *et al.*, 2011a). This applies not only to forestry, but also mining, hydropower, wind power, tourism and rural development. Local adaptation to the mix of different ownership categories is also crucial. Larsen (2009) stressed the need to recognize multiple world views and the differing philosophical foundations that guide understanding of both social and ecological systems, ranging from positivism to constructivism.

Equally important is the need to understand how sustainability is influenced by different strategies for regional development. For instance, the core–periphery strategy for regional development that has been applied in Sweden since the beginning of the industrialization has drained rural regions of inhabitants, natural resources and private ownership. From a rural perspective, this strategy has hampered social and cultural sustainability, as well as local economic sustainability. A prerequisite for a sustainable development of these dimensions in rural settings is conditions that promote opportunities for people to earn a living in a societal context where there is a public sense of belonging and empowerment in the sustainable management of the local landscape. Place-based and community-led strategies are more likely than the core–periphery strategy to provide such conditions in remote and rural regions (Dubois and Roto, 2012).

20.8.2 Landscapes with different histories: using space for time substitution

There is growing consensus about the need to develop sustainable use of forest resources, services and values, but there is little agreement about how to interpret and implement this goal (Lidskog *et al.*, 2013). By comparing possible actions across a range of landscapes with different histories, we can see which strategies might be effective where. Studies of countries and regions with different histories (Lehtinen, 2006; Angelstam *et al.*, 2011b, 2013b,d) are one way of understanding how forest landscape use and its consequences develop over time, and also the effects of different institutional contexts on sustainability. Comparisons of regions and countries with a shorter and longer history of the use of natural resources than in Bergslagen, and with different governance arrangements (Angelstam *et al.*, 2013b), can validate results from single case studies.

Forest landscape history provides the essential temporal context to run alongside spatial knowledge in attempts to implement sustainable forest management. By learning from the past, we can contribute to more sustainable forest landscapes across Europe in the future.

References

Agnoletti, M. (ed.) (2013) *Italian Historical Landscapes. Cultural Values for the Environment and Rural Development*. Springer, Dordrecht, The Netherlands.

Ahti, T., Hämet-Ahti, L. and Jalas, J. (1968) Vegetation zones and their sections in northwestern Europe. *Annales Botanici Fennica* 5, 169–211.

Andersson, K., Angelstam, P. Axelsson, R., Elbakidze, M. and Törnblom, J. (2013a) Connecting municipal and regional level planning: analysis and visualization of sustainability indicators in Bergslagen, Sweden. *European Planning Studies* 21, 1210–1234.

Andersson, K., Angelstam, P., Elbakidze, M., Axelsson, R. and Degerman, E. (2013b) Green infrastructures and intensive forestry: need and opportunity for spatial planning in a Swedish rural–urban gradient. *Scandinavian Journal of Forest Research* 28, 143–165.

Andrews, C.C. (1872) *Report on the Forests and Forest-Culture of Sweden*. Government Printing Office, Washington, DC.

Angelstam, P. (1997) Landscape analysis as a tool for the scientific management of biodiversity. *Ecological Bulletins* 46, 140–170.

Angelstam, P. and Bergman, P. (2004) Assessing actual landscapes for the maintenance of forest biodiversity – a pilot study using forest management data. *Ecological Bulletins* 51, 413–425.

Angelstam, P., Wikberg, P.E., Danilov, P., Faber, W.E. and Nygrén, K. (2000) Effects of moose density on timber quality and biodiversity restoration in Sweden, Finland and Russian Karelia. *Alces* 36, 133–145.

Angelstam, P., Persson, R. and Schlaepfer, R. (2004a) The sustainable forest management vision and biodiversity – barriers and bridges for implementation in actual landscapes. *Ecological Bulletins* 51, 29–49.

Angelstam, P., Mikusinski, G. and Fridman, J. (2004b) Natural forest remnants and transport infrastructure – does history matter for biodiversity conservation planning? *Ecological Bulletins* 51, 149–162.

Angelstam, P., Andersson, K., Axelsson, R., Elbakidze, M., Jonsson, B.-G. and Roberge, J.-M. (2011a) Protecting forest areas for biodiversity in Sweden 1991–2010: policy implementation process and outcomes on the ground. *Silva Fennica* 45, 1111–1133.

Angelstam, P., Axelsson, R., Elbakidze, M., Laestadius, L., Lazdinis, M., Nordberg, M., Pătru-Stupariu, I. and Smith, M. (2011b) Knowledge production and learning for sustainable forest management: European regions as a time machine. *Forestry* 84, 581–596.

Angelstam, P., Elbakidze, M., Axelsson, R., Dixelius, M. and Törnblom, J. (2013a) Knowledge production and learning for sustainable landscapes: seven steps using social–ecological systems as laboratories. *Ambio* 42, 116–128.

Angelstam, P., Grodzynskyi, M., Andersson, K., Axelsson, R., Elbakidze, M., Khoroshev, A., Kruhlov, I. and Naumov, V. (2013b) Measurement, collaborative learning and research for sustainable use of ecosystem services: landscape concepts and Europe as laboratory. *Ambio* 42, 129–145.

Angelstam, P., Andersson, K., Isacson, I., Gavrilov, D.V., Axelsson, R., Bäckström, M., Degerman, E., Elbakidze, M., Kazakova-Apkarimova, E.Yu., Sartz, L. *et al.* (2013c) Learning about the history of landscape use for the future: consequences for ecological and social systems in Swedish Bergslagen. *Ambio* 42, 150–163.

Angelstam, P., Roberge, J.-M., Axelsson, R., Elbakidze, M., Bergman, K.-O., Dahlberg, A., Degerman, E., Eggers, S., Esseen, P.-A., Hjältén, J. *et al.* (2013d) Evidence-based knowledge versus negotiated indicators for assessment of ecological sustainability: the Swedish Forest Stewardship Council standard as a case study. *Ambio* 42, 229–240.

Axelsson, R., Angelstam, P. and Svensson, J. (2007) Natural forest and cultural woodland with continuous tree cover in Sweden: how much remains and how is it managed? *Scandinavian Journal of Forest Research* 22, 545–558.

Axelsson, R., Angelstam, P., Elbakidze, M., Stryamets, N. and Johansson, K.-E. (2011) Sustainable development and sustainability: landscape approach as a practical interpretation of principles and implementation concepts. *Journal of Landscape Ecology* 4, 5–30.

Axelsson, R., Angelstam, P., Degerman, E., Teitelbaum, S., Andersson, K., Elbakidze, M. and Drotz, M.K. (2013a) Social and cultural sustainability: criteria, indicators and verifier variables for measurement and maps for vizualisation to support planning. *Ambio* 42, 215–228.

Axelsson, R., Angelstam, P., Myhrman, L., Sädbom, S., Ivarsson, M., Elbakidze, M., Andersson, K., Cupa, P., Diry, C., Doyon, F. *et al.* (2013b) Evaluation of multi-level social learning for sustainable landscapes: perspective of a development initiative in Bergslagen, Sweden. *Ambio* 42, 241–253.

Bass, S. (2004) Certification. In: Burley, J., Evans, J. and Youngquist, J.A. (eds) *Encyclopedia of Forest Sciences*. Elsevier, Oxford, UK, pp. 1350–1357.

Bergdahl, E., Isacson, M. and Mellander, B. (1997) *Bruksandan – Hinder Eller Möjlighet?* Ekomuseum Bergslagens Skriftserie 1. MediaPrint, Uddevalla, Sweden.

Bindler, R., Segerström, U., Pettersson-Jensen, I.M., Berg, A., Hansson, S., Holmström, H., Olsson, K. and Renberg, I. (2011) Early medieval origins of iron mining and settlement in central Sweden: multiproxy analysis of sediment and peat records from the Norberg mining district. *Journal of Archaeological Science* 38, 291–300.

Birkne, Y., Rydberg, D. and Svanqvist, B. (2013) *Skogens Sociala Värden – en Kunskapssammanställning*. Skogsstyrelsen Meddelande 9, Skogsstyrelsen, Jönköping, Sweden.

Björn, I. (2000) Takeover: the environmental history of the coniferous forest. *Scandinavian Journal of History* 25, 281–296.

Bladh, G. (2008) Selma Lagerlöf's Värmland. In: Jones, M. and Olwig, K. (eds) *Nordic Landscapes: Region and Belonging on the Nordic Edge of Europe*. University of Minnesota Press, Minneapolis, Minnesota, pp. 251–282.

Boonstra, W.J., Ahnström, J. and Hallgren, L. (2011) Swedish farmers talking about nature – a study of the interrelations between farmers' values and the socio-cultural notion of naturintresse. *Sociologia Ruralis* 51, 420–435.

Bradshaw, R. and Hannon, G. (1992) Climatic change, human influence and disturbance regime in the control of vegetation dynamics within Fiby Forest, Sweden. *Journal of Ecology* 80, 625–632.

Bradshaw, R.H.W. and Zackrisson, O. (1990) A two thousand year history of a northern Swedish boreal forest stand. *Journal of Vegetation Science* 1, 519–528.

Brookes, A. (1988) *Channelized Rivers: Perspectives for Environmental Management*. Wiley, Chichester, UK.

Brumelis, G., Jonsson, B.G., Kouki, J., Kuuluvainen, T. and Shorohova, E. (2011) Forest naturalness in northern Europe: perspectives on processes, structures and species diversity. *Silva Fennica* 45, 807–821.

Brynte, B. (2002) *C.L. Obbarius. En Nydanare i Bergslagens Skogar vid 1800-talets mitt* [C.L. Obbarius. An Innovator in Bergslagen's Forests during the mid-19th century]. Totab, Hållsta, Sweden.

Dahlström, N. and Nilsson, C. (2005) Influence of woody debris on channel structure in old growth and managed forest streams in central Sweden. *Environmental Management* 33, 376–384.

Dahlström, N., Jönsson, K. and Nilsson, C. (2005) Long-term dynamics of large woody debris in a managed boreal forest stream. *Forest Ecology and Management* 210, 363–373.

Davis, P. (2011) *Ecomuseums – A Sense of Place*. Continuum International Publishing Group, London.

Degerman, E. and Nyberg, P. (2002) The SILVA project – buffer zones and aquatic biodiversity. In: *Sustainable Forestry to Protect Water Quality and Aquatic Biodiversity*. Kungl. Skogs- och Lantbruksakademiens Tidskrift 7, 107–112.

Degerman, E., Sers, B., Törnblom, J. and Angelstam, P. (2004) Large woody debris and brown trout in small forest streams – towards targets for assessment and management of riparian landscapes. *Ecological Bulletins* 51, 233–239.

Dubois, A. and Roto, J. (2012) *Making the Best of Europe's Sparsely Populated Areas. On Making Geographic Specificity a Driver for Territorial Development in Europe*. Nordregio Working Paper 15, Nordic Centre for Spatial Development, Stockholm.

Duncker, P.S., Raulund-Rasmussen, K., Gundersen, P., Katzensteiner, K., De Jong, J., Ravn, H.P., Smith, M., Eckmüllner, O. and Spiecker, H. (2012) How forest management affects ecosystem services, including timber production and economic return: synergies and trade-offs. *Ecology and Society* 17(4): 50.

Edman, T., Angelstam, P., Mikusiński, G., Roberge, J.-M. and Sikora, A. (2011) Spatial planning for biodiversity conservation: assessment of forest landscapes' conservation value using umbrella species requirements in Poland. *Landscape and Urban Planning* 102, 16–23.

Elbakidze, M. and Angelstam, P. (2007) Implementing sustainable forest management in Ukraine's Carpathian Mountains: the role of traditional village systems. *Forest Ecology and Management* 249, 28–38.

Elbakidze, M., Angelstam, P., Andersson, K., Nordberg, M. and Pautov, Yu. (2011) How does forest certification contribute to boreal biodiversity conservation? Standards and outcomes in Sweden and NW Russia? *Forest Ecology and Management* 262, 1983–1995.

Elbakidze, M., Andersson, K., Angelstam, P., Armstrong, G.W., Axelsson, R., Doyon, F., Hermansson, M., Jacobsson, J. and Pautov, Yu. (2013a) Sustained yield forestry in Sweden and Russia: how does it correspond to sustainable forest management policy? *Ambio* 42, 160–173.

Elbakidze, M., Hahn, T., Mauerhofer, V., Angelstam, P. and Axelsson, R. (2013b) Legal framework for biosphere reserves as learning sites for sustainable development: a comparative analysis of Ukraine and Sweden. *Ambio* 42, 174–187.

Emanuelsson, M. and Segerstrom, U. (2002) Medieval slash and burn cultivation: strategic or adapted land use in the Swedish mining district? *Environment and History* 8, 173–196.

European Commission (2006) *The Leader Approach. A Basic Guide*. Office for Official Publications of the European Communities, Luxembourg.

European Commission (2013a) *A New EU Forest Strategy: For Forests and the Forest-Based Sector*. Publication No. COM(2013) 659, Brussels.

European Commission (2013b) *A New EU Forest Strategy: For Forests and the Forest-Based Sector*. Commission Staff Working Document SWD(2013) 342, Brussels.

Forest Europe, UNECE and FAO (2011) *State of Europe's Forests 2011. Status and Trends in Sustainable Forest Management in Europe*. Jointly prepared by Forest Europe (The Ministerial Conference on the Protection of Forests in Europe) Liaison Unit Oslo, the United Nations Economic Commission for Europe and the United Nations Food and Agriculture Organization of the United Nations. Forest Europe Liaison Unit, Oslo, Norway.

Foster, N.W., Beall F.D. and Kreutzweiser, D.P. (2005) The role of forests in regulating water: the Turkey Lakes Watershed case study. *The Forestry Chronicle* 81, 142–148.

Fransson, S. (1965) The borderland. *Acta Phytogeographica Suecica* 50, 167–175.

Geijerstam, J. and Nisser, M. (2011) *Swedish Mining and Metalworking – Past and Present*. Vol. 24, National Atlas of Sweden. Norstedts Förlagsgrupp, Stockholm.

Haberl, H., Winiwarter, V., Andersson, K., Ayres, R.U., Boone, C., Castillo, A., Cunfer, G., Fischer-Kowalski, M., Freudenburg, W.R., Furman, E. *et al.* (2006) From LTER to LTSER: conceptualizing the socioeconomic dimension of long-term socio-ecological research. *Ecology and Society* 11(2): 13.

Hæggström, C.-A. (2000) The age and size of hazel (*Corylus avellana* L.) stools of Nåtö Island, Åland Islands, SW Finland. In: Agnoletti, M. and Anderson, S. (eds) *Methods and Approaches in Forest History*. CAB International, Wallingford, UK, pp. 67–77.

Hånell, B. (2006) *Dikad Skogsmark och Myr med Djup Torv som Resurser för Uthålligt Torvbruk i Sverige*. Rapport, Institutionen för Skogsskötsel, Sveriges Lantbruksuniversitet, Umeå, Sweden.

Heckscher, E. (1935–1949) *Sveriges Ekonomiska Historia från Gustav Vasa*. Albert Bonniers Förlag, Stockholm.

IMFN (2008) *Model Forest Development Guide*. International Model Forest Network Secretariat, Ottawa.

Jutila, E. (1985) Dredging of rapids for timber-floating in Finland and its effects on river-spawning fish stocks. In: Alabaster, J.S. (ed.) *Habitat Modification and Freshwater Fisheries. Proceedings, Symposium on Habitat Modification and Freshwater Fisheries, Aarhus (Denmark), 23 May 1984*. Published for Food and Agriculture Organization of the United Nations, Rome by Butterworths, London, pp. 104–108.

Keskitalo, E.C.H. and Lundmark, L. (2009) The controversy over protected areas and forest-sector employment in Norrbotten, Sweden: forest stakeholder perceptions and statistics. *Society and Natural Resources* 23, 146–164.

Lagerås, P. (2007) *The Ecology of Expansion and Abandonment: Medieval and Post-medieval Land-Use and Settlement Dynamics in a Landscape Perspective*. National Heritage Board, Stockholm.

Länsstyrelsen i Västmanlands Län (2014) *Regionalt Utvecklingsprogram (RUP) för Västmanlands Iän 2014–2020*. Västerås, Sweden.

Larsen, G.L. (2009) An inquiry into the theoretical basis of sustainability. In: Dillard, J., Dujon, V. and King, M.C. (eds) *Understanding the Social Dimension of Sustainability*. Routledge, New York and London, pp. 45–82.

Lehtinen, A.A. (2006) *Postcolonialism, Multitude, and the Politics of Nature. On the Changing Geographies of the European North*. University Press of America, Lanham, Maryland.

Lehtonen, H. and Huttunen, P. (1997). History of forest fires in eastern Finland from the fifteenth century AD – the possible effects of slash-and-burn cultivation. *The Holocene* 7, 223–228.

Lidskog, R., Sundqvist, G., Kall, A.-S., Sandin, P. and Larsson, S. (2013) Intensive forestry in Sweden: stakeholders' evaluation of benefits and risk. *Journal of Integrative Environmental Sciences* 10, 145–160.

Lowenthal, D. (2000) Nature and morality from George Perkins Marsh to the millennium. *Journal of Historical Geography* 26, 3–27.

Magis, K. and Shinn, C. (2009) Emergent principles of social sustainability. In: Dillard, J., Dujon, V. and King, M.C. (eds) *Understanding the Social Dimension of Sustainability*. Routledge, New York and London, pp. 15–44.

Mansourian, S., Vallauri, D. and Dudley, N. (eds) (2005) *Forest Restoration in Landscapes, Beyond Planting Trees*. Springer, New York.

Manton, M.G., Angelstam, P. and Mikusinski, G. (2005) Modelling habitat suitability for deciduous forest focal species – a sensitivity analysis using different satellite land cover data. *Landscape Ecology* 20, 827–839.

Marsh, G.P. (1864) *Man and Nature; or, Physical Geography as Modified by Human Action*. Charles Scribner, New York.

Mathisen, K.M. and Skarpe, C. (2011) Cascading effects of moose (*Alces alces*) management on birds. *Ecological Research* 26, 563–574.

McDonald, G.T. and Lane, M.B. (2004) Converging global indicators for sustainable forest management. *Forest Policy and Economics* 6, 63–70.

Mikusinski, G., Angelstam, P. and Sporrong, U. (2003) Distribution of deciduous stands in villages located in coniferous forest landscapes in Sweden. *Ambio* 33, 520–526.

MOE (2013) *The Swedish Environmental Objective System*. Information Sheet, Ministry of the Environment, Government Offices of Sweden, Stockholm.

Montelius, S. (1953) The burning of forest land for the cultivation of crops. 'Svedjebruk' in Central Sweden. *Geografiska Annaler* 35, 41–54.

Nelson, H. (1913) En Bergslagsbygd. *Ymer* 33, 278–352.

Nilsson, C., Lepori, F., Malmqvist, B., Törnlund, E., Hjerdt, N., Helfield, J.M., Palm, D., Östergren, J., Jansson, R., Brännäs, E. and Lundqvist, H. (2005) Forecasting environmental responses to restoration of rivers used as log floatways: an interdisciplinary challenge. *Ecosystems* 8, 779–800.

Nordberg, M., Angelstam, P., Elbakidze, M. and Axelsson, R. (2013) From logging frontier towards sustainable forest management: experiences from boreal regions of North-West Russia and North Sweden. *Scandinavian Journal of Forest Research* 8, 797–810.

Östlund, L., Zackrisson, O. and Axelsson, A.-L. (1997) The history and transformation of a Scandinavian boreal forest landscape since the 19th century. *Canadian Journal of Forest Research* 27, 1198–1206.

Pettersson, Ö. (1999) *Population Changes in Rural Areas in Northern Sweden 1985–1995*. CERUM Working Paper No. 13, CERUM (Centre for Regional Science), Umeå, Sweden.

Putnam, R.D. (2001) Social capital: measurement and consequences. In: Helliwell, J.F. (ed.) *The Contribution of Human and Social Capital to Sustained Economic Growth and Well-being*. Human Resources Development Canada, Ottawa, pp. 117–135.

Ramberg, L. (1976) Effects of forestry operations on aquatic ecosystems. In: Tamm, C.O. (ed.) *Man and the Boreal Forest*. *Ecological Bulletins* 21, 143–149.

Ranius, T. and Roberge, J.-M. (2011) Effects of intensified forestry on the landscape-scale extinction risk of dead wood dependent species. *Biodiversity Conservation* 20, 2867–2882.

Rannikko, P. (1999) Combining social and ecological sustainability in the Nordic forest periphery. *Sociologia Ruralis* 39, 394–410.

Reed, M.G. (1999) 'Jobs Talk': retreating from the social sustainability of forestry communities. *The Forestry Chronicle* 75, 755–763.

Rosenström, U., Mickwitz, P. and Melanen, M. (2006) Participation and empowerment-based development of socio-cultural indicators supporting regional decision-making for eco-efficiency. *Local Environment* 11, 183–200.

Royal Swedish Academy of Agriculture and Forestry (2009) *The Swedish Forestry Model*. Stockholm. Available at: http://www.skogsstyrelsen.se/Global/myndigheten/Skog%20och%20miljo/ENGLISH/retrieve_file.pdf (accessed 9 December 2014).

Sarmela, M. (1987) Swidden cultivation in Finland as a cultural system. *Suomen Antropologi – Journal of the Finnish Anthropological Society* 4/1987, 241–262.

Singh, S.J., Haberl, H., Chertow, M.R., Mirtl, M. and Schmid, M. (2013) *Long-Term Socio-Ecological Research (LTSER): How to Study Society-Nature Interactions across Spatial and Temporal Scales*. Springer, New York.

Snow, C. (1959) *The Two Cultures*. Cambridge University Press, Cambridge, UK.

Söderberg, T. (1932) *Stora Kopparberget under Medeltiden och Gustav Vasa*. S Victor Petterssons Bokindustriaktiebolag, Stockhom.

Sporrong, U. (2008) The Swedish landscape: the regional identity of historical Sweden. In: Jones, M. and Olwig, K. (eds) *Nordic Landscapes: Region and Belonging on the Nordic Edge of Europe*. University of Minnesota Press, Minneapolis, Minnesota, pp. 141–156.

Stephens, M.B., Ripa, M., Lundström, I., Persson, L., Bergman, T., Ahl, M., Wahlgren, C. H., Persson, P-O and Wickström, L. (2009) *Synthesis of the Bedrock Geology in the Bergslagen Region, Fennoscandian Shield, South-Central Sweden*. Publication Ba 58, Swedish Geological Survey, Uppsala, Sweden.

Storaunet, K.O., Rolstad, J. and Groven, R. (2000) Reconstructing 100–150 years of logging history in coastal spruce forest (*Picea abies*) with special conservation values in central Norway. *Scandinavian Journal of Forest Research* 15, 591–604.

Streyffert, T. (1957) *Influence of Ownership and Size Structure on Forest Management in Sweden. A Study of Fundamentals*. Bulletin of the Royal School of Forestry No. 23b, Stockholm.

Stridsman, M. (2013) Nytänk behövs för att möta globala utmaningar. *SkogsEko* 3, 3.

Suominen, O. (1999) Impact of cervid browsing and grazing on the terrestrial gastropod fauna in the boreal forests of Fennoscandia. *Ecography* 22, 651–658.

Tillväxtverket (2011) *Genuint Sårbara Kommuner. Företagandet, Arbetsmarknaden och Beroendet av Enskilda Större Företag*. Rapport 0112, Tillväxtverket, Stockholm.

Törnblom, J., Degerman, E. and Angelstam, P. (2011) Forest proportion as indicator of ecological integrity in streams using Plecoptera as a proxy. *Ecological Indicators* 11, 1366–1374.

Törnlund, E. and Östlund, L. (2002) The floating of timber in northern Sweden: construction of floatways and transformation of rivers. *Environment and History* 8, 85–106.

Vail, D. and Hultkrantz, L. (2000) Property rights and sustainable nature tourism: adaptation and maladaptation in Dalarna (Sweden) and Maine (USA). *Ecological Economics* 35, 223–242.

Wieslander, G. (1936) Skogsbristen i Sverige under 1600 – och 1700 – talen. *Sveriges Skogsvårdsförbunds Tidskrift* 34, 593–633.

Zackrisson, O. (1977) Influence of forest fire on the north Swedish boreal forest. *Oikos* 29, 22–33.

Zaremba, M. (2012) *Skogen vi Ärvde*. Svante Weyler Bokförlag, Stockholm.

Part V

Lessons From the Past for the Future?

In this final section of the book, we start by considering the history of woodland conservation and how that is now applied across Europe (Chapter 21). Woodland history, through attempting to catalogue ancient woodland, has proved a valuable guide to which sites might be given priority and has been a way of attracting public attention and support (Chapter 22).

History can also help us to address some of the major threats (such as tree diseases) to the future of our woods (Chapter 23).

Lastly, we reflect on the major ideas and trends to emerge from the last two decades of interest in European woods and forests, and suggest some of the areas where work might be needed in the next two decades (Chapter 24).

21 The Development of Forest Conservation in Europe

James Latham*

Bryn Ffynnon, Llanddona, Anglesey, UK

21.1 Introduction

Much of the forest cover of Europe is 'protected' for a variety of interests and purposes, and hence nature may also be conserved in some way, but protection and conservation do not always go hand in hand. Forests may be protected for practical reasons, such as to ensure timber supplies or for the physical protection of villages in mountainous regions, or for aesthetic, cultural or political reasons. Nature conservation interests may sometimes be damaged by management for these other interests. Equally, the designation of areas as important for nature conservation may mean that some other services such as wood production or grazing are curtailed; and if the special conservation interest is a non-woodland habitat or species, then trees and woods may be cleared for conservation reasons.

Protection and conservation are often viewed largely as passive processes – the prevention of damaging activities. However, some kind of intervention may be needed to achieve the desired goals, not least because of the cultural nature of much of the European landscape. The management of woodland for conservation and the integration of this with other forestry management objectives has recently been reviewed by Götmark (2013) for temperate forests and by Kraus and Krumm (2013) for central Europe.

This chapter draws particularly on an overview of forest conservation in Europe, the 'Protected Forest Areas in Europe' project (Latham *et al.*, 2005; Frank *et al.*, 2007; and data available online for the European COST E27 project 'Protected forest areas in Europe – analysis and harmonisation (PROFOR)' at http://www.efi.int/portal/virtual_library/information_services/cost_c27/). A second strand looks at the state of the habitats and species identified under the European Union (EU) 1992 Habitats and Species Directive, whose condition is reported on every 6 years (see http://bd.eionet.europa.eu/article17/reports2012/).

21.2 Why Conserve Forests?

21.2.1 As a spiritual place

Concern for the conservation of trees and forests is not simply a modern or utilitarian idea: Welzholz and Johann (2007) found that the earliest activities that could be considered to

---

*E-mail: jimstardrift@googlemail.com

be some sort of protection had a religious motivation and were universally reported across Europe. A recurring theme is the idea of a 'tree of life', beautifully epitomized by the Norse Yggdrasil – an ash tree stretching from deep in the earth to the heavens and holding all life within its branches. Welzholz and Johann (2007) emphasize the etymological connection between words such as sanctuary and sacred. 'Holy groves' existed throughout Europe. Tacitus, Roman senator and historian, provides an early written account of such 'groves' in his 98 AD work *Germania*:

> Woods and groves are the sacred depositories; and the spot being consecrated to those pious uses; they gave to that sacred recess the name of the divinity that fills the place which is never profaned by the steps of men. The gloom fills every mind with awe, revered at the distance, and never seen but with the eye of contemplation.

Mythic associations with forests appear to have overlapped and coexisted for a while with Christianity before being replaced or, perhaps more often, absorbed. Welzholz and Johann (2007) comment on the symbolic association of church pillars and towers with high forest structure – 'the sacred forest made stone'.

In Lithuania, where Christianity was not formally adopted until the 14th century, and even then was not forcefully imposed, a rich tradition persists, including the wearing of giant masks symbolizing pagan gods on feast days (Rowell, 1994). A widespread theme that is widely thought to relate to this transition is that of the 'Green Man'. Depictions of humans and animals spouting or being transformed into foliage are widespread in churches in Britain and western Europe. Their strict meaning is unknown, but they are often interpreted as representing some sort of fusion of Christian and pagan ideas (Hayman, 2003; Hooke, 2010).

The spiritual dimension of the conservation of forests, which encourages protection, continues to the present day (UNECE, 2011). Some continuity of past traditions may remain in Christmas trees, maypoles and the festive use of mistletoe and other greenery. Concern for the conservation of trees and woods is also a recurring theme in modern fiction, from Astrov in Uncle Vanya and the Ents in Lord of the Rings, to Dr Seuss's Lorax and the Tree of Souls in the film Avatar. Saving tropical forests and tree-planting campaigns are common among the activities of 'green' movements in Europe, with the associated ethics epitomized by Jean Giono's 1953 book, '*The Man who Planted Trees*'.

21.2.2 As a place for the chase

Another reason for early forest protection was to reserve areas for hunting for the nobility for the purposes of status and recreation (Fletcher, Chapter 9). These remain common modern objectives for woodland management.

From the early Middle Ages, there was widespread protection of 'forests' for hunting interests. Not all of these were necessarily dense forest in the modern sense, but many did have extensive tree cover. Within these large areas, the rights of the common people to fell trees, hunt, fish or graze livestock were often curtailed and controlled. Welzholz and Johann (2007) suggest that this was connected with social changes in Europe: as the disruption of the migrations and invasions that characterized the Dark Ages came to a close, more stable and influential systems of royalty became established that allowed this sort of land use appropriation to develop. The first evidence of these practices were reported from the 10th to the 12th centuries from Germany, Ireland and the UK (although there are indications that they were established earlier under Frankish monarchs; see Fletcher, Chapter 9); the practice was well established in Austria, Lithuania, the Netherlands, Italy, Romania and Sweden at the height of the interest in royal hunting forests from the 14th to 17th centuries.

The royal reservation of these forests generally waned in the 18th century through a combination of the broad social, economic and political changes of the enlightenment, and the reduction in the power of absolute monarchy (Watkins, 2014). However, many important modern forest reserves are derived from royal hunting forests; examples include Windsor Great Park (in the UK), Fontainebleu (France), Reichswald (Germany), and Białowieża (Poland)

(Fletcher, Chapter 9; Latałowa *et al.*, Chapter 17). Reservation for hunting also contributed to the survival of particular species long after their disappearance through the countryside generally; examples include bison (*Bison bonasus*) in Lithuania and Poland, beaver (*Castor fiber*) in Poland and ibex (*Capra ibex*) in Austria.

21.2.3 As a source of raw materials and a barrier against the elements

Concern about a perceived loss of forest and a consequent future reduction in timber production led to many countries seeking to limit rates of harvesting, forest clearance for farming and the practice of wood-pasture (Williams, 2002; Johann *et al.*, 2012; Hartel *et al.*, Chapter 5). Modern equivalents are regulations that govern felling and the clearance of woodland.

The protection and conservation of trees and woods as productive resources became increasingly important from the 16th century onwards (Box 21.1) This concern would also lead to the development of planned forest management (Savill, Chapter 7). There was also a realization that forests in some situations should be kept because they provide protection for people and their property from floods, landslips and avalanches. In the alpine areas of Austria, 'protection forests' to prevent avalanches falling on villages were established in the early 16th century (Frank *et al.*, 2005).

Box 21.1. Examples of early laws and customs promoting forest protection. (From sources in Latham *et al.*, 2005, and Welzholz and Johann, 2007.)

- 7th century, Portugal, the *Visigothic Code* – protection of cork oaks and pines for the services (food, shelter) they provided.
- 8th century, Ireland, the *Brehon Law*, classified trees into four classes in relation to value and included sanctions for damaging them.
- 13th century, Belgium, *Keurboeck van Soniën*, an ordinance regulating the use of the Zoniënwoud south of Brussels.
- 1355, Czech Republic, *Code of laws for the Czech Crown Lands* outlining the need to protect forests.
- 1398, Lithuania, the *Salynas Treaty*, which refers to the 'saint place' near the Nevèlis River where hunting, forest felling and even walking were restricted.
- 14th century, Romania, the so-called *Letter of the forbidden forest* restricting rights to use forests for timber, grazing, hunting, etc. without the owners' permission.
- 14th century onwards, Switzerland, 'banning letters' (*Bannbriefe*) were issued, prohibiting or restricting certain kinds of use of forests within defined areas.
- 1511, Austria, regulations for the *Vienna Forest* prohibiting grazing and clearance for grazing.
- 1517–1518, Austria, protection forests established in Tyrol and Carinthia to protect villages from avalanches.
- 1524, Austria, the Archbishop of Salzburg signed a law to ensure the sustainable use of forests to supply the mining industry.
- 1558, Sweden, beech and oak declared *the King's property* because of their use in shipbuilding and also their importance for pig foraging.
- 1014, Slovenia, the *Annex to the Ortenburg Forest Regulation* indicates protection of forests for their role in military defence.
- 1669, France, the first national forest law *Ordonnance portent règlement général pour les Eaux et Forêts* providing general protection and regulation.
- 1748, Spain, the *Ordenanza para el fomento, cultivo y conservacion des montes* was passed, the first regulations on forest development, cultivation and protection.
- 1771, Slovenia, *The Forest Ordinance for Carniola and Istria* was passed, regulating silvicultural use.
- 1852, Austria, *Reichsforstgesetz* – protection of forests for public welfare.
- 1854, Belgium, *The Belgium Forest Act*, regulating forest exploitation.
- 1874, Transylvania, the 'Hungarian' forest law, protection of forests to prevent soil erosion and avalanche.
- 1883, Bulgaria, the *1st Forest Act*, with forest and land protection as a responsibility of the state.

21.2.4 For a new form of communing with the forests

From about 1800, a set of new motivations for forest conservation made their appearance. These related to: the visual landscape; the appreciation of the picturesque; a more explicit recognition of woods and forests as places for recreation and public physical well-being; and the use of forests for scientific research and for wildlife protection. The last of these, nature conservation in its various forms, appears as only of minor importance in the first half of the 19th century, but it then becomes more prominent, with its significance rapidly increasing during the second half of the 20th century (Welzholz and Johann, 2007).

There may here be an element of relabelling what were once religious values as an appreciation and awe of nature; there is also an element of 'hunting' with the camera, rather than with gun or spear. Moreover, in the modern world, an increasing separation from working on the land and being obviously connected to trees and woods perhaps allows people more opportunity to step back and appreciate that nature is indeed at risk.

21.3 Type and Extent of Protected Forest Areas

It is hard to establish the numbers of sites and total area of forest included across Europe. There is no standard definition of a 'protected forest', and interpretations vary between countries, particularly in terms of what types of activity are allowed within protected areas. In addition, protection may happen without any sort of formal designation of the forests concerned.

In practice, most countries have several types of protected area designation (nature reserve, national park, etc.) which overlap to varying degrees with each other. They may be managed or regulated by a variety of different bodies (Vodde, 2007). For example, in England an area of woodland may be under protection by statutory designations, including those as Site of Special Scientific Interest, Special Area of Conservation, National Nature Reserve and Local Nature Reserve; it may also be covered by policies derived from it being ancient woodland (Goldberg, Chapter 22), or be subject to a tree preservation order. Certain damaging activities such as excessive felling are regulated by felling licences. Separately, the woodland might be protected through ownership by a voluntary conservation body (i.e. a non-governmental organization, or NGO) (Fig. 21.1; Kirby, 2003). While the obvious way to increase woodland protection might be seen as through increasing the extent of statutorily protected sites (arrow a in Fig. 21.1), in some circumstances, changing overall forestry policies (arrow b), or the area managed for conservation on a voluntary basis (arrow c), might prove more cost-effective.

About 14% of the forest area of European countries is reported as 'protected', in amounts ranging from none (in Malta, Luxembourg) to more than half of the wooded area in Liechtenstein and Slovakia (Fig. 21.2) (Eurostat, 2011). The significance of nature conservation as the prime motivation for protection differs. Slovakia reported that 57% of its forest area was protected, but that only 4% was protected for the conservation of biodiversity; in other countries, such as the UK, the countries' respondents have interpreted forest protection as synonymous with conservation of biodiversity, such that the figure is 100%, or in some cases more!

Over 260 types of protected areas have been reported from 25 countries, with each state typically reporting about ten types of protected areas, but with some reporting more than 20 types (Latham *et al.*, 2005). There was a gradation between strict, legally protected areas where nature conservation is the prime aim and little management activity occurs, to much broader areas where the protection may be less and apply to only certain aspects of the forest system, be covered by general policy rather than strict laws or even be voluntary.

'National Parks' were reported from every country except Belgium (where one has since been established). In some countries, this designation carries a high level of ecosystem protection with strict controls on human intervention (France, Portugal), in line with the International Union for Conservation of Nature (IUCN) definitions (Dudley and Phillips, 2006). In others, there is more an

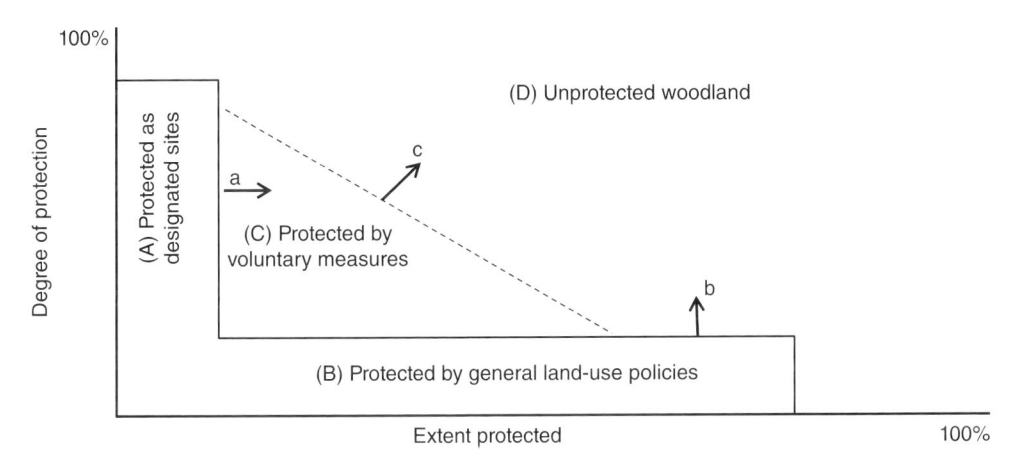

Fig. 21.1. Schematic view of woodland protection in England. (From Kirby, 2003.) Key: Zone A – protected sites with a high level of statutory protection (although not absolute), but of limited extent; Zone B – more limited protection through general statutory land-use policies and regulations over the majority of the woodland resource; Zone C – varying levels of voluntary protection afforded by, for example, ownership by non-governmental organizations (NGOs), interest of owner, certification schemes, etc.; Zone D – zone where no effective protection applies. Arrows a–c represent the way that the boundaries of the zones might shift to achieve different conservation outputs.

emphasis on nature protection within the context of cultural values, which may emphasize aspects such as historic features, aesthetics and recreation (the UK, Lithuania).

In part, this difference in the interpretation of what a national park is reflects differing values placed on the conservation of habitats and species that are – or are strongly associated with – cultural landscape forms such as coppice woods or wood-pastures. These require significant intervention to maintain them and their associated species, as opposed to more natural forests and landscapes where the emphasis is often upon allowing natural processes to operate.

Tensions develop where a desire or pressure to rely on 'natural processes' leads to loss of species and features from sites because they would have depended on management to sustain them. Particular examples of this are the loss of open-space butterflies from many British coppice woods following the neglect of these woods, and the loss of old trees and open habitat mosaics from eastern European wood-pastures where domestic livestock grazing has been banned (Hartel et al., Chapter 5; Buckley and Mills, Chapter 6).

Branquart et al. (2007) explored differences between types of protected forest and other characteristics of a country's forest cover using a multivariate approach. At a European level, protected forest types could broadly be separated according to the overall level of restriction placed on activity within them, which correlated well with an independent assessment of their placement within the various IUCN categories (Dudley and Phillips, 2006).

This overall 'strength of protection' was negatively correlated with the position of countries along the main axis of a principal components analysis (PCA) and with a number of independent national forestry statistics (MCPFE, 2003), notably two economic variables: gross domestic product per capita (GDP) and the economic value of national wood production. Thus, countries with a relatively high GDP and a high importance of forestry within that tended to have a relatively low level of restrictions on forestry; these included Austria, Switzerland, Germany, France, Norway and Sweden. Countries with relatively low GDP or low forestry contribution showed the reverse (Romania, Cyprus).

Where the forest industry is particularly important to the economy of a country, so too will be the opportunity costs of restrictions in protected areas. However, protected areas may

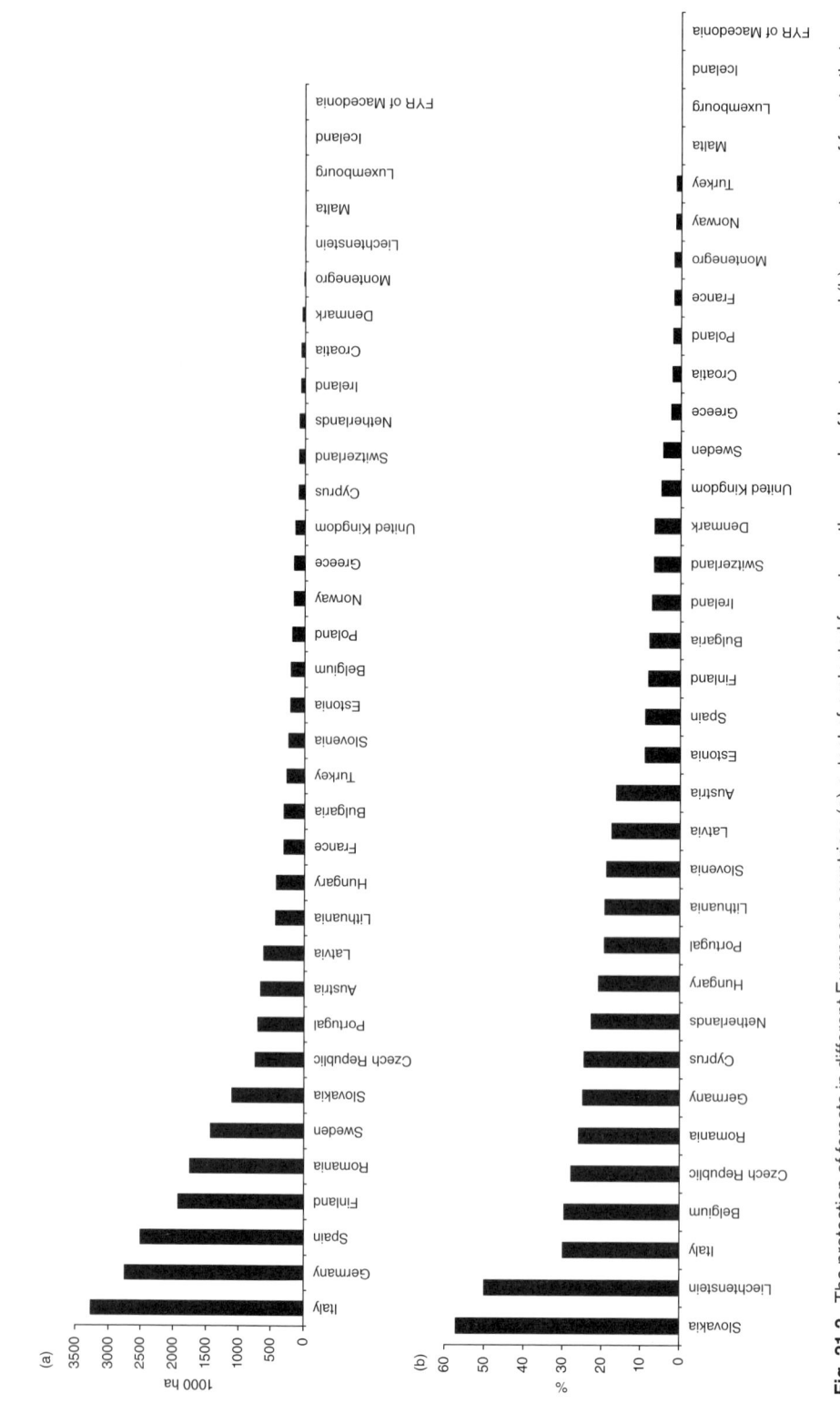

Fig. 21.2. The protection of forests in different European countries: (a) extent of protected forests as thousands of hectares; and (b) percentage of forests that are protected. (Based on Eurostat, 2011.)

also be under less pressure, because there is likely to be an extensive forest area and a well-developed forest culture, such that a statutory high level of protection for protected areas is less critical.

The secondary axis of variation shows a separation between types of protected forests in different countries, with types of restrictions. This was also correlated with how much forest cover a country has. At one end are countries where the restrictions largely relate to maintaining forest integrity and production, while at the other are those where restrictions are more about access and more specific interferences (e.g. hunting, berry collection). So countries with a high forest cover, for example Finland or Sweden, will tend to rely heavily on forests economically; forests are such an important part of the country's well-being that the question of threats to it simply do not arise. Rather, restrictions tend to be general and relate to actions to maintain the forest infrastructure. Countries with a relatively low forest cover, say the UK or the Netherlands, tend to rely less on forests for their GDP, and there may be greater competition for land and, hence, greater threats to forest habitats. In turn, restrictions tend to be stricter and more diverse, with tight limitations on many activities.

21.4 Selection of Protected Areas

In most countries, the selection of protected sites has been a somewhat ad hoc accumulative process. An initial selection is made on the basis of the best sites known to the ecologists/natural historians of the day, which is generally biased towards sites that are not a priority for other productive purposes, such as farming. Later, more formal criteria are drawn up to help with the process of evaluating candidate sites to fill gaps in the series and to ensure that the selection has a more objective basis (Box 21.2).

Further refinement has involved using gap analysis to identify where key habitats or species are under-represented across the suite of sites, often judged against potential natural vegetation or remotely sensed habitat maps (Nilsson and Götmark, 2003; Araujo et al., 2007; Branquart et al., 2008; Rosati et al., 2008). There has also been increasing concern that the relationships between sites and the surrounding countryside be considered, partly because of recognition that many species may exist as metapopulations across the landscape, but also because with climate change, species may need to move to new locations (Angelstam et al., 2003; Bruinderink et al., 2003; Gaston et al., 2006; Vos et al., 2008; Lawton et al., 2010).

21.5 Developing a European Perspective

Historically the priorities for protected areas in terms of habitats and species have been determined by individual countries. Nationally

---

Box 21.2. Development of the protected area system (Sites of Special Scientific Interest, SSSIs) in Britain.

In the guidelines for the selection of protected areas Derek Ratcliffe (1989) wrote that:

An inherited knowledge of the best places to find and collect rare and local species developed during Victorian times and became part of a common fund of knowledge amongst naturalists ... When, in 1915, Charles Rothschild compiled a list of desirable nature reserves ... he drew on the opinions of many leading figures throughout the country. Thirty two years later the listings of the Society's Nature Reserves Investigation Committee ... represented the distillation of collective knowledge from a large body of informed opinion. A similar sifting was applied to the choice of SSSIs when these became a statutory category.

Woodland surveys carried out from the 1960s onward, but particularly during the early 1980s, allowed for a more systematic assessment of which sites should be brought into the protected site system. Ten evaluation 'criteria' set out by Ratcliffe (1977) – from size and diversity to intrinsic appeal, were converted into more-or-less elaborate scoring procedures (Goodfellow and Peterken, 1981; Kirby 1993) to make the selection of sites more consistent.

rare woodland types and species might be accorded high status even though they are abundant elsewhere in Europe, while nationally common, but Europe-wide rare features have received less attention. International conventions and protocols have been used to try to address this issue and to make the conservation process more consistent across the continent.

Specific commitments on the conservation of species and habitats are included in the Bern Convention (1979) (Council of Europe, 2013) but, particularly for the EU, in the Habitats and Species Directive (1992) and the Birds Directive (1979). The Habitats and Birds Directives (European Commission, 2013a,b) are built around two pillars: the Natura 2000 Network of protected sites; and a strict system of species protection. Member States are required to select sites to cover habitats and species identified as of community importance for nature conservation (Latham et al., 2005). Inevitably though, there have been differences between countries in the way that this aspect of the Directive has been interpreted and hence in the number and extent of sites selected by different countries.

The 69 forest types included (Table 21.1) in the sites that have been listed for protection under the EU Habitats and Species Directive include examples from across the

Table 21.1. Woodland types listed for protection under the European Union Habitats and Species Directive.[a]

Type[b]	Habitat
Broadleaved evergreen forest	6310 Dehesas with evergreen *Quercus* spp.
	9320 *Olea* and *Ceratonia* forests
	9330 *Quercus suber* forests
	9340 *Quercus ilex* and *Quercus rotundifolia* forests
	9360 Macaronesian laurel forests (*Laurus, Ocotea*)
	9370 Palm groves of *Phoenix* sp.
	9380 Forests of *Ilex aquifolium*
	9390 Scrub and low forest vegetation with *Quercus alnifolia*
Coniferous forests of the Mediterranean, Anatolian and Macaronesian regions	9520 *Abies pinsapo* forests
	9530 (Sub) Mediterranean pine forests with endemic black pines
	9540 Mediterranean pine forests with endemic Mesogean pines
	9550 Canary Island endemic pine forests
	9560 Endemic forests with *Juniperus* spp.
	9590 *Cedrus brevifolia* forests
Thermophilous deciduous forests	91B0 Thermophilous *Fraxinus angustifolia* woods
	91H0 Pannonian white oak woods
	91I0 Euro-Siberian steppic woods with *Quercus* spp.
	91M0 Pannonian-Balkanic turkey oak forests
	9230 Galicio–Portuguese oak woods with *Quercus robur* and *Q. pyrenaica*
	9240 *Quercus faginea* and *Quercus canariensis* Iberian woods
	9250 *Quercus trojana* woods
	9280 *Quercus frainetto* woods
	9310 Aegean *Quercus brachyphylla* woods
	9350 *Quercus macrolepis* forests
	93A0 Woodland with *Quercus infectoria* (Anagyro foetidae-Quercetum infectoriae)
Beech forests	9110 Acidophilous (*Luzulo-Fagetum*) beech forests
	9120 Atlantic acidophilous beech forests with *Ilex* and sometimes *Taxus*
	9130 *Asperulo-Fagetum* beech forests
	9150 Medio-European limestone beech forest of the *Cephalanthera-fagion*
	91K0 Illyrian *Fagus sylvatica* forests (*Aremonio-Fagion*)
Mesophytic deciduous forests	9160 Sub-Atlantic and medio-European oak or oak–hornbeam forest of the *Carpinion betuli*
	9170 *Galio-Carpinetum* oak hornbeam forests
	9180 *Tilio-Acerion* forests of slopes, screes and ravines
	91G0 Pannonic woods with *Quercus petraea* and *Carpinus betulus*
	91L0 Illyrian oak–hornbeam forests

Continued

Table 21.1. Continued.

Type[b]	Habitat
Flood plain forest	9030 Natural forests of primary succession stages of land upheaval coasts 91E0 Alluvial forests with *Alnus glutinosa* and *Fraxinus excelsior* 91F0 Riparian mixed forests of *Quercus robur, Ulmus laevis* and *Ulmus minor, Fraxinus excelsior* or *F. angustifolia*, along great rivers (*Ulmenion minoris*) 92A0 *Salix alba* and *Populus alba* galleries 92B0 Riparian formations on intermittent Mediterranean watercourses with *Rhododendron ponticum, Salix* and others 92C0 *Platanus orientalis* and *Liquidambar orientalis* woods 92D0 Southern riparian galleries and thickets (*Nerio-Tamaricetea* and *Securinegion tinctoriae*)
Mountain beech forests	9110 Acidophilous (*Luzulo-Fagetum*) beech forests 9120 Atlantic acidophilous beech forests with *Ilex* and sometimes *Taxus* 9130 *Asperulo-Fagetum* beech forests 9140 Medio-European subalpine beech forests with *Acer* and *Rumex arifolius* 91K0 Illyrian *Fagus sylvatica* forests (Aremonio-Fagion) 9270 Hellenic beech forests with *Abies borisii-regis*
Alpine coniferous forest	91Q0 Western Carpathian calcicolous *Pinus sylvestris* forests 91R0 Dinaric dolomite Scots pine (*Pinus sylvestris*) forests 9410 Acidophilous *Picea* forests of the montane to alpine levels 9420 Alpine *Larix decidua* and/or *Pinus cembra* forests 9430 Subalpine and montane *Pinus uncinata* forests
Acidophilous oak and oak–birch forest	9190 Old acidophilous oak woods with *Quercus robur* on sandy plains 91A0 Old oak woods with *Ilex* and *Blechnum* in the British Isles
Non-riverine alder, birch and aspen forest	9030 Natural forests of primary succession stages of land upheaval coasts 9040 Nordic subalpine/subarctic forests with *Betula pubescens* ssp. *czerepanovii*
Hemiboreal forest and nemoral coniferous and mixed broad-leaved–coniferous forest	9020 Fennoscandian hemiboreal natural old broadleaved forests rich in epiphytes 9070 Fennoscandian wooded pastures 91C0 Caledonian forest 91T0 Central European lichen Scots pine (*Pinus sylvestris*) forests 91U0 Sarmatic steppe pine forest
Mire and swamp forest	91D0 Bog woodland
Boreal forest	9010 Western taiga 9050 Fennoscandian herb-rich forests with *Picea abies* 9060 Coniferous forests on or connected to glaciofluvial eskers

[a]See: http://ec.europa.eu/environment/nature/legislation/habitatsdirective/index_en.htm#interpretation
[b]The order reflects broadly the distribution of these types across the continent from the Mediterranean to boreal zones.

range of European forest types. However, they do not necessarily represent all the woodland variation that is considered important within Member States: for example, the mixed deciduous woodland of lowland Britain with its characteristic vernal aspect of bluebells, *Hyacinthoides non-scripta*, is not represented.

It should be noted that the designation and protection of the Natura sites represent only a part of the conservation obligations required under the Directive. Member States are expected to bring these habitats and species into a favourable conservation status (Table 21.2), which requires consideration of the occurrence of the habitat and species outside the Natura series as well. Under Article 17 of the Directives, Member States must report on the conservation status of the habitats and species every 6 years (European Commission, 2014); the evaluation matrix used for habitats is presented in Table 21.2. For further detail see http://bd.eionet.europa.eu/activities/Reporting/Article_17/reference_portal and https://circabc.europa.eu/faces/jsp/extension/wai/navigation/container.jsp.

The most recent round of reporting was being analysed at the time of writing, but results from the 2006 round are available at http://bd.eionet.europa.eu/article17/habitats-progress.

The majority of habitat types for which an overall assessment was made were classed as 'unfavourable – bad' or 'unfavourable – inadequate' (Fig. 21.3a). This is not surprising given that poor condition contributed to their original inclusion in the Directive. There were, however, differences between the status recorded for different groupings, as illustrated in Fig. 21.3b. While all but one of the six assessments of Alpine larch/*Pinus cembra* forests was judged to have good structures and function, this was the case for only seven of the 25 assessments for Sub-Atlantic/medio-European oak–hornbeam woodland assessments, and so the overall assessment for the latter was very low. Among species, 13 out of 25 national wolf (*Canis lupus*) assessments showed rising population trends, as did 11 out of 20 brown bear (*Ursus arctos*) studies (see Hearn, Chapter 14); whereas the stag beetle (*Lucanus cervus*) had a declining trend in 17 of 32 national assessments, and only one rising trend.

The reporting process is still being refined, but over time should provide a broad measure of the effectiveness of the Directives. The European Commission is also empowered to take infraction proceedings against Member States that fail to implement the Directives adequately. This potentially opens the way for voluntary conservation groups and others to lobby for action where they feel that Member States have not taken sufficient action to address conservation problems.

The Natura series of protected sites is a major contribution to European biodiversity conservation but can only reflect those habitats and species that have been agreed as of European significance. There have been concerns expressed by some conservation NGOs that the current lists are incomplete, but at present there seems little likelihood that there will be any major revisions to them. In addition, many forests and the species that they contain fall outside the scope of the Directive and indeed of other nature conservation sites. Their long-term protection depends on integrating their needs with other objects of management, such as wood production.

21.6 Forest Protection and Conservation as Part of Land-use Practice

Within the EU, forestry is primarily a responsibility for the Member States. Most countries have some sort of regulatory framework governing the treatment of forest areas, and this may require the integration of conservation measures alongside wood production, forest recreation management, etc. General principles for this have been set out at various times, for example by Peterken (1977). Lindenmayer *et al.* (2006) proposed five broad requirements:

- maintenance of connectivity;
- maintenance of landscape heterogeneity;
- maintenance of stand structural complexity;
- maintenance of aquatic ecosystem integrity; and
- use of natural disturbance regimes to guide human disturbance regimes.

The potential for integrating biodiversity with forest management from the stand to landscape scales has also recently been reviewed by Kraus and Krumm (2013), particularly for central Europe.

Frequently, these measures are supported by grants or subsidies, recognizing that the conservation is generally a shared non-market benefit from the forests. Within the EU, much of this support is through the rural development programmes for 2014–2020 (http://ec.europa.eu/agriculture/rural-development-2014-2020/index_en.htm). Voluntary forestry management certification schemes – notably those run by the Forest Stewardship Council (https://ic.fsc.org/) and the Programme for the Endorsement of Forest Certification (http://www.pefc.org/) – may also require that conservation measures be taken.

The success or otherwise of such measures should be addressed by the development

Table 21.2. Assessing the conservation status of a habitat type by biogeographical region within a European Member State (MS).

Parameter	Conservation status			
	Favourable ('green')	Unfavourable – Inadequate ('amber')	Unfavourable – Bad ('red')	Unknown (insufficient information to make an assessment)
Range with biogeographic region	Stable (loss and expansion in balance) or increasing and not smaller than the favourable reference range	Any other combination	Large decrease: equivalent to a loss of more than 1% a year within period specified by MS or More than 10% below favourable reference range	No or insufficient reliable information available
Area covered by habitat type within range	Stable (loss and expansion in balance) or increasing and not smaller than the favourable reference area' and without significant changes in distribution pattern within range (if data available)	Any other combination	Large decrease in surface area: equivalent to a loss of more than 1% a year (indicative value MS may deviate from this if duly justified) within period specified by MS or With major losses in distribution pattern within range or More than 10% below 'favourable reference area'	No or insufficient reliable information available
Specific structures and functions (including typical species)	Structures and functions (including typical species) in good condition and no significant deteriorations/ pressures	Any other combination	More than 25% of the area is unfavourable in its specific structures and functions (including typical species)	No or insufficient reliable information available
Future prospects (for range, area covered and specific structures and functions)	The prospects of the habitat for its future are excellent/ good, with no significant impact from threats expected; long-term viability assured	Any other combination	The prospects of the habitat are bad, severe impact from threats expected; long-term viability not assured	No or insufficient reliable information available
Overall assessment of conservation status	All 'green' or three 'green' and one 'unknown'	One or more amber' but no 'red'	One or more 'red'	Two or more 'unknown' combined with 'green' or all 'unknown'

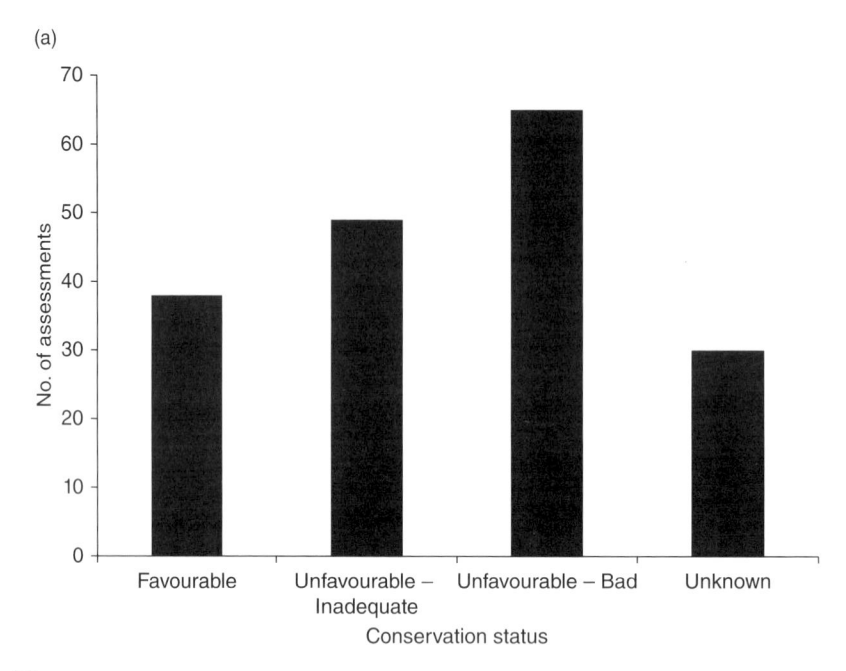

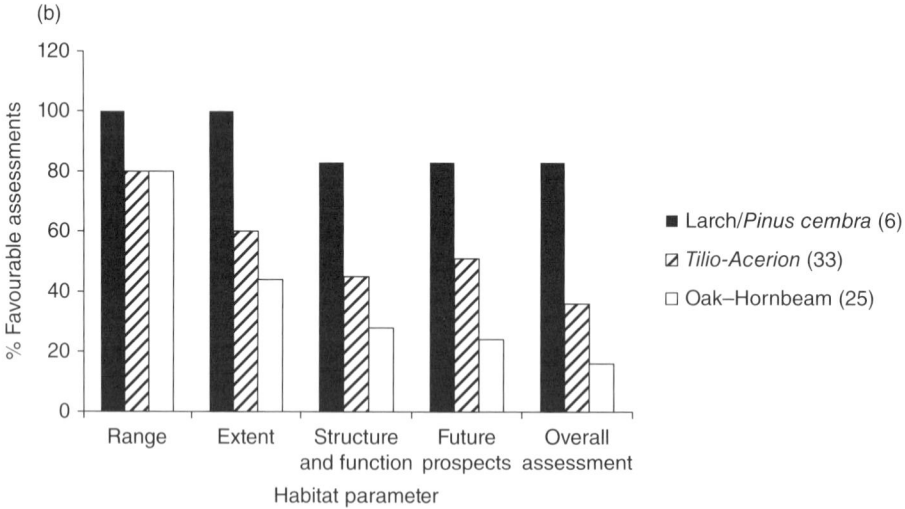

Fig. 21.3. (a) Overall assessment of conservation status for European Union (EU) Habitats Directive forest types; for the 2001–2006 period, assessments were made for every forest type in each biogeographic zone in each EU country. (b) Percentage of assessments judged to be favourable for three forest types: Alpine larch (*Larix decidua*)/*Pinus cembra* woodland (six assessments); *Tilio-Acerion* forests (33 assessments); and Sub-Atlantic/medio-European oak–hornbeam woods (25 assessments).

of appropriate monitoring programmes (Hurford and Schneider, 2006), but at present these are not consistently applied across the Continent. Hence, there has been recent work to try to simplify these in the Pan-European SEBI (Streamlining European Biodiversity Indicators) initiative (see http://biodiversity.europa.eu/topics/sebi-indicators). Many forest monitoring systems are based on indirect (structural) indicators of biodiversity or on habitats rather than on taxonomic data. There may also be a discrepancy between the taxa currently monitored (butterflies, birds, vascular plants) and the taxa that may be most at risk (e.g. saproxylic insects). The monitoring of rare/threatened species needs to be combined with looking at groups that are typical for forests and/or are potentially affected by changes in silvicultural practices (or other natural trends). More comparisons of managed and unmanaged forests under equivalent conditions would allow an improved assessment of the effects of forest management on biodiversity (Götmark, 2013).

21.7 Rewilding and Forest Conservation

A tension in European forest conservation discussions can be between 'naturalness' and 'biodiversity' conservation, because of the cultural nature of many landscapes. However, there is interest in trying to explore what more natural forests might be like (Bastrup-Birk, 2014). Minimum intervention forest reserves in Europe have been widely used for the long-term monitoring of forest dynamics as controls against which to judge the effects of interventions – whether for production or conservation purposes (Parviainen *et al.*, 2000). Implicitly or explicitly, there has often been an assumption that these reserves are in a near-natural state or might come to resemble in some ways the original natural forests that covered the Continent.

There has also been much debate in Europe about the consequences of allowing much larger areas to revert to nature – to 'rewild'. In places, this is happening by default (Navarro and Pereira, 2012; Hearn *et al.*, 2014), but often in the absence of some of the key players, notably large wild herbivores. These would in the original landscapes, Vera (2000) argues, have helped to create a shifting vegetation mosaic that would have allowed the coexistence of species assemblages of closed forest, forest edge and open habitats. Hence, there are projects to re-establish such large herbivores so that people can 'enjoy the benefits from ecosystems and landscapes, inhabited and shaped by viable populations of all large herbivores of the region, living in the wild'.

The long-term outcomes of rewilding are generally unpredictable (Hughes *et al.*, 2011; Lorimer and Driessen, 2014). None the less, rewilded areas are likely to be interesting, diverse and give people a sense of wilderness, even if they are totally novel landscapes (Plate 8). They may also be potentially easier and cheaper to maintain than trying to reinstate cultural landscape management against prevailing economic conditions.

21.8 From the Past to the Future

The motivation for forest protection has varied over time. Currently, nature conservation is seen as important, but will it have as high a value in future, or will there be a shift back to a more utilitarian view of forests and the goods and services that they provide? In the UK guidelines for the selection of protected sites, Ratcliffe (1989) stated that:

> The primary objective of nature conservation is to ensure that the nation's heritage of wild flora and fauna and geological and physiographic features remains as large and diverse as possible, so that society may use and appreciate its value to the fullest extent.

There are two key challenges in this statement. First, the reference to society indicates an appreciation that nature conservation is intimately associated with human society, and not something separate from it. Secondly, how large and how diverse do we want the future conservation heritage to be, whether viewed at a country or a Europe-wide scale?

21.8.1 Conservation for people?

The need to integrate nature conservation with wider societal use of the environment was recognized by the Convention on Biological Diversity (CBD) that arose from the Earth Summit in Rio in 1992. The Ecosystem Approach takes this association further, and emphasizes the functional importance of biodiversity for ecosystems to deliver 'ecosystem services' on which people rely on for survival and well-being. In this way, nature conservation can be seen as practically, as well as morally, important.

This way of looking at the relationships between society and nature has gathered political momentum. The Millennium Ecosystem Assessment (2005) was a global assessment of the status and value of ecosystems through their delivery of services that was carried out by a collaboration of international bodies. The UK National Ecosystem Assessment (2011) provides a more local analysis of broad habitat types, their status and their contribution to ecosystem services.

Forest protection and conservation will continue to be important within this approach, but there may be more pressure for biodiversity conservation to be integrated with the delivery of other goods and services. Nature conservation may become less focused on discrete sites and individual species populations, but take account of the landscape more generally. Protected sites are, however, likely to continue to have a role within the landscape as core biodiversity hotspots and as stepping stones for organisms as they adjust their distributions in response to climate change (Thomas et al., 2012).

Nevertheless, there are risks for nature conservation objectives from putting an emphasis on ecosystem services – what happens, for example, if society values other goods and services more? There is also a specific conflict with much species protection legislation that focuses on threats to the individual organism. It may be difficult to permit a project that has overall benefits in ecosystem service terms, but would at the same time lead to the death of some individuals of a protected species.

21.8.2 What sorts of woods and forests will be conserved in future?

The current composition of European forests (or that which existed in the recent past) forms the template for most conservation designations and programmes, but that composition is likely to change. Past forest clearance and management may mean that sites are carrying an extinction debt and will continue to lose species even if the environment stays the same (Siitonen and Ranius, Chapter 11; Hermy, Chapter 13).

Furthermore, current climate change projections indicate that forest species will change their distributions and that the current assemblages may not exist in future (Lindner et al., 2010). Climate change may also trigger changes in farming practices, thereby leading to shifts in the extent and distribution of forests. The effects of chronic air pollution, in particular increasing levels of nitrogen, are still working their way through the system (Bobbink and Roelofs, 1995; Bobbink et al., 2003). New threats from pests and diseases may take out major tree and shrub components of the forest systems over quite short periods (Potter, Chapter 23).

Future tree and woodland conservation needs to recognize different sets of templates:

- For the short term, it is essential that we maintain the range of habitats and species assemblages as they have existed in the past. This is the approach epitomized under the EU Habitats and Species Directive. In the longer term, this is unlikely to be successful because of environmental change.
- Secondly, we can look to manage woodland and forest landscapes to make them more resilient in the face of future economic, social, biological and physical environments. This may involve letting species go extinct in some areas, as long as they thrive elsewhere, or accepting changes in species assemblages (including those currently not considered to be native to a region) if the resulting structure and composition is likely to survive better under future conditions. Future woods and forests may therefore lack historical analogues.
- Thirdly, there is likely to be an increasing role for areas left to 'natural development/rewilding' in Europe. These will change in response to different natural or near-natural processes, and their ultimate state is unpredictable. Thus they are a complement to, rather than a substitution for cultural landscape conservation.

References

Angelstam, P.K., Butler, R., Lazdinis, M., Mikusinski, G. and Roberge, J.-M. (2003) Habitat thresholds for focal species at multiple scales and forest biodiversity conservation – dead wood as an example. *Annales Zoologici Fennici* 40, 473–482.

Araujo, M.B., Lobo, J.M. and Moreno, J.C. (2007) The effectiveness of Iberian protected areas in conserving terrestrial biodiversity. *Conservation Biology* 21, 1423–1432.

Bastrup-Birk, A. (2014) *Developing a Forest Naturalness Indicator for Europe*. European Environment Agency, Luxembourg.

Bobbink, R. and Roelofs, J.M. (1995) Nitrogen critical loads for natural and semi-natural ecosystems: the empirical approach. *Water, Air, and Soil Pollution* 85, 2413–2418.

Bobbink, R., Ashmore, M., Braun, S., Flückiger, W. and van den Wyngaert, I.S.J.J. (2003) Empirical nitrogen critical loads for natural and semi-natural ecosystems: 2002 update. In: Achermann, B. and Bobbink, R. (eds) *Empirical Critical Loads for Nitrogen. Expert Workshop, Berne, 11–15 November 2001. Proceedings*. Environmental Documentation No. 164, Swiss Agency for the Environment, Forests and Landscape (SAEFL), Berne, Switzerland, pp. 43–170. Available at: http://www.iap.ch/publikationen/nworkshop-background.pdf (accessed 11 December 2014).

Branquart, E., Latham, J., Lier, M. and Saudyte S. (2007) A general analysis of protected forest area types in Europe. In: Frank, G., Parviainen, J., Vandekerkhove, K., Latham, J., Schuck, A. and Little, D. (eds) *COST Action E27 – Protected Forests in Europe – Analysis and Harmonisation (PROFOR): Results, Conclusions and Recommendations*. Federal Research and Training Centre for Forests, Natural Hazards and Landscape (BFW), Vienna, pp. 7–16.

Branquart, E., Verheyen, K. and Latham, J. (2008) Selection criteria of protected forest areas in Europe: the theory and the real world. *Biological Conservation* 141, 2795–2806.

Bruinderink, G.G., van der Sluis, T., Lammertsma, D., Opdam, P. and Pouwels, R. (2003) Designing a coherent ecological network for large mammals in northwestern Europe. *Conservation Biology* 17, 549–557.

Council of Europe (2013) *Convention on the Conservation of European Wildlife and Natural Habitats* [Bern Convention]. Brussels. Available at: http://www.coe.int/t/dg4/cultureheritage/nature/bern/default_en.asp (accessed 11 December 2014).

Dudley, N. and Phillips, A. (2006) *Forests and Protected Areas: Guidance on the Use of the IUCN Protected Area Management Categories*. International Union for Conservation of Nature, Gland, Switzerland and Cambridge, UK.

European Commission (2013a) The Habitats Directive. Brussels. Available at: http://ec.europa.eu/environment/nature/legislation/habitatsdirective/ (accessed 11 December 2013).

European Commission (2013b) The Birds Directive. Brussels. Available at: http://ec.europa.eu/environment/nature/legislation/birdsdirective/index_en.htm (accessed 11 December 2014).

European Commission (2014) Habitats Directive reporting. Article 17 reporting. Brussels. Available at: http://ec.europa.eu/environment/nature/knowledge/rep_habitats/index_en.htm (accessed 11 December 2014).

Eurostat (2011) *Forestry in the EU and the World: A Statistical Portrait*, 2011 edn. Eurostat European Commission, Brussels.

Frank, G., Schwarzl, B., Johann, E., Hackl, J., Schweinzer, K.-M. and Hauk, E. (2005) Country Report – Austria. In: Latham, J., Frank, G., Fahy, O., Kirby, K., Miller, H. and Stiven, R. (eds) *COST Action E27 – Protected Forest Areas in Europe – Analysis and Harmonisation (PROFOR) – Reports of Signatory States*. Federal Research and Training Centre for Forests, Natural Hazards and Landscape (BFW), Vienna, pp. 7–26.

Frank, G., Parviainen, J., Vandekerkhove, K., Latham, J., Schuck, A. and Little, D. (2007) *COST Action E27 – Protected Forests in Europe – Analysis and Harmonisation (PROFOR): Results, Conclusions and Recommendations*. Federal Research and Training Centre for Forests, Natural Hazards and Landscape (BFW), Vienna.

Gaston, K.J., Charman, K., Jackson, S.F., Armsworth, P.R., Bonn, A., Briers, R.A., Callaghan, C.S.Q., Catchpole, R., Hopkins, J., Kunin, W.E., *et al.* (2006) The ecological effectiveness of protected areas: the United Kingdom. *Biological Conservation* 132, 76–87.

Goodfellow, S. and Peterken, G.F. (1981) A method for survey and assessment of woodland or nature conservation using maps and species lists: the example of Norfolk woodlands. *Biological Conservation* 21, 177–195.

Götmark, F. (2013) Habitat management alternatives for conservation forests in the temperate zone: review, synthesis and implications. *Forest Ecology and Management* 306, 292–307.

Hayman, R. (2003) *Trees, Woodland and Western Civilization*. Bloomsbury, Hambledon and London, London.

Hearn, R., Watkins, C. and Balzaretti, R. (2014) The cultural and land-use implications of the reappearance of the wild boar in north-west Italy: a case study of the Val di Vara. *Journal of Rural Studies* 36, 52–63.

Hooke, D. (2010) *Trees in Anglo-Saxon England: Literature, Lore and Landscape*. Boydell and Brewer, Woodbridge, UK.

Hughes, F.M.R, Stroh, P.A., Adams, W.M., Kirby, K.J., Mountford, O. and Warrington, S. (2011) Monitoring and evaluating large-scale, 'open-ended' habitat creation projects: a journey rather than a destination. *Journal of Nature Conservation* 19, 245–253.

Hurford, C. and Schneider, M. (2006) *Monitoring Nature Conservation in Cultural Habitats*. Springer, Dordrecht, The Netherlands.

Johann, E., Agnoletti, M., Bölöni, J., Erol, S.Y., Holl, K., Kusmin, J., Latorre, Jesús G., Latorre, Juan G., Molnár, Z., Rochel, X. *et al.* (2012) Europe. In: Parrotta, J.A. and Trosper, R.L. (eds) *Traditional Forest-Related Knowledge: Sustaining Communities, Ecosystems and Biocultural Diversity*. World Forests, Vol. 12, Springer, Dordrecht, The Netherlands, pp. 203–250.

Kirby, K.J. (1993) Assessing nature conservation values in British woodland. *Arboricultural Journal* 17, 253–276.

Kirby, K.J. (2003) Woodland conservation in privately-owned cultural landscapes: the English experience. *Environmental Science and Policy* 6, 253–259.

Kraus, D. and Krumm, F. (eds) (2013) *Integrative Approaches as an Opportunity for the Conservation of Forest Biodiversity*. European Forest Institute, Joensuu, Finland.

Latham, J., Frank, G., Fahy, O., Kirby, K., Miller, H. and Stiven, R. (2005) *COST Action E27 – Protected Forest Areas in Europe – Analysis and Harmonisation (PROFOR) – Reports of Signatory States*. Federal Research and Training Centre for Forests, Natural Hazards and Landscape (BFW), Vienna.

Lawton, J. (ed.) Brotherton, P.N.M., Brown, V.K., Elphick, C., Fitter, A.H., Forshaw, J., Haddow, R.W., Hilborne, S., Leafe, R.N., Mace, G.M. *et al.* (2010) *Making Space for Nature: A Review of England's Wildlife Sites and Ecological Network*. Report Submitted to the Secretary of State, the Department for Environment, Food and Rural Affairs (Defra) on 16 September 2010. Defra, London. Available at: http://archive.defra.gov.uk/environment/biodiversity/documents/201009space-for-nature.pdf (accessed 11 December 2014).

Lindenmayer, D.B., Franklin, J.F. and Fischer, J. (2006) General management principles and a checklist of strategies to guide forest conservation. *Biological Conservation* 131, 433–445.

Lindner, M., Maroschek, M., Netherer, S., Kremer, A., Barbati, A., Garcia-Gonzalo, J., Seidl, R., Delzon, S., Corona, P., Kolström, M., Lexer, M.J. and Marchetti, M. (2010) Climate change impacts, adaptive capacity, and vulnerability of European forest ecosystems. *Forest Ecology and Management* 259, 698–709.

Lorimer, J. and Driessen, C. (2014) Wild experiments at the Oostvardersplassen: rethinking environmentalism for the Anthropocene. *Transactions of the Institute of British Geographers* 39, 169–181.

MCPFE (2003) *State of Europe's Forests 2003 – The MCPFE Report on Sustainable Forest Management in Europe*. Ministerial Conference on the Protection of Forests in Europe (Forest Europe), Liaison Unit, Vienna.

Millennium Ecosystem Assessment (2005) *Ecosystems and Human Well-Being: Synthesis*. Island Press, Washington, DC.

Navarro, L.M. and Pereira, H.M. (2012) Rewilding abandoned landscapes in Europe. *Ecosystems* 15, 900–912.

Nilsson, C. and Götmark, F. (2003) Protected areas in Sweden: is natural variety adequately represented? *Conservation Biology* 6, 232–242.

Parviainen, J., Bucking, W., Vandekerkhove, K., Schuck, A. and Päivinen, R. (2000) Strict forest reserves in Europe: efforts to enhance biodiversity and research on forests left for free development in Europe (EU-COST-Action E4). *Forestry* 73, 107–118.

Peterken, G.F. (1977) General management principles for nature conservation in British woodlands. *Forestry* 50, 27–48.

Ratcliffe, D.A. (1977) *A Nature Conservation Review*. Cambridge University Press, Cambridge, UK.

Ratcliffe, D.A. (ed.) (1989) *Guidelines for Selection of Biological SSSIs*. Nature Conservancy Council, Peterborough, UK.

Rosati, L., Marignani, M. and Blasi, C. (2008) A gap analysis comparing Natura 2000 versus National Protected Area network with potential natural vegetation. *Community Ecology* 9, 147–154.

Rowell, S.C. (1994) *Lithuania Ascending: A Pagan Empire within East-Central Europe, 1295–1345*. Cambridge University Press, Cambridge, UK.

Thomas, C.D., Gillingham, P.K., Bradbury, R.B., Roy, D.B., Anderson, B.J., Baxter, J.M., Bourn, N.A.D., Crick, H.Q.P., Findon, R.A., Fox, R. *et al.* (2012) Protected areas facilitate species' range expansion. *Proceedings of the National Academy of Sciences of the United States of America* 109, 14063–14068.

UK National Ecosystem Assessment (2011) *UK National Ecosystem Assessment Technical Report. Understanding Nature's Value to Society* UNEP-WCMC (United Nations Environment Programme World Conservation Monitoring Centre), Cambridge, UK. Available at: http://uknea.unep-wcmc.org/Resources/tabid/82/Default.aspx (accessed 11 December 2014).

UNECE (2011) *State of Europe's Forests 2011*. United Nations Economic Commission for Europe, Oslo.

Vera, F.M.W. (2000) *Grazing and Forest History*. CAB International, Wallingford, UK.

Vodde, F. (2007) Organisations involved in the establishment and maintenance of protected forest areas. In: Frank, G., Parviainen, J., Vandekerkhove, K., Latham, J., Schuck, A. and Little, D. (eds) *COST Action E27 – Protected Forests in Europe – Analysis and Harmonisation (PROFOR): Results, Conclusions and Recommendations*. Federal Research and Training Centre for Forests, Natural Hazards and Landscape (BFW), Vienna, pp. 41–49.

Vos, C.C., Berry, P., Opdam, P., Baveco, H., Nijhof, B., O'Hanley, J., Bell, C. and Kuipers, H. (2008) Adapting landscapes to climate change: examples of climate-proof ecosystem networks and primary adaptation zones. *Journal of Applied Ecology* 45, 1722–1731.

Watkins, C. (2014) *Trees, Woods and Forests. A Social and Cultural History*. Reaktion, London.

Welzholz, J.C. and Johann, E. (2007) History of protected forest areas in Europe. In: Frank, G., Parviainen, J., Vandekerkhove, K., Latham, J., Schuck, A. and Little, D. (eds) *COST Action E27 – Protected Forests in Europe – Analysis and Harmonisation (PROFOR): Results, Conclusions and Recommendations*. Federal Research and Training Centre for Forests, Natural Hazards and Landscape (BFW), Vienna, pp. 17–40.

Williams, M. (2002) *Deforesting the Earth: From Prehistory to Global Crisis*. University of Chicago Press, Chicago, Illinois.

22 The UK's Ancient Woodland Inventory and its Use

Emma Goldberg*
Natural England, Peterborough, UK

22.1 Introduction

One approach to woodland conservation relies on designating large natural areas where forest cover has been unbroken and where the full range of biodiversity, from genetic to ecosystem processes, can be maintained. This approach has its roots in the North American preservationist idea (Leopold, 1949), and it was developed by MacArthur and Wilson (1967) into a discussion on how large such reserves need to be for the species populations within them to be sustainable. However, what do we mean by 'natural'? Is any forest in Europe large enough, and with so little evidence of past or present human intervention, for this approach to be practical? Even large forest systems such as those of Białowieza in Poland (Nilsson, 1997; Latałowa *et al.*, Chapter 17) or Fontainebleau in France (Pontailler *et al.*, 1997) show evidence of human activity; they are ancient forests but they cannot be claimed as wholly 'natural', 'primary' or 'virgin' forests (Kirby and Watkins, Chapters 3 and 4).

The preservationist approach is particularly inappropriate in the smaller, more densely populated and less forested European countries, such as Belgium and the UK. The surviving small, fragmented patches of woodland have long been isolated and also heavily managed as part of the local economy, and much of their characteristic fauna and flora results from this management (Verheyen *et al.*, 1999; Rackham, 2003).

Not all patches of forest are the same though: some have a long woodland history while others have grown up or been created more recently. The concept of 'ancient woodland' began partially as a quest to find any vestiges of the pre-Neolithic forest cover – the former 'virgin forest' in the UK (Watkins, 1988). Indeed, Peterken (1974) initially referred to 'primary' woodland indicators in his comparison of old versus young woodland in Lincolnshire. Since then, ancient woodland has become a rather different concept, and in the exploration of ancient woodland, we have uncovered a great deal about our cultural as well as our biological heritage.

The Ancient Woodland Inventory, initiated in 1981, has become an important tool for woodland conservation (Goldberg *et al.*, 2007, 2011), but its initial aims and objectives were not quite the same as the purposes to which it is now put. This can lead to misconceptions as to what inclusion (or not) of a site in the inventory means and misunderstandings in how that information is used. Developments

---

*E-mail: emma.goldberg@naturalengland.org.uk

in technology have made some of the original methods and maps look very primitive. Tensions inevitably arise when trying to extend approaches carried out with limited resources for research purposes to the real and very messy world of conservation policy and practice.

22.2 Developing the Ancient Woodland Concept

Through the 1970s, Oliver Rackham and George Peterken concluded, based primarily on field work in lowland England, that the sites judged to be most important for conservation were those with a long continuity of woodland cover (Peterken, 1983; Rackham, 2003). Thus, it was useful to distinguish woods that had existed since medieval times from plantations on open ground and from the recent secondary woodland that started to become common in the early 19th century and expanded substantially in the 20th century.

At the start of the 'inventory process', the definition of ancient woodland used was an area of woodland that had had continuous tree cover since 1600 (see Box 22.1). Peterken (1977) selected this threshold date because it reflected the start of the availability of accurate maps, at least for some parts of the country, and because it seemed to represent a point before which relatively little planting of new woods had occurred. The use of this (or any other) threshold date meant that debates as to whether the woods were primary, virgin, natural, etc., were circumvented.

Studies elsewhere in Europe subsequently confirmed this distinction of ancient woodlands based on their historical origin applied more generally. Furthermore, many species appear more likely to be found in ancient woods, e.g. Wulf (1997), Brunet and Von Oheimb (1998), Hermy et al. (1999), Verheyen et al. (1999). The date used to define ancient woodland in these studies does vary though, suggesting that the effect of age is perhaps more of a continuous response than a threshold effect.

It is one thing to identify that ancient woods are, in principle and in general, worth conserving (Box 22.2); it is quite another to produce a consistent country-wide list of where they are and then to get it accepted as part of forestry policy. Oliver Rackham noted in 1980 that it would not be possible even if 'at most a sixth of what is green on the map (i.e. a little over 1% of the total land area) survives as ancient woodland … [this] is still too much to be completely catalogued on a national scale'. George Peterken described the Ancient Woodland Inventory as having been 'an expensive act of faith' in its time because there was no way of knowing how much it would be used (Goldberg et al., 2011).

Box 22.1. Definitions of different types of woodland. (Derived from Peterken, 1977.)

Primary Woodland
  Woodland is considered 'primary' if it has existed continuously since before the original forests in that district were fragmented. It is, in a sense, hypothetical because the status of any one example can never be proved – one cannot prove that a break in continuity has never occurred.

Secondary Woodland
  'Secondary' woods are those that have originated on previously unwooded land. The status of secondary woods can be proved positively by using evidence which is independent of the ecosystem, such as earthworks, cultivation remains, maps and historical records (Rackham, 1971, 1976).

Ancient Woodland
  As a matter of practical convenience, it is valuable to have a category of 'ancient' woodland (Rackham, 1971), or 'Medieval' woodland (Peterken and Harding, 1974), whose status can be proved. This type of woodland is contrasted with recent secondary woodland, and distinguished simply by a threshold date; an origin before this date qualifies a wood to be 'ancient'. Ancient woodland therefore comprises both primary woodland and secondary woods originating before the threshold. The threshold itself can for convenience be placed at about 1600, before which time secondary woods were rarely created by planting.

Recent Woodland
  Areas of land that have become wooded since the threshold date.

> **Box 22.2.** Importance of ancient semi-natural woodland, based on Peterken (1983).
>
> Ancient semi-natural woods are far more important for nature conservation than other woodland types because of the following factors.
>
> **1.** They include all primary woods. These are the lineal descendants of Britain's primeval woodland, whose wildlife communities, soils and sometimes structure have been least modified by human activities over the millennia. Their tree and shrub communities preserve the natural composition of Atlantic forests. Once destroyed, they cannot be recreated.
> **2.** They provide baselines (or 'controls' in scientific parlance) against which to measure the effects of human practices on soils, on the productivity of woodland communities and on food webs, etc.
> **3.** Their wildlife communities are generally, but not invariably, richer than those of recent woods. The latter tend to contain only weed species or impoverished plant forms of heathland, meadow or bog, according to the prior land use.
> **4.** They contain a very high proportion of rare and vulnerable wildlife species, in other words those species most in need of protection if all species are to survive in Britain. Many of these species require the stability afforded by the continuity of suitable woodland; they cannot colonize newly created woodland, or do so only slowly.
> **5.** Where large, old trees have been present for several centuries, these provide refuges for the characteristic inhabitants of primeval woodland, such as lichens and beetles.
> **6.** They contain other natural features that rarely survive in an agricultural setting, such as streams in their natural watercourses and micro-topographical features formed under periglacial conditions.
> **7.** They are reservoirs from which the wildlife of the countryside has been maintained (and could be restored). For example, hedges and recent woods in the vicinity of ancient woods are more likely to contain woodland species than sites isolated from such woods.
> **8.** They have been managed by traditional methods for centuries. Not only are they ancient monuments whose value to historians and to village community consciousness is arguably as great as that of the older buildings in a parish, but, where traditional management continues or can be revived, they are living demonstrations of conservation in the broader sense of a stable, enduring relationship between humans and nature.

However, the inventory proved to be what was required to turn around the conservation of ancient woodland. It anticipated the current fashion for 'evidence-based' policy, in showing where our most important woodland sites were and why they were under threat through estimates of recent loss, and provided a focus for shifts in forestry and planning policies that would improve their protection. Thus, Rackham (2003) was able to write 'one aspect in which I am gratified to be proved wrong is the compilation of a register of ancient woodland: a formidable task which I thought impossible in 1980'.

22.3 The Creation of the Ancient Woodland Inventory

The Ancient Woodland Inventory project was initiated in 1981 by George Peterken of the then Nature Conservancy Council (the British government wildlife service). The inspiration came from a forestry conference in the 1970s where a delegate observed that if landowners only knew where ancient woods were, then 'of course they would do their best to look after them'. A trial in Norfolk showed that it was possible to produce a provisional county list from readily available maps and other sources relatively quickly and cheaply (Goodfellow and Peterken, 1981). So a start was made in extending this approach to the rest of the country (Kirby et al., 1984).

The inventory was seen as having four possible applications. It would:

- create a factual basis for a site-specific forestry policy directed at ancient woods, thereby not burdening all woods with the silvicultural specifications needed to protect and maintain ancient woods;
- form an efficient basis for conservation surveys, i.e. efforts could be concentrated on woods that were most likely to be wildlife rich;

- enable Sites of Special Scientific Interest (SSSIs) (the legally protected sites) to be selected and justified more easily because we would know more about the wider population of wildlife-rich woodland; and
- establish a baseline for monitoring any further losses of ancient woods, hence reflecting concern about the reduction in extent and conservation quality that had been going on since 1945.

The methods adopted had to be quick and simple in order to be able to cover the thousands of woods scattered across the country. They also had to be largely desk based, as field surveys were not available for most of the woods considered. Ordnance Survey 1:25,000 maps of 1880–1960 (but mostly revised for woodland c.1930s) were used to draw up an initial selection of candidate sites. This period corresponded more or less with the historic low point in woodland cover at around the beginning of the 20th century. It reduced the number of sites to be considered by ruling out old woods that had by this time been cleared, as well as most of the 20th century afforestation that had occurred after the formation of the Forestry Commission in 1919 (Quine, Chapter 15). Only sites of over 2 ha on the base maps were included for further study. Establishing the occurrence of smaller sites on the historic maps that were readily available in the 1980s was difficult, and their inclusion would have greatly inflated the number of woods involved disproportionately to the extra area of ancient woodland identified.

Candidate sites were checked against earlier, mostly 19th century, maps, for example the Ordnance Survey 1st Edition maps (surveyed 1805–1873; scale 1:63,360). Features that might indicate ancient woodland, including wood names, boundaries and context within the landscape were noted at the same time.

Older 17th and 18th century maps were used where they were available, although their accuracy was not always known. Woods might be shown in only a highly stylized form, or only if they were well-known landmarks. Small woods might be omitted by the surveyors or during the simplification that occurred in the engraving of the copper plates used to print the maps. In areas of complex topography, there may have been more cartographic errors and the hachuring (finely spaced lines) used to depict slopes often covered the woodland symbols. The apparent absence of a wood from an older map was therefore not necessarily proof that it was not there. Sites absent from all the older maps were deleted from the inventory only where there was a clearly depicted alternative land use, or supporting evidence for a recent origin from other sources. Equally, some sites present on old maps were removed where they were interpreted as likely to have been planted as 18th century landscaping on large estates (Spencer and Kirby, 1992).

Sites or parts of sites identified as ancient were also categorized by the current composition and structure of the tree and shrub layer. The division was into site-native/'semi-natural' stands (including, for example, coppice, as well as high forest stands), or plantations, which were usually composed of non-native trees. This categorization was derived from a variety of sources, including field surveys already undertaken, aerial photographs, Forestry Commission and private estate forestry maps, and information held by other parties, such as local councils, voluntary wildlife organizations, forest owner groups. Aerial photos were also used to provide information on the extent and likely cause of woodland loss, i.e. where candidate sites no longer existed. This categorization as semi-natural/plantation/lost was a crude, but useful, measure of conservation priority and threat at the time, but covered a wide variation of conditions within each category.

The results were summarized for each site on a database, collated as 'county' reports e.g. Whitbread (1986), and also summarized by country within the UK (Roberts et al., 1992; Spencer and Kirby, 1992) (Table 22.1).

22.4 Developing and Using the Inventories

22.4.1 England: the 'Red Queen' dilemma

Like the Red Queen in Lewis Carroll's *Through the Looking Glass*, who had to run as

Table 22.1. Estimates of ancient woodland cover in the UK.

Country	Estimate for 1992 ('000 ha)	Latest estimate ('000 ha)
England	341[a]	364[c]
Wales	56[a]	95[d]
Scotland – ancient, of semi-natural origin	136[b]	120[e]
Scotland – long established, of plantation origin	172[b]	187[e]
Scotland – other on the General Roy map (the Roy Military Survey of Scotland, 1747–1755)	Captured as ancient woodland origin	16[e]
Northern Ireland – ancient, probably ancient or possibly ancient, of semi-natural origin	Not part of the original inventory project	4[f]
Northern Ireland (long established)	Not part of the original inventory project	6[f]

[a]Spencer and Kirby (1992); [b]Roberts et al. (1992); [c]Personal communication, Natural England (August 2014); [d]Forestry Commission (2011); [e]Scottish Natural Heritage (2014); [f]Woodland Trust (2007).

fast as she could just to stand still, the inventory needed to be upgraded before it was complete. The original compilation of information for each site was done using tracing paper and coloured pens, but as geographic information systems (GIS) became readily available, users of the inventory came to expect that it would be available in digital form. In England, an initial attempt was made to create a digitized version in the late 1990s.

The availability of the digital data meant that people could download and interrogate the data more easily and compare it with their own data. We rapidly learned that errors had crept into the data: some polygons were mislabelled, semi-natural as plantation and vice versa; in other cases, the polygon boundaries did not match with the corresponding underlying base-map features. In part, this was because the world had moved on, both technologically and literally.

The natural shift in magnetic north meant that the raster images of the woods as shown on the base maps were lying in the wrong position according to modern positioning. Technical advances in GIS allowed boundaries to be defined to a much higher level of precision than in the first digitization phase. Consequently, when users 'zoomed in' on a site, the boundaries and relevant map features no longer coincided.

The boundaries were updated in 2003 to the new Ordnance Survey Mastermap to give greater accuracy within the GIS layer. The stand categorization (semi-natural or plantation) was revisited using the Forestry Commission's (2001) forest cover map based on recent aerial photographs. Even so, there remained some obvious misclassifications where, for example, areas of broadleaved semi-natural woodland were classed as coniferous/mixed because a dense understorey of the evergreen holly (*Ilex aquifolium*) was interpreted as underplanting with conifers.

The GIS technology could define woodland boundaries to within a few metres, but it was not possible from aerial photographs and old maps alone to judge the edges of an ancient woodland to this level of accuracy. There might be a narrow band of recent woodland where the wood expanded outward beyond the medieval wood bank, for example, which would not be detectable except by detailed field survey; or if the GIS line were drawn at the outer limits of the observed tree canopy this could include several metres of adjacent field below the shade of the trees.

These differences are not important if the aim is simply to show the location of an ancient wood, but they do affect the calculation of the total area of ancient woodland when this is added up across c.32,000 woodland polygons. In 2003, the total area of ancient woodland in England, calculated from the digital data set, was greater than the original estimates (Spencer and Kirby, 1992) (Table 22.1), despite known losses of woodland in the intervening period! Small differences in boundaries are also critical in discussions on the loss of woodland to roads or housing, because whether a particular strip is ancient or not might affect whether permission is given for that development to go ahead.

As well as the changes arising from the digitization process, there were more systematic reviews, between 1995 and 2005, of the data for particular counties (for example, Hampshire, Dorset and Kent) and for the whole of the Forestry Commission estate. Particularly significant was a pilot study in 2004 that identified ancient woods of less than 2 ha in Wealden District, East Sussex, in south-east England (Westaway, 2005). South-east England is heavily populated, as well as being heavily wooded, so there is a high level of threat to the woods from house and road building. Hence, it was particularly important that ancient woods had been identified correctly and to establish the importance of the smaller areas (below 2 ha) that had not been included in the original work. Geo-rectified historic maps had by this time become available as a data layer, which greatly simplified comparisons between old and new maps. This approach to inventory revision was subsequently extended to other counties in south-east England (Westaway et al., 2007a,b; Hume et al., 2010; Bernsteade-Hume and Morris, 2012; Sansum et al., 2012).

The number of sites of ancient woodland increases almost exponentially in heavily wooded areas as the size threshold is lowered, but the increase in total area of ancient woodland captured is much less (typically a 10–20% increase in area for a doubling of the number of sites). The revision of the inventory in the Weald (the centre of Wealden District), for example, tripled the number of ancient woodland sites, but the overall area only went up by about 30%. In one region, the revision identified 523 ancient woods with an average area of 1.9 ha, compared to 240 sites on the original inventory of average area 13.3 ha. To date, resources have not allowed for this type of revision to be extended to the rest of England. There is also a question as to how cost-effective it would be to extend this approach relative to the number of woods that might come under threat in less densely populated parts of the country.

22.4.2 Wales

The Welsh Ancient Woodland Inventory was developed along similar lines to that in England and the results were available in digital form by 2004. However, the number of woods potentially obscured by hachuring on the early 19th century Ordnance Survey maps was much greater because Wales has more hills and mountains than England. The whole data set was therefore revised using 1:10,560 scale, mid-19th century maps that had become available in digital form. This larger scale allowed for the consideration of woods down to an area of 0.5 ha.

The revised Ancient Woodland Inventory, completed in 2011, identified around 95,000 ha of ancient woodland, compared with the 62,000 ha digitized in 2004. The revision also included two additional categories: Restored Ancient Woodland Sites to cover where a plantation has been recently restored to semi-natural woodland; and Ancient Woodland Sites of Unknown category (AWSU), which include young trees or clear-felled areas. The new data set has formally redefined the baseline for ancient woodland in Wales from 1600 to the date of the earliest usable 19th century maps, c.1830 (Forestry Commission, 2011).

22.4.3 Scotland

The 1981 ancient woodland project started in England and Wales because major surveys of semi-natural woods had already been undertaken in Scotland in the 1970s, but by 1984, Nature Conservancy Council staff in Scotland felt it would be worth applying the historical inventory approach there as well. The threshold date for defining ancient woodland was initially set at 1600, but soon effectively changed to 1750 because of the availability of maps made about this time as part of a military survey led by General William Roy (Scottish

Natural Heritage, 1997; Smout *et al.*, 2004). There were concerns as to the accuracy of Roy's maps, and particularly the black and white photocopies used for the inventory, which led to the inclusion of some semi-natural woods that were missing from them, but clearly present on mid-19th century maps as part of the 'ancient' category (Roberts *et al.*, 1992). The results were then digitized, as in England and Wales, in the 1990s.

Loss of ancient woodland to development is less of a threat in Scotland than in England and Wales, especially when compared with other causes of loss, such as high grazing pressure. A study carried out using the new Native Woodland Survey of Scotland (Forestry Commission, 2014) showed that around 9% of the woodland originally included in the Ancient Woodland Inventory had since been lost to development compared with *c.*12.5% to open ground.

Ancient woodland has not received the same weight in conservation and planning terms in Scotland as in England and Wales, partly because of greater uncertainty as to its consistent identification. The less intensive land use in Scotland means that ancient woodland features and species may survive more often beyond the mapped boundary of the individual fragments of ancient woodland, which have expanded and contracted in response to changing grazing patterns. More recently, the Scottish Native Woodland Inventory, a detailed habitat survey of the whole woodland resource (Scottish Natural Heritage, 2014), has replaced the Ancient Woodland Inventory as the key reference data set for conservation planning.

22.4.4 Northern Ireland

The Nature Conservancy Council had no remit in Northern Ireland and so this country was not included in the original Ancient Woodland Inventory. However, in 2006, the Woodland Trust completed the provisional Ancient Woodland Inventory for Northern Ireland, taking advantage of lessons from the British work and advances in GIS technology. Woods down to 0.25 ha were included because the total extent of ancient woodland is very small. The Woodland Trust was also more systematic in the use of field evidence, for example through trying to define the number of ancient woodland indicator plants needed per site to qualify a wood as ancient where documentary evidence was missing or ambiguous (Woodland Trust, 2007).

22.5 Testing the Limits of the English Inventories

Two classes of problem raised by Spencer and Kirby (1992) when describing the Ancient Woodland Inventory remain critical today: 'those caused by incomplete information for a site, and those which arose because the wood fell between or outside the categories used'.

22.5.1 Uncertain evidence

The original inventory took a precautionary approach and included as ancient woodland many sites where the evidence was sparse or ambiguous. It was a 'provisional' list, and inclusion on the list made relatively little difference to what the owner of the wood might do. Since then, the importance of ancient woodland has been more recognized, and planning and forestry policies in England now refer to such woodland (Forestry Commission, 2005; ODPM, 2005; DCLG, 2012). The fact that a site is ancient has more implications for the owner; hence the basis for determining that a site is ancient is more likely to be challenged. The sites that prove most contentious are those that have conflicting map/documentary evidence, or where the historical and field evidence seem to suggest different conclusions (see Box 22.3).

22.5.2 What is a wood?

Wood-pasture and parkland (see Hartel *et al.*, Chapter 5) were recognized as a form of ancient woodland by Peterken (1977), but were not always included in the original inventories. Where recognizable stands of trees were

Box 22.3. Examples of recent challenges to the inclusion of sites in the ancient woodland inventory.

- *Four Acre Wood, Bolnore, Sussex.* This small area (about 2 ha) is on the edge of a much larger ancient woodland block and was proposed as part of a housing development. It had a well-documented history as woodland up to the 1950s when it was cleared, with some soil disturbance from stump removal. During the 1980s, it began to grow up again as woodland. Field surveys showed a high presence of ancient woodland indicators. Although the site had been cleared of trees for 30 years, much of the rest of the flora had survived and it was in effect a woodland glade, but on the edge of the woodland. It was argued successfully at public inquiry that it should be treated as part of the adjacent ancient wood (Asquith, 2007).
- *Downlands Farm, Sussex.* Housing development was proposed that would affect an area of woodland. Part was absent from mid-19th century maps but there was well documented ancient woodland on either side and ancient woodland plants occurred within the disputed section. Further investigation showed that the central zone was not present on an 18th-century map and that the distribution of the key ancient woodland plants within it was consistent with spread in from the adjacent, well-documented ancient woods during the last 100 years. The disputed area was removed from the ancient woodland inventory (Dobson, 2008).
- *Broughton Woods, Lincolnshire.* Part of this wood was proposed for clearance to create a golf course. At the time, most of it was not listed as ancient woodland. However, a review of the map evidence suggested that this was an oversight, and documentary records also supported the case that the area was likely to be ancient. The flora was rather poor with few ancient woodland species over large areas. Nevertheless, the planning inspector ruled that there was sufficient evidence that it was ancient to justify refusing the golf course development (Cullingford, 2010).
- *Oaken Wood, Kent.* There is clear map and documentary evidence that this site was ancient, but it had a low diversity of plants over quite large areas of the woodland. The planning inquiry accepted the case that the site was ancient, but none the less approved the quarry proposal (Watson, 2013).
- *Racecourse Plantation, Norfolk.* The documentary evidence suggests that the site was formerly heathland, but the wood that is now on the site contains a relatively high number of ancient woodland indicators. The current hypothesis is that these may have survived among the heathland created in medieval times or been introduced recently, for example, on forestry machinery. There is not enough of a case from the flora to counter the documentary evidence.

shown on old maps and were still present, these could be mapped as ancient woodland. In contrast, sites with more scattered trees were omitted because their tree density (on maps or aerial photographs) was too low for them to be mapped as 'woodland'. Many important areas of veteran trees have, therefore, not been listed in the Ancient Woodland Inventory, or listed only in part. To address this problem, there has been extensive work to look at ways of developing a separate wood-pasture habitat inventory (Kirby and Perry, 2014).

22.5.3 How small can an ancient wood be?

The inventory originally included only woods that were above 2 ha on the base maps (generally 1:25,000 scale, from the 1930s) for pragmatic, and not ecological, reasons. It provided a common baseline to the 1979–1982 National Forestry Census and it avoided the problems of trying to identify consistently very small woods on early Ordnance Survey maps (scale 1:63,360). Restricting consideration to woods of 2 ha or more also helped to limit the work involved.

Subsequently the Welsh revision reduced the threshold area to 0.5 ha, and the south-east revision in England to 0.25 ha. Is there a minimum size below which the historical origin of the wood ceases to be a significant factor? Woods below 2 ha in area can show ancient woodland features such as wood banks, old coppice stools and ancient woodland indicator plants. Even individual veteran trees may support significant lichen or saproxylic species. However, small woods are more subject to edge effects from adjacent open ground (Kirby, 2004), for example, from the possible impacts of spray drift (Gove *et al.*, 2007). There tends to be more overlap between the number of indicator species present in small ancient and recent woods (Hill, 2003), so the significance of the flora as a guide to history is more difficult to interpret.

From one perspective, the majority of the species in such tiny fragments may be viewed as having unsustainable populations anyway, so is it worth using scarce conservation resources on identifying them? The contrary view is that small patches can form the nuclei from which species populations can expand back out into adjacent recent woods or other semi-natural habitats (see point 7 in Box 22.2). They may also act as links and stepping stones for species moving between larger blocks of woodland.

22.6 Conclusion

The Ancient Woodland Inventory remains 'provisional' which has caused some confusion at times, with the question being posed, 'when is the final version coming out?'. The strength of this approach is that the data are clearly labelled as 'live' and open to modification, if new evidence or interpretations become available.

The inventory has increased awareness of the value of ancient woodland among the conservation sector, but also among the general public, regulatory bodies and planning authorities. The term 'ancient woodland' is now widely accepted and its identification frequently adds capital value to woodland. Many more sites are being protected and better managed than would otherwise be the case, but not all threats have been defused.

Ancient woods are still damaged or destroyed by development, but this is more likely to be in the glare of public attention rather than without notice. The situation differs in Scotland, where the study of ancient woodland (Forestry Commission, 2014) has shown that as much as one eighth of Scottish ancient woodland may have been lost because of high herbivore pressure preventing regeneration, with few people noticing it.

The different countries have each moved to slightly different bases for setting the date thresholds for ancient woodland, and only England nominally holds to the original 1600 used in Peterken's definition. Even in England, the practical threshold date for available map evidence is often the 18th century, with map interpretation and other field evidence used to back up ancient woodland status. Research in the UK, as well as in other European countries, suggests that the ecological response to the influence of site history is graded rather than necessarily being a step change at a particular date (Stone and Williamson, 2013).

No quality or size threshold was ever attached to a site being classed as ancient; the lower size limits were purely pragmatic in order to make the creation of an inventory feasible. In theory, there must come a point where there is too little of the woodland system left, either because it is so small or because it is so degraded for so long a period that it has lost the special qualities of ancient woodland. This lack of a minimum 'quality' element could become increasingly important because the treatment of an area depends on whether or not it attracts the 'ancient woodland' label. Otherwise, the anomalous position could be reached where small degraded areas of ancient woodland receive greater protection than larger, more diverse and important areas of other habitats.

The original binary classification of woods into semi-natural or plantation was useful for broad-scale policy work, but it is overly simplistic for dealing with the complexities that can arise as semi-natural woods are brought back into management using native species and conifer plantations are restored to native broadleaved stands.

Conservation priorities have changed since the 1980s. The emphasis then was on identifying and protecting sites that were threatened with either clearance for agriculture or felling and conversion to plantations. Now, the need to increase habitat and species cover within a changing environment and with increasing threats from novel or emerging pests and diseases is a priority. The Ancient Woodland Inventory will remain an important conservation tool in this new world as a means of identifying critical conservation areas. Protecting the remaining areas of ancient woodland will remain a priority, but how we use that information will change.

References

Asquith, P.J. (2007) *Report on Appeals by Crest Nicholson (South) Limited Relating to Bolnore Village Phases 4 and 5, Haywards Heath, West Sussex. Appeal refs: APP/D3830/A/05/1195897-98 and APP/D330/A/06/1198282-83*. The Planning Inspectorate, Bristol, UK.

Bernsteade-Hume, V. and Morris, J. (2012) *Ancient Woodland Inventory for the Chilterns*. Chilterns Conservation Board, Chinnor, UK.

Brunet, J. and Von Oheimb, G. (1998) Migration of vascular plants to secondary woodlands in southern Sweden. *Journal of Ecology* 86, 429–438.

Cullingford, D.R. (2010) *Report on Land Containing Woodland and Golf Courses beside Forest Pines Hotel Golf and Country Club, Ermine Street, Broughton, Lincolnshire DN20 0AQ. Appeal Ref: APP/Y2003/A/09/2101852*. The Planning Inspectorate, Bristol, UK.

DCLG (2012) *National Planning Policy Framework*. Department for Communities and Local Government, London.

Dobsen, P.E. (2008) *Report to the Secretary of State for Communities and Local Government. Wealden District Council – Inquiry into Conjoined Section 78 Appeals Against the Refusal of 3 Outline Planning Applications. A) Downlands Farm, Snatts Road, Uckfield (appeal ref. 2046982) B) Bird In Eye North, Framfield Road, Uckfield (appeal ref. 2042597) C) Bird In Eye South, Framfield Road, Uckfield (appeal ref: 2053422)*. Planning Inspectorate, Bristol, UK. Available at: http://www.crawley.gov.uk/pub_livx/groups/operational/documents/plappdn/int165107.pdf (accessed August 2014).

Forestry Commission (2001) *National Inventory of Forests and Trees – England*. Forestry Commission, Edinburgh, UK. Available at: http://www.forestry.gov.uk/pdf/niengland.pdf/$FILE/niengland.pdf (accessed 12 December 2014).

Forestry Commission (2005) *Keepers of Time: A Statement of Policy for England's Ancient and Native Woodland*. Forestry Commission, Cambridge, UK.

Forestry Commission (2011) *Ancient Woodland Inventory 2011*. Available at: http://www.forestry.gov.uk/forestry/INFD-8VPJFD (accessed February 2014).

Forestry Commission (2014) *Scotland's Native Woodlands Results from the Native Woodland Survey of Scotland*. Forestry Commission, Edinburgh, UK.

Goldberg, E.A., Kirby, K.J., Hall, J. and Latham J. (2007) The ancient woodland concept as a conservation tool in Britain. *Journal of Nature Conservation* 15, 109–119.

Goldberg, E.A., Peterken, G.F. and Kirby, K.J. (2011) Origin and evolution of the Ancient Woodland Inventory. *British Wildlife* 23, 90–96.

Goodfellow, S. and Peterken, G.F. (1981) A method for survey and assessment of woodland or nature conservation using maps and species lists: the example of Norfolk woodlands. *Biological Conservation* 21, 177–195.

Gove, B., Power, S.A., Buckley, G.P. and Ghazoul, J. (2007) Effects of herbicide spray drift and fertilizer overspread on selected species of woodland ground flora: comparison between short-term and long term impact assessments and field surveys. *Journal of Applied Ecology* 44, 374–384.

Hermy, M., Honnay, O., Firbank, L., Grashof-Bokdam, C. and Lawesson, J.E. (1999) An ecological comparison between ancient and other forest plant species of Europe, and the implications for forest conservation. *Biological Conservation* 91, 9–22.

Hill, A. (2003) Plant species as indicators of ancient woodland in the Malvern Hills and Teme Valley Natural Area. PhD thesis, Coventry University, Coventry, UK.

Hume, V., Crose, M., Sansum, P., Westaway, S. and McKernan, P. (2010) *A Revision of the Ancient Woodland Inventory for West Sussex*. Sussex Biodiversity Record Centre, Henfield, UK.

Kirby, K.J. (2004) Balancing site-based protection versus landscape-scale measures in English woodland. In: Smithers, R. (ed.) *Landscape Ecology of Woodland and Trees*. ialeUK (International Association for Landscape Ecology – UK Region), Cirencester, UK, pp. 263–270.

Kirby, K.J. and Perry, S.C. (2014) Institutional arrangements of wood-pasture management: past and present in the UK. In: Hartel, T. and Pleininger, T. (eds) *European Wood-Pastures in Transition: A Social–Ecological Approach*. Earthscan from Routledge (imprint of Taylor & Francis), Abingdon, UK, pp. 254–270.

Kirby, K.J., Peterken, G.F., Spencer, J.W. and Walker, G.J. (1984) *Inventories of Ancient Semi-natural Woodland*. Focus on nature conservation 6, Nature Conservancy Council, Peterborough, UK.

Leopold, A. (1949) *A Sand-County Almanac*. Oxford University Press, New York.

MacArthur, R.H. and Wilson, E.O. (1967) *The Theory of Island Biogeography*. Princeton University Press, Princeton, New Jersey.

Nilsson, S. (1997) Forests in the temperate–boreal transition: natural and man-made features. *Ecological Bulletins* 46, 61–71.

ODPM (2005) *Planning Policy Statement 9: Biodiversity and Geological Conservation*. Office of the Deputy Prime Minister, London.

Peterken, G.F. (1974) A method for assessing woodland flora for conservation using indicator species. *Biological Conservation* 6, 239–245.

Peterken, G.F. (1977) Habitat conservation priorities in British and European woodlands. *Biological Conservation* 11, 223–236.

Peterken, G.F. (1983) Woodland conservation in Britain. In: Warren, A. and Goldsmith, F.B. (eds) *Conservation in Perspective*. Wiley, Chichester, UK, pp. 83–100.

Peterken, G.F. and Harding, P.T. (1974) Recent changes in the conservation value of woodlands in Rockingham Forest. *Forestry* 47, 109–128.

Pontailler, J., Faille, A. and Lemée, G. (1997) Storms drive successional dynamics in natural forests: a case study in Fontainebleau forest (France). *Forest Ecology and Management* 98, 1–15.

Rackham, O. (1971) Historical studies and woodland conservation. *Symposium of the British Ecological Society* 11, 563–580.

Rackham, O. (1976) *Trees and Woodlands in the British Landscape*. Dent, London.

Rackham, O. (2003) *Ancient Woodland*, rev. edn. Castlepoint Press, Dalbeattie, UK.

Roberts, A.J., Russell, C., Walker, G.J. and Kirby, K.J. (1992) Regional variation in the origin, extent and composition of Scottish woodland. *Botanical Journal of Scotland* 46, 167–189.

Sansum, P., Westaway, S. and McKernan, P. (2012) *A Revision of the Ancient Woodland Inventory for Maidstone Borough, Kent*. The Weald and Downs AONB (Area of Outstanding Natural Beauty) Unit, Robertsbridge, UK.

Scottish Natural Heritage (1997) *The Inventory of Ancient and Long-established Woodland Sites and the Inventory of Semi-natural Woodlands (Provisional)*. Information and Advisory Note Number 95, Scottish Natural Heritage, Edinburgh, UK. Available at: www.snh.org.uk/publications/on-line/advisorynotes/95/95.html (accessed February 2014).

Scottish Natural Heritage (2014) *A Guide to Understanding the Scottish Ancient Woodland Inventory*. Scottish Natural Heritage, Inverness, UK. Available at: http://www.snh.gov.uk/docs/C283974.pdf (accessed January 2014).

Smout, T.C., Macdonald, A.R. and Watson, F. (2004) *A History of the Native Woodlands of Scotland, 1500–1920*. Edinburgh University Press, Edinburgh, UK.

Spencer, J.W. and Kirby, K.J. (1992) An inventory of ancient woodland for England and Wales. *Biological Conservation* 62, 77–93.

Stone, A. and Williamson, T. (2013) 'Pseudo-ancient woodland' and the ancient woodland inventory. *Landscapes* 14, 141–154.

Verheyen, K., Bossuyt, B., Hermy, M. and Tack, G. (1999) The land use history (1278–1990) of a mixed hardwood forest in western Belgium and its relationship with chemical soil characteristics. *Journal of Biogeography* 26, 1115–1128.

Watkins, C. (1988) The idea of ancient woodland in Britain from 1800. In: Salbitano, F. (ed.) *Human Influence on Forest Ecosystems Development in Europe*. Pitagora Editrice, Bologna, Italy, pp. 237–246.

Watson, R. (2013) *Letter to Mr Hare, Detailing Planning Decision. DCLG*. Department for Communities and Local Government, London. Available at: https://www.gov.uk/government/uploads/system/uploads/attachment_data/file/237090/Hermitage_Quarry__Hermitage_Lane__Aylesford__ref_2158341__11_July_2013_.pdf (accessed August 2014).

Westaway, S. (2005) *Weald Ancient Woodland Survey: A Revision of the Ancient Woodland Inventory for Wealden District*. High Weald AONB (Area of Outstanding Natural Beauty) Unit, Flimwell, Sussex, UK.

Westaway, S., Grose, M. and McKernan, P. (2007a) *A Revision of the Ancient Woodland Inventory for Mid Sussex District, West Sussex*. High Weald AONB (Area of Outstanding Natural Beauty) Unit, Flimwell, East Sussex, UK.

Westaway, S., Grose, M. and McKernan, P. (2007b) *A Revision of the Ancient Woodland Inventory for Tunbridge Wells Borough, Kent*. High Weald AONB (Area of Outstanding Natural Beauty) Unit, Flimwell, East Sussex, UK.

Whitbread, A. (1986) *Herefordshire Inventory of Ancient Woodlands (Provisional)*. Nature Conservancy Council, Peterborough, UK.

Woodland Trust (2007) *Back on the Map, an Inventory of Ancient and Long Established Woodland in Northern Ireland. Preliminary Report*. The Woodland Trust, Bangor, Northern Ireland, UK. Available at: www.backonthemap.org.uk/NR/rdonlyres/09F70BD6-8E68-4328-90B7-05DFE9483550/0/070115Preliminaryreport.pdf (accessed February 2014).

Wulf, M. (1997) Plant species as indicators of ancient woodland in northwestern Germany. *Journal of Vegetation Science* 8, 635–642.

23 Tree and Forest Pests and Diseases: Learning from the Past to Prepare for the Future

Clive Potter*
Centre for Environmental Policy, Imperial College London, UK

23.1 Introduction

Invasive pests and diseases, many of them unknown to science a decade or two ago, pose a significant threat to Europe's woods and forests. The spread of *Chalara fraxinea* (ash dieback, now properly identified and renamed as *Hymenoscyphus fraxineus*) throughout north-western Europe, and its arrival in the UK in 2012, is just the latest in a series of pest and disease outbreaks that have swept through Europe's forests over the last 10 years (Boyd et al., 2013). Well-documented epidemics include those caused by the oak processionary moth (*Thaumetopoea processionea*), an insect pest now widespread throughout Belgium, the Netherlands and Germany, chestnut blight (*Cryphonectria parasitica*), a fungal pathogen that was first recorded in Italy in 1938 but which has been spreading steadily since that date, and the pine tree lappet moth (*Dendrolimus pini*), a native of continental Europe, Russia and Asia, where it has long caused periodic and large-scale damage to pine plantations.

The cumulative impact of these and various other pests and diseases is often dramatic. The loss of almost 30 million elms in the UK between 1970 and 1990 due to an epidemic of Dutch elm disease (DED) was undoubtedly a very significant environmental event, impoverishing many upland woodland communities in Scotland and Wales, and also removing the culturally valued 'elmscapes' of lowland England (Tomlinson and Potter, 2010). The current outbreak of *Chalara* in Denmark, Poland and Lithuania has been similarly damaging, with extensive loss of woodland areas traditionally dominated by ash and serious impacts on commercial forestry operations. In some locations, dieback has been so severe that up to 90% of standing trees have been lost (Kowalski, 2006). Other less pathogenic outbreaks can have equally serious long-term consequences. Fungal-like pathogens such as *Phytophthora alni* have been identified as a significant cause of long-term decline in the native alder communities in the UK, for instance, while a number of weaker invasive diseases are thought to be contributing to acute oak decline syndrome across much of Europe (Boyd et al., 2013).

Drawing on previously unpublished archival research, epidemiological modelling work and field interviews, this chapter charts the progress of three important but biologically different epidemics – DED, *P. ramorum* blight and ash dieback – assessing their environmental and cultural impact and reflecting

---

*E-mail: c.potter@imperial.ac.uk

© CAB International 2015. *Europe's Changing Woods and Forests: From Wildwood to Managed Landscapes* (eds K.J. Kirby and C. Watkins)

on comparisons that can be made and lessons that can be drawn.

23.2 Background

Daszak *et al.* (2000) consider that the introduction of alien pathogens has, until recently, been one of the most underestimated causes of global anthropogenic environmental change. The consequences of their introduction and spread are likely to be especially severe in woodlands. The rate at which new pests and diseases are being introduced and are managing to establish themselves is increasing, largely due to the globalization of the horticultural trade in plants and plant materials that has occurred over the last 20 years (Dehnen-Schmutz *et al.*, 2010). Plants are increasingly being shipped into the European Union (EU) from China, Japan, Australasia, Africa and South America, and although at least some of these might be from certified sources, the growing throughput of material greatly enhances the risk that known and unknown pathogens will be introduced into previously 'virgin' environments in which there are native trees with no co-evolved resistance to disease. The growing preference for 'instant woody landscapes' further exacerbates the problem because semi-mature or even mature trees with large root balls are much more likely to carry disease.

This complicated but highly lucrative 'plants for planting' trade pathway is the route through which many recent invasive pathogens have arrived in the EU, including the various *Phytophthora* spp. that are carried as spores on live plants and in root balls. Other pathways of disease introduction may become just as significant in future years. For example, the US Forest Service estimates that up to 50% of maritime shipments and 9% of air freight use solid wood as a packing material, and this can play host to insect pests such as the Asian longhorn beetle (*Anoplophora glabripennis*) and the pine wood nematode.

The Asian longhorn beetle was probably introduced into the USA in the early 1980s on the wood used to pack and transport the pipework manufactured in China for the refurbishment of New York city's sewage system (Haack, 1997). Sweet chestnut blight is thought to have first been brought to Europe during World War I, when infested chestnut wood from North America was used to pack ammunition being transported to Belgium and France (Biraghi, 1950). The westward spread of the great spruce bark beetle (*Dendroctonus micans*) over the last 30 years has been facilitated by trade in logs with bark. Wood chips imported to supply biomass for energy generation may also be infested with various insect pests, including the emerald ash borer (*Agrilus planipennis*), a particularly virulent insect pest that has devastated native ash populations in North America. The expansion of renewable energy generation in many EU member states has led to estimates of huge increases in imports if what seems likely to be an industrial-scale demand for wood chips is to be satisfied. The risk of new introductions can only increase proportionately.

Nevertheless, many will argue that invasive pathogens are nothing new, as they have been present and caused damage to native plant communities, woodlands and treescapes throughout history (Orwig *et al.*, 2002). The root pathogen *P. cinnamomi*, for instance, has been spreading around the world for at least the last 150 years and we know that an outbreak of oak processionary moth was recorded in Flanders in 1800 (Potter *et al.*, 2015). Attention is very much on present-day threats and consequences and on the need for better *predictions* of emerging risks. Notwithstanding, it is helpful to look back at the historical record in order to draw lessons from what Cornell *et al.* (2010) have called 'the disease archive of the past'.

Recent epidemics may be unprecedented in their scale and impact, but the sudden emergence of a tree pest capable of killing large numbers of trees is not new. An outbreak of chestnut blight in North America around 1910 destroyed most of North America's chestnut forests in the space of 30 years (Frankel, 2010), and twice in the 20th century, pandemics of DED have spread throughout North America, Europe and south-west Asia, killing elm trees (*Ulmus* spp.) in large numbers.

The tree disease outbreak archive is thus potentially very large, dating back to the earliest examples of tree pests and disease outbreaks.

Tracing the different trajectories of specific disease introductions and their spread and impact can be valuable in helping us to understand how biology and epidemiology interact with human behaviour and policy responses (or the lack of them) to determine eventual outcomes. Examining aftermaths, meanwhile, can show us the extent to which natural environments, communities and people are able to adapt to the loss of particular tree species on what is often a landscape scale; and whether management programmes have any real effect on long-term outcomes.

23.2.1 Dutch elm disease, ramorum blight and ash dieback

In this chapter we compare three outbreaks, of which one is long past its peak (DED), and the other two, ramorum blight and ash dieback, still very much in progress.

DED is a fungal pathogen with a long history of introduction and spread in Europe generally and in the UK especially. Carried on the bodies of scolytid bark beetles, the fungus has been causing elm death since the 1920s, when it was first identified. The disease was first described in north-west Europe in 1918, and was endemic in Scandinavia and southern Italy by the late 1930s. In the UK, it was first identified in Hertfordshire in 1927, and this marked the beginning of a first outbreak that would persist throughout the 1930s and 1940s. However, it was the second and much more virulent outbreak in the UK in the early 1970s that stands out as one of the most dramatic environmental events of the last 40 years, killing almost 30 million trees and restructuring the traditional 'elmscapes' of large parts of lowland England, Scotland and Wales.

P. ramorum, the causative organism of ramorum blight, is a similarly virulent fungal-like pathogen which has been traced to South-east Asia, where it is thought to have entered international trade pathways on exotic horticultural plants of various kinds. Before entering the UK in the late 1990s via the nursery trade, it had previously been identified in the USA, where the resulting epidemic of 'sudden oak death' infected millions of tan oaks and seriously depleted the coastal forests of California and Oregon (Cobb *et al.*, 2012). From an initial site of infection in south-west England, the pathogen has spread throughout the UK and is currently regarded as a high-risk outbreak by the UK's plant health authorities, threatening a range of tree species, including commercial timber species such as larch (*Larix* spp.) as well as core plant species making up lowland heathland (Defra, 2013b).

Ash dieback has been spreading westward throughout Europe for over a decade and its arrival in the UK in early 2012 attracted considerable media and public attention. Current predictions are that this pathogen will significantly deplete the UK's ash woodland over the next 5–10 years.

23.3 The Dutch Elm Disease Outbreak

As is often the case with major disease epidemics, the early stages of the UK DED outbreak of the 1970s were hard to detect and easily ignored. In 1960, Tom Peace, a forest pathologist based at the UK Forestry Commission's Alice Holt research station, had published an assessment of DED in Europe and the UK which concluded that while 'the disease may long continue to be a minor nuisance' it was unlikely to have significant long-term consequences (Peace, 1960). This became the conventional wisdom about DED until well into the 1970s and coloured the way that the UK authorities would respond to the first reports from foresters, estate managers and gardeners of extensive dieback in elms during the first stages of the new outbreak.

Forestry Commission (FC) officers and others have since confirmed that these early reports were largely dismissed on the grounds that this was merely a flare-up of an already established, but less virulent, form of the disease (Tomlinson and Potter, 2010). Following the Peace report, received wisdom and scientific advice maintained that DED was a manageable problem that did not justify extensive and costly intervention. With the benefit of hindsight, it is now clear that something had changed and subsequent research confirmed that the later observations were the first symptoms of a much more virulent strain of the disease imported into the UK from North America on a consignment of rock elm (*Ulmus thomasii*) (Brasier and Gibbs, 1973). For the time being, FC officials responded to public expressions of concern by referring to the previous outbreak and largely dismissing reports of dieback as evidence of another concerning, but probably geographically limited, flare-up.

As the outbreak continued to spread, plant health officials changed their view of the seriousness of the outbreak only slowly. However, by the end of the decade, the FC was coming under growing pressure from the local authorities most affected by the disease to deal with the growing number of standing dead trees along roadsides and in public spaces. Even so, the FC still rejected the case for compulsory sanitation felling in order to control the disease, and its first intervention in the outbreak, in 1970, was to recommend a programme of voluntary sanitation felling instead. Only after this had failed did the FC begin to contemplate legislative action.

The summer of 1971 saw protracted negotiations between the FC, the UK Ministry of Agriculture, Fisheries and Food and the Treasury about what should be done and how any control measures should be funded. The willingness of the UK government and its agencies to deal with the outbreak was heavily compromised by the anticipated high costs of control measures, together with considerable uncertainty as to whether these would prove effective. Eventually, in October 1971, it was agreed that the 1967 Plant Health Act would be invoked in order to give statutory powers to local authorities 'to enter land and inspect elms trees and to take steps – extending if need be to enforced destruction of trees without compensation – to prevent further spread of the disease'.

Tomlinson and Potter (2010) have shown how this decision to construct the DED problem as an issue of local amenity rather than a national emergency, and hence to devolve responsibility on to local authorities, was critical in deciding how the outbreak would subsequently be managed. Most local authorities, when confronted with the (uncompensated) costs of large scale felling, decided they could only justify sanitation felling in settings where standing dead elms posed risks to public safety. Across large swathes of rural England, little or no sanitation felling consequently took place.

Critics at the time argued that the plant health authorities had 'done too little, too late' and contended that central government should have been much better coordinated and quicker to act once the disease had broken out. In a widely publicized series of articles in the *New Scientist*, Jon Tinker argued that the compulsory felling policy eventually put in place by the FC was under-resourced and abandoned too early, and that more should have been done to contain the spread of the outbreak by restricting the movements of contaminated timber (Tinker, 1971).

In epidemiological terms, the epidemic followed a classic trajectory, building slowly at first during an extended lag phase (our work suggests an earlier than previously thought date of first introduction of 1962), before breaking out into an exponential phase of disease spread, and tailing off as the host population of live elms was eliminated. By the time of the first comprehensive FC survey of elms conducted in late 1971, 700,000 of the then estimated 18 million elms in southern England were dead or dying and a further 1.6 million showed evidence of disease. These figures were revised upwards as the outbreak unfolded and by 1980 there were only a few mature elms remaining in the UK.

The creation of a *cordon sanitaire* was briefly contemplated in 1972 in order to prevent the disease from reaching northern England and Scotland, but rapidly abandoned on grounds of cost. A ban on the movement of elm logs was instituted in 1974 following the realization that transportation of infected material was accelerating spread, but this proved difficult to enforce and by the middle of the decade the FC was advising ministers that '(we) believe there is no alternative to allowing the disease to take its course, with all this implies for the future of the common species. … The FC believes that all should be done to repair the damage done by the disease by a sustained programme of planting hardwood trees other than elms over the next few years' (Forestry Commission, 1972; quoted in Tomlinson and Potter, 2010).

Subsequent analysis of the management of the outbreak (Potter *et al.*, 2011) has revealed that there were undoubtedly delays in identifying the threat and a confused and poorly coordinated and resourced policy response. However, biology trumped policy at an early stage, given the great difficulty of containing the spread of this very virulent disease once it becomes established. In a counterfactual world, DED might have been prevented, but this would have required action to restrict imports of diseased timber at an even earlier stage than previously considered, based on there having been a full risk assessment of the disease system as it developed in North America during the early 1960s. Neither action was likely given the lack of advance intelligence and the embryonic nature of international biosecurity policies and protocols at the time.

The aftermath of DED in Britain is still with us. At the time, there was curiously little official retrospection on the outbreak and no formal commissions of enquiry. DED was largely consigned to the historical record as an unfortunate, but probably unique, disease event.

In Brighton and Hove and in some parts of the South Downs, mature elm trees survived owing to the protection afforded by topography and the coast, and East Sussex County Council instituted a vigorous DED Control Programme in order to protect the remaining trees. This programme continues to the present day. Elsewhere, the disease remains endemic and it continues to spread into the far north and west. In infected areas, young elms regenerate in hedgerows and on the edges of woodland until they become large enough to succumb to disease (Plate 8). Simulation studies suggest that this will continue to take place on a roughly 20 year cycle, with successive disease fronts building up and spreading out, before then subsiding as the available population of trees is killed off by the disease (Harwood *et al.*, 2011). Attempts are continuing to reintroduce elms by using strains that are disease resistant on a small scale. The Conservation Foundation, for instance, a UK charity with interests in biodiversity protection, has set up its 'Great Elm Experiment' to distribute micro-propagated, disease-resistant saplings to schools, local authorities and private landowners.

The scientific legacy of DED in the UK has arguably been significant; experience of the outbreak among a small but influential community of forest pathologists triggered further work on the causes and nature of tree diseases. For many scientists working in the field, the DED outbreak continues to be an important point of historical reference (Brasier, 1996).

The institutional legacy of DED is somewhat harder to trace. As MacLeod *et al.* (2010) observed, in the years following the outbreak, the safeguarding of plant health generally and tree health especially, received much less attention in the UK than did animal health concerns. While plant health has now been consolidated as a responsibility of government (Defra and Forestry Commission, 2011), this development has largely been driven by the need to manage the impact of pests and diseases on commercial agricultural and horticultural crops and products. Attempting to minimize the risks to *public goods* such as biodiversity, ecosystem services and amenity that might be posed by new invasions of tree pests and diseases is a relatively new policy concern, and one with which decision makers are only just beginning to come to terms.

23.4 'Sudden Oak Death' (Ramorum Blight) in the UK

The current outbreak of ramorum blight in the UK shares many of the features of DED, not least in terms of the difficulty of dealing with changing outbreak characteristics and a shifting risk profile. All the same, this more recent experience suggests that anticipating risk and taking precautionary action in the way that would have been required to prevent the DED epidemic is still just as difficult in a world of genetic mutation, expanding markets and the existence of powerful vested interests committed to the promotion of tree trade (Maye *et al.*, 2012; Potter, 2013).

Ramorum blight is thought to have arrived in the UK as a single introduction, probably on infected nursery stock originating from within the European Single Market via 'plants for planting' (i.e. live plants intended to remain planted, to be planted or to be re-planted) pathways. Almost certainly originating in South-east Asia, the highly damaging disease owes its common name – sudden oak death – to outbreaks in the USA in 2001, where it has killed millions of oaks and tan oaks along the coastal flanks of California and Oregon (Cobb et al., 2012).

The disease is not currently affecting native English oaks, but it does pose risks for a range of tree species, including beech, ash and larch (Grunwald et al., 2008). From an initial outbreak in Cornwall in 2002, the disease has expanded throughout the south-west, into southern Wales and western Scotland, particularly on larch. In common with DED, long-distance spread has been assisted by human means, in this case through movements of infected plants within the nursery trade. Indeed, during the early stages of the current epidemic, the pathogen was largely confined to, and largely seen as a problem for, the nursery trade. Measures were put in place by the Department of Food, Environment and Rural Affairs (Defra) acting through its agency, the Food and Environment Research Agency (Fera), to monitor and destroy any infected stock.

None the less, by 2002, ramorum blight was being found in woodland gardens in south-west England, typically where its principal UK host species, *Rhododendron ponticum*, was present as an understorey plant, and there were fears of further spread into surrounding semi-natural woodland. Certainly, there was evidence of the subsequent infection of nearby susceptible trees such as beech, ash, sweet chestnut and evergreen oaks.

Despite a comparatively swift initial containment effort from the plant health authorities, ramorum blight has continued to spread, with infections increasing in number and geographical range. Defra put in place an Emergency Programme in south-west England soon after the disease was confirmed in 2002, with powers given to plant health inspectors to enter and inspect woodland gardens in the region and to destroy any infected material found in nurseries and in woodland. Annual surveys of nursery stock were initiated, with a policy of destroying all infected plants found within 2 km of an identified infection.

In 2009, following a full science and policy review, a new programme was established with increased funding in order to achieve the goal of containing the spread of the disease. This included money for an extensive programme of clearance of rhododendron understorey in private woodland throughout the country in an attempt to remove one of the disease's main host species. However, the outbreak entered a new phase following the discovery in 2009 of the disease in Japanese larch – an important forestry tree that accounts for almost 10% of the conifer growing stock of Britain. A new programme of sanitation felling was agreed, with over 700 ha of plantation larch cleared to date (2013) in Scotland, for example. Here, the presence of extensive commercial larch plantations means that the incidence of the disease is now greatest, with areas such as Dumfries and Galloway particularly badly affected.

Despite these timely measures, and in a manner strongly reminiscent of DED, plant health inspectors and policy makers discovered that they were dealing with an epidemic that had unpredictable characteristics. The pathogen's ability to spread across an expanding host range means that the outbreak has kept ahead of attempts to contain it, jumping from ornamental shrubs like rhododendrons to woodland trees like beech and now to important commercial forestry species.

Forest pathologists agree that ramorum blight is a complex disease system that is hard to diagnose, and may be present for long periods without the host plants showing symptoms. Meanwhile, large numbers of spores may be being produced by infected plants. The authorities have implemented measures designed to prevent further importation of infected material, but because the fungus can infect plants without outward signs, port inspections based on purely visual evidence may be ineffective. Control has proved difficult because of a reluctance to report early signs of the disease in some private gardens and woodlands during the early stages of the outbreak. Other contributory factors have been an initial lack of clarity regarding who would bear the costs of clearance, the potential for continued spread on the footwear of garden visitors and walkers, and the special epidemiological features of ramorum blight as a disease system.

In summer 2013, the FC announced that eradication of the pathogen in the UK was no longer achievable, though efforts continue to slow its spread and contain its impact within designated control zones. It seems likely that, while the disease may be spreading more slowly than DED, its eventual, cumulative impact on landscapes, commercial forestry and the horticultural heritage may be just as significant. Indeed, modelling work suggests that the area now affected by ramorum blight is already so great that, even were effective management solutions available for all the landscapes at risk, they would be prohibitively expensive to implement (Harwood et al., 2010).

23.5 A Landscape Without Ash?

The ramorum experience in the UK demonstrates the ad hoc manner in which disease epidemics tend to be responded to and managed under conditions of uncertainty. Our final case study confirms how difficult it is to contain an introduced disease, even when there appears to be considerable political will and stakeholder support for doing so.

C. fraxinea is a fungal pathogen that causes dieback and extensive mortality of ash trees. It was discovered for the first time in the UK in a nursery in Buckinghamshire in February 2012 and confirmed as present in the wider environment when found in woodland in Norfolk in October of that year. The disease had been spreading westward from eastern European forests since the late 1990s, and its impact on ash woodland in countries such as Denmark and Norway is well documented in the scientific literature (Kowalski, 2006).

Although not listed on the UK's national risk register, *Chalara* had long been seen by pathologists as a potential threat (Kowalski, 2006), but until the time of its introduction, large consignments of potentially diseased young ash trees were still being imported into the UK, largely to supply domestic forest nurseries. A failure to stop these imports when it was known that the risks were high has been seen by some commentators as a significant mistake, albeit that the confusion surrounding the classification of the disease being observed on the continent and the extent of its pathogenicity did not help matters (Potter, 2012).

Once identified as present in the UK, however, the official response to the pathogen was both immediate and dramatic. Unlike the DED and ramorum blight outbreaks, where the case for intervention was assembled slowly and in reaction to different phases of an unfolding outbreak, the government moved quickly to put various management measures in place. The Cabinet Office Crisis Committee (COBR) met in special session during the early days of the outbreak, and this was followed by a decision by Defra to ban imports of (potentially infected) ash saplings. Emergency powers were invoked to prevent movements of infected ash around the country. On the ground, the relevant forestry agencies in England and in the devolved administrations undertook extensive ground survey and air surveillance work to identify infected trees in order to establish how far the disease had already spread. An interim *Chalara* Control Plan was put together and Ian Boyd, Defra's Chief Scientist, convened a Tree Health and Plant Biosecurity Expert Taskforce (Defra, 2013b) to look more generally at the mounting threats to tree health from other invasive pests and diseases, and to make recommendations for improving plant biosecurity (see further discussion below).

Chalara followed a series of other new tree pest and disease introductions into the UK, such as the oak processionary moth and Asian longhorn beetle, and the extent and depth of the official response is partly due to raised awareness of the threat to tree health in government circles generally. A Tree Health and Plant Biosecurity Action Plan had been published by Defra in 2011, and plant health was moving up the political agenda alongside the reform of the EU's Plant Health Regime (Defra and Forestry Commission, 2011).

Nevertheless, the intense media and public response to *Chalara* took many by surprise and it was arguably this sudden politicization of tree health as a policy issue that prompted the speedy government response. Over a period of weeks, tree health was promoted from being an issue of largely expert concern to a major focus of public debate and media coverage; questions were asked about institutional competence and broader critiques of biosecurity breaches in the live plant trade (Potter *et al.*, 2015). As Pidgeon and Barnett (2012) observed, by the end of 2012, *Chalara* appeared to have joined a list of rapidly developing risk cases such as bovine spongiform encephalopathy (BSE) and the MMR (measles, mumps and rubella) vaccination affair, where what had changed was not the risk itself so much as the way it is perceived and interpreted by the public.

However, having failed to prevent the introduction of the pathogen in the first instance, the impact of government actions on the course of the outbreak itself has so far been minimal. Based on experience in Denmark, future years will see very widespread tree mortality, albeit not quite as rapid or complete as in the case of DED. While the greater genetic variability of the ash genus means that not all trees will succumb, Danish experience suggests that the incidence of this natural resistance will be low, probably in the range of 1–4% of all trees. Modelling work undertaken for Defra, meanwhile, suggests that the disease will be widespread by 2017 and that the resulting depletion of the UK's standing hardwood stock will be significant.

The renamed '*Chalara Management Plan*' (Defra, 2013a) is really a strategy for adapting to loss. By putting the emphasis on the long-term development of resistant strains of ash and the need to restock the landscape on this basis, the plan implicitly accepts that large numbers of native trees will be lost in the meantime. New breeds of disease-resistant ash trees will undoubtedly emerge from laboratories and field experiments in coming years. Woodland may be restocked, but it is unlikely that there will be a replanting programme extensive enough to replace the widely scattered mature trees in open countryside that are such important (if often undervalued) landscape features in many parts of the country. The *Chalara Management Plan* makes some sensible suggestions for slowing the rate of spread in order to buy time (such as encouraging the removal of those recently planted stands of imported ash saplings most likely to be infected), but the strategy is the familiar one of adaptation to an outbreak and its consequences rather than one of control.

23.6 The Lessons from History

Managing an outbreak once it is established is a difficult enterprise, and although there are some examples of successful outcomes (such as the recently contained Asian longhorn beetle outbreak in the UK), the typical progression in each of the cases analysed above is from attempted control of a pest or disease to some sort of adaptation to its consequences. This is particularly true of the ramorum and *Chalara* outbreaks discussed above, where genuine efforts at eradication and control involving large numbers of biosecurity professionals and extensive engagement with stakeholders have failed to prevent continued spread (Potter *et al.*, 2015). We can expect further improvements in the way in which pests and diseases are monitored and in the chemical and other treatments available to control diseases, but these *ex post* responses will always be second best from a disease control point of view given the biological virulence and complex epidemiology of many tree pest and disease systems.

The alternative to better outbreak management – the prevention of new introductions into jurisdictions such as the EU and the UK in the first place – is also fraught with difficulties. There are powerful interests ranged against any tightening of import controls and the better regulation of the trade in plants that might prevent the entry of damaging pathogens into a country such as the UK. Prior to the formation of the European Single Market, many Member States had their own national standards. After the Single Market these were replaced with a Europe-wide system of standards, risk assessments and plant passports. Under this system, any producer wishing to move plants and plant material within the EU must be issued with a phytosanitary certificate (a plant passport) by the plant health authorities in their jurisdiction following an inspection to verify that the plants/material pose no threat to plant health.

As a recent review of the EU's Plant Health Regime (PHR) acknowledges, this system appears to work well for those trades that have a long record of compliance, but is philosophically and operationally flawed as a system for managing the risks posed by previously unknown invasive pathogens. Yet this is precisely the problem in relation to tree pests and diseases. Many recent invasive pathogens were unknown to science before they were identified as threats. According to Brasier (2008), sudden oak death, Dutch elm disease, and the proliferating range of phytophthora diseases affecting a range of tree species and native plant communities worldwide, did not appear on any international lists before they were identified. Moreover, it is estimated that only 7–10% of all fungal species with the potential to become pathogenic have so far been described.

There are also structural weaknesses in the PHR itself. Founded on a 'weakest link public good' principle, the effectiveness of this regime in preventing disease introductions from outside the EU can only be as good as that of the weakest member state. However, inspection and quarantining standards vary widely across the EU, with some Member States having much less rigorous procedures than others in their approach to inspection and diagnostic testing. As the UK's Tree Health and Plant Biosecurity Taskforce report puts it 'the UK's biosecurity from non-European threats is completely dependent on the level of biosecurity applied by other EU member states' (Defra, 2013b).

The emphasis on visual inspections of random samples of material present in consignments is increasingly problematic given that pathogens may be present as largely invisible propagules, mycelium or spores in the roots, leaves or substrate of imported plants. The consequences for the UK have been the series of new disease invasions reported above.

For critics like Brasier, the only guaranteed way to minimize the risk is to limit the movement of plant materials around the world, but this seems unlikely given the commercial interests at stake and the EU's concern to maintain open borders. It also implies adopting a more critical stance on the relationship between plant biosecurity and trade, an eventuality that organizations like the World Trade Organization (WTO) and many industry lobbyists would be anxious to resist. Globally, there is a fundamental conflict between the free trade agenda of organizations like the WTO and the ability of its members to enforce import bans and biosecurity controls in the interests of protecting their environments. Biosecurity protocols and safeguards do exist, but they are often poorly designed and not well suited to the regulation of the complex trade pathways through which diseases are being spread.

The extent to which Europe's forests, woodlands and treescapes can be protected from further damaging pest and disease outbreaks will depend both on the ability of plant health authorities to prevent new introductions and the effectiveness with which outbreaks can be managed to minimize their impacts where pests and diseases have become established. The lessons from history are not encouraging on either count.

References

Biraghi, A. (1950) La distribuzione del cancro del castagno in Italia. *L'Italia Forestale e Montana* 5, 18–21.
Boyd, I., Freer-Smith, P., Gilligan, C. and Godfray, C. (2013) The consequences of tree pests and diseases for ecosystem services. *Science* 342, 823–840.
Brasier, C. (1996) New horizons in Dutch elm disease control. In: Forestry Commission (Edinburgh) *Report on Forest Research 1996*. Her Majesty's Stationery Office, London, pp. 20–28.
Brasier, C. (2008) The biosecurity threat to the UK and global environment from international trade in plants. *Plant Pathology* 57, 792–808.
Brasier, C. and Gibbs, J. (1973) Origin of the Dutch elm disease epidemic in Britain. *Nature* 242, 607–609.
Cobb, R., Filipe, R., Meentemeyer, R., Gilligan, C. and Rizzo, D. (2012) Ecosystem transformation by emerging infectious disease: loss of large tanoak from California forests. *Journal of Ecology* 100, 712–722.

Cornell, S., Constanza, R., Sorlin, S. and van der Leeuw, S. (2010) Developing a systematic 'science of the past' to create our future. *Global Environmental Change* 20, 426–427.

Daszak, P., Cunningham, A.A. and Hyett, A.D. (2000) Emerging infectious diseases of wildlife – threats to biodiversity and human health. *Science* 287, 443–449.

Defra and Forestry Commission (2011) *Action Plan for Tree Health and Plant Biosecurity, October 2011*. Department for Environment Food and Rural Affairs (Defra), London. Available at: http://www.fera.defra.gov.uk/plants/plantHealth/documents/treeHealthActionPlan.pdf (accessed 15 December 2014).

Defra (2013a) *Chalara Management Plan, March 2013*. Department for Environment Food and Rural Affairs (Defra), London. Available at: https://www.gov.uk/government/uploads/system/uploads/attachment_data/file/221051/pb13936-chalara-management-plan-201303.pdf (accessed 15 December 2014).

Defra (2013b) *Taskforce on Tree Health and Plant Biosecurity, Final Report, 20th May 2013*. Department for Environment Food and Rural Affairs (Defra), London. Available at: https://www.gov.uk/government/publications/tree-health-taskforce-final-report.pdf (accessed 15 December 2014).

Dehnen-Schmutz, K., Holdenrieder, O., Jeger, M. and Pautasso, M. (2010) Structural change in the international horticultural industry: some implications for plant health. *Scientific Horticulture* 125, 1–15.

Frankel, S. (2010) *American Chestnut: The Life, Death and Rebirth of a Perfect Tree*. University of California Press, Berkeley, California.

Gibbs, J.N. and Howell, R.S. (1972) *Dutch Elm Disease Survey 1971*. Forest Record No. 82, Forestry Commission, UK. Her Majesty's Stationery Office, London.

Grunwald, N., Goss, E. and Press, M. (2008) *Phytophthora ramorum*: a pathogen with a remarkably wide host causing sudden oak death on oaks and ramorum blight on woody ornamentals. *Molecular Plant Pathology* 9, 729–740.

Haack, R. (1997) New York's battle with the Asian Longhorn Beetle. *Journal of Forestry* 95, 12–15.

Harwood, T., Tomlinson, I., Potter, C. and Knight, J. (2011) Dutch elm disease revisited: past, present and future management in the UK. *Plant Pathology* 60, 545–555.

Kowalski, T. (2006) *Chalara fraxinea* associated with dieback of ash in Poland. *Forest Pathology* 36, 246–270.

MacLeod, A., Pautasso, M., Jeger, M. and Haines-Young, R. (2010) Evolution of the international regulation of plant pests and challenges for future plant health. *Food Security* 2, 49–70.

Maye, D., Dibden, J., Higgins, V. and Potter, C. (2012) Governing biosecurity in a neoliberal world: comparative perspectives from Australia and the United Kingdom. *Environment and Planning A* 44, 150–168.

Orwig, D.A., Foster, D.R. and Mausel, L. (2002) Landscape patterns of hemlock decline in New England due to the introduced hemlock woolly adelgid. *Journal of Biogeography* 29, 1475–1487.

Peace, T. (1960) *The Status and Development of Elm Disease in Britain*. Research Information Note No. 252, Forestry Commission, Farnham, UK.

Pidgeon, N. and Barnett, J. (2012) *Chalara and the Social Amplification of Risk*. Department for Environment Food and Rural Affairs (Defra), London.

Potter, C. (2012) Saving Britain's trees: countering the growing threat from invasive pests and diseases. *ECOS* 34, 25–30.

Potter, C. (2013) A neoliberal biosecurity? The WTO, free trade and the governance of plant health. In: Dobson, A., Barker, K. and Taylor, S.L. (eds) *Biosecurity: The socio-politics of invasive species and infectious diseases*. Routledge (imprint of Taylor & Francis), Abingdon, UK and New York, pp. 123–136.

Potter, C., Harwood T., Knight, J., Tomlinson, I. (2011) Learning from history, predicting the future: the UK Dutch elm disease outbreak in relation to contemporary tree disease threats. *Philosophical Transactions of the Royal Society of London B: Biological Sciences* 366, 1966–1974.

Potter, C., Dandy, N., Marzano, M., Bayliss, H. and Porth, E. (2015) *The Stakeholder Landscape for Tree Health in the UK*. Research Report for Department for Environment Food and Rural Affairs (Defra), London. [In press.]

Tinker, J. (1971) Elms: a phoney war. *New Scientist* 30, 719–720.

Tomlinson, I. and Potter, C. (2010) Too little, too late? Science, policy and Dutch elm disease in the UK. *Journal of Historical Geography* 36, 121–131.

24 Reflections

Charles Watkins[1]* and Keith J. Kirby[2]

[1]*School of Geography, University of Nottingham, Nottingham, UK;*
[2]*Department of Plant Sciences, University of Oxford, Oxford, UK*

24.1 Introduction

We are more aware than ever before of the variety of forms that landscapes with trees, woods and forests can take across Europe, thanks to easier travel and the way that images, data and opinions can be easily found across the Web. This spatial heterogeneity is matched by temporal variety. People have valued and used trees and woods in different ways in different places, and at different times in the same place. Indeed, Europe's woods and forests have been providing ecosystem goods and services by different names throughout the last 10,000 years.

The wide range of potential benefits provided by woodland, including the production of timber, firewood and a range of non-timber forest products, as well as carbon sequestration, landscape and culture, wildlife and game conservation, public access and shelter make it a complex land use to understand and manage. There is a general feeling that woods and forests should be conserved, but which, where, what for and how?

The extent, depth and quality of our knowledge of the history of trees and woodland changes all the time, and beliefs that once seemed to have a solid basis in fact have to be qualified if they are to retain truth and validity. The current debates over the nature of the pre-Neolithic landscape are a case in point; others are the degree to which past shortages of wood were real or only perceived, and the relationship of forest clearance and erosion in Mediterranean regions.

24.2 Ways of Exploring and Understanding Woodland Histories

Studies on a single site such as the Białowieża Forest in Poland, or into how a group of species such as birds has changed over time, have a part to play in helping to resolve some of these issues. The influence of soil changes under different land uses over time would also seem to be particularly ripe for this type of narrowly focused research.

Such studies need to be complemented by interdisciplinary efforts to bring together historical, cultural and technical knowledge. The current composition and structure of a wood can easily contain features whose explanation depends on events and conditions from previous decades, centuries or millennia. They may be present or past environmental conditions, the accidents of past ownership or cultural

---

*E-mail: charles.watkins@nottingham.ac.uk

practice, or deep past events such as the location of glacial refuges.

Palynologists have made considerable progress in trying to understand how what is observed in the pollen record is affected by differential pollen production, dispersal and survival. This has increased the sophistication of the interpretation that can be placed on a particular level of pollen presence in terms of the surrounding landscape structure. Similar efforts are needed in other areas, such as the analysis of sub-fossil beetle remains.

This taphonomic issue also applies to human records: historical documents, even detailed maps, are not an unbiased representation of past conditions. Yet we are rapidly losing knowledge as the last of the generations who practised what we now call 'traditional management' die off. Their knowledge of how wood-pastures or coppices were treated should be captured through oral history projects to fill in the gaps in the 'official' accounts. We cannot judge the value or sustainability of past land uses such as these, including whether it would be desirable to revive them, unless we know what was actually done on the ground and whether it still works, for example, under higher levels of nitrogen deposition than in the past. Where traditional management is revived, for whatever reasons, we need to ensure that the experience of those (often volunteers) who do such work is also recorded as this, in turn, may throw light on past practices.

There is an increasingly important role in the study and analysis of woodland change for interested and skilled amateurs who have the time to explore and collate the wealth of data, information and photographs that are becoming available. Until the rise of modern university academic disciplines and research departments in the last century, this was, after all, how most botanical, archaeological and historical information was collected. We need to make more use of the potential of citizen science.

24.3 Issues for the Future Historian

History has not stopped. The woods and the landscapes around them are still changing and we cannot predict what the outcome of the current trends will be. Ancient woods are important, distinct and a priority for conservation because of rare plants and beetles, but many were themselves 'new' once. The development of recent woods, of plantations of non-native species and of non-traditional management systems, should not be ignored either; at the very least, their study may confirm our hypotheses (or prejudices) as to what they will contain, but from time to time there will be surprises.

Historical records suggest that the assemblages of species that make up our woods and forests were different in the past, and climate change and the spread of tree diseases more or less guarantee that they will be different in the future. There may be lags in species population responses to past change, such that future extinctions are already built into the landscapes that we are trying to conserve.

Many conservation priorities and practices are based around the conservation of defined forest types or of species within their native range. However, our definitions of what are native species and their range limits may need to change because of environmental shifts, or because species are reoccupying lands from which they were driven in the past by persecution. Will this lead to new forms of local distinctiveness (and at what scale), or to the homogenization of landscapes across large swathes of Europe, with the loss of the patterns that we currently value?

We need to understand better how quickly and under what conditions species have spread in the past in response to changing conditions. The early appearance of temperate species in northern Europe in the early Holocene might imply quite rapid spread as the climate warmed (though even so, not necessarily rapid enough to cope with current rates of warming and through a much more fragmented landscape). If many of these species spread back, not from southern Europe but from much nearer northern refugia, then our estimates of their capability to cope with future climate change will need to be rethought.

In some cases, the spread of species has been assisted by humans, although for every conservation reintroduction success story (such as that of the beaver) there are cases of species that became pests (the grey squirrel) or of diseases (ash dieback) that accidentally spread through the movement of animals and plants across borders. Which species should we move in future, which should we leave alone?

The speed of mechanization in the forest industry has been such that many forests have gone from being harvested by axe and crosscut saw, through chainsaws and to tree harvesters in less than one crop rotation. We cannot yet say what the long-term outcome will be in terms of the effect on the structure and composition of forests.

While mechanization has drastically reduced the numbers of people working in woods, this is partly offset by an increasing number of people visiting them for recreational purposes, such as to view wildlife. What will be the effect of this changing use on the woods?

24.4 From Cultural Landscapes Back to Wildwood?

We used the term 'wildwood' in the title of the book to conjure up an image of the forest landscape as it might have been before the spread of farming in the Neolithic period, albeit that research increasingly suggests that humans were affecting the European woods and forests long before that, so in one sense the term may be redundant, even misleading. However, it can also be used to suggest a new direction for some of our conservation efforts: to try deliberately and *knowingly* to step back from intervening in the management of the landscape, to let it 'rewild'.

Any deliberate rewilding project, or the de facto rewilding caused by large-scale abandonment of agricultural land, will change the nature of that land. There will be losses as well as gains in species, but there will in most cases certainly be a loss of the historical meaning of the landscape. We may decide that the expected benefits outweigh the drawbacks, but that assessment should be based on an understanding of the cultural nature of the particular landscape in question. What develops will not be a return to some past 'natural state' but actually a new form of cultural landscape in which the human input is much reduced, but can never be totally eliminated. Thus rewilding is a complement to current conservation practices, not a complete substitute.

24.5 Europe's Woods and Forests: The Future?

The past and future of European woods and forests cannot be understood in isolation from global issues. In addition to the important implications of the globalization of trees diseases, we need to consider broad issues of climate change and the potential value of growing crops of trees as a contribution to carbon sequestration.

European countries are significant players in the global wood trade; they are also active in international discussions on promoting sustainable forest management and conservation worldwide. How we treat our woods and forests matters not only in terms of what we can get from them, but also in terms of the messages it sends to others. If we cannot cope with the challenges posed by climate change, the spread of pests and diseases and the conservation of our big carnivores and rare plants, and if we cannot show how we are managing forests sustainably, or organizing markets for carbon sequestration in forests, then how can we expect the rest of the world to do this?

There is now the possibility that European demand for firewood may increase, stimulating a major revival of coppice. If so, the benefits and costs for biodiversity of re-coppicing large areas of woodland that have been unmanaged for many years need to be considered carefully. On the one hand, re-coppicing could have substantial benefits with the creation of much more of the mixture of open and closed woodland habitat that so many valued species prefer. On the other hand, it could lead to a significant downward trend in the amount of fallen dead wood.

Another important factor to consider is whether the European forest area will be maintained. The recent increases in the extent of European woodland are based on continued efficient production of agricultural crops in Europe, much supported by the Common

Agricultural Policy (CAP), combined with the import of cheap food supplies from across the world. This state of affairs may not continue. There may be demands to clear woodland growing on recently abandoned land to help to increase the production of food in Europe.

There may also be demands to increase the production of wood and timber from established plantations and to expand their extent. Modes and methods of management, such as the reintroduction or protection of wood-pastures, which benefit wildlife, may again be seen as unnecessarily luxurious.

There is increasing uncertainty as to what the future holds, and we need better ways of dealing with the changes that occur and a better understanding of the effects of policies we make. We are at a critical point for many types of wooded landscapes in Europe and the wildlife they contain. We have the knowledge to catch and restore many important cultural types of woodland management over the next few decades, before they are lost forever. We can greatly improve the integration of biodiversity and landscape conservation within modern productive forests to increase the range of benefits that they provide. We can set aside some large areas to reverse the historical trend of increasing human domination and direction on the natural world.

The process will not necessarily be straightforward; it will involve social and natural scientists working together, and politicians and practitioners agreeing on priorities. Crucial to success will be learning the lessons of history. Knowing how our woods have come to be as they are is the first step in planning how we make them what we want them to be in future.

Index

Note: italic page numbers indicate figures and tables.

Abies spp. *see* fir
Acer spp. *see* maple; sycamore
acid rain 53, 111
acidophilous oak/oak-birch forests 10, *12–13*
Aesculus hippocastanum 143
afforestation 8, 9, 11, 26, 48, 98–99, 101, 103, 207–209
    birds 166–167
    see also Atlantic spruce forests
agriculture 3, 8, *10*, 292–293
    afforestation 48
    EU 8, 72, 349–350
    intensification of 3, 66, 148, 165, 175, 207
    slash-and-burn 236, 249, 257, 283, *283*, 293
    woodland history 19, 26, 27, 38, 41
    wood-pastures 66, 141
    see also fodder
Agrilus planipennis 51, 338
Ailanthus altissima 11, 84
Alces alces 39, 116, 124, 252, 259
alder (*Alnus* spp.) 10, 11, 26, 246, 258
    A. glutinosa 24, 82, 270, 273, 316
        rotation of 101, 236–237
alnocoltura system 235–237, *236*
alpine coniferous forests 10, *12–13*
Alps 20, 31, 35, 68, 195, 200
    wood-pastures in 68, 69, 71
ancient trees 4, 25–26, 63, 72
    see also veteran trees
ancient woodland 4, 6, 23, 48, 49, 102, 103, *178*, 233, 234, 326
    semi-natural 328, 334
    understorey vegetation in 177–178
Ancient Woodland Inventory (UK) 326–334
    categories of woodland in 327–328
    creation of 328–329
    development/use of 329–332
    disputes with 333
    estimates of UK cover in *330*

    problems/uncertainty with 332–334
    responses to/future of 334
    woodland size for inclusion 333–334
Anemone nemorosa 36, 130, 174, 183
Anglo-Saxon period 77, 79, 118, 270
Anoplophora glabripennis 338, 344
Apennines 21, 83, 235, 237
apple (*Malus* sp.) 63, 70
Arbutus unedo 82
archaeology 18, 24–26, 41, 193, 243, 250–252
Ariège (France) 80
Artemisia 38, 249, 253
Arum maculatum 177, 179
ash (*Fraxinus* spp.) 10, 11, 27, 143, 253
    coppicing of 77, 79, 82, 84
    F. excelsior 22–24, 51, 66, 82, 101, 122, 270, 271–274, 316
    manna (*F. ornus*) 82
    pests/diseases of *see* ash dieback
    veteran trees 141, 144
ash dieback (*Hymenoscyphus fraxineus*) 14, 51, 165, 337–339, 349
    response to UK epidemic of 343–344
Asian longhorn beetle (*Anoplophora glabripennis*) 338, 344
aspen (*Populus tremula*) 11, 78, 83, 133, 177
Atlantic spruce forests 209–216
    dead wood in 212–216
    development stages 212–216
    diversification 210–211, 215
    dominance of Sitka spruce in 210, 212
    history of 209–210
    species richness 211–216, *211*, *212*, 218
    stand dynamics 212–216, *213*
aurochs (*Bos primigenius*) 36, 39, 40, 46, 116, 193–195, 202, 259
    ecological role of 194–195
    extinction of/rebreeding attempts 194

Austria 7, 8, *12*, 69, *81*, 197, 198
  close-to-nature forestry in 107
  forest conservation in 310, 311, 313, *314*
  timber production in 94, 101
  woodland history in 25
avalanches 6, 38, 96, 311
Aveto valley (Italy) 21, *228*, 233–234, *233*, 236–238

back-to-nature movement 110
Baden-Württemberg (Germany) 101, 102, 168
Balkan peninsula *10*, 29, 35, 37, 50, 143, 195, 197
Baltic states 50, 93, 123, 197
bats 9, 134
Bavaria (Germany) 26, 85, 119, 120, 132
bear, brown (*Ursus arctos*) 37, 119, 195–197, 229, 252, 293, 318
  range/population collapse of 197
beaver (*Castor fiber*) 38, 46, 51, 116, 124, 193, 197–199, 311
  environmental impact of 198
  population collapse of 198
  reintroduction of 198–199, 348
beech (*Fagus* spp.) 40
  coppicing of 77, 78, 82–85, 273
  F. orientalis 10
  F. sylvatica 10, 25, 27, 36, 51, *270*, 272, 276, 282, 288
  plantations of 103, 266, 274, 275
  selection systems for 96
  understorey vegetation 182
  veteran trees 145, *149*
  wildlife 155, 200
  in wood-pastures 63, 65
beech forests *10*, *12–13*, 35, 85, *316*
beekeeping 124, 252, 253, 255, *256*, 258
bees, solitary 135, 144
beetles (Coleoptera) 3, 6, 130, 132, 133, *175*, 211
  Asian longhorn (*Anoplophora glabripennis*) 338, 344
  great Capricorn (*Cerambyx cerdo*) 144, 147
  saproxylic *see* saproxylic insects
  see also click beetles; hermit beetle
Belarus 7, *12*, 198, 254–255
Belgium 7, *12*, 80, 99, *100*, *178*, 311, *314*, 337, 338
  wood-pasture in 63, 69, *70*
  see also Flanders
Belházy, Emil 66
Bergslagen (Sweden) 290, 292–301
  agriculture in 292–293, 298
  forest ownership/land use in 292–294, *293*, *300*
  iron/charcoal industry in 293, 298
  land cover dynamics in *294*
  recreation/tourism in 294, 298
  sustainable forest management in 294–301
    barriers to 294–299
    bridges towards 299–300
    ecological sustainability 294–297, *295*
    economic sustainability *295*, 297–298
    forest types 295–296
    implementation schemes 299–300
    landscape history 290, 292, 294, 301
    rural employment/development 298, 299
    social/cultural sustainability *295*, 298–299
    stakeholder engagement 300
    water management 296–297
  timber/pulp industry in 293–294, 298
Bern Convention (1979) 9, 196, 316
Betula spp. *see* birch
Białowieża Forest (BNP, Poland) 37–38, 46, 103, 158, 161, 243–260, 310, 326, 347
  archaeological record for 250–252
  archival record for 252–255
  beekeeping in 252, 253, 255, *256*, 258
  bór lado in 256, 258, 259
  charcoal/wood tar production in 253, 255, 258
  fire history of 255–258, 260
  forestry in 254, 255
  interplay of natural/cultural forces in 257–259
  in Iron Age 246–250, 257, 260
  iron production in 26, 253, 257, 258
  land-use changes in 255
  large herbivores in 259
  livestock grazing in 153–154
  paleoecological record for 245–250
    charcoal particles 245, 246, 249, 250, 258
    and forest fires 246–249
    four phases 246–250
    methods in study 245–246
    pollen records 245, *246*, 249, 253
    radiocarbon dating *248*, 252
  previous studies on 243–245, *244*
  timber extraction in 253, 254
biodiversity 6, 8, 11, 26, 52, 53, 61, 68, 71, 72, 102, 107, 111–113, 123, 124, 136, 150, 165, 167, 184, 186, 195, 199, 207, 208, 211, 212, 215, 216, 217, 218, 227, 231, 234, 237, 238, 287, 290, 291, 293, 294, 296, 297, 298, 299, 312, 318, 320, 321, 322, 326, 341, 349, 350
Biodiversity, Rio Conference on (1992) 14, 321
Biolley, Henry 107, 109–110
biomass 82, 86, 87, 177, 179, 186, 214, 217–218, 293, 338
birch (*Betula* spp.) *10*, 11, 24, 35, 249, 253, 271–273
  B. pendula 270, 282
  B. pubescens 22–23, *270*
  coppicing of 77, 83, 133, 237
birds 4, 49, 51, 131, 135, 154–168, *156–157*, 211
  and afforestation 166–167, 211, 214–217
  and agricultural/forestry intensification 165–166
  current population trends 164–165
  and current woodland management trends 167–168
  in early Holocene 154–161
    closed/open forest scenarios 155, 158–159, *160*
    effects of management on 161–163
    and fragmentation of woodland 159–161
  edge habitat 158–161, *160*
  high forest 164
  hunting 122–124
  understorey vegetation 131, 158, 159, 163, 164, 166, 167, *175*
  wood-pasture/coppice 159, 163–164
Birds Directive (EU, 1979) 9
bison (*Bison bonasus*) 39, 124, 161, 252, 259, 311

black cherry (*Prunus serotina*) 84, 182, 183
Black Death 80
black grouse (*Tetrao tetrix*) 155, *156*, 159, 166, 214, 215
black locust *see Robinia pseudoacacia*
Black Sea 34, 250
blackbird (*Turdus merula*) 131, 155, *156*, 164–167
blackcap (*Sylvia atricapilla*) 155, *156*, 158–160, 162–165, *162*
blackthorn (*Prunus spinosa*) 39, 122, 132, 177
blue tit (*Cyanistes caeruleus*) 155, *157*, 160, 163, 165, 167
bluebell (*Hyacinthoides non-scripta*) 27, 130, 174
Bohemia 48, 80, 197
Bonasa bonasia 154, *156*
Bonawe Ironworks (UK) 48
boreal forests 11–13, 25, 35, 294–296
    coniferous 3, 8
    fire in 52
Bos primigenius see aurochs
Bosnia-Herzegovina 7, *81*
bracken (*Pteridium aquilinum*) 36, 52, 134, 183
bramble (*Rubus fruticosus*) 39, 130, 132, 135, 164, 176, 183
brambling (*Fringilla montifringilla*) 155, *156*, 165
Britain (UK) 178, 310, *314*
    afforestation in 48, 208–212, 275
        see also Atlantic spruce forests
    ancient woodland in 266, 267, *267*, 276, 326, *330*
        loss of 332, 334
        see also Ancient Woodland Inventory
    birds in 154–157
    coppice in 77–80, *81*, *82*, 84, 85, 270–276
    DEFRA 342–344
    fauna in 33, 37, 51
    forest conservation in 310, *314*, 315, 322
        see also Ancient Woodland Inventory
    forest cover in 7, 8
    forest types in 12
    historical ecology in 230, 231
    hunting in 116, 118, 119, 121, 122
    parks in 118–119, 272, 274–275
    pests/diseases in 337, 339–345
        ash dieback 337–339, 343–344
        Dutch elm disease 337–341
        ramorum blight 337–339, 341–343
    plantations in 93, 94, 103, 266, 275
        introduced species in 99, *100*, 101
    wildlife in 132, 135, 195, 199, 201
    wooded commons in 272–273, 276
    woodland history in 22–26, 28, 35, 265–277
        Borderland zone 268, *269*, 272–277
        Boreal zone 268, *269*, 275
        by region/zone 267–268, 268–270, 276–277
        exploited wildwood 267, 269
        historical stages 266–267, *267*
        improved traditional management 273–275
        Lowlands zone 268–270, *269*, 271, 272, 276, 277

        original wildwood 267, 268–269
        plantations 275
        revival/restoration of native woodland 275–276
        South-east Lowlands zone 268, *269*, 270, 272–277
        traditional woodland management 269–272
        Western Uplands zone 268, *269*, 271–277
    wood-pasture in 62, 65, 68, 270–272, 274–277
    see also England; Northern Ireland; Scotland; Wales
broadleaved woods/forests 3, 8, *10*, 12–13, 35
    and fire 51–52
Bronze Age 26, 47, 79, 116, 265
bryophytes 52, 130, 133, *175*, 211, *211*, 212, 215, 216
Bulgaria 7, *12*, *100*, 311, *314*
    coppice in 78, *81*, 84
buntings 154, *156*, 159
butterflies/moths 130, 132, 135, *175*, 313

Calluna vulgaris 38, 52, 246, 247, 249, 251, 256
Campemoor (Germany) 25
canary pine (*Pinus canariensis*) *10*, 78
Canis lupus see wolf
capercaillie (*Tetrao urogallus*) 155, *156*, 158, 166, 167
Capreolus capreolus see roe deer
Caprimulgus europaeus 131, 216
carbon emissions 5, 53, 86
carbon sequestration 54, 87, 179, 276, 298, 347, 349
Carduelis spp.
    C. cannabina *156*, 163, 214
    C. spinus 155, *156*, 166
Carlowitz, Hans Carl von 94, 107
Carpathian Mountains 37, 195, 197, *317*
Carpinus spp. see hornbeam
Castanea sativa see sweet chestnut
Castor fiber see beaver
Catalonia (Spain) 80, 84
cattle 26, 27, 68, 70, 93, 116, 119, 123, 141, 147, 161, 237, 254, 255, 256, 257, 270, 271
Cerambyx cerdo 144, 147
Certhia familiaris 155, *156*, 160, 163
Cervus elaphus see red deer
chaffinch (*Fringilla coelebs*) 155, *156*, 163, 165
Chalara see ash dieback
Champagne (France) 24
charcoal 47, 48, 253
    analysis 27–28, 78, 280
    and coppicing 80, 82, 84, 280
chestnut see horse chestnut; sweet chestnut
chestnut blight (*Cryphonectria parasitica*) 337, 338
chiffchaff (*Phylloscopus collybita*) *156*, 158, 162, 164, 165
Chiltern forests (UK) 200, 273, 276
Chorthippus parallelus 37
Circus cyaneus *156*, 166, 214
clear-felling systems 96–97, *97*, 99, 102, 103, 208, 209
    reaction against 109, 110, 112

click beetles 23, 143, 146, *146*, 147
　　violet (*Limoniscus violaceus*) 11, 146, 148
climate change 3, 5, 25–26, 46, *47*, 48, 53–54,
　　148, 150, 211
　　and birds 165, 168
　　and close-to-nature forestry 112, 113
　　European policy on 8
　　fire risk 52
　　forest conservation 315, 322
　　postglacial 33, 34
　　understorey vegetation 175, 185
　　see also carbon sequestration
close-to-nature forestry 59, 77, 107–113
　　ecological implications of 112–113
　　and forest certification 111–112
　　natural disturbances mimicked in
　　　　77, 107, 111, 113
　　origins/development of 107, 109–110
　　politics of 111
　　principles of 110–111
　　regulations for 110
　　and single-stem management 109
　　terminology of 107–109
coastal habitat *10*, 195, 210, 339, 342
Coelodonta antiquitatis 39
Colbert's Ordinance (1669) 94, 284, *285*
commons, wooded 236, 270, 272–275
coniferization 275, 276
coniferous forests 3, 8, *10*, *12*–*13*, 50
conifers 94, 99–102
　　advantages of plantations of 101
　　coppicing of 77–78
　　monocultures of 107, 111, 112, 275, 276
connectivity 49, 135, 150, 165, 179, 217, 291, 294,
　　297, 318
conservation *see* forest conservation
continuous cover forestry 50, 85, 96, 102, 107, 110,
　　164, 167, 215, 274
coppice/coppicing 4, 6, *10*, 25, 50–51, 59, 61, 65,
　　77–87, 96, 119, 237
　　archaeological evidence for 77, 78
　　and carbon sequestration 87
　　and construction/fencing 79, 86
　　European extent of 81–82, *81*
　　evolution/mechanism of 78
　　for extraction industries 80
　　flora/fauna of 129–136
　　　　birds 131, 135, 163–164
　　　　and conservation strategies 135
　　　　in dead wood 133
　　　　diversity of 129–134
　　　　impacts of deer browsing on 134–135
　　　　invertebrates 130–132, 313
　　　　mammals 133–134
　　　　plants 130, *131*
　　future for 86–87, 136
　　genetic 'fixing'/'freezing' 78
　　high forest, conversion to *10*, 77, 82, 84–85, 87,
　　　　93–94, 99, 129, 130, 175, 266, 273–274
　　historic development of 78–80
　　methods of 82–84
　　　　cutting 83–84
　　　　three systems 82–83
　　regulations for 79, 80

rise/fall of 77, 80–81, 84–86
　　reinstatement of 85–86, 349
　　rotation lengths for 77, 79, 80, 84, 95, 164,
　　　　271–272
　　short-rotation (SRC) 82, 84, 86, 133, 135–136
　　species amenable/not amenable to 77–78, 82
　　standards/maiden trees 79, 83
　　weirs/fish traps 78–79
　　yields 85
coppice-with-standards 82–83, 96, 130, *131*,
　　280–282, *281*, 284, 288
CORINE database 69
cork 6, 50, 63, 69, 124
cork oak (*Quercus suber*) 50, 71, 82, 99, 123, 124,
　　132, *316*
Cornus spp. 78
Corsican pine (*Pinus nigra*) 24, 210, *316*
corvids *156*, 159, 164
Corylus avellana see hazel
COST (Cooperation in Science and Technology)
　　programme 14
Cotta, Heinrich 94, 107
Couvet Forest (Switzerland) *97–98*
crab apple (*Malus sylvestris*) 37
Crataegus monogyna 39, 132
Crete (Greece) 19–21
Croatia 7, *12*, 62, *81*, *100*, *314*
crossbill 154, 155, 166
Cryphonectria parasitica 337, 338
CSR (competitive, stress tolerant, ruderal)
　　model 177, *184*
cultural landscapes 61, 103, 176–177, 199, 209, 217,
　　227, 236, 293, 297, 313, 321, 349
Cyanistes caeruleus 155, *157*, 163, 165, 167
　　see also blue tit
Cyprus 101, 313, *314*
Czech Republic 7, *12*, 94, 311, *314*
　　coppice in 79, 85, 86, 132
　　woodland history in 23, 25
　　wood-pasture in 63, 68, 69, *70*

Dama dama 118, 119, 121, 123, 135
Danube region 37
Dartmoor (UK) 21, 38
Dauerwald Movement 107, 110, 113
dead wood 4, 6, 37, 112, 142, 144, 158, 163, 294
　　in Atlantic spruce forests 212–216
　　species associated with *see* saproxylic insects
Dean, Forest of (UK) 22, 26, 80, 266, 271–273
deer 50, 51, 62, 86, 116, 123, 161, 167, 193, 238, 276
　　fallow (*Dama dama*) 118, 119, 121, 123, 135
　　impact of 202
　　muntjac (*Muntiacus reevesi*) 135
　　range/distribution of 201–202
　　see also red deer; roe deer
deer parks 118–119, 121–122, 134–135
DEFRA (Department of Food, Environment and
　　Rural Affairs, UK) 342–344
dehesa system 6, 50, *62*, 123, 283, *316*
dendrochronology 25, 255–257
Dendrocopos leucotos 157, 158, 163
Dendrocopos major 157, 165, 167
Dendrocopos minor 157, 165–167

Denmark 99, *100*, 119, *178*, 195, 207–208, *314*, 337, 343, 344
  coppice in 78–79
  forest cover in 7, 8
  forest types in *12*
  veteran trees in *146*
  woodland history in 27, 38
Diptera 143, 144
dog's mercury (*Mercurialis perennis*) 130, 177, 183
Dolomiti Bellunesi National Park (Italy) 84
Domesday Book 23, 65, 79, 118, 270, 271
Douglas fir (*Pseudotsuga menziesii*) 111, 210, 266
'Dover Boat' 41, 47
downy oak (*Quercus pubescens*) 21, 82, 84, 134–135
drought *10*, 53, 85, 101, 150
Dryocopos martius 157, 164, 166
dunnock (*Prunella modularis*) 155, *156*, 158–160, 163, 167
Dürrenstein (Austria) 25
Dutch elm disease (DED) 38, 337–341, 344, 345
  management of UK epidemic 340–341

eastern hornbeam (*Carpinus orientalis*) 82
ecosystem conservation 228, 321–322
ecosystem services 6, 71–72, 179, 207, 217, 276, 347
edge habitat 49, 123, 130, 136, 158–161, *160*, 175, 176, 214, 333–334
EFI (European Forest Institute) 14
elder (*Sambucus nigra*) 132
elk (*Alces alces*) 39, 116, 124
elm (*Ulmus* spp.) *10*, *11*, 24, 27, 38, 143, 253
  diseases of *see* Dutch elm disease
  English (*U. procera*) 28, 177
  veteran trees 143, 144
  wych (*U. glabra*) 28, 82, 274
emerald ash borer (*Agrilus planipennis*) 51, 338
Ename Wood (Belgium) 27
England 4, 6, 37, 199, 200
  afforestation in 103, 208
  ancient woodland in 329–331, *330*, 334
  conservation in 9, 312, *313*
  coppice in 79, 80, 83, 135
  hunting in 116, 118, 121–122
  pests/diseases in 337, 339, 342
  woodland history in 21, 38, 48
English elm (*Ulmus procera*) 28, 177
English Lake District 22, 209, 218
Environmental Impact Assessment 11, 14
epiphytes 23, 49, 52, 130, 210, 277, 317
Equus spp.
  E. ferus 39, 40, 259
  E. hydruntinus 40
Erithacus rubecula 131, 155, *156*, 164–167
Estonia *100*, 141, 197, *314*
  forest cover in 7, 8
  forest types in *13*
Eucalyptus spp. 48, 52, 71, 167, 218
  coppicing of 79, 82, 95, 96
  E. camaldulensis 82
  E. globulus 82
  in single-aged stands 99, 101, 208

EUNIS (European Nature Information System) Habitat Classification 8
European bison (*Bison bonasus*) 9
European forestry policy 8–14
  see also under European Union
European Landscape Convention (Council of Europe, 2000) 11, 14
European Union (EU) 8–9, 12–13, 48, 101, 141, 165, 338
  Birds Directive (1979) 9
  carbon emissions targets of 86
  Common Agricultural Policy (CAP) 72, 349–350
  forest conservation *see* Habitats and Species Directive
  Forest Strategy 9, 12–13, 290–291
    2010 objectives 9
  pests/diseases 338, 343–345
  Plant Health Regime (PHR) 343, 345
European woods/forests 347–350
  categories of 3, 8, *10–11*
  forest cover of *see* forest cover
  future for 348–350
  public/private ownership of 8, 292–294, *293*, 300, 312
  range/zones of 3–4
  research programmes for 14
even-aged stands 4, *11*, 27, 50, 85, 93–103, 107, 129, 207–218
  and afforestation 207–209
    see also Atlantic spruce forests
  arising after fire 38, 96
  of conifers/introduced species 99–101, *100*, *102*, 210
    advantages of 101
    disadvantages of 107
  diversity in 208–211, 215–217
  economics of 94, 95, 99, 207, 209
  forest design 216–217
  future options for 217–218
  growth of 98–99
  landscape setting of 217
  optimum rotation 95
  origins/development of 93–95
  regulations 94, 102
  silvicultural systems for 95–98
exotic/introduced species *11–13*, 51, 59, 111
  see also invasive species
extinction debt 150, 175, 184, 322

Fagus spp. *see* beech
fauna of woods/forests 3, 4
  fragmentation of forests 49
  hunting of *see* hunting
  introduced species 59
  postglacial recolonization 34, 36–41
  Vera's shifting mosaic cycle 39–40, *39*
  woodland management 51
  see also birds; insects; mammals
Faustmann formula 94–95
Felix sylvestris 51
fencing 79, 80, 120, 167
ferns 36, 182
  bracken (*Pteridium aquilinum*) 36, 52, 134, 183

fertilizers 49, 71, 110
field maple (*Acer campestre*) 19, 82, 272, 273
Finland 14, 101, 124, 143, 155, 195, 198, *314*, 315
 forest cover in 6, 7, 8
 forest types in 13
 woodland history in 22–23, 35
fir (*Abies* spp.) *10*
 silver (*A. alba*) 25, 82, 96, 98, 122, 284, 287, 288
fire 6, *11*, 27–28, 33, 38, 40, 47, *47*, 51–53, 237, 246–249, 255–257
 and hunting 116
 species adapted to *10*, 84, 256
firewood/fuelwood 47, *70*, 93, 101, 167
 coppice 79, 81, 82, 84, 86
 fuel prices 48, 85, 94, 280
Flanders 27, 69, 176, *176*, 184, 338
flies (Diptera) 143, 144
flood plain forests *10*, *12–13*, 40, 141, 272, 274
floods *10*, 38, 40, 78, 155, 234, 276, 311
flycatchers 155, *157*, 158, 160, 161, 163, 167–168
fodder 19–21, 46, 52, *70*, 80, 234–236
Fontainebleau Forest (France) 22, 310, 326
forest certification 14, 111–112, 318
forest conservation 11, 14, 52, 102, 112, 309–322
 aesthetics/wildlife protection 312, 313
 ancient woods concept in 326–327
 of common land 232
 in conflict with other interests 309
 economics of 311, 313, 315
 in EU *see* Habitats and Species Directive
 future for 321–322, 348–349
 historical ecology *see* historical ecology
 nature conservation 312, 313
 origins of 309–312
  early laws 311
 as part of land-use practice 318–321
 productivity/protection from forests 311, 315
 rewilding 321, 322, 349
 royal hunting forests 310–311
 selection of protected areas 315
 spiritual dimension of 310
 type/extent of protected areas 312–315, *313*, *314*
  National Parks 312–313
 understorey plants 183–186
 wood-pastures 63, 66, 68, 69, *70*, 71
forest cover 4–6, *5*, 47–49
 increase in 5
 landscape ecology 49
 in postglacial landscape 38–41, *39*
 timber industry 48–49
 variation in 6–8, *7*
forest edges *see* edge habitat
'Forest History and Traditional Knowledge' (research grouping) 124
forest laws 4, 94, 102, 283–288, 311
 hunting 116–119, 124
forest litter 21, 23, 37, 47, 52, 85, 87, 129, 132, 135, 237, 283
Forest Stewardship Council (FSC) 14, 111, 291, 318
forestry 47, 52, 66
 close-to-nature 59
 development of 94
 German school of 94, 107, 113

historical ecology 230–231
 see also even-aged stands
Forestry Commission (UK) 48, 102, 211, 275, 276, 329, 339–340
fox (*Vulpes vulpes*) 122, 134
France *178*, 231, 234, 338
 close-to-nature forestry in 107, 111
 coppice in 79, 80, *81*, 82, 84, 133, 280–282, *281*, 284
 forest conservation in 311–313, *314*
 forest cover in 6, 7, 8
 Forest Ordinance (1669) 94, 284, *285*
 forest types in *12*
 grazing in 282, 283
 high forest in 164
 hunting in 117–121, 124
 plantations in 96, *97*, 102, 103
  introduced species in *100*, 101
 sartage system in 283, *283*, 284
 timber production in 94, 200, 282, 284, 287
 wildlife in 193, 197, 198
 woodland history in 22–24, 26–28, 279–285
  Basque agriculture 282–284, *282*
  forest cover 48
  forestry legislation 283–285, *285*
  Forestry School, Nancy 279–280, 284
  historical forest uses/management 280–283
 wood-pasture in 63, *70*
 see also Lorraine
Franconian Alb (Germany) 26
Fraxinus spp. *see* ash
Fringilla spp.
 F. coelebs 155, *156*, 163, 165
 F. montifringilla 155, *156*, 165
fruit/fruit trees 47, 63, 66, *70*, 71
FSC (Forest Stewardship Council) 14, 111, 291, 318
fungi 211
 see also pests/diseases
furniture industry 86, 93, 109, 273, 274

Garrulus glandarius 35, 155, *156*
Gayer, Karl 102, 107, 109
gean (*Prunus avium*) 78, 272
Genova (Italy) 233–238, *233*
Germany *178*, 337
 close-to-nature forestry in 107, 109, 110
 coppice in 79, 80, *81*, 85, 86, 132
 forest conservation in 310, 313, *314*
 forest cover in 7
 forest types in *12*
 foundations of forestry in 94, 107, 113
 Hauberg system in *236*, 237
 hunting in 117–120, 124
 plantations in 96, 99, 102, 103
  even-aged 208, 209
  introduced species in 99, *100*, 101
  regulations for 94, 102
 wildlife in 132, 167–168, 194, 195, 197, 198
 woodland history in 23, 25–28, 38, 48
 wood-pasture in 63, 69, *70*, 71
Geum urbanum 177, 180
GIS (geographical information systems) 18–19, 26, 27, 330, 332

glass industry 80, 185, 280
Global Forest Resources Assessment 2010 (FAO) 4, *81*
Gnorimus spp. 143
goats 19–21, 68, *70*, 87, 123, 234–235
goldcrest (*Regulus regulus*) 155, *156*, 166, 215
grasshopper (*Chorthippus parallelus*) 37
grassland 3, 26, 34, 63, 64, 103, *185*
grazing 3, 4, 6, *10*, 24, 27, 33, 47, *47*, 50, 61–62,
    153–154, 253–254
  and coppicing/pollarding 80, 86–87, 282–283
  and environmental degradation 93
  and hunting 123, 271
  rights 71, 80, 271, 272, 310, 313
  see also wood-pastures
great Capricorn beetle (*Cerambyx cerdo*) 144, 147
great tit (*Parus major*) 155, *157*, 161, 164, 165
Greece 7, *12*, 143, 229, 234, *314*
  afforestation in 48, 101
  coppice in 79, *81*, 82
  woodland history in 19, 28
  wood-pasture in 62, 63, 69, *70*
grouse *156*, 161
  black (*Tetrao tetrix*) 155, *156*, 159, 166, 214, 215
  hazel (*Bonasa bonasia*) 154, *156*
Gulo gulo 9, 195, 293
Gurnaud, Adolphe 107, 109

Habitats and Species Directive (EU, 1992) 9, 146,
    309, 316–318
  conservations status of sites under 317, *319*, *320*
  monitoring sites under 320–321
  woodland types listed under 316–317
Harz Mountains (Germany) 27–28, 197
Hatfield Forest (UK) *62*, *268*, 276
Hauberg system 236, 237
hawthorn (*Crataegus monogyna*) 39, 132
Hayley Wood (UK) 79, 239
hazel (*Corylus avellana*) 22–25, 39, 228, 270, 271,
    272, 276
  coppicing of 50, 78, 79, 82, 132, 273
heather (*Calluna vulgaris*) 38, 52
heathland 3, 38, 53, 69, 102
Heck cattle 194, 195
Hedera helix 130, 183
hedgehog (*Erinaceus*) 37
hedgerows 28, 68, 122, 179
hemiboreal forests 11–13
hen harrier (*Circus cyaneus*) 156, 166, 214
herbs 21, 33, 34, 36, 49, 180–181, 216, 237
hermit beetle 141, 143, 145–148
  Osmoderma eremita 11, 141, 143
high forest management 26–27, 112
  and coppice *see under* coppice/coppicing
  future for 103
  regulations 79
  silvicultural systems 95–98
historical ecology 227–239, 265
  classification of wooded systems 231–233, *233*
  conservation policies 227
  forestry/woodmanship 230
  in interdisciplinary conservation
    approach 231–232

Italian case studies *see under* Italy
  in management strategies 232–238
  micro-historical approach in 231, 232
  re-naturalization 228–229, 234
  role of 232
Hoge Veluwe National Park (Netherlands) 135
holly (*Ilex aquifolium*) 24, 272, *316*, 330
holm oak (*Quercus ilex*) 19, 22, 71, 82–86, 131, *316*
  veteran trees 147
Holocene epoch *see* postglacial forest landscape
honeysuckle (*Lonicera periclymenum*) 130,
    202–203, *202*
hop hornbeam (*Ostrya carpinifolia*) 19, 35, 82
hornbeam (*Carpinus* spp.) 133, 155, 245, 253, 258
  C. betulus 10, 19, 35, 36, *270*, 286, *316*
  coppicing of 77, 82, 86
  eastern (*C. orientalis*) 82
  in wood-pastures 63, *64*, 66
horse, wild (*Equus ferus*) 39, 40
horse chestnut (*Aesculus hippocastanum*) 143
Hortobágyi National Park (Hungary) 66, *67*
human population changes 80, 119
Hungarian oak (*Quercus frainetto*) 19, 82, 85, 134–135
Hungarian Plain 37, 99, 101
Hungary 7, *12*, 78, *81*, *178*, 194, *314*
  introduced species in 99, *100*, 101, 103
  wood-pasture in 63, 66, *67*, 69, *70*
hunting 6, 26, 47, 50, 59, 65, 116–124, 161, 198, 214, 235
  damage caused by 119–121
  early impacts of 116
  early modern parks for 118–119, 252, 310–311
  game numbers 120, 121
  land area enclosed for 123
  and land-use change 123–124
  medieval reserves 117–118
  modern 121–124
  and poaching 117, 119, 121
  and prestige/nobility 117–120, 122
Hyacinthoides non-scripta 27, 130, 174
Hymenoscyphus fraxineus see ash dieback

Iberian peninsula 5, 34, 35, 37, 123, 155, 208
  see also Portugal; Spain
Ice Age 33, 34
  see also postglacial forest landscape
Ilex aquifolium 24, 272, 316, 330
insects 130–132
  see also beetles; butterflies/moths; saproxylic
    insects
International Union for Conservation of Nature
    see IUCN
International Union of Forest Research
    Organizations (IUFRO) 14, 230
introduced species *see* exotic/introduced species
invasive species 84, 183
invertebrates 9, 50, 51
  see also insects
IPBES (Intergovernmental Platform on
    Biodiversity and Ecosystem Services) 231
IPCC (Intergovernmental Panel on Climate
    Change) 87
Ips typographus 6

Ireland 35, 37, 38, *178*, 195, 197
  afforestation in 48, 167, 209, 211, 212, 214, 218
  forest conservation in 310, 311, *314*
  forest cover in 7, 8
  forest types in *12*
  hunting in 116
  introduced species in 99, *100*
  plantations in 99
Iron Age 26, 47, 246–250, 257, 260
  coppicing in 79, 80
iron industry 80, 185, 253, 270, 292, 293
Italy 119, 143, *178*, 337, 339
  ancient woods in 233, 234
  coppice in 80, *81*, 82–85
  forest conservation in 310, *314*
  forest cover in 7
  forest types in *12*
  historical ecology in 228, 231–238
    catalogue of rural landscapes initiative 234
    classification of woods in 232–233, *233*
    litter collection 237
    re-naturalization 228–229, 234
    restoration of bogs 238
    soil fertility 236–237
    trees used for leaf fodder 234–236
  introduced species in 99, *100*
  plantations in 94
  veteran trees in 143, 145, *149*
  wildlife in 195, 197, 200
  woodland history in 21–22, *22*, 28, 29, 35
  wood-pasture in 63, 68, *70*, 233, *233*
IUCN (International Union for Conservation of Nature) 193, 227, 312, 313
  Red List 141, 144, 147, 193, 211, 228
IUFRO (International Union of Forest Research Organizations) 14, 230
ivy (*Hedera helix*) 130, 183

jay (*Garrulus glandarius*) 35, 155, *156*
Juglans regia 35
Juniper (*Juniperus* spp.) *10*, *316*
Jutland (Germany) 38
Jynx torquilla *157*, 159, 162, 165

kermes oak (*Quercus coccifera*) 19–20, 82
Kerry slug (*Geomalacus maculosus*) 11
Kielder Forest (UK) 218

Laetiporus sulphureus 143, 144
Lagopus muta 154, 155, *156*
Landsat data 4–5
landscape/amenity conservation 11
  see also recreation; tourism
landscape history 230, 234, 268, 290, 292, 294, 301
land-use changes 27, *47*, 68, 72, 123, 147, 184, 218
  intensification 71, 147, 175, 217–218, 237
  see also afforestation
Lány Game Park (Czech Republic) 23

larch (*Larix decidua*) *10*, 21, 210, 235, 266, *317*, *320*, 342
larks/pipits 154–155, 166
Latvia 7, 8, *13*, *314*
Les Landes (France) 103, 208
lichens 23, 52, *175*, 333
Lidar (light detection and ranging) 18, 26
Liguria (Italy) 21–22, *22*, 228, *228*–229, 233, 234, 237
  alnocoltura system in 236–237
Ligustrum vulgare 122
lime (*Tilia* spp.) 3, *10*, *11*, 24, 27, 245, 253, 258
  coppicing of 51, 77, 82
  large-leaved (*T. platyphyllos*) 27, 272
  silver (*T. tomentosa*) 82
  small-leaved (*T. cordata*) 27, 82, 144, *270*, 271, 272, 275
  T. alba 19
  veteran trees 141
Limoniscus violaceus 11, 146, 148
Lincolnshire (UK) 122, *268*, 270, 271, 326, 333
linnet (*Carduelis cannabina*) 156, 163, 214
Lisbon Meeting (1998) 8
Lithuania 7, *12*, *100*, 252, 310, 311, *314*, 337
livestock 6, 41, 46, 52, 54, 66, 68, 87, 120, 161, 201, 253
  predators on 185, 196
  see also cattle; goats; grazing; pigs; sheep
Loch Katrine (UK) 26
lodgepole pine (*Pinus contorta*) 210
logging 9, 46, 61, 141, 254, *256*
  residues from 133, 293
Lonicera periclymenum 130, 182–183, *182*
Lonkay, Antal 66
Lorraine (France) 279–280
  coppice-with-standards in 281–282, *281*
  Forestry School in 279–280, 284
  impact of French forestry policies in 285–288
Luscinia megarhynchos see nightingale
Luxembourg *81*, 312, *314*
lynx (*Lynx lynx*) 9, 49, 51, 197, 259, 293

Macaronesian region *10*, *316*
Macedonia, Former Yugoslav Republic of (FROM) *81*, *314*
  forest cover in 7
Malus sp. 63
  M. sylvestris 37
mammals 4, 193–202, 214
  large carnivores 193, 195–197, 202
    see also bear, brown; lynx; wolf
  postglacial megafauna 34, 36, 38–41, 116, 193–195
  threatened species 193
  see also specific mammals
mammoth (*Mammuthus primigenius*) 39
manna ash (*Fraxinus ornus*) 82
maple (*Acer* spp.) *10*, 27, 253
  A. criticum 19
  A. platanoides 144, 183
  A. pseudoplatanus see sycamore
  field (*A. campestre*) 19, 82, 272, 273
  veteran trees 143
maritime pine (*Pinus maritima*) 99
MCPFE (Ministerial Conferences on the Protection of Forests in Europe) 8, 112, 291

medicinal plants 69, 70, 71, 80
medieval period 4, 25, 40, 48, 50, 65, 194, 195, 198, 283–284
  coppicing in 79–80, 82, 84, 271
  hunting in 116–118
Mediterranean 3, 4, 6, 8, 28, 34, 69, 103, 197, *316*
  and climate change 53, 54
  coppice in 81, 82, 84–86, 133
  environmental degradation in 94, 347
  fire risk in 6, 52
  forest types in *10*, 12
Mercurialis perennis 130, 177, 183
Mesolithic period 77, 78
mesophytic deciduous forests 10, *12–13*
mice 133–135, 188
military surveys 66, *67*
Millennium Ecosystem Assessment (2005) 322
Milovicky Wood (Czech Republic) 132, 135
mire/swamp forests *11–13*
mistle thrush (*Turdus viscivorus*) 131, 155, *156*
mixed broadleaved–coniferous forest 11, 35, 109, *317*
Moldova 7, 8, 12
molecular genetic analysis 18, 28–29, 36–37
monocultures 94–95, 99, 101, 102, 107
Montenegro 7, *81*, 314
moose (*Alces alces*) 252, 259
mouflon (*Ovis musimon*) 123, 135
mountain beech forests 10, *12–13*, 83, 85
Muntiacus reevesi 135
Muscicapa striata 155, *157*, 165
mushrooms 69, *70*, 80

Natura 2000 protected sites 9, 235, 238, *238*, 316–318
naval timber *see* shipbuilding
nemoral coniferous forest 11, 82, 209, *317*
Neolithic period 24–25, 28, 46, 49–50, 141, 269, 349
  coppicing in 77–79
The Netherlands 99, *100*, 194, 310, *314*, 315, *337*
  coppice in 79, *81*, 135
  forest cover in 7, 8
  forest types in 12
  woodland history in 4, 47
  wood-pasture in 68
nettle (*Urtica dioica*) 52, 176, 177
New Forest (UK) 22, 39, 163, 266, 275
nightingale (*Luscinia megarhynchos*) 135, 155, *156*, 159, 162, 161 167
nightjar (*Caprimulgus europaeus*) 131, 216
non-riverine alder/birch/aspen forests *11–13*
non-wood forest products 6, 47, 69
Northern Ireland 209, 330, *330*, 332
Norway 13, 35, 99, *100*, 198, 313, *314*, 343
  forest cover in 7, 8
Norway spruce (*Picea abies*) 10, 11, 21, 25, 35, 36, 51, 82, 84, 158, 200, 210, 253, 256, 284
  selection systems for 96, *98*
  sustainable harvesting of 107
  yield of 101
nuthatch (*Sitta europaea*) 155, *157*, 158, 160, 163, 165, 166

oak (*Quercus* spp.) 10, 11, 24, 122, 245, 258
  archaeological study of 24–27, 40–41
  coppicing of 77, 79, 80, 82–86, 237, 273, 274
  cork (*Q. suber*) 50, 71, 82, 99, 123, 124, 132, *316*
  downy (*Q. pubescens*) 21, 82, 84, 134–135
  as fodder/grazing material 19–20, *20*, 51, 234
  kermes (*Q. coccifera*) 19–20, 82
  planting of 48, 274, 275
  Portuguese (*Q. faginea*) 82, *316*
  postglacial spread of 35, 37, 39
  Pyrenean (*Q. pyrenaica*) 79, 82, 85
  Q. cerrioides 83, 84
  Q. frainetto 19, 82, 85, 134–135
  Q. ilex see holm oak
  Q. petraea see sessile oak
  Q. robur 10, 19, 35, 63, 85, 270, 271, 273, 274, 282, *316*, 317
  Q. rubra 182
  rotation/yield of 95, 101
  Turkey (*Q. cerris*) 19, 35, 82, 83, 133, 134–135, 228–229, *316*
  understorey vegetation 182, *182*
  veteran trees 72, 141, 143, 144, *146*, 148
  wildlife 35, 132–135, 155, 164, 167
  in wood-pastures 62, 63, *64*, 65, 69, 228
oak processionary moth (*Thaumetopoea processionea*) 337
olive 3, 21, 123, 233, 234
orchards 47, 61, 141, 233, 234
Ordnance Survey maps 329–331, 333
Orkney Islands (UK) 38
ornamental species 51, 342
Oryctolagus cuniculus 51, 119–121
osier willow (*Salix viminalis*) 84
Ostrya sp. 10
  O. carpinifolia 19, 35
Ovis musimon 123, 135
owls 134, 155, *156*, 161
Oxalis acetosella 182, 183

paganism 310
palaeoecology 14, 243
palynological studies 39, 78, 82, 280, 348
  see also pollen analysis
pannage 65, 271, 272, 274
paper/pulp industry 82, 101, 298, *298*
parks 118–119, 121–122, 134–135, 141, 272, 274–275
partridge 122, 123, *154*
Parus major 155, *157*, 161, 164, 165
pear (*Pyrus* sp.) 63, *64*, 70
PEFC (Programme for the Endorsement of Forest Certification) 14, 111–112
peppered moth 53
pesticides 49, 160
pests/diseases 3, 6, 51, 99, 112, 150, 165, 201, 211, 334, 337–345
  and EU 338, 343–345
  future control of 344–345, 349
  and global plant/timber trade 338–340, 345, 349

pests/diseases (continued)
    introduction/spread of 338–339
    national biosecurity measures 341, 343–345
    see also ash dieback; Dutch elm disease;
        ramorum blight
pheasant (*Phasianus colchicus*) 122–123
Phillyrea latifolia 19
Philoscopus sibilatrix 156, 158, 159, 161–162, 163, 165
Phoenicurus phoenicurus 155, *156*, 159, 163
Phylloscopus collybita 156, 158, 162, 164, 165
Phylloscopus trochilus see willow warbler
Phytophthora ramorum see ramorum blight
Picea spp. 122, 258, *317*
    P. abies see Norway spruce
    P. sitchensis see Sitka spruce
    plantations of 98–99, 103, 209, 210
Picoides tridactylus 157, 158, 166
Picus viridis 131, *157*
pigs *62*, 65, 68, *70*, *71*, 123, 161, 201, 253, 274
pine (*Pinus* spp.) *10*, 25, 26, 27, 48, 167, 249, 258
    canary (*P. canariensis*) *10*, 78
    Corsican (*P. nigra*) 24, 210, *316*
    lodgepole (*P. contorta*) 210
    maritime (*P. maritima*) 99
    pests/diseases of 337, 338
    P. cembra *10*, *317*, 318, *320*
    P. halepensis *10*
    P. mugo *10*
    P. sylvestris see Scots pine
    plantations of 99, 103
    stone (*P. pinea*) 22, 28
    understorey vegetation 182, *182*
pine marten (*Martes martes*) 37
plane (*Platanus orientalis*) 35
plantations 4, 6, 8, *11–13*, 52, 94, 207
    regulations for 94, 102
    see also even-aged stands
Pleistocene era 33, 34, 116, 143, 193
Plenterwald system 107, *108*, 109, 110
Poland 3, 82, 94, 101, 158, *178*, 311, *314*, 337
    forest cover in *7*
    forest types in *12*
    wildlife in 194, 195, 197, 198
    see also Białowieża Forest
pollarding 4, 19, 46, 61, 65, 119, 234, 273, 275
    grazing 282–283
    veteran trees 147, *149*
pollen analysis 27, 33, 34, 36, 38, 40, 78, 235, 243, 274, 348
    pollen diagram *238*
pollution 23, 46, 47, 52–53, 277, 322
poplar (*Populus* spp.) *10*, *11*, 24, 101, 253
    coppicing of 78, 82, 96, 133
    P. nigra 29
    P. tremula *11*, 78, 83, 133, 177
    P. tremuloides 78
    veteran trees 143
Portugal 4, 5, 7, 8, *12*, 48
    coppice in *81*, 82, 95, 96
    forest conservation in 311, 312, *314*
    introduced species in 99, *100*
    wildlife in 132, 167, 195
    wood-pasture in 50, 63, 68, 69, 71, *71*

Portuguese oak (*Quercus faginea*) 82, *316*
postglacial forest landscape 33–41
    birds 154–157
    megafauna 34, 36, 38–41, 116, 193–195
    molecular genetic markers 36–37
    open/closed tree cover 38–41, *39*
    spread of canopy 35–36
    woodland ground flora 36
potash 245, 253, 255, *256*
Primula elatior 180–183
private forest owners 109, 111
privet (*Ligustrum vulgare*) 122
Pro Silva 110, 113
Protaetia spp. 143
    P. marmorata 141, 148
Prunella modularis 155, *156*, 158–160, 163, 167
Prunus
    P. avium 78, 272
    P. serotina 84, 182, 183
    P. spinosa 39, 122, 132, 177
Pseudotsuga menziesii 111, 210, 266, 288
ptarmigan (*Lagopus muta*) 154, 155, *156*
Pteridium aquilinum 36, 52, 134, 183
Pterocarya 33
pulpwood production 82, 95, 96, 101, 293–294, 298, *298*
Pyrenean oak (*Quercus pyrenaica*) 79, 82, 85
Pyrenees 29, 34, 83, 280
Pyrus sp. 63, *64*, *70*

Quercus spp. *see* oak

rabbit (*Oryctolagus cuniculus*) 51, 119–121
ramorum blight (*Phytophthora ramorum*) 165, 337–338, 339, 341–343, 344, 345
    control of 342–343
Rangifer tarandus 39
recreation *70*, *71*, *72*, 167, 218, 294, 298, 312
red deer (*Cervus elaphus*) 25, 39, 117, 120–123, 135, 252, 259
    management of 202
    range/distribution of 201–202
Red List 141, 144, 147, 193, 211, 228
redpolls 154, *156*, *157*, 166
redstart (*Phoenicurus phoenicurus*) 155, *156*, 159, 163
reforestation 71, 72, 98–99, 103, 212
    and understorey vegetation 174, 177–178
reindeer (*Rangifer tarandus*) 39
re-naturalization 228–229, 234
renewable energy 86, 186, 293, 338
rewilding 321, 322, 349
Rhododendron ponticum 51, 342
Rio Earth Summit (1992) 14, 111, 321
roads 49, 71
robin (*Erithacus rubecula*) 131, 155, *156*, 164–167
Robinia pseudoacacia *11*, 66, 79, 84, 99, 101, 103
roe deer (*Capreolus capreolus*) 39, 121, 123, 124, 135, 201, 259
Roman period 46–48, 117, 119, 177, 178, 194, 249, 250, 269, 310
    coppicing in 77, 79, 80

Romania 7, 12, 81, 310, 311, 313, *314*
  wood-pasture in *62, 63, 64, 65, 66, 68, 69, 70, 72*
rose chafer (*Protaetia marmorata*) 141, 148
Rottmanner, Simon 120
Rubus fruticosus 39, 130, 132, 135, 164, 176, 183
Russia/Russian Federation 3, 5, 6, 36, 197, 198, 291

Salix spp. *see* willow
salt industry 280
Sambucus nigra 132
saproxylic insects 33, 140–150, 158, 333
  diversity of 143
  and environmental factors 144–147
    adult/larvae requirements 146
    surrounding landscape 145–146
    tree characteristics 144–145
  microhabitats for 141–144, *142*
  preservation strategies for 147–150
  survey methods for 146–147
  threatened species 140–141, 144, 147
  wood-pastures/parks 141
Sarló-hát (Hungary) 78
sartage system 283, *283*, 284
savannah 34, 61, 123
Savernake Forest (UK) 26
sawmills 48–49, 101, 287, 298
Scandinavia 3, 8, 25, 35–37, 103, 111, 195, 339
  see also Denmark; Norway; Sweden
Sciurus spp. *see* squirrel
sclerophyllous forests 8, *10*, 155
Scotland 19, 37, 38, 48, 199, 201, 266, 271, 272, 274
  afforestation in 103, 167, 209, 216
  ancient woodland in *330*, 331–332, 334
  hunting in 116, 117
  pests/diseases in 337, 339, 342
Scots pine (*Pinus sylvestris*) 10, 11, 21, 24, *30*, 208, 210, 214, 252, 266, 270, 288, *317*
  clear-felling of 97
  postglacial spread of 35, 37
  and species richness 212
  yield of 101
scrub 8, *10*, 34, 163
seed dispersal 28, 34, 35
selection systems 83, 96, 112, 273, 288
sequoia (*Sequoiadendron sempervirens*) 78
Serbia 7, 12, *81*, 82
service (*Sorbus torminalis*) 272
sessile oak (*Quercus petraea*) 10, 19, 27, 35, 63, *82, 83, 85, 86*, 270, 272–274, 276, *282*, 286–288, *316*
  veteran 144
SFM *see* sustainable forest management
shade tolerance 27, 39, 78, 129–130, 177, 183, 215, *282*, 288
sheep *26*, 62, 65, 68, 70, 71, 87, 93, 116, 123, 234–235, 237, 274
shelterwood systems 96, 97, 109, 284–286, 288
Sherwood Forest (UK) 24–26
shipbuilding 48, 94, 148, 281, 284
shredding 19–20, 234, 237, 275
shrews (*Crocidura* spp./ *Sorex* spp.) 37, 133, 193
Sicily 118, 119, 121

silver fir (*Abies alba*) 25, 82, 96, *98*, 122, 284, 287, 288
silver lime (*Tilia tomentosa*) 82
single-stem management 109, 110
siskin (*Carduelis spinus*) 155, *156*, 166
Sitka spruce (*Picea sitchensis*) 48, 99, 101, 209, 210
  see also Atlantic spruce forests
Sitta europaea 155, *157*, 158, 160, 163, 165, 166
slash-and-burn cultivation 236, 249, 257, 283, *283*, 293
Slovakia 7, 12, *81*, 213, *314*
  introduced species in 99, *100*
Slovenia 7, 8, 12, *81*, 84, 311, *314*
  close-to-nature forestry in 107
  high forest management in 94
soil 27, 33, 109, 185, 292
  erosion 52, 63, 93–94
  microbes 85, 215
  moisture 11, 35, 51, 82, 85
  nutrients/fertility 5, 10, 17, 51, 52, 85, 178–180
  protection 6
  waterlogged 11, 101, 269
  and wood-pastures 63, 66, 68
Somerset Levels (UK) 24, 269
song thrush (*Turdus philomelos*) 155, *156*, 165, 166
Sorbus spp. 272
Spain 234, 311, *314*
  afforestation in 48
  coppice in 79, 80, *81*, 82, 84, 85, 95, 96
  forest cover in 5, 6, 7
  forest types in 12
  hunting in 123–124
  single-aged stands in 99
  wildlife in 131, 155, 167, 195, 198
  woodland history in 28, 29
  wood pasture in 50, 62, 63, 69, 71, *71*
spiders 132, 143, 211, 214
spotted flycatcher (*Muscicapa striata*) 155, *157*, 165
spruce *see* Picea
spruce bark beetle (*Ips typographus*) 6
squirrel (*Sciurus* spp.)
  grey (*S. carolinensis*) 3, 134, 199–200, 349
  red (*S. vulgaris*) 51, 134
  impact of grey squirrel on 199–200
SRC (short-rotation coppice) 82, 84, 86, 133, 135–136
Star Carr (UK) 78
steppe zone 34, 39, 40, 198, *317*
stone pine (*Pinus pinea*) 22
storms 6, 38, 65, 150
strawberry tree (*Arbutus unedo*) 82
Streptopelia turtur 131
sudden oak death *see* ramorum blight
Sus scrofa *see* wild boar
Suserop Skov (Denmark) 27
sustainable forest management (SFM) 9, 65, 72, 94, 102, 107, 294, 300–301
  certification 14, 112
  coppicing 77
  ecological 294–297, *295*
  economic *295*, 297–298
  European scale 290–291
  implementation of, requirements for 300–301
  regional scale 291–292
  social/cultural *295*, 298–299
  in Sweden *see* Bergslagen

Sweden 31, 82, 86, 119, *178*, 196, 198
    close-to-nature forestry in 110
    forest conservation in 310, 311, 313, *314*, 315
    forest cover in 6, 7, 8
    forest types in 13
    pests/disease in 6
    plantations in 96
        conifers in 99, *100*, 101
    SFM in *see* Bergslagen
    veteran trees in 141, 143, 145, 147, 148
    wood-pasture in 63, 68, 69, *71*
sweet chestnut (*Castanea sativa*) 10, 28, 35, 233–234, 237, 267
    coppicing of 79, 82, 83, 85, 86, 134, 273
    rotation of 101
Switzerland 21, *81*, 86, 200, 237
    close-to-nature forestry in 107, *108*, 109, 110
    forest conservation in 311, 313, *314*
    forest cover in 7
    forest types in 12
    plantations in 94, *98*
        introduced species in 99, *100*
    wood-pasture in 63, 68, 69, *71*
sycamore (*Acer pseudoplatanus*) 35, 77, 84, 183, 200, 266
Sylvia atricapilla see blackcap
Sylvia communis / S. curruca 156, 159, 163

tanning industry 80, 81, 237
Tetrao tetrix 155, *156*, 159, 166, 214, 215
Tetrao urogallus 155, *156*, 158, 166, 167
Thaumetopoea processionea 337
thermophilous deciduous forests 8, *10*, *12–13*
Thiérache region (France/Belgium) 184–185, *185*
thrushes 131, 155, *156*, 165, 166
Thuja plicata 200
Tilia spp. *see* lime
timber production 48–49, 51, 66, 68, *70*, 94, 101, 253, 280–281, 298
timber shortages 48, 94, 98, 99, 209
timber trade 93, 101, 349
    and pests/diseases 338
    see also even-aged stands; high forest management
tits 155, *157*, 158, 160, 163, 165, 167
tourism 21, 68, *71*, 124, 165, 294, 298
Transylvania (Romania) 62, 63, *64*, 65, 66, 311
tree of heaven (*Ailanthus altissima*) 11, 84
treecreeper (*Certhia familiaris*) 155, *156*, 160, 163
Troglodytes 131, 155, *156*, 159, 160, 165
Tronçais, forest of (France) 26–27
Tsuga heterophylla 210
Turdus merula 131, 155, *156*, 164–167
Turdus philomelos 155, *156*, 165, 166
Turdus viscivorus 131, 155, *156*
Turkey 28, 68, *81*, 101, 143, *314*
Turkey oak (*Quercus cerris*) 19, 35, 82, 83, 133–135, 228–229, 316
turtle dove (*Streptopelia turtur*) 131

Ukraine 7, 195, 198
Ulmus spp. *see* elm

understorey vegetation 36, 47, 50, 78, 129, 130, 174–186
    and agricultural intensification 175
    ancient/recent forests compared 177–178, *184*
    biodiversity of 175–177, *176*, 182–186
        see also under Atlantic spruce forests
    climate change 175, 185
    colonization of new forests by 178–181
        limitations on 179–181, *180*, 184
    conservation/forest expansion 183–186
    coppice 129, 130
    environmental changes 182–183, *182*
    forest management systems 181–182
    grazing 183
    invasive species 183
    seed banks in 178
    shade tolerance 177
    stress/disturbances 177
    wildlife in 131–133, 135, *175*
        birds 131, 158, 159, 163, 164, 166, 167
    woodland history 174–175
United States (USA) 40, 78, 178, *178*, 179, 183, 196, 338–340
Ursus arctos 9, 37, 119, 195, 293
Urtica dioica 52, 176, 177

Vaccinium myrtillus 182, 183, 185
Val d'Aveto (Italy) 21
Vera's shifting mosaic cycle 38–40, *39*, 160
veteran trees 4, 25, 68, 72, 118, 333
    decline of 141, 147
    and fungi 140, 142–144, 148
    habitat types for 141
    and insects *see* saproxylic insects
    microhabitats in 141–144, *142*
    preservation strategies for 147–150
    sap flows in 143–144
Vienna Meeting (2003) 8
violet click beetle (*Limoniscus violaceus*) 11, 146, 148
Vosges Mountains (France) 109, 286–288
Vosges region (Germany) 109, 197, 286–288
Vulpes 122, 134

Wales 199, 209, 215, 216, 266, 271, 274, *330*, 331, 337, 339, 342
walnut (*Juglans regia*) 35
warblers (*Sylvia*) 131, 135, 155, *156*, 162, 165, 214
water management 8, 68, 296–298
Weald (UK) 80, 266, 331
weevils 23
weirs 78–79
western hemlock (*Tsuga heterophylla*) 210
western red cedar (*Thuja plicata*) 200
wetlands 10, 38, 52, 79, 103, 161, 184, *185*, 238, 296–297
whitebeam (*Sorbus aria*) 272
Whitelee forest (UK) 19, 209
whitethroat/lesser whitethroat (*Sylvia communis/ S. curruca*) *156*, 159, 163
wild ass (*Equus hydruntinus*) 40

wild boar (*Sus scrofa*) 36, 39, 117, 120, 123, 124, 196, 238, 259
    decline/reintroduction of 200–201
    impact of 201
wild cat (*Felix sylvestris*) 51
wild cherry (*Prunus avium*) 78, 272
wild horse (*Equus ferus*) 39, 40
wild ox *see* aurochs
wildwood 1, 46, 116, 159–160, *160*, 162, 266, *267*, 268–270, 349
willow (*Salix* spp.) 10, 24, 246, *316*
    coppicing of 78, 79, 82, 96, 133
    osier (*S. viminalis*) 84
    veteran trees 143
    in wood-pastures *62*, 63
willow warbler (*Phylloscopus trochilus*) 150, 155, *156*, 162–167
Windsor Great Park (UK) *268*, 275, 310
windthrow 37, 52, 96, 101–102, 107, 130, 133, 210, 216, 268
wolf (*Canis lupus*) 3, 9, 51, 59, 116, 124, 161, 193, 259
    protection/reintroduction of 196, 318
    range/population collapse of 195–196
wolverine (*Gulo gulo*) 9, 195, 293
wood anemone (*Anemone nemorosa*) 36, 130, 174
wood banks 66, 330, 333
wood pellets 86
wood tar 253, 255, 258
wood warbler (*Philoscopus sibilatrix*) *156*, 158, 159, 161–163, 165
wooded meadows 234–235
woodland ground flora *see* understorey vegetation
woodland history 18–29, 347–349
    archaeology/palaeoecology 18, 24–27
    biological indicator 22–23, 177–178
    digitized photographs/drawings 18, 21–22, *22*
    future issues for 348–349
    historical records 23–24
    interdisciplinary approach to 18, 279–280
    Internet 18
    Lidar/GIS 18–19, 26, 27
    molecular genetic analysis 18, 28–29
    oral history 18, 19–21, 348
    pollen/charcoal analysis 27–28
    see also postglacial forest landscape
woodland management 6, 46–54, *47*
    changes to soil 52
    emergence of 46–47, 49–50
    forest cover 47–49

forest laws 4, 94
intensive/extensive 97, 109
oral history 21
wildlife conservation 9
woodmanship 230–231
wood-pastures 3, 4, 6, 26–27, 46, 50, 59, 61–73, 270–271, 274
    agriculture 66–68
    conservation 63, 66, 68, 69, 70, 71, *71*
    diversity of/terms for 61–62, *62*, 65, 69
    ecosystem services provided by 63, 64
    grazing in 61–63, 65, *70*
        in conflict with other activities 66–68
    historical development of 65–68
        regulation 65–66
    historical ecology 233, *233*
    hunting 118, 123
    management interventions in 63–64
    multiple functions of 69, *70*–71, 72
    national inventories of 69
    revival of 68–69
    soil 63, 66, 68
    sustainable management of 65, 66, 72
    threats to 69–72
    tree species used in 63, *64*
    veteran trees in 63, 68, 72, 141
    wildlife 68, 132, 159, 162–163, 165, 350
woodpeckers 131, *157*, 158, 162–163
    black (*Dryocopos martius*) *157*, 164, 166
    great spotted (*Dendrocopos major*) *157*, 165, 167
    green (*Picus viridis*) 131, *157*, 159, *162*
    lesser spotted (*Dendrocopos minor*) *157*, 165–167
    three-toed (*Picoides tridactylus*) *157*, 158, 166
    white-backed (*Dendrocopos leucotos*) *157*, 158, 163
woolly rhinoceros (*Coelodonta antiquitatis*) 39
wren (*Troglodytes troglodytes*) 131, 155, *156*, 159–161, 165
wryneck (*Jynx torquilla*) *157*, 159, *162*, 165
wych elm (*Ulmus glabra*) 28, 82, 272
Wye Valley (UK) 271, 272, 276
Wytham Woods (UK) 135

yew (*Taxus baccata*) 78, 272

Zagori region (Greece) 19

THE BEGINNER'S
— GUIDE TO —
RUGBY

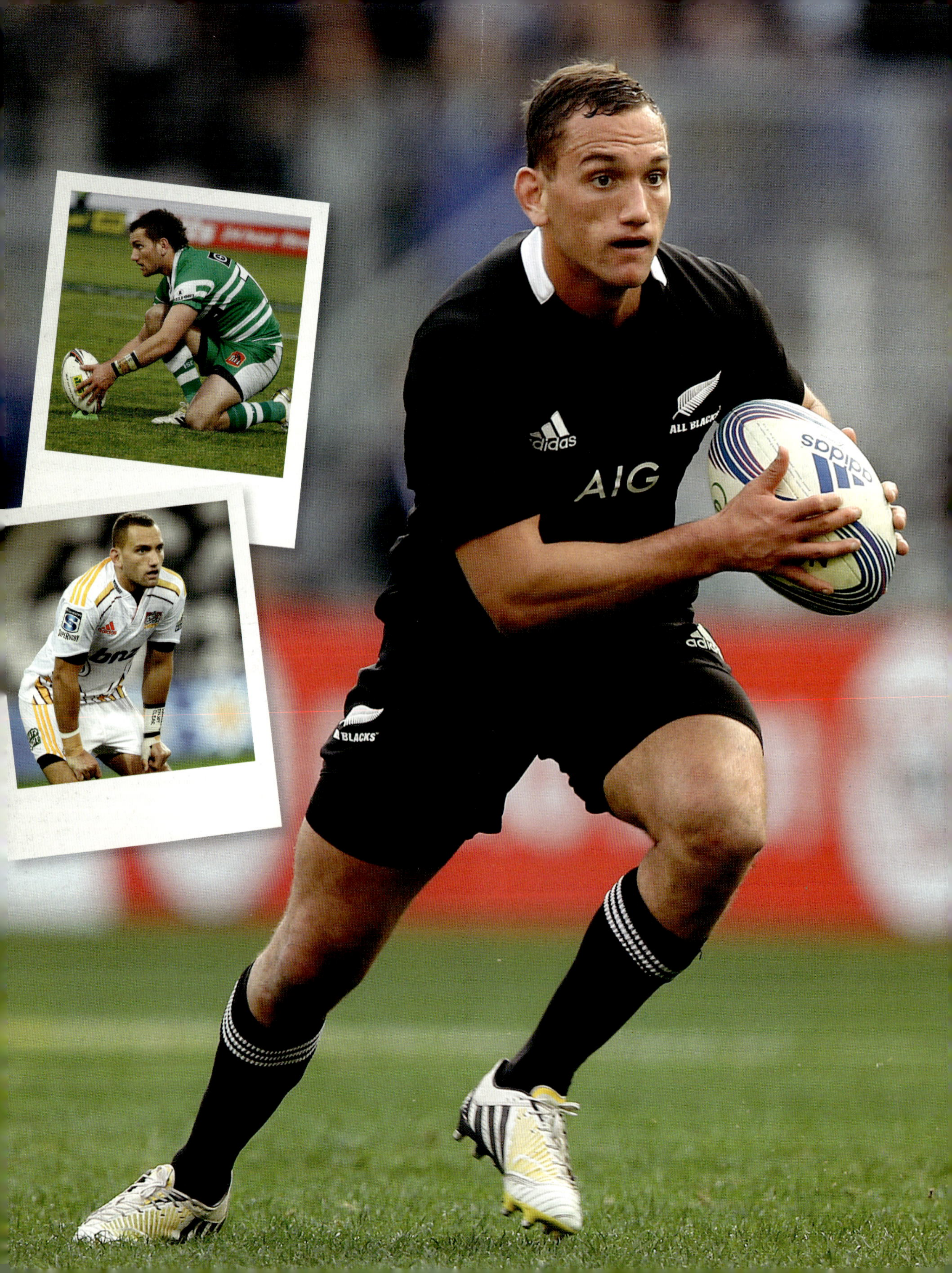
adidas
ALL BLACKS
AIG
BLACKS

THE BEGINNER'S
— GUIDE TO —
RUGBY

AARON CRUDEN

RANDOM HOUSE
NEW ZEALAND

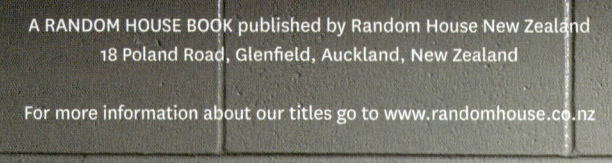

A RANDOM HOUSE BOOK published by Random House New Zealand
18 Poland Road, Glenfield, Auckland, New Zealand

For more information about our titles go to www.randomhouse.co.nz

A catalogue record for this book is available from the National Library of New Zealand

Random House New Zealand is part of the Random House Group
New York   London   Sydney   Auckland   Delhi   Johannesburg

First published 2015

© 2015 text Aaron Cruden, images Adrian Malloch unless otherwise credited on page 208

The moral rights of the author have been asserted

ISBN 978 1 77553 790 8

This book is copyright. Except for the purposes of fair reviewing no part of this publication may be reproduced or transmitted in any form or by any means, electronic or mechanical, including photocopying, recording or any information storage and retrieval system, without permission in writing from the publisher.

Design: Kate Barraclough
Front and back cover photographs: Adrian Malloch
Additional cover images: Shutterstock, Dorling Kindersley (Steve Gorton)

Printed in China by Everbest Printing Co. Ltd

DISCLAIMER

Rugby is a highly physical game with lots of contact. To enjoy it, you need to be physically and mentally prepared. You also need to understand how to play rugby safely; no physical activity should be engaged in without taking some precautions.

We recommend the use of a mouthguard in all practice situations. Qualified trainers should be present when partaking in any contact situation, like scrum and tackle activities.

The information in this book is presented for the purpose of educating and entertaining readers on the skills and basics of rugby. Furthermore, we make no claims about the safety or appropriateness of any information found in the content, and consequently cannot be liable for any resulting loss, damage or injury.

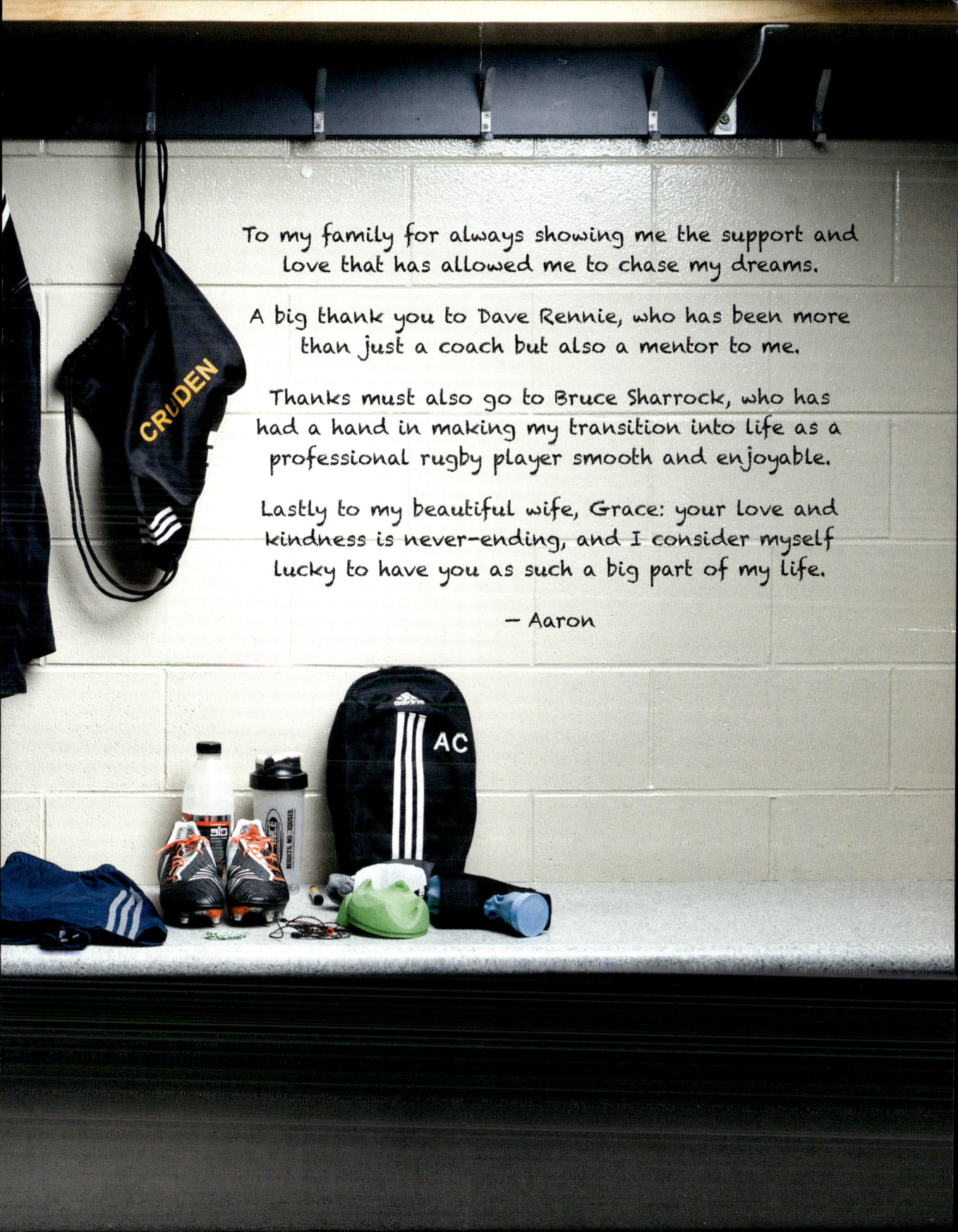

To my family for always showing me the support and love that has allowed me to chase my dreams.

A big thank you to Dave Rennie, who has been more than just a coach but also a mentor to me.

Thanks must also go to Bruce Sharrock, who has had a hand in making my transition into life as a professional rugby player smooth and enjoyable.

Lastly to my beautiful wife, Grace: your love and kindness is never-ending, and I consider myself lucky to have you as such a big part of my life.

— Aaron

CONTENTS

FOREWORD
BY DAVE RENNIE
9

AARON CRUDEN FAST FACTS
11

THE BOY FROM PALMY
13

MY JOURNEY TO THE ALL BLACKS
17

CHAPTER 1
WHAT IS RUGBY?
43

CHAPTER 2
HANDLING
81

CHAPTER 3
PASSING
99

CHAPTER 4
RUNNING
121

CHAPTER 5
KICKING
139

HYDRATION
161

CHAPTER 6
TACKLING
165

NUTRITION
179

CHAPTER 7
SET PLAY, PHASE PLAY, PENALTIES, ATTACK AND DEFENCE
183

LEADERSHIP
192

INJURY
194

AARON'S BALL TRICKS
196

FINAL WHISTLE

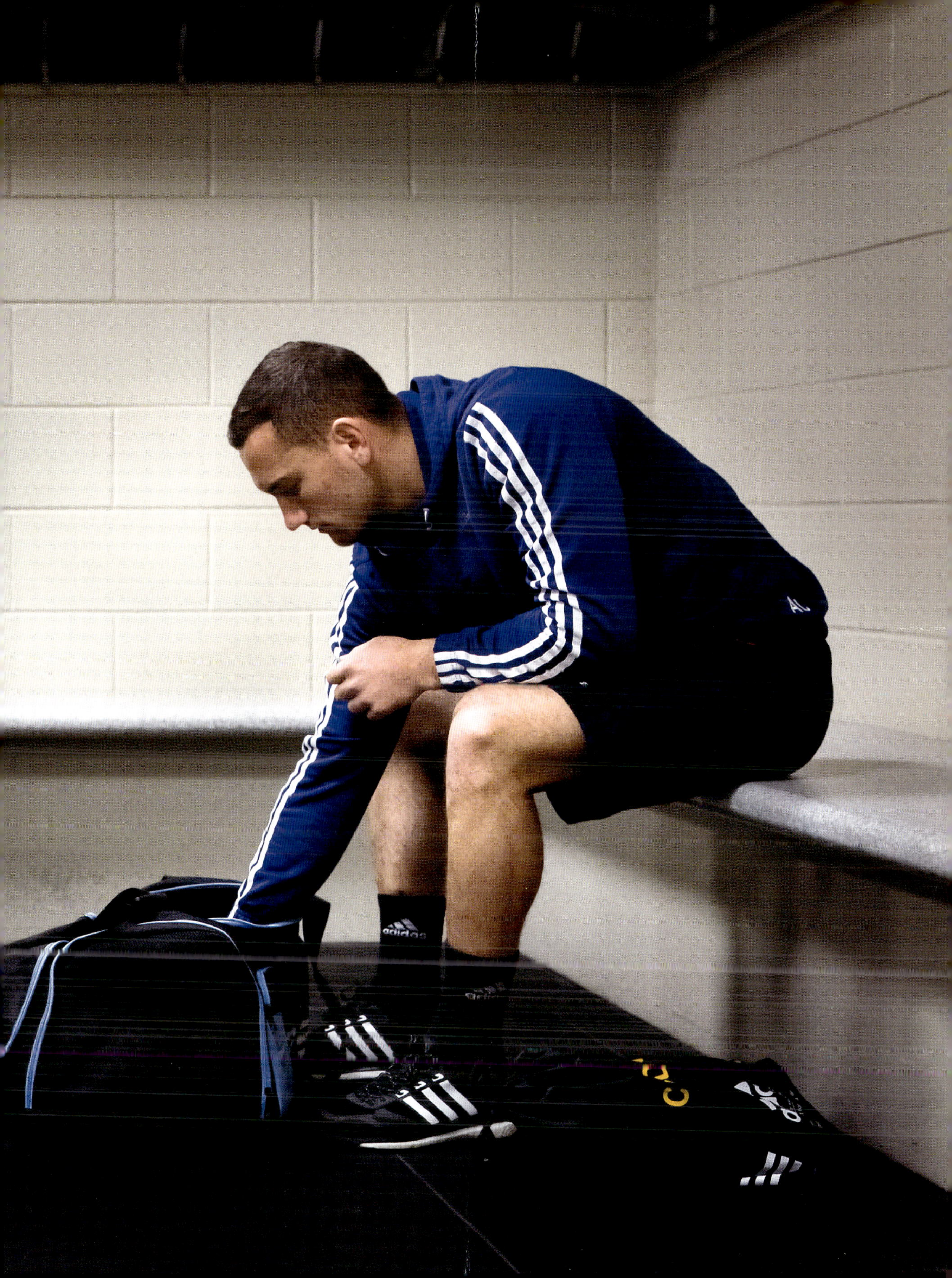
adidas

Watching a New Zealand Under-20 training match with Dave Rennie.

FOREWORD

I was impressed from the very first time I saw Aaron play. It was 2006, his final year at Palmerston North Boys' High. I loved the free spirit he played with, a mindset to attack from anywhere and the ability to create space for himself and others. A brave defender, he clearly enjoyed the responsibility of running the ship and gave confidence to those around him.

When I was coach of the Manawatu Turbos, I took the backs down to Boys' High for an opposed training and he carved us up a couple of times. It was clear from the start that he had the potential to be something special.

It was over the next few years that I learnt of the quality of the man. He was the best player in Manawatu club footy in 2007 but wasn't allowed to play in the NPC because he was too young.

His battle with cancer the following year has been well documented but it reflected the strength of character which has defined him. Within nine months of that diagnosis he had captained New Zealand to the 2009 IRB Junior World Championships title in Japan and been crowned U20 Player of the Year.

While he has enjoyed success in Super Rugby and on the international stage, he is still humble, approachable and generous with his time to charities and fans. He's a great team man who has high expectations of himself, demands the same from others and is an excellent role-model for our young players.

Dave Rennie
Head Coach, Chiefs Rugby Club

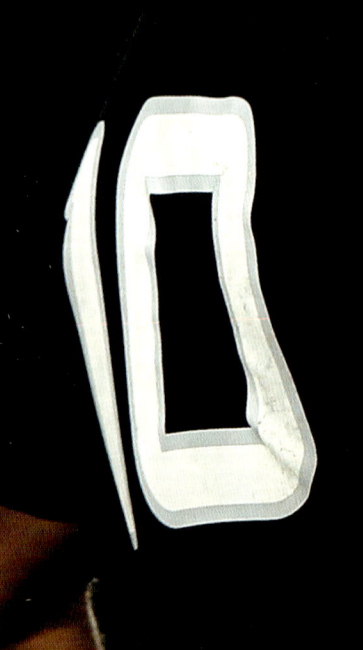

Playing for the All Blacks against France in June 2013.

BORN: 8 January 1989

HEIGHT: 1.78m

WEIGHT: 84kg

TEST DEBUT: v Ireland at New Plymouth 12 June 2010

PROVINCIAL TEAM: Manawatu Turbos

SUPER RUGBY TEAM: Chiefs

USUAL POSITION: First five-eighth

AARON CRUDEN FAST FACTS

Aaron made his provincial debut for Manawatu Turbos in 2008 and then captained New Zealand to the 2009 IRB Junior World Championship title in Japan, where he was named IRB Junior Player of the Year. In 2010 Aaron made his Super Rugby debut against the Brumbies.

On 20 February 2010, Aaron scored his first Investec Super Rugby points, when he converted a try in the Hurricanes' 47–22 win over Western Force at the Westpac Stadium.

On 24 April 2010, Aaron scored his first try for the Hurricanes in their 33–31 success against the Highlanders in Dunedin, and that same year he made his All Blacks debut against Ireland in New Plymouth.

AARON'S PLAYING CAREER

College Old Boys Juniors 1998–2001

Manawatu U11 1999

Manawatu U13 2000–01

Manawatu U16 2004

Palmerston North Boys' High School 2002–06 1st XV 2004–06

Hurricanes School 2006

College Old Boys Seniors 2007–CURRENT

Manawatu B 2007

Manawatu Turbos 2008–CURRENT

New Zealand U20 2009

Hurricanes 2010–11

Chiefs 2012–CURRENT

All Blacks 2010–CURRENT

Kicking for goal for Manawatu against Taranaki.

All Blacks against Ireland in November 2013.

THE BOY FROM PALMY

Most people would have looked at Aaron Cruden when he was growing up in Palmerston North and thought that his pocket-sized 1.78-metre frame would not be able to withstand the difficulties of test rugby.

Yet Aaron's agile movements and graceful poise, seen throughout the Square in Palmerston North, often on his trusty skateboard, hinted of talents that would serve him well on rugby's highest stage. He was in fact thinking about his trusty deck when a phone call came from All Blacks coach Graham Henry in 2011.

With the country gripped with World Cup fever, 22-year-old Aaron — initially not selected in the squad — unexpectedly saw his skateboarding become a national topic. With just six tests to Cruden's name, suddenly Henry was talking about the testicular cancer survivor in front of packed media conferences.

The long-serving Manawatu player, who started in the U11 grades before cracking the Turbos senior squad in 2008, was certainly a familiar face around Palmerston North Boys' High School. Now, after groin injuries to Dan Carter and Colin Slade, he was suddenly the premier All Blacks first five-eighth while the team were in the midst of a World Cup campaign.

'Keeping Aaron Cruden off his skateboard has been a major,' Henry dryly noted at a media conference during the tournament. 'Last week he was skateboarding around Palmerston North, having a couple of beers and watching us play. Now he's the number one No. 10 in the country.'

As it turns out, Aaron, now with almost 40 tests to his name, has become a much better rugby player than a skateboarder. After all, if that mistimed ollie or kick-flip before being called up to the camp had been more serious, the graze on his knee might have stopped him becoming an unlikely star in the biggest sporting event in New Zealand's history.

Aaron had established a fine rugby record before taking part in the 2011 Rugby World Cup. Captain of his school's First XV, he received his first Manawatu jersey in 2008, and captained New Zealand to the 2009 IRB Junior World Championship, earning the Junior Player of the Year title in the process. With a devastating turn of speed and a sharp running game, many thought that Aaron was potentially one of the best ball-in-hand attacking players New Zealand had ever produced at No. 10.

He made his Hurricanes debut aged 21, and in July 2011, just before the World Cup tournament, the Chiefs' new head coach Dave Rennie announced they had signed Aaron for the 2012 season.

It was an interesting time for first-fives around the country. Daniel Carter's big boots were being filled by Aaron more and more often. A young sensation called Beauden Barrett was also attracting attention, Colin Slade was putting his injuries behind him and getting more game time, while an entire new generation led by the likes of Ihaia West were waiting in the wings.

However, 2012 was to be Aaron's year. He would make his Chiefs debut in a loss against the Highlanders, but would achieve five goals from five attempts from the kicking tee.

While his sparkle when attacking the line was well known, Aaron would continue to improve his game, notably becoming a fine kicker, with his punting distance and accuracy shining throughout the season. Eventually, his boot would strike over 251 points, leading Super Rugby that season, while Aaron would win his first championship in the competition.

As for the All Blacks, Aaron would feature in 11 tests in 2012, putting in a mammoth performance when New Zealand defeated the Springboks 21–11 in Dunedin.

Injuries have played a part in his career but having survived cancer at 19, he certainly has a fighting spirit. A knee injury cut short Aaron's World Cup campaign in the final against France in 2011, and less than a year later, after playing a leading role in the All Blacks' 60–0 rout of Ireland in Hamilton, he would again limp from the field.

However, in 2013 the star put aside his minor injuries in another magnificent season, winning his second Super Rugby

Making a break against Argentina in the quarterfinal of the 2011 Rugby World Cup.

medallion before adding another nine internationals to his ledger. He would sign off a fantastic year by kicking the match-winning conversion against Ireland, which would confirm the first perfect season in the professional era.

There have been lots of match-winning moments, even if 2014 saw the Chiefs denied in their quest for three straight titles. Many believed that Aaron's broken thumb, suffered in the Chiefs' 43–43 draw with the Cheetahs, was a critical reason for the Chiefs' inability to make it three in a row.

However, Aaron began his 2014 test campaign in stellar fashion, taking a quick tap in the final moments against England in Eden Park, which would lead to Conrad Smith scoring the match-winning try.

Aaron's remarkable rugby journey continues. It looks like he will cement his place as one of the All Blacks' premier first five-eighths and if for some reason it doesn't all work out — well, we can guarantee that inside him will be one small kid from Palmy who will just try even harder.

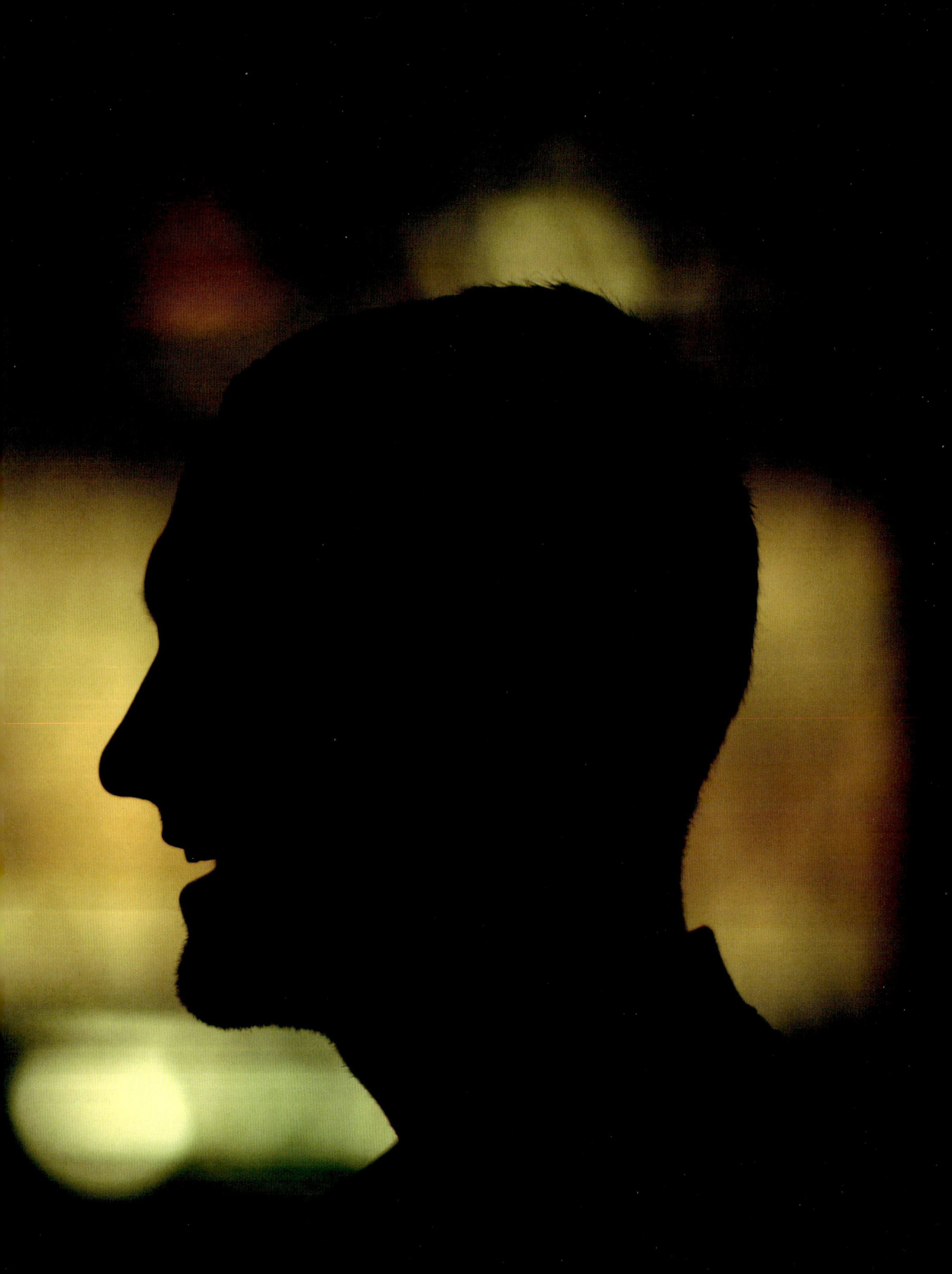

MY JOURNEY TO THE ALL BLACKS

STARTING OUT SMALL

When you're an All Black, one of the questions that people often ask is when you first dreamed of putting on the black jersey. People seem to think that you're born with the hunger to play for the All Blacks.

It wasn't like that for me. It wasn't until my late teens that I dared to believe that being an All Black might be a realistic goal. That's when I began to chase my dream, to put in all the hard work needed to make it come true.

One thing I have learned is that the harder you work, the more your dreams come true.

Dad played at second-five, lock and loose forward for Wairarapa Bush, at fullback for Taranaki and on the blindside flank for Manawatu. Uncle Snow (John Cruden) played for the Manawatu Turbos and Uncle Bruce (Bruce Hemara) played for NZ Maori, so you would think playing rugby was an obvious choice for me, especially with so many family members there to give me advice. Unfortunately when I was six years old my parents split up. That's tough on any kid, but they both kept supporting me and always encouraged me as I became more interested in sport.

Mum always made it clear that we had to make our own decisions about which sporting path each of us took. The great thing about our family was that there was never any pressure to play one particular

Posing for a photo with my team-mates after a regional tournament.

sport. After all, it's easier to get hooked on a sport if it is your passion and not another person's. You're always better at doing things if you enjoy them.

Mum was the odd one out, because we were a family of boys. Kurt was the oldest of my brothers, Jarrad is six years younger than me and I have a half-brother, Stewart, who is about 12 years younger. Kurt and I played a lot of sports and games at a very young age: touch rugby, bullrush, backyard rugby and games like that — all useful parts of my rugby education — but also soccer, cricket and skateboarding. We grew up in a typical Palmy house on a quarter-acre in a quiet street, a few shrubs out the front, then came the house — weatherboards and a corrugated-iron roof — and out the back a sleep-out, a shed and a fence around a good-sized yard with a veggie garden, a lemon tree and a rotary clothes line. Like backyards all over Palmy and the rest of the country, ours was the scene of some epic sporting encounters!

When I was six years old, I wasn't the biggest kid around, and since Kurt was playing soccer I followed his lead. At this very young age, we played soccer the way that kids still play it — we formed a big bunch around the ball and hacked away at it. It didn't look much like real soccer, and even less like rugby, unless you're talking about the ruck!

But like a lot of other sports, soccer can teach you core rugby skills, too. I like to think I was a quick learner, and besides the obvious, basic ability to kick a ball, soccer also gave me an idea of how to look for opportunities to create space.

I didn't make any rep teams, but because I was in my older brother's team, I was playing a couple of grades above my age. Kurt was faster than me, but I think I had a little more skill on the ball. Kurt might not agree!

As a little kid, I worshipped my older brother, and wanted to be just like him. Most people try to copy someone they admire, so there's nothing wrong with looking up to someone else. You shouldn't be afraid to follow in someone else's footsteps, at least at first! It's not quite the same as copying, and you'll eventually want to find your own place and strengths in your chosen activity.

When I was about nine years old, Kurt switched to rugby and I decided to do the same. I often followed him and his mates down to the rugby ground, which was close to home. I guess that this is where my love of rugby really started.

Of course, coming from such a strong rugby family, rugby had always been there in the background. My first rugby memory is of watching the 1995 Rugby World Cup game when Jonah Lomu ran over the top of English player Mike Catt —something my brother did to me lots of times in the backyard. I was six. Like a lot of Kiwi kids, we would regularly get up in the early hours of the morning with the family to watch the games. Mum would come and wake us up, and we would trudge sleepily into the lounge. We would slowly wake up as we watched the haka and all that — and we would get a hot Milo and biscuits as a treat! It was probably a few years before I saw a whole game through, as I usually fell asleep, but these were great family rugby moments. It didn't matter that I was playing soccer at the time.

My first memory of actually playing rugby was running around in bare feet in the cold at Paneiri Park by the rubbish dump in Awapuni. My feet would take a little while to warm up but once they did, playing with all my mates in the mud, rain and wind — it was great! Running on such heavy surfaces might even have helped my agility. My first boots were a second-hand pair of Mizunos. I think they might have been a bit too big and I was puzzled by the laces, which seemed incredibly long. But I had grown up watching dad lacing on his boots, so getting my very own pair was pretty cool.

I played juniors for College Old Boys — the same club as my dad and Uncle Snow. By now, I had that typical rugby bedroom with posters of my idols on my wall — players like Jonah Lomu, Zinzan Brooke, Sean Fitzpatrick, Christian Cullen and Jeff Wilson. I began to add team photos of the representative sides that I was selected for.

Friends would come around home in those days and there was always a soccer or rugby ball lying about, so we would play

With my brothers Kurt (left) and Jarrad (in front of me), and my cousin Ryan (right).

a bit of both. If one team was thrashing the other team in a game of soccer, we would toss the round ball away and play with the oval one instead. Or we might just bounce on the trampoline or pick up the cricket bat to have a few overs. I'm sure the variety was good for us. We never got sick of or bored with anything, least of all rugby.

College Old Boys Junior Rugby team photo. The Crudens have been playing for College Old Boys for years now.

I remember going to the park to watch my dad play footy, but my memories aren't really of him running or kicking or tackling. What I remember is all of us young kids running around on the sideline or on the next field kicking our own ball and playing our own game. I guess we looked up every now and then if the crowd roared or if there was a try, but our game was much more important!

When you are 10 years old, you tend

College Old Boys Junior Rugby
Under 13's – 2001
Sponsored by T-Market Fresh
Played 13, Won 13, Points For: 612 - Against: 67.

BACK ROW: Tim Bryson, Reece Sutton, Ryan Sayer, Sam Howard, Jonathan Phillips, Shaun Bradley.
MIDDLE ROW: Mike Sayer (Coach), Nick Dorn, Mason Hanly, Beau Borell, Tipene Wehipeihana, Kayne Dunlop, John Cruden (Coach).
FRONT ROW: Ryan Bowater, Karl Mehlhopt, Aaron Cruden, Ryan Cruden, Cameron Smith.
ABSENT: Christian Loketti.

A Few of My Heroes

JEFF WILSON

JONAH LOMU

CHRISTIAN CULLEN

to run around and imitate your rugby heroes. When we played in the backyard or at the local park, I would practise the chip-and-chase that I had seen Jeff Wilson do in test matches. I couldn't believe the control and ease with which he did those things. Christian Cullen was another hero. I'd practise doing the dives that Wilson and Cullen did when they scored tries. Meanwhile, my older brother was working on being Jonah, so I had to do a lot of tackling. He was pretty successful at running over me, too!

Dad had played all over the field but was mostly a flanker. I had no desire to be in the forwards, but I think it was only because there was a spare spot in the midfield that I started out playing at second-five and centre. My first rugby coaches were John Cruden and Mike Sayer. They were outstanding, and I really owe them. They worked us hard but they also kept everything fun, which I reckon is really important when you are playing sport at a young age. I didn't get any favours because my uncle was the coach — both Mike and Uncle Snow had sons playing and they didn't get any special treatment either, that's for sure. I got Player of the Day on a few occasions but we shared it around a bit. And after the games, our coaches would give us Coke and lollies. You didn't mind the coaches being tough if they also rewarded you.

Just as important as the coaches were the mums. They would take turns bringing along a nice tray with cut-up oranges for halftime. We would put the orange quarters in our mouths like mouthguards and flash big, orange grins at one another.

After the games we would all hang out together. A stop at the dairy was always on the cards. If it was hot, we'd get an ice cream, but usually with the cold Palmy weather we'd get a hot pie. The drive home was always really positive, with Mum telling me about all the team's good plays. It was like a gentle version of a coaching debrief! As I got older my parents would also point out some of the negative things, but they must have been good at keeping it positive, because it's the supportive things that I mostly remember.

Kurt and I didn't get any money for the number of tries we scored like the kids in some families, but my mates and I had little competitions about how many tries we had all scored over the season.

It was pretty classic, Kiwi-kid rugby time.

HIGH-SCHOOL DAYS

I was playing club rugby while I went to primary school — which was all at Central Normal School in Palmerston North, apart from a year when I went to Hillneath Primary in Wairoa out there on the East Coast. After primary school I went to Palmerston North Intermediate Normal School, but played rugby for College Old Boys. At that stage I usually played in the centres. This was also when I was picked in my first rep team, the Manawatu Under-11s, for the province I still play for every now and then to this day.

I was allowed to choose which high school I went to. Kurt was going to Awatapu College in Palmy, but I decided to go to Palmerston North Boys' High. The reason was rugby. I was pretty passionate about the game by that stage, and Palmerston North Boys' was top dog in footy in our area.

It was a big school and a little intimidating at first but I soon made the most of all the activities it had on offer. I was small back then in the third and fourth form (now called years 9 and 10), and maybe a little chubby as well. Don't think just because you are small that you can't put on the pounds! But there was lots of sport available. I played social cricket in the summer. Basketball was also the go on Friday nights. I'm no one's idea of the ideal build to play basketball, but it was always fun. Like rugby, you don't have to be a big lad to be competitive. I played a lot of touch rugby at high school, which was really good for my catching and passing skills as well as developing some early hand-eye coordination. I even went to the touch nationals three times. They usually named a tournament team at the end of each year and a New Zealand side. I was selected for this more than once, but often when the touch team went on tour, I had other, more important rugby commitments.

Rugby was the main focus for me when I was at Palmy Boys', as it was for most of the kids there. The school has a house system and I was in Gordon House. All of

Making the First XV is an exciting achievement for any schoolboy.

Evading a defender, with good friend Tim Bryson watching my back.

the inter-house sporting fixtures were hard-fought, but none were quite as full on as the rugby games. When there was a footy game coming up, it was always a hot topic among the boys as to who would win those mate-on-mate battles. There were several of these games each year, and I remember them fondly. Murray House, with all the boarders in it, was pretty strong. They won the competition a few years in a row in my early years there.

I was determined to make the school First XV, just as Dad had done. The Palmerston North Boys' High First XV coach when I was there was Rhys Archibald. I knew who he was but I wanted him to notice who *I* was. It wasn't easy, but I was determined and prepared to work hard. I ran and swam and played any and every sport that came my way, and I practised and practised the skills you need to play rugby — not just the fancy stuff, but the basics, too. Never neglect the basics.

I did more than just show up to training. We worked on our skills all the time. In those days, my mates and I used to walk to school with a rugby ball and stand on opposite sides of the street. When there was a break in the traffic, we would kick the ball to each other, or put in a long spiral pass. As we got better, we didn't worry too much about the breaks in traffic. We would kick the ball to each other over the cars. Palmy was a rugby-mad town, so no one seemed to mind too much!

This might have helped me later on with my cross-kicks to tall forwards loitering over by the touchline. You have to be careful, though. A car is about as unforgiving as a pack of forwards. You have to give them just about as much respect as you do the big boys up front.

Besides the basic skills — running, kicking, catching and passing — we did practise some fancy stuff. We were always throwing behind-the-back passes, as they were becoming quite popular in games we saw on TV. Spinning the ball on your finger the way basketball players do was also on our 'to-do list'. It's still on the list for me! But one trick I did manage to master was where the ball is on the ground and you flick it up with your feet from behind your back and into your hands. Give it a try — it looks awesome!

In my first year at PNBHS, I was in the U14A grade. They had three A teams and the rest of the players made up teams under that. From there, I made the U15 Colts and spent a couple of years playing there. By the time I turned 15, I was picked in the First XV. I was younger and smaller than most, which made it tough at times, but playing at that level of footy for three years definitely did me a lot of good.

The pressure did get to me a bit at first. The crowds were pretty overwhelming, with the whole school out there watching you, everyone in uniform and lined up behind a strip of white tape really close to the sideline. Then they would do a chant or the haka. It was a thrill, but it was terrifying at the same time. I have very fond memories of those early days, and I think learning to handle the pressure was an important part of my development.

My strength was running and evading players — all those years getting squashed whenever Kurt caught me in the backyard had seen to that! At first the coach played me at second five-eighth, but the following year, in sixth form (year 12), he moved me to first five-eighth. I had done a bit of goalkicking, but while I was playing in the midfield, I hadn't been taking it too seriously. But now I decided I needed to work on the goalkicking side of my game. There wasn't much structured time for goalkicking practice, so my mates and I did it ourselves. We would keep practising long after everyone had gone home after the usual Tuesday and Thursday training. We'd keep going until it got too dark to see what you were kicking, let alone catching. And whenever we had some spare time, we would go to the park near home and kick some more.

The key to all that practice was making it fun! We kept it interesting by running little competitions amongst ourselves.

We didn't have proper kicking tees, but we made our own. We used a flat marker cone at first. Then we got inventive and put a bit of plastic pipe in the centre of the cone to lift the ball up a bit more. Then we sawed the end off on a slight angle so the ball would sit just the way we liked it. When we forgot to bring the tee, there was always dirt to make a mini-mountain, just like in the olden days.

One of the drills we did after the normal shot-at-goal competition was to place the ball on the tryline near the corner and see who could hit the upright with a placekick. If you got it first, the other guys had to buy you a Coke or a packet of lollies.

As well as goalkicking competitions, we played a lot of 'forceback' and a game we called 'kicking tennis'. In kicking tennis, you had an area on the field (the court) and a halfway mark (the net) and you had to kick

Shaking hands and showing sportsmanship is an important part of all sports.

the ball into the opposition's court. If the ball fell short of the halfway line or out of bounds, it was a point to them. But if you got it to land on the grass without them catching it, it was a point to you.

In sixth form (year 12), we got into the semifinals of the national schoolboy rugby competition and we played against Christchurch Boys'. They had a star-studded line-up, with Colin Slade, Nasi Manu, Tim Bateman and Matt Todd. We lost 16–9 and they went on to win the championship while we fronted up against Wesley College in the play-off for third and fourth. We won the bronze medal, and while that was encouraging, we were disappointed not to have won gold. That feeling that you could have done better can be a good thing. When you are close to something you really want and then fall short, you have to get back on the horse and try even harder the next time. I guess that's why I went back to school for seventh form (year 13).

Those school rugby days were great, with trips away to Wellington, Hawke's Bay and Taranaki. When we travelled, we stayed in motels or were billeted with other families. When visiting teams came to Palmy, we could usually billet a couple of players in the sleep-out at home. That made us a popular billet, as it was much more comfortable to be billeted along with a mate rather than staying alone with a bunch of strangers. I loved the travel side of it. I met so many different people and I certainly saw a bit more of the country. I came to realise what a huge part rugby plays in New Zealand culture.

At high school I was also lucky enough to go on a couple of rugby trips to Australia. Rugby has taken me to a lot of places around the world. Besides Aussie, I've been to South Africa, England, Wales, Ireland, Scotland, Italy, France, Argentina, the USA and Japan. But before my first trip to Australia, I hadn't even been to the South Island. Both of the trips to Australia were to Sydney. It was a bit of an eye-opener. Sydney seemed huge, even compared to Auckland. The grounds were hard and didn't have the lovely lush grass that our pitches have. And rugby over there is a posh game — not like here, where everyone plays it, whether your dad wears white gumboots or toecaps or a three-piece suit. The schools were flash too, and when we got billeted the places we stayed were pretty nice!

I still remember some of the scores from my schoolboy games, as our coach made us a highlights DVD at the end of each season. Now we spend hours going over the footage from our professional games, but when I used to watch those DVDs, I didn't analyse my game or anything like that. Still, I was always interested to see how I went, what I did well and what I could have done better.

I was always impressed playing at the Gully Ground at New Plymouth Boys' High. It has a terraced bank that completely surrounds the rugby pitch and it makes you feel like you are playing in a large hole. Their whole school gets out there to do the haka and it's awesome, even if it's meant to fire up their team, not you.

And speaking of sideline action, we had some bizarre moments, too. Somehow the Palmy boys — it was mostly the boarders, I think — started parking a rotary hoe on the sideline and firing it up when we were playing. The feral roar of the motor revved

I'll always remember my time on the field for Palmerston North Boys' High.

us up as well. More recently, we have had a chainsaw roaring into action when the Chiefs franchise plays. Maybe the Palmy boys were ahead of their time!

My mates from those days are scattered all over New Zealand, and the rest of the world. One of the advantages of being a professional rugby player is that I get to travel, and I catch up with mates from my high-school days when I can. That's part of what makes rugby such a great game: it makes for some lasting friendships.

Lots of guys, like myself, Andre Taylor, Kurt Baker and Ma'afu Fia strived to play for our school First XVs, but it's interesting that a lot of the guys we played with haven't gone on in the game. I sometimes wonder if it's the guys like me who had to struggle to make the First XV who work a little harder and who have the work ethic to play footy at a higher level when they leave school.

If nothing else, First XV rugby taught me the truth of our school motto: *Nihil boni sine labore* — that's Latin for 'nothing is achieved without hard work'. No one is going to hand you what you want on a plate. You have to work hard and go and get it. I think I have followed this motto for most of my life!

I was lucky enough to be asked to captain the First XV in my seventh-form year. It was a huge honour, and I reckon Mum and Dad were pretty proud. It was also a responsibility, and I took it just as seriously as I was taking my playing skills. I asked lots of questions of my parents, Uncle Snow and my coaches. At first-five, part of your role is to make decisions and boss the backline around, so you soon learn to back yourself. But being team captain gives you a whole new set of things to think about.

Towards the end of my time at high school, I set some new goals. Rhys, our coach, was quite good friends with Dave Rennie (Renns), who had just become the Manawatu coach. I began to think that maybe if I worked hard enough, I might catch Renns's eye, too. And if I could get selected for Manawatu, who knows how far I could go?

At last, I had started to dream the dream. Maybe I could become an All Black.

OUT IN THE BIG WIDE WORLD

When I left school, I was presented with the opportunity to do an apprenticeship for six to eight months to become a builder. I was pretty keen, but professional rugby got in first. I had started playing senior club rugby for College Old Boys — we won the Hankins Shield that year — and in 2007 Dave Rennie said he was looking at me. Around this time Manawatu Rugby also sent me to the International Rugby Academy of New Zealand (IRANZ) in Palmerston North. It was a three-week camp which focused on skills and also mental toughness and leadership. There were some great teachers at the academy including a couple of my rugby heros, Grant Fox and Jeff Wilson.

They were great role-models, and I think a lot of what I know about backline play and goalkicking is down to them. Foxy taught me to go about kicking as though I was a machine, doing everything the same way. He also showed me it's important to pause before kicking and relax, to imagine your foot striking the ball and the ball sailing through the posts, and to tell yourself that, when you do things, do them exactly the way you've practised them. That way the pressure never gets to you.

Later that year I got approached by player agent Bruce Sharrock, who talked to me about how rugby as a professional career could work for me. It would mean a whole lot of hard work — there certainly wouldn't be time to carry on with my building apprenticeship — but it sort of appealed. I was offered a development contract and I signed on the dotted line.

To top up my income over the summer, I got a job in a café for six months making coffees. I got quite good at it. I couldn't do all the fancy designs on top of the froth you see in the big city centres, but the customers seemed to think I made a good

Going through my goalkicking process for Manawatu.

cup of the warm black stuff. Occasionally I regret that I didn't get my building qualification, but maybe I will fall back on coffee in years to come!

I was still quite small a year out from school, and Renns thought it would be too much of a risk for me to play for the Turbos straight away. So I turned out for Manawatu B instead. It was my first experience playing with and against adults. I loved it! They were bigger than me — some of them *much* bigger — but I quickly found out I had natural advantages, too: speed off the mark, agility, guile, all the stuff I'd been using for years in schoolboy rugby and before that, to get away from Kurt. I had a good season and by the beginning of the following season, in 2008, I was selected for the Manawatu Turbos.

Not long after that I began suffering pains in my groin, and when I visited the doctor, he sent me off for a scan. It was a bit freaky, but nothing like when Mum and I were sitting in the doctor's room and he told us the results. I heard the words 'testicular cancer' and I just went numb. Then questions started flooding in. How will this affect my rugby? Why me? Crikey — will I die? Here I was — a typical Palmy lad, socialising with friends and having a good time with my footy, fully on the right track to a decent career and who knows, maybe even working my way toward that black jersey — what the heck was this? You never expect to hear you have cancer at the age of 19. And after a few more scans, the news was even worse. The cancer had spread to my lungs. The doctors — there were a few of them on my case by now — reckoned there was a good chance they could beat it, but there was always that other unspoken possibility hanging there in the air.

Life will try and knock you down. The big part is getting back on the horse.

I had to stand up in the dressing room at my next footy training and tell the boys that I was sick and wouldn't be able to see out the season. I didn't tell them what was wrong, but I cried, and I reckon most of them worked out it was pretty serious. Some of them had tears in their eyes, too.

I had surgery, and more than three months of chemotherapy. The chemo was horrible: my hair started falling out, and I had no energy. I felt sick most of the time after each lot of pills and all sorts of strange things happened to my body shape. But in November 2008, Mum and I were sitting in the doctor's room again and he gave me a big grin and told me I was in remission: the cancer had gone away. Mum had taken the news of my cancer amazingly calmly, but the news that I was getting better was too much for her. It was probably the first time I ever saw her cry.

Behind everyone who ever gets to the

Celebrating beating England in the final of the IRB Junior World Championship in 2009.

top level in sport, there's a huge team: family, coaches, mates. In my case, there were also doctors, nurses and radiotherapists. They were all legends. And every team needs a captain: that was Mum, no question.

I bounced right back. As my energy levels returned to normal (and my hair grew back), I knew what I wanted — to enjoy life, play rugby and maybe even play some of it in a black jersey. Just over a year after the shock of the diagnosis, I got to pull on that jersey. In 2009 I was selected for the U20 rugby team to represent New Zealand at the Junior World Championship in Japan. I received the further honour of being made captain. I had been the Palmy Boys' First XV captain, but there was extra pressure here. New Zealand had won the tournament the previous two years, and I didn't want to be in charge of the team that lost it!

To cut a long story short, we had a great tournament and managed to defend our title. For me it was a surprise and an honour after the final to receive the Junior Player of the Year award. Each team puts in a nomination for the player of the tournament and then the International

Rugby Board make their pick.

I was pretty proud, and quite choked up. I gave my playing jersey to Mum, to thank her for everything she had done over the last rollercoaster year, and for all the rest. In all my life I'd never seen her cry, and now that was twice in 18 months!

Looking back, I think the horrible experience of cancer and the treatment made me stronger and more mature as a person far earlier than if it hadn't happened. Now that I've played at the highest level, I can tell you that having the biggest tight forward running straight at you is nothing compared to going through cancer treatment. The whole ordeal made me realise how short life can be. I'm determined to make the most of every opportunity.

I don't really get into rituals before rugby games like a lot of players do, but I do tape my wrist and write a few things on it in the dressing room. What I write changes from game to game. I might write goals I have for a particular game, such as 'composure, control'. I might write my role, or reminders of team goals. It's always something I can quickly look down at and read and remember.

But one thing I write never changes. I always write GEJ0342 on the strapping on my wrist. It's my patient number from hospital. I used to see it so often when I was getting my treatment that it just stuck with me. Seeing it there helps me keep things in perspective.

SUPER DUPER

I got my first Super Rugby contract for the 2010 season, when I got the call-up for the Hurricanes. It was a big moment, but it also seemed just the logical next step in my natural progression. On the one hand, I told myself I had worked hard enough to deserve it. But on the other hand, I couldn't help but realise what it meant. I was rubbing shoulders with All Blacks, and playing the game on the same stage as them. I could measure myself against them, and I might even turn out to be good enough to play not only against them, but with them.

I made my debut against the Brumbies in a pre-season game at Porirua at the end of January 2010, when I came on as a replacement for Willie Ripia just after halftime. I converted Charlie Ngatai's try, the first of two conversions I got in the 26–5 win. I was a bundle of nerves until I lined up my first kick, but I told myself to trust my processes and became calm, because I was doing what I knew. In the following

In my first competition game of Super Rugby for the Hurricanes, against the Blues in February 2010.

My debut test for the All Blacks, against Ireland on 12 June 2010. I'm standing between Zac Guildford (left) and Sam Whitelock.

game, my first real competition game, I came on as a substitute against the Blues at Albany. Unfortunately with my first touch of the ball I was monstered in a tackle from Benson Stanley which left me winded and on all fours. What a welcome! In the next game, against the Western Force, I managed to set up Ma'a Nonu for a try, and while I missed the conversion, I got my first real Super Rugby points when I converted Tyson Keats's try a few minutes later. My professional rugby career was underway.

I was still pretty giddy with the fact I was a Super Rugby professional when, one day later that same year, I was travelling from Napier back to Palmerston North with Grace — my girlfriend then, my wife now, and always amazing — when she took a call on my cellphone.

'You'd better pull over and take this,' she said to me.

It was the phone call all Kiwi rugby kids dream of getting. It was Ted (Sir Graham Henry). He told me I was going to be named in the All Blacks.

It took a little while to sink in. It still hadn't sunk in when Mum rang later to tell me she had heard my name announced in the team. It was very special but I was very surprised.

Walking into the All Blacks' changing room for the first time was intimidating and scary — I couldn't believe it. Although I had played with some of these guys and against all of them, to be in a room with some of the best players in the world, at that moment — yeah, that was an 'oh boy' moment!

In the All Blacks you put on a few jerseys for promotional photos and head shots and so on, but it wasn't until I threw on the real jersey, which has your name embroidered on the hem and the name of the opposition on the sleeve — that I had one of the many 'wow' moments my career has given me. It was a dream come true. Or nearly. It wasn't true until I stepped across the touchline, coming on as a substitute for Dan Carter in the fifty-third minute of the test against Ireland in New Plymouth, that I knew I really was an All Black. I was nervous sitting there on the bench, although I knew the plan was to get me on with time remaining in the second half. The butterflies in my stomach were really going for it when I got the word to stand up to do warm-ups before heading onto the park. But as soon as I was on, looking at the way the Irish defence was set, thinking about the game plan and making my first call, my nerves were gone. I was reasonably pleased with my first 27 minutes of international rugby, and we won handsomely, 66–28. I couldn't wait to get out there again.

My next chance was against Wales a week later, in the last ever test at Dunedin's Carisbrook rugby ground, but replacing Dan Carter late in the game I didn't have much impact. We won again, 42–9, coming home strongly after the Welsh started well.

It looked like I wasn't going to leave my mark on the second test against Wales at Waikato Stadium, either. I came on late, and quietly did my bit. Time was up on the clock, but we were keeping play going. A ruck formed just inside the Welsh 22, in line with the uprights. Piri Weepu cleared to me on the right. I had two players outside me — Cory Jane and Rene Ranger were nice and wide. There were three defenders, but they were close in. I drifted a little wide and considered my options. If I shovelled it wide to Cory, he would likely have drawn one, maybe two tacklers, but the fullback, Lee Byrne, could probably have picked up Rene. I could cut out Cory and give Rene a chance to get on the outside — and that's what the defence was expecting me to do, as they hung off me a bit. I ran up to the line, did a little shimmy to align myself with a small gap in the wall in front of me and then I dropped the ball onto my right boot. Byrne seemed to have read it, and he was moving across to gather. I chased hard, because you always do, but I was pretty sure he had it covered. Then the ball did what rugby balls do and bounced awkwardly for him. He slipped, and I saw it appear beyond him over the tryline. It was a simple matter of running past him, diving on the ball and scoring my first try for the All Blacks. I threw the ball high, stayed on one knee and pointed to the crowd. Piri Weepu kicked the conversion. I dimly heard the ref's whistle go to signal full-time: we had won, and I had scored a try! The surge of adrenaline was amazing. I was still buzzing days later.

I played three more tests in 2010, coming on as a replacement against the Springboks and in the first test against the

Wallabies. I was honoured with my first start in the second test in Sydney, but like the rest of the team, I had a pretty scrappy day. I was subbed at 60 minutes by Colin Slade, playing his first game for the All Blacks. We were lucky to win that one.

By the end of the home season, I hadn't had much game time for the All Blacks. I got another call from Ted, this time to say my name wasn't going to be on the list for the end-of-year tour. This was the first time I had ever been dropped from a rugby team. The highs and lows of team selection! I was disappointed for most of that day, but once I shrugged it off I realised that it had given me the opportunity to keep getting better, rather than being satisfied with where I was. Looking back, I think I was like a lot of players who make a team for the first time. You are happy and delighted just to get there, and you think you've made it. But if I'm honest, I didn't really feel I was fully part of the All Blacks team. Perhaps I didn't even feel as though I deserved to be there. After I was dropped, I realised I badly wanted to get back in, and to feel I belonged there.

Deep down, you usually know what you have to work on. You don't need too many people to tell you. I knew I needed more strength and a few kilos to help with kicking. I got to work on that, and I worked harder than ever on the road, in the pool and in the gym, and on my skills on the field. I was determined to get into the All Black side to contest the 2011 Tri-Nations Trophy, because that would give me the inside running for a spot in the side playing in the Rugby World Cup.

But when the World Cup kicked off, I was just another fan watching the pool games. I had missed out on the Tri-Nations squad — the selectors preferred Colin Slade — and when they read out the names for the World Cup squad itself, I wasn't there either. It had been a long and tough season, more of the rollercoaster I'd been riding since 2009. I decided I deserved a break, and booked a holiday. I was all set to go to Disneyland in Los Angeles before the end of the World Cup. I was going to ride real rollercoasters!

I watched the All Blacks put away Tonga, Japan and France with Dan Carter wearing No. 10 and Colin Slade on the bench. Then on Saturday 1 October, when the All Blacks were supposed to have had a quiet captain's run ahead of their pool game against Canada on the Sunday, I got a phone call. It was Ted, and he sounded worried. Dan had picked up a bad groin injury during kicking practice, and Ted wanted me to come into the squad as back-up to Colin. I felt awful for Dan — we all did — but I was stoked to be back in black. I had a sore knee after spilling off my longboard when my crew and I were cruising Palmerston North just the week before but I knew it wouldn't hold me back.

We prepared for the quarterfinal against Argentina on Sunday 9 October the way we prepare for every test. A week out, we gathered and talked in a pretty low-key way about the challenge ahead. The next day, Wednesday, we began our real build-up, letting the excitement of the occasion seep in. Thursday was a day off. I reckon the coaches let us have a day in the middle of our preparations so that we're fresh and ready to ramp it up the following day, which we did on that Friday. The training is hard and physical, as tough as we can make it without injuring ourselves. Saturday, we had the captain's run, putting the finishing touches on all our preparations. There's a bit of a ritual we do the evening before a game, where we gather for a light stretching session. I went to bed and tried to visualise my role in the game. I was expecting to come on for Colin late in the game, all going well. I tried to imagine what that would be like — the nerves, the excitement of playing in front of a sell-out crowd at Eden Park.

Game day was surreal. Two weeks before, I had been eating chips and skating in Palmy. Now here I was on the team bus, I had my sounds on and as we headed from the Heritage Hotel to Eden Park, we found ourselves passing wave after wave of

Kicking for goal in the 2011 Rugby World Cup quarterfinal against Argentina.

New Zealanders on the footpaths and roads waving flags and chanting, faces painted black and many wearing All Blacks jerseys. They were looking up at us — at me — with the same expression on their faces that I must have worn when I looked up at the posters of my heroes on my bedroom wall back in primary-school days.

It made me realise what all this meant to the country. I knew what it meant to me to be in the team again but it wasn't until that moment that I realised what it meant to everybody else — yet another 'wow' moment.

It all happened very quickly after that. Only 30 or so minutes had gone before Colin didn't get up after taking a kick. The physio went out and word quickly came back that I'd better warm up. It looked like poor Colin had strained his groin, just like Dan Carter. Sure enough, as Colin limped off, I ran on.

I was up for this one. I kicked one conversion in our 33–10 victory, but more importantly, I felt I was combining pretty well with Piri Weepu and the other backs. But with Colin's injury ending his tournament and Mils Muliaina also out with a shoulder injury, the boys were dropping like flies! I had gone from the number three first-five in the country to number one, and there weren't too many behind me.

I got a start against the Wallabies. We finished them off, thanks to a wonderful performance by our forwards and amazing work by Israel Dagg and Cory Jane under high kicks. I had a small personal moment late in the first half when I decided we ought to try to turn some of the territorial dominance we'd had into points. My granddad had once told me to keep practising my drop goals, as you never know when you might need to try to score one. I called it and dropped back into the pocket, Piri spun me a beautiful pass and from the moment I struck it, I knew it was over the posts.

Then I was faced with the prospect of starting in the No. 10 jersey in the World

Watching my drop kick fly over the posts against Australia in the 2011 World Cup semifinal.

The crowd at Eden Park celebrates as the referee blows the final whistle of the Rugby World Cup.

Cup final. We tried to tell each other that this was just another game, but we all knew that it was far more than that. While most people reckon the All Blacks are the number one rugby team in the world, we hadn't been able to win a World Cup since that first one in 1987. People badly wanted it to be this year, this team. The team felt the weight of everyone's hopes, and we wanted it just as badly for ourselves.

It was a thrill to run out onto Eden Park for the final. The noise was unbelievable! Eden Park had been refurbished and its capacity expanded especially for this moment, and there were over 60,000 people in there all yelling their heads off. It was a tough game right from the start, and while we got on the board quite early, we couldn't really dominate the French team. In the thirty-fourth minute, I was tackled and came down awkwardly. My lower leg was trapped while my upper body kept on going, and my knee bent the wrong way. It hurt like anything, and I knew straight away that it was serious. I couldn't play on. Stephen Donald, who had been called into the squad when Colin's injury happened, replaced me and became an All Black legend when he kicked a goal in the fourth minute of the second half that ended up scoring the winning points. We won the World Cup! I was stoked to have played a part in the journey, but bitterly disappointed my knee had stopped me doing more.

Injuries are always frustrating but when you play a contact sport like rugby you have to be aware of the potential risks. Like anything in life, when things set you back, you have to stay positive and put all your efforts into getting back on track as quickly as possible. And to have been part — even a small part — of the winning All Blacks team in the 2011 Rugby World Cup was very, very special.

Since then, I have played a lot more test rugby, partly thanks to Dan Carter's run of injuries, and the break he's had away from international rugby. By the end of 2014 I'd played 37 tests for the All Blacks and had scored 280 points, 25 of them from my five tries. I've also been lucky enough to be part of a Super Rugby champion team not once, but twice.

By the time I played in the World Cup, I already knew I was shifting from the Hurricanes to the Chiefs. It wasn't a decision I took lightly — and I copped a bit of stick for it — but I felt it would do me good to follow Renns (Dave Rennie) to Waikato. I think it's worked out pretty well. We picked up the Super Rugby title in 2012 and 2013, the first Chiefs side to win the competition, and we did it back-to-back!

I think I have always believed in myself, but without the support of my team I couldn't have made it this far. My team isn't just the players that I play with — it's my family, my brothers, my wife, my coaches and those that support me. I've had advice from some of the greats of the game, but some of the greatest advice is from the people you care about. When I dropped that goal at a pivotal moment against the Wallabies in the World Cup semifinal in 2011, I said a quiet thanks to Granddad as I turned to run back to halfway. I still get text messages of encouragement from Mum before kick-off, and critiques from Dad after the final whistle. It's true what they say: rugby is a team game.

Celebrating after beating the Brumbies 27–22 in the 2013 Super Rugby final.

A happy bunch of boys after a hard-fought win against Taranaki at FMG Stadium in Palmerston North.

We won the World Cup!

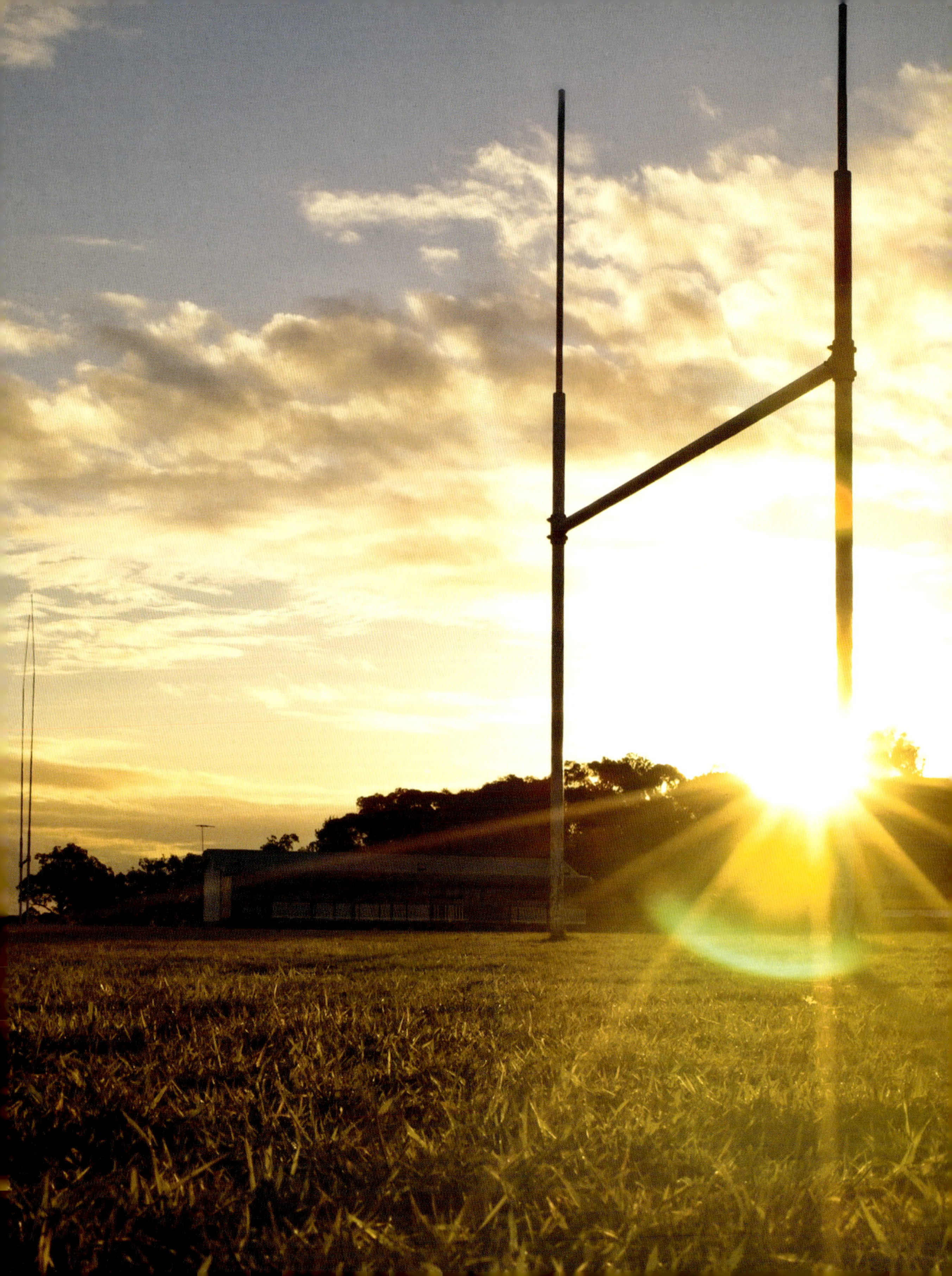

CHAPTER 1
WHAT IS RUGBY?

Rugby Union is a simple game, really. It takes 80 minutes to play at the highest level, divided into two periods of 40 minutes each. It's played on a field that is 100 metres long x 70 metres wide with a set of H-shaped posts at each end, and it's played with an oval ball. You can kick or pass the ball or run with it, but you're not allowed to pass the ball forward to another team member.

Your team is trying to get the ball down the field and place it over the opponent's **goal line** (often called the '**tryline**') to score a **try**, which is worth five points. But wait! There's more! You're then allowed to have a crack at getting a **conversion** by kicking the ball from a position on the field of play in line with where the try was scored. If the ball is kicked through the goal between the posts and over the crossbar, the try is said to have been converted and an extra two points are awarded to your team.

If one team breaks the rules, the referee can award a **penalty** to the other team, and you can score points from these, too. A penalty kick is just like a conversion. You kick it at the goal from the spot where the rules were broken, and if the ball goes between the posts and over the bar, your team gets three points. And if any player **drop kicks** the ball — dropping it onto the ground and kicking it at the moment it lands — through the posts and over the bar during play, that's worth three points to their team, too.

At the end of the game each team's points are totalled up and the one with more points wins the game. No surprises there!

Rugby is supposed to have started when some kids were playing soccer at an English school called Rugby in 1823. They reckon a kid called William Webb Ellis caught the soccer ball in the middle of the game and ran with it in his hands. I guess everyone else chased him to get it back, and enjoyed themselves so much that they decided to turn it into a new game with its own set of rules. But I've also heard that the story about William Webb Ellis was written 50 years after it is meant to have happened, so maybe it's just that: a story. Who knows? But the Rugby World Cup is named after him, and even if the story is just a myth, winning the Webb Ellis Cup in 2011 does make it a bit more special.

Things must have happened pretty fast, though, because in 1845 the very first rugby laws were written by pupils at Rugby School. Those first rules went through quite a few changes over the years, but they were reasonably settled when the first game of rugby was played in New Zealand in 1860. By the early years of the twentieth century, rugby was played all over the world — or, at least, all over the bits of it that the British were involved with. It was a global game. But it wasn't until 1995 that rugby became one of the last global sports to turn professional and start paying its top players.

They reckon rugby is now played in over 110 countries. It's one of the fastest-growing sports on the planet and the Sevens version of it (played between teams of seven players) has been re-introduced to the Olympics in 2016. Funnily enough, the reigning Olympic champions in the 15-a-side game are the United States, who won it in 1924, the last year it was played before being dropped from the list of Olympic sports.

GETTING STARTED

It's easy for kids to get into the wonderful game of rugby in New Zealand. There are differences in the rules and laws of the game that kids play at different ages. These have been worked out to help you gain the skills and understanding of the game while keeping it safe and fun for everyone, no matter what shape, size and age you are. It's called Small Blacks Rugby, and it's a great system. It's designed to give you all the skills you'll need when you start playing proper rugby — with the same rules that adults play — when you reach high school.

One of the things the Small Blacks programme does really well is look after kids who are a bit nervous about the tackling and **contact** side of the game. If you're small like me, or too shy or kind to feel comfortable about knocking over another kid, it sort of eases you into it. In fact, when you first start out, there's no tackling at all. Instead, you 'tackle' someone by ripping a flag from a belt that they wear and holding it up. That's why the first level of rugby is called 'Rippa Rugby'. It's heaps of fun, and lots of boys and girls play it all over the country. There's also touch rugby, where instead of tackling someone to the ground, you just have to touch them. People of all ages play touch rugby: I've been involved in touch rugby for most of my career in tackle rugby.

By the time you get to be a teenager, you start playing something pretty close to the game that adults play. But even at high school, there are a few variations, mostly in the scrum, where everyone is keen to protect the growing bones of the players. These rules make sure that the game isn't only lots of fun, it's also safe!

THE RUGBY PITCH

The rugby field can look pretty big, even to grown-up players — especially when they get tired. In all U10 grades — where the players are under 10 years old — you play on half the full-sized field, usually playing cross-wise with the **sidelines** as the goal lines and the real goal line and 10-metre line as the **touchlines** (the 'out' or sidelines). Once you're older than 10, you'll be playing on a full field.

The full-sized playing field (pitch) can't be bigger than 100 metres long by 70 metres wide. As well as the sidelines, there is an **in-goal area** at each end of the field, between the goal line and the '**dead ball line**'. These vary in size, but aren't supposed to be deeper than 22 metres.

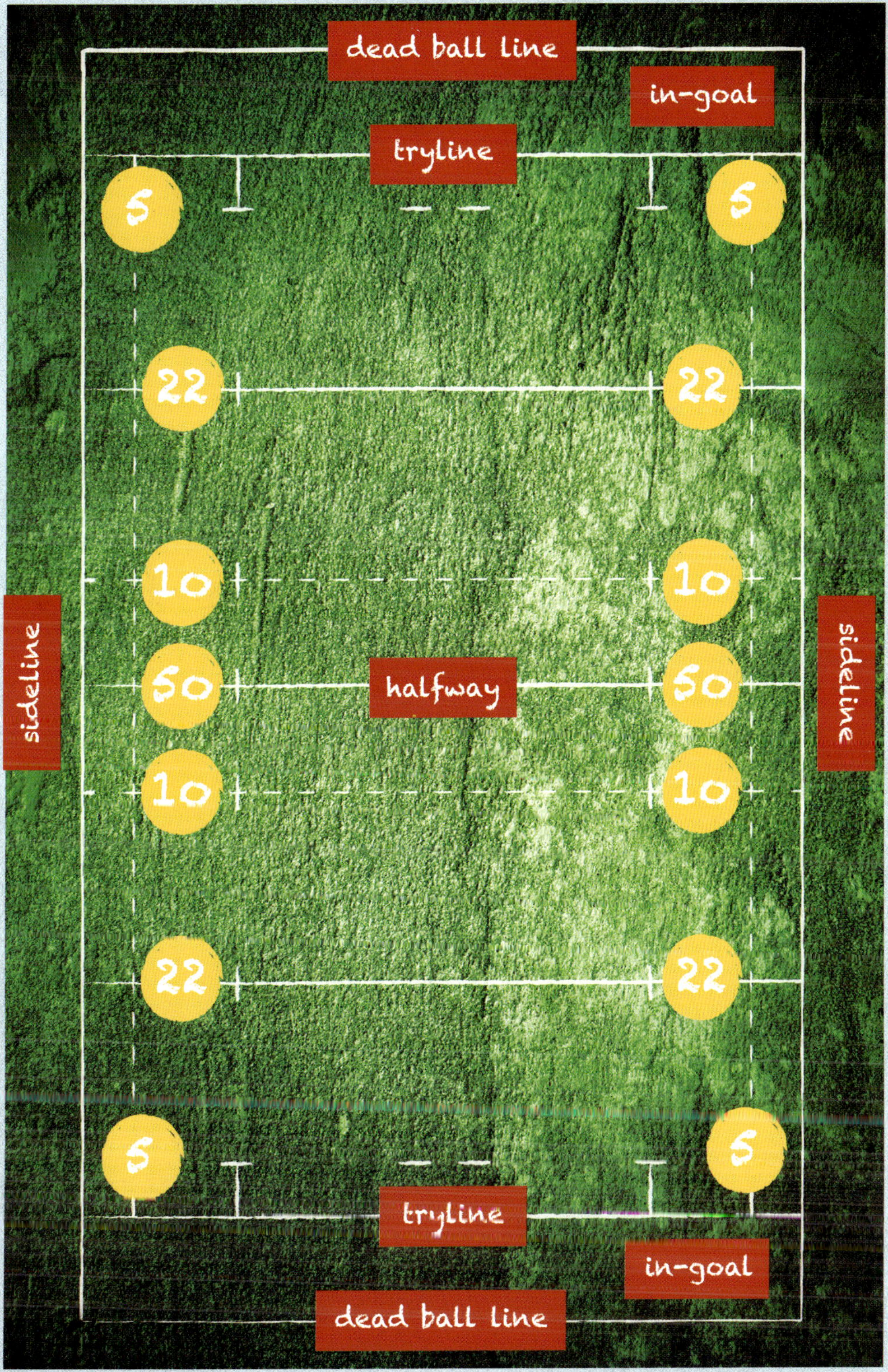
dead ball line
in-goal
tryline
5
5
22
22
10
10
sideline
50
halfway
50
sideline
10
10
22
22
5
5
tryline
in-goal
dead ball line

There are other lines, too:
- The **halfway line** at 50 metres from each tryline.
- A dotted **10-metre line** on each side of the halfway line, which is used to work out whether the kick-off has gone 10 metres. More about this soon.
- A solid **22-metre line** marked 22 metres from each tryline, which is used for several of rugby's rules. We'll get to those, too.
- Two dotted lines set at 5 and 15 metres, marked parallel to each sideline (or touchline). These lines are used mostly to identify the areas for **lineouts** (we'll learn about lineouts soon!).
- A 5-metre line marked parallel to the goal line. These lines are used mostly to identify the areas where a **scrum** or penalty can be set (we're getting to scrums and penalties).

The goalposts are found on the tryline and are 5.6 metres apart, with a crossbar set 3 metres above the ground. The height of the posts varies. In the backblocks of the Manawatu where I live, the posts are often stumpy little things. But in other parts of the world, especially South Africa, I've seen posts that are really tall and make you feel like a midget when you're lining up a kick.

DID YOU KNOW?

The length of a full-sized rugby ball is between 28 and 30cm, its end-to-end circumference is 76 to 79cm, its width between 58 and 62cm. It must weigh between 400 and 440 grams and the air pressure is supposed to be between 0.67 and 0.70kg/square cm.

THE RUGBY BALL

Probably the most important piece of equipment in the game of rugby is the ball. It might seem a bit weird if you're used to the round balls you use for most other sports, but you soon get used to it.

Did you know rugby balls were originally made from all sorts of materials, including pigskin and cowhide? My dad used to play with a leather ball that got really soggy when it was wet, making it both slippery and really heavy. Most rugby balls are made from a waterproof synthetic material these days, and I'm pretty pleased about that.

If you look closely at the four panels the ball is made out of, you'll see the surface has got bumps all over it, as though it's got a bad case of pimples. This helps you grip the ball, even when it's wet and muddy.

Until you're 13, you'll play with a smaller version of the standard rugby ball. U11 Small Blacks play with a size 3 ball, U13 Small Blacks play with a size 4 and everyone else after that plays with a size 5 ball. This makes sense. Imagine if you were a kid and you had to play cricket with a grown-up's bat!

RUGBY GEAR

HEADGEAR is optional. It's been worn for a long time, but modern headgear is nothing like what players used to wear. Some players who wear it think it can protect your head from getting bumped and you getting concussion, but the doctors don't think so. But headgear does give people confidence and it can stop some cuts and bruises and helps keep your ears from getting squashed in the scrum.

RUGBY JERSEYS come in all shapes and sizes. Some of the guys like the body-hugging types because they're harder to grab. Others (mainly props in the scrum) like them a bit baggier. As in most sports, jerseys can be all sorts of different colours and patterns. They can be striped, coloured and have pictures or writing on them. There are jerseys with hoops of colour and jerseys all the same colour, with the most famous jersey being our one, which is solid black.

RUGBY BOOTS aren't used in all grades, as some junior teams (like Rippa Rugby teams) play in bare feet. But as you get older, you'll wear boots with **sprigs** on the soles, so that you can get a grip in muddy conditions. Most boots have six to eight sprigs. Sometimes you can change the sprigs around on your rugby boots, and sometimes the boots have the sprigs moulded into the soles.

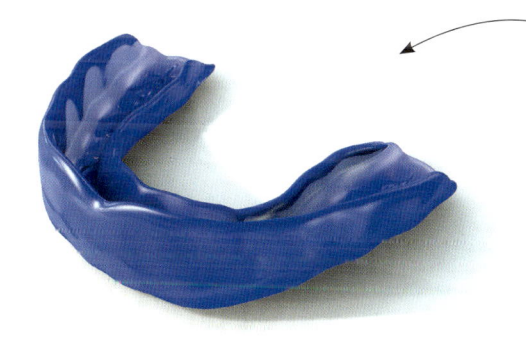

MOUTHGUARDS come in all sorts of sizes, shapes and colours. You have to wear one in games and in any training where body contact is involved. They're not just to protect your teeth: they also help stop a bump under the chin from causing concussion.

KICKING TEES are a tool I use when I'm taking a **shot at goal**. Players used to make a hole in the field with their heel to place the point of the ball in, and later in rugby's history they made mini-mountains out of sand which was brought onto the field in a bucket. How weird is that? There are all sorts of different kicking tees available, and it's a matter of finding one that suits your style of kicking.

MY GEAR BAG

rugby jersey

personalised bag
(keep training hard and
one day you might earn
one of these!)

shower in a can!

trackpants

mouthguards

undershorts

drawstring bag

CRUDEN

protein shake

marker I use to write down things I need to focus on

(clean!) rugby boots

boot bag

sports drink

shorts

socks

ice pack

good-luck bracelet

earphones

spare tees

POSITIONS

As I mentioned earlier, U7 Small Blacks play a seven-a-side version of the game called Rippa Rugby, where there is no tackling — you rip flags off your opponents instead of tackling. It's really simple, and there aren't any positions. The emphasis is on having fun!

When you move into the U10 grades (between the ages of 7 and 9), you have 10 players in a team, five in the **forwards** and five players in the **backs** (we'll talk about forwards and backs very shortly).

From the moment Small Blacks move into the U11 grade, it's 15-a-side all the way, and the positions you play are the same as the positions we play in the All Blacks.

A rugby team is simply separated into forwards and backs.

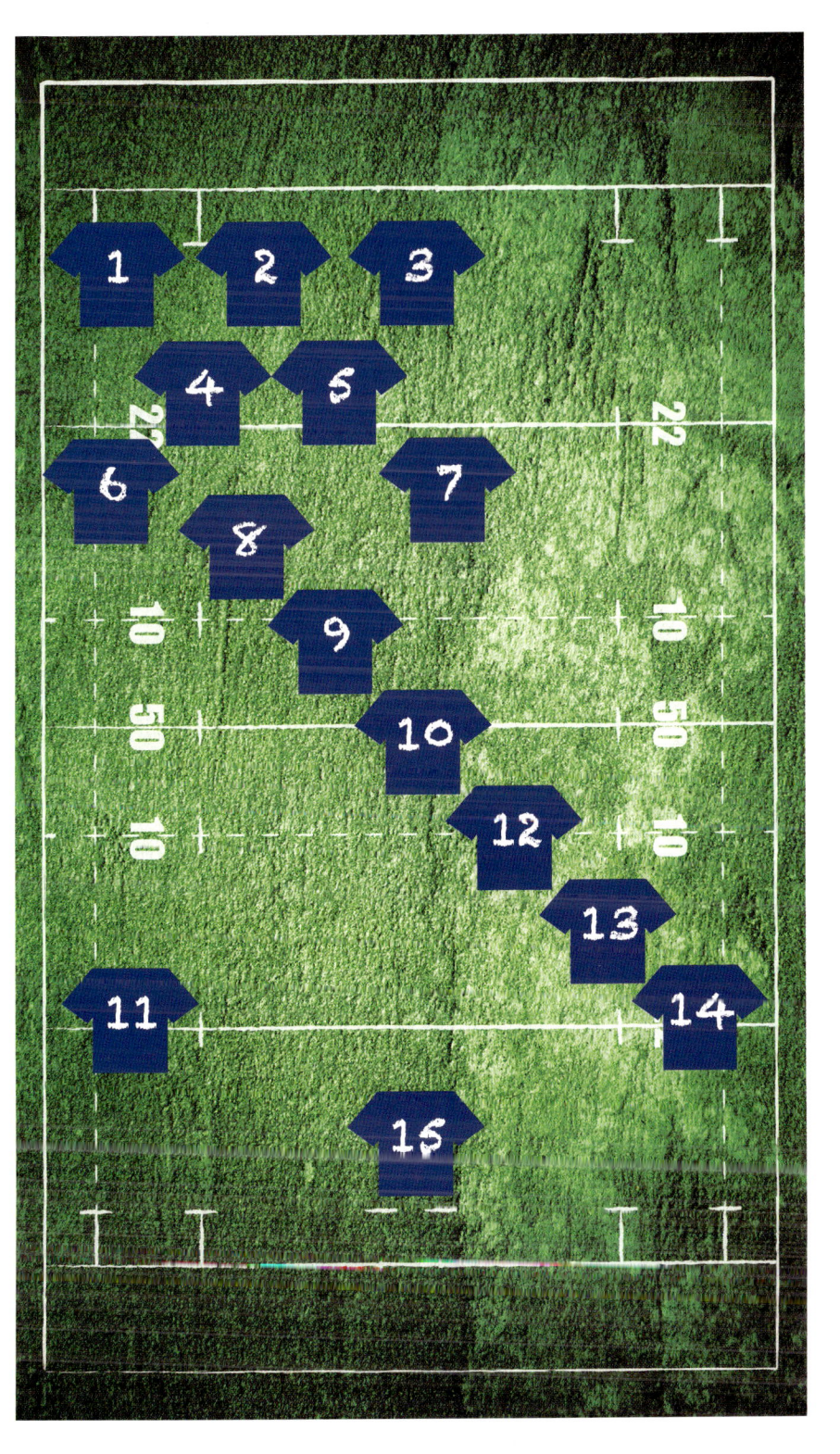

FORWARDS

1: Loosehead prop

2: Hooker

3: Tighthead prop

4: Lock

5: Lock

6: Blindside flanker

7: Openside flanker

8: No. 8

BACKS

9: Halfback

10: First five-eighth

11: Left wing

12: Second five-eighth

13: Centre

14: Right wing

15: Fullback

FUN FACT

There have been over 1000 All Blacks since 1884. A prop called James Allan was All Black number one. I'm All Black number 1105!

A South African scrum against the All Blacks in August 2010.

Forwards (numbered 1 to 8) are often the bigger, sturdier players on the team and their main job is to win the ball in the scrums or lineouts, or when a player carrying the ball has been tackled.

The backs (numbered 9 to 15) are often smaller, quicker and more agile than the forwards. After the forwards have won the ball, it's often delivered to the backs to use it, ideally to score points!

The formation of the players (that is, where they stand) when a scrum is happening is designed to make sure that when players are passing the ball to one another, it goes backwards rather than forwards between team-mates. Remember, passing the ball forward is against the rules.

WHAT ARE THE POSITIONS?

PROPS

BIG AND BEEFY. Props are often the muscle men of the team. You'll sometimes hear them called 'the fatties', usually by TV commentators safely tucked up in the commentary box well away from where the Franks brothers or Tony Woodcock might hear them. Like all props, these guys are big and tough, with the strongest legs and upper bodies in the team. Props have to be strong in the upper body, as their job in the scrum is to prop up the hooker (which is where their name comes from). The loosehead prop packs down in the scrum on the left-hand side of the hooker, and the tighthead prop packs down in the scrum on the right-hand side. They're also often called upon to lift other players receiving kick-offs or in the lineout. Technically, they have to have a low body position (that is, they have to bend in the middle and get their heads as low as their hips) for the work they do in scrums, rucks and mauls. This isn't easy! They provide a lot of the pushing power in the scrum (in grades where pushing is allowed), which is why they've got such big, strong legs. But even though they're so big

Tighthead prop Charlie Faumuina (left), hooker Keven Mealamu (centre) and loosehead prop Wyatt Crockett pack down the front row of an All Blacks scrum against England.

and strong, they're often fast and agile, too. Some of the props I've played with have been faster over the first 10 metres than the backs, which is another reason why you want to be careful calling them fatties!

HOOKER

WIRY AND WILY. The hooker is the player that packs down in the middle of the front row of the scrum, which might explain why old hookers have ears that look like your baby sister made them out of Play-Doh. They use their feet to 'hook' the ball back on their own side when it is rolled into the scrum. It's also usually the hooker's job to throw in the ball at lineouts (though my dad remembers when it was a winger's job to do the throw-ins). The hooker is expected to do a lot of the work in rucks and mauls (we'll come to those shortly), and to be one of the best tacklers in the side. You can see why hookers have to be exceptionally fit. If you look at players like Dane Coles and Keven Mealamu, you'll see that they aren't the biggest guys on the field. Their job is to be just about the most skilful.

LOCKS

TALL AND RANGY. Locks are often lanky, because one of their main jobs is to reach

Tall lock Brodie Retallick wins a lineout, playing against Australia.

high above everyone else in the lineouts to catch the ball. Even though you're allowed to lift players in the lineouts these days, it helps if your locks are good jumpers. The springiness, the catching ability and the timing that make a good lock also make your locks a natural choice when it comes to deciding who catches the kick-off for your team.

Being really tall is obviously an advantage if you're a lock, but you still have to be strong and durable, because locks are involved in the scrums, rucks and mauls. Like props, locks need to be able to achieve a nice, low body position when they're pushing in the scrums just behind the hooker, and if this is hard for props, it's even harder for the 'tall timber'! Like hookers, locks can often get 'cauliflower ears' after years of having their ears squashed in the scrums and banged about in the rucks and mauls. Lots of old locks are proud of their ears, but don't let them hear you making jokes about them!

While the locks' first job is in the scrums and lineouts, it helps if you are good around the field, too. You need to cover big distances quickly so that you can be one of the first players to help when one of your team-mates has been tackled to the ground so it's useful if you're fast and fit, too.

One of the best locks I've played with is Brodie Retallick, who plays with me in

Flankers Justin Collins (left) and Jerome Kaino (right) make a tackle for the Blues against the Cheetahs in 2008.

the Chiefs and the All Blacks. Brodie is a man I really look up to — he's 2.04m tall, after all — and at 120kg he weighs nearly as much as one and a half Aaron Crudens! He is good at tackling, lineouts and running with the ball because of his strength, determination and skills.

FLANKERS

TOUGH AND ATHLETIC. Your flankers are two of what are called the **loose forwards**. Like the props, hooker and locks (the **tight five**), flankers are part of the scrum. But unlike the tight forwards, the 'loosies' can break away quickly as soon as the ball comes out and get involved in play.

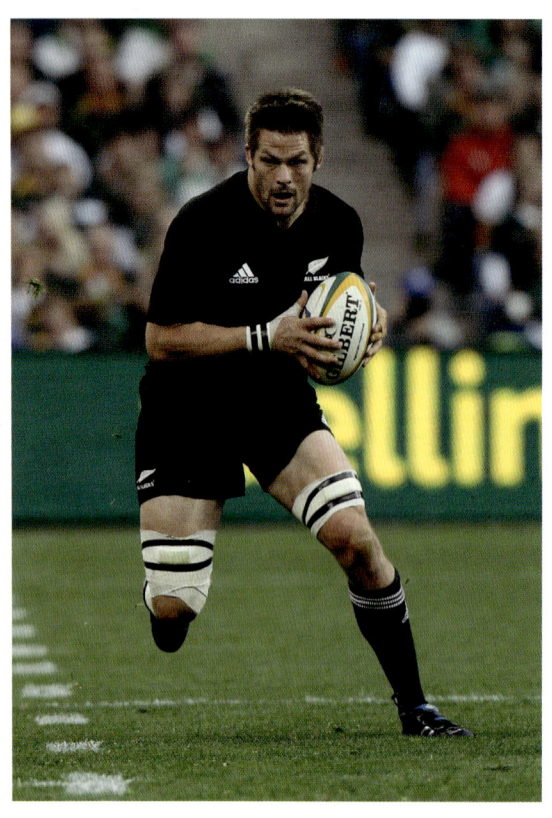

Flanker Richie McCaw runs with the ball in two hands while playing against South Africa in 2010.

This means that the flankers are important in defence, as they can form a screen to tackle players who are trying to run close to the scrum.

In New Zealand and many other parts of the world, the flanker on the side of the scrum closest to the sideline is called the 'blindside' flanker. Their main role is to defend the narrow space between the scrum and the sideline, and so they are usually quite big, strong players. The 'openside' flanker is on the side of the scrum furthest from the sideline, where they can quickly break away from the scrum and try to catch the halfback or the first-five (that's me!) before the ball can get out into the backs. This requires speed, and openside flankers are usually reasonably small and fast.

The explosive acceleration and physical presence that make for a good flanker mean that flankers are usually among the most athletic players in a team. They are useful as jumpers in the lineout and as runners in support of someone carrying the ball, ready to receive a pass. That means they're often the first there when play breaks down (that means their team-mate who is carrying the ball is tackled). So one of the main jobs of the faster flanker, the openside, is to be very good at winning the ball when it is on the ground after a tackle.

Flankers work really hard because they're into everything. One minute they're driving the ball-carrier backwards. The next they are up on their feet trying to get their hands on the ball. A few seconds later they're **out wide** in support of a speedy back running the ball, or they're there — head down, driving — in the middle of a maul. That's why flankers are said to be the players in the team with the biggest engine. This means they can just keep going, and I guess there isn't a better example than Richie McCaw.

Not only can Richie tackle hard and win the ball for our team off the ground but he is a good runner with the ball, he's always there in support and has good hands (catching skills). This guy would be one of the best I have ever played with and against, and he just doesn't stop.

Part of what makes Richie McCaw and other flankers so good is that they must be able to understand how a game of rugby is structured, so that they can run the right lines both in attack and defence. They have to have great vision and awareness, and they have to understand the laws of the game, especially the rules around the tackle, rucks and mauls.

NUMBER EIGHT

RANGY AND RUGGED. The number eight is the third loose forward, and has a lot of the same qualities as the flankers. But they also have their own job to do. They're an important part of the pushing power in a scrum, so they're usually big and strong. But because they have to break away from the scrum very quickly, they're also fast and athletic. They pack down at the back of the scrum, usually binding onto the two locks, and they control the way the ball comes out of the scrum. They can hold it at the back of the scrum with their toe until the halfback is ready for it, or they can pick it up and run with it. Number eights love doing this, and if my halfback or flanker misses him, guess who gets to tackle him! Me! Luckily,

I learned at an early age that little guys like me can cut down big guys by tackling them low — or at least we can grab them and hang on until help arrives.

Kieran Read is one of the world's best number eights and when I'm playing with

Number eight Kieran Read tries to break through the Sharks' defence in July 2014.

Halfback Aaron Smith passes the ball during a Super Rugby match between the Highlanders and the Western Force in 2011.

him, I'm certainly glad he is on my side. He is very fit, extremely skilful and one of the best tacklers in the side.

HALFBACK

SMALL AND NUGGETY. I guess I could have been a halfback because these players are often the smallest in the side. They're also often the cheekiest, because even though they're small, they have to boss the big forwards around. The halfback is the link between the forwards and the backs, and they are expected to have a fast, long and accurate pass on both sides. At scrum time, the halfback rolls the ball into the tunnel between the two front rows and then scoots around the back to collect it when it comes out. At the lineout, they're there ready to receive the ball from the forwards and pass it to the backs, and at rucks and mauls, they're right there, barking orders to the forwards and spinning the ball out to the backs when the moment is right. Like the openside flanker, you need your halfback to be exactly where the ball is at all times. They cover a lot of ground, and they have to be fit!

As well as passing, halfbacks need to be able to kick with their left and their right boots, as they're often the best-placed person to kick the ball downfield when they are close to their own tryline. And because they're often in the way when the opposition halfback and loosies break from the scrum with the ball in hand, they need to be good tacklers, too.

It's important if you play in this position that you establish a good relationship with your first-five, and I think this is true off the field as well. Aaron Smith, the All Blacks halfback, is a good buddy of mine and we have played a lot together, especially in our Palmerston North club and Manawatu representative rugby days. He has a very fast pass and an accurate boot (kick). But he also has great vision, runs the ball well and is one of the bravest players you'll ever see.

FIRST FIVE-EIGHTHS

TRICKY AND SKILFUL (if I do say so myself!). As a first five, I have to make the tactical decisions during the game, on whether to kick the ball to gain space and a tactical advantage, move it to my outside backs, or to run with the ball myself. The first five's special skill is kicking, but running is important, too. If you are willing and able to run with the ball, it keeps the opposition guessing about what you're going to do. So not only do I need to be able to pass, kick and run, I also have to have the judgement

DID YOU KNOW?

In New Zealand we call number 10 the first five-eighth, but in Europe they call it 'fly-half'.

to know which one is best in each situation. And because (like most first-fives) I'm quite small, I often find opposition loosies lining me up as the weak point in our defensive line. I've had to become a good tackler, too. I'm also the team's goalkicker, which has its own set of pressures.

First-five is a position of responsibility. You can't hide when you're in that position, so if you have a weakness or two in your game, you might want to look at playing a different position or work hard to improve your weaknesses. You have to understand the game tactically and be able to make quick decisions, and you have to be able to communicate effectively, because it helps if everyone in your team knows what you're planning to do. At first, I was a fairly quiet first-five but I worked on this part of my game and now I feel confident both in making my decisions and in telling everyone else what I need them to do to make the plan work.

There's heaps of pressure on a first-five's shoulders, but I wouldn't have it any other way. You have to believe in yourself. I remember a game for the All Blacks against England where I decided to take a tap penalty rather than kick for touch or take a shot at goal. It took the opposition by surprise, we scored and everyone reckoned it was a great decision! If I hadn't been confident in my play then it wouldn't have been a success and everyone would have been telling me what a stupid decision it was. Work hard on your skills, and believe in yourself.

SECOND FIVE-EIGHTHS

CRASH AND BANG. The second-five is a '**midfielder**', positioned outside me and inside the centre in a standard backline formation. I played second-five through most of my primary and secondary school years. At second-five, you are expected to pass if there is enough space for your outside backs to do something with it, but you often need to hang onto the ball and take the tackle, to 'straighten the attack' by running straight down the field, or to make the ball available through a ruck or maul so that it can be passed out to the backs again. I preferred to try to beat my opposite number rather than to be tackled, by sidestepping, feinting or swerving. Maybe I should have passed more!

Second-fives have to have an accurate pass and be very strong tacklers. Ma'a Nonu has been one of the best All Black second-fives in recent years. He's much bigger than me (most second-fives are, which is one of the reasons I moved into first-five), and he possesses all the great qualities of a midfielder. He has great vision — if there's a gap, he'll find it — but he's also really strong, so that if he's tackled he can keep moving forward and often even get a pass away. This is one of his best skills. And, needless to say, you know you've been tackled if it's Ma'a who's hit you!

If you want to play in this position, you need to be fit and strong and have a good feel for attacking play.

CENTRE

STRONG AND SELFLESS. The other midfielder is the centre, who is positioned outside the second-five and inside the wing in a standard backline formation. The centre's main role is to provide time and

Second-five Ma'a Nonu attempts to fend off the tackle of South African Victor Matfield with centre Conrad Smith (right) outside in support.

space for the players outside them, and there isn't a better player in this position than Conrad Smith.

Conrad's first 'black' New Zealand jersey was an All Blacks one, which just goes to show that you don't need to be in all the representative teams along the way to achieve the highest of honours.

Centres need to be good tacklers as well as good passers. They need to support the ball-carrier all the time, and this is where Conrad excels. He is always 'popping up' in the game where you least expect him to appear, which happens because he just seems to know how play is going to unfold and because he also happens to be the fittest player in the All Blacks.

If you want to be a good centre, get fit, make sure you can pass well from both sides and be sure to communicate really well with the players outside and inside you, both on attack and defence.

WINGERS

FAST AND FANCY. The players at the end of the chain of backline players are the wings, and they're usually the fastest players in the team. You can tell when a backline is working well, because the wings score most of the tries. That's why they get called 'finishers' — because they finish backline movements by scoring. It's that kind of fancy-pants, glory-hogging stuff that sees the fatties call the wings 'pretty boys'. But as Julian Savea or Cory Jane will tell you, wings actually do a lot more than just run fast and score tries. Along with the fullback, they're part of what gets called 'the **back three**', and their role is often interchangeable with the fullback's. Wings have to tackle well, be safe under the high ball — that is, be able to catch high kicks — and it's a good idea to be able to kick well from close to their own tryline, too.

Traditionally, the idea was that the forwards and the rest of the backs created space for the speedy player to have a clear run at the tryline. But as teams have improved their defence, different ideas about how the wing should play have seen different physical types taking the wing role. Some are still tall and lean and fast. Others are slower, but better at deception — at swerving and dodging. Some of the best — like Cory Jane — are good at both. Cory is 1.83m tall and weighs 85kg, he's really fast and light on his feet, and has the best **fend** (the ability to push a tackler away with his hand) in the game. But you get other wings who are all about size and power, like Julian Savea. Julian is fast as well, but he's 1.92m tall, weighs 108kg and is just as good at squashing tacklers as running around them.

Winger Julian Savea makes a break for the Hurricanes in 2013.

FULLBACK

BRAVE AND SKILLED. The fullback is often called the 'last line of defence' and is traditionally positioned behind all their team-mates. It's funny, though, because in the normal run of things, your fullback should be doing far less tackling than most of the other positions. The trouble is, by the time your fullback is lining up a tackle, the opposition have got past everyone else,

Fullback Israel Dagg jumps to catch a high ball.

so the fullback's tackle is often the most important one — or the one everyone notices if they miss!

As well as the obvious need to be a very good tackler, the fullback needs to be very good at catching high balls and must have a good kick themselves. Good fullbacks 'read' a game well — they can tell where the ball is going to go, and they get into position before it happens. On attack, the fullback often runs the ball back at the opposition after catching a high kick or a kick through, and is often used as an extra man in the backline.

Israel Dagg shows a lot of courage at fullback. He's not only very good at catching the ball when it comes to him, but his skill in running forward and leaping high into the air to take a ball falling where there are lots of other players waiting has become legendary. But Izzy's also one of the most elusive and deceptive runners in the game — a quality that makes him a very good fullback. Vision, an eye for the right line to run and timing — knowing just when to run into the backline — are all useful skills to have in your toolbox.

Wingers these days do so much of the fullback's role that they can often play very well in that position if they have to. Look at Cory Jane and Ben Smith. And first-fives

DID YOU KNOW?

Players wear numbers on their jersey numbered from 1 to 15, starting with the loosehead prop. However, when rugby first started in New Zealand, the fullback wore number 1 and the lock wore number 15.

can often slot into the fullback position because they have a lot of skills in common. Beauden Barrett and Colin Slade are both first-fives, but both have played very effectively at fullback as well.

DID YOU KNOW?

The first international rugby test match was between Scotland and England in 1871. New Zealand's first test match was against Australia in 1903.

BASIC SKILLS

To play rugby at any level, you need to work hard on the skills needed to play your position, but you should also work on basic skills that will help you succeed in the game, both individually and as a member of a team. Some of it is physical, of course, but a lot of what makes rugby players good — even great — is all in your head, and it starts with your attitude.

As in every sport, winning isn't the most important thing when you're learning rugby. It's enough to want to play well and to have the will to succeed. That means trying your hardest, both in your games and also at training. Set high standards for yourself: be honest with yourself and take responsibility for what you do. Get in the habit of working hard. If you and your team-mates all do this, then you'll probably succeed. But if you're beaten by a better team, don't try to make excuses. Be a good sport and congratulate them on playing well. Above all, enjoy yourself!

Books about rugby sometimes leave out the need for players to understand the game. This is pretty basic if you want to be a good rugby player. Understand the rules, understand how to play the game, understand what a game plan is and understand what rugby can do for you! Some of it you will find out for yourself, but don't be afraid to ask other players what they know — older players like your dad or your brothers or your coach, and younger players like your team-mates, and even your opponents!

MY THOUGHTS ON THE GAME

AT THE START OF THE MATCH

If my team wins the toss, the responsibility of starting the game usually falls to me. The match always starts with a kick-off, then the receiving team collects the ball and generally tries to move it downfield to score.

The team without the ball (the defending team) tries to stop this, usually by tackling, and tries as hard as it can to get the ball. Believe me, rugby is a lot more fun when you have the ball!

TACKLING

You can only **tackle** the ball-carrier (the person who has the ball in their hands). Tackling can be done by anyone, no matter what position they are playing.

In a lot of other games — like rugby league and American football, for example — everything stops when a tackle has been completed. Rugby is different. When you're tackled, you can play the ball immediately. That means you can throw it to a team-mate (backwards, of course) or, if you're close enough to the goal line, you can reach out and place it over for a try. But if you can't play the ball immediately, you have to release it, that is, let go of it. Ideally, you should release the ball in such a way that your team stands a good chance of getting their hands on it. You're allowed to place the ball behind you. Your team-mates should be there to pick it up, and the game carries on. We'll talk about this skill in more detail later on.

RUGBY'S KEY ELEMENTS

SCRUM

The scrum is a formation used to restart play. The referee may order a scrum for a number of reasons: the most common reasons are a **knock-on**, where the ball has been knocked forwards, or a **forward pass**. The scrum is a unique part of our game and when it's executed well it's poetry in motion — 16 bodies all doing their best to out-muscle the opposition, both physically and mentally. When your scrum is on top, you tend to be on top in the game itself, so coaches put a lot of importance on getting scrummaging right.

Eight forwards on each side **bind** into a unit and the two units come together, with the front rows interlocking to leave a tunnel between them.

If the other team made the mistake that led to the scrum, your halfback gets to 'feed' the ball into the tunnel midway between the front rows and the hookers try to rake it back with their heels. It should then roll between the locks and arrive at the back of the scrum, where the No. 8 controls it with their toe.

The scrum ends when the ball comes out of the tunnel at the back of the scrum, no matter who picks it up.

LINEOUT

The lineout is how rugby players put the ball back into play after it has gone out (that is, over the sidelines or '**in touch**', as it's often called).

The rules say that there must be at least two players, one from each team, standing where the ball went out of play for a lineout to be formed. There's actually no restriction on the number of players from each team who can be involved in the lineout. The team that is throwing the ball in decides how many will participate, and the opposition has to match that number. It's usually just the forwards who are involved.

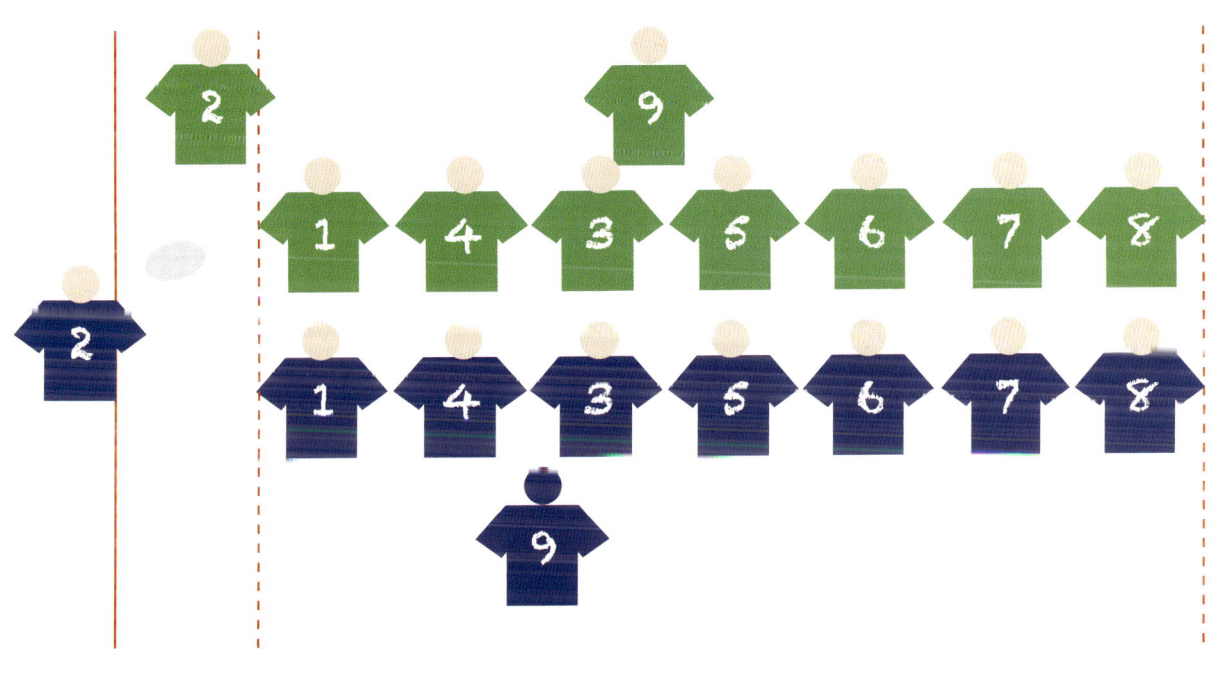

The two sets of forwards line up alongside each other a metre apart facing the sideline; if the other team put the ball into touch, then one of your players (usually the hooker) throws the ball between the two lines. The ball is meant to go straight down the middle, and it has to travel 5 metres from the touchline for it to be a legal throw. Players from both teams can jump or be lifted to get the ball. Rugby teams usually have a few different plans for lineouts, and they use a code so that everyone from your team who is involved in the lineout knows where the ball is going and who is meant to receive it.

Sometimes when the ball goes over the touchline, you don't need to have a lineout. If no one from the opposition has arrived to form a lineout, then the player who has collected the ball can make a quick throw-in to a team-mate (or even to themselves). This speeds up the game and can sometimes catch the opposition napping.

Lifting in the lineout was against the law for most of the history of rugby. They allow it these days at all levels after you get to high school. It involves 'support players' gripping your legs or your shorts as you jump, to give you a boost — and making sure you come down safely, too.

MAUL

If you're carrying the ball and someone from the other team grabs you and hangs on in such a way that you're both still on your feet, then it's a 'maul'. Other players from both teams can join the fun by 'binding' on with one or both arms. They have to join from behind their team-mates who are involved in the maul. If the maul stops moving and the ball doesn't come out within five seconds, play will restart with a scrum. The team who didn't have the ball when the maul formed gets to feed the ball into the scrum.

RUCK

Suppose you get tackled. You're lying on the ground and a couple of other players — one of your team-mates and one of your opponents — arrive. They can form a 'ruck' over the ball by binding onto one another. Once a ruck is formed, no one can touch the ball with their hands until it is out; you have to use your feet to rake (ruck) it back on your side. Other players can join in, as long as they arrive from behind their team-mates who are involved in the ruck.

My dad remembers what he calls 'the good old days', when you were allowed to rake the sprigs of your boots over players who were lying on the ball in the ruck. Funnily enough (Dad reckons), the ball used to come out of the ruck pretty quickly in those days. But there's a rule against rucking people on purpose in the modern game. Dad and his mates grumble about that, but I reckon it's a pretty good rule!

DID YOU KNOW?

There are about 158 rucks in the average game of professional rugby, but only 12 mauls.

KICK-OFF / RESTARTS / DROP-OUT

FUN FACT

The Ranfurly Shield is one of the oldest rugby trophies in the world. The first Shield match took place in 1904 when Wellington beat Auckland.

All of these kicks must be a **drop kick** (when you drop the ball from your hands and kick it just as it touches the ground). **Restarts** happen at the beginning of each half of the game, and after points have been scored. These always happen from the exact centre of the pitch.

There's another kind of restart, which is taken from anywhere along the 22-metre line. This happens after the ball has '**gone dead**' (that is, it has rolled over the dead ball line, or been forced over the goal line by the defending team). A '**22 drop-out**' can be as long or as short as you want: the ball only needs to cross the 22-metre line.

At both kinds of restart, everyone from the kicking team must be behind the kicker.

REFEREE SIGNALS

When a referee blows their whistle, they always make a signal to explain why they have stopped play. Players do lots of their own signalling when the referee blows his whistle: everyone has their own opinion. But the only signal that counts is the opinion of the person with the whistle. Here are some of the most common signals.

Advantage

Free kick

Try or penalty try

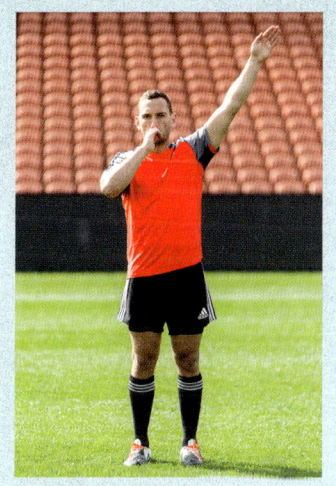

Penalty kick	Throw forward/Forward pass	Knock-on
Not releasing ball immediately in the tackle	Tackler not releasing tackled player	Handling ball in ruck or scrum
Throw in a lineout not straight	High tackle (foul play)	Scrum awarded

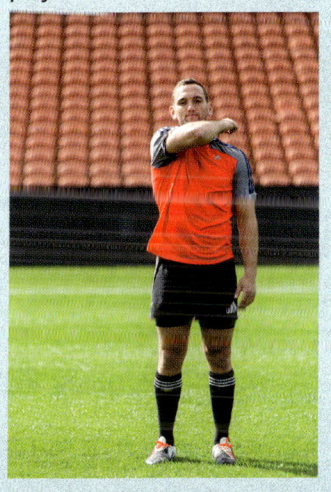

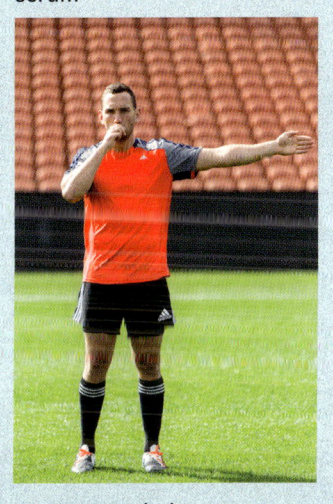

RUGBY WORLD CUP: DID YOU KNOW?

- The William Webb Ellis Cup is made of silver coated with gold.
- The tournament is held every four years.
- The highest number of points scored in a RWC game was 162, in the match where New Zealand beat Japan 145–17 in 1995.
- The same whistle is used to start the opening game of every Rugby World Cup.

1987

- The All Blacks won the 1987 Rugby World Cup tournament with a 29–9 victory over France in the final at Eden Park.
- David Kirk, who had taken over the leadership of the side when original captain Andy Dalton was ruled out (he was injured at training before the tournament even started), was the first rugby player to hold aloft the William Webb Ellis Cup.
- The All Blacks scored three tries in the final, on their way to the 1987 title.

GLOSSARY

10-METRE LINE a dotted line 10 metres on each side of the halfway line, which is used to work out whether the kick-off has gone 10 metres

22 DROP-OUT a method of restarting play when the ball has gone over a team's dead ball line

22-METRE LINE a solid line marked 22 metres from each tryline, that marks a team's defensive zone

BACK THREE the two wingers and the fullback

BACKS players wearing jersey numbers 9 through 15, forming the backline

BIND using your arms and hands to hold another player in the scrum, ruck or maul

BREAKDOWN area where there has been a tackle or the ball is loose on the ground

CONTACT where two players come together, usually in a tackle situation

CONVERSION kicking the ball from a position on the field of play in line with where the try was scored. If kicked between the posts and over the crossbar, the team scores 2 points

DEAD BALL LINE line at the end of the rugby pitch

DROP KICK kicking the ball after it has been dropped from the player's hands and strikes the ground. Worth 3 points if it goes over the crossbar between the goalposts

FEND pushing a tackler away with your hand

FORWARD PASS a pass thrown by a player in front of themselves. This is against the rules

FORWARDS players wearing jerseys 1 through 8

GOAL LINE (TRYLINE) the line at each end of the field, with the goalposts on it

GONE DEAD when the ball has rolled over the dead ball line, or been forced over the goal line by the defending team

HALFWAY LINE the line across the middle of the field

IN TOUCH when the ball is outside the playing area on the field

IN-GOAL AREA the area between the goal line and dead ball line

KNOCK-ON when the ball has bounced forward after touching a player's hands, arms, or upper body

LINEOUT the two teams line up perpendicular to the sideline to contest a ball being thrown back onto the field

LOOSE FORWARDS (LOOSIES) the two flankers and the No. 8

MIDFIELDER a general name used to describe the second-five and centre

OUT WIDE playing near the edge of the field

PENALTY KICK a kick awarded to one team after a serious breaking of the rules by the other team

PHASE a period of open play (that is, between rucks and mauls) where the ball remains in the possession of one team

RESTART to start play again

SCRUM a formation used to restart play. Eight forwards from each team bind into a unit then join together to contest the ball

SHOT AT GOAL to try to kick the ball between the goalposts

SIDELINES (TOUCHLINES) lines indicating the two long edges of the pitch

SPRIGS metal bumps on the bottom of rugby boots to help players grip the ground

TACKLE the act of bringing a player to the ground while they are carrying the ball

TIGHT FIVE (TIGHT FORWARDS) two locks, two props and the hooker

TRY a try is scored when the ball is touched down in the other team's in-goal area, and is worth 5 points

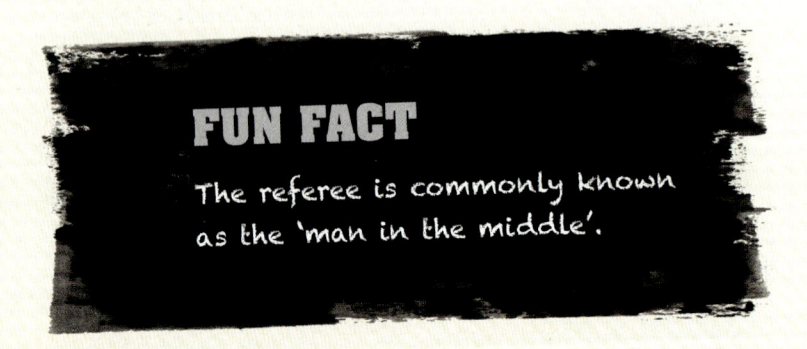

FUN FACT

The referee is commonly known as the 'man in the middle'.

AARON'S SKILL TIPS

The next chapters talk about the key skills you need to become a really good rugby player. There are some general things to remember, too.

For you to perform well at any level of rugby, you must develop a number of skills. Becoming a skilled player begins with mastering the basics. Over time, your skills and techniques will progress, allowing you to apply those skills in increasingly competitive situations.

Here are five principles for your skill development:

- Make sure you keep the fun in rugby.
- Train and compete at a level that's right for your age.
- Develop a foundation of solid techniques — passing, catching, kicking — and continually work on these.
- Keep competition in perspective. Look at each game as a step on a journey and don't worry too much about the result.
- Don't set your heart on playing any particular position, because sometimes you end up playing — and being best at — a position you never dreamed you'd play.

Offloading the ball against Ireland in June 2012.

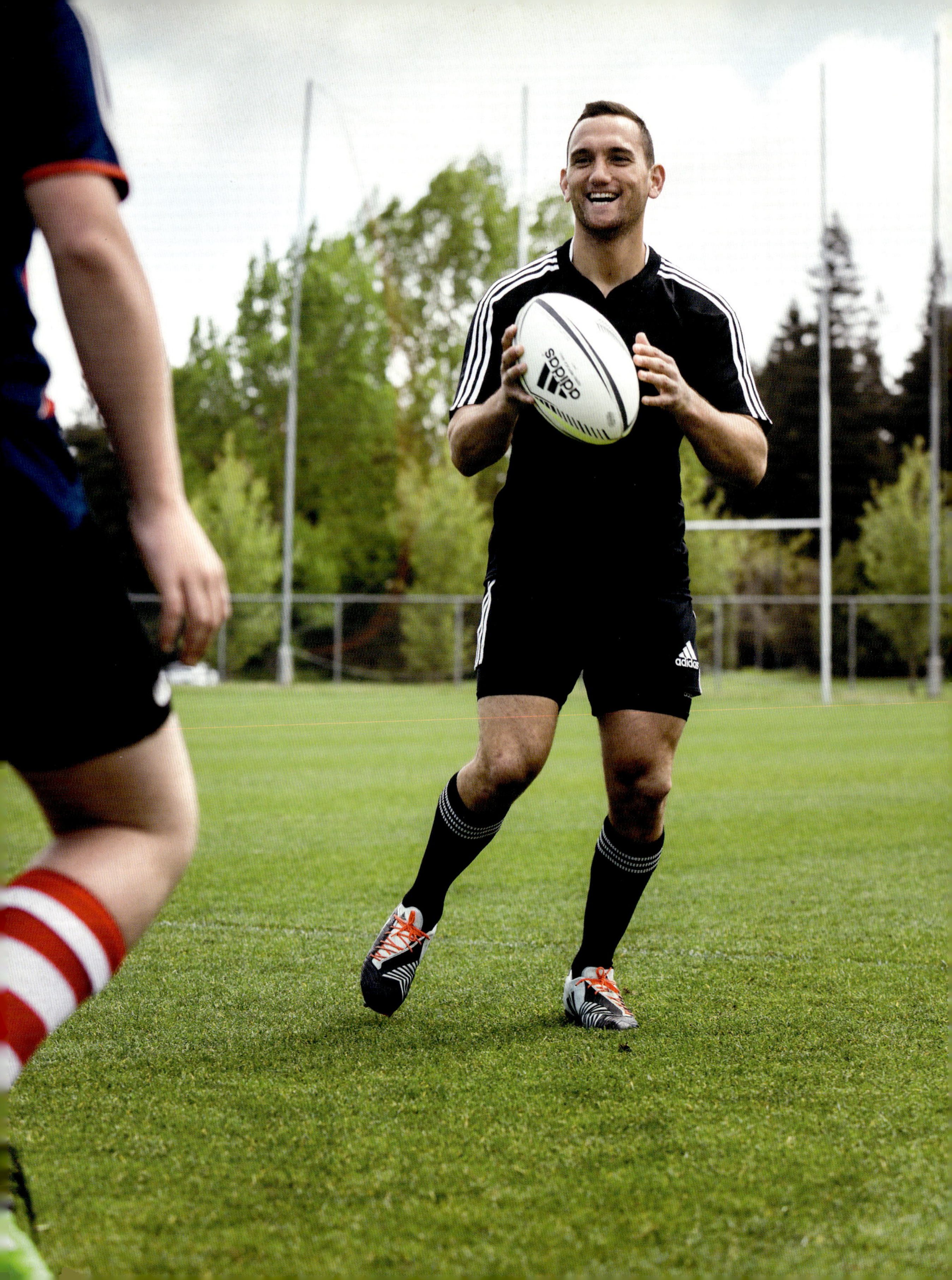
adidas

CHAPTER 2
HANDLING

Handling is everything in rugby — from how you hold the ball before you kick, to the millions of different ways you can pass the ball, through to catching the ball in general play or in the air. The more you understand the ball and how to handle it, the more skilled you'll be as a player.

The most distinctive thing about rugby — the thing that makes it different and great — is the oval ball we use. I love coming up with different ways of using the ball. You'll discover most of these in your own backyard. It's very important to become familiar with the shape and size of the ball, the way it flies through the air and the weird things it can do when it bounces.

You can get to know the rugby ball as you're walking to school, mooching around the house (mind Mum's precious stuff!), or even lying in bed.

Throwing passes or kicking to your mates, even just tossing the ball in the air and catching it — the more you handle a rugby ball, the better a player you will become.

AARON'S FLASH RUGBY TRICKS

In a lot of games, I've found myself having to juggle the ball, catch it in a strange position or from a strange angle, or dribble it along the ground with my foot. You never know what will be necessary! So becoming really familiar with the ball is super important. And best of all, you can use these tricks to impress your mates!

CHECKLIST

- Move the ball around your body using both hands. If you drop it, try to catch it before it hits the ground.

82

- Move the ball around your legs in a figure eight — around the outside of your right leg, inside your right leg, around the outside of your left leg, around the inside of your left leg, then start again. Now change direction. And when you've mastered both, try to do it with one hand!
- Throw the ball into the air and catch it higher and higher each time.
- A slightly trickier throw is to chuck the ball in the air with one hand and catch it with the same hand.
- Try dribbling the ball along the ground like a soccer player, using just your feet. I like to lean over the ball so my head is close to being directly over the ball, but be careful of using this technique if you're still working on balance — you don't want to trip over the ball and do a face-plant! Dribbling is an old skill, but a goodie. When rugby was first invented, dribbling was one of the main methods of getting the ball down the field. It's faded out, but you never know when you might have to control the ball like a soccer ball in a game.
- If you want to show off, try spinning the ball up in the air so that it lands on its point and bounces back to me. This

THE CRUDEN CLUE

Bounce the ball on the ground as you walk, as you need to understand the different ways the ball bounces. Not near traffic though!

trick is a bit of an Aaron special.
- To finish up, throw the ball up and over your head and catch it behind your back.

If you can do all these tricks, you're on your way to becoming a rugby trick expert.

WHEN IT GOES BAD!

Get your technique correct right from the start. If you don't, you could be practising bad habits. And doing the skill or trick once doesn't mean you have mastered it. Do it over and over again! Coaches say it all the time — it gets boring listening to them, but they're right: practice makes perfect.

Don't make the skill too simple just to feel good about what you can do. Test yourself, push yourself. This will help your skills grow stronger and stronger.

Practising and experimenting with your handling skills really sets you up, because if a strange bounce or wonky pass happens in a match, you will have practised your response. Your instincts will take over.

ACTIVITIES

Try these activities at home on your own to get to know the rugby ball:

First, take the ball and move it around your head. Then move the ball around your waist, your knees and your ankles. Try to keep control as you move it around these body parts.

Now throw the ball into the air, taking one step forward, and catch it behind your back.

Try holding the ball between your knees with one hand in front and one hand around the back. Then change hands quickly, before the ball falls. Do it again, keep doing it, and see how fast you can go.

Think you've got it sussed? Try doing it while you're walking, or jogging, or running hard. And try bouncing the ball in front of you as you run so that it bounces back to you. When you get good at this, it can look as though you've got the ball on a bit of string. And along the way to getting good at this, you'll see the ball bounce in just about every way imaginable — all good training!

EXTRA FOR EXPERTS

My favourite trick is hooking the ball between the instep of one foot and the heel of the other and flicking it up off the ground from behind me and catching it in front. Slick! Soccer players call this 'the rainbow', and it's my go to trick with a rugby ball.

CATCHING ON BOTH SIDES

When I'm about to receive a pass, I tend to extend my hands (fingers pointing at the ball, almost like a piano player) in the direction the pass is coming from. This gives the passer a nice big target to aim at, and it means I can catch the ball earlier, which gives me more time to decide what to do next.

I also like to catch the ball with 'soft hands' (hands that have a little 'give' in them). Lightly punch one hand with the other and curve the 'receiving hand' on impact, a bit like a baseball mitt closing around the ball. Practise moving the receiving hand with the punch as well.

Bingo! Soft hands! This helps to make your catching in any ball sport more secure. If your hands are 'hard', the ball can simply bounce off them. And if I have started with my hands pointing at the ball, the movement of softening them as I catch also sets me up to pass the ball on immediately if I have to.

CHECKLIST

- Keep your head up! I tell myself this to remind myself not to drop my chin on my chest (as you would if you were checking to see if your bootlaces were tied) or tilt

my head back (as if you were looking for seagulls).
- Keep your eyes open! It sounds basic, but lots of people learning to catch don't do it. Sometimes, frowning at the ball — really giving it the evils — will help you focus on it.
- Always have your hands up and pointing at where the ball is coming from. Position them as though you're about to play the piano.
- In rugby, it's best if you're running when you receive a pass. To do this, you need to come from **depth** so that you run on to the ball. What do I mean by 'depth'? Imagine a line from the passer straight across the field (in rugby, he's not allowed to pass forward of this line). The further behind the line you start, the 'deeper' we say you are when you start to run to receive the pass.
- Your hips should be square (facing towards the tryline) as you receive a pass. If not, you're closing off one of your passing or receiving sides. And you should open your shoulders toward the pass. Pointing both hands at the passer will make you do this naturally.
- Finally, it's important to catch the ball first before doing what you're planning to do next. If you're thinking too far ahead, you might forget to do the most important thing first: catch the ball!

THE CRUDEN CLUE

Keep your eyes on the ball right up to the moment the ball goes into your hands.

WHEN IT GOES BAD!

You have to have your hands up so they present a target for the ball-carrier to pass to. If you put your hands down by your side, there's a good chance they'll throw the ball at your tummy button. Try it! Get a mate and say 'Pass me the ball' with your hands down low or up high. I bet they pass to where your hands are.

Rugby is all about getting your hands on the ball first, so reach toward the oncoming ball so that you get it before your opponent does.

One of the biggest mistakes in rugby is taking your eyes off the ball. It can be tempting to have a quick look at the huge, sweaty No. 8 who is running up to smash you. Don't! Watch the ball. The ball is everything.

Don't catch with hard hands. Soft hands will accept the ball evenly and comfortably.

Practise, practise, practise! At first-five, you get the ball a lot, so this is the one skill I can't be shabby on. I can't afford to drop the ball or juggle it. But anyone playing rugby needs to be able to catch the ball when it's passed to them.

ACTIVITY

This activity sounds really simple — and I guess it is — but what it does give you is a sound base to work from. Get these passes and catches right, and you'll never drop a ball again.

Next time you watch the All Blacks warming up before a test you'll notice that they use a lot of these simple activities. You never get so good at the game that you can afford to ignore the basics.

STAGE 1

You and a friend stand separated by around one arm's length, side on to each other and facing the same way. Stand still and simply pass the ball back and forth. See how many you can do without dropping the ball. Such short passes require the soft hands we've talked about. You can swap sides to practise passing to the left and the right.

STAGE 2

Now move out till you're about 1.5 metres apart, but keep passing the ball back and forth.

STAGE 3

Increase the distance between you to about 3 metres, and keep passing.

STAGE 4

Time to start moving! Try some **lateral** passing (this means passing the ball to your side) at a walking pace for about 30 metres.

STAGE 5

Increase your pace to jogging, and keep up the lateral passing for about 30 metres.

STAGE 6

Now it's time to run! Keep up the lateral passing at a varied running pace for about 30 metres.

EXTRA FOR EXPERTS

Get close to a wall — not an inside wall, unless you've got a rumpus room! Pass against the wall and catch the ball as it comes off the wall on different angles.

You can also have two mates facing you and you turn your back. When you hear them shout 'Go' you turn around and one of them passes the ball to you. You have to catch the ball from whoever passes it.

Develop that quick instinct. The more you do this, the more time you'll have for your next move.

PICKING UP / FALLING ON THE BALL ON BOTH SIDES

You'll hear rugby players talk about 'the **loose ball**'. Sometimes the ball can be just sitting there on the ground in the open. Sometimes it may have a lot of bodies around it or it could be skidding along the ground after a kick.

The first step is to get the ball quickly and safely. Then you can consider what's happening around you and decide on the best available option.

I like to approach the loose ball with my body position low (bent at the hips, knees slightly bent). I keep my eyes on it all the time, and I get side on (facing the touchline).

I try to get my front foot ahead of the ball so that I'm straddling it. This helps your balance.

Try to get your hands on the ball as quickly as possible without knocking it on. Pick the ball up by sweeping your hands from behind the ball.

And remember: watch the ball!

CHECKLIST

- The number one rule when the ball is on the ground is to make the decision to go for it.
- When you're on your feet, straddle the ball, bend your knees and put both hands on the ball.
- Once you have the ball, look up, so that you can consider your next move.
- If you don't have time to pick up the ball from a standing position, fall or slide

THE CRUDEN CLUE

Always secure the ball with both hands. You can use one hand to scoop up the ball, but it is tricky and risky.

onto it. Make sure you slide to the side of the ball and get both hands on it. I often have to do this when I'm running back to tidy up a loose ball that's spilled behind the backline or has been kicked through. It takes practice to get the technique right! Get a strong grip on the ball before you stand up, because you might have to wrestle with opposition forwards for it the moment you do.

WHEN IT GOES BAD!

If you lift your head too early — sneaking a peek at that wall of tight forwards thundering toward you — you might juggle or, even worse, knock the ball on.

Approaching the ball front on seems like the right thing to do, but you're off balance. Take the time to step ahead of the ball and pick it up with the sweep movement.

Keep your hands soft when picking up the ball. The harder your hands, the more likely you are to drop it.

ACTIVITY

You have a ball and your friend runs behind you and follows you.

You place the ball on the ground and your friend picks it up as smoothly as possible. They then run ahead of you and place it for you to pick up. Keep taking turns as you run.

EXTRA FOR EXPERTS

Your friend rolls the ball along the ground. You run and slide/fall onto the ball, grab it and get back to your feet. Return the ball to your mate and repeat.

CATCHING THE HIGH BALL

When a big bomb has gone up — someone has kicked the ball high into the air — I love the challenge of out-jumping the opposition to catch it. It doesn't matter if you're tall or short in this game: catching the high ball is all about technique.

You need to watch the ball and call for it, because if everyone is watching the ball and running towards where it will land — well, you can imagine the crashes that can happen.

Get your hands up early, above eye level. Rather than trying to catch it in your hands, you're trying to grab the ball in your arms in a big hug against your chest. This reduces the chances of the ball bouncing out.

Lots of players like to turn so they are side on (facing the touchline), so that they are leading with their shoulder and if they drop the ball it will go sideways and not forward.

Practise catching the high ball with both feet on the ground at first. As you get more confident of your catching technique, try jumping so that you catch the ball while you're in the air. The laws of rugby say you can't be tackled if your feet are off the ground.

As you get good at jumping and catching, try running forward and jumping to catch the ball. We call this 'attacking possession'. It takes lots of practice and lots of courage, because there are often opposition players attacking possession, too!

CHECKLIST

- Once you've seen that the ball is in the air and that you'll be able to reach it, make the commitment and then call out loudly that you are going to catch it. 'Mine!' is a common call.
- You must get under the ball. It sounds obvious, but you need to move your feet: do whatever it takes — run, shuffle — to get there early, then steady yourself in the position where you think the ball will land.
- Even when you think you have it right, don't stand there like a statue. Keep your feet ready so you can adjust your position.
- Stand in a slight side-on body position so you don't knock the ball forward.
- Have your hands high, pointing at the ball.
- If you're running to receive the ball, bring your front leg up as you jump. This is for protection, but it also helps turn your momentum into height for your jump.

THE CRUDEN CLUE

Communication is vital to prevent team-mates all trying to catch the ball. Shouting 'my ball' works for me.

WHEN IT GOES BAD!

If you take your eyes off the ball — give up! It's that simple.

Get side on. If you're front on, you're an easy target for the tackle, and if you do drop the ball it will go forward toward the opposition. Knock-on!

Call! If a heap of players jump or run for the same ball, injuries can happen. Calling lets everyone on your team know who has it covered.

Some of the funniest moments on a rugby field are when players misjudge and don't get under the high ball. Don't be the one who provides the fun! Getting your jump wrong can also be funny to watch — remember, timing is everything, and you'll only learn this through practice.

If you catch the ball below your eye level you can be distracted by your opponents coming toward you and it makes it tricky to press the ball into your chest.

ACTIVITY

With a friend, throw the ball accurately to one another, high into the air. The throw should be made so that it can be caught on the full (before it bounces). You can get funky with your throws. Spin some, throw wobbly ones — it gets players used to the ball coming down awkwardly, which it often does!

DID YOU KNOW?

In some situations, you can call 'mark' when you catch the ball on the full. Play stops, and you are allowed a free kick of the ball from the spot where you caught it. These days, you can only call a 'mark' when you are inside your own 22-metre zone. In the olden days you could call a 'mark' anywhere on the field if you caught the ball on the full.

EXTRA FOR EXPERTS

Do the same as above but this time with a bigger distance between you and your friend. You may have to use a kick instead of a throw. For a real challenge, stand with your back to your friends and kick over your head to them without looking.

RIPPING THE BALL

Robbing the opposition of the ball in a standing situation sounds like something only the strongest players can do, but it's amazing what you can do when you want it enough. For someone like me, who has always enjoyed playing rugby with the ball more than without it, ripping the ball off people comes naturally. And being a small guy, I need this skill to hang onto the ball once I've got it. You can have a lot of fun with a mate practising this — see who is the 'takeaway king'!

Being strong in the upper body helps, but technique usually wins overall. It's also useful when your team-mates are there to help.

CHECKLIST

- When you go to rip the ball off someone, get in close to the ball-carrier and keep your eye on the ball.
- Latch onto the ball like a limpet with your hands above and below the ball.
- Push up first then pull down hard (not side to side), using your body weight an the momentum of your approach to add force.
- Alternatively, wrap both arms around the ball and use a washing-machine-type action to rip their grip off the ball. This can also be a useful defence against rippers, giving them a moving target to try to grab. Even the biggest players have lost the ball when I've had a go at them. Sometimes being small allows you to sneak in close and use that washing-machine action.

FUN FACT

There are four ways of scoring in a game of rugby. They are from a penalty kick, drop kick, try or conversion.

THE CRUDEN CLUE

First, bend down below the height of the ball.

WHEN IT GOES BAD!

If you have a poor body position it won't matter how hard you try, that ball is going nowhere.

Commit yourself to it. If you don't jerk the ball up and down, or get that washing machine going like you mean it, the ball-carrier will just grin at you like a playground bully at a little kid who wants his ball back.

ACTIVITY

You and a team-mate are both on your knees. Your friend has the ball.

With your team-mate using only a little bit of resistance at first, work on the technique of getting the ball from their grasp. Get them to increase the resistance as you get better at it.

EXTRA FOR EXPERTS

Using a friend or team-mate, try to get the ball off them while they are resisting as hard as they can.

GLOSSARY

ATTACK your team's play when it has the ball

DEPTH the amount by which the player is behind the ball or player. The further behind the ball you start, the deeper you are when you start to run to receive the pass

LATERAL to the side of the player, sideways

LOOSE BALL when the ball is in the open

RUGBY WORLD CUP: DID YOU KNOW?

1991

- The second Rugby World Cup was hosted by England.
- Australia won the 1991 Rugby World Cup, beating England in the final.

FUN FACT

There is an invitation team made up of players from all around the world called the Barbarians (or 'Baa-Baas'). They have a uniform of black-and-white hooped jerseys, but they don't have a home ground, or a clubroom!

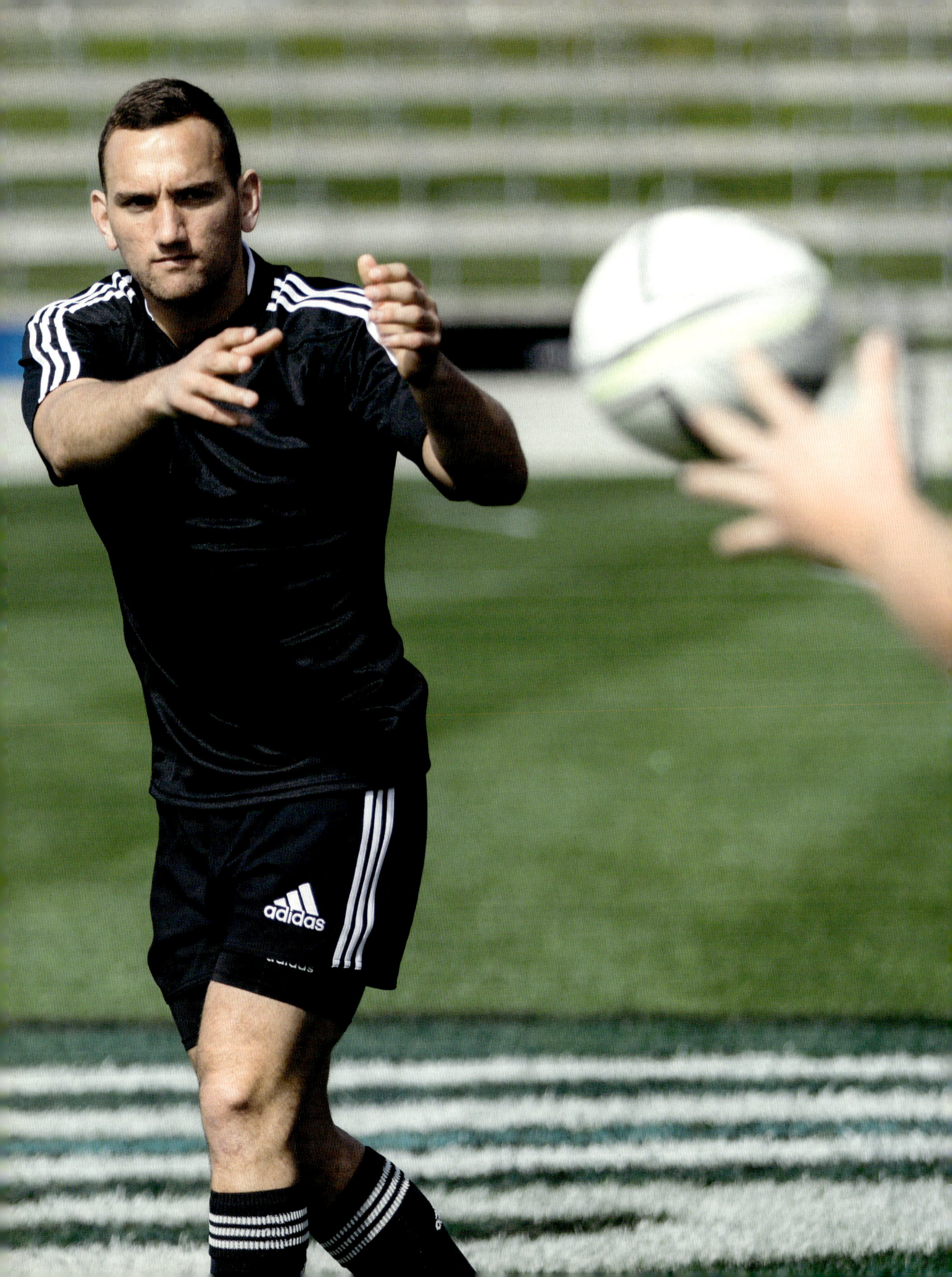
adidas

CHAPTER 3
PASSING

PASSING TO BOTH SIDES

I like to run straight down the field or towards the closest defender to try to 'draw' the defence. If an opposition player **commits** himself to tackling you, that reduces the number of players there are to tackle your team-mates. As soon as you see that an opposition player is committed to tackling you, your goal is to pass the ball as soon as possible.

You need to be able to pass off either hand — to both sides — at about waist height for shorter passes or at chest height for longer passes. All of your passes should be aimed just in front of the person you're passing to (the receiver) so that they can run on to the ball.

There are lots of different sorts of passes. I use many types of passes in a game, and some players seem to be able to get the ball to their team-mates in all kinds of weird ways and in all sorts of unlikely situations, but it all starts with the basics.

There are two basic sorts of pass: I'll call them the standard pass and the **spiral** pass.

CHECKLIST

- Always look at the player you want to pass the ball to, and then commit to making the pass.
- To be sure of your control of the ball, you need to use both hands to pass. Only try one-hand passes once you feel your skill levels are high enough.

- Your fingers should be placed down on the seams of the ball for a basic pass.
- Now all you have to do is swing your arms and let go of the ball so that it sails to the **receiver**, arriving at about chest height. Follow through with your fingers toward the target. When you've finished the motion of passing, your fingers should be pointing at where the ball has gone.
- To learn how to do a spiral pass, think of the ball as a bullet you're sending in the intended direction, pointed end first. To make this happen, use one hand to provide the power and the spin, and the other to guide the pass.
- Hold the ball so that your fingers are across the seams (when your elbows are bent, the points of the ball will be aiming up and down). Stand side on to the direction you want to pass.
- To pass to your left, aim your left foot toward your receiver. Move your legs apart, put your weight on your right foot and hold the ball above your right hip.

THE CRUDEN CLUE

A pass is only as good as the ease with which it can be caught. After all, passing is a two-step process: pass and catch.

- Now transfer your weight to your left foot, and as you do it, sweep your arms in the direction you want to pass. Your left hand guides the ball toward the target, while your right hand comes upward, making the ball spin. Follow through: when you've completed the pass, your fingers should be pointing at the target.
- Passing to the right is the exact opposite of passing to the left!
- Practise spinning the ball by tossing it straight up in front of you, spinning it in the opposite direction each time. You'll get more power as you get the hang of what each hand is supposed to do. Spiral passes can travel much further than a standard pass, but they are also harder to catch and tend to dip toward the ground as they reach the end of their travel.

WHEN IT GOES BAD!

Look at the target area (your receiver should be showing you their hands) and aim your pass there. If you don't identify your target, the ball could go anywhere!

If you don't have a straight follow-through, there is a good chance the ball will be passed downwards or way up into the air. That's about as much use as passing it behind or too far in front of the receiver — no use at all!

ACTIVITY

Do the same activity as for catching (see page 88) but this time, use three people.

The person who is practising their passing goes in the middle so they can pass the ball to both sides. You can swap positions around.

EXTRA FOR EXPERTS

Once you're comfortable with the basics, try cut-out passing, lob passing, pop passing and overhead passing.

A **lob** is a high, loopy pass that drops to the receiver rather than goes straight to them. You use it to get the ball over the head of an opponent who is trying to intercept the ball, or to make the ball 'hang' in the air for a receiver who has started their run a little too far back.

A **cut-out pass** is when you miss one of your team-mates by passing to the player beyond them in the chain. Sometimes the player who is missed out does their best to make it look as though they're going to catch the ball — they act as a '**decoy**' runner.

A **pop pass** is a tiny little flick of the wrists that puts the ball into the air right beside you. You use this pass if you run right up to an opponent and a team-mate is running up close behind you. You 'pop' the ball into the space, and as you take the tackle, your team-mate runs through the gap at your elbow.

Finally, the **overhead pass** is like a netball or basketball pass. You have to get your hands way up high so that you can flick the ball over the head of an opponent standing between you and the receiver.

DID YOU KNOW?

There can be around 250 passes in a professional rugby game.

PASSING FROM THE GROUND TO BOTH SIDES

Sometimes, you just have to pass the ball quickly off the ground from a crowded area. This isn't just a skill for the halfbacks: it's a skill to be mastered by the whole team. I prefer not to play halfback — not because I'm scared of those big, sweaty forwards, but because my job is to be the person who catches the first pass after a ruck or maul, and then makes the decisions. But every now and then, the halfback gets caught up in the ruck or maul and because I'm close by, it makes sense that I do the job.

Because first-fives like to stand as far away as possible from those piles of sweaty forwards, the person who throws the pass must be able to pass quickly to a receiver some distance away. This is usually done with a spiral pass. But every now and then, you'll need to use the halfback's trick of 'popping' a little pass to someone running close to the ruck or maul. It helps if you know how to do this to both sides.

CHECKLIST

- You have to get close to the ball (pretend you're trying to read what's written on it in really small writing).
- Place one foot next to the ball and the other foot further away, pointing to the target.
- Bend your knees, stoop and grip the ball.

- Sight the target, then pass from the ground in one motion by transferring your weight from the foot closest to the ball onto the other one.
- If you follow through with your arms to the target point, you'll improve your accuracy.

WHEN IT GOES BAD!

As usual, you'll run into problems if you don't keep your eyes on the ball. Perhaps you're watching what happens to the halfback when the loose forwards come around the ruck. Perhaps you're looking at your options. Perhaps you're just admiring the scenery. Remember, if you take your eyes off the ball, there's a good chance you'll muck it up.

Sometimes reaching too far for the ball rather than getting the passing foot into position next to it causes problems. Reaching can put your body off balance and means the pass will be inaccurate.

You might be front on instead of side on to the receiver. You'll never get any speed in the pass that way! Or you might have a 'backswing' before you pass, which robs you and your receiver of precious time. If your feet are placed correctly, you'll be side on and the power you need for a nice, fast pass will come from your transfer of weight from one foot to the other — no need for a backswing!

ACTIVITY

Place three cones in a line, with the first and last cones 3 metres apart, and four balls on the ground next to the middle cone. Grab a mate, and ask him to stand a few metres away off to the side of the line of cones.

Start from the first cone. Run to the middle cone and pass the ball off the

THE CRUDEN CLUE

Stay nice and low throughout the whole movement.

ground to your mate standing to your right. Run around the third cone and come back to the middle. Pass the ball off the ground to your mate (who is now on your left-hand side). Run around the cone in front of you, and so on.

EXTRA FOR EXPERTS

Put the ball on the ground at the bottom of a wall and pass from there. This will make sure your elbows stay close and tight and you don't use a backswing.

Try this technique to help get your body down low. Instead of placing your foot next to the ball, put your knee on the ground and your other foot pointing to the target. Pass with two hands. When you have mastered that pass on both sides, do it again, this time passing the ball using one hand. Keep your guiding hand out of it. You wouldn't do this in a game — you want to move quickly to the next phase — but it helps develop the muscles and the skills you need for passing off the ground.

RECORD
The most rugby passes in one minute by two players is 59. Now there's something to aim for!

DUMMY PASS

As you can guess from the name, a dummy pass is one you throw to fool the opposition. Everything is the same as throwing a real pass apart from the bit about letting the ball go. I love selling dummies! It works quite well for first-fives, because everyone expects you to pass the ball. A dummy pass often makes the defence drift wide of you and a nice, big gap opens up for you to run through with a little shimmy, a wiggle of your hips.

Make sure your action is convincing. Do everything as you normally would. If the defender is watching your face, make eye contact and wear a determined expression as if you mean to draw and pass. Throw the pass, but not the ball, and then bring your arms back so that you're carrying the ball in front of you again. Accelerate, change direction if you need to and head through that juicy gap!

CHECKLIST

- Conviction (looking like you mean business) and vision are the main ingredients of a classic dummy pass. Some players watch the ball — they don't study you or your face — as you run it up to them. When I see this, I know the dummy is on!
- Basically, to sell a dummy, you make a

THE CRUDEN CLUE

You must 'read' the defence before playing a dummy pass. Try to pick out a player you think will fall for it.

perfectly normal pass without releasing the ball, so don't forget to make all your usual movements (wrists, elbows) that make it look like a pass is about to be made.

- The movement of your pass can be quite short. Sometimes even waving the ball in the direction of the next player in the chain — **'showing the ball'**, as it's called — can be enough to convince a tackler that you're going to pass.
- It's a quick action and then *you* have to be quick — take off before they've noticed you still have the ball.

WHEN IT GOES BAD!

If you don't look convincing, you will probably get smashed, as the defence will see the move coming. When you try to sell a dummy and no one's buying it — that's when it *does* go bad!

Also, if the person you're pretending to pass to isn't believable, the opposition aren't going to buy it. A dummy pass must have a dummy target, so there's no use pretending to throw a pass if there's no one there to pass to.

If the ball is not out in front of your body, the opposition can guess that you aren't going to pass. Carry it out in front of you at every opportunity. That not only makes it easier to pass in both directions, but it also means you can *pretend* to pass in both directions.

ACTIVITY

Use a small area with a tryline at the end.

Two of you run along with a ball and a friend defends the line. The defender can only move sideways, not forwards or backwards.

The aim for you and your attacking mate is to score at the other end of the line. Either pass or use a dummy pass to score.

Mix up the attackers and defenders after the attackers have been up and down the area twice — or make it into a competition, so that you only change when someone is caught.

EXTRA FOR EXPERTS

Walk around your area dummy passing to family members, pets, street signs and obstacles. A crowded footpath is also a great place to practice!

DIVE PASS

If you really need to clear the ball from a busy area quickly, I prefer the dive pass. You could also call it the 'superhero pass', as you need to move dramatically as you throw the ball.

It's one skill we're seeing less and less of these days, because the laws around 'clearing out' players mean you're much better protected from defenders coming around and grabbing or tackling you when you're working behind the rucks and mauls. Throwing a dive pass also puts you out of play for a bit because you end up on the ground when the pass has been completed.

Halfbacks should practise this skill more than anyone else as it's more than likely they will have to use it during a game.

CHECKLIST

- With the dive pass, you must keep your weight over the ball and place both feet close to the ball.
- Hold the ball on both sides to keep it secure.
- A hard leg drive is necessary, to propel you and the ball toward the target, like Superman launching himself off a tall building and into the air.
- Like most passes, throw your hands toward the target (follow through) and

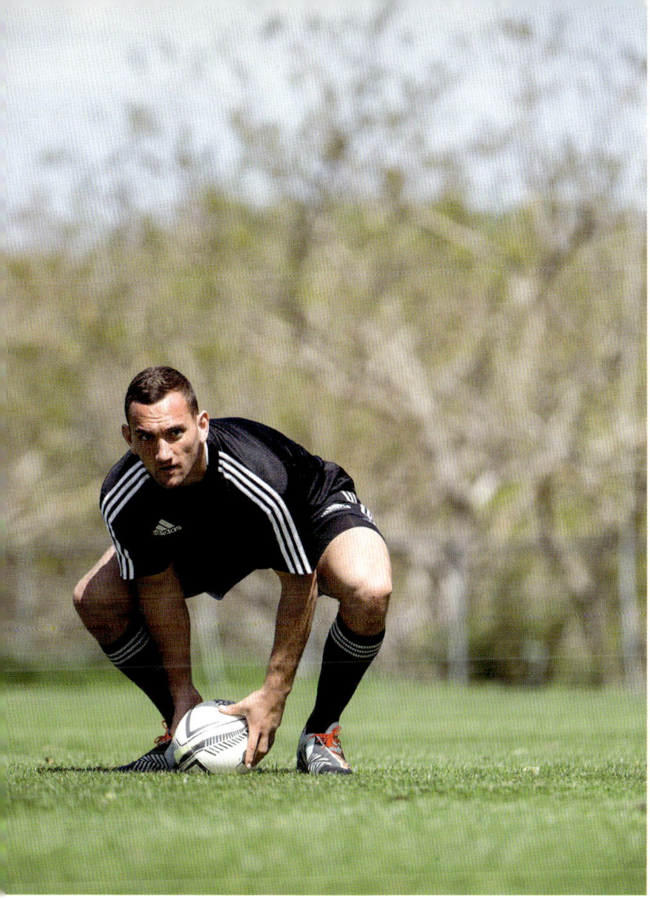
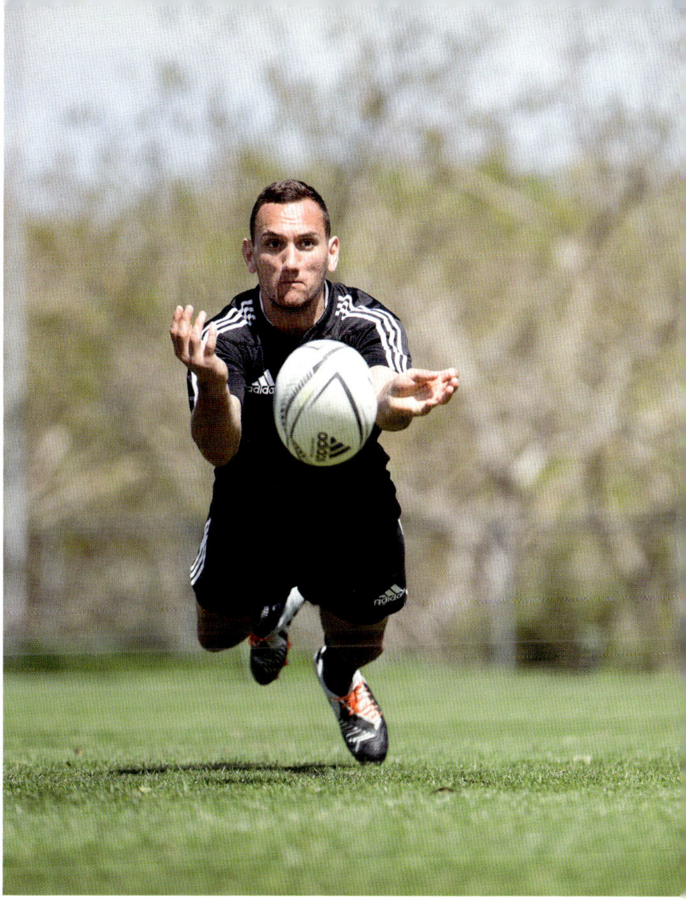
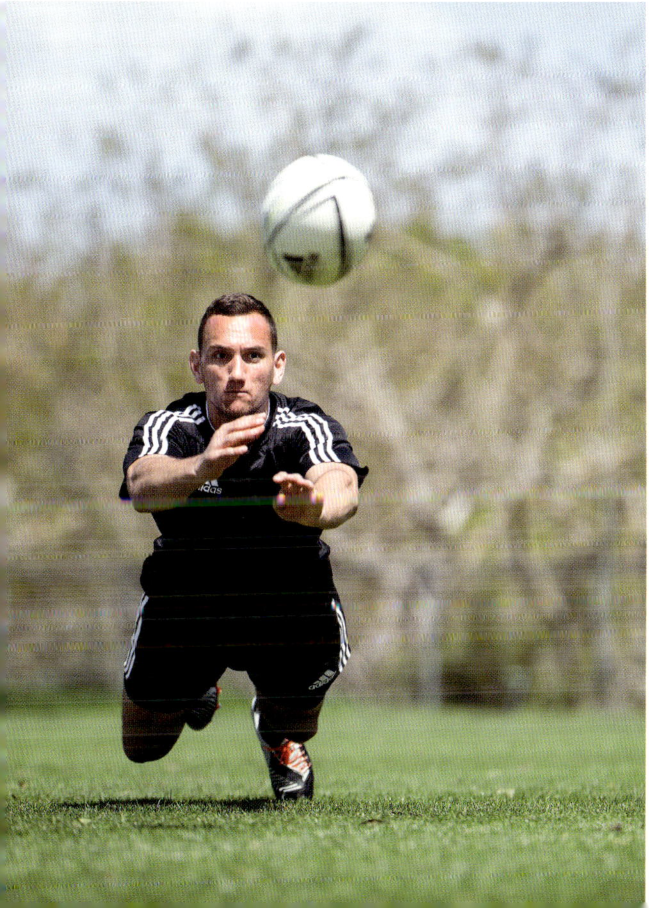
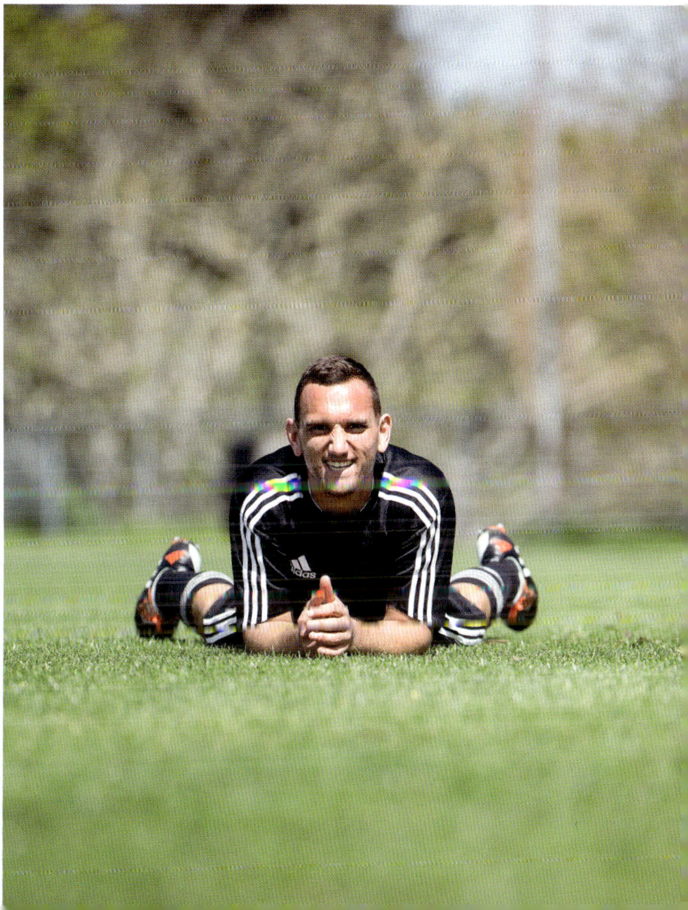

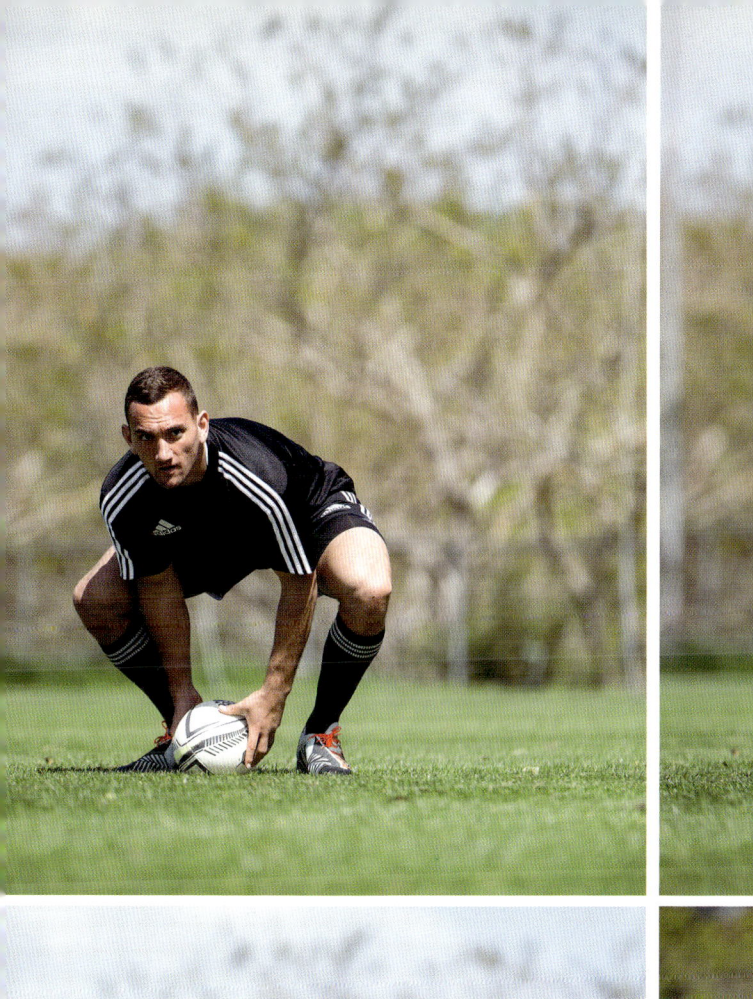

flick the wrists to impart a little more force.

- Get to your feet as quickly as possible and follow the ball, because it's time to support your team member. If they are tackled, you might have to do it all over again!

WHEN IT GOES BAD!

If you stand up too early, before passing the ball, the pass isn't going to have any power. This also happens if there's no follow-through. It's just like tennis, golf and cricket — you must follow through on the swing.

THE CRUDEN CLUE

For a more accurate pass, throw your hands towards the target.

ACTIVITY

Get a mate to stand ready to receive the pass, starting off close, then moving further away. Practise the pass and what you need to do to get to your feet again once you've thrown it.

If any of your other mates are hanging around, invite them to try reaching you before you before you've cleared the ball. Vary the distance as you get better at it.

THE OFF-LOAD

THE FLIP PASS CHECKLIST

This is a skill that has come into rugby from its cousin code, rugby league, where the need to **keep the ball alive** in the tackle is very important. As the ball-carrier is tackled, and often even as they are falling to the ground, they free one or both hands and 'off-load' the ball to a support player who arrives at their shoulder (we call this 'running off' the tackled player). Properly done, an off-load is one of the coolest acts in rugby, but it's a high-risk one, too.

There are a number of different ways to off-load the ball, but we'll look at the one-hander and the risky 'flip' options.

- Lean slightly in towards the tackler.
- As you enter contact (that is, you're tackled), use one hand to flip the ball across your midriff and upwards, as though you were about to flick the ball back up over your own shoulder.
- Use your wrist (as your elbow comes up) and extend your arm toward the target for more accuracy. The idea is to just flip the ball into the space beside you as you are tackled, for a team member to run on to.

WHEN IT GOES BAD!

If you run into heavy traffic and there are lots of bodies around, the flip is not really an option. You'll be more likely to bounce it off someone and lose it.

If you're being lined up for a **ball-and-all tackle**, where your opponent folds you up in a great big hug so that the ball is squashed between the two of you, you're not going to be able to execute this pass.

Off-loads usually go wrong if your support is too slow in arriving. You need someone running near you just as you flip the ball into that space.

THE ONE-HANDER

If you're going to use the one-hander, lead into the tackle with the arm that's not carrying the ball. The idea is to try to get your upper body into space beyond the tackler and extend your arm through the tackle and well in front of you as you fall.

Hook the ball around and backwards (using a swinging action) to your support player.

Use the wrist to give the ball an extra push and extend your arm toward your target for more accuracy.

WHEN IT GOES BAD!

If there is not enough space between your arm and the tackler, don't let the ball go, as it's likely to hit something it's not supposed to.

THE CRUDEN CLUE

Make sure you 'create the space' to make this pass.

It's common to lose the ball if you try to pass it before the tackle.

Remember, if your support from team-mates is too slow there's no point making the pass as there will be no one there to collect it.

I don't use off-load passes very often in games because they're so risky, but when I do it's usually at a critical moment. If we're desperate to break the defensive line or to go for a try, these off-loads can be the difference between winning and losing. You'll be a hero if it works, but you'd better hide somewhere if it doesn't come off, as your coach will wish you'd tried a conventional pass. It helps a lot if the people playing around you know that you are likely to try it. This is one of the reasons I like playing with Sonny Bill Williams so much. He's not only good at giving off-load passes: he's always looking for them too and arrives at exactly the right time to receive them.

ACTIVITY

These are great passes to practise with your mates, as you're not likely to be encouraged to practise them at team trainings. The off-load is an instinctive pass that you do when the moment seems right.

Start simply by walking or running and flicking the ball to a team-mate. Then add in a tackler to make sure you get the techniques right. You could even try these passes on your knees first!

SCISSORS OR SWITCH PASS

It's natural for players to run a little bit cross-field as the ball is passed along the backline. This can mean your speedy players, the wings, have no space by the time the ball reaches them. That's why it's a good idea to '**straighten**' **your attacking lines** by changing the direction of your pass. The unexpected change of direction can also be a very effective tactic, catching out the defence. A 'scissors' pass is a pass designed to change the direction of an attack or baffle an opponent.

CHECKLIST

- The ball-carrier has to run at an angle to the defence, to try to drag a defender across the field with them so that a space opens for the team-mate they're going to pass to.
- It can be a good idea to screen the ball from the opposition, which means that you present it towards the people to your outside (and away from the person who is actually the target of your pass).
- The receiver of the pass runs at an angle to your path. They might run straight (while you're running across-field), or they might run slightly across-field in the opposite direction to you. Either way, they will cross close behind you so that your path and their path will make an X like an open pair of scissors (that's where this play gets its name). As your mate passes behind you, you transfer the ball to them: a little pop pass is best here. What you're both hoping is that a gap will have opened as the defence drifts across to try to catch you, and your mate will slice nicely through it.
- After the pass, your job is to support the runner and your other team-mates.

WHEN IT GOES BAD!

Communication and timing are everything in the scissors. Both you and the runner need to know it's on. If you begin the move too early, it will be ineffective. If you or your runner are too slow, it won't be any better.

The receiver has to run the right angle to make the most of the gap you create.

And of course, the scissors requires you to execute the transfer — the pass — perfectly.

ACTIVITY

In your backyard or on a field, you and a mate run zigzagging across the area, performing a scissors pass each time you cross.

THROW INTO LINEOUT

I bet Aaron Smith was surprised to find himself throwing the ball into the lineout against England at Twickenham in November 2014. He's a halfback, and it's usually the hooker who throws into the lineout these days. I've never had to do it in a game, but I have quite enjoyed practising it — you can challenge yourself to see how accurate you can be. You may or may not want to be a hooker, but when you're starting out in rugby, you never know what position you'll end up playing. Best to start practising those throws now, just in case!

I've played in a number of games where the lineout throw has been the most important feature of the match. If we hadn't won our lineout we would have lost the game, or if our lineout had gone better, we could have won it. If you can master this skill, you will be a much sought-after player for any team.

CHECKLIST

While everyone develops their own style of throwing, there are a few basics you'll need to master.

Understand what your team plans to do from the lineout. Is the ball going to be caught and thrown to the halfback straight away (what we call '**off the top**')? Do your forwards intend to take the ball into a maul and try to drive downfield? Are you planning a sneaky move around the front of the line? How you throw and where you aim will depend on the lineout option.

You'll need to know how much force you need to throw to the front, middle and back of the lineout. The type of throw you use will also depend on the style preferred by your jumpers. Some like to receive a flat, hard throw, and some prefer a high, loopy throw.

Every team has its own codes for communicating the lineout call to the jumper and thrower. Numbers are popular, and all sorts of complicated systems get used. I have found that simple is best, so long as it's not so simple that the opposition can work it out.

'One, two, seven' could mean throw the ball to the second jumper. 'Three, four, eight' could be a throw to number four. You can also use words like 'footy, gorilla, rabbit' — this pass is going to the front, as the first word starts with 'f'. Where do you think this throw is going? 'Mouse, bottle, tree.' Yes, you've got it — it's going to the middle because the first word started with the letter 'm'.

Try this one. 'One, four, five, penguin'? Or how about 'two, eight, five, buffalo'? Well, the first call says the throw is going

THE CRUDEN CLUE

Find a throwing stance that feels right for you!

to number two in the lineout because a penguin is a bird and has two legs. And the second call says the throw is going to number four in the lineout because a buffalo has four legs. Crazy, I know! You can have fun coming up with your own calls.

Now to learn about the throw itself. The art of throwing starts with your fingers spread wide on the ball across the seams.

Hold the ball directly above your head.

Your foot placement is a matter of personal choice, but if you're having problems with your throw, changing the way you plant your feet may be the solution. Some people like to plant both their feet about a shoulder-width apart; others like one foot placed right in front of the other.

Push your elbows forward as you throw so the power comes from your wrist and forearm.

Throw accurately and extend your hands toward the target with a good follow-through. Your hands will end up pointing toward where the ball needs to be.

Once you've thrown the ball in, you can take up position inside the 5-metre line to prevent opponents from moving around the front, or you can get ready to help form a maul and drive the ball.

WHEN IT GOES BAD!

You might do a perfect throw, but it will be wasted if you haven't understood who you're throwing to.

If your feet are poorly positioned, the ball may go in a surprising direction.

The biggest problems at lineout time can arise when you're trying to be too complicated. Like many things in rugby and in life — keep it simple!

ACTIVITY

This drill can be done almost anywhere — I like to try to stand on the tryline facing the goalposts. Use cones to space out the distance of each throw.

Aim to hit the upright posts at various heights.

Progressively set targets for your practice. For example, aim to hit the goalposts three times in a row, then five in a row, and 10 in a row at each distance. Your muscles get used to doing what you want them to do, and that helps you to become consistent.

Once you've mastered technique and accuracy, you need to work on timing. Once your throwing has reached a suitable standard, you can then start working with your jumpers.

RUGBY WORLD CUP: DID YOU KNOW?

1995

- Jonah Lomu scored four tries against England in the All Blacks' 45–29 semifinal win in Cape Town, South Africa, in what some say is the best individual Rugby World Cup performance of all time.
- In the final, it was New Zealand versus South Africa at Ellis Park. The game went all the way to extra time.
- Joel Stransky made South Africa world champions with a drop goal to achieve a 15–12 win.

GLOSSARY

BALL-AND-ALL TACKLE where your opponent folds you up in a great big hug so the ball is squashed between the two of you

COMMIT to give yourself entirely to a cause or action

CUT-OUT PASS when you miss one of your team-mates by passing to the player beyond them in the chain

DECOY a faked pass or kick, designed to trick the opposing player

KEEP THE BALL ALIVE keeping the game going and not breaking any rules

LOB a high, loopy pass

OFF THE TOP catching the ball in the lineout and throwing it to the halfback straight away

OVERHEAD PASS similar to a netball or basketball pass

POP PASS a tiny flick of the wrists that puts the ball into the air right beside you. You 'pop' the ball into space and your team-mate runs through the gap at your elbow

RECEIVER the player who gets the ball

SHOWING THE BALL pretending you're going to pass to another player by waving the ball in their direction

SPIRAL a ball that is spinning as it goes through the air

STRAIGHTEN THE ATTACK running straight towards the tryline, not across the field

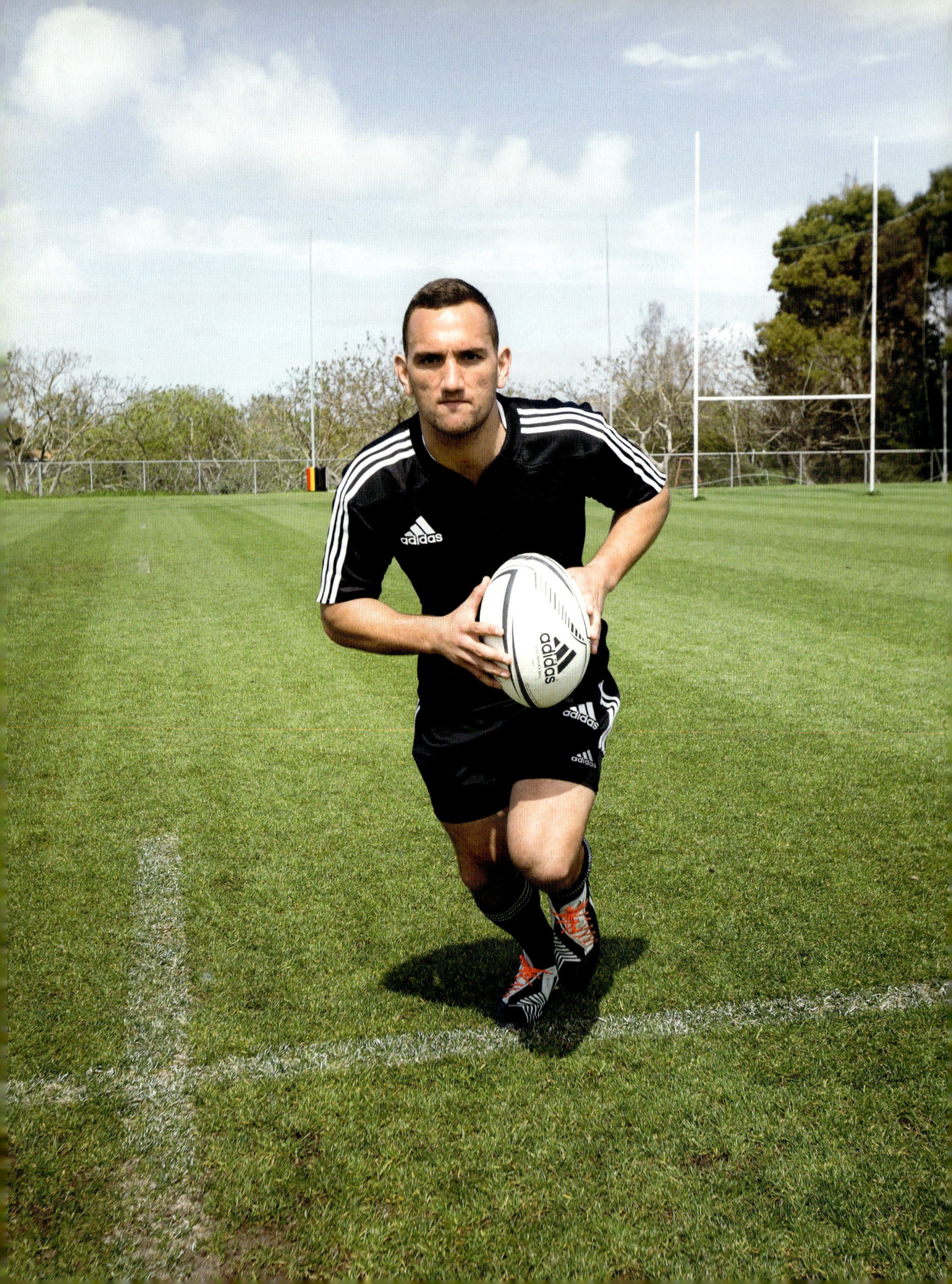
adidas
adidas
adidas
adidas

CHAPTER 4
RUNNING

RUNNING WITH THE BALL (IN ONE OR TWO HANDS)

When running with the ball in two hands, I can evade players, draw a defender or go through a gap.

Holding the ball in two hands keeps the opposition guessing what I'm about to do. I might be about to pass (in either direction), sell a dummy, kick or run. If I run with the ball in one hand — best reserved for those big lads with big hands — or tucked under one arm, my options are much more limited.

It's important to practise running with the ball using two hands most of the time. You can also practise moving the ball from your left hand to your right and back again — again, this can cause confusion to the opposing team.

CHECKLIST

- It may sound way too simple, but before you try anything else, run with the ball. You need to get the feel of the ball in your hands.
- Hold the ball in two hands, then move the ball to your left side (don't pass, but practise the motion), to your right side, and then hold it in the middle, keeping two hands on the ball at all times.
- Practise scoring tries with two hands. Make sure the ball is nice and secure as you force it down.
- Always carry the ball tightly (with soft hands). Keep your fingers in the same direction as the seams for the standard pass.

THE CRUDEN CLUE

Don't always carry the ball under the same arm. Keep the opposition guessing!

- If you're carrying the ball in one hand, keep it secure by pressing it to part of your chest and upper arm. Imagine carrying a baby in one arm while running. (Don't try this at home with your baby brother or sister!)
- Now try carrying the ball on the other side. Try moving it from one side to the other while changing direction or running on an angle.
- I set up Julian Savea for a try against Australia in 2014 simply by holding the ball out in front of me in two hands and moving it a bit to the left and right to confuse the opposition. That created space, which I was able to slip through and pass to give Julian a free run to the line.

WHEN IT GOES BAD!

Running with the ball sounds like a simple skill. Why would you bother to practise it? But this is one of the most important skills in our game — I see way too many young players tucking the ball under their arm when running or evading. You should reserve the one-handed tuck for when you need to fend or when you're trying to trick your opponent about your intentions.

Keep the ball in front of your chest. If you carry the ball too high or too low, it won't be secure.

Sometimes players find carrying the ball in two hands a little uncomfortable when they are starting to run, as they feel they can't get speed up without swinging their arms. But I know plenty of loose forwards who lick their lips when they see a player who tucks the ball as soon as they catch it! Easy meat!

ACTIVITY

Make a grid 40 metres square with a cone every 10 metres.

Sprint out 10 metres carrying the ball in two hands in front of you. Then tuck the ball under your right arm for the second 10 metres. Switch it back in front of you for the third 10 metres and under your left arm for the last 10 metres.

Pretend the cones are tacklers. Keep the ball away from them as you go around them.

EXTRA FOR EXPERTS

Practise scoring tries using both hands as you reach a cone. Score with the ball under your right arm and then under your left.

EVASION

I started out as a second-five, and I love to evade opponents while running. My ability to dodge people might have something to do with being a small guy in a game played by big guys! I'm quite good at stepping off both feet. I know my footwork will get me out of trouble and also beat the defenders. But you can overdo it, and if you do, you'll find yourself cut off from your support (isolated) and the opposition will probably get the ball.

Not good. Remember, you're not the only player on the field!

I believe evasion is a natural skill set, and all rugby players should be encouraged to use their flair and ability. You can get better by practising changing direction and changing speed.

DID YOU KNOW?

The ball is only in play for around 38 minutes in an 80-minute rugby match.

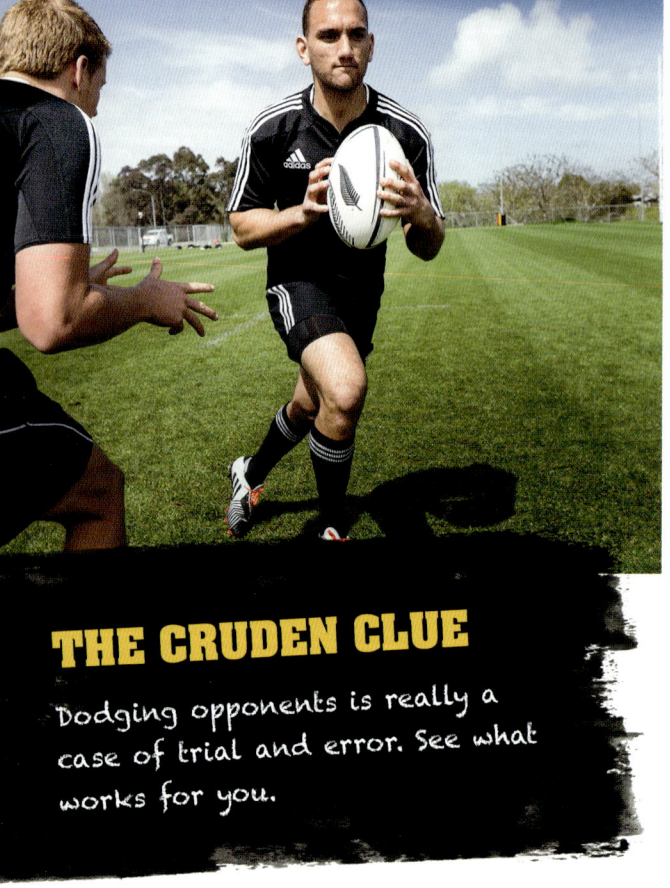

CHECKLIST

- Getting past defenders is harder than you'd think (that's why they invented passing), but the keys to successful evasion are changing direction and accelerating. Do this slowly at first, as it can stress your ankles and other joints.
- You can confuse defenders with a number of evasion techniques: sidestep, feint, swerve, spin, shuffle, stop and start, goose-step, dummy . . . Don't be afraid to develop a 'pet move'. I like to accelerate fast after a step, often with a dummy thrown in.
- Experiment with different ways to beat a defender to find what works for you. Sometimes the best practice can be dodging your way down a crowded street.

SIDESTEP CHECKLIST

- Ideally, carry the ball in two hands. If you hold the ball in one hand or under one arm, you might lose it.
- Run toward the defenders but picture two points in your mind: the defender and the space near them.
- Shorten or lengthen your stride (depending on your preference) for timing and good balance and then

THE CRUDEN CLUE

Dodging opponents is really a case of trial and error. See what works for you.

change direction close to your opponent by pushing sideways powerfully off your right foot to shift left and your left foot to go right. Plant your stepping foot to get a good grip.
- Always position the ball away from the defender so it is protected, and then accelerate if the gap you're eyeing up appears.

THE GOOSE-STEP

The goose-step is an evasion manoeuvre made famous by Australian rugby winger David Campese. You run until the defender is lining you up from the side. Then you slow down slightly to make them hesitate. At this point, you do a 'stutter step' — flick your legs quickly forward to baffle their sense of your speed — and then accelerate off, leaving the defender with their feet planted in the wrong position.

WHEN IT GOES BAD!

If you are not running towards the defender, there is not a lot of point trying to dodge them. It's called running evasion, not total evasion!

If you change direction too early or too late, you'll get into trouble.

Make sure you accelerate forward immediately after the sideways movement, otherwise the dodge is not as effective, as you'll soon be caught or the space you have created will close.

> **DID YOU KNOW?**
> There are usually around eight clean **line breaks** in a professional rugby game.

ACTIVITY

Give a friend the ball and then follow behind them as they run evasively at speed. Try to stay as close as 1 metre away.

Continue for 5–15 seconds — see how long you last. Allow the same amount of time to catch your breath, then repeat the activity.

EXTRA FOR EXPERTS

Play a game of bullrush with a number of friends — like all Kiwi kids! If you're the bull, play hard but very fair: no high or dirty tackles.

THE FEND

The fend helps you to create space while you're running by pushing off a defender using the palm of your hand. I have used the fend: it's great to give you an extra push as you waltz through a gap. There are some great fenders in the All Blacks — Cory Jane probably has the best fend in the game.

Fending is only effective if the tackler gets within an arm's length. This allows you either to push the tackler away or knock away their arms as the tackle is attempted.

CHECKLIST

- Before you bring out the big old fend, you have to get ready to sidestep or swerve around the opposing player.
- You then transfer the ball to the arm further away from the tackler. If the tackler comes within arm's length as you're sidestepping or swerving, use an open hand to push him away. Easy, eh?
- To generate extra force, try to lean away from the tackler as they approach and then toward them as you fend.

WHEN IT GOES BAD!

Raising your arm too early and allowing the tackler to drive underneath just makes it easy for them to grab you.

Your arm needs to be a bit bent as you fend, because you need to push away on contact.

Finally, if you're not focused on the bit of the defender you're going to plant your hand on and push, you'll probably miss an opportunity to get through. Remember, fending is not allowed in the U8 grades for our Small Blacks.

ACTIVITY

A good starting point is to walk about pretending to fend at objects. That way you'll get the action right and also get into the good habit of transferring the ball from one side of your body to the other.

Add the fend to your practice when you're running through the evasion activities described earlier with a mate.

THE CRUDEN CLUE

I like to push downwards if the tackler is coming at me low.

SCORING A TRY

Scoring a try is one of the greatest feelings in the game of rugby.

Because rugby is a team game, scoring a try is usually a cause for the whole team to celebrate. The All Blacks don't go in for the big celebrations that a lot of teams do, but we feel just as good about it.

Try-scoring is something we don't often practice. This surprises me, because scoring a try is worth five points.

CHECKLIST

- A try is scored when you apply downward pressure to the ball on the ground over the tryline. The ball only needs to touch the tryline to be considered 'over'.
- The number one thing to remember when you get the ball over the line is to make sure it is secure. This means having two hands on the ball if you can. If you have the ball under one arm, make sure you hold it tight and hard up against your chest.
- Diving or sliding over the line is recommended, but simply 'dotting' the ball down with two hands is also very safe.
- Sometimes you see people celebrating before they score a try — pointing to the heavens, or their mum in the stand, or

THE CRUDEN CLUE

Hold the ball tight with two hands as you dot it down.

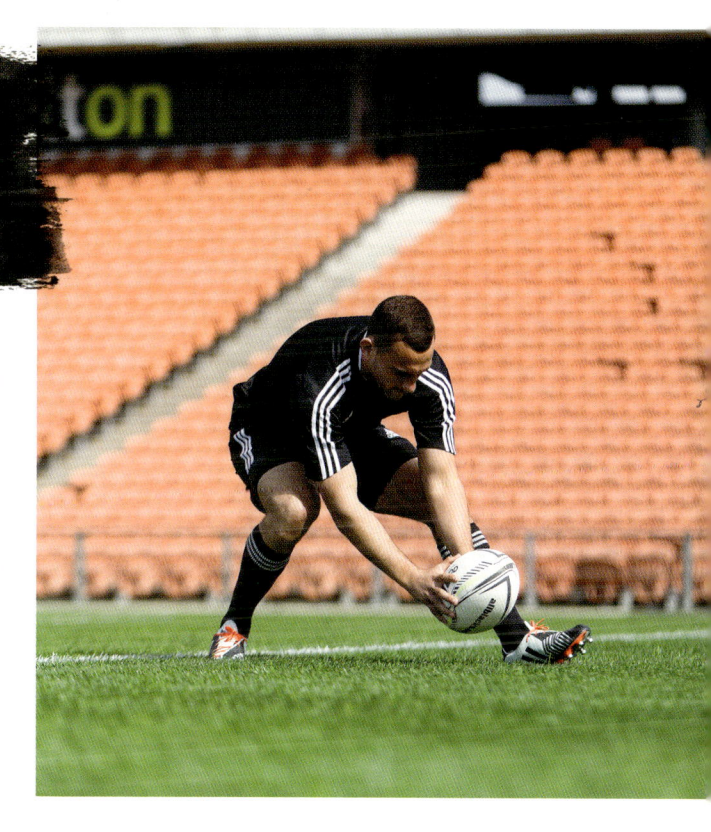

the TV camera. My number one rule is to make sure you ground the ball correctly: score first, celebrate later.

WHEN IT GOES BAD!

I've scored a lot of tries over the years, for a number of teams, but I have only scored a few for the All Blacks. It's important to make sure you don't miss these great moments by making a silly mistake.

When you go to press the ball to the ground using only one hand, it could fall or be hit away by an opponent.

One of the biggest boo boos that I see is when a player dives for a try with the ball tucked under the wrong arm. Remember, if there is a tackler coming from the right, carry the ball on the left side. It's all about keeping the ball safe!

a little uncomfortable at first as you'll probably favour one side more than the other.

When you've mastered this, see how many tries you can score past a mate who is defending.

ACTIVITY

Start by standing near a line and practise diving for the line with the ball tucked under each arm in turn. This will feel

RECORD

The fastest try in a rugby match was scored just 7.24 seconds after the referee blew his whistle to start play.

GETTING READY

Warming up before playing rugby is really important to me and it should become something you always do — get into the habit!

Warming up prepares your heart, muscles, joints and your mind for any match or training session. It improves your performance, helps you get mentally prepared, and it's a big factor in avoiding injury.

Usually after a game, all you want to do is hang out with your mates or go home. Take the time to cool down first. Helping the muscles to relax again is just as important as warming up.

WARM-UP

Here's a warm-up that works well for me. I believe a rugby warm-up should be broken into these important segments:

1. To start with, it's best to do 4–5 minutes of easy exercise such as jogging, skipping and lateral movements. Lateral movements are sideways skips or leg-crossing activities.
2. Use the next 3–5 minutes to work through low-intensity or half-paced skill activities, such as passing along the line to team-mates and 'quick hands' drills (passing along a line as quickly as you can) to warm up the body and mind for the next stage of training.
3. Now it's time to move on to dynamic stretching, which means warming up your muscles by slowly moving your arms and legs in the ways you need them to move during the game.
4. Gradually increase each movement as you repeat it. Do not force your muscles to stretch beyond their normal range.

ACTIVITIES

CALF EXERCISE

Get into a sprint-race starting position and stretch your calves by swapping heels on the ground (pause for one or two seconds). Complete four to six movements for each leg, rest, and repeat the set.

SIDE EXERCISE

Stand up straight and then take a big step forward so you form a lunge position.

Place one hand on your hip and reach the other hand over the top of your head.

Hold for a few seconds and then stand up straight again.

Complete five movements each side, rest, and then do another set.

HURDLING BACK AND FORWARD

With your hands on your hips, raise one leg in front of you and then turn your knee around 90 degrees. (The lower leg should be hanging down in a natural state.)

Now rotate that leg around to the side, a little behind your body, and then place your foot onto the ground.

Do five of these hurdles each side, moving backwards as you go, rest, then repeat. Then repeat the activity with your feet moving in a forward direction.

DID YOU KNOW?

The largest muscle in the human body is the gluteus maximus. The gluteus maximus is located in your buttocks.

ROTATING CORE

Lie on the ground for this exercise.

Place your arms out so they are extended and your palms are facing upward. Keep your legs together as you raise your knees to around 90 degrees.

Slowly but gently, lower your legs to one side, pause for a couple of seconds and then go to the other side. Make sure your shoulders stay flat on the ground.

Do five movements from side to side, then have a rest.

HAMSTRING STRETCH

Sit down with your legs apart and flat on the ground. Lean forward with a nice straight back and hold for 30 seconds. Rest, then repeat.

UPPER BODY WINDMILLS

Keeping your back straight and holding your arms straight out in front of you, rotate your arms slowly around like a windmill. Repeat in the opposite direction.

CALF STRETCHING

Place the ball of your foot against a wall or rugby goalpost and your heel on the ground. (Your front leg should be straight.)

Now put your hands on the wall or goalpost and push your hips forward to feel the stretch along the entire calf muscle. Don't push so hard that it hurts. Hold for 30 seconds, rest, then repeat on the other side.

POSITION-SPECIFIC EXERCISES

You should always work with your teammates to do the sorts of activities that will prepare you for the requirements of your position on the field.

Outside backs should do swerves, sidesteps, one-on-one **beating the defender**, and inside backs should do ball handling, running onto the ball, **box kicks** for the halfbacks, and **line kicks** for first-fives.

Loose forwards should do wrestling for the ball, running off players for short passes, and getting down and up off the ground, whereas tight forwards should do lineouts, throws, lifting, and one-on-one scrums.

Remember, you need to be ready (warmed up) for the game and the tasks you will have to perform.

Then you can slowly introduce body contact and gradually increase the intensity.

UPPER BODY SWINGS

Keeping your back straight, swing your left and right arms to your right side (your left arm will swing across your chest). Then swing your arms back to the left side.

QUAD STRETCH

Kneel down on the ground with your front knee at a 90 degree angle. Keep your back straight, chest out and push forward.

Hold for 30 seconds, rest, then repeat the exercise.

COOL-DOWN AND STRETCHES

Cooling down after playing or training lets your body **recover** more quickly, so you are better able to front up for the next week's activities. It's also the best time to improve your flexibility — to get bendier — so learn how to do it correctly.

Cooling down and stretching after rugby should last for 10–15 minutes. It should consist of aerobic exercise — slow jogging or brisk walking around the field with a ball is one of the best ways to cool down.

You should also do static stretching, which means you should stretch on the spot for 10 minutes after your light jog.

If you want to achieve greater flexibility, hold these stretches for 45–60 seconds during the cool-down period, as this will help your muscles relax.

ON TOUR

Being on tour is fun, but it does depend on what country you are in and how much time you actually get to look around.

The tour I have enjoyed the most so far was a short tour to South Africa in 2013. I enjoyed it because of how we played rather than what I saw or experienced away from rugby! It was a huge challenge to play in such a big, intimidating environment as Ellis Park.

RUGBY WORLD CUP: DID YOU KNOW?

1999
- The All Blacks defeated England at Twickenham 30–16 and Scotland 30–18 in the quarterfinals, but then were unexpectedly beaten 43–31 by France in a semifinal in London.
- Australia beat France in the final to become the first two-time Rugby World Cup champions.

2003
- The All Blacks lost to Australia in the semifinal.
- In the final, Australia played England, and the scores were level at 17–17 at full-time. Jonny Wilkinson kicked a last-minute drop goal in extra time to break the deadlock, and England earned the northern hemisphere its first Rugby World Cup title with its 20–17 victory.

GLOSSARY

BEATING winning against your opposition (beating a team) or evading a tackler (beating a defender)

BOX KICK a kick taken from behind a scrum or breakdown, usually by the halfback. The kicker turns to face their own tryline and kicks the ball back over their shoulder, into the 'box' of space behind the breakdown

LINE BREAK when a player with the ball runs through the opposition's defensive line without being tackled

LINE KICK a kick for the sideline

RECOVER catch your breath, return to your natural state, or to get something back (like the ball!)

Stadiums I've been lucky enough to play at over the years

Kings Park, Durban, South Africa

Stadio Olimpico, Rome, Italy

Aviva Stadium, Dublin, Ireland

Twickenham, London, England

ANZ Stadium, Sydney, Australia

136

Murrayfield, Edinburgh, Scotland

Stade de France, Paris, France

Soldier Field, Chicago, USA

FNB Stadium, Johannesburg, South Africa

Millennium Stadium, Cardiff, Wales

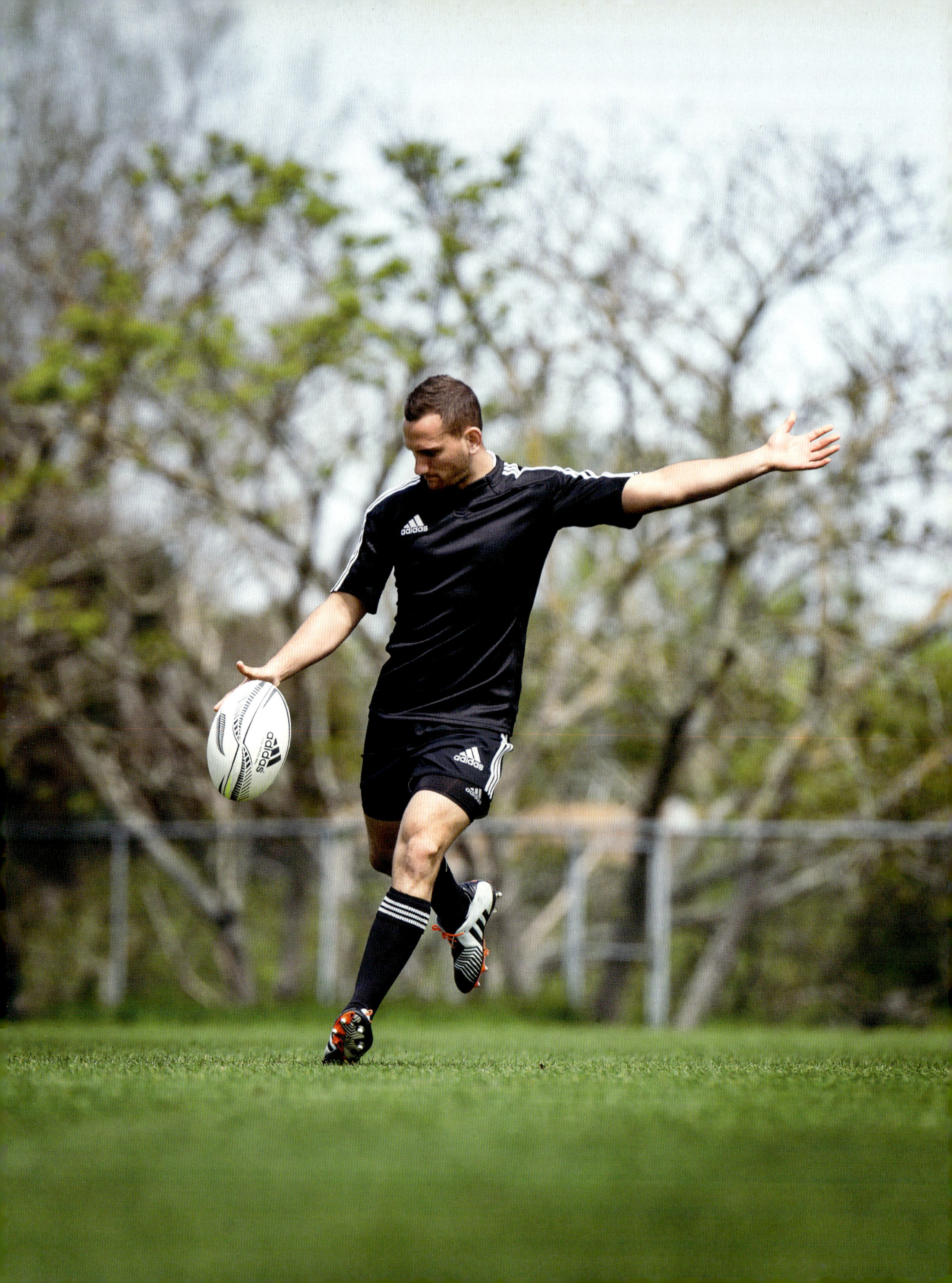
adidas
adidas
adidas

CHAPTER 5
KICKING

After passing, kicking is the second most important skill for a rugby player to have. To be honest, at times I think it is the most important.

If you think you have to practise a lot to get your passing right, that's nothing compared to practising when it comes to putting boot to ball.

KICKING OFF EITHER FOOT

All players need to be able to **kick off either foot**. As a first-five, I have to kick a lot. One of the objects in rugby is to advance the ball towards the opposition's goal line. Kicking is one of the most effective ways of doing this, whether it's by kicking into touch a long way downfield so that play restarts closer to your opponent's goal line (which is a great way of relieving pressure, if your opponent is hot on attack), or kicking behind the defence so that they have to turn around and run back. This is a great way of applying pressure!

Most basic kicks these days are about putting the ball into space or down the field, or relieving pressure.

One of the most elegant kicks in rugby is a long spiral punt. It's not used as much these days, but it's wonderful to see when it does get used!

THE CRUDEN CLUE

Experiment with the ball to find your 'sweet spot'. My sweet spot is just above the point of the ball.

- Lean slightly forward so you are standing 'over' the foot that you intend to kick with.
- Step forward into the kick (but don't rush). Plant your left foot (if you're kicking with your right), point the toes of your kicking foot and swing it to make contact with the ball with the 'bootlaces', or your upper foot. Your kicking foot should follow through in the direction of the target, and your head should initially stay down.
- Try to return to a nice, balanced state, as this will help you balance out the whole movement — a bit like landing after a jump or flip in gymnastics.

CHECKLIST

- The first thing to do before you even think about kicking is to control the ball with both hands. Your fingers should be around the middle of the ball and across the seams.
- The key to being a confident kicker is to drop the ball consistently. If you're kicking with your right foot, release your left hand (this allows your right foot to follow through to your left).

WHEN IT GOES BAD!

Like most things in our game — and many other games — if you lift your head too early, you'll miss controlling the all-important moment of contact with the ball.

Turning your shoulders at the point of contact with the ball will reduce the power and affect the direction that the ball will fly in. Stay straight and square.

One of the biggest mistakes I see in our

game is when players try to kick the ball too far, losing timing, balance and power. Remember, it's all about technique. Like a golf shot, it all works better if the timing is right, not just the force.

Younger players or people who haven't kicked a lot of rugby balls tend to throw the ball into the air to kick. This doesn't help to guide the ball onto the right place on your foot, and also can make the contact between the ball and the foot happen too early or too late. The ball will go straight up into the air or roll along the ground — neither one is a good look!

THE CRUDEN CLUE

In a game I can kick the ball about 50-60 metres but accuracy is more important to me than distance.

ACTIVITY

With a friend standing just 5 metres away, start with trying to kick front on and accurately to one another. The kicks should be a little 'snap' action so they can be caught on the full (leaving out the full-on follow-through you would use if you were going for distance). Turn about 45 degrees to the side and repeat.

Next, kick the ball accurately to one another 10 metres away, then 20, until you reach your maximum distance.

EXTRA FOR EXPERTS

Play a game of kicking golf. Nominate a target and kick the ball to it. See how many kicks it takes to land the ball on the target.

CHIP KICK

The chip kick is sometimes called the most dangerous kick in rugby. It's a little kick you pop just over the head of the defenders, so that either you or one of your team-mates can run through and re-gather it on the other side. It's dangerous, because you can use it to take a whole line of defenders out of play. But if you get it wrong, it can be dangerous for your own team, too!

I like to move onto the ball and I'm always looking for space behind the defence. But remember, when using this kick, you have to be pretty sure you're going to get the ball back, whether by re-gathering it yourself or by letting your team-mates know you're going to try it. Communication is essential. I like using chip kicks because I can accelerate quite quickly off the mark. If I've told my team-mates it's on, they're right there for me to pass to once I've re-gathered the ball.

DID YOU KNOW?

There are about 40-50 kicks in general play [excluding placekicks and restarts] in an average match.

CHECKLIST

- Control the ball by having both hands on it. Keep your head down and your eyes on the ball.
- This time, have a short follow-through with the kicking foot — it's just a little snap of the foot rather than a full-throttle swing. The less time your foot is in contact with the ball, the better.
- If you raise your toe (instead of pointing it) you get a little backspin. This is good, as the ball should bounce back into your waiting arms.
- Always keep your head and shoulders still and always chase the kick, especially if you chose to chip because you intended to collect it!

WHEN IT GOES BAD!

Chip kicks don't work that often, to be honest!

Kicking the ball too hard is the most common fault. First, If you kick it too hard, you won't get there in time to collect it. Second, if it goes too high, others will have time to get under it as well. It's a fine line between technique and force!

Kicking the ball straight to an opposition player might sound like something too silly to list as a thing that can go wrong, but believe me, I've done it! If you kick the ball too close to the defender, they can charge it down. Worse, the person whose head you're trying to kick the ball over might catch it and just take off. Fail!

Failing to chase hard enough, and failing to tell your team-mates that in a few seconds you'll be on the other side of the defence gathering up the bounce and looking for support, are other ways to muck up your chip kick.

THE CRUDEN CLUE

Make sure you put the ball into space. You're not aiming for the defenders!

ACTIVITY

With a friend standing a few metres away, run toward them and gently kick the ball over their head. Recover the ball (aim to take it on the full), return and kick again.

If you can't re-gather the ball, it's their turn and you are the defender. It's a bit like piggy in the middle but you only need two players.

EXTRA FOR EXPERTS

Practise on your own and see if you can get the ball to pop back to you as you chip kick (with a bit of backspin).

DROP KICK

My granddad always said that you need to be able to do this kick, as you never know when it will win you a game! He was right. As a first-five, I need the drop kick for a whole lot of other reasons, too. I use it at restarts, and an accurate drop kick enables our team to get underneath the ball, win it and go forward. I use it at 22 drop-outs, and quite a lot of the time I'm using it to go a long way downfield in this situation. And I have it up my sleeve as a way of scoring points from the field of play.

Basically, a drop kick is a kick where the ball is actually dropped onto the ground for your foot to strike as it swings through. For low kicks, strike the ball close to the ground — these are good for restarts when you have a kick in mind that goes straight to a chasing team-mate. For high kicks, allow a fraction of a second's pause after the ball has hit the ground to let it bounce higher. These are good if you want a number of team-mates under the ball, as this gives them more time to get there.

I usually have a call for these different kicks. Sometimes we use visual signals such as holding up a number of fingers hidden from the opposition behind the ball before I kick. For example, 1 finger is to go short, 2 is for a flat kick and 3 is to go long.

I get side on to my target, and I like to drop the ball to the ground to the side of my planted foot and in front of my kicking foot.

CHECKLIST

- The most important thing is the drop of the ball. It's that simple. Drop the ball

THE CRUDEN CLUE

Keep your hands on the ball for as long as possible to guide it to the ground.

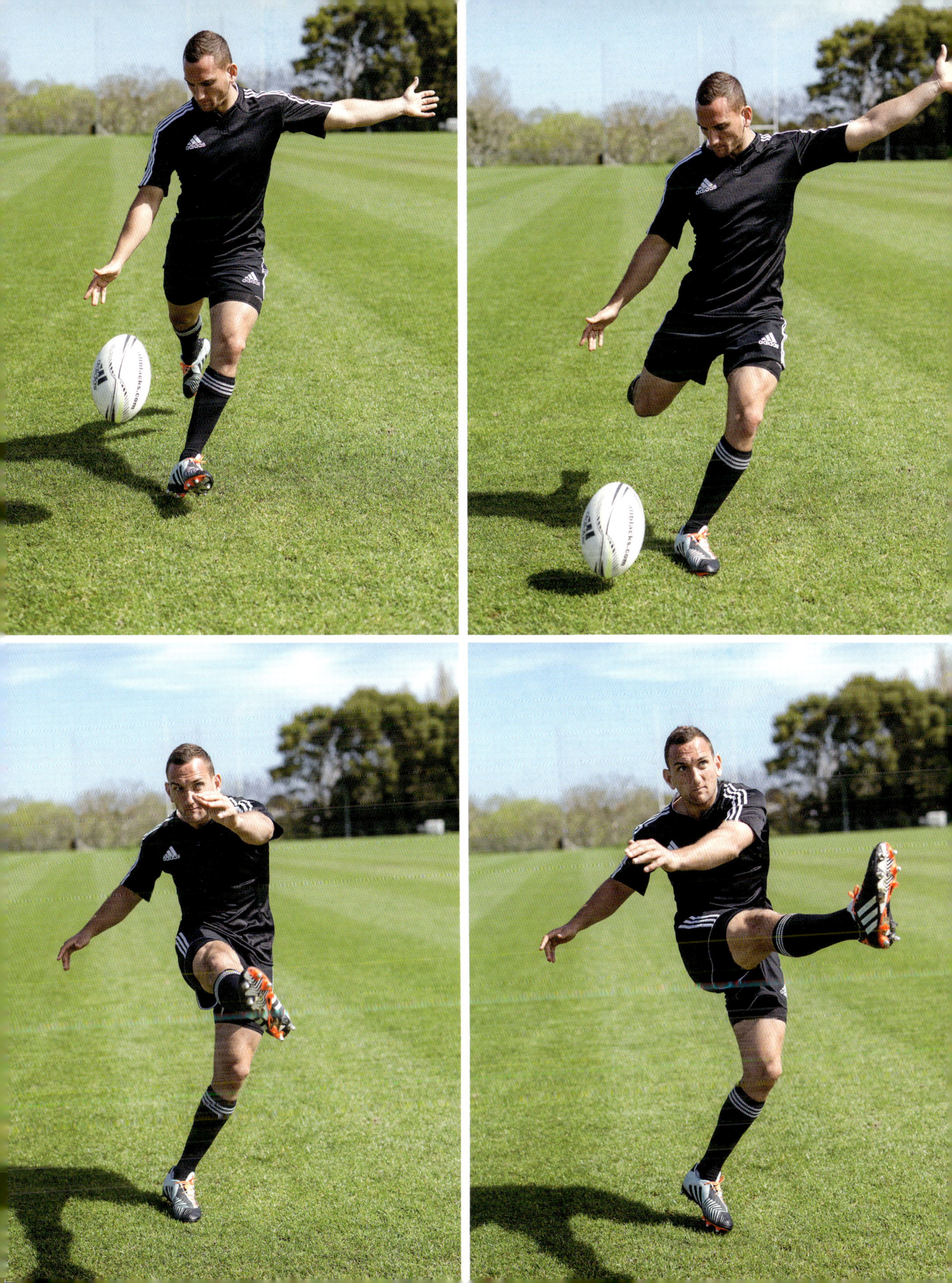

onto its point. Actually, I should say guide the ball onto the point, as this is the most important part of the drop kick.
- When the ball hits the ground, it's time to kick. Timing is everything.
- Kick with the instep of your foot, the area between the arch of your foot and your bootlaces.
- You then transfer your weight through with a little skip or hop. This comes quite naturally, so there's no need to try to introduce it. But if things aren't going too well, then practise this motion continually.
- Always have a big follow-through. Yes, even in the drop kick!
- If you're trying to do a high restart, let the ball bounce a little before your foot strikes it so that you can get underneath it more. This is how you get the 'hang time' that lets your team-mates get under a short restart.
- If you want it to travel low — which we sometimes do with restarts or drop-outs from the 22 — then hit it close to the ground the moment it lands.

DID YOU KNOW?
The longest recorded successful drop goal is 77.7 metres — three-quarters of the length of the field! It must have been a windy day!

WHEN IT GOES BAD!

There are a lot of things that can go wrong with a drop kick, because the ball is bouncing off a piece of grass before you kick it.

If the ball topples as it falls, it's not going to end well. The same goes for if you don't hit it with the right part of your foot.

The other problems apply to most things that we've already covered, such as lifting your head too early, taking your eyes off the ball, rotating your shoulders as you kick, trying to kick the ball too hard and just generally losing the timing.

ACTIVITY

On a rugby field, see if you can drop kick the ball so it lands exactly 10 metres away. Make sure you practise this going both to the left and to the right.

EXTRA FOR EXPERTS

Using the goalposts, see if you can get the ball over the posts from a number of different angles and distances. If the wind is blowing, study the effects it has on your kick and adjust accordingly.

GRUBBER KICK

Like a chip kick where you apply backspin, the grubber is one of those kicks that uses the oval shape of the ball to your advantage. The grubber is intended to roll end over end along the ground. When you get skilled in performing grubber kicks, you can make the ball run toward an opponent like a puppy dog, only to leap unexpectedly over their head.

Sometimes I need to apply pressure by putting a grubber kick through a gap in the opposition's defensive line. Sometimes, I aim the kick so that it sits up nicely for me or for a team-mate to run onto and re-gather it.

There are actually a lot of different ways to achieve this, so have some fun experimenting with the way you strike the ball, and make sure you watch carefully to see how it behaves.

Unless you intend the kick to go into touch, you should aim for only enough distance so that your team-mates can reach it.

CHECKLIST

- Your hand position on the ball should be controlled and secure, as most grubber kicks are done on the run.
- You need to position your body weight

> **THE CRUDEN CLUE**
> When you make contact with the ball, the knee on your kicking leg should be bent.

over the top of the ball, because you have to push the ball forward onto the ground.
- Your toe should be pointed to the ground and you should use a short, stabbing motion of your lower leg. What you're actually doing is kicking the ball into the ground in front of you. It's a bit like skipping a stone on a pond.
- Follow through and then chase the kick down the field.

WHEN IT GOES BAD!

The most common mistake made by people trying a grubber is kicking the ball into an opponent's legs. This happens a lot and what usually happens next is it bounces favourably for the opposition. Not a good thing!

If you kick the ball too far, the grubber kick becomes ineffective. You'll only look to the grubber for distance if you're close to the sideline and aiming for territory.

Spiralling the ball rather than hitting it so that it goes end over end robs the grubber of its natural advantage, because the spiral will make the ball behave unpredictably off the ground.

If the ball contacts the ground too far in front of you — more than 2 metres — it will start hopping and bounding and might not do what you want it to do.

Remember to point your toe. This is one kick you don't want to give any air.

ACTIVITY

With a friend, grubber kick a ball accurately to one another 5 metres away, then 10 metres, then from some distance, and finally running towards one another (for pressure!).

EXTRA FOR EXPERTS

Kicking golf again. In your backyard, grubber kick the ball to try to hit or stop on an object. Pick nine sturdy objects (not the pot plants!). See if you can better your score each time with the same objects. Challenge a friend and see if they can beat your score.

THE UP-AND-UNDER (THE BOMB)

A great way to apply pressure to the opposition is to put the ball high into the air, allowing your team-mates to get underneath and battle for possession.

In the olden days, this was called a 'Garryowen' kick, because it was named after the Garryowen Football Club, which was famous for using the tactic. Don't ask me how a rugby club came to be called Garryowen!

I like to use this option when I am at the back (that is, where the fullback usually stands). I make it so that either I can chase and contest the ball myself, or so my team-mates have a chance to get the ball. Sometimes I remain in the **pocket** (right behind my team-mates) after hoisting a bomb in case my kick is returned.

Try to make contact with the ball just behind the point, to make it rotate end over end in the air. Raising your toes on contact will help with this. This enables support players to catch the ball in the air more easily.

This is another kick you'll want to experiment with. Sometimes you can get the ball to float or fly differently so that it becomes harder for the opposition to catch — good on a windy day, I reckon!

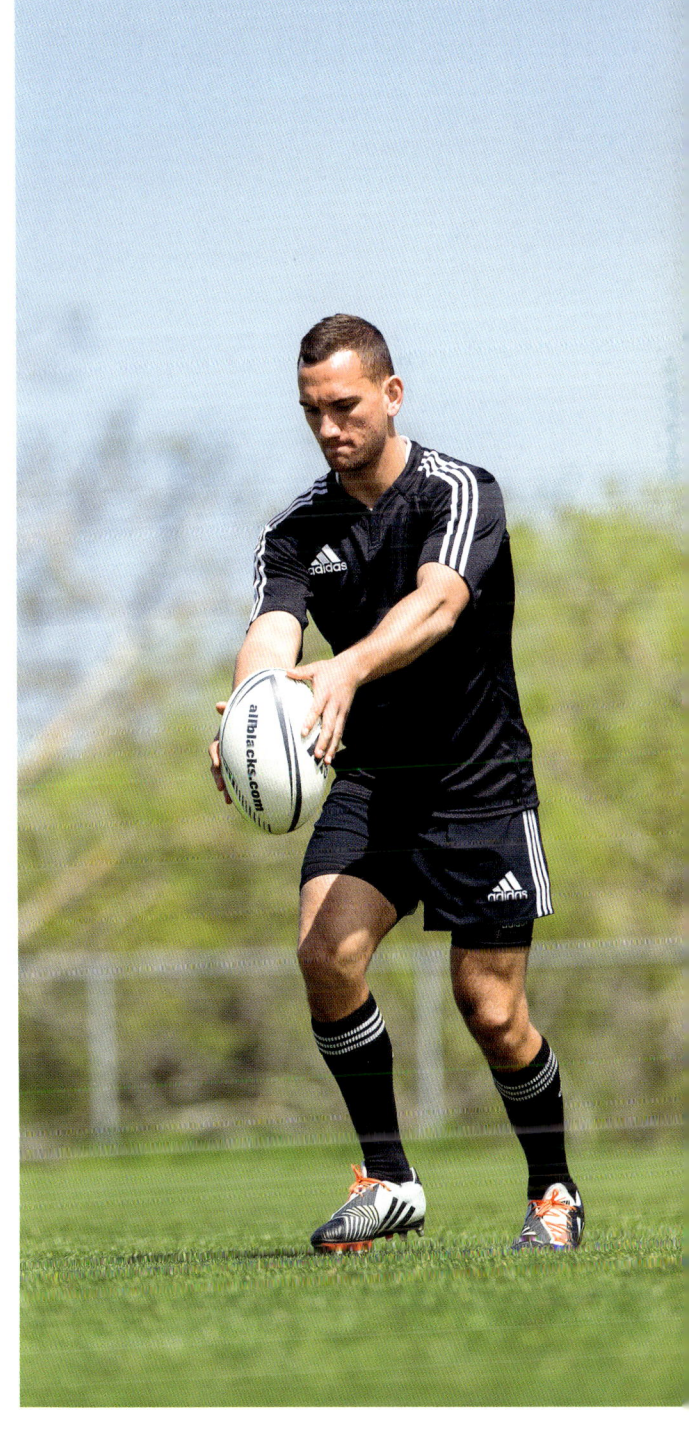

THE CRUDEN CLUE

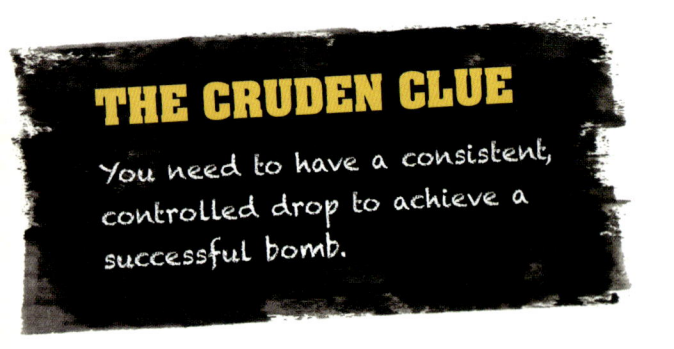

You need to have a consistent, controlled drop to achieve a successful bomb.

CHECKLIST

- We all love doing bombs! It's a great skill to practise because you don't need a big area. All you need is a lot of sky!
- Hold the ball in an upright position and, as always, keep your eyes on the ball and your head over the top of it.
- Now for the tricky bit. Turn slightly side on to protect the ball. (This is just a slight adjustment.) Swing your hips into the kick and make contact with the ball just behind and under its point as you drop it.
- Raise your toes on contact and make sure you have a high follow-through. Give it heaps! You want this one to come down through the clouds.
- Make sure the kick will land where your team-mates have got just as much of a chance of reaching it as your opponents, so they can 'contest possession'.

WHEN IT GOES BAD!

The kick might go too far or not far enough. Remember, this is a pressure kick — you want it to put pressure on your opponents, not on your own team.

Don't lift your head too early (a common kicking error). Make sure the ball is dropped correctly onto your foot.

In an up-and-under, if your follow-through isn't high enough, you won't get any height at all. We talk about follow-through a lot, and this is one occasion where it definitely makes a big difference.

ACTIVITY

With a friend, bomb kick a ball accurately to one another from 10–15 metres apart. Every time somebody drops the ball, the kicker gets a point. First to 10 is the winner!

EXTRA FOR EXPERTS

Do the same activity, but this time follow the kick and put pressure on your friend to catch it.

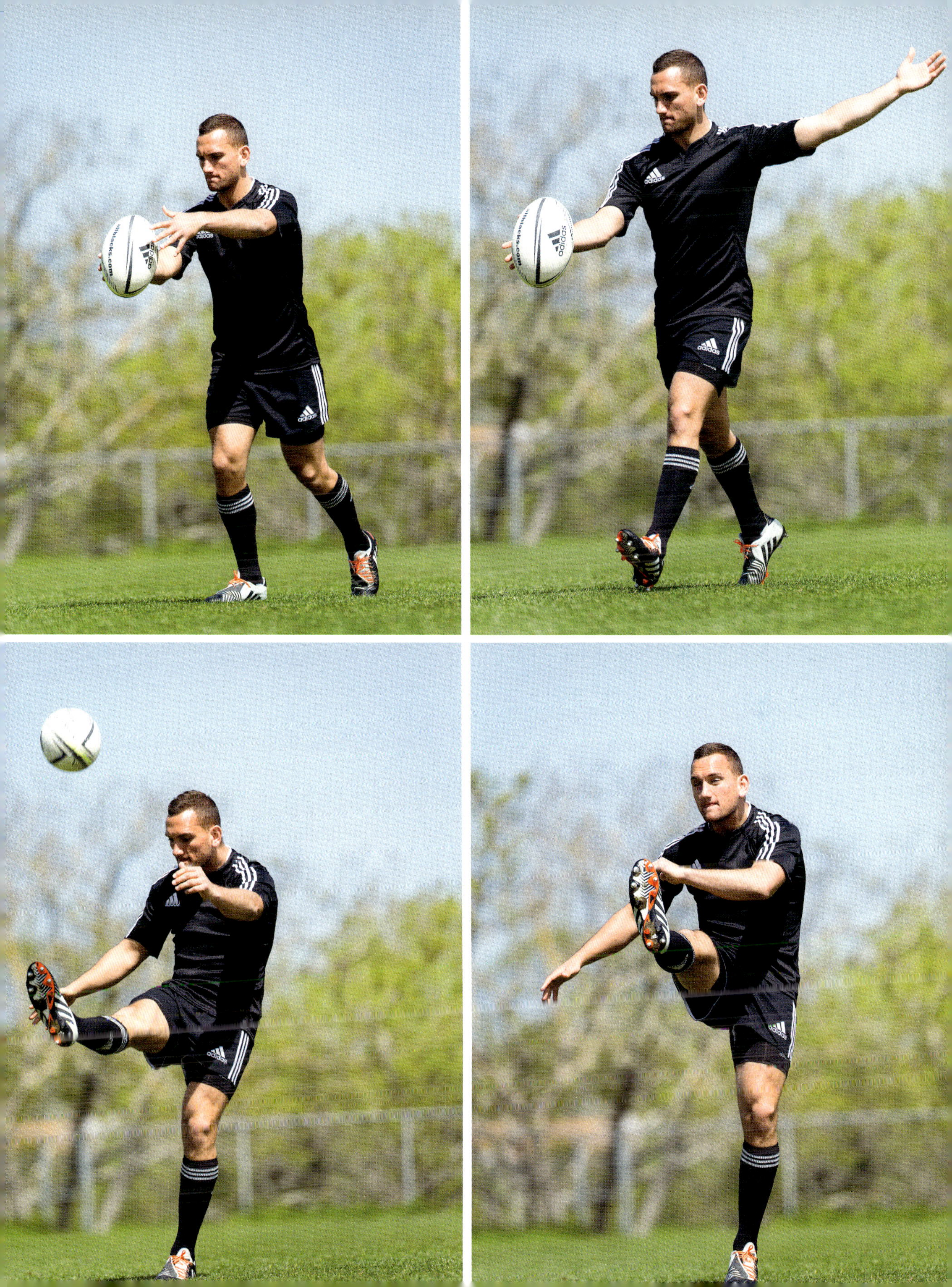
adidas

PLACEKICK

One of the most enjoyable tasks I have on a rugby field is kicking at goal. Kicking at goal involves kicking a placekick.

To kick a ball accurately through the posts from where it has been placed on the ground is a big challenge, no matter what grade you're playing. But that's what you need to do to score points from a penalty or to convert a try. Sometimes the easy ones are just as challenging as the difficult ones.

Goalkicking is a very personal skill. People have their own styles of goalkicking, and none of it is very important as long as the ball goes over. You only need to look at the following comments on how to do a goalkick when something is going wrong. When it does, you'll usually find you've missed a step along the way.

The key to a successful kick in my opinion is the placement and alignment of the non-kicking foot. It must be hip-width from the ball and aligned to the target.

CHECKLIST

- I like to lean the ball slightly forward to expose the 'sweet spot', but a lot of players now lean the ball right over so they are looking right down on top of the sweet spot.
- Before I move back, I stand over the ball in the kicking position, with my non-kicking foot close to the ball.
- Now I just move back to a comfortable position. You can do whatever you feel is best. I turn 45 degrees then take three steps directly back on that angle.

- Then I focus my eyes on the 'sweet spot' on the ball again.
- I look at the target, which could be an imaginary bull's eye set in the crowd and in the middle of the posts. Then I bring my focus back to the ball.
- When I pick my spot in the crowd, I then draw a line in my mind to the ball from that spot. It's my visual graphic, just like you see on TV!
- Relax, relax, and relax! Breathing is crucial.
- Focus. You could use some key words to help with this. I tell myself: 'Process. Trust the middle (goalposts). Get through the ball.'
- I then visualise the successful kick (I play a little film in my head in which I kick the goal successfully). I do a little shuffle before I go into my approach. Yeah, I know — this has sometimes made the opposition charge my conversion attempts too early.
- Most kickers approach the ball by running in an arc but I come straight at the ball.
- I then turn my non-kicking shoulder side on to the target, place my non-kicking foot in line with the target and swing through the strike zone, the middle of which is the ball.

DID YOU KNOW?

There are usually about 6 penalty goals in a match, about 3 for each team.

foot, and then I return to a balanced stance and watch the ball go through the posts (of course). Hopefully, yours will too!

I'll never forget the 2013 All Blacks test against Ireland, because the score was 22–22 and time was up on the clock when I had a chance to kick the winning conversion. I missed! But that little shuffle I do meant that the Irish players had rushed the conversion too early. The ref gave me another chance. I had to refocus on my own processes and, luckily, the second attempt went straight through the posts for a 24–22 victory. That's what they call a happy ending!

WHEN IT GOES BAD!

Lots of things can go wrong with placekicks, but the main ones are quite simple: lifting your head, planting your non-kicking foot too far from the ball, and having your leading shoulder too open (you need to keep it hunched with the follow-through).

Again, if you have a poor or non-existent follow-through and an unbalanced finishing position, things can get ugly.

- I keep my head over the ball and down after the initial strike, just like in a golf shot.
- I like to make contact with the ball using the top of my **instep**, and then rise onto the toes of my non-kicking foot.
- Finally, I follow straight through. This makes me do a hop on my non-kicking

ACTIVITY

This is the routine I did when I was at school.

Standing right in front of the posts (as close as you like), take 5–10 shots at goal. Then move 15 metres to the left of the centre and take 5–10 shots. Finally, move 15 metres to the right of the centre and take 5–10 shots.

If you can, move right over near the touchline on both sides (5 metres from touch) and have two or three shots. Then move back from the 22 and have one or two long shots.

EXTRA FOR EXPERTS

Increase the number of shots, increase angles and increase distance. Maybe even practise when it's raining because, as you know, rugby is a winter sport and often played in the wet. The wet ground and a wet ball can affect your mental preparation for a kick, besides making it physically difficult. It's a lot easier in a game if you know you can do it from your training.

THE CRUDEN CLUE
Find a rhythm that suits you!

BANANA KICK

The banana kick curls in the air off your boot, and you can control the direction it curls. It's not a standard tactical kick — I only really use the banana kick when I have to kick the ball into touch and I am close to the touchline.

CHECKLIST

- Hold the ball directly in front of you with a pointy end in each hand.
- Aim with the pointy end of the ball, by lining up one end with where you want the ball to go (like a gun sight).
- If your target is on your left, line up the pointy part of the ball with your target and kick the back of the ball, a little toward the right-hand end, with the top of your right foot. Instead of following through to the left (as in a normal kick) the follow-through will be straight, or even a little to the right. You're trying to put lots of spin on the ball.
- Do the opposite if you are kicking to a target on the right-hand side.
- Only take a couple of small steps forward and guide the ball to your foot.
- Did I mention you must follow through?

WHEN IT GOES BAD!

Banana kicks are very difficult. It doesn't take much for a banana kick to turn into a banana split, and then everything else can turn to custard!

Don't take too long a stride into the

kick, as this kick is more about precision than power. The hit on the ball is the most important thing here.

Kicking the ball in the middle — near its waist — will also have a drastic effect, as it won't bend at all.

Finally, use this kick only when you have decided you must get the ball out of play. Concentrate on hitting it right.

ACTIVITY

As with all kicking activities, practice makes perfect. A mate is useful to fetch the ball for you and return it so that you don't need to go and get it each time.

As with all kicking, there are lots of ways you can hold, drop and strike the ball. Each makes the ball behave a bit differently. This is where practice and experimentation help.

Get your mate to move around so you have to try different styles.

EXTRA FOR EXPERTS

Kicking golf again. In your backyard, banana kick the ball to try to hit or stop on an object. You can deliberately 'banana' the ball along the ground to fool your opponents or get it around an obstacle.

THE CRUDEN CLUE

If you want the ball to curl to the right, kick the ball on the left-hand side!

TAP AND GO

The laws of rugby let you restart play after a penalty is awarded by just touching it with your foot — it has to actually leave your hands as you tap it — then running the ball or passing it. When a penalty has been awarded, everyone needs to concentrate just in case your team or the opposition decides to take a quick tap and go. Don't turn your back on play.

It might seem like a simple skill, but like everything, it's only when you've practised to the point where you know exactly what you're doing that you can be confident you'll get it right every time.

CHECKLIST

- It's best to place the ball on the ground where the mark has been made by the referee. That way, you know you're taking the tap kick from the right place. Otherwise, all your good work in restarting play quickly will be undone when the referee calls you back to do it again. I also prefer to do the tap kick with the ball on the ground, because doing the little kick to yourself in the air is a lot riskier.
- Move the ball slightly forward using the side of your foot, pick up the ball with two hands and run or make an accurate pass.

WHEN IT GOES BAD!

Lots of things can go wrong, but I guess the best bit of advice I can give is to back yourself. I've done a quick tap and go in a couple of test matches now and both moves have ended up with my team scoring. That said, one of the worst things you can do is take a quick tap kick when your team could easily have scored points another way, and something goes wrong. You'll feel pretty stink, but take responsibility for your decision. Hopefully your team-mates and coach will understand!

You can knock the ball on or drop it if you do your quick tap from your hands. Place the ball on the ground instead.

If the opposition sees what you are up to, you can run straight into an opponent. Or, just as bad, if your own team aren't awake, you can make a good break and find there's no one supporting you.

Remember, if you take a tap kick from the wrong place or the ball doesn't leave your hands, the referee will drag you back.

ACTIVITY

Start with the ball on the ground, move it forward with your foot, pick it up and repeat.

Try a difficult one by running up to a mark, tapping the ball by dropping it onto your boot and doing a little kick to yourself. Take off! Then do it again, and again.

CROSS-FIELD KICK

The cross-field kick has been in the game a long time but it has probably become a more common attacking weapon in the last few years as defensive lines have become harder to breach.

The kick is usually done by the first-five and it's usually received by a big tall forward or winger who is waiting way out towards the sideline, hopefully beyond the last person in the defensive line.

CHECKLIST

- Catch the ball from the halfback first! Then turn so your body is pointing toward where you want the ball to end up.
- I like to kick this ball quite low, as your receiver doesn't have to hang around waiting for it, and it doesn't give the opposition time to adjust and get under the kick as well.
- Kick the ball and watch it go into your mate's hands for a glorious try!

WHEN IT GOES BAD!

You can 'overcook' these kicks, which means you kick the ball too hard and it goes out over the sideline, so don't do a big follow-through.

Cross-kicks are another 'hero or zero' kick. If they work, you're a hero. If they don't, your total out of 10 will be a big fat zero!

Needless to say, everyone — especially your receiver — needs to know what's coming up, so communication is important. Your receiver needs to time their run so that they don't get ahead of the kick, or arrive too late for it. It's something you and the people who might receive cross-kicks need to practise together.

If you don't hit a cross-kick hard enough, the opposition can grab it and take off down the field. This is bad, because your team is set up to play with the ball, not to defend.

If there is too much hang time on the ball, the opposition will get under the ball to compete or arrive to tackle your poor, lonely receiver.

ACTIVITY

Using the full length of a rugby field, jog up and down with a mate and take turns kicking the ball across the field to each other.

Make sure you keep running forward so it's not just a basic to-and-fro kicking practice.

To make it fun, practise these kicks on your own, pretending you have just received the ball from your halfback. Kick with imaginary targets in mind.

HYDRATION

I like to always keep a water bottle handy.

A good test to see if you're hydrated or not is the colour of your wee. If it's clear, your hydration is good. If it's a little yellow then you probably need a bit more water.

Hydration is definitely important for your performance, and is something to keep in mind before, during and after any sports, especially rugby.

I usually drink a Powerade just over an hour before kick-off to get my energy levels up. During breaks in the game, I like to take little sips of water. After a game, I like to have a protein shake, usually made from protein powder, but when you're younger a banana will do just fine.

Dehydration (lack of fluids) can make you tired, affect your decisions on the field, cause cramps (that's where a tired muscle bunches up and pretty much ties itself into a knot: it hurts like heck!), heat stress or even heatstroke. Thirst is not a good gauge of your water needs during rugby. For you to be at your best, you should follow the following guidelines:

- 2 hours before the game — drink 300–500ml (1–2 glasses)
- 15 minutes before the game — drink 300ml
- During the game or training session — drink 100–150ml every 10–15 minutes
- Afterwards — drink 500–750ml per hour for 3 hours

Did you know you will continue sweating for quite a while after you stop exercising? You also lose more fluid in hot and humid conditions. So you need to increase your fluid intake, too.

You can drink water that is flavoured and sports drinks (4–8% carbohydrate) during activity lasting longer than one hour. Don't overdo it just because you like the drink — those sugary drinks are strictly for hydration and recovery.

Avoid caffeinated energy drinks immediately after exercise, and don't share drink bottles, because infections can spread between players. As my mum always said: 'Don't share your germs, son.'

AARON'S TRAINING SCHEDULE

DAY	SMALL BLACKS	TEENAGER	SUPER RUGBY
MONDAY	At school looking forward to playtime and lunchtime where we would play lots of games.	Training mainly during breaks in school time.	Review game. Gym work. On-field skill-based work.
TUESDAY	Rugby practice — lots of skills and fun games.	Practice with lots of skill development.	Team training in morning and gym work in the afternoon.
WEDNESDAY	School time: lots of games with mates. Practise my favourite Jeff Wilson and Christian Cullen moves.	Go to the park and play touch rugby with mates.	Day off.
THURSDAY	Rugby practice — fun skills and a bit of a team run.	Rugby practice with emphasis on team training.	Team clarity work in the morning which is low-tempo, role-focused stuff. Afternoon is a full-on team training session.
FRIDAY	More fun at school during breaks.	Nothing but school work. No rugby. Aw, man!	Captain's run, which sees the team at the ground going through a few low-tempo activities.
SATURDAY	Game day and more fun with mates afterwards!	Game day!	Game day, usually with a sleep in the afternoon if it's a night game.
SUNDAY	More running around with mates.	With mates kicking the ball around.	Recovery day, which could be time in the pools or stretching.

RUGBY WORLD CUP: DID YOU KNOW?

2007
- This tournament was hosted by France. The All Blacks lost to France in the quarterfinals in Cardiff.
- South Africa were the eventual winners. They won 15-6 against England.

RECORD
The largest rugby scrum ever recorded was in England in 2014. It consisted of 1008 people!

GLOSSARY
INSTEP arch of the foot

KICK OFF EITHER FOOT using either your right or left foot to kick the ball

POCKET the space behind the bunch or line of opposing players

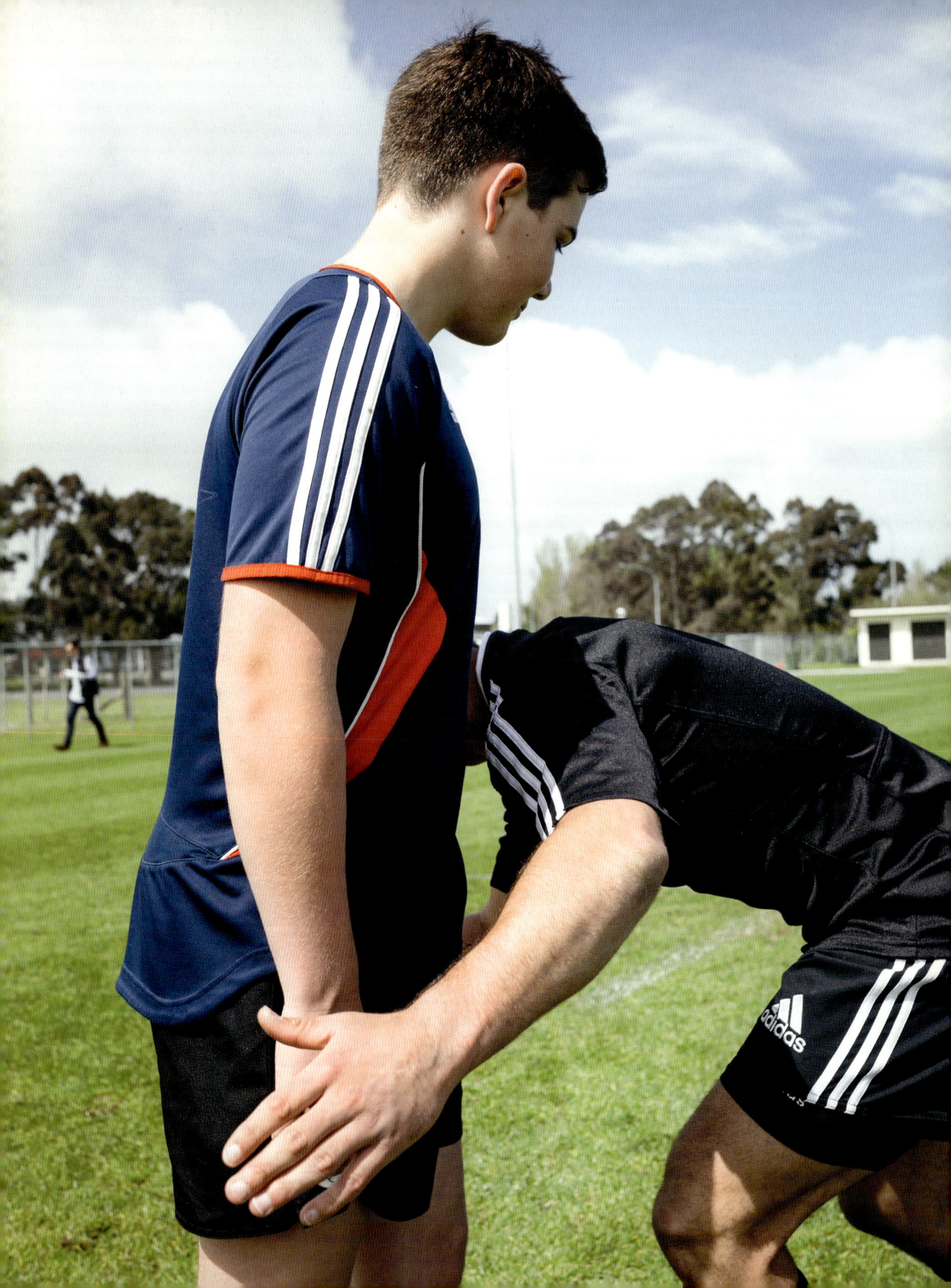
adidas

CHAPTER 6
TACKLING

Tackling is what rugby is all about, and it's what makes our great game different from lots of other games. You don't see players pulling one another to the ground in soccer or netball — or at least, you're not supposed to! Tackling — even being tackled — is fun, as long as you have a good technique.

Being able to tackle effectively and safely is the key to enjoying rugby. Making a try-saving tackle can be nearly as satisfying as scoring a try. Most people are afraid of this important part of the game when they start out. But after you've practised it a lot, even little guys like me find a method that suits us.

If you follow this checklist and the extra tips I've given you below, you'll find yourself looking forward to that next tackle!

CHECKLIST

- To start with, position yourself to one side of the ball-carrier and focus on the area you're going to hit (coming from front on gives you fewer areas to target). Coming from the side also gives the ball-carrier fewer options and importantly, it lets you get your head on the right side of the tackle.

DID YOU KNOW?

There are 258 tackles in a typical game. The openside flanker makes the most, with around 15, and the fullback makes the least — about 4.

THE CRUDEN CLUE

Get in as close as possible and get your shoulder on them first, then wrap your arms around.

- Keep your face up! Keep your eyes open and your back straight, and put your hands and arms in the ready position. You're about to grab someone like a monster!
- Now it's time to prepare yourself for the impact — the fun part.
- Keep your eyes open! I know I've mentioned it, but if you close your eyes (which is your natural **instinct** when you're about to crash into something), you could end up with a serious injury because you're not looking at what's coming towards you.
- I think it's best to focus on the ball-carrier's core — the area in a line around their middle at the height of their tummy button. Sometimes you'll want to go a bit higher and hit them where they're carrying the ball.
- It's time to get in close, but keep your feet alive. Move your feet quickly and keep adjusting so that your leading foot is close to the ball-carrier. Remember never to '**plant**' your feet — good tacklers keep moving. If you're stuck in one spot, the ball-carrier is likely to move or change their angle and you'll be

left standing there, looking silly.
- Avoid swinging your arms into the tackle, as this is dangerous for both you and the ball-carrier. Always make firm contact with your shoulder into their core, keeping your head behind the player you're tackling. Coaches love to say 'cheek to cheek' and that's the best way to explain it — your face cheek against your opponent's bum cheek.
- Whip your arms around the ball-carrier. Wrap, grip and squeeze with your arms. A firm grip is best or else the ball-carrier might break through.
- Remember to continue driving and pumping your legs to make the tackle count. March! March! March!
- Finally, and if everything has gone well, you'll both finish on the ground with you on the top. Let the player go (the referee might be yelling 'release' to remind you to do this, and to remind the tackled player that they have to let the ball go). This is your cue to get to your feet and pluck the ball off the ground.
- Because I'm a little lad, I like to 'chop tackle' bigger players. This means I aim quite low, sometimes even at their ankles. This 'chops' the player down like a tree, and I always remember the advice I was given as a kid: 'the bigger they are, the harder they fall'. Even the big boys have small ankles.

WHEN IT GOES BAD!

Waiting for the ball-carrier to come to you gives the opposition the chance to do all that stuff we were talking about in the evasion section. *You* want to have the options, not them. Go into the tackle with momentum.

Injuries happen to tacklers when they close their eyes, drop their head or get their head in front of the player they're tackling. Head position is everything.

When you're young and feel bullet-proof, diving into a tackle with no control — front on, feet leaving the ground — seems like a great thing to do. But it's not a good idea at all.

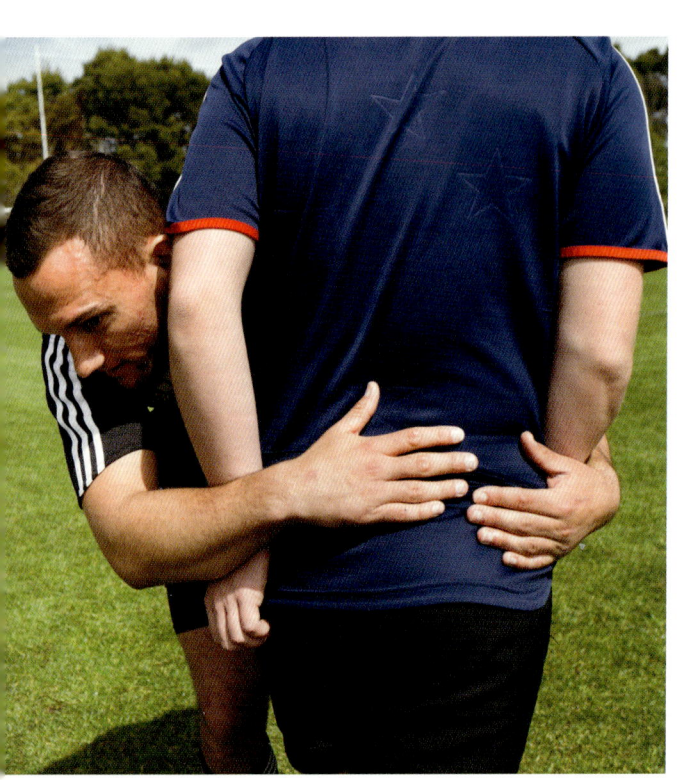

Swinging your arms into the tackle is not only dangerous, it's also ineffective. Shoulders bring the giants down, not the arms. But that said, you do have to use your arms to wrap up the player once you've hit them with your shoulder, otherwise you will be penalised for a dangerous shoulder charge.

Remember, your job isn't finished just because you've made a great tackle. A tackle is only as good as your ability to get to your feet quickly and contest the ball. Getting the ball is what it's all about.

ACTIVITY

You and a friend are on your knees facing each other with about a 2 metre gap between you. Your friend is the attacker and has a ball, and you are the tackler. (Mouthguards in, please, folks!)

Your friend with the ball runs on their knees to your right and you make the tackle! Then your friend with the ball runs on their knees to your left and you make the tackle! Swap roles and do it again.

EXTRA FOR EXPERTS

Do the same, but this time get to your feet and jog slowly.

MORE ON THE TACKLE

The type of tackle you use depends upon the size of the tackler, the starting gap between the attacker and tackler, and what the tackler is trying to accomplish (which can change, depending on field position).

Are you trying to tackle only or do you want to tackle and get the ball?

Is it an ordinary tackle, or is this a last-gasp tackle that you're making to save a certain try?

Even though rugby is a team game, it often boils down to a one versus one moment.

SIDE-ON

The basic side-on tackle generally occurs when there is enough space between the attacker and defender for you to control the attacker's run so they can only run one way. This is the most common type of tackle, especially these days where teams tend to use a 'drift' defence (tacklers approach their targets from the inside toward the sidelines, to encourage the ball to be shifted wide where the attack will run out of room).

With a side-on tackle you can bring your target down more easily, as you're not presenting yourself as an obvious obstacle to be evaded or bumped off.

THE PASSIVE LOW TACKLE

The checklist is similar to your basic side-on tackle but the difference is that the tackler isn't trying to dominate the hit (by driving with the legs). Rather, you're trying to fall with the tackled player so that you end up on top and in a good position to get to your feet quickly to contest the ball.

The passive low tackle is a great way to achieve a turnover (when you take the ball from the other team).

FRONT-ON

The front-on tackle has become much more common these days, because rugby has turned into a far more explosive and direct game.

Front-on tackling includes the 'smother' or 'ball-and-all' tackle, in which you target the upper body where the ball is being carried. The approach to the attacker is more upright and requires good timing. Quite frequently, you're not trying too hard to get the player on the ground: it's enough to have stopped them and prevented them from moving the ball before your team-mates arrive to form a maul.

While you're not going to be able to

get your cheek pressed against their bum cheek as in the tackles described previously, the mechanical aspects of front-on tackling are similar. Most importantly, you have to get your head to the side of the body of the person you're tackling and not directly in line with the body.

The smother tackle requires the defender to hit their shoulder on the chest area and wrap both arms around the ball and the player. You should come from slightly low and drive up and forward as your shoulder hits them. This momentum gives you the advantage: if you're not carrying any momentum into the tackle, you might well get bumped off.

The smother tackle forces the tackled player to remain upright and often leads to a turnover.

The low front-on tackle requires you to step out to the side before the hit so that you — and especially your head — are not directly in line with the run of the attacker.

The low tackle can also be separated into two types. There is the aggressive hit, where you push with your legs to try to stop and even drive back the attacker. There is also the passive hit (which works well for smaller defenders). The passive hit requires the tackler to be strong but prepared to fall with the attacker still going forward and ending up on top.

Low tackles allow players of any size to take down players of any size.

ACTIVITIES

These can be done with a hit shield (a big, spongy pad) to prevent the body taking too much of a pounding in training. Most can be done with just one team-mate. The ball-carrier shouldn't run at full speed, and you shouldn't make the tackle as hard and mean as possible, because you're just trying to get the technique right.

GETTING THE SHOULDER ON

Two players are kneeling facing each other about 2 metres apart.

The attacker moves towards the defender, who is also moving forward and targeting one side of the attacker.

When you are close enough, the attacker will propel themselves forward using their hips and knees. Hit the attacker with your shoulder, wrap their arms up tightly to their body, and keep driving through the tackle. Check that your head position is behind the falling player.

Repeat five times on each shoulder.

FOOT IN, SHOULDER ON

Two players are standing facing each other 3 metres apart.

The attacker walks forward towards the defender. The defender moves forward, shuffles and crouches, takes a big step into the attacker with the leading leg (feet close and weight going forward), hits with the shoulder on the attacker's core (don't wrap your arms around at this stage) and keeps walking forward.

Practise five times on each side.

OPENING UP ONE SIDE

Two players are standing facing each other 4 metres apart.

The attacker walks directly at the defender. The defender shuffles forward and just before contact (still in direct line with the attacker) steps sideways onto their outside leg (the leg farthest away from the point of contact) and then drives inwards off that leg into the tackle.

Drive the arms through and squeeze; keep driving through the tackle.

> **RECORD**
> The most rugby tackles made in one hour is 4130 but don't worry, it wasn't in a real match!

EXTRA FOR EXPERTS

MANIPULATING THE ATTACKER

Two players stand opposite one another on a corner at each end of a 5 metre x 5 metre grid.

The defender walks straight forward as the attacker starts off. The first three steps are important. Go forward, not to the side where the space is.

The defender should stay on the inside as the attacker tries to score. This will push the attacker towards the sideline.

If the attacker steps back inside they should be tackled easily by powering across, changing the leading foot and shoulder with one step.

MANIPULATING THE ATTACKER (2)

Two players are facing each other at either end of a 10 metre x 10 metre grid, on opposite corners. The defender can be 1 metre forward of their corner.

The attacker can score a try on either side of the grid which they are facing (on the left or right of the defender).

The defender will move forward to position themselves on the outside of one of the attacker's shoulders to encourage them to run to the side they are leaving open (this is called 'showing the attacker the sideline').

Move forward quickly and then move in towards the attacker and cut down their time and space. Keep moving in on the attacker even if they try to put you off with fancy footwork.

As you get more familiar with the angles of approach and engagement (when you make contact with the person you're tackling), make the grid bigger and do the drill at faster and faster speeds, until you're eventually doing it at full pace.

FALLING IN THE TACKLE

No matter what position you play, one thing is guaranteed: you will get tackled!

The first thing you need to think about is your own safety. You also need to consider holding onto the ball and presenting it to your team-mates as best as you can.

There are so many different ways a tackle can unfold, depending on the angles, the force, speed, size, intentions and technique of the tackler and the tackled player. You'll never be fully prepared for all the ways you'll fall over in a tackle.

But the checklist below is a good starter, and if you practise this it can become second nature.

CHECKLIST

- When you get tackled, you can't always control how you fall. I've found that if I keep to these basic principles on how to fall in a tackle, when I get tackled, my instincts kick in and I tend to do what I've practised.
- As you are getting tackled, make sure you protect the ball by holding it in both hands. I like to position the ball away from the point of contact and turn toward my support during the tackle if possible.

THE CRUDEN CLUE

Falling on top of the opposition is the best result. They can't play much rugby if they're underneath you!

- Sometimes staying on your feet as long as possible is a good option, but only if the support is a little slow getting to you. The opposition is going to try to keep you upright so that the ball is trapped in the maul and they get the feed to the scrum.
- If you're going down, make sure your grip on the ball is firm, tuck your shoulder under and roll onto your upper back, positioning your body between the ball and the opposition. This is so they can't get to it easily and your team-mates can.
- Finally, reach out and place the ball as far back on your own team's side as you can, especially if the tackler has jumped to his feet and is ready to have a go at getting the ball.

WHEN IT GOES BAD!

Your team-mates will have problems if you fall facing the opposition. Don't panic if it happens, because you're allowed one movement to place the ball in the tackle, and this can be rolling over to deliver the ball to your own side, as long as it's done immediately.

Don't ever put one hand down as you fall. Your hand and arm aren't great shock absorbers, as any skateboarder or snowboarder finds out the hard way!

If the support from your team-mates is too slow, you're in a bit of trouble, because you must release the ball immediately. If you hang on while the tackler or another opposition player who has arrived on the scene are tugging at the ball, you'll be penalised.

ACTIVITY

You can practise this on your own. Start in a standing position, drop to your knees, then roll onto your hip, then shoulder.

Place the ball 'long' to your pretend team-mates. Repeat this so it is comfortable to do and you'll get the feel of the rolling action you need with the ball in two hands.

Maybe even put down a mattress and get a friend to tackle you onto this so you can get your technique right.

THE JACKAL

You probably know that jackals are dog-like animals that arrive on the scene to gnaw away at the dead body of an animal that another animal has killed. That's what a jackal in rugby is named after: they arrive to challenge for the ball just after a tackle has been made. There's an art to this, and it pays to know what you're doing, because there are usually lots of big forwards arriving on the scene at the same time. But a properly executed turnover can be an important moment in a match, because it lets your team go onto attack before the opposition can organise their defence. For this reason, it's a good idea if everyone in your team is good at getting turnovers.

When you arrive at a tackle, the jackal is one of two options available to you. It's best for the jackal to be the first player to the tackle, besides the tackler and the tackled ball-carrier. If an opposition player has got there first, or if another team member is already there, your job is clearing out the opponent or supporting your team-mate.

CHECKLIST

- Approach from your side of the tackle (otherwise you'll be penalised). Get yourself low: bend your knees, bend at the waist and make sure you plant your feet wide apart to give yourself a nice, stable base. It's best if your feet are parallel to one another, although you can place them as though you're taking a big stride, especially if the tackled player has placed the ball a long way back.
- Be prepared to be tackled by the opposition, as they want the ball, too! I like to 'turtle' my neck down into my shoulders so I have added protection (like a shrug movement).
- Reach out with strong arms and put your hands on the ball.
- If you can pick up the ball cleanly, do it, and allow yourself to fall towards your support (it's pretty rare that all this is going on with only three players involved). The opposition will probably help out with this, as they'll be trying to push you away from the ball. Get the ball to your support as soon as you can.
- If the tackler isn't letting go, make it obvious that you have your hands on the ball and are contesting it. You (the jackal) have the rights to the ball here and the referee will award you a penalty.

WHEN IT GOES BAD!

If your head is up looking for oncoming opposition, there is a good chance that you will sustain an injury. Keep your noggin down and out of harm's way.

Don't bend so far that your shoulders are below your hips, as this is just asking to fall over face-first. As soon as the referee sees you aren't supporting your own body weight (supporting it with the tip of your nose doesn't count), you'll be penalised.

ACTIVITY

Using a couple of team-mates, have one player lie on the ground placing the ball as though he's been tackled, and the other two competing to get the ball (like jackals).

You can vary the distance between the two challenging players and the tackled player.

CLEARING OUT

There are lots of jackals about in the modern game, so it's often necessary to remove them. The act of driving the opposition away from the tackled ball is called 'clearing out'.

It's usually thought of as a job for the big forwards, but the backs are quite often closest to the tackled ball. Every now and then, even well-groomed outside backs will find themselves having to roll up their sleeves and clear out an opponent or two. You'll be surprised what you can achieve if your technique is right, and if you're my build, it can be really satisfying to see the look on their faces when they see who has cleared them out!

- The best place to contact the opposition player is under their shoulder as they are leaning over the tackled player or the ball.
- You have to be lower than the player you mean to clear, so a low stance makes it easier to drive an opposing player away. I've got a natural advantage here, being quite low to the ground to start with!
- I aim for the chest and outreaching arms, wrapping my arms around the arms of the opposition and squeezing to make sure that the movement looks legal.
- Finally, I drive through and up with my legs and power them away, keeping my feet moving forward and driving hard!

CHECKLIST

- I like to approach the opposition with my hands up around my chest and my weight on the balls (the front part) of my feet.
- The laws of rugby mean you have to approach a tackle 'through the gate', the imaginary area on your own side of the tackled ball area, not the side or from the opposition's side.
- I like to focus on one opponent.
- I think 'face up, eyes open and chin off chest'.

WHEN IT GOES BAD!

The number one rule is safety, both your own and that of others. You must pay attention both to the rules and to good technique, or people will get hurt.

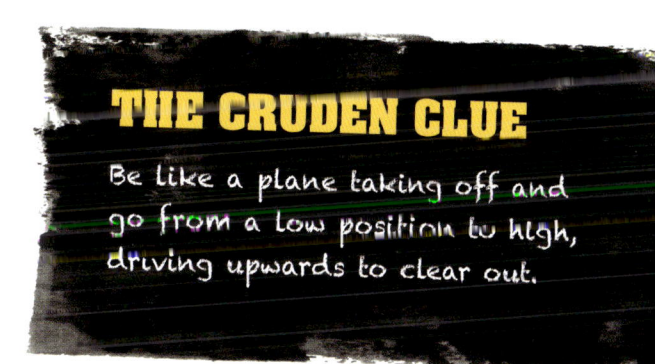

THE CRUDEN CLUE

Be like a plane taking off and go from a low position to high, driving upwards to clear out.

If you fall over in the act of clearing out, you'll be penalised. You must stay up and stay strong!

Driving from high to low just doesn't work — the opposition player won't be going anywhere. Go in low to start with.

Things change fast around the tackled ball, so don't make your mind up too early about what you're going to do. You might be going in as a jackal and at the last moment decide to clear out an opponent who has arrived with the same idea.

Remember, if a player is bigger and heavier than you, then technique is the only sure way to move them away from the ruck.

ACTIVITY

Using a mate who is reaching over to grab a ball on the ground, move in (slowly!) and remove them (carefully!).

After you've practised clearing them out as they reach for the ball, try doing it when they have their hands on the ball. Slowly and carefully!

Get your technique and action 100% correct before you start using force.

Introduce the option of clearing out to the drill described for the jackal, so that you can practise making quick decisions about your best option on arrival at a tackled ball situation.

NUTRITION

Just as important as hydration is the need to ensure you're eating the right foods to keep up your energy levels. Your body is like a car — it needs fuel!

If you have any doubts about what you're eating — let's say you're vegetarian or don't eat high-energy foods — I recommend that you talk to a dietician to make sure you are meeting all your energy requirements. Even if you are a young player this is beneficial, and it goes for whatever sport you want to play.

Weight restrictions apply to lower-grade rugby. I am strongly against young players losing weight in order to make weight-graded teams. You see people do crazy stuff just before weigh-in — wearing plastic rubbish bags under their clothes while they go for a run to make them lose weight through sweat, taking pills to make them go to the toilet more, going without food for days beforehand — it's mindless. These techniques can lead to dehydration and undernourishment in the short term, and it can harm your mental and physical performance, to say nothing of doing actual physical damage.

In the long term, these techniques may affect normal growth and development and lead to serious health consequences.

Make sure over half your food **intake** comes from carbohydrate-based foods. That's yummy things like potatoes, pasta, bread, rice, cereals and bananas. Lots of players like to eat six smaller meals spaced through the day rather than three big ones. You'll soon figure out what works for you.

Increase the amount of carbohydrate-based foods a few days before playing, but ensure your food intake contains at least 15% protein — eggs, fish, meat, chicken and shakes (OK, not chocolate thickshakes: I mean fruit smoothies and things like that). Your body needs that protein to build muscle and to repair itself after training and matches.

There are lots of diet supplements around these days. The word 'supplement' means 'extra', and supplements are exactly that — they're no substitute for a good, balanced diet.

Finally, eat some protein and carbohydrates directly following training to improve your recovery and — yes, the one you already know — avoid foods that are high in fat before and during exercise.

AARON'S SPECIAL RECIPES

I like to eat lots of different foods, so let's make something that tastes great, is healthy and that you can make as a treat for the whole family.

THE BIG BOY

This recipe makes six burgers.

½ onion
500g lean beef mince
tomato sauce or BBQ sauce
1 egg
6 fresh wholemeal buns
lettuce
tomatoes
cucumber

1. Start by finely chopping the onion. If you struggle with knives, maybe an adult can help you out with this part. Rugby is best if you have all your fingers!

2. Combine the mince, two squirts of sauce, the egg and two tablespoons of water.

3. Mix this together really well.

4. With wet hands, make six patties by rolling the mixture into balls and then flattening them. This is the fun bit!

5. Grill the patties in an oven or on a barbecue or fry in a pan, until they're cooked through. Make sure the patties are brown all the way through. If they're still pink inside, they need a bit longer.

6. When the patties are almost done, put the buns on the grill for a couple of minutes to toast them slightly. Well, this is how I like it.

7. Prepare the vegetables — I like lettuce, tomatoes and cucumber — by washing and cutting them. You can really add anything you like here.

8. Then stack it all up and dig in.

THE BANANA KICK SHAKE

A lot of the guys in the team use this recipe to rebuild energy after a hard game. It's a banana shake: delicious and nutritious! You'll need a blender or food processor for this one. Make sure the lid's on properly or you could be cleaning up a mess!

2 bananas
2 cups milk
1 cup low-fat natural yoghurt
1 tsp lemon juice

1. Chuck the bananas into a blender or food processor and pour in the milk, yoghurt and lemon juice.

2. Whizz on high speed until smooth and pour into a glass. Enjoy!

3. Instead of bananas, you can experiment and use berries, like strawberries and blackberries.

RUGBY WORLD CUP: DID YOU KNOW?

2011

- The 2011 Rugby World Cup was the largest sporting event ever held in New Zealand. New Zealand won! The All Blacks beat France 8–7 in the final and joined South Africa and Australia as teams that have won the Rugby World Cup twice.

GLOSSARY

INSTINCT a natural behaviour

INTAKE the process of taking food into the body through the mouth

MANIPULATE influence or control what is in front of you

PLANT positioning your foot suddenly

Practising my restarts with the All Blacks in 2012.

CHAPTER 7
SET PLAY, PHASE PLAY, PENALTIES, ATTACK AND DEFENCE

Rugby coaches talk about 'set' and 'phase' play. Set play is what happens after a restart (a set piece is always announced by the ref's whistle). Phase play is what happens the rest of the time, including the tackles, rucks and mauls. If you understand how set play and phase play unfold during a game and the opportunities each presents, you'll go a long way to being successful and having fun on the rugby field.

I like to think of phase play as the part that joins together the set pieces. Your approach to phase play determines how you and your team will go.

WHAT ARE THE SET PIECES?

The very first set piece in a game of rugby is when the ball is kicked off to start the game. Each restart after points have been scored is also a set piece. The person taking the kick — eek! me! — worries about these, as the length of the kick is all-important.

The other set pieces that are used to restart play are scrums, lineouts and penalties.

A scrum is obviously more than a restart of play; it is the thing that defines the game. It requires strength, perfect technique, timing and unity amongst all the players involved.

Front rowers (the hooker and props who

play in the front row of the scrum) are very proud of what they do in the scrum, and to be honest, I don't know all their tricks. It's like a secret society, and I've always thought it was best to leave it to them. I'm reliably told that new props and hookers soon learn everything they need to know about the dark arts of front-row play.

A lineout is all about gaining possession, so the forwards must operate as a combined unit to control the ball in the air and take it forward. This can be achieved in a number of ways, but they all have the common aim to be the team that gets its hands on the ball first.

Our hookers and jumpers spend hours and hours getting this right. You'll often see hookers and locks practising in the car park before a game so they get their timing right.

WHAT IS 'SECOND-PHASE' PLAY?

Simply put, second-phase play is everything that happens straight after a tackle, ruck or maul. There can be several passages of second-phase play. You'll often hear a commentator on TV say there have been 8–10 phases: each 'phase' is one of these periods of second-phase play, counting from the last restart.

WHAT HAPPENS AT A 'BREAKDOWN'?

When players are running wild and free with the ball in their hands and the wind in their hair, the game is going smoothly. As soon as they're tackled, play 'breaks down', which is why a tackle is often called a 'breakdown'. Players have to enter the breakdown 'through the gate' — from directly behind the ball. As a team, your job at the breakdown is to make sure your team keeps the ball or grabs the opposition's ball. This is where players like the flankers excel and there is none better than Richie McCaw. He has this amazing ability to get the ball from the breakdown before anyone else.

When the ball is on the ground and players from each team are in contact around it, the breakdown becomes a ruck. Once a ruck has been formed, the ball cannot be played with the hands, even to pick it up cleanly.

In a ruck, players tend to bind with each other and drive directly over the ball so that the player or players following on behind them are able to pick up the ball when it pops out the back and get on with running and passing. The ball is considered to be out of the ruck as soon as it goes past the last foot in the ruck, or when a player behind the ruck (who isn't bound to it) reaches in and puts their hands on it.

A maul is similar to a ruck, except that the ball is not on the ground.

WHY ARE TEAMS REWARDED OR PENALISED?

A team is rewarded by gaining territory, retaining possession and scoring points. It's not all about tries. The reward for superior skill is possession of the ball and the time and space to use it.

The reverse can be said of a team who is poor at the skills of running and passing. The penalty they suffer is that they no longer have the ball — without it you cannot win the game.

But like every sport, rugby has rules designed to penalise bad play. When a rule is broken, the referee punishes the infringing team by giving possession of the ball to their opponents. Sometimes this can lead to points being scored, too.

ATTACK

You are the attacking team if you have the ball, so your job is to try to get to the opposition's goal line — to attack it.

A rugby attack is a magnificent thing, whether it be a simple running and passing movement, or a complicated plan with players running this way and that on numerous lines of attack so that the other team doesn't know which way to shift, whether to come up hard or sit back on their heels and wait for the runners to arrive.

All the moves you see at a game or on telly are all about trying to attack the defence, to get through and score. Building an effective attack takes time, and that's why combinations with your fellow players are important. It's a bit like Aaron Smith and me or TJ Perenara and Beauden Barrett — we've played together long enough to instinctively know how to work together and trust each other on attack.

The job on attack is to look for space or weakness in the opposition, such as a gap where the defence hasn't come right up or a tackler has run out of the line, leaving a big hole. That's an invitation to you to take the ball forward.

If there are more attacking players than defending players — as happens when some of the opposition are slow to get organised, or they're caught in a ruck or maul — there's going to be space there. You naturally want to shift the ball into the space as quickly as you can. The All Blacks are always looking to create this situation. We'll spread the ball wide to one wing, then we'll go wide back the other way to try to 'stretch' the opposition's defence.

We sometimes talk about it in terms of 'town' and 'country'. In the breakdown area, there are lots of people: it's like being in town, with not a lot of room to move. In the country — out wide — there's lots of open space and room to really stretch out. So in the All Blacks, we're always looking to move from town to country.

Players with the ball need to look to pass the ball to a team-mate who is not guarded ('marked', as we call it), so this player can catch the ball, go forward and beat the defenders to the goal line.

Team-mates should support all the time, giving the player with the ball as many options — choices — as to what they do with the ball as possible. If the opposition only has to concentrate on one ball-carrier — a runner with no support — they'll sort them out pretty quickly.

SUCCESSFUL BASIC ATTACK

Initially, the attacking team must identify the situation in front of them and communicate the best option available to one another.

The attacking team's formation relies on the team's skill level, the running speed of the players and also the phase of play they find themselves in.

Generally, you should run straight (directly toward the tryline, not drifting towards the sideline) to commit the opposition — make them tackle you — as this creates space. If you run on an angle, the opposition are usually quite happy to let you do it. They just herd your backline to the side of the field where there's nowhere to go. Coaches tell kids that the shortest way to the tryline is straight ahead: I got told this heaps, and it still makes sense!

When the backline is set up, everyone should be deep enough so that they can run on to the ball as it is passed in front of them. Passes should be made close enough to the opposition to commit the defence, but everyone needs to maintain that depth.

The speed of both feet and hands will determine your team's ability to break through to score.

Players can break through a defence by passing, running evasively or kicking.

AARON'S ATTACK PRINCIPLES

Gain possession of the ball through your individual and team skills. Do well in scrums, lineouts, kick starts, tackles, ruck and mauls, and securing loose ball.

Once you have the ball, you attack by going forward — it's as simple as that. Make sure the opposition is on the back foot and too disorganised to prevent you and your team heading for the tryline.

The way to keep an attack going against a determined defence is for everyone to support their team-mates. This means telling the ball-carrier what their best option is, and being there to receive a pass or secure the ball if your team-mate goes to ground in a tackle.

If you can develop skills that give you options when you are tackled other than just going to the ground — they call this ability 'keeping the ball alive' — you'll be a big help to your team.

If the defence tries to stop your attack, your team needs to organise itself to keep the ball going forward. This sets the

DID YOU KNOW?
There are on average 16 scrums and 25 lineouts in a professional rugby game.

defence back and creates space. You need to be ready to use this space.

If the ball can't be kept alive in the tackle, then take it into a ruck or maul. What this does is makes sure the opposition have to stay behind the feet of the last player on their side of the ruck or maul. That gives you time and space to set up another attack.

When your team has the ball and is making progress toward the opposition's tryline, it's called pressure, and pressure is what we want. Pressure is what it's all about! Pressure, pressure, pressure until you score!

DEFENCE

It follows that if attacking is what you're doing when you have the ball, defending is what you're doing if your team hasn't got the ball.

The defence's main aim is to press forward and take away the time and space available to the attack. It's not just attackers who can put pressure on their opponents: tackling is the main weapon in defence. By closing in on the attack and tackling effectively, you can force them to hurry what they're doing and make a bad decision, or perform a skill poorly.

When the attack makes a mistake like this, the defence will strike, making every effort to regain the ball and launch an attack of their own.

No matter how crisp the training run, how polished the team's passing game, and how superb the team have been in previous matches, a well-drilled defensive wall playing at its best will shut down every play.

One of the best midfield defensive combinations is Conrad Smith and Ma'a Nonu. They've played together for so long for Wellington, the Hurricanes and the All Blacks, they instinctively know how to work together, trust each other and know each other's strengths and weaknesses. They can both read the opposition's attacking play and it's like they each know what the other is thinking and doing without having to look. It also helps that they can both tackle really effectively!

SUCCESSFUL BASIC DEFENCE

Players in defence must move forward together, to avoid leaving gaps.

Usually one player organises the defence and makes sure communication goes right through the whole team. In the All Blacks, we all take responsibility. You'll see if you watch closely that we are all pointing and telling the players around us who each of us is lining up in defence. It's a good habit to get into.

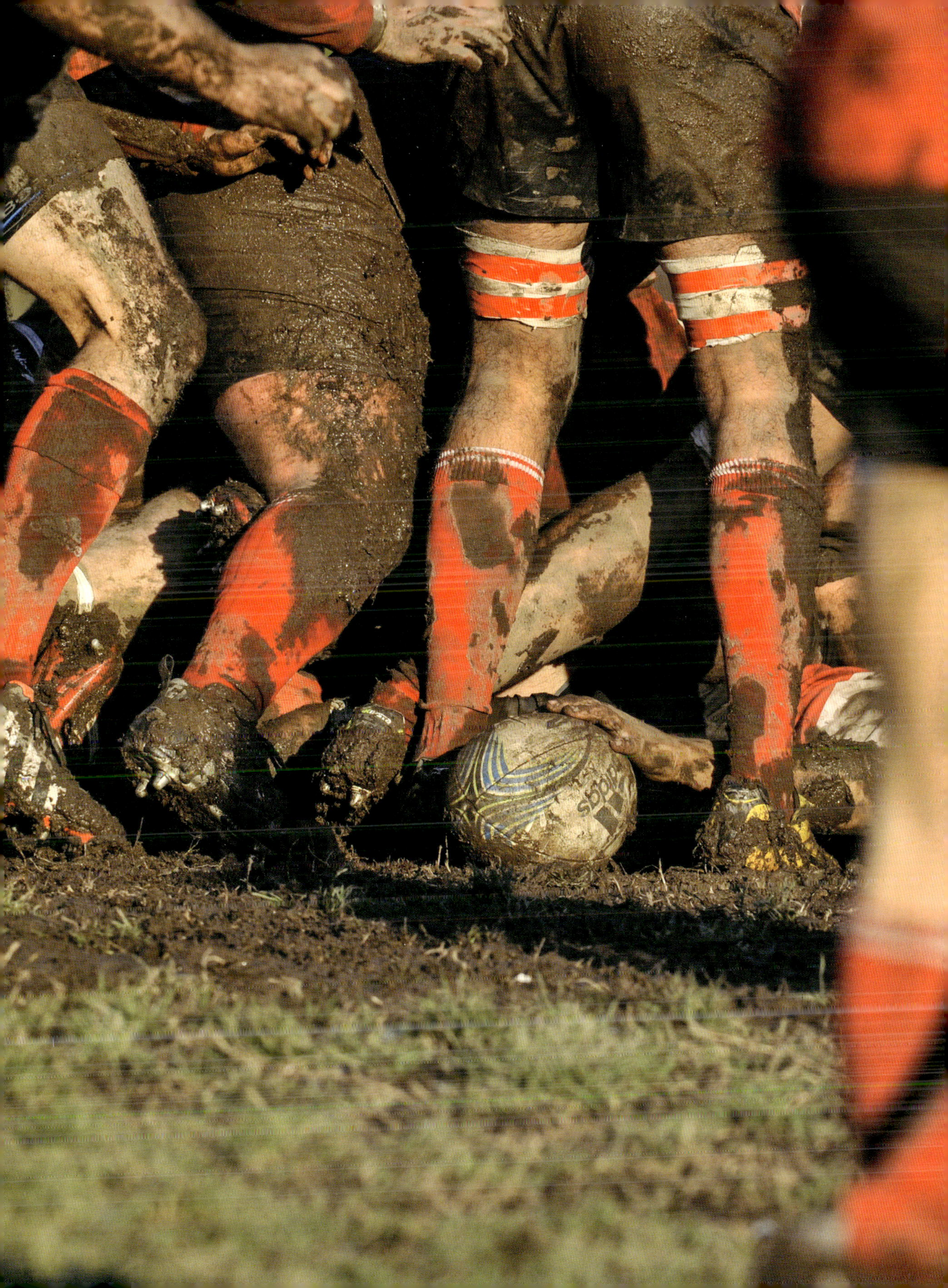

Tackling Zac Guildford in the 2013 Super Rugby semifinal.

We usually know whose fault it was when someone gets through, but you'll also notice, we don't yell at them or put them down. We encourage them by letting them know it's all right and everyone will do better next time.

Defence should always tackle with the intention of turning the ball over or tying the ball up. This is the responsibility of every player in the team. If you're not required to make the initial tackle, then provide immediate support to the tackler and other defenders outside you.

When you're defending, the rules help you out by making it so that the tackled player has to let the ball go. That means the defence has a good chance of getting their hands on it. That's why everyone should be interested in what happens straight after a tackle is made.

BASIC MAN-ON-MAN DEFENCE

If the attacking backline is set up and you have a defender to match each of them, you can defend man-on-man.

The defending team's halfback puts pressure on the opposition halfback, or drops back to help the wing standing on the blindside.

The first pair of hands to receive and pass the ball (usually the halfback) is supposed to be tackled by your flanker.

The second pair of hands is tackled by your first-five.

The third pair of hands is tackled by your second-five.

The fourth pair of hands is tackled by your centre.

The fifth pair of hands is tackled by your winger.

If an extra person — the opposition's blindside wing or the fullback — enters the line and there is a sixth pair of hands, then normally your fullback will make the tackle.

BASIC 'ONE-OUT' DEFENCE

If the opposition bring in extra players (for example, their forwards are standing in the backline, or the fullback and/or the blindside wing), 'one-out' defence can work well. It's called 'one-out', because it's the same as man-on-man except everyone shifts one position out to defend.

The first pair of hands — the halfback — is tackled by your halfback while your flanker goes for their first-five. Or, if your flanker takes care of their halfback, your halfback goes for their first-five.

The third pair of hands, usually the second-five, is tackled by your first-five.

The fourth pair of hands, usually the centre, is tackled by your second-five.

The fifth pair of hands is tackled by your centre.

This leaves a sixth pair of hands to be tackled by the wing.

The fullback is available to tackle a seventh attacker out wide if needed.

BASIC DRIFT DEFENCE

Most teams use a drift defence these days, although it's pretty complicated. The idea is that each defender aligns himself on the inside shoulder of his opponent and aims to chase them wide. The point of this is to make the attack run out of space.

AARON'S PRINCIPLES ON DEFENCE

Going forward is important in attack and it is just as important in defence. It reduces the opposition's time and space, and therefore the options available to them.

Your team-mates must understand their role within the team's defensive pattern, and tell everyone else what they're doing. Point at who you're marking or the area you're going to defend so everybody knows and no one is doubling up.

The defensive team can apply pressure to their opposition when they get forward and make effective tackles.

The next step is to get supporting players there to win the ball from the breakdown.

LEADERSHIP

The most obvious leader of a rugby team is the coach, especially when you're younger. I've been very lucky with my coaches. A good coach needs to have the respect of the whole team. They will be the person who will always think about the little things, they will be trying to make each of you a better player, they will be trying to get players wanting to be part of the team — and they'll have to deal with the hopes, dreams and opinions of the parents on the sidelines. So listen to your coach, suck up their knowledge, ask them questions and cut them some slack every now and then. They are developing just as much as you are.

Captaincy is an honour and a big responsibility. I captained the PNBHS First XV in my seventh form (year 13), the NZ U20 side and Manawatu after they had a few injuries.

I guess as a first-five, you are in control of moving the team around the field, and I believe I am fairly composed and have a cool tactical head — that is, I make good decisions under pressure.

These are good qualities in a first-five, and good qualities in a captain or leader, too. It's probably a surprise that not many first-fives captain their teams.

I try to make sure everyone is happy in their roles and make sure all the messages on the field are clear.

I don't think there is only one way to be a leader.

When I was younger, I would get frustrated and hot-headed. That used to put me off my game. I had to practise controlling myself and concentrating on doing each task well, regaining my cool and then continuing with the plan.

I think it's important to start with small successes when you are the leader of a team. That means focusing on no more than a couple of parts of the game and practising them carefully. Success can be measured by how well your team does in these aspects rather than a victory on the scoreboard — although a win is always nice!

Obviously one of the best leaders is All Blacks captain Richie McCaw. You would think it was all about following his example on the field, but it is so much more than that. He is as tough and professional in his preparation off the field as he is when he's playing.

I think his strength is that he makes every team member stick up their hand and take responsibility. You can only do this if you are leading the way, and that's what Richie does. He sets a high standard and he sets the example. He won't demand of others what he is unwilling to give himself, and you'll never find a man with more

Having already captained the New Zealand Under-20 team and the Manawatu Turbos, in 2014 I was lucky enough to be named co-captain of the Chiefs.

energy, enthusiasm and knowledge.

But as a leader, you also need to surround yourself with good people. Even in the All Blacks, where we have a captain like Richie, we also have a leadership group of a number of players who contribute to both on-field and off-field decision making.

The key to success is having the best people possible working alongside you and motivating you. Good teamwork will make each member better. Learn from each other, as well as learning from teams that you play.

DID YOU KNOW?

All Black Colin Meads was named Player of the Century at New Zealand Rugby's awards in 1999. He played 133 games for the All Blacks. Only one player has beaten this record so far — Richie McCaw.

If you're captain, make sure you hear others out before making final decisions. It is crucial that each member of your team believes that they are part of the decision-making process. Even if you're captain, you're not the whole team. I suppose it's a bit like the old saying: 'There is no "I" in TEAM!'

INJURY

TREATING INJURY

Believe it or not, the most common injury in rugby is a soft-tissue injury. That's all the soft bits that connect, hold or enclose your bones and the organs of your body.

If you get a soft-tissue injury — a sprain, strain or even just a bruise — you should remember the R.I.C.E.D. procedure.

R is for rest. Rest stops further damage to the injury. Avoid as much movement of the injured part as possible and don't put any weight on the injured part.

I is for ice. Ice cools off the tissue and reduces the swelling, pain and bleeding (which is what bruising is). Wrap ice in a damp towel or use an ice bag and place it on the injured area. A nice, big packet of frozen peas works well, too. Apply the ice for 20 minutes every two hours for the first 48 hours. Timing is really important. The earlier you get the ice on the better.

C is for compression. Between all the ice treatments that you'll be doing, as a general rule you should bandage the injury. A firm bandage helps to reduce bleeding and swelling. Make sure that the bandaging is not too tight or it could cut off circulation, causing tingling or pain past the bandage.

E is for elevation. Always raise the injured area to stop bleeding and swelling. For comfort and support, raise the injured area on a pillow, but a box or chair will do if that's all you've got. The ideal is to raise the injured area above your heart, but this is sometimes impossible.

D is for diagnosis. See a medical expert such as a doctor or physiotherapist if you are worried about the injury, or if it gets worse. If the pain or swelling hasn't gone down within 48 hours, it's time to see a doctor.

THINGS TO AVOID

Heat increases the bleeding in injured tissues, so avoid showers and hot baths (even though that's often all you feel like doing), hot-water bottles, saunas, liniments and heat packs.

Running, or exercising the injured part, will cause further damage, so don't resume exercise within 72 hours of the injury unless a doctor or physio says it's all right to do so.

Rubbing or massage can also make an injury swell and bleed more, so avoid it, as either will affect your recovery time.

DID YOU KNOW?

By far the most injuries in rugby happen during the tackle.

CONCUSSION

Concussion is a serious injury and can happen when you receive a blow to the head or body that causes your brain to bounce around inside your skull.

If you're knocked out or lose consciousness, you have obviously sustained a concussion, but it's important to know that a person can be concussed without being knocked out.

If you're concussed, you're often the last person to know and all you want to do is keep playing. Your coach, referee or parent must take responsibility and decide whether it is safe for you to play or not — and if they have any doubt, you should stop. You've got a lifetime of rugby ahead of you — missing one game isn't the end of the world!

Even if you don't seem to have been concussed at the time, problems with concussion can show up later. If you've had a knock to the head, it's important someone keeps a close eye on you and checks on you regularly for the first four hours after the injury.

Once it's been decided you have a concussion, you have to stand down — not play any rugby — for at least 23 days.

All of this might sound like a big deal, but that's concussion for you. It has to be taken seriously, or the effects can be very bad.

AARON'S BALL TRICKS

I believe these ball tricks will help you to develop a range of ball skills and make the business of learning fun. Although they're a bit zany, they do give you a lot of ball awareness and that is key to your skill development.

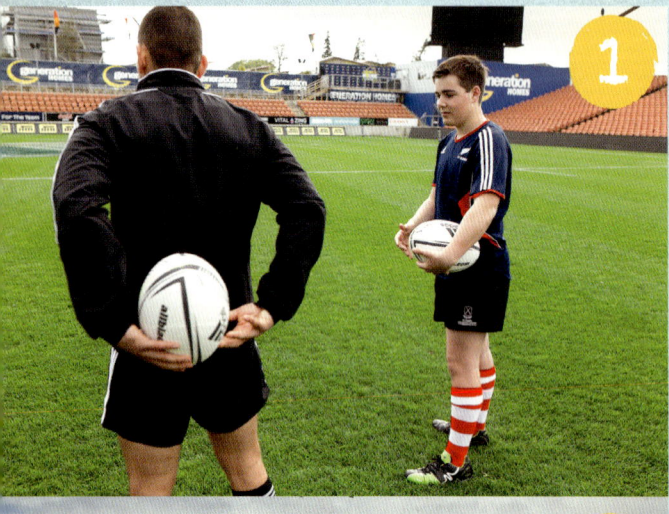

1. CHURNING BUTTER
Rating

Objective Move the ball around your body.

Start by holding the ball in two hands, then move the ball slowly around your middle, keeping it close.

As you get better, keep the ball away from your body by just using your hands.

By shifting it quickly from hand to hand, it looks even cooler, and you can change direction for a bit of variety.

Extension Pass the ball when it's behind your back rather than just moving the ball from one hand to the other.

2. THE BOUNCE Rating

Objective Pump the ball forward using the inside of the elbow.

Hold the ball with a bent arm about head height and then let the ball drop.

Straighten your arm sharply to let the ball bounce forward off your middle arm (inside elbow)

It's handy to have someone to catch the ball.

Extension Try 'the fake' by stopping short of hitting the ball with the inside of your elbow, and catch between your upper and lower arm.

3. THREAD THE NEEDLE
Rating 🏉🏉

Objective Move the ball back and forward between your legs while moving forward.

First get the hang of the 'in and out' skill, and then begin to walk.

Once you've learned this, start moving faster, even running.

Extension To master this, run quickly while moving the ball in and out between your legs.

4. 888 Rating 🏉🏉

Objective Move the ball in a figure eight between your legs.

Start by moving the ball between your legs and then behind your right leg.

Bring the ball back to the front and repeat by putting the ball between your legs, but this time go behind your left leg. Now repeat this figure eight movement over and over.

As you learn this skill, you can work on going as fast as you can.

Extension Keep the ball between your legs but move your hands from in front of your legs to behind your legs. This leaves the ball stationary but your hands and arms moving.

5. THE FUNNY FROG
Rating 🏉🏉

Objective Flick the ball up from the ground.

Stand upright with the ball between your

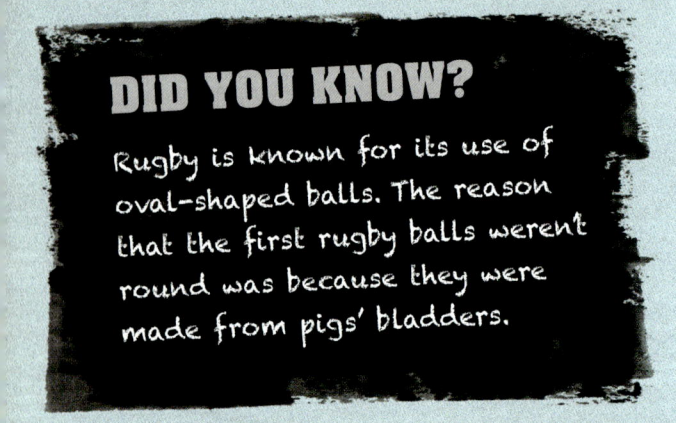

feet and squeeze the ball firmly with your ankles. (Don't worry if the ball is lengthways or across your feet.)

Squat slightly, then jump, flicking the ball upwards behind you and catch it!

Extension To really impress, flick the ball high enough so that you catch the ball behind your neck.

6. FLIP CATCH Rating 🏉🏉

Objective Throw the ball over your head and catch it behind your back.

Hold the ball in two hands in front of you, then throw the ball over your head with a nice toss and catch the ball in two hands behind your back.

Extension Flick it back the other way and catch it in front of you.

7. TURTLE NECK Rating 🏉🏉

Objective Keep the ball settled on the back of your neck.

First, place the ball on the back of your neck. This isn't as easy as you might think!

Bend from the waist until you're doubled over, let the ball go and keep it balanced on the back of your neck.

Get the feel of the ball being there, but keep the rest of your body balanced.

When this feels secure, walk around.

DID YOU KNOW?

Rugby is known for its use of oval-shaped balls. The reason that the first rugby balls weren't round was because they were made from pigs' bladders.

Extension Let the ball slide down your back for a Blind Donkey heel kick (see page 202).

8. BILLY BOUNCER Rating 🏈🏈

Objective Throw the ball to the ground so it bounces back to you.

Throw the ball to the ground using two hands, but make sure it lands on the point.

Notice which way the ball bounces when it lands on different points of the ball.

Now try to do it so the ball comes back to you every time.

It should be on a 45 degree angle so the ball comes back toward you. Catch and repeat.

Extension Do this one-handed or while walking around.

9. FOOTY FEET Rating 🏈🏈

Objective Keep the ball off the ground by kicking the ball from one foot to the other.

Start by dropping the ball onto one foot and catching it, then do two kicks and catch it.

Always use alternate feet, so you are in full control of the ball. Keep the ball up off the ground by kicking it from one foot to the other.

Extension Use your knee as well to keep the ball off the ground.

10. FILLED ROLL Rating 🏉🏉🏉

Objective Get the ball to run from one hand to the other.

Throw the ball in the air with a spinning action and as it comes down raise a hand toward the ball so it begins to spin along your extended arm.

Make a circle or barrel shape with your arms, chest and hands so the ball rolls toward the other hand.

Watch the ball roll around your first arm and as you get more control, let the ball move across your chest onto your other arm and into that hand.

Extension Get the ball to roll twice around your arms and chest from one spin. You need to circle your arms for this.

11. SQUASH THE BUG Rating 🏉🏉🏉

Objective Flick the ball up by rolling the ball up onto your foot.

Place your foot on the ball long-ways and put a little pressure on the ball.

While keeping the pressure on the ball, roll your foot toward you. As the ball begins to spin toward you, hook your toe under it, flick it up with a bent knee and catch it.

Extension Spin the ball so it spins or rolls up your leg into your hands.

12. FINGER SPIN Rating 🏉🏉🏉🏉

Objective Spin the ball on your finger.

Stand with your weight balanced evenly on each leg.

Hold the ball about 50cm in front of your face, with your arms bent at a 90 degree angle, and one hand on either side of the ball.

Now throw the ball up a little, with a quick snap of the wrists, while rotating your hands.

Cross one arm over the other and follow through with your fingers pointing in opposite directions.

Spin the ball a few times in the air with your hands, and catch it. This will help you increase your speed and steadiness with the rugby ball.

Speed is the most important factor of a spin, so work on increasing the speed by increasing the opposing force with your hands.

Once you've got the hang of spinning it, give it an extra-hard spin. Don't throw it too high. This time, don't catch it. Instead, put

RECORD

The longest rugby game was 24 hours 51 minutes. The game was a charity match and the two sides kept swapping reserves and players for more than a day. The score was 1742-828.

your back as you drop it.

As you drop it, flick your lower leg upward, which should make the ball strike your heel and propel the ball over your head.

Now do it so you can catch it!

You may need to practise the dropping of the ball before you get to the actual kicking part.

Extension Do a Foot Flicker (below) to get the ball into position instead of dropping the ball.

14. FOOT FLICKER
Rating 🏉🏉🏉🏉

Objective Flick the ball up from the ground and catch it.

Start by holding or catching the ball between your feet. Then one foot (usually the dominant one) rolls the ball up your support leg.

With your support leg, bring the heel up and flick the ball up. The ball should flick up over your shoulder or head where you catch it.

Extension Instead of catching it, trap it with your front foot and start again.

15. DROP THE EGG
Rating 🏉🏉🏉

Objective Drop the ball onto the top of

your index finger right in the middle under the spinning ball. Try to get it spinning on the tip of your finger.

Extension Transfer the ball, still spinning, to the other hand.

13. BLIND DONKEY
Rating 🏉🏉🏉🏉

Objective Heel-kick the ball over your head and catch it.

Start by holding the ball behind your back in two hands.

Then give it a small flick away from

14

15

your foot and hold it there.

Hold the ball in front of you, then drop the ball onto your foot.

See if you can catch the ball on your foot.

The move can be done more easily if you point your toes upward so you clamp the ball between your toes and your shin.

Extension Flick the ball up to your hands again and repeat.

16. ROLY POLY Rating 🏉🏉🏉

Objective Throw the ball in the air and catch it after you have achieved a forward roll.

Hold the ball in front of you and then throw it into the air so it will land a couple of metres from where you threw it.

Launch into a safe forward roll and stand up in the one movement.

Catch the ball!

Extension Throw the ball backwards and complete a backward roll before finishing by catching the ball.

16

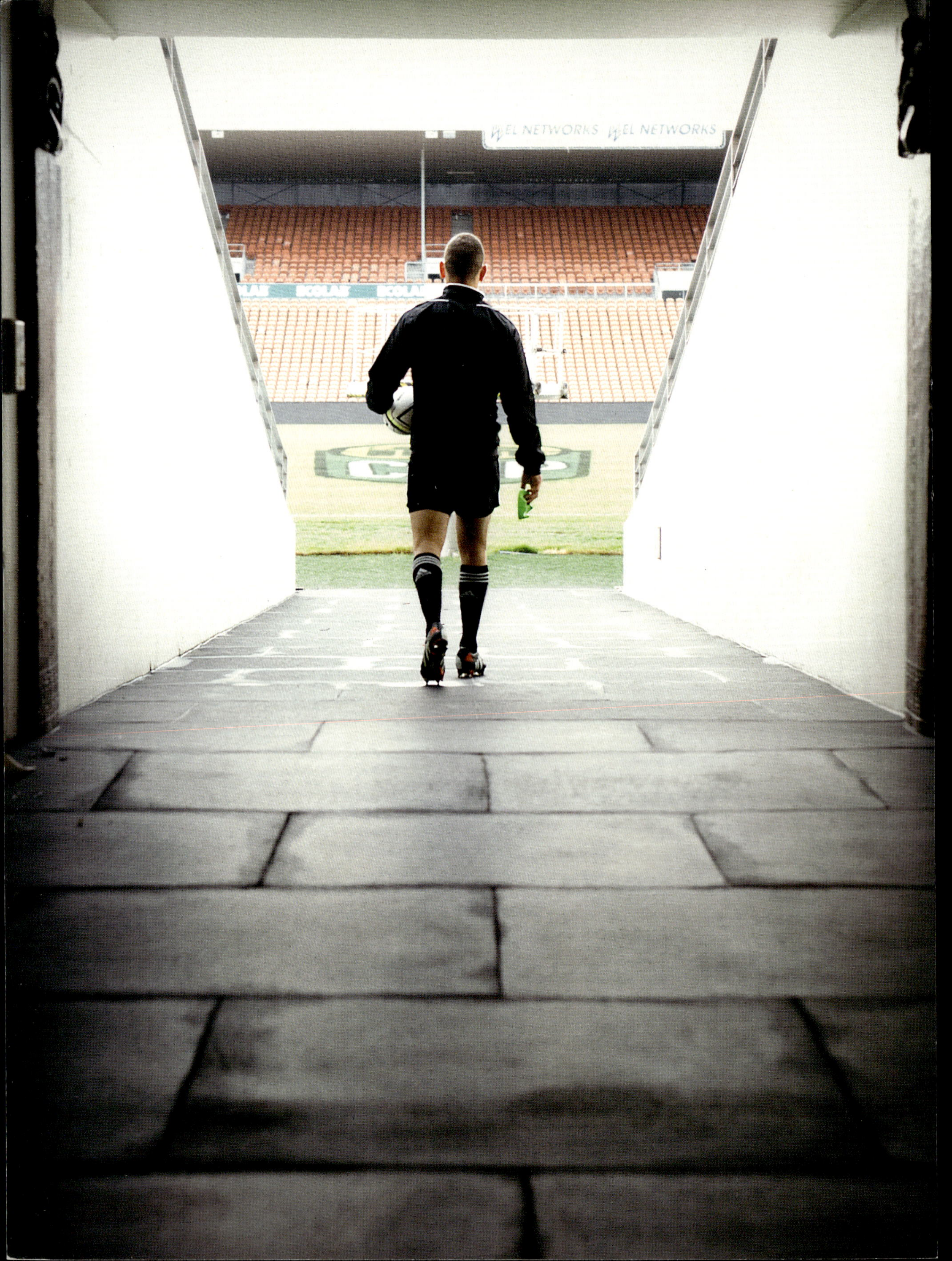
EL NETWORKS EL NETWORKS

FINAL WHISTLE

My first experience of rugby was a world away from becoming a professional rugby player or an All Black.

I wasn't watching my idols play rugby. I was watching my mates in the backyard. The All Blacks were untouchable, like gods, but then again, I guess it was natural that any passionate Kiwi kid would want to follow in the footsteps of their heroes. Rugby is New Zealand's national game!

Often the first full game of rugby a child plays is just down the road on the local field, against another team whose players live nearby, and a rugby side's first competition or tournament may only feature a small handful of teams. But there's no rush. There will be plenty of games for you to play as you go from being a child to teenager to adult.

My parents and first coaches made it simple: the way to get good enough to make dreams come true was to practise, practise and practise. So I did, doing lots of sessions a week, building a set of skills that made me a good player even though I was a 'tiny kid' — good enough eventually to be an All Black number 10.

Working and helping your team-mates, sacrificing your own goals, letting someone else take the kicks or promising to set someone else up for a try — these are the things that make you a team player, the biggest part of the art of rugby.

I believe it all starts at home and in the backyard with your mates.

The more support you get at home, the better you become. Have a set-up that lets you train at home, whether it's kicking the ball over the clothesline out back or evading the tackles of the shrubs out front.

The biggest names in the game, and all their silky skills, all started out small.

The friendship and the feeling of belonging that I get playing rugby has been a big part of my life, not just as a rugby player, but for me as a person as well.

Rugby isn't just about the game, the tackle or the scrum. For me, it has been about playing with my mates, developing confidence and the discipline that has taken me to the highest level. That's what I'll need if I want to continue to represent my country.

I love showing off my skills on the rugby field. I hope I have helped you to do the same.

ACKNOWLEDGEMENTS

The publishers would like to thank the following people:

Adrian Malloch, John McCrystal, Ree Davidson, Bruce Sharrock, John Knowles, Dave Rennie, Natalie Barrott, Andrew Flexman, Sean Harris, Jeff Andersen, Bill Heslop, Shona Pinny, Kate Barraclough, Tracey Wogan, Mike Wagg, Sarah Ell, Lindsay Calton and Palmerston North Boys' High School, NZ Rugby, Lachie Wright, Hamish Wright, Dave Dillon and Dave Morris.

IMAGE CREDITS

Photosport: 2 (main image), 8, 13, 28, 33, 34, 65, 182 (main image), 190, 193

Manawatu Rugby Union: 2 (insert above), 29, 41 (above), 182 (both inserts)

Getty Images: 2 (insert below), 10–11, 12, 15, 16, 21 (above and below), 31, 37, 38, 39, 40, 41 (below), 57, 58, 59, 60, 61 (below), 62, 66, 67, 79, 163

Cruden family: 18, 19, 20

photonewzealand/Corbisimages: 21 (middle), 45

Shutterstock: 22, 42, 47, 48, 49 (below), 50 (above and middle), 51 (middle and below), 55, 56 (above, Luke Schmidt), 56 (below), 61 (above), 63, 73 (above and below), 74 (above), 88 (Dario Vuksanovic), 94, 97, 111 (above), 135 (Paolo Bona), 136 (middle, Pavel L Photo and Video), 136 (below left, Neil Balderson), 137 (above right, Frederic Legrand – COMEO), 137 (middle, iofoto), 137 (below left, Capture Light), 176 (Daniel Goodings), 185, 189 (Paolo Bona)

Palmerston North Boys' High School: 23, 24, 26, 27

Museum of New Zealand Te Papa Tongarewa: 43, 50 (below)

Alexander Turnbull Library: 44 (1/2-194322-F)

Thinkstock: 71 (above)

123RF: 76, 172, 174

Dreamstime.com: 136 (above right, Instinia), 137 (below right, Ratmandude)

iStock: 136 (above right), 136 (below right, kokkai), 137 (above left)

Dorling Kindersley: 150 (Steve Gorton)